Geophysical Monograph 207

Geophysical Monograph Series

Geophysical Monograph 207

Magnetotails in the Solar System

Andreas Keiling
Caitríona M. Jackman
Peter A. Delamere
Editors

This Work is a co-publication between the American Geophysical Union and John Wiley and Sons, Inc.

American Geophysical Union

WILEY

This Work is a co-publication between the American Geophysical Union and John Wiley & Sons, Inc.

Published under the aegis of the AGU Publications Committee

Published by John Wiley & Sons, Inc., Hoboken, New Jersey
Published simultaneously in Canada

Library of Congress Cataloging-in-Publication Data is available.

ISBN: 978-1-118-84234-8

Cover image: Illustration of Earth's magnetosphere by C.S. Arridge (Mullard Space Science Laboratory, University College London, U.K.). (inset, top left) Illustration of Ganymede's magnetosphere by Xianzhe Jia (Department of Atmospheric, Oceanic and Space Sciences, University of Michigan, Ann Arbor, Michigan). (inset, top right) Illustration of the Jovian magnetosphere by John Spencer (Southwest Research Institute, Boulder, Colorado). (inset, bottom) Comet Hale-Bopp showing a dust tail (white) and an ion tail (blue) (from Chapter 10, and http://www.tivas.org.uk/solsys/tas_solsys_comet.html).

Printed in the United States of America

10 9 8 7 6 5 4 3 2 1

CONTENTS

CONTRIBUTORS

C. S. Arridge
Mullard Space Science Laboratory, Department of Space and Climate Physics, University College London, Surrey, United Kingdom
and
Now at Department of Physics, Lancaster University, Lancaster, United Kingdom

Nicolas Aunai
Heliophysics Science Division, NASA Goddard Space Flight Center, Greenbelt, Maryland

Joachim Birn
Los Alamos National Laboratory, Los Alamos, New Mexico
and
Space Science Institute, Boulder, Colorado

D. A. Brain
Laboratory for Atmospheric and Space Physics, University of Colorado, Boulder, Colorado
and
NASA's Lunar Science Institute, NASA Ames Research Center, Moffett Field, California

P. C. Brandt
Applied Physics Laboratory, Johns Hopkins University, Laurel, Maryland

J. F. Carbary
Applied Physics Laboratory, Johns Hopkins University, Laurel, Maryland

P. A. Delamere
Geophysical Institute, University of Alaska Fairbanks, Fairbanks, Alaska

M. K. Dougherty
The Blackett Laboratory, Imperial College London, London, United Kingdom

E. Dubinin
Max-Planck-Institut für Sonnensystemforschung, Göttingen, Germany

J. P. Eastwood
The Blackett Laboratory, Imperial College London, London, United Kingdom

M. Fraenz
Max-Planck-Institut für Sonnensystemforschung, Göttingen, Germany

K.-H. Glassmeier
Institute of Geophysics and Extraterrestrial Physics, Technical University of Braunschweig, Braunschweig, Germany

Tamas I. Gombosi
Department of Atmospheric, Oceanic and Space Sciences, University of Michigan, Ann Arbor, Michigan

Gerhard Haerendel
Max Planck Institute for Extraterrestrial Physics, Garching, Germany

J. S. Halekas
Department of Physics and Astronomy, University of Iowa, Iowa City, Iowa
and
NASA's Lunar Science Institute, NASA Ames Research Center, Moffett Field, California

Fred Hall IV
Geophysical Institute, University of Alaska Fairbanks, Fairbanks, Alaska

D. C. Hamilton
University of Maryland, College Park, Maryland

Michael Hesse
Heliophysics Science Division, NASA Goddard Space Flight Center, Greenbelt, Maryland

M. Holmström
Swedish Institute of Space Physics, Kiruna, Sweden

G. B. Hospodarsky
Department of Physics and Astronomy, University of Iowa, Iowa City, Iowa

Min-Shiu Hsieh
Geophysical Institute, University of Alaska Fairbanks, Fairbanks, Alaska

Caitríona M. Jackman
Department of Physics and Astronomy, University of Southampton, Southampton, United Kingdom

Xianzhe Jia
Department of Atmospheric, Oceanic and Space Sciences, University of Michigan, Ann Arbor, Michigan

Jay R. Johnson
Princeton Plasma Physics Laboratory, Princeton University, Princeton, New Jersey

L. Kepko
NASA Goddard Space Flight Center, Greenbelt, Maryland

S. A. Kiehas
Space Research Institute, Austrian Academy of Sciences, Graz, Austria

and

Department of Earth, Planetary, and Space Science, University of California, Los Angeles, California

S. M. Krimigis
Applied Physics Laboratory, Johns Hopkins University, Laurel, Maryland

Elena Kronberg
Max-Planck-Institut für Sonnensystemforschung, Göttingen, Germany

Norbert Krupp
Max-Planck-Institut für Sonnensystemforschung, Göttingen, Germany

W. S. Kurth
Department of Physics and Astronomy, University of Iowa, Iowa City, Iowa

Masha Kuznetsova
Heliophysics Science Division, NASA Goddard Space Flight Center, Greenbelt, Maryland

B. H. Mauk
Applied Physics Laboratory, Johns Hopkins University, Laurel, Maryland

David J. McComas
Southwest Research Institute, San Antonio, Texas

and

Department of Physics and Astronomy, University of Texas at San Antonio, San Antonio, Texas

Robert L. McPherron
Department of Earth, Planetary, and Space Sciences, University of California Los Angeles, Los Angeles, California

D. G. Mitchell
Applied Physics Laboratory, Johns Hopkins University, Laurel, Maryland

Antonius Otto
Geophysical Institute, University of Alaska Fairbanks, Fairbanks, Alaska

C. Paranicas
Applied Physics Laboratory, Johns Hopkins University, Laurel, Maryland

W. R. Pryor
Central Arizona College, Coolidge, Arizona

Aikaterini Radioti
Laboratoire de Physique Atmosphérique et Planétaire, Institut d'Astrophysique et de Géophysique, Université de Liège, Liège, Belgium

Geoffrey D. Reeves
Space Science and Applications Group, Los Alamos National Laboratory, Los Alamos, New Mexico

J. A. Slavin
Department of Atmospheric, Oceanic and Space Sciences, University of Michigan, Ann Arbor, Michigan

D. J. Southwood
The Blackett Laboratory, Imperial College, London, United Kingdom

T. Sundberg
School of Physics and Astronomy, Queen Mary College, University of London, London, United Kingdom

and

Center for Space Physics, Boston University, Boston, Massachusetts

Vytenis M. Vasyliūnas
Max-Planck-Institut für Sonnensystemforschung, Göttingen, Germany

M. Wiltberger
High Altitude Observatory, National Center for Atmospheric Research, Boulder, Colorado

Simon Wing
Applied Physics Laboratory, The Johns Hopkins University, Laurel, Maryland

Seiji Zenitani
National Astronomical Observatory of Japan, Tokyo, Japan

PREFACE

Spacecraft observations have established that all magnetized planets in our solar system (Mercury, Earth, Jupiter, Saturn, Uranus, and Neptune) interact strongly with the solar wind and possess well-developed magnetotails. Magnetotails are the site of many dynamic processes critical to the transport of mass, momentum, energy, and magnetic flux. The great differences in solar wind conditions, planetary rotation rates, ionospheric conductivity, and physical dimensions from Mercury's small magnetosphere to the giant magnetospheres of Jupiter and Saturn provide an outstanding opportunity to extend our understanding of the influence of these factors on magnetotail processes and structure.

It is not only the strongly magnetized planets that have magnetotails. Mars and Venus have no global intrinsic magnetic field, yet they possess induced magnetotails. Comets have a magnetotail that is formed by the draping of the interplanetary magnetic field. In the case of planetary satellites (moons), the magnetotail refers to the wake region behind the satellite in the flow of either the solar wind or the magnetosphere of its parent planet. The largest magnetotail of all in our solar system is the heliotail, the "magnetotail" of the heliosphere.

In the last decade there has been an abundance of new observations from spacecraft missions, such as MESSENGER, Geotail, Cluster, THEMIS, ARTEMIS, Galileo, New Horizons, Cassini, IBEX, and Deep Space 1. In addition, there have been great advances in the capabilities of simulations to allow modeling of many of the relevant processes. Hence, the time was ripe to address recent achievements of observational, theoretical, and modeling studies in a Chapman conference on "Fundamental Properties and Processes of Magnetotails," held in Reykjavik, Iceland, during 10–15 March 2013. The conference stimulated cross-disciplinary discussions among scientists specializing in all aspects of magnetotail physics.

To collect the discussions and results from the conference for the broader space science community, we endeavored on making this monograph, resting assured that the expertise would come from the expert attendees. While in the past the different types of magnetotails have been largely treated in separate books, our intention with this monograph was to take an integrated approach by treating the magnetotail as a fundamental phenomenon and incorporating the phenomenology and physics for all types of magnetotails in one book. As a result, this book should appeal to a broad community of space scientists. In fact, it should also be of interest to astronomers who are looking at tail-like structures in our galaxy and beyond.

This book is largely, but not exclusively, based on reviews and tutorials presented at the above-mentioned Chapman conference. It is organized into three sections: Introduction, Tutorials, and Specialized Topics. To set the stage, the first section exposes with an excellent discussion an "unsolved" fundamental problem in magnetospheric physics: the magnetotail. The specific question asked is: "How is the configuration of highly stretched magnetic field lines created and maintained?" The problem is addressed via global stress balance but ultimately points to the lack of any quantitative theory, as of yet, to answer this question. We believe that this chapter is a "must-read" for anyone interested in magnetotails.

The second section covers the large range of magnetotails in our solar system. Each tutorial chapter focuses on one magnetotail, reviewing various phenomena specific to the magnetotail at hand. These chapters are ideal for those who want to become familiar with a specific magnetotail. With respect to the current literature, we find this to be quite a unique collection of tutorials.

Each magnetotail has its own individual and unique characteristics. By taking a comparative approach to a number of specific science questions, one can develop a greater understanding of magnetotail dynamics in these different environments. This is the approach followed in the third section, where each chapter focuses on a special region of the magnetotail or a specific phenomenon while comparing it among different magnetotails. Many facets of magnetotail physics are covered in this section.

We are truly grateful to the many people who made the Chapman conference a success and the book a reality. All the conference attendees brought the energy and enthusiasm that made the meeting fun and rewarding. The other program committee members (Jin-Bin Cao, Colin Forsyth, Mervyn Freeman, Denis Grodent, Xianzhe Jia, Larry Kepko, Norbert Krupp, James Slavin, Michelle Thomsen, and Tsugunobu Nagai) contributed greatly to the science aspect of this conference. AGU staff (Cynthia Wilcox, Shermonta Grant, and Lynn Erwin) and Beverly Smith-Keiling provided outstanding support throughout the organization of and during the conference. The authors of the 22 chapters provided the substance for this book, and the many volunteers (listed below) who reviewed those chapters have given the book scientific credibility. The Icelandic hospitality welcomed us and made the field trips a true pleasure, visiting some of the local geological sites and, last but not least, a visit to the world-renowned Blue Lagoon. The National Science Foundation provided funding for many students to attend the conference. Additional funding, which helped with the costs of the conference center HARPA and the various field trips, came from the Institute of Geophysics and Planetary Physics (Los Alamos National Laboratory), the Space Sciences Laboratory (University of California, Berkeley), the Laboratory for Atmospheric and Space Physics (University of Colorado, Boulder), and the Geophysical Institute (University of Alaska, Fairbanks).

Andreas Keiling
University of California, Berkeley

Caitríona Jackman
University of Southampton, Southampton

Peter Delamere
University of Alaska, Fairbanks

Reviewers (in alphabetical order):

Otto Antonius, Chris Arridge, Fran Bagenal, Stanley Cowley, Tom Cravens, Peter Delamere, Dominique Delcourt, Rob Ebert, Stefan Eriksson, Karl-Heinz Glassmeier, Erika Harnett, Suzie Imber, Peter Israelevich, Mervyn Freeman, Herbert Funsten, Mary Hudson, Jing Liao, Andreas Keiling, Larry Kepko, Margaret Kivelson, Philippe Louarn, John Lyon, Barry Mauk, Robert McPherron, Steve Milan, Donald Mitchell, Rumi Nakamura, Takuma Nakamura, Masaki Nishino, Steven Petrinec, Aikaterini Radioti, Joachim Raeder, Dan Reisenfeld, John Richardson, Elias Roussos, Menelaos Sarantos, George Siscoe, Victor Sergeev, James Slavin, Marc Swisdak, Michelle Thomson, Vytenis Vasyliūnas, Marissa Vogt, Martin Volwerk, Dan Welling.

Section I: Introduction

1

Magnetotail: Unsolved Fundamental Problem of Magnetospheric Physics

Vytenis M. Vasyliūnas

1.1. INTRODUCTION

Planetary magnetospheres are observed to extend in the antisolar direction to distances orders of magnitude larger than the magnetosphere's characteristic linear dimension on the dayside. That the supersonic solar wind interacting with a planet's magnetic field should create an extended wake along its direction of flow can be readily understood. What is remarkable about the observed planetary magnetotails is that they do not consist merely of sundry perturbations in the magnetic field and plasma but display a well-ordered structure: highly stretched-out, oppositely directed magnetic fields, separated by regions of enhanced plasma pressure (sketched in Figure 1.1). This configuration has by now become so familiar, from numerous papers on observations and model descriptions, that it is easy to overlook the underlying physical question: How is such a configuration created and maintained? In this chapter I show that, at the level of physical understanding as distinct from mere empirical representation, explaining the magnetotail constitutes a fundamental problem in magnetospheric physics that still remains unsolved, despite extensive empirical data (mainly at Earth but increasingly at other planets as well) and diverse qualitative theoretical concepts: There is as yet (to my knowledge) no predictive first-principles theory, comparable to the Chapman-Ferraro model which is the basis for understanding the dayside magnetosphere.

Section 1.2 summarizes the principal properties of magnetotails and identifies those that lead to the problem of global stress balance, which is described quantitatively in section 1.3. The various proposed ideas of what maintains a magnetotail are reviewed in section 1.4.

Time variations, interesting and important as they are, are not dealt with; dynamical processes are mentioned only insofar as they illuminate some aspect of the basic problem. For simplicity, descriptions and illustrations generally presume that the planetary magnetic dipole is quasi-perpendicular to solar wind flow, which is the case for all magnetospheres observed to date (Earth, Mercury, Jupiter, Saturn) with the exception of Uranus and Neptune.

1.2. ESSENTIAL PROPERTIES

Evident in Figure 1.1 is the striking difference in the shape of magnetic field lines between the dayside and the nightside. Near the planet, the field lines are (nearly) dipolar at all local times. As one proceeds outward, on the dayside the field lines become more and more compressed as one approaches the magnetopause; on the nightside, to the contrary, they become stretched out and nearly aligned with the solar wind flow direction, the magnetic field direction reversing across the current sheet in the central region of enhanced plasma pressure and density (plasma sheet). In the lobes of the magnetotail above and below the plasma sheet, the magnetic field is approximately uniform, and the plasma pressure and density are very much lower than in adjacent regions.

Seemingly, there are processes that act to compress the magnetic field of the planet on the dayside and to pull it out on the nightside. The dayside process was identified already, for the magnetosphere of Earth, by Chapman and Ferraro in the 1930s: exclusion of the planet's magnetic field from the highly conducting plasma of the (then hypothetical) solar wind and its compression by the dynamic pressure of solar wind flow, a concept that by

Max-Planck-Institut für Sonnensystemforschung, Göttingen, Germany

Magnetotails in the Solar System, Geophysical Monograph 207. First Edition.
Edited by Andreas Keiling, Caitríona M. Jackman, and Peter A. Delamere.

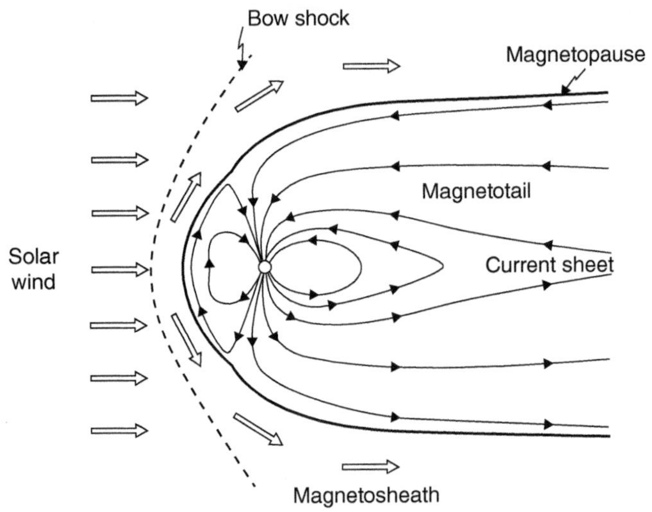

Figure 1.1 Schematic view of a magnetosphere, cut in the noon-midnight meridian plane. Open arrows: solar wind bulk flow. Solid lines within magnetosphere: magnetic field lines. The figure continues to the right out to distances in general much larger than the width of the figure. (In all figures of this chapter: sunward direction to the left; unless otherwise stated, magnetic field line directions appropriate for Earth or Mercury, reversed for Jupiter or Saturn.)

the 1960s had been developed into a well-defined mathematical theory [see, e.g., *Siscoe*, 1988, for review and references], capable of predicting quantitatively the location and shape of the magnetopause and the configuration of magnetic field lines. These predictions are mostly (with some exceptions, primarily during intense magnetic storms) in reasonable agreement with what is observed on the dayside of Earth's magnetosphere, but they disagree completely with what is observed on the nightside and especially in the extended magnetotail [see, e.g., *Vasyliūnas*, 2011, for a more details discussion]. The root of the disagreement clearly is the assumption in the Chapman-Ferraro model that pressure is the only force acting across the solar-wind/magnetosphere interface.

The concept of magnetic field lines from Earth being pulled back by the solar wind to form a (transient or permanent) magnetotail was suggested by *Parker* [1958] and *Piddington* [1960]; it acquired a concrete form with the proposal of the magnetically open magnetosphere [*Dungey*, 1961]. The more general concept of an (unspecified) tangential drag force across the solar-wind/magnetosphere interface was introduced by *Axford and Hines* [1961] and invoked to discuss a magnetotail by *Axford et al.* [1965]. A quasi-permanent (and very long) magnetotail created by internal hydromagnetic-wave or plasma pressure was proposed by *Dessler* [1964]. Before they could be developed into some reasonably coherent quantitative theory, these early theoretical ideas were soon overshadowed by the observational identification of Earth's magnetotail [*Ness*, 1965] and subsequent detailed studies of its properties that continue, on the basis of ever-expanding data sets, to this day.

One consequence has been to deflect the attention of theorists from the basic problem of understanding what creates the magnetotail itself to the applied task of modeling and calculating its various aspects (such as local stress balance between plasma and magnetic field, dynamical developments, etc.), taking for granted that the magnetotail exists and has its empirically determined parameter values. The basic problem arises primarily as the result of two general properties of observed magnetotails:

1. To first approximation, magnetic fields are nearly aligned (in opposite senses on the two sides of the central current sheet) with the direction of solar wind flow, which is also, more or less, the direction of the magnetotail axis.
2. In the lobe regions, magnetic stresses are dominant over plasma mechanical stresses (pressure or flow).

1.3. GLOBAL STRESS BALANCE PROBLEM

The two properties listed above imply that across any cross section perpendicular to the magnetotail axis there exists a net magnetic tension force directed approximately along the axis. As pointed out explicitly by *Siscoe* [1966], this force must be exerted ultimately on the massive planet; moreover, something external to the magnetotail must be exerting this force. Understanding the origin of this global force constitutes the fundamental (and in my view still unsolved) problem that the magnetotail poses for magnetospheric physics.

1.3.1. Plasma Momentum Equation

Quantitative discussion of stress balance begins with the plasma momentum equation in standard conservation form (partial time derivative of density of conserved quantity plus divergence of flux density of conserved quantity equals zero):

$$\partial \rho \mathbf{V} / \partial t + \nabla \cdot (\rho \mathbf{V}\mathbf{V} + \mathbf{P} - \mathbf{T}) = 0, \tag{1.1}$$

where ρ, $\mathbf{V}$, and $\mathbf{P}$ are the plasma mass density, bulk velocity, and pressure tensor, respectively, and $\mathbf{T}$ is the total stress tensor: Maxwell stress tensor $\mathbf{T}_M$ plus stress tensors of any other forces that might be present. Equation 1.1 has been written with the usual approximation of charge quasi-neutrality, nonrelativistic bulk flow, and Alfvén speed $V_A^2 \ll c^2$, neglecting electric field terms both in the momentum density and in the Maxwell stress tensor,

$$\mathbf{T}_M = \mathbf{B}\mathbf{B}/4\pi - \mathbf{I}\left(B^2/8\pi\right) \tag{1.2}$$

(Gaussian units, $\mathbf{I}$ = unit dyad δ_{ij}). The divergence of any stress tensor equals the corresponding force density, in particular,

$$\nabla \cdot \mathbf{T}_M = \mathbf{J} \times \mathbf{B}/c \tag{1.3}$$

($\mathbf{J}$ = current density). Gravity can be treated by including in $\mathbf{T}$ a gravitational stress tensor

$$\mathbf{T}_G = -\mathbf{gg}/4\pi G + \mathbf{I}\left(g^2/8\pi G\right) \tag{1.4}$$

($\mathbf{g}$ = gravitational acceleration, G = Newtonian gravitational constant), satisfying

$$\nabla \cdot \mathrm{T}_G = \rho \mathbf{g}, \tag{1.5}$$

but it is almost always simpler to just put the term $\rho\mathbf{g}$ on the right-hand (RH) side of equation 1.1. In any case, for most magnetospheric problems direct gravitational effects are so small that they need not be explicitly included.

1.3.1.1. Integrated Forces. The divergence term in equation 1.1 integrated over any fixed volume gives the net force acting on the contents of the volume, expressed as a surface integral over the boundary of the volume,

$$\mathbf{F} = \int d\mathbf{A} \cdot \left[\mathbf{BB}/4\pi - \mathrm{I}\left(B^2/8\pi\right) - \mathrm{P} - \rho\mathbf{VV}\right], \tag{1.6}$$

where $d\mathbf{A} = \hat{\mathbf{n}}\, dA$, with $\hat{\mathbf{n}}$ the outward normal to the boundary (throughout this chapter, $\hat{a}$ designates unit vector); the force can thus be determined just from values of physical quantities on the boundary, without the need to inquire into what goes on in the interior. With the force taken as acting on the volume *interior* to the (closed) bounding surface, the surface integral in equation 1.6 contains *exterior* quantities, and vice versa. If $\hat{\mathbf{n}}$ is reversed (i.e., "interior" and "exterior" are interchanged), the force $\mathbf{F}$ is also reversed; this is a statement of Newton's third law.

The component of $\mathbf{F}$ in a fixed direction $\hat{\mathbf{u}}$, assuming gyrotropic pressure, is

$$\begin{aligned} F_u = \int dA \{ & \left[B_n B_u (1-\xi)/4\pi\right] \\ & - \left[\hat{\mathbf{n}} \cdot \hat{\mathbf{u}}\left(B^2/8\pi + P_\perp\right)\right] - \left[\rho V_n V_u\right]\} \end{aligned} \tag{1.7}$$

where the first bracket [] is magnetic tension modified by pressure anisotropy $\xi \equiv 4\pi(P_\parallel - P_\perp)/B^2$, the second is total pressure (magnetic + plasma), and the third is bulk flow stress ("dynamic pressure," which can also be described as "negative tension"). If the direction of $\mathbf{B}$ or $\mathbf{V}$ or both is reversed, $\mathbf{F}$ remains unchanged.

With the volume integral of the time-derivative term included, the integrated equation 1.1 becomes

$$\mathbf{F} = (d/dt)\int d^3r\, \rho \mathbf{V}, \tag{1.8}$$

i.e., force (applied to a given volume) equals rate of change of linear momentum (within the volume): Newton's second law. With M the total mass enclosed within the volume and $\mathbf{R}$ its center of mass defined by

$$M \equiv \int d^3r\, \rho, \qquad M\mathbf{R} \equiv \int d^3r/\rho\, r, \tag{1.9}$$

the rate of change can be written as

$$\begin{aligned} (d/dt)\int d^3r\, \rho \mathbf{V} = & M\, d^2\mathbf{R}/dt^2 + 2(dM/dt)(d\mathbf{R}/dt) \\ & + \int d\mathbf{A} \cdot (\partial \rho \mathbf{V}/\partial t)(\mathbf{r} - \mathbf{R}) \end{aligned} \tag{1.10}$$

[*Vasyliūnas,* 2007a]. The last term can be shown to be $O(\delta\mathbf{R}\, d^2M/dt^2)$, where $\delta\mathbf{R}$ is a length sufficiently short to fit within the volume. For magnetospheric volumes of global scale, the enclosed mass in most cases changes, if at all, very slowly in comparison with time scales of interest; the terms in the second line of equation 1.10 can then be neglected, and equation 1.8 reduces to the more conventional form

$$\mathrm{F} \simeq M\left(d^2\mathbf{R}/dt^2 - \mathbf{g}_{\mathrm{ext}}\right) \tag{1.11}$$

where gravity has now been taken into account, with $\mathbf{g}_{\mathrm{ext}}$ the gravitational acceleration due to matter external to the volume (the enclosed mass cannot by self-gravity produce a net acceleration of itself).

The integrated equations (1.6)–(1.11) apply to any arbitrarily chosen volume, provided the integrals have been calculated over the corresponding volume or surface. An important distinction between volumes in the magnetosphere that do and those that do not include the planet follows from equation 1.11. Consider a magnetospheric closed surface 1 that surrounds the planet and nested inside it another closed surface 2 just outside the planet; the volume enclosed by surface 1 consists of the magnetosphere (or part thereof) plus the planet, and that enclosed by surface 2 is just the planet. The forces on the two volumes, calculated as integrals over the respective bounding surface, are

$$\mathbf{F}_{\mathrm{Pl+MS}} \simeq \left(M_{\mathrm{Pl}} + M_{\mathrm{MS}}\right)\left(d^2\mathbf{R}/dt^2 - \mathbf{g}_{\mathrm{ext}}\right), \tag{1.12}$$

$$\mathbf{F}_{\mathrm{Pl}} \simeq M_{\mathrm{Pl}}\left(d^2\mathbf{R}/dt^2 - \mathbf{g}_{\mathrm{ext}}\right). \tag{1.13}$$

Subtracting equation 1.13 from equation 1.12 gives the force on the magnetospheric volume with the planet excluded,

$$\mathbf{F}_{\mathrm{MS}} \simeq M_{\mathrm{MS}}\left(d^2\mathbf{R}/dt^2 - \mathbf{g}_{\mathrm{ext}}\right), \tag{1.14}$$

which can be calculated from integrals on bounding surface 1 minus surface 2. Now ($d^2\mathbf{R}/dt^2 - \mathbf{g}_{\mathrm{ext}}$) is essentially the same for the planet and for the magnetosphere, hence the force on the magnetosphere is smaller than that on the planet by the ratio of the respective masses:

$$|\mathbf{F}_{\mathrm{MS}}|/|\mathbf{F}_{\mathrm{Pl}}| \sim (M_{\mathrm{MS}}/M_{\mathrm{Pl}}), \tag{1.15}$$

which is a very small number indeed (e.g., of order 10^{-21} for Earth). It follows that a net (nonzero) force can be exerted only on the massive planet; *the force exerted on any magnetospheric volume is transferred to the planet, and the net force on a magnetospheric volume that does not include the planet is essentially zero*. This principle was first explicitly enunciated by *Siscoe* [1966] and further discussed by *Carovilano and Siscoe* [1973], *Siscoe* [2006], and *Vasyliūnas* [2007a].

1.3.1.2. Linear Momentum Transport. An alternative view of the momentum equation 1.1 is in terms of transport of linear momentum. The stress tensor **T** represents the flux density of linear momentum, i.e., the tensor component T_{un} is the amount per unit area of linear momentum component along direction $\hat{\mathbf{u}}$ being transported in the direction $\hat{\mathbf{n}}$ (or, equivalently, component along $-\hat{\mathbf{u}}$ transported in direction $-\hat{\mathbf{n}}$). The surface integrals in equations 1.6 or 1.7 represent the net flux of linear momentum (or specified component thereof) across the entire surface. The total linear momentum contained inside a volume can be changed only by transport across the enclosing surface, because linear momentum is a conserved quantity.

From this point of view, it is also meaningful to discuss partial forces, across a segment of a bounding surface, by applying equation 1.6 or 1.7 to an unclosed surface; the calculated **F** is then the linear momentum flux through (or, equivalently, the force exerted across) that part of the bounding surface. In the special case $\hat{\mathbf{u}} = \pm\hat{\mathbf{n}}$, of particular interest for magnetotails (see section 1.3.2), equation 1.7 becomes

$$\begin{aligned} F_u = \pm\int dA\Big\{&\left(B_n^2 - B_t^2\right)/8\pi \\ &-P_\perp\left(B_t^2/B^2\right) - P_\parallel\left(B_n^2/B^2\right) - \rho V_n^2\Big\} \end{aligned} \tag{1.16}$$

or for isotropic pressure

$$F_u = \pm\int dA\left\{\left(B_n^2 - B_t^2\right)/8\pi - P - \rho V_n^2\right\} \tag{1.17}$$

where $B_t \equiv \sqrt{B^2 - B_n^2}$ is the magnetic field tangential to the surface; $F_u > 0$ represents a net tension force (outward through the bounding surface) and $F_u < 0$ a net pressure force (inward through the bounding surface). In the surface integrals, tension is contributed only by the normal magnetic field; all other terms (tangential magnetic field, plasma pressure, flow stress) contribute pressure.

1.3.2. Implications for Magnetotail

The force between the magnetotail and the dayside magnetosphere can be estimated most simply by choosing the bounding surface shown by the double vertical dotted line in Figure 1.2: cross section perpendicular to the magnetotail axis, located at a distance near the interface between the magnetotail and the dayside magnetosphere, where the magnetic field is still nearly aligned with the solar wind flow but is beginning to turn toward a dipolar configuration. With this choice of surface and with the force component of primary interest being that along the magnetotail axis (or solar wind flow direction or Sun-planet line), taken as $\hat{\mathbf{x}}$ sunward in standard solar magnetospheric coordinates, the surface normal $\hat{\mathbf{n}} = -\hat{\mathbf{x}}$ and the force direction $\hat{\mathbf{u}} = \hat{\mathbf{x}}$ are aligned, and hence equation 1.16 or 1.17 can be applied. The first of the two magnetotail properties listed at the end of section 1.2 implies that, over much of the cross-sectional area, $B_n^{\,2} \gg B_t^{\,2}$; the second implies that $B^2/8\pi \gg$ plasma terms. The surface integral over the interface can then be estimated as $F_x = -F_{\mathrm{MT}}$ with

$$F_{\mathrm{MT}} \approx \left(B_T^2/8\pi\right)(1-\delta)A_T, \tag{1.18}$$

where B_T is the magnetic field strength in the magnetotail lobes, A_T is the cross-sectional area, and δ is a correction term for plasma sheet contribution [*Siscoe*, 1972; *Vasyliūnas*, 1987; see the Appendix]. The global magnetotail force is thus a net tension force; this is a direct result of two magnetotail properties, the stretched-out magnetic field configuration (what is universally called a "tail-like" field), and the extended lobes where plasma stresses are very small in relation to the magnetic field magnitude.

A bounding surface that encloses a volume containing the dayside magnetosphere (and the planet) can be constructed by taking the magnetotail cross section at the interface shown by the vertical dotted lines in Figure 1.2 and joining it to a surface just outside the dayside magnetopause. Similarly, a bounding surface that encloses a volume containing the magnetotail can be constructed by taking the same magnetotail cross section (now viewed as

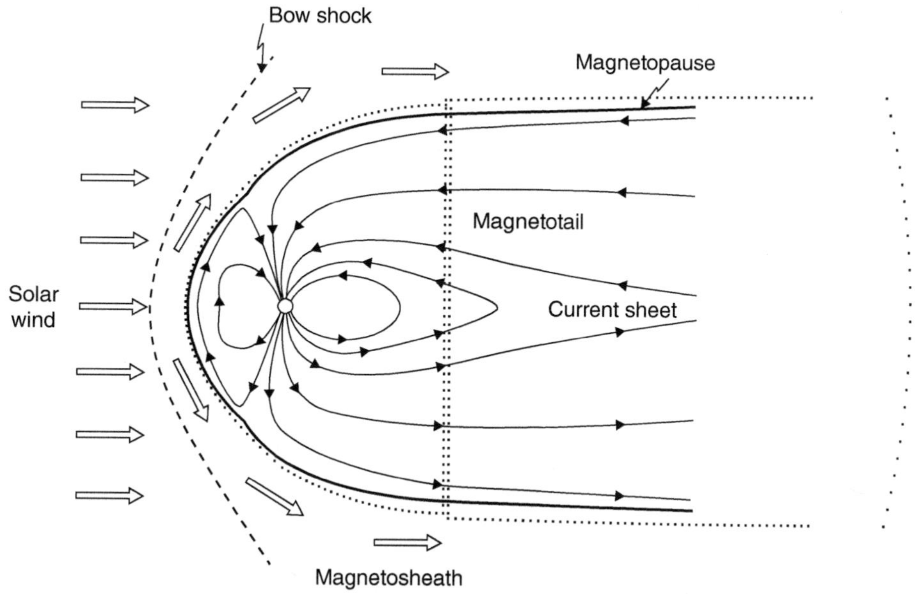

Figure 1.2 Same as Figure 1.1, with dotted lines added to show integration surfaces for discussion of stress balance.

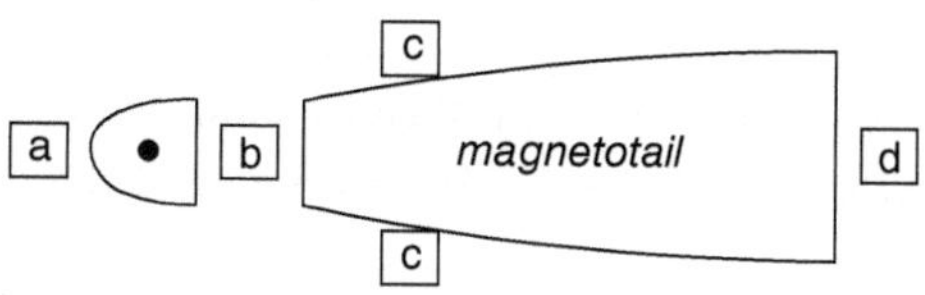

Figure 1.3 "Box model" representation of Figure 1.2 (see text for description).

the planet-facing boundary of the magnetotail) and joining it to a surface just outside the flanks and the top/bottom of the magnetotail, extended in the antisunward (solar-wind downstream) side as far as necessary to join it to the distant boundary where the magnetotail ends. (The reason for choosing the surfaces just outside the magnetic boundaries is discussed in section 1.4.) Both surfaces are indicated by dotted lines in Figure 1.2. To further simplify the discussion of global forces on enclosed volumes, Figure 1.3 shows these two volumes represented as boxes (planet surrounded by dayside magnetosphere on the left, magnetotail on the right), with the various interface/boundary surfaces labeled by letters:

a. Dayside magnetosphere boundary
b. Magnetosphere/magnetotail interface
c. Magnetotail flank boundaries
d. Magnetotail termination boundary

Surface (b) is the bounding surface introduced previously for the calculation of the force between the magnetotail and the dayside magnetosphere, equation 1.18. Surface (d) is the end of the stretched-out magnetic field configuration, discussed in section 1.3.3. The magnetotail problem is primarily about the $\hat{\mathbf{x}}$ components of the global forces acting on these volumes.

Forces acting on the dayside magnetosphere volume are F_a, the well-understood antisunward pressure force on the dayside magnetosphere exerted by the solar wind, and F_b, the antisunward tension force discussed above and given by $-F_{MT}$ of equation 1.18, exerted by the magnetotail. There is thus a net antisunward force on the volume, or equivalently a continual transport into the volume of antisunward linear momentum, both from the solar wind on the dayside and from the magnetotail on the nightside. In accordance with the principle stated in section 1.3.1.1, this force is applied to the massive planet and adds antisunward linear momentum to it, ultimately balanced by solar gravity (the quantitative effect on the planet's orbit is, of course, utterly negligible; Earth's orbit is displaced by something like 0.1 mm).

Forces acting on the magnetotail volume include the same force $F_b = +F_{MT}$, now viewed as the sunward tension force exerted by the dayside on the magnetotail (or, equivalently, as transport of sunward linear momentum into the magnetotail from the dayside) across its planet-facing boundary, and in addition forces F_c and F_d across the sides and the distant boundary of the magnetotail, respectively. These are far from being understood; the very concept indeed of surface (d), the magnetotail termination boundary, is still somewhat nebulous (see section 1.3.3). Unlike the dayside magnetosphere, the magnetotail volume does not include the planet; its center of mass, however, although different from that of the planet, is not accelerated relative to the planet: The magnetotail as a structure is not blown away or otherwise removed by the forces acting on it, even though some of the plasma within it might be. By the already-mentioned principle of section 1.3.1.1, the total force on the magnetotail volume must therefore be essentially zero, which imposes a far-reaching constraint on F_c and F_d:

$$F_b + F_c + F_d \simeq 0; \qquad (1.19)$$

equivalently, the sunward linear momentum that is transported into the magnetotail through its planet-facing boundary must be transported out through its boundaries elsewhere. Left unbalanced, the force F_b would collapse the magnetotail at roughly the Alfvén speed given by lobe magnetic field strength and plasma sheet density.

1.3.3. Termination of Magnetotail

Where and how the magnetotail ends in the downstream (antisunward) direction from the planet has not yet (to my knowledge) been observed unambiguously at any planet. Theoretically, there are several ways of defining the termination of the magnetotail and correspondingly several different definitions for the related concept of the length of the magnetotail. (*Caution*: It is not uncommon for papers to discuss "length of the magnetotail" without specifying which definition is meant.)

1.3.3.1. Closed Magnetosphere. The simplest case is that of a magnetically closed magnetosphere, where all the magnetic field lines that leave the planet return to the planet. The magnetic flux of the stretched-out field lines in one half of the magnetotail cross section must connect to that of the oppositely directed field lines in the other half, as one proceeds antisunward; since the total flux is finite, at some distance all of it has been connected and the magnetotail ends. (*Note:* "Connection" here refers to a property of the field configuration, "reconnection" to a physical process.) Although the length of the magnetotail is thus unambiguously defined, it is not uniquely determined: Its value obviously depends on the profile of the normal magnetic field component in the field reversal region. At one extreme, *Johnson* [1960], in his famous "teardrop" model, assumed that the magnetotail fields are confined within a surface that closes behind the planet at approximately the solar wind thermal speed, neglecting all interior stresses; this gives a distance from the planet to the end of the magnetotail about 5–20 times the distance to the dayside magnetopause. At the other extreme, *Dessler* [1964] assumed that the magnetotail fields, although closed in the sense of not connecting to an exterior medium, are (by the pressure of hydromagnetic waves generated in the solar wind interaction) kept open in the sense of not closing across the field reversal region; the magnetotail then extends indefinitely and ends only with the termination of the solar wind by interaction with the interstellar medium (thought then to occur at a heliocentric distance of some 20–50 AU but known now to be much farther out).

1.3.3.2. Open Magnetosphere. The case of an open magnetosphere is more complicated. A stretched-out field line can now connect either with the oppositely directed field in the other half of the magnetotail or with the magnetic field in the solar wind. The topology of an open magnetotail when the interplanetary magnetic field is parallel to the planet's magnetic dipole is illustrated schematically in Figure 1.4 for the entire system and in Figure 1.5 more specifically for the magnetotail. In both figures, the presentation is simplified: In Figure 1.5, the near-planet region is shown foreshortened in comparison to the rest (the segment to the left of the X line corresponds in reality to more than the full extent of Figure 1.1), and the deformation of the magnetic field by magnetosheath flow is not shown (outside the magnetotail the field lines are in reality tilted to the right, due to field line draping in the magnetosheath, and become vertical only after they cross the bow shock); Figure 1.4, intended to illustrate just the three-dimensional topology of magnetic field and plasma bulk flow, makes no pretense to realism in either scale or shape.

The most widely used definition for the length of an open magnetotail, proposed by *Dungey* [1965] and shown as $\mathscr{L}_{\rm MT}$ in Figure 1.4 and as $\mathscr{L}_c$ ("connected tail") in Figure 1.5, is given by solar wind speed multiplied by flow time across the open field line region (at Earth generally estimated from ionospheric plasma flow across the polar cap). It is often viewed as the distance to the last field line in the magnetotail connected to the planet (sometimes the length of the magnetotail is understood as the distance to the last *closed* field line in the magnetotail, which is the distance to the nightside X line in both figures). The magnetotail length defined by *Dungey* [1965], however, clearly is equal to the length of the open field line region *projected following magnetic field lines into the undisturbed solar wind*; the projected region is approximately a rectangle of length $\rm L_{MT}$ and width $\rm L_X$ (Figure 1.4), with $\mathrm{L_{MT}L}_X\times$ interplanetary magnetic field = open magnetic flux. At Earth, empirical estimates give $\mathrm{L}_X << \mathrm{L_{MT}}$, implying a highly elongated rectangle described by *Stern* [1973] as "a long narrow 'window' along the geomagnetic tail." Furthermore, a stretched magnetic field configuration is possible even when the field lines are no longer connected to the planet, as shown over the distance L_d ("disconnected tail") in Figure 1.5.

For the purposes of this chapter, which discusses the problem posed by the tension force in the magnetotail, termination of the magnetotail is appropriately defined

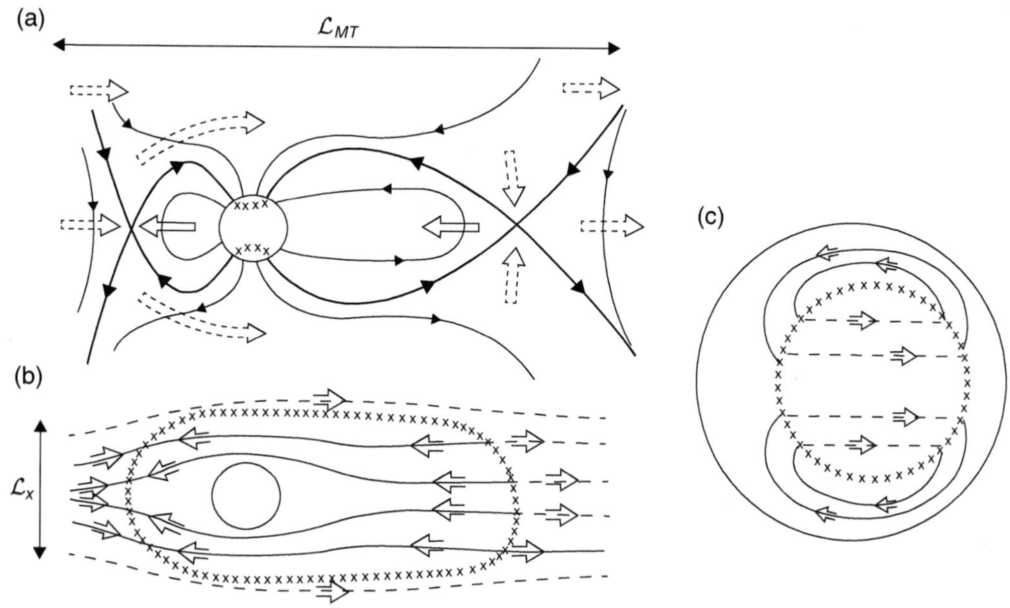

Figure 1.4 Schematic topological view of a magnetically open magnetosphere. (a) Noon-midnight meridian plane (solid lines: magnetic field lines, open arrows: plasma bulk flow directions). (b) Equatorial plane (lines: plasma flow streamlines, line of x's: magnetic X line = closed/interplanetary field line boundary). (c) Projection on ionosphere (lines: plasma flow streamlines, line of x's: open/closed field line boundary = projection of magnetic X line = polar cap boundary) [*Vasyliūnas*, 2011].

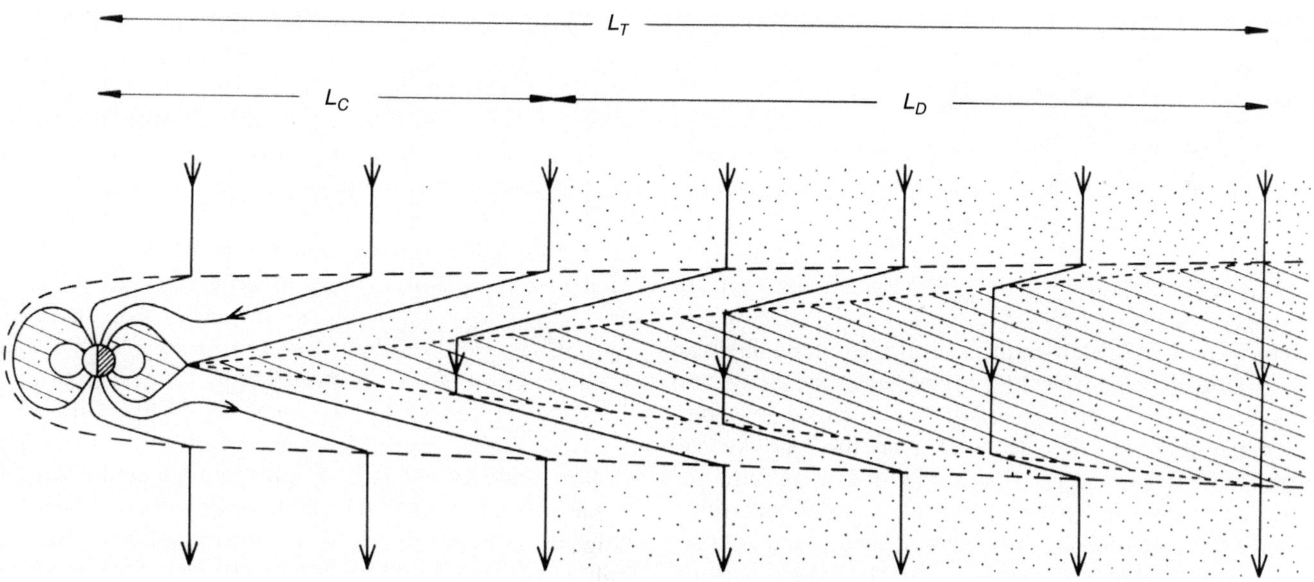

Figure 1.5 Sketch of the open magnetosphere and magnetotail in the noon-midnight meridian, showing magnetic field lines (solid lines), magnetopause (dashed line), and plasma sheet (dashed area) [from *Cowley*, 1991].

as the distance where that tension force is no longer present and no further antisunward transport of linear momentum need be accounted for. The tension force estimate of equation 1.18 applies for a cross section of the magnetotail not only at its interface to the dayside but also at all larger distances antisunward, as long as the twin properties of stretched antialigned ("tail like") magnetic fields and negligible plasma stresses hold. The magnetotail thus ends where (as, e.g., at distance L_d in Figure 1.5) the field components along the tail axis no longer are dominant, or the plasma stresses have become important over the entire cross section, or both.

1.4. WHAT MAINTAINS A MAGNETOTAIL?

The magnetotail problem in a nutshell is this: If the magnetotail is to be maintained as an enduring structure, neither collapsing nor being blown away, the net sunward magnetic tension force exerted on the magnetotail across its planet-facing boundary must be balanced by an equal antisunward net force across the remaining boundaries taken together. Alternatively, the sunward linear momentum transported into the magnetotail across its planet-facing boundary must be transported out (or, equivalently, an equal amount of antisunward linear momentum transported in) across the other boundaries: Linear momentum is a conserved quantity, and there is not sufficient mass in the magnetotail for storing it.

The necessary (but not sufficient) condition for maintaining a magnetotail is, from equations 1.18 and 1.19,

$$F_c + F_d \approx -\left(B_T^2/8\pi\right)(1-\delta)A_T. \qquad (1.20)$$

With the use of equation 1.7 to express the force $F_c + F_d$ as an integral over the surfaces (c) and (d), noting that, with $\hat{\mathbf{u}} = \hat{\mathbf{x}}$, $\hat{\mathbf{n}} \cdot \hat{\mathbf{u}} = -1$ for surface (d), and with the assumption (for simplicity) that pressure is isotropic, condition (20) may be rewritten as

$$\begin{aligned}\left(B_T^2/8\pi\right)(1-\delta)A_T = &+\int_{b+c+d} dA\{\rho V_n V_x\} \\ &+\int_c dA\left\{\hat{\mathbf{n}}\cdot\hat{\mathbf{x}}\left(P + B^2/8\pi\right) - B_n B_x/4\pi\right\} \\ &-\int_d dA\left\{P + \left(B_t^2 - B_n^2\right)/8\pi\right\}.\end{aligned} \qquad (1.21)$$

In equation 1.21 the left-hand (LH) side (for convenience written positive by reversing all signs) is the magnetic tension force that is required to be balanced by what is on the RH side. The first term on the right is the bulk-flow stress integrated over the complete (closed) bounding surface of the magnetotail volume, including surface (b) on which the bulk flow stress is usually considered negligible but turns out to be significant in some models; the remaining terms are the pressure and magnetic tension on surfaces (c) and (d).

Equation 1.21 can be considerably simplified. The magnetic tension force on surfaces (c) and (d) may be assumed negligible: on (d) by definition of the surface and on (c) by choice of surface just outside the flanks of the magnetotail (a deliberate choice, as noted in section 1.3.2: accounting for the tension force of stretched field lines by ascribing it to magnetic tension at the outer boundaries of the volume would not solve the magnetotail problem but would merely displace it elsewhere). The bulk-flow-stress integral can be written, by partitioning the complete bounding surface into inflow ($V_n<0$) and outflow ($V_n>0$) segments, as the sum of two integrals:

$$\int dA\{\rho V_n V_x\} = S_{\text{out}}\langle V_x\rangle_{\text{out}} - S_{\text{in}}\langle V_x\rangle_{\text{in}}, \qquad (1.22)$$

where

$$S_{\text{out}} \equiv \int_{V_n>0} dA|\rho V_n|, \quad S_{\text{in}} \equiv \int_{V_n<0} dA|\rho V_n| \qquad (1.23)$$

and $\langle V_x\rangle_{\text{out}}$, $\langle V_x\rangle_{\text{in}}$ are the (suitably weighted) averages of V_x over the corresponding segments. By conservation of mass,

$$S_{\text{in}} - S_{\text{out}} = dM/dt, \qquad (1.24)$$

where M is the magnetotail total mass; since M is small and slowly varying, dM/dt can be neglected and $S_{\text{in}} \simeq S_{\text{out}} \equiv S$ assumed as a reasonable approximation. Then

$$\int_{b+c+d} dA\{\rho V_n V_x\} = S\,\Delta V_x, \qquad (1.25)$$

where $\Delta V_x \equiv \langle V_x\rangle_{\text{out}} - \langle V_x\rangle_{\text{in}}$. The quantity S (>0 by definition) is the total plasma mass flow through the magnetotail and its region of interaction with the solar wind and ΔV_x is the average change of velocity between inflow and outflow ($\Delta V_x>0$ means net slowdown of antisunward or speedup of sunward flow).

Equation 1.21 is now now simplified to

$$\begin{aligned}\left(B_T^2/8\pi\right)(1-\delta)A_T \simeq &+S\Delta V_x \\ &+\int_c dA(\hat{\mathbf{n}}\cdot\hat{\mathbf{x}})\,P - \int_d dA\,P,\end{aligned} \qquad (1.26)$$

which states that what can balance the magnetic tension force (on the left) is (on the right) an appropriate velocity change of a mass flow through the magnetotail, plus the (external) pressure on the flanks as long as the magnetotail is flaring outward ($\hat{\mathbf{n}}\cdot\hat{\mathbf{x}} > 0$); the pressure on the distant outer boundary, to the contrary, acts in the same direction as the tension force at the inner boundary and hence cannot balance it, likewise the pressure on the flanks if $\hat{\mathbf{n}}\cdot\hat{\mathbf{x}} < 0$. (Internal pressure is discussed in Section 1.A.2.)

Solar wind pressure on the flaring flanks as a way to balance the magnetic tension force of the magnetotail was suggested by *Wentworth* [1967]. For any reasonable model of flaring, however, the effect is relatively small in comparison to what is needed; thus it is convenient to take it into account, along with all the other pressure terms, simply by redefining the correction term δ on the RH side [*Carovillano and Siscoe*, 1973]. Equation 1.26 is then further simplified to

$$\left(B_T^2/8\pi\right)\left(1-\delta\right)A_T \simeq S\Delta V_x \qquad (1.27)$$

as the basic global stress balance condition.

1.4.1. Flow Patterns to Maintain Magnetotail

The requirement for maintaining the magnetotail has now been reduced to having a plasma flow *through* the magnetotail (input ≃ output), inflowing and outflowing plasma having different average velocities, with the amount of mass flowing through and the velocity change between inflow and outflow corresponding to a linear momentum change rate that balances the magnetic tension force, in accordance with equation 1.27. Three types of flows which have been discussed in this context are sketched in the three panels of Figure 1.6. Each panel represents the magnetotail box of Figure 1.3, and the solid arrows at the boundaries show inflow and outflow, with flow speed indicated qualitatively by lengths of arrows. Dotted lines inside the box, meant to suggest flow lines within the magnetotail, are for orientation only: The linear momentum change within the entire volume is determined completely by inflow and outflow values on the enclosing boundary surface. The three flow systems are:

1. Inflow from magnetosheath along open magnetic field lines (Dungey cycle)
2. Boundary layer flow on closed field lines (Axford-Hines circulation/magnetospheric convection)
3. Plasma outflow from sources internal to the magnetosphere (planetary wind, Vasyliūnas cycle).

Sections 1.4.1.1, 1.4.1.2, and 1.4.1.3 describe each of the flows in more detail:

a. The underlying physical process and what can be said about its quantitative aspects
b. Transport of linear momentum (where does it come from, and where does it go?)
c. Questions of magnetic field topology and magnetic flux transport.

1.4.1.1. Inflow on Open Field Lines. Magnetotail formation is most easily understood in the case of a magnetically open magnetosphere [*Dungey*, 1961]. Reconnection of the planet's magnetic field with the interplanetary magnetic field at the dayside magnetopause forms flux tubes of open magnetic field lines, which are then pulled

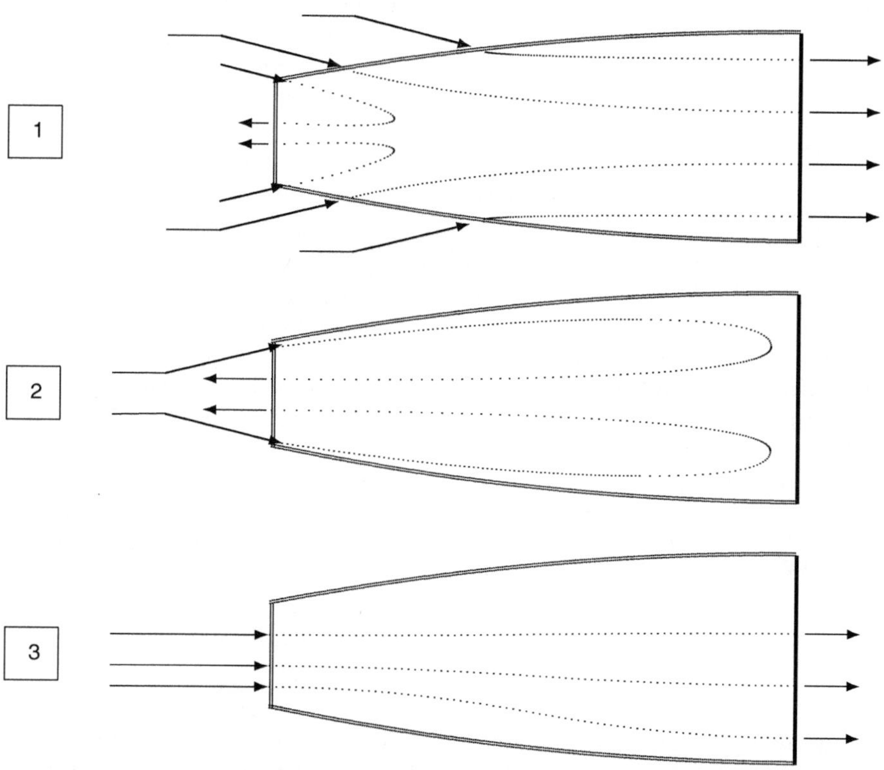

Figure 1.6 Possible flow patterns that provide stress to maintain magnetotail, described in text.

downstream by solar wind flow while remaining attached to the planet, thereby producing the highly stretched field configuration, until at some distance on the nightside the oppositely directed fields from the two hemispheres of the planet reconnect, forming on one side (sunward of the reconnection distance) flux tubes of closed field lines that flow back toward the planet, and on the other side flux tubes of interplanetary field lines that continue flowing with the solar wind. The topology has been sketched here in Figures 1.4 and 1.5.

The associated plasma flow system which accounts for stress balance is shown schematically in panel 1 of Figure 1.6; the view is in the midnight meridian. Magnetosheath (solar wind) plasma flows into the magnetotail along open magnetic field lines, predominantly at the top and bottom boundaries (plasma mantle), and then flows out, most of it downstream with the solar wind but part of it sunward. The boundary between the magnetosheath and the magnetotail in the plasma mantle region can be described to first approximation as rotational discontinuity plus slow-mode expansion fan [*Siscoe and Sánchez*, 1987; *Sánchez et al.*, 1990; *Siscoe et al.*, 1994]. The mass inflow $S = \int dA \rho V_n$ can then be related to the open magnetic flux $\Phi_{open} = \int dA\, B_n$ by noting that the plasma flow component normal to a rotational discontinuity equals the Alfvén speed calculated from the normal component of the magnetic field:

$$S = \int dA\, \rho\, B_n \left(4\pi\rho\right)^{-1/2} \approx \Phi_{open} \left(\rho / 4\pi\right)^{1/2} \qquad (1.28)$$

The velocity decrease across the boundary is given approximately by the Alfvén speed calculated from the interior (magnetotail) field B_T and the exterior density $\rho(\sim \rho_{sw}$ in order of magnitude):

$$\Delta V_x \approx B_T (4\pi\rho)^{-1/2}. \qquad (1.29)$$

Equations 1.28 and 1.29 show that the quantitative contribution from solar wind plasma inflow along open field lines to magnetotail stress balance is

$$S \Delta V_x \approx \Phi_{open}\, B_T / 4\pi, \qquad (1.30)$$

and comparison with equation 1.27 implies that this is adequate to maintain the magnetotail if

$$\Phi_{open} \approx B_T A_T (1-\delta)/2. \qquad (1.31)$$

The required amount of mass inflow, estimated from equations (1.29)–(1.31) and normalized to solar wind mass flux through area A_T, is

$$S / \rho_{sw} V_{sw} A_T \sim \xi\, B_T \left(8\pi\rho_{sw} V_{sw}^2\right)^{-1/2} \qquad (1.32)$$

[*Vasyliūnas*, 1987], where

$$\xi \equiv \left(\rho_{sw}/\rho\right)^{1/2} (1-\delta)/\sqrt{2} \sim O(1). \qquad (1.33)$$

The transport of linear momentum in this process is straightforward. Antisunward linear momentum of magnetosheath flow is carried into the magnetotail, predominantly through its top and bottom boundaries, by the inflowing plasma; reduction of the velocity corresponds to transfer of some antisunward momentum from bulk flow to tension of magnetic field lines, which then carry the antisunward momentum into the dayside magnetosphere and ultimately to the planet.

This process of maintaining the magnetotail is well understood qualitatively; its theory has been developed to the point of providing quantitative relations such as those discussed above and at least at Earth has proved quite useful for interpreting data and consistent with much that is observed. For this reason and also because Dungey's open-magnetosphere model is supported by many other observations (particularly the dependence of geomagnetic activity on the interplanetary magnetic field), the open-magnetosphere concept of the magnetotail is widely considered as the most likely or even as *the* explanation. Nevertheless, it is subject to some important limitations:

a. The process relies on reconnection between magnetic fields on the two sides of the magnetopause, which depends strongly on the (highly variable) orientation of the interplanetary magnetic field relative to the planet's magnetic dipole. It is not entirely clear how the magnetotail is to be maintained during a prolonged period of unfavorable relative orientation.
b. Although the theory of this magnetotail model is arguably more advanced than that of any other so far, it is to my knowledge still not capable of predicting from first principles even approximate values for key parameters of the open magnetosphere, such as open flux Φ_{open} or magnetotail length $\mathscr{L}_{MT}$ (whereas the Chapman-Ferraro theory *can* predict, e.g., the approximate distance to the dayside magnetopause). There is no well-established answer even to the basic question whether the configuration of the open magnetosphere is uniquely determined by solar wind parameters in a steady state [*Vasyliūnas*, 2011].

Once an open flux tube has been formed by reconnection at the dayside magnetopause, how far can it be carried downstream by the solar wind before it reconnects in the magnetotail with its partner from the other hemisphere? In the conventional quasi-steady-state view of

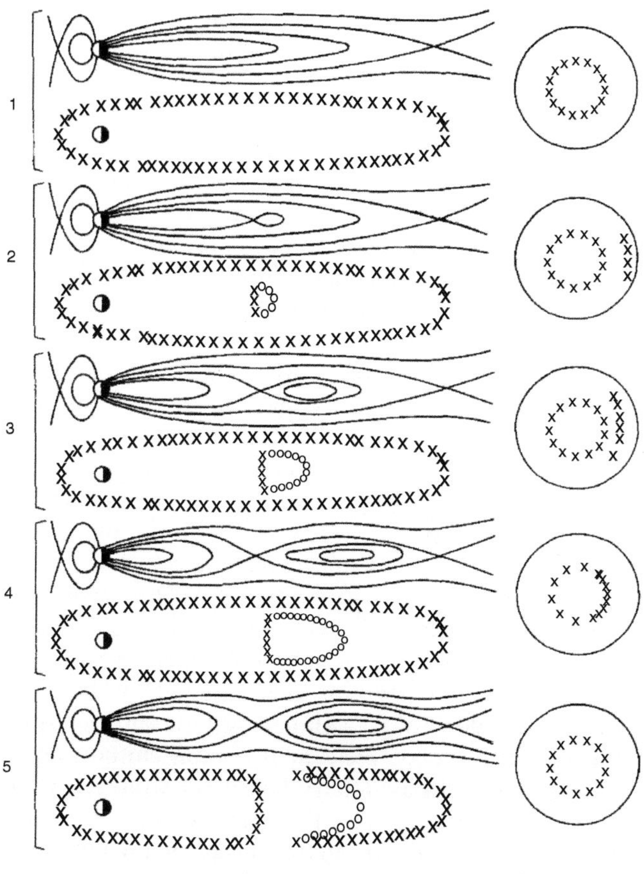

Figure 1.7 Possible changes of the magnetic field topology in the magnetotail of an open magnetosphere [*Vasyliūnas*, 1976]. Each panel has the format of Figure 1.4.

the Dungey cycle (as in Figure 1.4), magnetotail reconnection occurs at an X line, the location of which, although difficult to predict, is assumed to remain more or less fixed. It is, however, conceivable that the solar wind might continue carrying the entire distant magnetotail configuration (X line and all) downstream with itself, stretching the magnetotail indefinitely. Consideration of what might happen in such an extreme limit was what led to the concept of topological changes ("plasmoid" evolution) shown in Figure 1.7 [*Vasyliūnas*, 1976], although the concept was published from the outset as a model for magnetospheric substorms [see historical note in *Vasyliūnas,* 2011]; particularly as reformulated by *Hones* [1976, 1977], it became one of the most widely discussed (and disputed!) models for the magnetospheric substorm at Earth, subsequently applied extensively also to interpret analogous/similar dynamical events in the magnetospheres of other planets [e.g., *Syrjäsuo and Donovan*, 2007; *Hill et al.*, 2008; *Jackman et al.*, 2011].

Magnetotail dynamical events per se are beyond the scope of this chapter, but the plasmoid model does suggest an alternative form of the Dungey cycle, in which (differently from the conventional steady-state picture of magnetic flux and plasma circulating together between dayside and nightside reconnection regions) there is no direct return of the magnetic flux: Once reconnected at the dayside magnetopause, open flux tubes are carried away by the solar wind flow, total magnetic flux of the planet being conserved by creating, at a newly formed near-planet X line (Figure 1.7), the necessary return flux together with an equal amount of oppositely directed flux that flows away with the solar wind. What one effectively then has, on the average, is circulation of magnetic flux that need not be coupled to circulation of plasma; the amount of plasma in the return flow can in principle be as small as desired. Such a model may be relevant to, e.g., magnetospheric so-called sawtooth events at Earth [e.g., *Huang et al.*, 2009, and references therein] as well as to the controversial question of plasma return flow at Jupiter [*McComas and Bagenal*, 2007, 2008; *Cowley et al.*, 2008]. For this chapter, its main interest lies in the application of the concept to models (e.g., section 1.4.1.3) in which plasma return flow may be excluded.

1.4.1.2. Boundary Layer Flow. Almost simultaneously with the proposal of the open magnetosphere by *Dungey* [1961], *Axford and Hines* [1961] proposed a more general concept of a *viscous-like interaction* (also called *tangential drag*) at the magnetopause. In their own words: "By this we mean simply that some of the momentum of the solar wind is transferred across the boundary of the magnetosphere to the ionization within. The nature of this momentum transfer is, for present purposes, of minor importance; its existence, or the existence of an equivalent mechanism, is crucial." The essential new result put forward by *Axford and Hines* [1961] was a pattern of plasma circulation, universally known as magnetospheric convection. Of its possible causes, viscous-like interaction was only one of several mentioned, including the open field line model of *Dungey* [1961]; in subsequent comments on their paper, *Hines* [1974, 1993] emphasized that they had committed themselves neither to the viscous drag mechanism nor to the closed magnetosphere as the only options. Some years later, *Axford* [1969] had come to view magnetic reconnection rather than viscous drag as the dominant mechanism.

One signature of magnetospheric convection driven by viscous-like interaction (defined by *Axford* [1969] as "momentum transfer without field-line reconnection") in a closed magnetosphere is antisunward plasma flow in regions just inside the magnetopause, connected to sunward flow deeper inside the magnetosphere. By contrast, in the geometrically simplest open magnetosphere, plasma flow is sunward everywhere within the closed field line region, antisunward flow driven by tangential stress of reconnected field lines occurring only on open field lines (see Figure 1.4); more complicated geometries which contain antisunward flows in some closed field line regions may be envisaged [*Vasyliūnas*, 1984], but they require placing reconnection at unconventional sites. The observational identification at Earth of the so-called low-latitude boundary layer [*Hones et al.*, 1972, *Eastman et al.*, 1976, 1985; *Haerendel et al.*, 1978, and others] adjacent to and inside the magnetopause (on what are presumed to be, in all probability, closed field lines), with pronounced antisunward flow and with plasma properties intermediate between those of magnetosheath and magnetosphere, has revived interest in viscous-like interaction as an alternative to magnetic reconnection for, or at least a contributor to, driving magnetospheric convection and maintaining the magnetotail [*Rostoker*, 1987]. More recently, it has been argued that viscous-like interaction may be important in the magnetosphere of Jupiter [*Delamere and Bagenal*, 2010, and references therein].

The plasma flow system that could account for stress balance in the magnetotail of a magnetically closed magnetosphere is that sketched in panel 2 of Figure 1.6; the view is now in the equatorial plane. Antisunward-flowing plasma of the magnetopause boundary layers enters the magnetotail volume at the flanks, is eventually turned around, and flows back to the dayside magnetosphere in the middle. The transport of linear momentum in this process follows a path similar to that in the open magnetosphere (section 1.4.1.1). Antisunward linear momentum of magnetosheath flow is carried across the magnetopause, in this case by the tangential drag process and predominantly through the flanks, going into the antisunward flow of the plasma in the boundary layers; deceleration and ultimate reversal of the velocity correspond to transfer of all the antisunward momentum from bulk flow to tension of magnetic field lines, which then carry the antisunward momentum into the dayside magnetosphere and ultimately to the planet (the sunward linear momentum of the return flow implies an even larger magnetic tension but does not change the net amount of linear momentum going into the dayside magnetosphere).

The antisunward flow speed in the boundary layer may reasonably be equated to a significant fraction of magnetosheath flow speed (since the linear momentum is assumed to be supplied directly across the magnetopause, albeit by an unspecified mechanism), and since the flow is ultimately reversed, the velocity change ΔV_x may be assumed to be of a similar order of magnitude. For a given mass flow S, boundary layer flow is thus more efficient in maintaining stress balance than inflow along open field lines, for which ΔV_x given by equation 1.29 is in general considerably smaller than the flow speed in the exterior medium. On the other hand, since the boundary layers are (almost by definition) much smaller than the cross-sectional area of the magnetotail, S here is likely to be considerably smaller than the open-magnetosphere S from equation 1.28. Furthermore, S in the boundary layer eventually flows into the dayside magnetosphere and contributes to its mass source, whereas most of the open-magnetosphere S can flow out into the solar wind again, with only a small fraction contributing to the mass budget of the dayside magnetosphere.

Obtaining theoretical estimates of mass flow S in the boundary layer presupposes having at least a rough semi-quantitative model of the viscous-like interaction. At present, however, there is no agreement even on what physical process constitutes the mechanism of tangential drag. Possibilities suggested at various times include cross-field diffusion by scattering from waves, diffusion by MHD turbulence, localized direct penetration, Kelvin-Helmholtz instability of the magnetopause (which has recently received particular attention [*Delamere and Bagenal*, 2010]), and others. A reliable estimate of $S\Delta V_x$, to determine the contribution of boundary layer flow to magnetotail stress balance, is thus extremely difficult.

Magnetic field topology should not present any difficulties in this process: Field lines remain closed as they are stretched in the boundary layer and contract in the return flow. Constraints on the maximum extent of this magnetotail may be imposed, however, by pressure changes from the expanding and contracting flux tube volumes [*Erickson and Wolf*, 1980; *Kivelson and Spence*, 1988, and others].

1.4.1.3. Outflow from Interior Source. Magnetospheres of the giant planets (Jupiter and Saturn) differ from those of the terrestrial planets (Earth and Mercury) in two significant respects [see, e.g., review by *Bagenal*, 2009, for a comparative survey]:

1. Rotational effects are more important, because of both the faster rotation of the planet and the larger size (in kilometers as well as in planetary radii) of its magnetosphere, posing the problem of what radial force (if any) can sustain the centripetal acceleration of corotating plasma.
2. Mass input to the magnetosphere is primarily from sources deep within the magnetosphere, specifically from moons of the planet (Io at Jupiter, Enceladus at Saturn), posing the problem of where does the added mass go.

Hill et al. [1974] and *Michel and Sturrock* [1974] proposed that, at a rapidly rotating planet, magnetospheric plasma can always be maintained in (at least approximate) corotation on the dayside by the constraining action of solar wind pressure, but on the nightside, once the magnetic field has become too weak to exert a sufficiently large radial tension force, the plasma can flow out down the magnetotail, forming what they called the "planetary wind." Although proposed before the interior (planetary-moon) mass sources in the magnetosphere were discovered, this concept provides an obvious mechanism for the added mass to go away: It can flow out with the planetary wind. (Plasma transport in the inner magnetosphere, from the source region to the distance where the planetary wind begins, is a separate problem, difficult and much studied but beyond the scope of this chapter.)

This plasma flow system can be invoked to account (at least in part) for stress balance of the magnetotail, as sketched in panel 3 of Figure 1.6; the view is in the equatorial plane. Antisunward-flowing plasma of the planetary wind enters the magnetotail volume through the planet-facing boundary and eventually exits, with reduced speed, through the distant boundary (or possibly through the flanks). The transport of linear momentum is now somewhat different from that in sections 1.4.1.1 and 1.4.1.2: The antisunward linear momentum no longer comes from the solar wind flow but from the planet, which acquires an equal amount of sunward linear momentum (through the force that accelerates to corotation the plasma that is then lost to the planetary wind; it is a force and not just a torque because outflow is predominantly on the dusk side, as suggested in the figure). Deceleration of the outflow corresponds to transfer of part of the antisunward momentum from bulk flow to tension of magnetic field lines; these then carry the antisunward momentum into the dayside magnetosphere and ultimately to the planet, where it subtracts from the sunward momentum acquired by the planet in the flow acceleration process mentioned above. The net sunward linear momentum transported into the planet from the magnetotail equals the antisunward momentum remaining in the outflow after it has exited the magnetotail (simple physical analog: think of the planet-plus-magnetotail system as the rocket and the distant outflow as the rocket's exhaust).

The amount of mass supplied to the magnetosphere by a planetary-moon source is determined almost entirely by the properties of the moon and by its interaction with the surrounding medium, taking into account both physical and chemical processes. For Io at Jupiter and Enceladus at Saturn, the input rates are reasonably well determined by a combination of empirical data and modeling [see, e.g., review by *Bagenal and Delamere*, 2011, and references therein]. Under the assumption that most of the plasma ultimately gets out through the magnetotail, the value of S is given thereby. The value of ΔV_x, however, remains unknown: At present, it is difficult enough to determine from observations the mean value of the outflow speed, let alone its change with distance, and theory for ΔV_x is virtually nonexistent. Determining the relative contribution to magnetotail stress balance is thus even more difficult for the interior-source outflow than for the boundary layer flow of section 1.4.1.2.

Planetary wind outflow presents a problem of magnetic topology. At both Jupiter and Saturn, the source of plasma lies deep within the magnetosphere, in the region of essentially dipolar magnetic field lines, and mass ouflow carries net magnetic flux with it; furthermore, the outflow represents a net loss from the system, not a circulation. While the mass loss is negligible in relation to the total mass, the accompanying removal of magnetic flux is not: e.g., at the estimated present plasma source rate from Io, a fraction $\sim O(10^{-3})$ of Io's mass would be removed over the age of the solar system, but the equivalent of the entire magnetic flux from one hemisphere of Jupiter's surface would be removed in roughly a year. Clearly, magnetic flux must return even as plasma does not. The simplest topological configuration that accomplishes this was suggested by *Vasyliūnas* [1983], in a much-cited figure shown here as Figure 1.8. The basic model is essentially that discussed already in connection with the Dungey cycle in section 1.4.1.1 and illustrated in Figure 1.7; the same topological sequence is shown as function of local time in panels 1–4 of Figure 1.8, and as a function of elapsed time in panels 1–4 of Figure 1.7 (this is in fact how Figure 1.8 was actually derived).

1.4.2. Hybrid Models

One may speak of hybrid models in two senses: combining different models, i.e., having the magnetotail maintained by several processes acting simultaneously, or modifying one model by including in it features from another.

Combination of flow patterns 1 and 2 (inflow on open field lines, and boundary layer flow) is possible without much complication and has long been discussed for Earth's magnetosphere. The relative importance of the two in accounting for the magnetotail depends partly on the question, what fraction of the magnetic flux in the lobes of the magnetotail is open flux created by dayside reconnection and what fraction is closed flux transported into the tail by boundary layers?

Combination of flow patterns 1 and 3, i.e., inflow on open field lines (Dungey cycle) and outflow from interior sources (Vasyliūnas cycle), is also possible but somewhat more complicated and subtle [*Cowley et al.*, 2003; *Vasyliūnas*, 2007b].

Combination of flow patterns 2 and 3, i.e., boundary layer flow driven by viscous-like interaction on closed field lines and outflow from interior source, faces the problem of how to reconcile the return flow, which is an essential part of pattern 2, with the outward flow, which is an equally essential part of pattern 3.

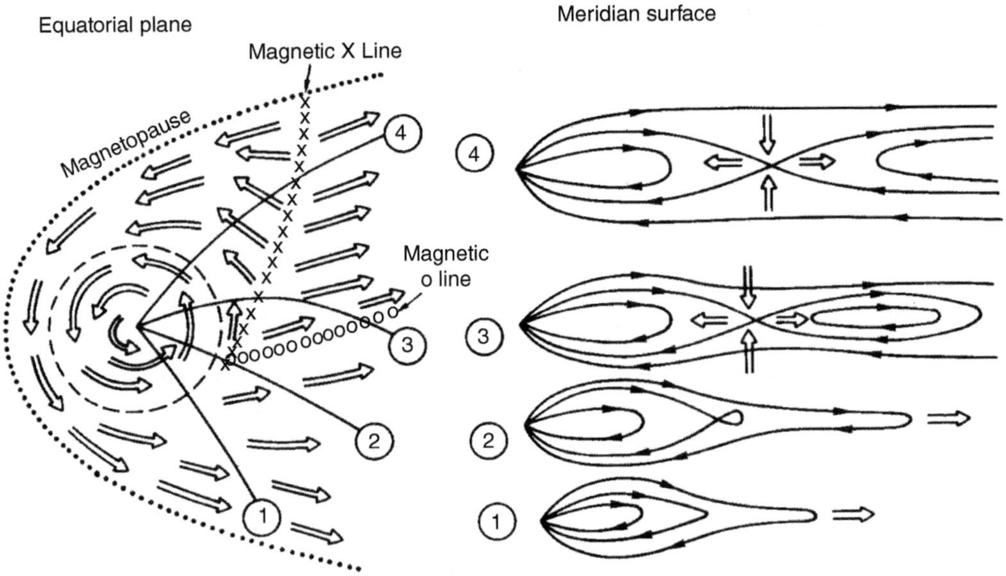

Figure 1.8 Qualitative sketch of planetary wind flow and magnetic topology [*Vasyliūnas*, 1983]. Magnetic field directions are appropriate for Jupiter.

It is possible, at least in principle, to modify the boundary layer model so as to remove the return flow: Assume that boundary layer flow carries too much linear momentum to be turned back by magnetic field tension and continues to flow antisunward indefinitely, breaking open the closed magnetic field lines the same way as outflow from an interior source, by the topological sequence of Figures 1.7 and 1.8. With this modification, flow pattern 2 could be combined with 3. Particularly for Jupiter, an interesting possibility might be: on the dawn side, boundary layer outflow driven by tangential drag from the solar wind; on the dusk side, rotationally driven outflow from the internal (Io) source.

1.5. CONCLUSION

The fundamental problem of the magnetotail is how to maintain the configuration of highly stretched magnetic field lines within low-density lobes, separated by a relatively thin field reversal region and plasma sheet. Such a configuration implies a magnetic tension force at the inner (planet-facing) boundary of the magnetotail, exerted ultimately on the massive planet, which must be balanced by a tension-like force at the outer boundary, and the question is, what is the nature and origin of this force? Equivalently, the force at the inner boundary represents sunward linear momentum flowing into the magnetotail, which (to conserve momentum) must be balanced by inflow of antisunward linear momentum, represented by the force at the outer boundary.

Comparing terms in the integrated stress balance equation shows that the required force can only be provided by the velocity change of an appropriate mass flow-through. There are several qualitative conceptual models for this: (1) solar wind plasma entry on open field lines, slowed down (slightly) by magnetic force, (2) boundary layer flow, produced by (unspecified) tangential drag from the magnetosheath, with field lines pulled out until the flow is stopped and reversed, and (3) outflowing plasma from a source in the interior of the magnetosphere, slowed down by stretched field lines.

Concept 1 is generally considered the dominant explanation of Earth's magnetotail (at least during periods of significant geomagnetic disturbances) and has been widely applied, with or without adequate justification, to interpret observations of other magnetotails. It depends sensitively on orientation of the interplanetary magnetic field relative to the planetary magnetic dipole; the expected dependence is in agreement with many observations at Earth but, for lack of suitable observations, is difficult to check at other planets.

Concept 2 has been proposed for Earth, particularly during prolonged periods with a northward component of the interplanetary magnetic field, and recently has been receiving increasing consideration for Jupiter. As a potential explanation of magnetotails, however, concept 2 has been little studied (although the concept itself has received much attention as an alternative to magnetic reconnection), and concept 3 hardly at all.

Theoretical understanding is more advanced for concept 1 than for the others, but none have yet been developed into a quantitative theory. Quantitative properties of the open magnetosphere and magnetotail (open flux, length of reconnection line, effective length of magnetotail) have been related empirically to solar wind parameters, and quantitative properties of other models (boundary layer thickness, mass transport rates, flow profiles) have been empirically estimated, but for none of these are there as yet any real theoretical predictions (first-principles or otherwise).

APPENDIX: SOME QUESTIONS ABOUT INTERNAL PRESSURE

1.A.1. Plasma Sheet Effects on Global Tension Force

Plasma pressure plays a direct role in the local stress balance of the magnetotail [e.g., *Rich et al.*, 1972; *Siscoe*, 1972]. For the global stress integrals taken over the full cross section of the magnetotail, equation 1.16 or 1.17, the effects of the plasma pressure as well as the reduction of magnetic field within the plasma sheet from its value in the lobes are in general relatively small compared to the total. Hence it is useful to represent the integral as calculated from lobe values only, minus a correction term δ. Some sample estimates for δ, for isotropic pressure:

a. Ratio $\beta = 8\pi P/B^2$ assumed constant over the thickness of the plasma sheet [*Carovillano and Siscoe*, 1973]:

$$\delta = 2\left(A_{ps} / A_T\right)\beta / (1+\beta)$$

b. Magnetic field within plasma sheet assumed to decrease linearly from B_T at the plasma sheet boundary to zero at the field reversal:

$$\delta = 8A_{ps}/3A_T$$

In the above, A_{ps} and A_T are the cross-sectional areas of the plasma sheet and of the entire magnetotail, respectively.

1.A.2. Can Internal Pressure Maintain a Magnetotail?

In this chapter, I have repeatedly emphasized the exclusive role of the external forces in the global stress balance of the magnetotail. It might be asked: Why can't internal pressure maintain a magnetotail, simply by inflating the magnetic field? Don't some papers speak of internal pressure breaking the field open to make a magnetotail?

The problem of making a magnetotail by inflating the field is that the magnetotail exerts a tension force on the planet, and by Newton's third law the planet then exerts an *equal* force on the magnetotail, but the mass of the planet is enormously larger than the mass of the magnetotail, and hence the magnetotail will be accelerated right into the planet! In the absence of other forces, such collapse can be avoided only if the pressure in the magnetotail impacts *directly* on the planet's surface (or if the inflation is axially symmetric, as in a ring current, so that the forces from all directions cancel).

What is really described when one talks about making a magnetotail by "breaking open" the magnetic field as a result of internal pressure is not *maintaining* the magnetotail but *creating* it in time: An initial high pressure, in the process of breaking the field lines, creates a flow, which is what subsequently maintains the magnetotail.

ACKNOWLEDGMENTS

I am grateful to Fran Bagenal, Alex Dessler, and George Siscoe for useful discussions and to one referee for very careful proofreading that significantly improved the text.

REFERENCES

Axford, W. I. (1969), Magnetospheric convection, *Rev. Geophys.*, *7*, 421–459.

Axford, W. I., and C. O. Hines (1961), A unifying theory of high-latitude geophysical phenomena and geomagnetic storms, *Can. J. Phys.*, *39*, 1433–1464.

Axford, W. I., H. E. Petschek, and G. L. Siscoe (1965), Tail of the magnetosphere, *J. Geophys. Res.*, *70*, 1231–1236.

Bagenal, F. (2009), Comparative planetary environments, in *Heliophysics: Plasma Physics of the Local Cosmos*, edited by C. J. Schriver and G. L. Siscoe, pp. 360–398, Cambridge Univ. Press, New York.

Bagenal, F., and P. A. Delamere (2011), Flow of mass and energy in the magnetospheres of Jupiter and Saturn, *J. Geophys. Res.*, *116*, A05209, doi:10.1029/2010JA016294.

Carovillano, R. L., and G. L. Siscoe (1973), Energy and momentum theorems in magnetospheric dynamics, *Rev. Geophys. Space Phys.*, *11*, 289–353.

Cowley, S. W. H. (1991), The structure and length of tail-associated phenomena in the solar wind downstream from the Earth, *Planet. Space Sci.*, *39*, 1039–1043.

Cowley, S. W. H., E. J. Bunce, T. S. Stallard, and S. Miller (2003), Jupiter's polar ionospheric flows: Theoretical interpretation, *Geophys. Res. Lett.*, *30*(5), 1220, doi:10.1029/2002GL016030.

Cowley, S. W. H., S. V. Badman, S. M. Imber, and S. E. Milan (2008), Comment on "Jupiter: A fundamentally different magnetospheric interaction with the solar wind" by D. J. McComas and F. Bagenal, *Geophys. Res. Lett.*, *35*, L10101, doi:10.1029/2007GL032645.

Delamere, P. A., and F. Bagenal (2010), Solar wind interaction with Jupiter's magnetosphere, *J. Geophys. Res.*, *115*, A10201, doi:10.1029/2010JA015347.

Dessler, A. J. (1964), Length of magnetospheric tail, *J. Geophys. Res.*, *69*, 3913–3918.

Dungey, J. W. (1961), Interplanetary magnetic field and the auroral zones, *Phys. Rev. Lett.*, *6*, 47–48.

Dungey, J. W. (1965), The length of the magnetospheric tail, *J. Geophys. Res.*, *70*, 1753.

Eastman, T. E., E. W. Hones, Jr., S. J. Bame, and J. R. Asbridge (1976), The magnetospheric boundary layer: Site of plasma, momentum and energy transfer from the magnetosheath into the magnetosphere, *Geophys. Res. Lett.*, *3*, 685–688.

Eastman, T. E., L. A. Frank, and C. Y. Huang (1985), The boundary layers as the primary transport regions of the Earth's magnetotail, *J. Geophys. Res.*, *90*, 9541–9560.

Erickson, G. M., and R. A. Wolf (1980), Is steady convection possible in the Earth's magnetotail? *Geophys. Res. Lett.*, *7*, 897–900.

Haerendel, G., G. Paschmann, N. Sckopke, H. Rosenbauer, and P. C. Hedgecock (1978), The frontside boundary layer of the magnetosphere and the problem of reconnection, *J. Geophys. Res.*, *83*, 3195–3216.

Hill, T. W., A. J. Dessler, and F. C. Michel (1974), Configuration of the Jovian magnetosphere, *Geophys. Res. Lett.*, *1*, 3–6.

Hill, T. W., et al. (2008), Plasmoids in Saturn's magnetotail, *J. Geophys. Res.*, *113*, A01214, doi:10.1029/2007JA012626.

Hines, C. O. (1974), Preamble to paper 42, in *The Upper Atmosphere in Motion*, pp. 933–935, Geophys. Monogr. 18, AGU, Washington, D.C.

Hines, C. O. (1993), The ionosphere then and now, in *The Axford Symposium*, edited by E. Keppler, pp. 15.1–15.8, Rep. MPAE–W-100-93-29, Max-Planck-Institut für Aeronomie, Katlenburg-Lindau, Germany.

Hones, E. W., Jr. (1976), The magnetotail: Its generation and dissipation, in *Physics of Solar Planetary Environments*, edited by D. J. Williams, pp. 558–571, AGU, Washington, D.C.

Hones, E. W., Jr. (1977), Substorm processes in the magnetotail: Comments on "On hot tenuous plasmas, fireballs, and boundary layers in the Earth's magnetotail" by L. A. Frank, K. L. Ackerson, and R. P. Lepping, *J. Geophys. Res.*, *82*, 5633–5640.

Hones, E. W., Jr., J. R. Asbridge, S. J. Bame, M. D. Montgomery, S. Singer, and S.-I. Akasofu (1972), Measurements of magnetotail plasma flow made with Vela 4B, *J. Geophys. Res.*, *77*, 5503–5522.

Huang, C.-S., A. D. DeJong, and X. Cai (2009), Magnetic flux in the magnetotail and polar cap during sawteeth, isolated substorms, and steady magnetospheric convection events, *J. Geophys. Res.*, *114*, A07202, doi:10.1029/2009JA014232.

Jackman, C. M., J. A. Slavin, and S. W. H. Cowley (2011), Cassini observations of plasmoid structure and dynamics: Implications for the role of magnetic reconnection in magnetospheric circulation at Saturn, *J. Geophys. Res.*, *116*, A10212, doi:10.1029/2011JA016682.

Johnson, F. S. (1960), The gross character of the geomagnetic field in the solar wind, *J. Geophys. Res.*, *65*, 3049–3051.

Kivelson, M. G, and H. E. Spence (1988), On the possibility of quasi-static convection in the quiet magnetotail, *Geophys. Res. Lett.*, *15*, 1541–1544.

McComas, D. J., and F. Bagenal (2007), Jupiter: A fundamentally different magnetospheric interaction with the solar wind, *Geophys. Res. Lett.*, *34*, L20106, doi:10.1029/2007GL031078.

McComas, D. J., and F. Bagenal (2008), Reply to comment by S. W. H. Cowley et al. on "Jupiter: A fundamentally different magnetospheric interaction with the solar wind", *Geophys. Res. Lett.*, *35*, L10103, doi:10.1029/2008GL034351.

Michel, F. C., and P. A. Sturrock (1974), Centrifugal instability of the Jovian magnetosphere and its interaction with the solar wind, *Planet. Space Sci.*, *22*, 1501–1510.

Ness, N. F. (1965), The Earth's magnetic tail, *J. Geophys. Res.*, *70*, 2989–3005.

Parker, E. N. (1958), Interaction of the solar wind with the geomagnetic field, *Phys. Fluids*, *1*, 171–187.

Piddington, J. H. (1960), Geomagnetic storm theory, *J. Geophys. Res.*, *65*, 93–106.

Rich, F. J., V. M. Vasyliūnas, and R. A. Wolf (1972), On the balance of stresses in the plasma sheet, *J. Geophys. Res.*, *77*, 4670–4676.

Rostoker, G. (1987), Contribution to "Dialog on the relative roles of reconnection and the 'viscous' interaction in providing solar-wind energy to the magnetosphere," in *Magnetotail Physics*, edited by A. T. Y. Lui, pp. 409, 411, and 413, Johns Hopkins Univer. Press, Baltimore, Md.

Sánchez, E. R., G. L. Siscoe, J. T. Gosling, E. W. Hones Jr., and R. P. Lepping (1990), Observations of rotational discontinuity–slow expansion fan structure of the magnetotail boundary, *J. Geophys. Res.*, *95*, 61–73.

Siscoe, G. L. (1966), A unified treatment of magnetospheric dynamics with applications to magnetic storms, *Planet. Space Sci.*, *14*, 947–967.

Siscoe, G. L. (1972), On the plasma sheet contribution to the force balance requirements in the geomagnetic tail, *J. Geophys. Res.*, *77*, 6230–6234.

Siscoe, G. L. (1988), The magnetospheric boundary, in *Physics of Space Plasmas (1987)*, edited by T. Chang, G. B Crew, and J. R. Jasperse, pp. 3–78, Scientific Pub., Cambridge, Mass.

Siscoe, G. L. (2006), Global force balance of region 1 current system, *J. Atmos. Solar-Terr. Phys.*, *68*, 2119–2126, doi:10.1016/j.jastp.2006.09.001.

Siscoe, G. L, and E. R. Sánchez (1987), An MHD model for the complete open magnetotail boundary, *J. Geophys. Res.*, *92*, 7405–7412.

Siscoe, G. L., L. A. Frank, K. L. Ackerson, and W. R. Paterson (1994), Properties of the mantle-like magnetotail boundary layer: GEOTAIL data compared with a mantle model, *Geophys. Res. Lett.*, *21*, 2975–2978.

Stern, D. P. (1973), A study of the electric field in an open magnetospheric model, *J. Geophys. Res.*, *78*, 7292–7305.

Syrjäsuo, M., and E. Donovan (2007), *Proceedings of the Eighth International Conference on Substorms (ICS-8)*, Univ. of Calgary, Alberta, Canada.

Vasyliūnas, V. M. (1976), An overview of magnetospheric dynamics, in *Magnetospheric Particles and Fields*, edited by B. M. McCormac, pp. 99–110, D. Reidel Pub. Co., Dordrecht, Holland.

Vasyliūnas, V. M. (1983), Plasma distribution and flow, in *Physics of the Jovian Magnetosphere*, edited by A. J. Dessler, pp. 395–453, Cambridge Univ. Press,New York.

Vasyliūnas, V. M. (1984), Steady state aspects of magnetic field line merging, in *Magnetic Reconnection in Space and Laboratory Plasmas*, edited by E. W. Hones, Jr., pp. 25–31, Geophys. Monogr. 30, AGU, Washington, D.C.

Vasyliūnas, V. M. (1987), Contribution to "Dialog on the relative roles of reconnection and the 'viscous' interaction in providing solar-wind energy to the magnetosphere," in *Magnetotail Physics*, edited by A. T. Y. Lui, pp. 411–412, Johns Hopkins Univ. Press, Baltimore, Md.

Vasyliūnas, V. M. (2007a), The mechanical advantage of the magnetosphere: Solar-wind-related forces in the magnetosphere-ionosphere-Earth system, *Ann. Geophys.*, *25*, 255–269.

Vasyliūnas, V. M. (2007b), Topology of plasma flow: Intertwining of the Dungey and the Vasyliūnas cycles, *Eos Trans. AGU*, *88*(52), Fall Meet. Suppl., Abst. SM44A-10.

Vasyliūnas, V. M. (2011), Physics of magnetospheric variability, *Space Sci. Rev.*, *158*, 91–118, doi:10.1007/s11214-010-9696-1.

Wentworth, R. C. (1967), The geometry of the magnetosphere, *J. Geophys. Res.*, *72*, 4582–4586.

Section II: Tutorials

2

Mercury's Magnetotail

T. Sundberg[1,2] and J. A. Slavin[3]

2.1. INTRODUCTION

Observations of Mercury's magnetosphere by the Mariner 10 and the MErcury Surface, Space ENvironment, GEochemistry, and Ranging (MESSENGER) spacecraft have revealed a small but yet complex planetary environment, in many ways different from any other in our solar system. Mercury's radius is only about a third that of Earth (1 R_M = 2440 km), but it is an extremely dense object; the core of the planet has a radius of ~2000 km and likely contains more than 60% iron. The planet's silicate regolith surface also provides a continuous source of heavy ions to Mercury's space environment, in particular sodium ions, and significant amounts of neutral sodium, calcium, and magnesium are present in the extended exosphere [e.g., *McClintock et al.*, 2008, 2009; *Vervack et al.*, 2010]. Mercury supports a magnetic field that is strong enough to stand off the solar wind on the dayside of the planet, with the magnetopause (the boundary between the magnetosphere and the solar wind) on average 0.45 R_M above the subsolar surface [*Winslow et al.*, 2013]. It follows an eccentric orbit around the Sun with a heliocentric distance that varies between 0.31 AU at perihelion and 0.47 AU at aphelion, which leads to large annual variations in the average solar wind properties. The average interplanetary magnetic field (IMF) and solar wind density vary with a factor of ~2 over the Mercury year, as shown in Table 2.1.

The MESSENGER spacecraft was launched in 2004 and subsequently performed three flybys of the planet (hereafter labeled M1, M2, and M3) before the orbit insertion in March 2011 [*Solomon et al.*, 2001, 2008]. During its first year at Mercury, MESSENGER followed a highly elliptical orbit with a 12-h period, shown in Figure 2.1. The initial periapsis was ~200 km, the inclination ~82°, and the apoapsis ~15,200 km. In the figure, the MSO subscript refers to Mercury solar orbital coordinates, where the *X* axis is directed from the center of the planet toward the Sun, the *Z* axis is normal to Mercury's orbital plane, positive toward the north celestial pole, and the *Y* axis is positive in the direction contrary to orbital motion. The orientation of the orbit is fixed in a heliocentric reference frame, which means that the spacecraft samples all different local times at two different altitudes during a Mercury year, equivalent to 88 Earth days. After the primary mission, the apoapsis was lowered and the orbital period was shortened to 8 h. For details on the flyby trajectories, see, for example, *Slavin et al.* [2012a]. The instrument suite onboard MESSENGER was designed to answer numerous questions about Mercury's composition, history and environment, and three of these instruments are of particular interest here: the magnetometer (MAG) [*Anderson et al.*, 2007], the energetic particle spectrometer (EPS), and the fast imaging plasma spectrometer (FIPS) [*Andrews et al.*, 2007]. The EPS measures high-energy electrons and ions with energies from ~35 keV to ~1 MeV, and the FIPS measures thermal and low-energy ions with energy per charge ratios between <50 eV/*q* and 13 keV/*q*. The typical sampling periods for the MAG, EPS, and FIPS are 0.05, 3, and 8 s, respectively.

This review covers results collected from the Mariner 10 and MESSENGER flybys and MESSENGER's first year at Mercury (the high-apoapsis orbit), and it provides a follow-up on previous reviews by *Slavin* [2004], *Baumjohann*

[1]Center for Space Physics, Boston University, Boston, Massachusetts

[2]School of Physics and Astronomy, Queen Mary College, University of London, London, United Kingdom

[3]Department of Atmospheric, Oceanic and Space Sciences, University of Michigan, Ann Arbor, Michigan

Magnetotails in the Solar System, Geophysical Monograph 207. First Edition.
Edited by Andreas Keiling, Caitríona M. Jackman, and Peter A. Delamere.

Table 2.1 Interplanetary conditions at Mercury orbit.

Phase	R (AU)	B (nT)	Spiral Angle	N_p (cm^{-3})	P (nPa)	Beta	M_A	M_s
Perihelion	0.31	46	17	73	26.5	0.5	3.9	5.5
Aphelion	0.47	21	25	32	11	0.9	5.7	6.1

Note: The table gives the heliocentric distance of Mercury (R), the average magnetic field (B), the IMF spiral angle, and the solar wind proton density (N_p), pressure (P), beta, and Alfvénic and sonic Mach numbers. From *Slavin and Holzer* [1981] and *Fujimoto et al.* [2007].

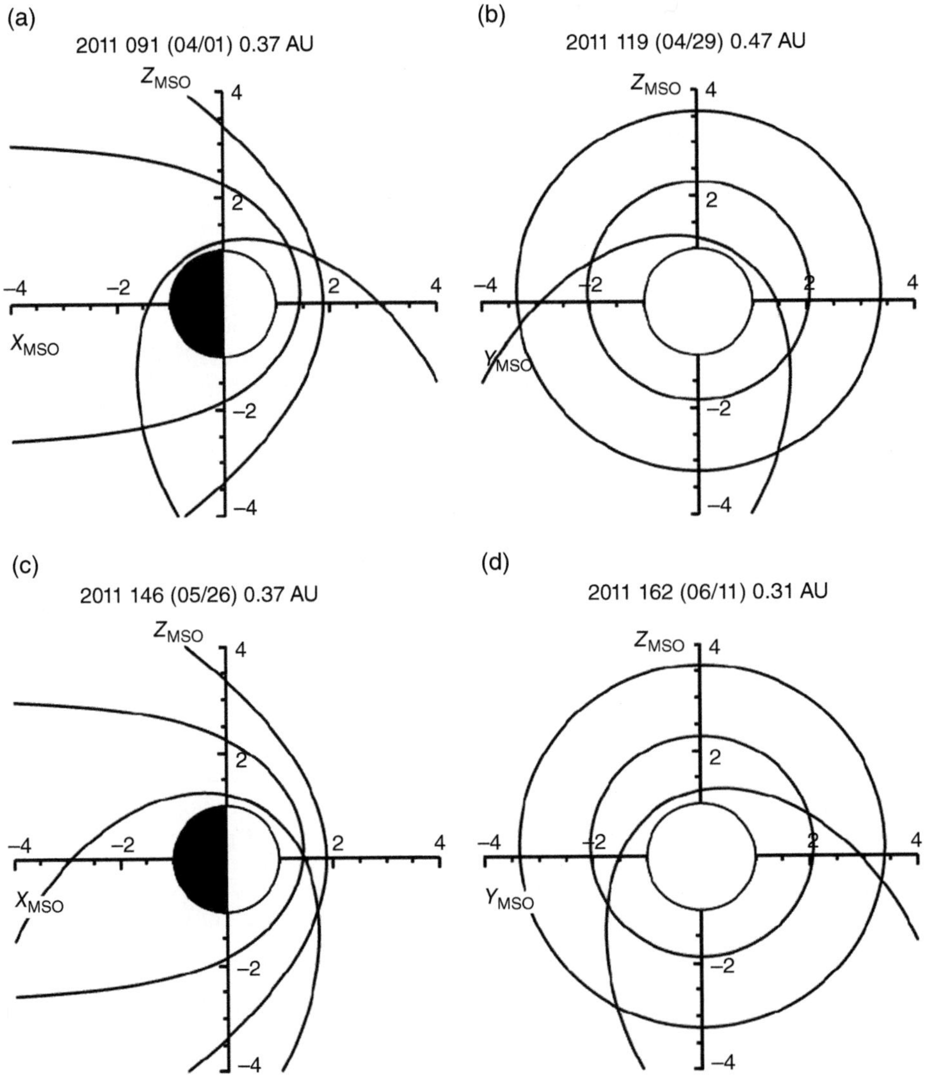

Figure 2.1 Overview of MESSENGER's orbit during the primary mission: (a) short tail orbit, (b) dusk side orbit, (c) long tail orbit, and (d) dawn side orbit. Lines mark the approximate location of the magnetopause and the bow shock for each viewing geometry. From *Boardsen et al.* [2012].

et al. [2006], and *Fujimoto et al.* [2007]. The MESSENGER mission continues to provide new and more comprehensive results, but the information gathered here should cover the most fundamental aspects of the magnetosphere in general and the magnetotail in particular.

2.2. PLANETARY MAGNETIC FIELD

The structure of Mercury's intrinsic magnetic field and its driving dynamo has been a topic of great debate in the past, and it is only in the last few years that we have been able to characterize its magnetic field configuration in detail [e.g., *Anderson et al.*, 2008, 2010, 2011a, 2012; *Alexeev et al.*, 2008, 2010]. The planetary field is best described by a 190–215 nT R_M^{-3} dipole moment aligned with the planetary axis of rotation and with the same polarity as that of Earth but shifted 479 ± 6 km northward of the planetary center. The field is predominantly dipolar, and higher order moments have only a low influence on the overall magnetic field characteristics [*Anderson et al.*, 2012]. The dipole offset is most evident when viewed in a cylindrical coordinate system along the Z_{MSO} axis, as shown in Figure 2.2. Here, the trace of the zero-inclination point of the magnetic field, i.e., the point where the radial component, $Z_{\rho 0,}$ goes to zero, is a direct measure of the dipole offset, assuming an axially aligned dipole. The results of this analysis are shown in Figure 2.3. The data show a slight systematic variation in $Z_{\rho 0}$ as a function of longitude (which is given in a body-fixed coordinate system) and/or local time, which indicates a small residual field not captured by the dipole representation. This deviation is generally best arranged in solar-orbital coordinates, which indicates that the primary source of the residual field is most likely of magnetospheric origin rather than from asymmetries in the magnetic dynamo, such as, for example, variation in the location and shape of the magnetopause [*Anderson et al.*, 2012; *Johnson et al.*, 2012].

The offset dipole also leads to different magnetic environments in the northern and southern hemispheres. The surface field at the northern pole is about 3.4 times larger and the size of the polar cap differs with a factor of ~4, which imposes significant hemispheric differences in the solar wind precipitation to the surface [*Anderson et al.*, 2012].

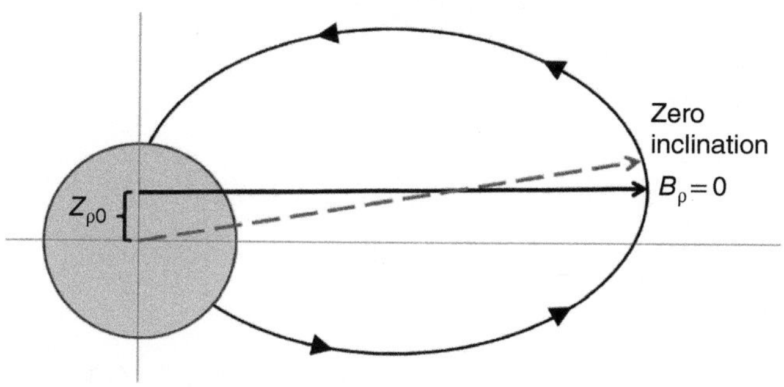

Figure 2.2 Mercury's offset dipole. The schematic shows how the zero-inclination offset can be used to determine the dipole position, given that the magnetic field and rotational axes are aligned. From *Anderson et al.* [2011a]. Reprinted with permission from AAAS.

2.3. MAGNETOSPHERE

Overall, Mercury's space environment bears many similarities with that of Earth, albeit on much smaller scales. A schematic overview of the system is presented in Figure 2.4. Mercury's intrinsic magnetic field forms a magnetic cavity around the planet, the magnetosphere, which protects it from the solar wind. The solar wind dynamic pressure compresses the magnetosphere on the dayside, and it is stretched out into a long tail on the nightside, likely around 125 R_M on average [*Milan and Slavin*, 2011]. The interaction with the solar wind also drives large-scale plasma convection cycles within the magnetosphere, primarily through magnetic reconnection at the dayside magnetopause, a process through which the planetary magnetic field interconnects with the IMF. This leads to a magnetic field in the cusps and the polar caps that are open to solar wind precipitation, with the cusps being main points of entry for solar wind plasma. Other regions familiar from Earth include the low-latitude boundary layer (LLBL), which is a region of relatively cold and dense plasma just inside the magnetopause, and the plasma sheet in the equatorial magnetotail. The plasma sheet extends from Mercury's surface to the nightside reconnection line marked in the central magnetotail in Figure 2.4, often referred to as the near-Mercury neutral line (NMNL), and it is contained between the two tail lobes.

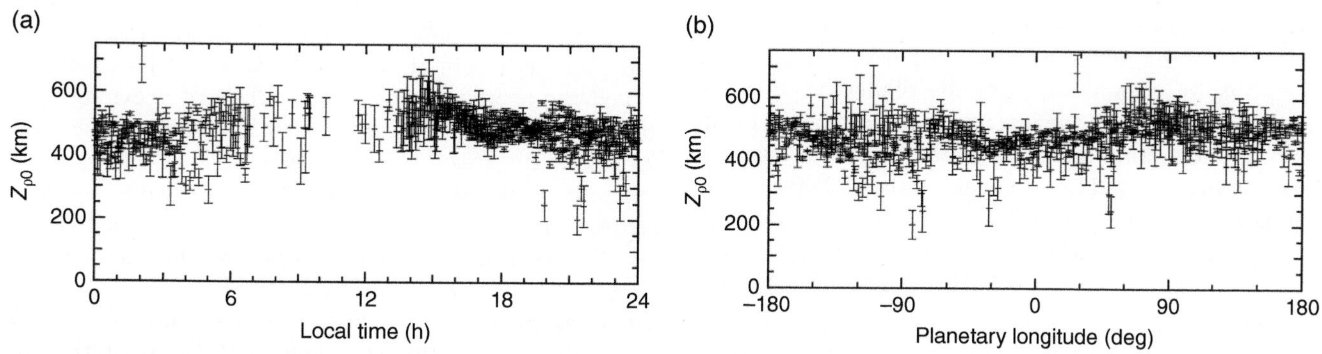

Figure 2.3 Trace of the dipole offset vs local time and planetary longitude. From *Anderson et al.* [2012].

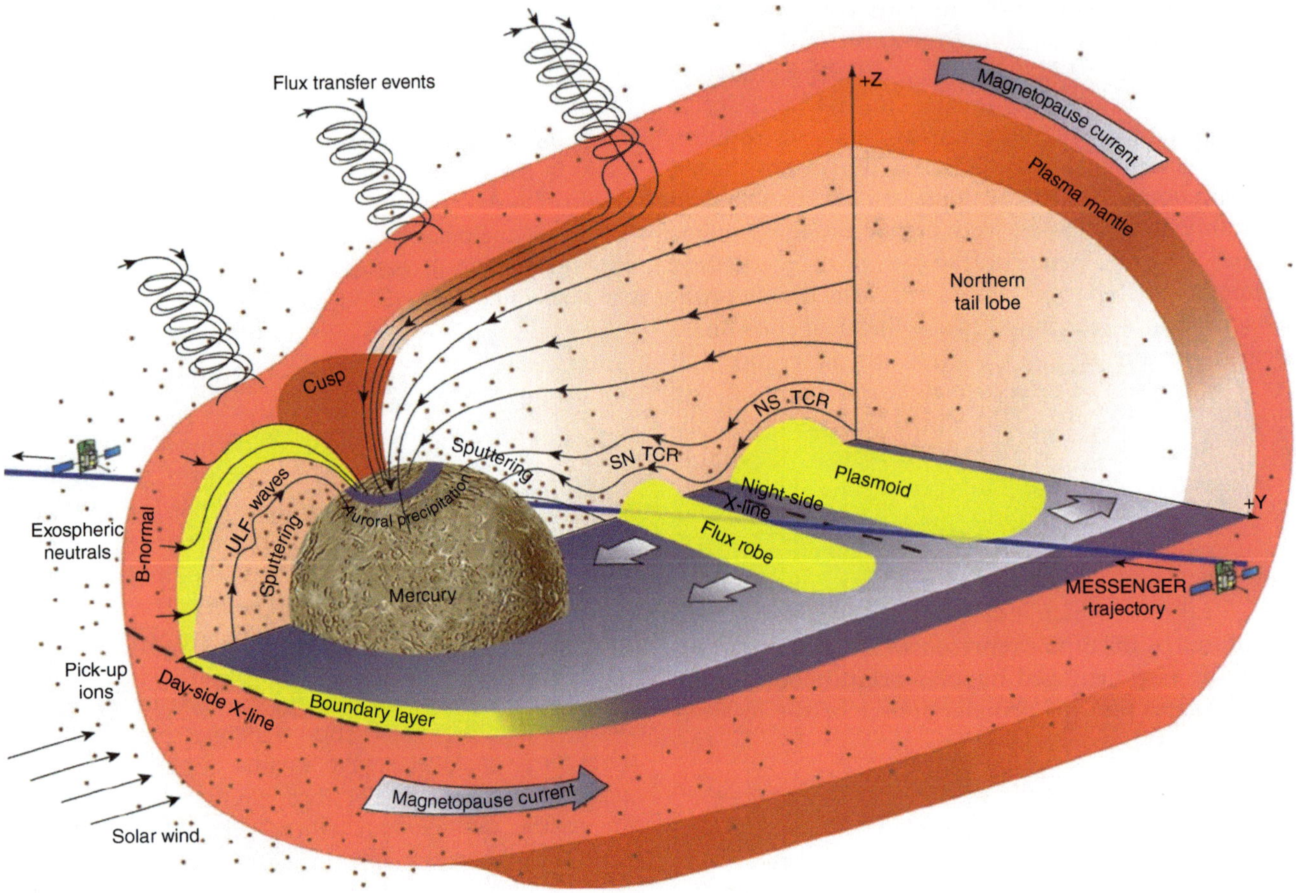

Figure 2.4 Schematic overview of Mercury's magnetosphere, including features observed during the M1 and M2 flybys. From *Slavin et al.* [2009b]. Reprinted with permission from AAAS.

We will primarily use the Mercury solar magnetospheric (MSM) coordinate system in this review. The MSM system has its origin at the center of the planetary dipole moment but is otherwise equivalent to the MSO coordinates described earlier.

2.3.1. Magnetopause and Bow Shock

Mercury's magnetosphere is on the order of ~7–8 times smaller than the terrestrial, but as the planet itself scales only by a factor of 2.6, the planet occupies an important fraction of the internal magnetosphere. As at Earth, the location and shape of the magnetopause are primarily determined by a pressure balance between the magnetic pressure in the magnetosphere and the magnetic and the dynamic pressure in the solar wind. The resulting geometry can be described by the Shue model,

$$\sqrt{\left(X'^2_{\mathrm{MSM}} + R'^2_{\mathrm{MSM}}\right)} = r_0\left[2/(1+\cos\theta)\right]^{\alpha}, \quad (2.1a)$$

where r_0 is the stand-off distance of the subsolar magnetopause, α is the flaring parameter, and $R'_{\mathrm{MSM}} = \sqrt{\left(Y'_{\mathrm{MSM}^2} + Z'_{\mathrm{MSM}^2}\right)}$ is the cylindrical component [*Shue et al.*, 1997; *Slavin et al.*, 2009a; *Winslow et al.*, 2013]. The primed coordinates refer to an abberated coordinate system which is rotated so that X'_{MSM} is in the upstream direction of the solar wind. Due to Mercury's relatively high orbital velocity this normally leads to an aberration of 5–6°, depending on the solar wind velocity and Mercury's orbital phase. *Winslow et al.* [2013] determined the average best-fit parameters of the magnetopause to $r_0 = 1.45\ R_M$, and $\alpha = 0.5$; however, the true stand-off distance of the magnetopause varies as a function of the solar wind ram pressure with instantaneous values between 1.55–1.35 R_M for pressures in the 8.8–21.6 nPa range. On the nightside, the magnetotail approaches its cylindrical limit much faster than what would be expected from the terrestrial analogy; The final cross-tail radius is on average around 2.7 R_M, although it also varies extensively depending on internal and external conditions. This will be discussed in more detail in section 2.5.

The average shape of the bow shock is approximately that of a conic of revolution located around half a Mercury radii upstream of the subsolar magnetopause.

$$\sqrt{\left((X'_{\mathrm{MSM}} - X_0)^2 + R'^2_{\mathrm{MSM}}\right)} = p\varepsilon / (1 + \varepsilon \cos\theta), \quad (2.1b)$$

where X_0 is the focal point on the X axis, p is the focal parameter, and ε the eccentricity [*Slavin et al.*, 2009a]. *Winslow et al.* [2013] also determined the best-fit parameters of the bow shock, which were 0.5 R_M, 2.9 R_M, and 1.02, for X_0, p, and ε, respectively. The subsolar distance of the bow shock tends to vary with the upstream Alfvén Mach number, with higher Mach numbers leading to a more compressed shock geometry. As the Mach numbers are generally lower at Mercury orbit than at Earth and the outer planets, Mercury's bow shock is also generally less pronounced and provides less energy dissipation, which can, for example, be seen by the general lack of shock ramps and overshoots [*Masters et al.*, 2013]. Nevertheless, bow shock processes still dictate much of the solar wind–magnetosphere coupling and it can also at times be a direct source of wave activity in the magnetosphere [*Sundberg et al.*, 2013] and a wealth of plasma waves in the upstream region [*Le et al.*, 2013].

2.3.2. Composition

Mercury's surface provides a continuous source of neutral sodium, potassium, calcium, and other constituents for the exosphere through several different source processes, for example, photodesorption by solar UV flux at the dayside surface, ion sputtering, thermal desorption, and micrometeoroid impact [e.g., *Potter and Morgan*, 1985; *Hunten and Sprague*, 2002; *Killen et al.*, 2001, 2007]. Additional magnetospheric processes may also directly or indirectly lead to increased emission rates, either through surface sputtering or by helping fresh sodium diffuse to the surface [e.g., *Miura et al.*, 2009; *Burger et al.*, 2010]. The sodium exosphere can be observed as a long bright tail that extends up to 100–1000 R_M from the planet in the antisunward direction as the atoms are accelerated away from the Sun by the radiation pressure [e.g., *Potter et al.*, 2002; *Baumgardner et al.*, 2008]. Electron stimulated desorption and sputtering by solar wind electrons and ions also constitutes a source of heavy planetary ions for the magnetosphere [e.g., *Elphic et al.*, 1991; *McLain et al.*, 2011; *Dukes et al.*, 2011], and together with charge exchange and photoionization of the exospheric neutrals, this gives an abundance of heavy planetary ions within the magnetosphere. Apart from protons and solar wind originating He^{2+}, Na^+ is by far the most common of the heavier ions, in particular at the cusp and within the central plasma sheet [*Zurbuchen et al.*, 2011; *Raines et al.*, 2013]. The average observed densities of He^{2+} and Na^+ by FIPS are 3.9×10^{-2} and $5.1 \times 10^{-3}\,\mathrm{cm}^{-3}$, respectively, averaged over all altitudes and local times [*Raines et al.*, 2013]. Abundance ratios of less common ions relative to Na^+, derived from the M1 observations, are given by *Zurbuchen et al.* [2008].

The sources, transport, and evolution of the heavy planetary ions are not understood. For example, the enhanced count rates of Na^+ ions in the cusps may suggest sputtering by solar wind ions as the primary source process, funneled into the cusp along open field lines, and theoretical models have long predicted the cusps to be important for solar wind impact on the surface [*Massetti et al.*, 2003; *Sarantos et al.*, 2007]. Planetary ions are sputtered from the open surface at the cusps with low initial energies, on the order of 0.1–1 eV [e.g., *Wurz et al.*, 2010; *Delcourt*, 2013], but they can subsequently be accelerated by centrifugal forces to energies above 100 eV, which is well beyond that required to overcome Mercury's gravity [e.g., *Delcourt et al.*, 2002, 2012]. However, these energies are still lower than the characteristic energies of 2–4 keV reported by *Raines et al.* [2013], which brings doubt to whether the ions observed are indeed local to the cusp or, alternatively, whether they may have been picked up in the magnetosheath or the outer magnetosphere and injected into the cusp along open field lines. In contrast to the Na^+ observations, the He^{2+} ions measured by MESSENGER are relatively uniform and, therefore, more likely to originate from a more uniform sputtering of a helium-enriched surface, and it indicates a global rather than regional source process [*Zurbuchen et al.*, 2011].

Sodium ions are observed not only with greatly increased count rates in the cusps and the plasma sheet but commonly also within the dayside magnetosphere and in the LLBL [*Raines et al.*, 2013]. This mass loading of the magnetosphere may have important effects on the system both by increasing the plasma density and by introducing finite Larmor radius effects, as the ion gyro motion about the magnetic field is generally nonnegligible compared to the scale size of the system. This is in particular true for localized regions with large gradients such as the plasma sheet, the boundary layers, and the cusp, and ion-kinetic turbulence may potentially be important within these regions [*Uritsky et al.*, 2011]. Estimates of typical ion Larmor radii and gyro periods are given in Figure 2.5.

2.3.3. Magnetic Cusps

The magnetic cusps, which separate the dayside magnetosphere from the tail lobes, are important entry points for solar wind particles, and they are two of the most

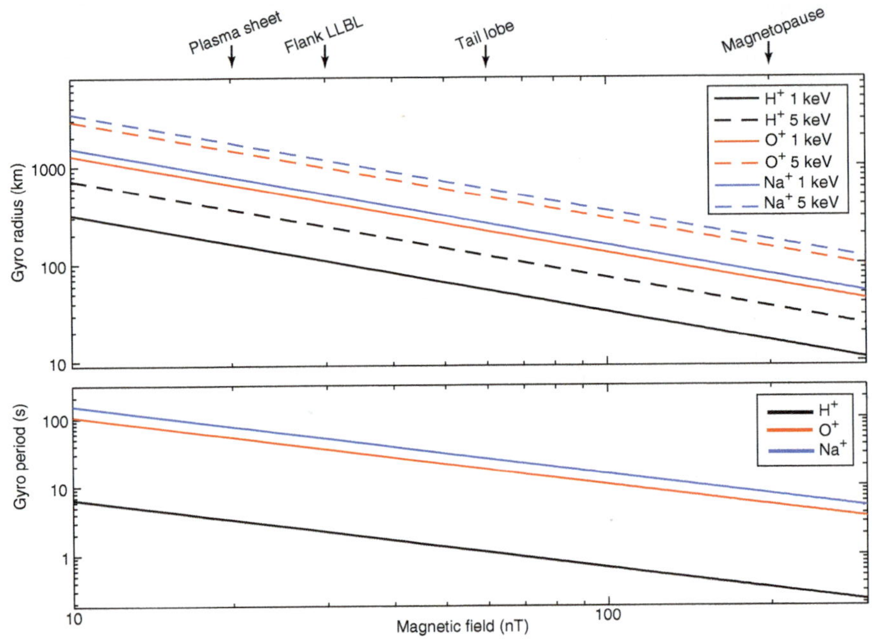

Figure 2.5 Ion gyro-radii and gyro periods within Mercury's magnetosphere. The black, red and blue lines mark the expected properties of H^+, O^+, and Na^+, respectively, with the solid lines representing 1 keV ions and the dashed lines 5 keV. Approximate magnetic fields in different significant areas of the magnetosphere are indicated at the top.

dynamic regions of the magnetosphere. These are characterized by an increased particle flux and decreased magnetic pressures, they are frequently very dynamic in their appearance, and they have an average extent of roughly 11° in latitude and 4.5 h in local time, although with large day-to-day variations [*Winslow et al.*, 2013]. The average particle flux in the northern cusp was estimated by *Winslow et al.* [2013] to ~1.1×10^{24} protons/s, and the particle pressure to ~1.8–1.9 and ~1.5–2.1 nPa for low-altitude (~400 km) and high-altitude (~700) orbits, respectively, with positive IMF B_X components yielding ~40% higher pressures on average (where B denotes the magnetic field). This B_X dependence in the particle influx is in agreement with the predictions by *Sarantos et al.* [2001]: When the B_Z component is negative and there is reconnection at the dayside magnetopause, the B_X component controls which of the two hemispheres is connected to the upstream (sunward) leg of the IMF. Given the solar wind flow direction, the hemisphere connected to the upstream region will be more susceptible to solar wind particles traveling antisunward along the field line, which will increase the particle impact rate. No measurements are available from the southern cusp due to the spacecraft orbit; however, as the expected magnetic field strength at the surface differs with almost a factor of 4 between the southern and the northern cusps [*Anderson et al.*, 2012], the particle flux impacting the southern hemisphere should be larger than that measured at the northern cusp. The high particle pressures indicated by the magnetometer also correspond to increased count rates of planetary ions, in particular Na^+, which even becomes dominant over He^{2+} within this region by a factor of ~2 [*Zurbuchen et al.*, 2011; *Raines et al.*, 2013].

2.3.4. Low-Latitude Boundary Layer

MESSENGER data collected at both the first and second Mercury flybys revealed a thick boundary layer inside the prenoon magnetopause with enhanced proton populations and depressed magnetic fields [*Slavin et al.*, 2008, 2009a; *Anderson et al.*, 2011b]. (At the third flyby, MESSENGER went into the safe mode during the second half of the flyby, and no data were collected in the dayside magnetosphere.) Both observations showed that the LLBL was contained within an inner and an outer current sheet, resembling a "double magnetopause" [*Slavin et al.*, 2008]. These dual current sheets are indeed unique to Mercury, as no other magnetosphere within our solar system shows such an abrupt inner boundary with a second current sheet inside the magnetopause proper. Due to their equatorial trajectories, the MESSENGER flybys have given the best view of this feature to date, and statistical studies are yet to be presented from orbit. An overview of the M1 LLBL measurements is given in Figure 2.6. The total widths of the layer were on the order of 1000 and 1400 km for M1 and M2, respectively,

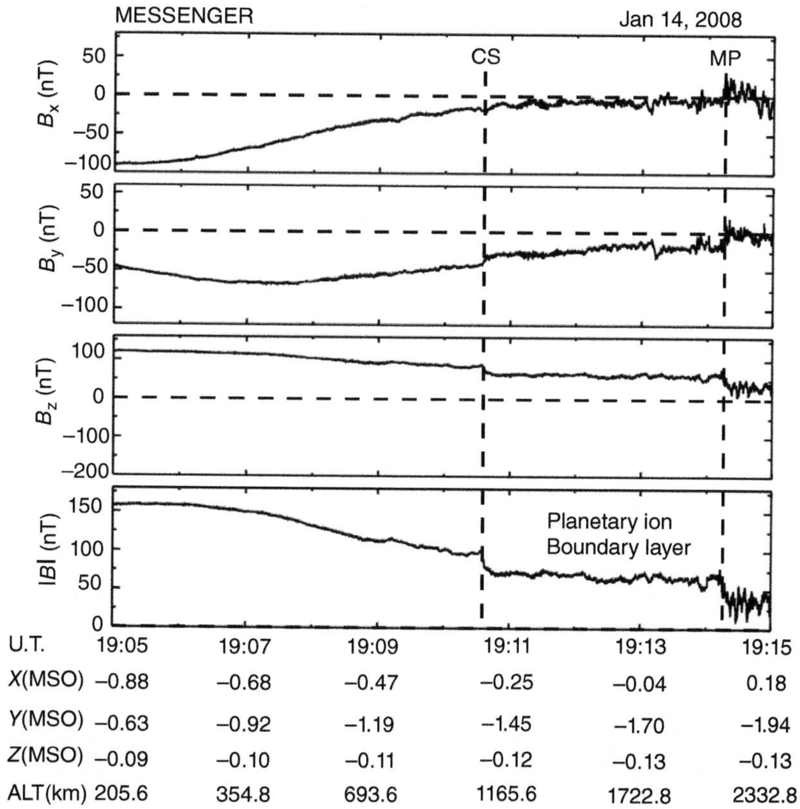

Figure 2.6 MESSENGER observations from the dayside low-latitude boundary layer observed during the M1 flyby. The two dashed lines mark the inner current sheet (CS) and the magnetopause proper (MP). Adapted from *Slavin et al* [2008]. Reprinted with permission from AAAS.

and the magnetic pressure decreased by ~1.6 nPa on both occasions, from 4.1 to 2.5 nPa for M1 and 4.9 to 3.3 for M2 [*Slavin et al.*, 2008; *Anderson et al.*, 2011b]. The magnetic field also showed an increase in kinetic fluctuations in this region, indicative of a more thermalized plasma and a higher plasma beta, the ratio between the thermal and magnetic field pressures [*Uritsky et al.*, 2011]. Both observations also showed a steplike increase in the proton count rate simultaneous with the decrease in the magnetic field [*Anderson et al.*, 2011b; *Raines et al.*, 2011]; however, the thermal pressure exerted by the protons is insufficient to explain the diamagnetic decrease in full, falling short by up to 1.2 nPa for M1 and 0.6 nPa for M2, assuming pressure equilibrium at the inner boundary. (These values were derived from the FIPS measurements by *Raines et al.* [2011] using a forward modeling technique, assuming a thermalized Maxwell-Boltzmann ion distribution.) *Slavin et al.* [2008] argued this to be an effect of heavy planetary pick-up ions accelerated to high velocities by the fast flow velocities in the magnetosheath that possibly would have gone undetected by FIPS [*Raines et al.*, 2011]. In this scenario, the thickness of the layer is strongly influenced by the penetration depth of the Na^+ ions, approximately equivalent to an ion gyro radius. For a typical magnetosheath flow velocity of 300 km/s, this would equal ~1000 km, which matches well with that observed. Hybrid simulations by *Wang et al.* [2010] and *Müller et al.* [2012] also show the existence of a thick magnetopause that may be consistent with the MESSENGER observations. In particular, *Müller et al.* [2012] interpret the second current sheet as a consequence of a density gradient within the LLBL, toward higher altitudes, caused by solar wind protons injected into the magnetosphere along the open field lines in the polar cap. The resulting pressure gradient feeds a current, **J**, which in turn exerts a $\mathbf{J} \times \mathbf{B}$ force at the inner edge of the boundary layer that will enhance the initial imbalance in the density distribution and thus narrow and increase the current in the current sheet, thus forming the second current sheet at the inner edge of the LLBL.

Despite the very different IMF conditions for the two flybys, the boundary layers observed were remarkably similar: The magnetosphere was relatively quiet during M1, with northward IMF and no clear signs of magnetic reconnection at the magnetopause, whereas the IMF was southward and the dayside magnetosphere was extremely active during M2 [*Slavin et al.*, 2009b]. *Anderson et al.* [2011b] thus concluded that the source of the layer must be independent of the reconnection rate and that the repopulation of this layer must also be fast in order to sustain a sufficient population, given the short Dungey time scales and the fast antisunward flows expected for southward IMF.

The LLBL not only covers the dayside magnetosphere, but clear boundary layers have also been observed at the dusk side flank. A thick boundary layer, around 2000 km (0.8 R_M) wide, was present at the inbound leg of M1 and somewhat thinner layer was also detected during M3; however, it was completely absent for M2 [e.g., *Slavin et al.*, 2012a]. These results are in concert with terrestrial measurements, which generally show a thicker boundary layer when the IMF is northward [e.g., *Mitchell et al.*, 1987], and as they differ from the dayside observations, this indicates that a few different mechanisms may be involved in the mass transfer into the LLBL and/or that the mass transfer rate is asymmetric. Heavy pick-up ions may be one source of asymmetry: As the pick-up ions become energized by the motional electric field in the magnetosheath, they are accelerated in opposite directions for northward and southward IMF [*Slavin et al.*, 2008; *Sarantos et al.*, 2009], leading to a preferential mass loading of the dawn or the dusk side, respectively. Alternatively, the layer might also form due to proton drift from the cusp, where protons injected into the open field lines at the cusp would drift toward the dawn side, forming a thicker boundary on the morning side than on the evening side, independent of the IMF direction [*Slavin et al.*, 2008; *Anderson et al.*, 2011b]. Kelvin-Helmholtz waves at the magnetopause could also be an important source process for the LLBL, especially on the dusk side flank [*Sundberg et al.*, 2012a]. This will be discussed in more detail in section 2.4.2.

2.3.5. Plasma Sheet

The hot and dense plasma found in Mercury's plasma sheet clearly separates it from the surrounding tail lobes [*Raines et al.*, 2011, 2013]. As the plasma sheet is in pressure balance with the tail lobes, this change in plasma composition can also easily be identified in the magnetic field by a depressed field magnitude compared with that in the lobes and by an increase in kinetic scale fluctuations [*Korth et al.*, 2011; *Uritsky et al.*, 2011]. By using the magnitude of these depressions as a proxy of the plasma pressure, *Korth et al.* [2012] traced out the overall characteristics of Mercury's plasma sheet using four Mercury years of magnetic field measurements. Their results show a plasma that extends on the order of 2 R_M into the nightside magnetotail from the planetary surface, with an average thermal pressure of ~1.45 nPa, centered at the magnetic equator. Although the measurements show a significant dawn-dusk asymmetry in the raw data [*Korth et al.*, 2011], much of this asymmetry can in fact be attributed to orbital effects; studies at Earth have shown that the particle density in the terrestrial plasma sheet correlates well with that of the solar wind, and given Mercury's eccentric orbit, such a dependence would imply a yearly variation in the average plasma sheet density. Due to the orientation of MESSENGER's orbit, this would lead to higher observed densities in the postmidnight compared to premidnight plasma sheet [*Korth et al.*, 2012, and references therein]. After normalization, a small asymmetry still persists in the data, with an ~0.05 nPa/h dusk-to-dawn gradient with local time. This is less significant than would be expected from terrestrial observations where the difference in drift trajectories at dawn and dusk lead to significantly different particle densities. Within the plasma sheet, the first adiabatic invariant should generally not be conserved for the ions, as stochastic effects would heavily influence their expected motion and prevent closed ion drift paths to form within the equatorial plasma sheet, like those shown by *Schriver et al.* [2011a]. Such drift paths are likely only stable for very quiet conditions and primarily for sub-keV energies [*Korth et al.*, 2012]. The heavy ion population on the other hand shows a consistent enhancement toward the dusk side that is not easily explained by drift models but potentially by nonadiabatic processes affecting their drift motions [*Raines et al.*, 2013].

The equatorial projections of mean recorded pressure deficits are shown in Figure 2.7a, normalized to a heliocentric distance of 0.39 AU. Figure 2.7b shows the same data, but also normalized by the ratio of orbits with positive detections. This reduces the influence of a limited set of uncommon detections and thus gives a better overview of the general extent of plasma sheet. On an orbit-to-orbit basis, the plasma sheet density can vary by up to an order of magnitude. In terms of plasma composition, ion moments inferred from FIPS observations during the M1 and M2 flybys showed plasma densities of ~4–12 cm^{-3} and plasma betas of ~1–2. Typical characteristic temperatures are likely in the ~2–5 keV range [*Zurbuchen et al.*, 2011; *Raines et al.*, 2011]. Heavy ion densities are in general also elevated within the plasma sheet, with average observed densities of $5.1 \times 10^{-3} cm^{-3}$ for Na^+, $8 \times 10^{-4} cm^{-3}$ for the O^+ group, and 3.4×10^{-4} and ~$3.9 \times 10^{-2} cm^{-3}$ for He^+ and He^{2+}, respectively [*Raines et al.*, 2013].

2.4. EXTERNAL DRIVING

Mercury's magnetosphere is heavily driven by the solar wind, much more so than any other magnetosphere in our solar system. The reconnection rates are much larger than at Earth, and there are almost always signs of reconnection taking place at the dayside magnetopause even for the most unfavorable IMF orientations, as when the magnetospheric

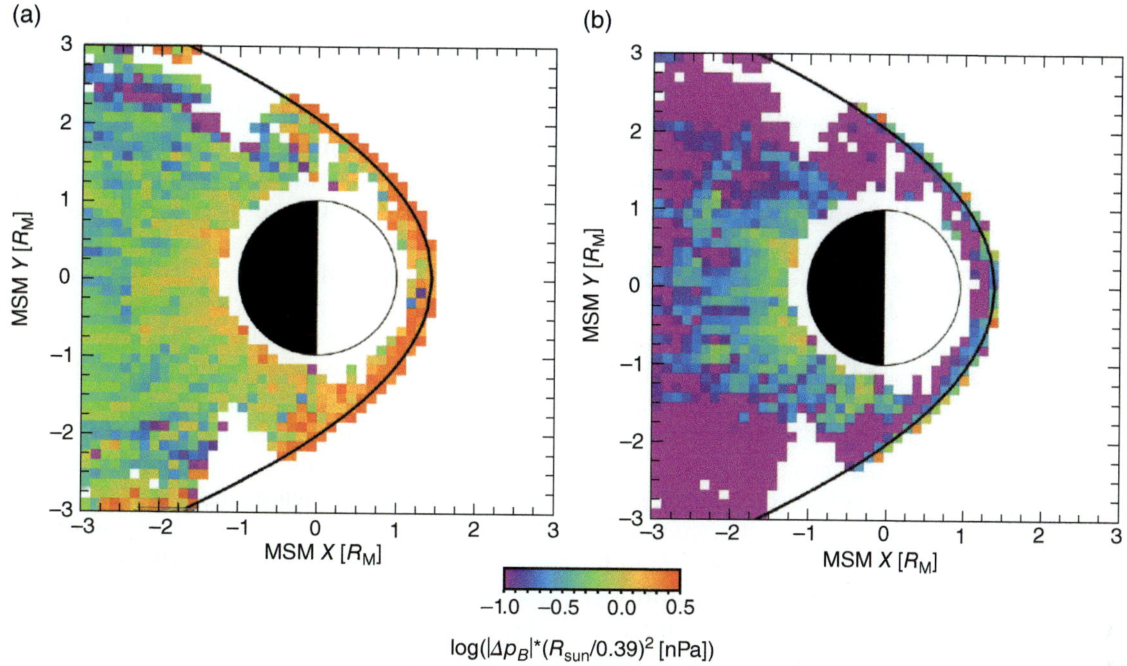

Figure 2.7 Equatorial plasma sheet pressures inferred from the magnetic pressure deficit: (a) distribution normalized to a heliocentric distance of 0.39AU: (b) each bin has also been normalized by the ratio of contributing orbits, in order to reduce the influence of outliers in the data. Adapted from *Korth et al.* [2012].

field and the IMF are nearly parallel. Flux transfer events (FTEs) are also both larger and much more frequent than what we are used to from the other planets [*Slavin et al.*, 2009b, 2012b; *DiBraccio et al.*, 2013].

2.4.1. Reconnection

One way to estimate the reconnection rate is to measure the ratio between the inflow and outflow velocity at the boundary or, equivalently, the ratio between the normal and absolute components of the magnetic field

$$\alpha = V_{in}/V_A = B_N/B_{MP}, \quad (2.2)$$

where α is a dimensionless reconnection rate, V_{in} is the inflow velocity, V_A is the local Alfvén velocity (which defines the outflow velocity), B_N is the magnetic component normal to the magnetopause, and B_{MP} is the magnitude of the magnetic field just inside the magnetopause. Although MESSENGER is not well suited to measure the plasma flow velocity, α can readily be estimated using magnetic field measurements. By this method, *DiBraccio et al.* [2013] showed that the reconnection rate in the dayside subsolar region is on average three times larger than what is normally seen at Earth, with a typical ratio of 0.15–0.2 compared to an average of 0.05 at Earth [*Mozer and Retinò*, 2007]. Their derived reconnection rates were also in principle independent of the magnetic field shear angle at the magnetopause (i.e., the angle between the magnetosheath and magnetospheric magnetic fields), without the expected predominance for higher reconnection rates in the antiparallel configuration. This unusual behavior could potentially be explained by the solar wind conditions at Mercury orbit: the relatively strong magnetic fields compared to the plasma density in the inner heliosphere give rise to low Mach numbers both in the solar wind and in the magnetosheath, conditions which are believed favorable for reconnection over a larger range of shear angles [*Swisdak et al.*, 2003]. *DiBraccio et al.* [2013] reported a weak observational dependence of the reconnection rate on the plasma beta, with lower betas being more favorable for reconnection, although further statistics are necessary to fully clarify their dependence. They also estimated the average dawn-dusk potential over the cusp to 29kV, a value which is in good agreement with that estimated from the M2 flyby and consistent with a Dungey cycle (the circulation of plasma and magnetic flux from the dayside magnetopause to the near-Mercury neutral line) time scale of ~2min [*Slavin et al.*, 2009b].

Evidence for reconnection is not only seen at the magnetopause proper; signatures of large FTEs are frequently seen within the magnetosheath and inside the magnetopause. These FTEs form through torsion on newly reconnected flux tubes, which causes the magnetic field lines to wrap up into flux ropes, giving rise to strong core fields surrounded by an azimuthal outer shell. These can be observed as a local peak in the magnetic field intensity along the core direction of the FTE together with a negative-positive or positive-negative excursion in the azimuthal component, depending on the helicity and propagation direction of the structure. The large-amplitude FTEs observed near the dayside magnetopause [*Slavin et al.*, 2010a] were found to contain between 0.001 and 0.2 MWb of magnetic flux. The largest of these events contain a significant fraction of the magnetic flux required to drive the Dungey cycle at Mercury, and they could potentially contribute with as much as ~2 MWb of open magnetic flux to the magnetotail over a period as short as 1–2 min [*Slavin et al.*, 2010b]. A much more extensive study of these large FTEs at the dayside magnetopause found a mean FTE magnetic flux content of 0.06 MWb, which could account for about 30% of the total magnetic flux circulated by the Dungey cycle [*Imber et al.*, 2014].

FTE observations are not only limited to the dayside region, but they can also be observed at high latitudes and in the tail when the IMF is northward and there is reconnection poleward of the cusp. As these FTEs are pulled across the magnetopause by the solar wind, they locally compress the tail, thereby giving rise to regions of higher magnetic field pressure within the tail lobes, so-called traveling compression regions (TCRs), traveling antisunward through the magnetotail. For this reason, the southern magnetotail is particularly well suited for long-term monitoring of the reconnection process with MESSENGER, as the spacecraft passes through this region much slower than it does near the subsolar point, and the impact of FTEs can be monitored over a larger volume. Figure 2.8 shows one such magnetopause crossing from 11 April 2012, analyzed in detail by *Slavin et al.* [2012b]. On this occasion, MESSENGER observed signs of 163 FTEs over a period of 25 min as it traveled southward through the southern tail lobe and the magnetosheath, approximately $X_{MSM} = -1.6\ R_M$ into the tail. These FTEs were partly direct observations of flux ropes within the magnetosheath and partly TCRs within the magnetosphere. Each FTE had duration of ~2–3 s, with an average recurrence rate of ~8–10 s. This long period of repeated, regular, flux rope structures, referred to as an FTE shower by *Slavin et al.* [2012b], shows that the generation of FTEs can be a quasi-periodic process at Mercury. The FTEs were found to contain very large amounts of magnetic flux, ~1 MWb, however, it is not

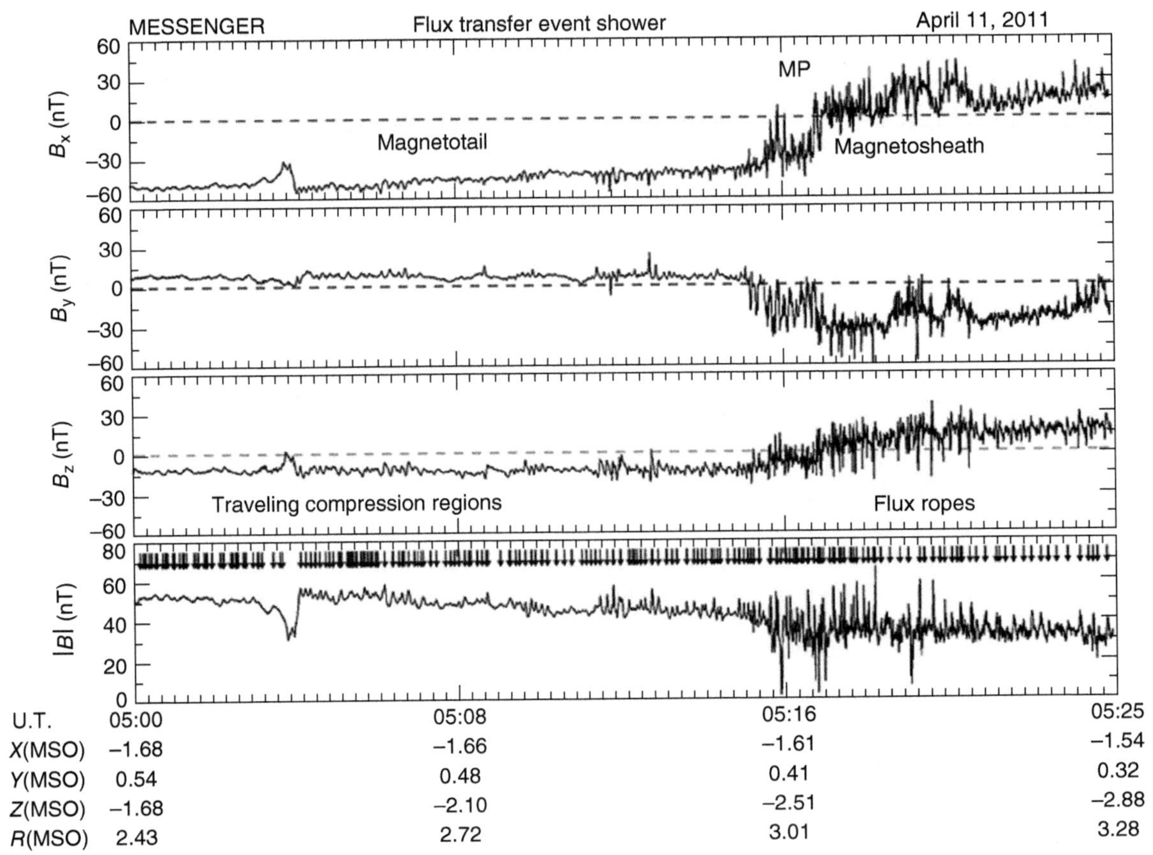

Figure 2.8 MESSENGER observation of an FTE shower. During a period of ~25 min, MESSENGER observed 163 FTEs and TCRs in the southern magnetotail and magnetosheath, each marked by an arrow in the bottom panel. From *Slavin et al.* [2012b].

clear whether the large flux content of these lobe FTEs relative to those on the dayside is due to differences in the modeling technique used in the two studies or intrinsic differences in the FTEs. Further study is required to bring light on this issue.

2.4.2. Kelvin-Helmholtz Instability

The Kelvin-Helmholtz (KH) instability is a surface instability that can develop at the boundary between two fluids of different streaming velocities, much like a strong wind over the ocean. This instability is of particular importance at the magnetopause, where the fast streaming solar wind plasma in the magnetosheath meets the stagnant magnetospheric plasma on the inside. Small perturbations along the boundary grow into large-scale vortices, transporting momentum, energy, and matter across the magnetopause [e.g., *Otto and Faifield*, 2000]. The instability primarily develops at the equatorial flanks of the magnetotail where the planetary magnetic field and plasma characteristics are favorable and the magnetosheath velocity begins to approach that of the solar wind [e.g., *Foullon et al.*, 2008]. At Mercury, the first proper identification of KH waves was made during the M3 flyby, where a series of magnetopause crossings were observed in the magnetic field data, indicative of a nonlinear wave train propagating along the boundary [*Boardsen et al.*, 2010, *Sundberg et al.*, 2011]. Subsequent observations from orbit have shown that such waves are common, and that they have large-scale implications on the magnetosphere, both in terms of solar wind mass transfer and as a source for ultralow frequency (ULF) wave activity [*Sundberg et al.*, 2012a]. Figure 2.9 shows one such observation taken at the postnoon magnetopause, close to the equatorial plane. On this occasion, MESSENGER observed a recurrent wave pattern just inside the magnetopause, with peak-to-peak variations of up to 40 nT in B_X, clear sawtooth-like oscillations in B_Y, and

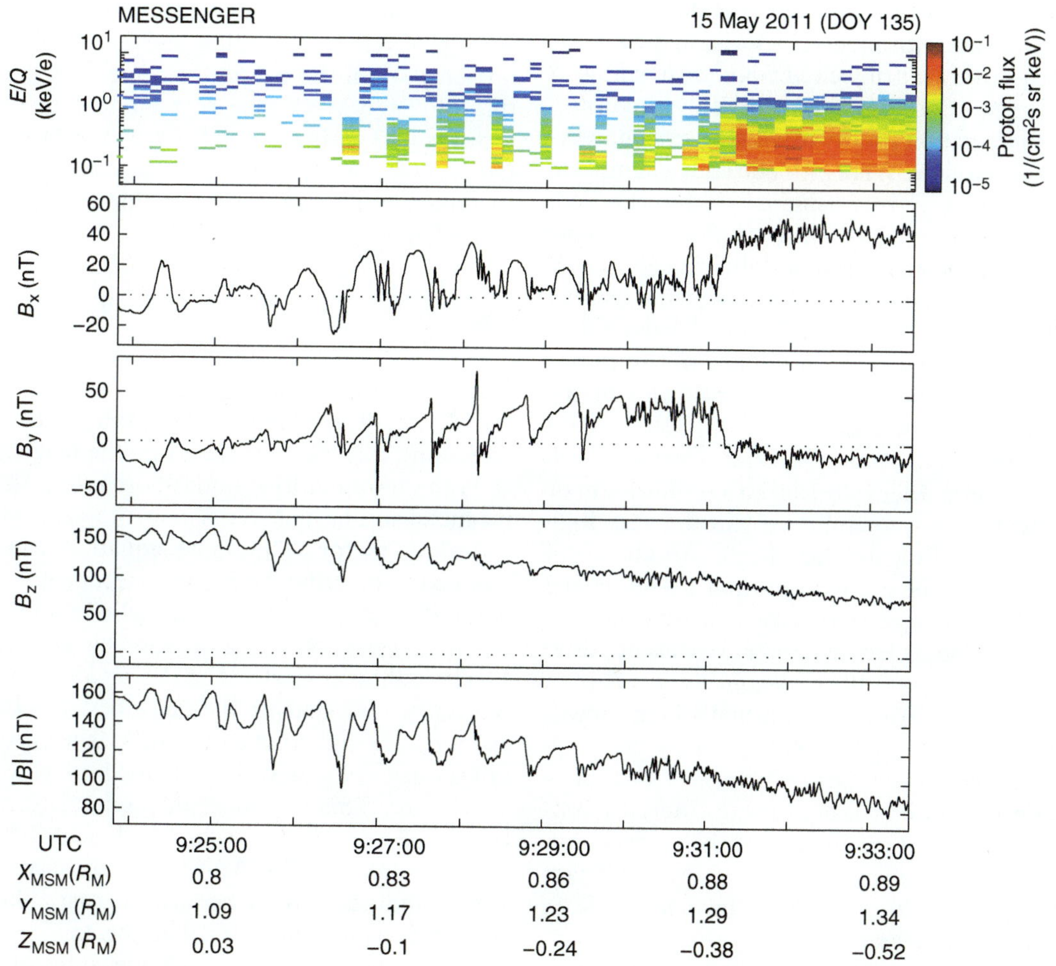

Figure 2.9 MESSENGER observations of a KH wave train. The top two panels show the FIPS energy over charge spectrogram and Na^+ count rate, and the bottom four panels show the *X, Y, Z*, and absolute components of the magnetic field. Adapted from *Sundberg et al.* [2012a].

a double-peak structure in the absolute field. In addition, FIPS measurements recorded periodic encounters with magnetosheath-like plasma in phase with the magnetic field waves, present well inside the magnetopause proper. MESSENGER also observed strong wave activity deep inside the magnetosphere with the same periodicity, likely on near-conjugate field lines with the boundary waves. These observations show that the KH instability not only causes perturbations within the LLBL but also may be an important source of wave energy within the magnetosphere.

All KH wave encounters reported by *Sundberg et al.* [2012a], six in total, were made either at the postnoon magnetopause or on the dusk side flank, and all during periods of northward IMF. There are several possible explanations for this bias, with finite Larmor radius effects as the most prominent. In contrast to the larger scale sizes at Earth and Saturn, the KH instability at Mercury is anticipated to be in the kinetic domain, which can have important implications on its characteristics. The opposite vorticity of the waves on the dusk side and dawn side of the magnetosphere (clockwise and counter-clockwise, respectively, when viewed from the north) lead to asymmetric conditions where the vortices are corotating with the ion gyration on the dawn side and counter-rotating on the dusk side. What influence this ion motion has on the instability yet remains to be clarified. Gyrokinetic simulations [*Nakamura et al.*, 2010] show that the ion gyration primarily has a stabilizing effect as it increases the width of the velocity shear layer and thereby lowers the instability growth rate. This stabilizing effect will be increased on the dawn side as the ion gyro radius is extended due to the convection electric fields in the magnetosheath and the boundary layer and decreased on the dusk side. These results are in good agreement with observations, however, they only lead to a stabilization of the LLBL and they do not fully explain the very high growth rates observed on the dusk flank. An analytical approach shows that the ion gyro motion would indeed partly destabilize the boundary layer for intermediate wavelength, but the predicted asymmetry is likely opposite that observed [e.g., *Nagano* 1979; *Glassmeier and Espley*, 2006; *Sundberg et al.*, 2010]. The unusually high growth rates observed are instead best explained by the high resistivity of Mercury's regolith and the lack of a conducting ionosphere. In contrast to Earth, Mercury provides no direct means for dissipation of the wave energy, and the instability can thus grow without any significant dampening at the low-altitude end of the field line [e.g., *Miura and Kan*, 1992; *Miura*, 1996].

An alternative explanation to the dawn-dusk asymmetry observed may be an asymmetric plasma composition within the LLBL: When the IMF is northward and the conditions are most favorable for the KH instability, the prenoon sector of the LLBL may in fact be significantly more mass loaded than the postnoon sector, both due to heavy pick-up ions and a dawnward drift of solar wind plasma that enters into the magnetosphere through the cusps. This influx of heavy ions could substantially broaden the velocity shear layer and thereby dampen the growth rates on the dawn side.

2.5. TAIL DYNAMICS

The strong solar wind driving at Mercury leads to a very active magnetotail with short time scales. Dayside reconnection feeds open flux into the tail at a relative rate that is much larger than that seen at Earth, and the average Dungey cycle is on the order of a few minutes compared to ~1 h at Earth.

2.5.1. Loading-Unloading and Steady Convection

Variations in the input rate of open magnetic flux in the tail and the closure of magnetic flux in the plasma sheet give rise to a loading and unloading of the magnetotail. When the dayside reconnection rate supersedes that at the NMNL, the flux content in the tail lobes grows, and it shrinks when tail reconnection is dominant. On average, the magnitude of the tail lobe field can be described by the following equation [*Slavin et al.*, 2012a, and references therein]:

$$B(x) = A|x|^{G} + B_0, \tag{2.3}$$

where G is the flaring parameter (which determines the radial rate of expansion of the tail in the antisunward direction), x is the distance along the magnetotail in R_M, B_0 is the terminal lobe intensity required to balance the magnetic and particle pressure in the solar wind, and A is a scaling factor. The same equation can equally be adapted to describe the plasma sheet profile. As the magnetotail quickly approaches its cylindrical limit a few R_M down the tail, this implies a much larger G value at Mercury than at Earth. The flaring also depends heavily on the current state of the tail lobes, and the values are significantly higher when the IMF is northward and the tail is quiet compared to the loaded tail when the IMF is southward. This is evident in Figure 2.10, which shows the fit of equation (2.3) for all three MESSEGER tail crossings and the first Mariner 10 passage, separated by IMF conditions. Both northward cases show G values of ~ −5.5, which indicates that the tail quickly reached the terminal width, and both measurement sets follow the equation nicely. The M2 and M3 flybys, on the other hand, which took place during southward IMF, show large differences when compared to each other.

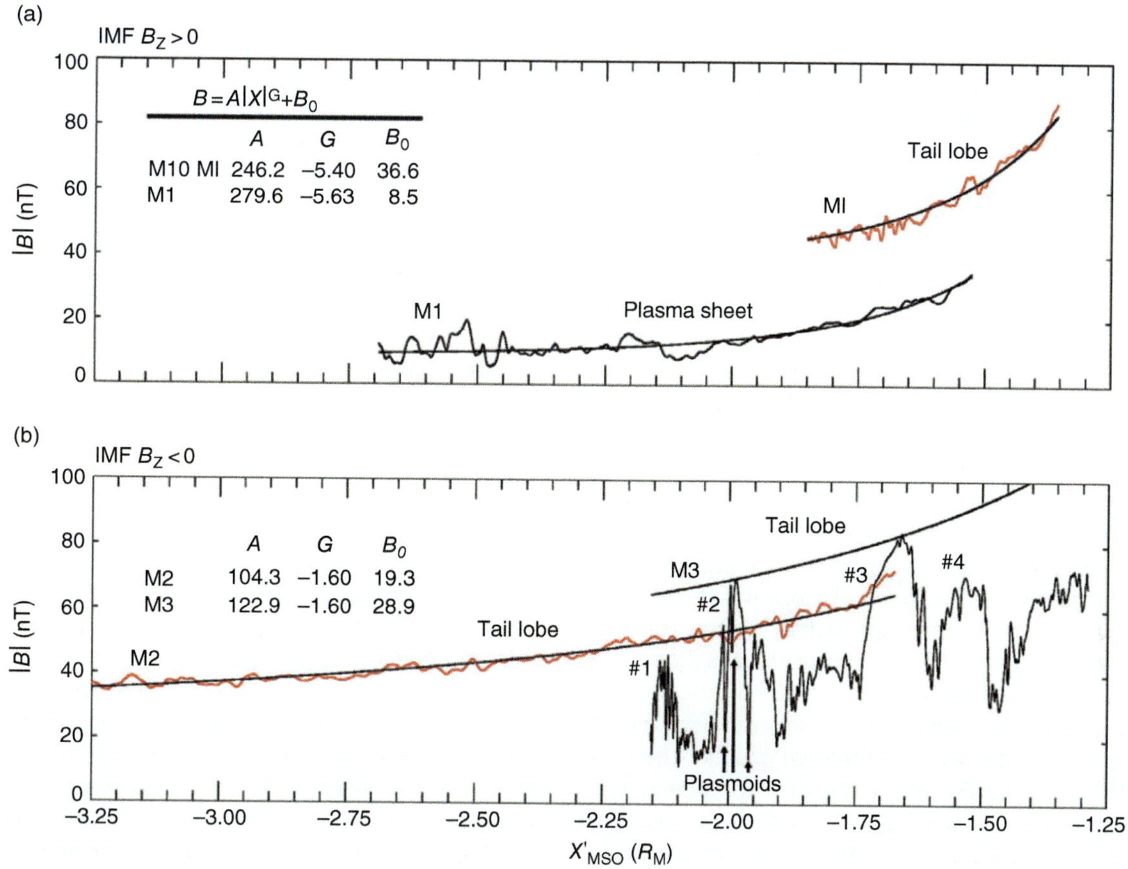

Figure 2.10 Magnetotail magnetic field measurements for the first Mariner 10 and the three MESSENGER flybys. The top and bottom panels show events with northward and southward IMF, respectively. For each measurement set, the fit to equation (2.3) is overlain on the data. From *Slavin et al.* [2012b].

The magnetic field magnitude measured for M2 follows that predicted all throughout the tail crossing with only minor fluctuations about the equilibrium, with a G value of −1.6. The field trace for the M3 flyby, on the other hand, is far from equilibrium. The data show four clear occasions of loading–unloading of the tail lobe where the field magnitude increases with 20–40 nT over a period of a few minutes before it drops back to the background level. *Slavin et al.* [2012a] interpreted these data sets as two different states of the magnetotail. The M2 data are reminiscent of the steady magnetospheric convection events seen at Earth [*Sergeev et al.*, 1996]: When the IMF is southward and the reconnection rate on the dayside remains moderate and reasonably continuous, the magnetotail can at times assume a quasi-steady behavior where the input and closure rates of open magnetic flux are in balance. In this scenario, the return of magnetic flux through the plasma sheet is achieved through a steady convection without the sudden bursts of tail reconnection that give rise to substorms. This is also in agreement with the measured IMF during the flyby, which showed steady southward IMF both before and after the magnetosphere transition. For the M3 data, however, the loading-unloading cycle is obvious; it both is more intense than we have previously observed at Earth and takes place on much shorter time scales. The relative increase in the lobe field on this occasion was close to a factor of 2, compared to the standard case of ~1.3 reported from terrestrial measurements [e.g., *Slavin et al.*, 2012a, and references therein]. Each unloading cycle is also associated with either sunward or antisunward traveling plasmoids or TCRs, both associated with bursts of reconnection. TCRs and plasmoids were also observed during M2, but with much smaller impact on the magnetosphere.

2.5.2. Transient Events

The nonsteady nature of magnetotail reconnection introduces many transient features in the plasma sheet and the tail lobes, in particular giving rise to plasmoids, TCRs, and dipolarizations. Plasmoids are essentially flux ropes or magnetic islands that form in the central plasma sheet as a result of reconnection. The exact nature of their formation is still somewhat of an unresolved issue, but theories suggest that they are related either to reconnection within a sheared magnetotail [e.g., *Hughes and Sibeck*, 1987] or between multiple reconnection X lines [*Slavin et al.*, 2003]. The first scenario can occur when there is a significant B_Y component present in the IMF. After reconnection, the IMF exerts an opposite force on the two tail lobes along the Y axis due to magnetic tension within the open magnetic field. For a negative B_Z and positive B_Y, this tension skews the northern tail lobe toward the dawn side and the southern tail lobe toward dusk [*Cowley*, 1981], resulting in a tilt of the current sheet away from the equatorial plane. Subsequent reconnection between the two tail lobes at the NMNL can then generate twisted flux rope structures, traveling both planetward and tailward. The magnetic tension in the outer wraps of the flux rope tends to shorten the magnetic field lines and gives the plasmoid a more circular cross section. Because the plasma sheet material from which the plasmoid forms is relatively incompressible, the result is that the north-south height of the plasmoid exceeds that of the plasma sheet and compresses the lobes as it moves sunward or antisunward. For a spacecraft positioned within the magnetotail, these will be observed as a south-north or a north-south oscillation in the field for planetward and tailward propagating structures, respectively. The alternative formation scenario with multiple reconnection lines could potentially occur whenever the magnetotail current sheet is unstable to reconnection. Multiple NMNLs can form simultaneously at different distances from the planet, and each such pair forms a closed loop of magnetic flux, either traveling tailward or planetward depending on the local conditions. The sense of the magnetic perturbation in this case will be identical to that above.

MESSENGER first detected tailward then planetward convecting plasmoids and TCRs in both the M2 and M3 flybys, and the separation between these two polarities gives a hint of the location of the NMNL on these occasions. For M2, the change between tailward and sunward structures was observed at $X = -2.8\ R_M$ and at $-1.8\ R_M$ for M3. This gives reliable evidence for the formation of a NMNL at $X = \sim -2$ to $-3\ R_M$, i.e., a distance of around 1 R_M to 2 R_M from the nightside surface. The periodic recurrence of the plasmoids, approximately every 10–30 s, also led *Slavin et al.* [2012a] to speculate that these formed at multiple reconnection sites, rather than by a twisted magnetotail.

Other features commonly associated with tail reconnection are dipolarization events, and they are believed to be intimately associated with particle acceleration and substorm processes [*Baker et al.*, 1986, 1996]. At Mercury, these are observed within the central plasma sheet, characterized by ~1 s increases in the B_Z component of the magnetic field, followed by a gradual decrease back to preonset values over ~10 s [*Christon et al.*, 1987; *Sundberg et al.*, 2012b]. This increase in the northward magnetic field component is a sign of a quick reconfiguration of the magnetotail where the initially stretched field lines in the tail temporarily returns to a more dipolar configuration. These events are likely also associated with bursts of fast flowing plasma, so called bursty-bulk flows [*Angelopoulos et al.*, 1994], which could be initiated either by bursty reconnection in the tail or by kinetic structuring within the plasma sheet itself. Once initiated, the front of the dipolarization region is subject to instabilities due to its lower entropy compared to the surrounding plasma and imbalances in the **J** x **B** and plasma pressure forces, which cause it to grow in amplitude as it travels inward through the magnetotail. *Sundberg et al.* [2012b] reported increases of 20–70 nT, comparable to ~4 times the background field. The events were also both observed as isolated and quasi-periodic features in the magnetometer data. An example of the latter is shown in Figure 2.11, where a train of 10 dipolarizations is seen within one 4-min plasma sheet crossing, each separated in time with 10–50 s. An analysis of the current sheet orientation at the dipolarization front showed evidence that not all of these were crossed head on, but a large fraction of the events were crossings of the flank side of a localized convection channel, as the current sheet normal clearly deviated from the X_{MSM} direction. This localized nature means that these features also must be intimately connected to field-aligned currents in order to balance the discontinuities in the gradient and curvature-drift currents in the plasma sheet. The exact nature of this current closure is still undetermined, and the time scales involved in the process may be too short for a stable current system to be set up, as the transit time required for a current-carrying Alfvén wave to reflect of the surface and back to the plasma sheet may exceed to the lifetime of the dipolarization [*Sundberg et al.*, 2012]. The transient nature of this coupling is further discussed in Chapter 21.

2.5.3. Energetic Electrons

Bursts of highly energetic electrons and protons were reported during the Mariner 10 flybys [*Simpson et al.*, 1974]. These were later reinterpreted as energetic electrons only [e.g., *Armstrong et al.*, 1975; *Christon et al.*, 1987], and MESSENGER measurements have now confirmed that such electron bursts are commonly occurring in the magnetosphere [*Ho et al.*, 2011a]. The most intense electron bursts take place in primarily two different regions of the magnetosphere: above the cusp and along all latitudes of local midnight, as shown in Figure 2.12a. Both of these groups are similar in their observed intensities and spectra, and they are primarily perpendicular to the background

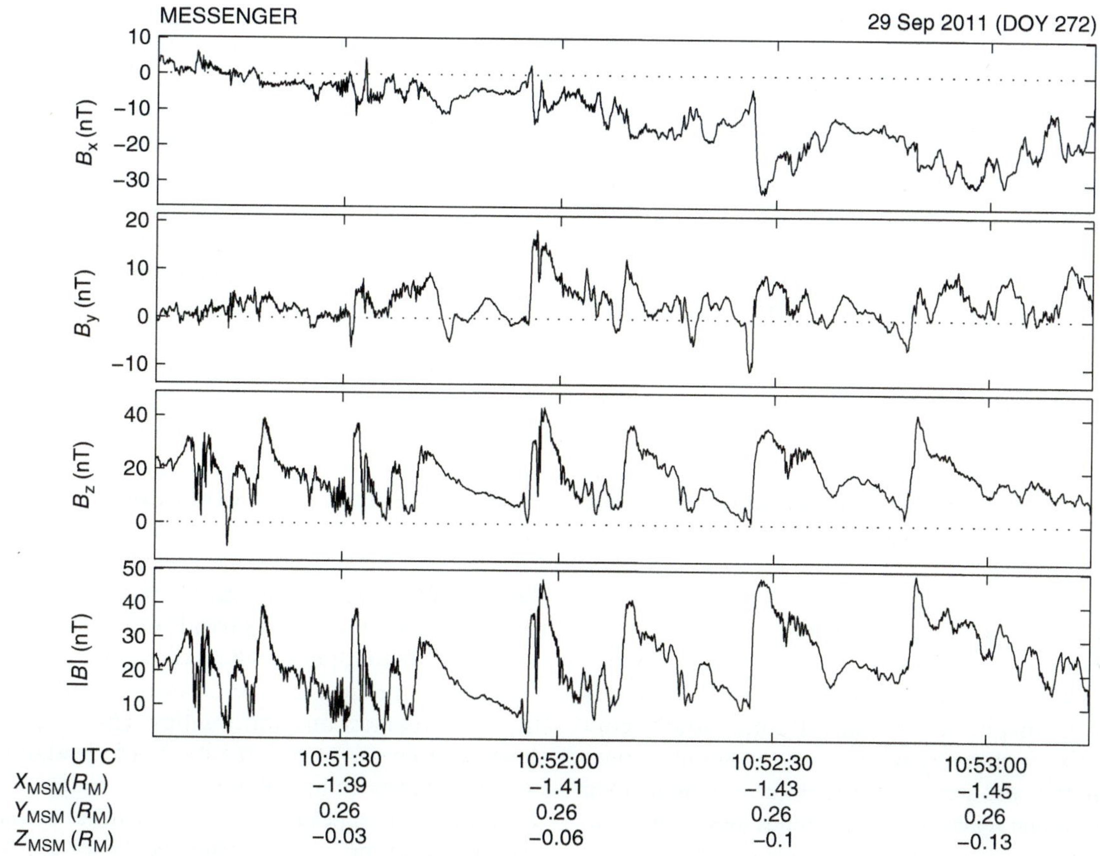

Figure 2.11 Series of dipolarization events observed inside the plasma sheet. Each dipolarization front is marked by a rapid increase in the *Z* component of the magnetic field.

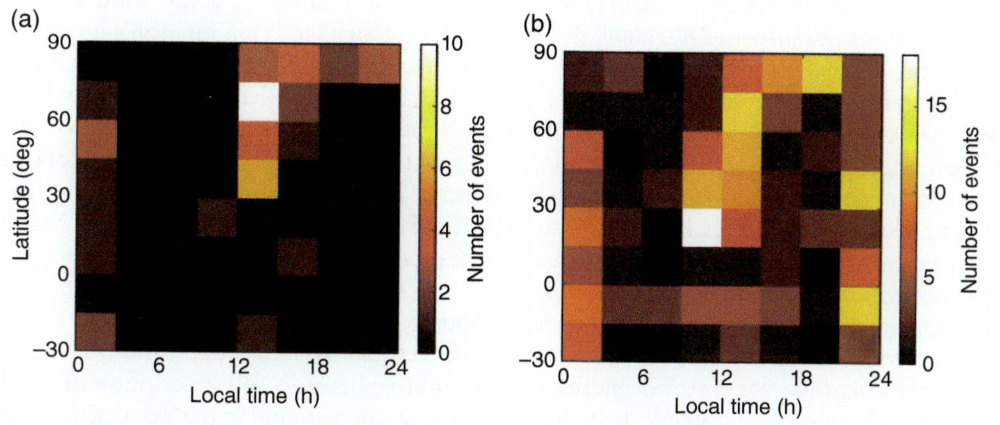

Figure 2.12 Distribution of energetic electron events in Mercury's magnetosphere: (a) location of the most intense bursts of electrons; (b) both moderate- and high-intensity events. From *Ho et al.* [2012].

magnetic field. *Ho et al.* [2012] interpreted this as evidence for both groups being accelerated by similar mechanisms, likely by reconnection in the tail and at the dayside magnetopause. A third group of energetic electrons was also found upstream of the bow shock, directed upstream, and these are more likely attributed to acceleration at the shock. For this group, the required acceleration (~100keV) could potentially be achieved by shock drift of solar wind electrons when the bow shock is quasi-perpendicular, i.e., the local IMF is perpendicular to the bow shock normal, although not all the events covered in the analysis were classified as quasi-perpendicular shocks.

In addition to the high-intensity electron bursts, events with lower intensity (2–10 times the average background) were also seen at all local times, albeit constrained to the magnetic equator. These are visible in Figure 2.12b. The X-ray spectrometer (XRS) also regularly reports indirect observations of bursts of moderate-energy electrons below the energy threshold of the EPS instrument. As electrons in the 1–10 keV energy range impact the instrument, they excite X-rays which can be picked up by the detector [*Ho et al.*, 2011b; *Starr et al.*, 2012]. These moderate-energy bursts are possibly from the same source as the low-intensity bursts reported by *Ho et al.* [2012], with the EPS measurements covering only the high-energy tail of the population. Some of the bursts in the XRS have also been observed simultaneously with X-ray fluorescence from the nightside surface, likely caused by electrons impacting the regolith. Energetic electrons could thus potentially sputter material from the surface and provide an additional source of exospheric atoms and ions [e.g., *McLain et al.*, 2011; *Schriver et al.*, 2011b].

2.6. SUMMARY

Recent results from Mercury orbit have revealed a very dynamic magnetosphere, often under continuous reconfiguration. Although many of the features observed have terrestrial analogues, they generally differ greatly in their properties and time scales: Reconnection is more dominant than at Earth, KH waves are ubiquitous at the dusk side magnetopause, and tail dynamics in the form of plasmoids, TCRs, and dipolarizations can be seen with a recurrence period as short as a few seconds. In addition to what has been already covered in the review, Mercury is also host to a wealth of ULF pulsations [*Boardsen et al.*, 2009, 2012] that have characteristics different from those observed at Earth, potentially due to the warm nature of the plasma.

New discoveries are continuously made through the MESSENGER mission, yet many issues still remain unresolved. The importance of the heavy planetary ions in moderating magnetospheric dynamics is still an open question, as is their extent within the magnetosphere and their life cycle. The nonconducting regolith may also have important effects on substorm processes, and it can in part explain the high reconnection rates and the large KH growth rates as neither the surface nor the ionosphere can provide any significant line tying which would help slow down the processes. In addition, induction currents within the conducting core may play a great role to moderate the solar wind–magnetosphere interaction [e.g. *Grosser et al.* 2004; *Glassmeier et al.* 2007]. The presence of field-aligned currents (FAC) is also an unresolved issue. Although *Slavin et al.* [1997] drew attention to a potential set of FAC in the Mariner 10 data, no encounters with FAC have yet been reported from the MESSENGER mission, and the typical R1/R2 current systems that we are used to from Earth do not seem to be present. The short time scales may be a limiting factor, and the currents could potentially in part be carried by kinetic Alfvén waves rather than stationary currents [*Lyatsky et al.* 2010]. Measurements and modeling combined [e.g., *Trávníček et al.*, 2009; *Benna et al.*, 2010] will doubtless help shed light on many of these aspects. Modeling efforts are essential to help resolve and understand many of the unique aspects of Mercury's magnetosphere, and they can provide important details on not only the solar wind forcing [e.g., *Delcourt et al.*, 2011; *Baker et al.*, 2013] but also surface-exosphere-magnetosphere interactions [e.g., *Kallio and Janhunen*, 2003; *Leblance et al.*, 2007; *Orsini et al.*, 2007; *Wang and Ip*, 2011] and internal acceleration process [e.g. *Delcourt et al.*, 2005, 2007; *Zelenyi et al.*, 2007]. Finally, next decade the BepiColombo mission, a joint effort by the European Space Agency and the Japan Aerospace Exploration Agency, will bring major new advances in our understanding of Mercury's space environment. With its two orbiters, BepiColmbo will provide our first opportunity to separate magnetopheric dynamics that are driven by changes in the upstream solar wind from those due to internal dynamics and instabilities. The measurement of these external solar wind drivers should lead to important new discoveries concerning Mercury's magnetic dynamo, the nature of magnetospheric substorms and charged particle acceleration, the feedback between changes in the magnetospheric magnetic field and induction currents in the planetary interior, and the role of cusp dynamics, KH waves, flux transfer events, and dayside magnetopause reconnection in direct solar wind impingement on the planetary surface. This mission's advanced instrumentation, for example the Mercury Plasma Particle Experiment and the Plasma Wave Instrument, will provide the first high-resolution, three-dimensional charged particle distribution functions, exospheric and energetic neutral particles, electric fields, and plasma waves. These measurements are critical for progress in understanding Mercury's magnetosphere because they will allow us to process beyond MHD to reveal the kinetic effects (e.g., large ion gyro-radii relative to current sheets) that are expected to be more important at Mercury than for any of the other planetary magnetospheres. A full description of the mission and its payload can be found in the work of *Benkhoff et al.* [2010].

REFERENCES

Alexeev, I. I., E. S. Belenkaya, S. Yu. Bobrovnikov, J. A. Slavin, and M. Sarantos (2008), Paraboloid model of Mercury's magnetosphere, *J. Geophys. Res.*, *113*, A12210, doi: 10.1029/2008JA013368.

Alexeev, I. I., et al., (2010), Mercury's magnetospheric magnetic field after the first two MESSENGER flybys, *Icarus*, *209.1*, 23–39.

Anderson, B. J., M. H. Acuña, D. A. Lohr, J. Scheifele, A. Raval, H. Korth, and J. A. Slavin (2007), The Magnetometer instrument on MESSENGER, *Space Sci. Rev., 131*, 417–450.

Anderson, B. J., M. H. Acuña, H. Korth, M. E. Purucker, C. L. Johnson, J. A. Slavin, S. C. Solomon, and R. L. McNutt, Jr. (2008), The structure of Mercury's magnetic field from MESSENGER's first flyby, *Science, 321*, 82–85.

Anderson, B. J., et al. (2010), The magnetic field of Mercury, *Space Sci. Rev., 152* (1–4), 307–339.

Anderson, B. J., C. L. Johnson, H. Korth, M. E. Purucker, R. M. Winslow, J. A. Slavin, S. C. Solomon, R. L. McNutt, Jr., J. M. Raines, and T. H. Zurbuchen (2011a), The global magnetic field of Mercury from MESSENGER orbital observations, *Science, 333*, 1859–1862.

Anderson, B. J., J. A. Slavin, H. Korth, S. A. Boardsen, T. H. Zurbuchen, J. M. Raines, G. Gloeckler, R. L. McNutt, Jr., and S. C. Solomon (2011b), The dayside magnetospheric boundary layer at Mercury, *Planet. Space Sci., 59*, 2037–2050.

Anderson, B. J., C. L. Johnson, H. Korth, R. M. Winslow, J. E. Borovsky, M. E. Purucker, J. A. Slavin, S. C. Solomon, M. T. Zuber, and R. L. McNutt, Jr. (2012), Low-degree structure in Mercury's planetary magnetic field, *J. Geophys. Res., 117*, E00L12, doi:10.1029/2012JE004159.

Andrews, G. B., et al. (2007), The Energetic Particle and Plasma Spectrometer Instrument on the MESSENGER spacecraft, *Space Sci. Rev., 131*, 523–556.

Angelopoulos, V., C. F. Kennel, F. V. Coroniti, R. Pellat, M. G. Kivelson, and R .J. Walker, C. T. Russell, W. Baumjohann, W. C. Feildman, and J. T. Gosling (1994), Statistical characteristics of bursty bulk flow events, *J. Geophys. Res. 99*, 21, 257–21,280, doi:10.1029/94JA010263.

Armstrong, T. P., S. M. Krimigis, and L. J. Lanzerotti, (1975), A reinterpretation of the reported energetic particle fluxes in the vicinity of Mercury, *J. Geophys. Res., 80*(28), 4015–4017, doi: 10.1029/JA080i028p04015.

Baker, D. N., J. A. Simpson, and J. H. Eraker (1986), A model of impulsive acceleration and transport of energetic particles in Mercury's magnetosphere, *J. Geophys. Res., 91*, 8742–8748, doi:10.1029/JA091iA08p08742.

Baker, D. N., T. I. Pulkkinen, V. Angelopoulos, W. Baumjohann, and R. L. McPherron (1996), Neutral line model of substorms: Past results and present view, *J. Geophys. Res., 101*, 12,975–13,010, doi:10.1029/95JA03753.

Baker, D. N., et al. (2013), Solar wind forcing at Mercury: WSA-ENLIL model results, *J. Geophys. Res., 118*, doi:10.1029/2012 JA018064.

Baumgardner, J., J. Wilson, and M. Mendillo (2008), Imaging the sources and full extent of the sodium tail of the planet Mercury, *Geophys. Res. Lett., 35*, L03201, doi:10.1029/2007GL 032337.

Baumjohann, W., et al. (2006), The magnetosphere of Mercury and its solar wind environment: Open issues and scientific questions, *Adv. Space Res., 38*(4), 604–609, doi:10.1016/j.asr.2005.05.117.

Benkhoff, J., J. van Casteren, H. Hayakawa, M. Fujimoto, H. Laakso, M. Novara, P. Ferri, H. R. Middleton, and R. Ziethe (2010), BepiColombo—Comprehensive exploration of Mercury: Mission overview and science goals, *Planet. Space Sci., 58*, 2–20, doi:10.1016/j.pss.2099.09.020.

Benna, M., et al. (2010), Modeling of the magnetosphere of Mercury at the time of the first MESSENGER flyby, *Icarus, 209*, 3–10.

Boardsen, S. A., J. A. Slavin, B. J. Anderson, H. Korth and S. C. Solomon (2009), Comparison of ultra-low-frequency waves at Mercury under northward and southward IMP, *Geophys. Res. Lett., 36*, L18106, doi:10.1029/2099GL039525.

Boardsen, S. A., T. Sundberg, J. A. Slavin, B. J. Anderson, H. Korth, S. C. Solomon, and L. G. Blomberg (2010), Observations of Kelvin-Helmholtz waves along the dusk-side boundary of Mercury's magnetosphere during MESSENGER's third flyby, *Geophys. Res. Lett., 37*, L12101, doi:10.1029/2010GL043606.

Boardsen, S. A., J. A. Slavin, B. J. Anderson, H. Korth, D. Schriver, and S. C. Solomon (2012), Survey of coherent ~ 1 Hz waves in Mercury's inner magnetosphere from MESSENGER observations, *J. Geophys. Res., 117*, A00M05, doi:10.1029/2012JA017822.

Burger, M. H., et al. (2010), Monte Carlo modeling of sodium in Mercury's exosphere during the first two MESSENGER flybys, *Icarus, 209*, 1, 63–74

Christon, S. P., J. Feynman, and J. A. Slavin (1987), Dynamic substorm injections—Similar magnetospheric phenomena at Earth and Mercury, in *Magnetotail Physics*, edited by A. T. Y. Lui, pp. 393–400, Johns Hopkins Univ. Press, Baltimore, Md.

Cowley, S. W. H. (1981), Magnetospheric asymmetries associated with the Y-component of the IMF, *Planet. Space Sci., 29*, 79–96.

Delcourt, D. C. (2013), On the supply of heavy planetary material to the magnetotail of Mercury, *Ann. Geophys., 31*, 1673–1679.

Delcourt, D. C., T. E. Moore, S. Orsini, A. Millilo, and J.-A. Sauvaud (2002), Centrifugal acceleration of ions near Mercury, *Geophys. Res. Lett., 29*(12), doi:10.1029/2001GL013829.

Delcourt, D. C., K. Seki, N. Terada, and Y. Miyoshi (2005), Electron dynamics during substorm dipolarization in Mercury's magnetosphere, *Ann. Geophys., 23*, 3389–3398.

Delcourt, D. C., F. Leblanc, K. Seki, N. Terada, T. E. Moore, and M.-C. Fok (2007), Ion energization during substorms at Mercury, *Planet. Space Sci., 55*, 1502–1508, doi:10.1016/j.pss.2006.11.026.

Delcourt, D. C., T. E. Moore, and M.-C. H. Fok (2011), On the effect of IMF turning on ion dynamics at Mercury, *Ann. Geophys., 29*, 987–996, doi:10.5194/angeo-29-987-2011.

Delcourt, D. C., K. Seki, N. Terada, and T. E. Moore (2012), Centrifugally stimulated exospheric ion escape at Mercury, *Geophys. Res. Lett., 39*, L22105, doi:10.1029/2012GL054085.

DiBraccio, G. A., J. A. Slavin, S. A. Boardsen, B. J. Anderson, H. Korth, T. H. Zurbuchen, J. M. Raines, D. N. Baker, R. L. McNutt Jr., and S. C. Solomon (2013), MESSENGER observations of magnetopause structure and dynamics at Mercury, *J. Geophys. Res., 118*, 997–1008, doi:10.1002/jgra.50123.

Dukes, C. A., W. Y. Chang, M. Famá, and R. A. Baragiola (2011), Laboratory studies on the sputtering contribution to the sodium atmospheres of Mercury and the Moon, *Icarus, 212*, 463–469, doi:10.1016/j.icarus.2011.01.027.

Elphic, R. C., H. O. Funsten, B. L. Barraclough, D. J. McComas, M. T. Paffet, D. T. Vaniman, and G. Heiken (1991), Lunar surface composition and solar wind–induced secondary ion mass spectrometry, *Geophys. Res. Lett.*, *18*, 2165–2168, doi:10.1029/91GL02669.

Foullon, C., C. J. Farrugia, A. N. Fazakerley, C. J. Owen, F. T. Gratton, and R. B. Torbert (2008), Evolution of Kelvin-Helmholtz activity on the dusk flank magnetopause, *J. Geophys. Res.*, *113*, A11203, doi:10.1029/2008JA013175.

Fujimoto, M., W. Baumjohann, K. Kabin, R. Nakamura, J. A. Slavin, N. Terada, and L. Zelenyi, (2007), Hermean magnetosphere-solar wind interaction, *Space Sci. Rev.*, *132*, 529–550, doi:10.1007/s11214-007-9245-8.

Glassmeier, K.-H., and J. Espley (2006), ULF waves in planetary magnetospheres, in *Magnetospheric ULF Waves: Synthesis and New Directions*, Geophys. Monogr. Ser., vol. *169*, edited by K. Takahashi et al., pp. 341–349, AGU, Washington, D. C.

Glassmeier, K. H., J. Grosser , U. Auster, D. Constantinescu, Y. Narita, and S. Stellmach (2007), Electromagnetic induction effects and dynamo action in the Hermean system, *Space Sci. Rev.*, *132*(2–4), 511–527.

Grosser, J., K. H. Glassmeier, and A. Stadelmann (2004), Induced magnetic field effects at planet Mercury, *Planet. Space Sci.*, *52*(14), 1251–1260.

Ho, G. C., et al. (2011a), MESSENGER observations of transient bursts of energetic electrons in Mercury's magnetosphere, *Science*, *333*, 1866–1868, doi:10.1126/science.1211001.

Ho, G. C., R. D. Starr, R. E. Gold, S. M. Krimigis, J .A. Slavin, D. N. Baker, B. J. Anderson, R. L. McNutt, Jr., L. R. Nittler, and S. C. Solomon (2011b), Observations of suprathermal electrons in Mercury's magnetosphere during the three MESSENGER flybys, *Planet. Space Sci.*, *59*, 2016–2025.

Ho, G. C., S. M. Krimigis, R. E. Gold, D. N. Baker, B. J. Anderson, H. Korth, J. A. Slavin, R. L. McNutt, Jr., R. M. Winslow, and S. C. Solomon (2012), Spatial distribution and spectral characteristics of energetic electrons in Mercury's magnetosphere, *J. Geophys. Res.*, *117*, A00M04, doi:10.1029/2012JA017983.

Hughes, W. J., and D. G. Sibeck (1987), On the 3-dimensional structure of plasmoids, *Geophys. Res. Lett.*, *14*(6), 636–639, doi:10.1029/GL014i006p00636.

Hunten, D. M., and A. L. Sprague (2002), Diurnal variation of sodium and potassium at Mercury, *Meteor. Planet. Sci.*, *37*, 1191–1196.

Imber et al. (2014), MESSENGER observations of large dayside flux transfer events: Do they drive Mercury's substorm cycle? *J. Geophys. Res. Space Physics*, *119*, 5613–5623, doi:10.1002/2014JA019884.

Johnson, C. L., et al. (2012), MESSENGER observations of Mercury's magnetic field structure, *J. Geophys. Res.*, *117*, E00L14, doi:10.1029/2012JE004217.

Kallio, E., and P. Janhunen (2003), Solar wind and magnetospheric ion impact on Mercury's surface, *Geophys. Res. Lett.*, *30*, 1877, doi:10.1029/2003GL017842, 17.

Killen, R. M., A. E. Potter, P. Reiff, M. Sarantos, B. V. Jackson, P. Hick, and B. Giles (2001), Evidence for space weather at Mercury, *J. Geophys. Res.*, *106*, 20, 509–20, 525.

Killen, R., et al. (2007), Processes that promote and deplete the exosphere of Mercury, *Space Sci. Rev.*, *132*(2–4), 433–509.

Korth, H., B. J. Anderson, J. M. Raines, J. A. Slavin, T. H. Zurbuchen, C. L. Johnson, M. E. Purucker, R. M. Winslow, S. C. Solomon, and R. L. McNutt, Jr. (2011), Plasma pressure in Mercury's equatorial magnetosphere derived from MESSENGER Magnetometer observations, *Geophys. Res. Lett.*, *38*, L22201, doi:10.1029/2011GL049451.

Korth, H., B. J. Anderson, C. L. Johnson, R. M. Winslow, J. A. Slavin, M. E. Purucker, S. C. Solomon, and R. L. McNutt, Jr. (2012), Characteristics of the plasma distribution in Mercury's equatorial magnetosphere derived from MESSENGER Magnetometer observations, *J. Geophys. Res.*, *117*, A00M07, doi:10.1029/2012JA018052.

Le, G., P. J. Chi, X. Blanco-Cano, S. Boardsen, J. A. Slavin, and B. J. Anderson (2013), Upstream ultra-low frequency waves in Mercury's foreshock region: MESSENGER magnetic field observations, *J. Geophys. Res.*, *118*, 2809–2823, doi:10.1002/jgra.50342.

Leblanc, F., et al. (2007), Mercury's exosphere origins and relations to its magnetosphere and surface, *Planet. Space Sci.*, *55.9*, 1069–1092.

Lyatsky, W., G. V. Khazanov, and J. A. Slavin (2010), Alfven wave reflection model of field-aligned currents at Mercury, *Icarus*, *209*, 40–45, doi:10.1016/j.icarus.2009.11.039.

Massetti, S., S. Orsini, A. Milillo, A. Mura, E. De Angelis, H. Lammer, and P. Wurz (2003), Mapping of the cusp plasma precipitation on the surface of Mercury, *Icarus*, *166*(2), 229–237.

Masters, A., J. A. Slavin, G. A. DiBraccio, T. Sundberg, R. M. Winslow, C. L. Johnson, B. J. Anderson, and H. Korth (2013), A comparison of magnetic overshoots at the bow shocks of Mercury and Saturn, *J. Geophys. Res.*, *118*, doi: 10.1002/jgra.50428.

McClintock, W. E., E. T. Bradley, R. J. Vervack, R. M. Killen, A. L. Sprague, N. R. Izenberg, and S. C. Solomon (2008), Mercury's exosphere: Observations during MESSENGER's first Mercury flyby, *Science*, *321*, 92–94.

McClintock, W. E., R. J. Vervack, Jr., E. T. Bradley, R. M. Killen, N. Mouawad, A. L. Sprague, M. H. Burger, S. C. Solomon, and N. R. Izenberg (2009), MESSENGER observations of Mercury's exosphere: Detection of magnesium and distribution of constituents, *Science*, *324*(5927), 610–613.

McLain, J. L., A. L. Sprague, G. A. Grieves, D. Schriver, P. Travinicek, and T. M. Orlando (2011), Electron-stimulated desorption of silicates: A potential source for ions in Mercury's space environment, *J. Geophys. Res.*, *116*, E03007, doi:10.1029/2010JE003714.

Milan, S. E., and J. A. Slavin (2011), An assessment of the length and variability of Mercury's magnetotail, *Planet. Space Sci.*, *59*(15), 2058–2065, doi:10.1016/j.pss.2011.05.007.

Mitchell, D. G., F. Kutchko, D. J. Williams, T. E. Eastman, L. A. Frank, and C. T. Russell (1987), An extended study of the low-latitude boundary layer on the dawn and dusk flanks of the magnetosphere, *J. Geophys. Res.*, *92*(A7), 7394–7404, doi:10.1029/JA092iA07p07394.

Miura, A. (1996), Stabilization of the Kelvin-Helmholtz instability by the transverse magnetic field in the magnetosphere-ionosphere coupling system, *Geophys. Res. Lett.*, *23*, 761–764.

Miura, A., and J. R. Kan (1992), Line-tying effects on the Kelvin-Helmholtz instability, *Geophys. Res. Lett.*, *19*, 1611–1614.

Miura, A., et al. (2009), The sodium exosphere of Mercury: Comparison between observations during Mercury's transit and model results, *Icarus*, *200*(1), 1–11.

Mozer, F. S., and A. Retinò (2007), Quantitative estimates of magnetic field reconnection properties from electric and magnetic field measurements, *J. Geophys. Res.*, *112*, A10206, doi:10.1029/2007JA012406.

Müller, J., S. Simon, Y.-C. Wang, U. Motschmann, D. Heyner, J. Schüle, W.-H. Ip, G. Kleindienst, and G. J. Pringle (2012), Origin of Mercury's double magnetopause: 3D hybrid simulation study with A.I.K.E.F, *Icarus*, *218*, 666–687, doi:10.1016/j.icarus.2011.12.028.

Nagano, H. (1979), Effect of finite ion Larmor radius on the Kelvin-Helmholtz instability of the magnetopause, *Planet. Space Sci.*, *27*, 881–884, doi:10.1016/0032-0633(79)90013-8.

Nakamura, T. K. M., H. Hasegawa, and I. Shinohara (2010), Kinetic effects on the Kelvin-Helmholtz instability in ion-to-magnetohydrodynamic scale transverse velocity shear layers: Particle simulations, *Phys. Plasmas*, *17*, 042119, doi:10.1063/1.3385445.

Orsini, S., L. G. Blomberg, D., Delcourt, R. Grard, S. Massetti, K. Seki, and J. Slavin (2007), Magnetosphere–exosphere–surface coupling at Mercury, *Space Sci. Rev.*, *132*(2–4), 551–573.

Otto, A., and D. H. Fairfield (2000), Kelvin-Helmholtz instability at the magnetotail boundary: MHD simulation and comparison with Geotail observations, *J. Geophys. Res.*, *105*, 21,175–21,190, doi:10.1029/1999JA000312.

Potter, A. E., and T. H. Morgan (1985), Discovery of sodium in the atmosphere of planet Mercury, *Science*, *229*, 651–653.

Potter, A. E., R. M. Killen, and T. H. Morgan (2002), The sodium tail of Mercury, *Meteor. Planet. Sci.*, *37*, 1165–1172, doi:10.1111/j.1945-5100.2002.tb00886.x.

Raines, J. M., J. A. Slavin, T. H. Zurbuchen, G. Gloeckler, B. J. Anderson, D. N. Baker, H. Korth, S. M. Krimigis, and R. L. McNutt Jr. (2011), MESSENGER observations of the plasma environment near Mercury, *Planet. Space Sci.*, *59*, 2004–2015, doi:10.1016/j.pss.2011.02.004.

Raines, J. M., et al. (2013), Distribution and compositional variations of plasma ions in Mercury's space environment: The first three Mercury years of MESSENGER observations, *J. Geophys. Res.*, *118*, 1604–1619, doi:10.1029/2012JA018073.

Sarantos, M., R. M. Killen, and D. Kim (2007), Predicting the long-term solar wind ion-sputtering source at Mercury, *Planet. Space Sci.*, *55*(11), 1584–1595.

Sarantos, M., J. A. Slavin, M. Benna, S. A. Boardsen, R. M. Killen, D. Schriver, and P. Trávníček (2009), Sodium-ion pickup observed above the magnetopause during MESSENGER's first Mercury flyby: Constraints on neutral exospheric models, *Geophys. Res. Lett.*, *36*, L04106, doi:10.1029/2008GL036207.

Sarantos, M., P. H. Reiff, T. W. Hill, R. M. Killen, and A. L. Urquhart (2001), A *Bx*-interconnected magnetosphere model for Mercury, *Planet. Space Sci.*, *49*, 1629–1635, doi:10.1016/S0032-0633(01)00100-3.

Schriver, D., et al. (2011a), Quasi-trapped particle population around Mercury, *Geophys. Res. Lett.*, *38*, L23103, doi:10.1029/2011GL049629.

Schriver, D., et al. (2011b), Electron transport and precipitation at Mercury during the MESSENGER flybys: Implications for electron-stimulated desorption, *Planet. Space Sci.*, *59*, 2026–2036, doi:10.1016/j.pss.2011.03.008.

Sergeev, V. A., R. J. Pellinen, and T. I. Pulkkinen (1996), Steady magnetospheric convection: A review of recent results, *Space Sci. Rev.*, *75*, 551–604, doi:10.1007/BF00833344.

Shue, J.-H., J. K. Chao, H. C. Fu, C. T. Russell, P. Song, K. K. Khurana, and H. J. Singer (1997), A new functional form to study the solar wind control of the magnetopause size and shape, *J. Geophys. Res.*, *102*, 9497–9511, doi:10.1029/97JA00196.

Simpson, J. A, J. H Eraker, J. E. Lamport, and P. H. Walpole (1974), Electrons and protons accelerated in Mercury's magnetic field, *Science*, *185*, 160–166.

Slavin, J. A. (2004), Mercury's magnetosphere, *Adv. Space Res.*, *33*(11), 1859–1874, doi:10.1016/j.asr.2003.02.019.

Slavin, J. A., and R. E. Holzer (1981), Solar wind flow about the terrestrial planets, 1. Modeling bow shock position and shape, *J. Geophys. Res.*, *86*, 11,401–11,418, doi:10.1029/JA086iA13p11401.

Slavin, J. A., J. C. J. Owen, J. E. P. Connerney, and S. P. Christon (1997), Mariner 10 observations of field-aligned currents at Mercury, *Planet. Space Sci.*, *45*(1), 133–141, doi:10.1016/S0032-0633(96)00104-3.

Slavin, J. A., R. P. Lepping, J. Gjerloev, D. H. Fairfield, M. Hesse, C. J. Owen, M. B. Moldwin, T. Nagai, A. Ieda, and T. Mukai (2003), Geotail observations of magnetic flux ropes in the plasma sheet, *J. Geophys. Res.*, *108*(A1), 1015, doi:10.1029/2002JA009557.

Slavin, J. A., et al. (2008), Mercury's magnetosphere after MESSENGER's first flyby, *Science*, *321*, 85–89, doi:10.1126/science.1159040.

Slavin, J. A., et al. (2009a), MESSENGER observations of Mercury's magnetosphere during northward IMF, *Geophys. Res. Lett.*, *36*, L02101, doi:10.1029/2008GL036158.

Slavin, J. A., et al. (2009b), MESSENGER observations of magnetic reconnection in Mercury's magnetosphere, *Science*, *324*, 606–610.

Slavin, J. A., et al. (2010a), MESSENGER observations of large flux transfer events at Mercury, *Geophys. Res. Lett.*, *37*, L02105, doi:10.1029/2009GL041485.

Slavin, J. A., et al. (2010b), MESSENGER observations of extreme loading and unloading of Mercury's magnetic tail, *Science*, *329*, 665–668.

Slavin, J. A., et al. (2012a), MESSENGER and Mariner 10 flyby observations of magnetotail structure and dynamics at Mercury, *J. Geophys. Res.*, *117*, A01215, doi:10.1029/2011JA016900.

Slavin, J. A., et al. (2012b), MESSENGER observations of a flux-transfer-event shower at Mercury, *J. Geophys. Res.*, *117*, A00M06, doi:10.1029/2012JA017926.

Solomon, S. C., et al. (2001), The MESSENGER mission to Mercury: Scientific objectives and implementation, *Planet. Space Sci.*, *49*, 1445–1465, doi:10.1016/S0032-0633(01)00085-X.

Solomon, S. C., et al. (2008), Return to Mercury: A global perspective on MESSENGER's first Mercury flyby, *Science*, *321*, 59–62.

Starr, R. D., D. Schriver, L. R. Nittler, S. Z. Weider, P. K. Byrne, G. C. Ho, E. A. Rhodes, C. E. Schlemm II, S. C. Solomon, and P. M. Trávníček (2012), MESSENGER detection of electron-induced X-ray fluorescence from Mercury's surface, *J. Geophys. Res.*, *117*, E00L02, doi:10.1029/2012JE004118.

Sundberg, T., S. A. Boardsen, J. A. Slavin, L. G. Blomberg, and H. Korth (2010), The Kelvin-Helmholtz instability at Mercury: An assessment, *Planet. Space Sci.*, *58*, 1434–1441, doi:10.1016/j.pss.2010.06.008.

Sundberg, T., S. A. Boardsen, J. A. Slavin, L. G. Blomberg, J. A. Cumnock, S. C. Solomon, B. J. Anderson, and H. Korth (2011), Reconstruction of propagating Kelvin-Helmholtz vortices at Mercury's magnetopause, *Planet. Space Sci.*, *59*, 2051–2057.

Sundberg, T., S. A. Boardsen, J. A. Slavin, B. J. Anderson, H. Korth, T. H. Zurbuchen, J. M. Raines, and S. C. Solomon (2012a), MESSENGER orbital observations of large-amplitude Kelvin-Helmholtz waves at Mercury's magnetopause, *J. Geophys. Res.*, *117*, A04216, doi:10.1029/2011JA017268.

Sundberg, T., et al. (2012b), MESSENGER observations of dipolarization events in Mercury's magnetotail, *J. Geophys. Res.*, *117*, A00M03, doi:10.1029/2012JA017756.

Sundberg, T., et al. (2013), Cyclic reformation of a quasi-parallel bow shock at Mercury: MESSENGER observations, *J. Geophys. Res.*, *118*, 6458–6464, doi:10.1002/jgra.50602.

Swisdak, M., B. N. Rogers, J. F. Drake, and M. A. Shay (2003), Diamagnetic suppression of component magnetic reconnection at the magnetopause, *J. Geophys. Res.*, *108*, 1218, doi:10.1029/2002JA009726, A5.

Trávníček, P. M., P. Hellinger, D. Schriver, D. Herčík, J. A. Slavin, and B. J. Anderson (2009), Kinetic instabilities in Mercury's magnetosphere: Three-dimensional simulation results, *Geophys. Res. Lett.*, *36*, L07104, doi:10.1029/2008GL036630.

Uritsky, V. M., J. A. Slavin, G. V. Khazanov, E. F. Donovan, S. A. Boardsen, B. J. Anderson, and H. Korth (2011), Kinetic-scale magnetic turbulence and finite Larmor radius effects at Mercury, *J. Geophys. Res.*, *116*, A09236, doi:10.1029/2011JA016744.

Vervack, R. J., et al. (2010), Mercury's complex exosphere: Results from MESSENGER's third flyby, *Science*, *329*, 672–675.

Wang, Y.-C., and W.-H. Ip. (2011), Source dependency of exospheric sodium on Mercury, *Icarus*, *216*, 387–402.

Wang, Y.-C., J. Mueller, U. Motschmann, and W.-H. Ip (2010), A hybrid simulation of Mercury's magnetosphere for the MESSENGER encounters in year 2008, *Icarus*, *209*, 46–52, doi:10.1016/j.icarus.2010.05.020.

Winslow, R. M., C. L. Johnson, B. J. Anderson, H. Korth, J. A. Slavin, M. E. Purucker, and S. C. Solomon (2012), Observations of Mercury's northern cusp region with MESSENGER's Magnetometer, *Geophys. Res. Lett.*, *39*, L08112, doi: 10.1029/2012GL051472.

Winslow, R. M., B. J. Anderson, C. L. Johnson, J. A. Slavin, H. Korth, M. E. Purucker, D. N. Baker, and S. C. Solomon (2013), Mercury's magnetopause and bow shock from MESSENGER Magnetometer observations, *J. Geophys. Res. Space Phys.*, *118*, 2213–2227, doi:10.1002/jgra.50237

Wurz, P. J. A. Whitby, U. Rohner, J. A. Martín-Fernández, H. Lammer, C. Kolb (2010), Self-consistent modelling of Mercury's exosphere by sputtering, micro-meteorite impact and photon-stimulated desorption, *Planet. Space Sci.*, *58*(12), 1599–1616.

Zelenyi, L., M. Oka, H. Malova, M. Fujimoto, D. Delcourt, and W. Baumjohann (2007), Particle acceleration in Mercury's magnetosphere, *Space Sci. Rev.*, *132*, 593–609, doi:10.1007/s11214-007-9169-3.

Zurbuchen, T. H., J. M. Raines, G. Gloeckler, S. M. Krimigis, J. A. Slavin, P. L. Koehn, R. M. Killen, A. L. Sprague, R. L. McNutt, Jr., and S. C. Solomon (2008), MESSENGER observations of the composition of Mercury's ionized exosphere and plasma environment, *Science*, *321*, 90–92, doi:10.1126/science.1159314.

Zurbuchen, T. H., et al. (2011), MESSENGER observations of the spatial distribution of planetary ions near Mercury, *Science*, *333*, 1862–1865.

3

Magnetotails of Mars and Venus

E. Dubinin and M. Fraenz

3.1. INTRODUCTION

Mars and Venus do not have a global magnetic field and as a result their magnetic tails are different from tails of planets possessing intrinsic magnetospheres. In contrast, the magnetospheres of Mars and Venus are induced and are formed due to direct interaction of solar wind with their atmospheric/ionospheric shells. The existence of an induced magnetosphere with a magnetic tail around Venus was already observed in the early space missions Venera 9 and 10 [*Yeroshenko*, 1979; *Dubinin et al.*, 1978; *Dolginov et al.*, 1981; *Podgornyi et al.*, 1982]. The induced origin of the magnetotail on Mars was established by the Phobos-2 measurements [*Yeroshenko et al.*, 1990], although some observational features have also indicated a contribution from internal sources [*Dubinin et al.*, 1983, 1994; *Smirnov and Israelevich,* 1983]. Finally, this issue was resolved by the Mars Global Surveyor observations, which have shown that, at present, Mars does not possess a global intrinsic magnetic field but rather has strong magnetic anomalies of crustal origin [*Acuña et al.*, 1998]. At that time, a lot of information about properties of induced magnetospheres was already obtained by the Pioneer Venus Orbiter (PVO) that has explored the Venus space for almost 14 years [*Russell et al.*, 2006].

As a result of the direct interaction with the ionospheres of Mars and Venus, the magnetic flux tubes carried by solar wind are mass loaded by planetary photoions and draped around the planets forming the long magnetotails consisting of two lobes of opposite polarity separated by a plasma sheet [*Russell and Vaisberg,* 1983]. The resulting geometry of a magnetotail is essentially the same as that proposed by *Alfven* [1957] for cometary tails. The length of an induced magnetotail is approximately determined by the time which the magnetic flux tube loaded by the ionospheric plasma takes to slide around the planetary obstacle. According to estimates made by *Yeroshenko* [1979] and *Saunders and Russell* [1986] for Venus, this transit time is about 10–30 min, implying a magnetotail length of $50R_V$–$100R_V$. The tail-side boundary layer on Venus was well defined, at least up to $12R_V$, the most distant space sampled by PVO [*Saunders and Russell*, 1986] and up to $\sim 10R_M$ on Mars by Phobos-2. It is interesting to note that the Solar and Heliospheric Observatory (SOHO) spacecraft passing in 1996 through the Venus wake at a distance of 45×10^6 km has observed unusual high fluxes (2–4×10^3 cm^{-2}/s) of singly ionized oxygen ions which were attributed to the ions streaming from Venus [*Grünwaldt et al.*, 1997].

There are many similarities between the solar wind interaction at Mars and Venus, although some differences appear due to the stronger gravity on Venus and finite Larmor radius effects on Mars. Strong local crustal magnetic fields at Mars also add complexity and variability to the Martian magnetic field topology at the near-planet tail.

This chapter focuses mainly on recent results obtained from observations on the Mars Express (MEX) and Venus Express (VEX) spacecraft (see, e.g., *Dubinin et al.*, 2011, 2013a; *Bertucci et al.*, 2011; and references therein). MEX was inserted into an elliptical orbit around Mars in 2004. This eccentric elliptical orbit has a periapsis altitude of about 275 km, an apoapsis of 10000 km, an orbital inclination of 86° and a period of 6.75 h. VEX was inserted in 2006 into a highly elliptical polar orbit around Venus with a 24 h period and pericenter and apocenter of 180–350 and 66000 km, respectively. The plasma packages (ASPERA-3 and ASPERA-4) on both spacecraft are almost identical [*Barabash et al.*, 2006,

Max-Planck-Institut für Sonnensystemforschung, Göttingen, Germany

Magnetotails in the Solar System, Geophysical Monograph 207. First Edition.
Edited by Andreas Keiling, Caitríona M. Jackman, and Peter A. Delamere.

2007a]. There is no magnetometer on MEX. Magnetic field measurements on VEX are performed by the magnetometer MAG [*Zhang et al.*, 2006].

3.2. GENERAL FEATURES OF MAGNETOTAILS ON MARS AND VENUS

Since the orientation of the tail lobes and current sheet is controlled by the clock angle of the upstream interplanetary magnetic field, the relevant coordinate system used for the description of induced magnetospheric configurations has the X^* axis antiparallel to the upstream solar wind flow and the Y^* axis along the cross-flow magnetic field component of the IMF such that the IMF is in the ($X^* Y^*$) plane and B_y^* is always positive. Then the motional electric field $-V_{SW} \times B_{IMF}$ is along the Z^* axis.

Figure 3.1a shows the magnetic field polarity of the Venus tail in the Y^*Z^* plane based on analysis of 160 VEX orbits crossing the tail current sheet from August 2006 to January 2011. All field values within the tail ($-3R_V \le X \le -0.5R_V$) are combined. A clear separation between two magnetic lobes is observed while in the Venus Solar Orbital coordinates the distribution is random. A similar distinct separation was observed by PVO in the distant Venus tail ($-12R_V \le X_{VSO} \le -10R_V$) [*Saunders and Russell,* 1986]. The magnetic tail is also well displayed by plotting the projections of the field vectors onto the $X^* Y^*$ plane (Figure 3.1b). A shift of the current sheet ($\sim 0.2R_V$) in the $+Y^*$ direction probably appears due to asymmetry of the draping related to the X component of the IMF.

Spectrograms of electron and ion (O^+) fluxes in Figure 3.2 illustrate the morphology of plasma distribution in the Martian tail [see also *Fedorov et al.*, 2006, 2008]. The induced magnetosphere confined within the magnetospheric boundary (MB) is surrounded by the boundary layer adjacent to the magnetosheath filled by shocked solar wind. The boundary layer is a site where the momentum carried by solar wind is directly transferred to the planetary plasma that consists mainly of oxygen ions. The mean ion energy in the boundary layer gradually decreases from several hundred to tens of electron-volts when approaching the lobes. The lobes are filled by low-energy atomic and molecular oxygen ions, the polar wind. It is worth noting that the measured energies of ions in the lobes are shifted upward by the negative spacecraft potential. The spacecraft potential can often be reasonably evaluated from a downward shift of the characteristic energy peaks of CO_2 photoelectrons [see, e.g., *Fraenz et al.*, 2010], which are good tracers of polar wind on Mars and Venus [*Dubinin et al.*, 2006; *Frahm et al.*, 2006; *Coates et al.*, 2008]. Ambipolar electric fields arising along the magnetic field lines threading the dayside planetary ionospheres at the periods of solar minimum extract and accelerate the ionospheric ions to the lobes.

In the center of the tail, ions are energized up to several hundreds of electron-volts forming the plasma sheet separating the magnetic lobes. The magnetic normal and tangential stresses push the planetary plasma tailward, acting like a jet engine. Note that although there is no magnetometer on MEX a crossing of the current sheet can be

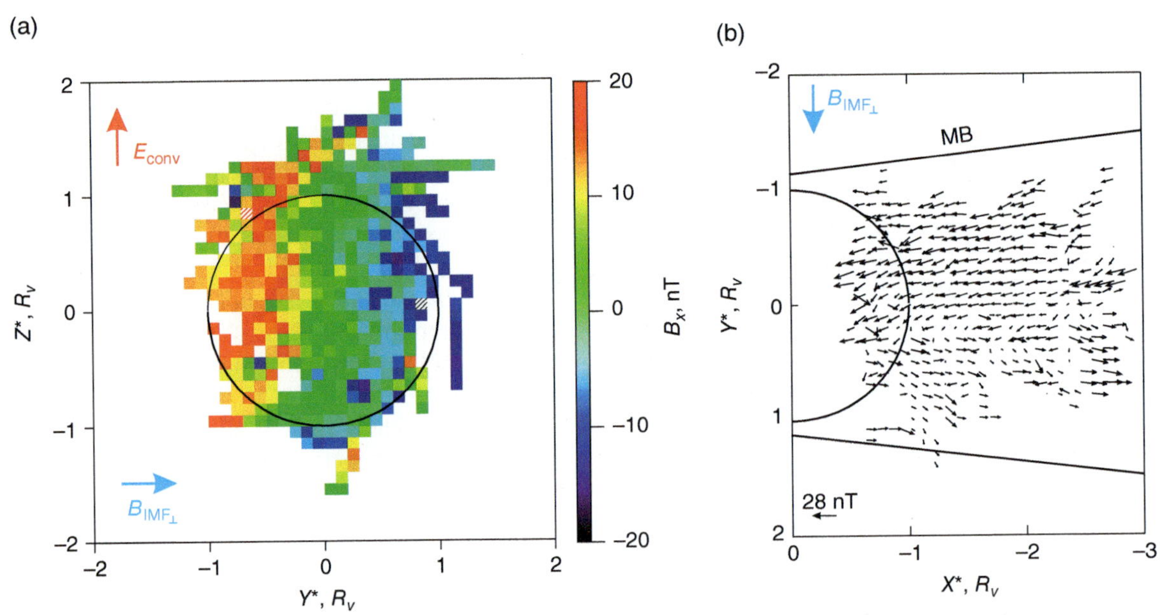

Figure 3.1 (a) Magnetic field polarity (the B_x component) of the Venus tail in the Y^*Z^* plane (magnetic coordinates) based on analysis of 160 VEX orbits crossing the tail current sheet from August 2006 to January 2011. (b) Magnetic field vectors in the X^*Y^* plane.

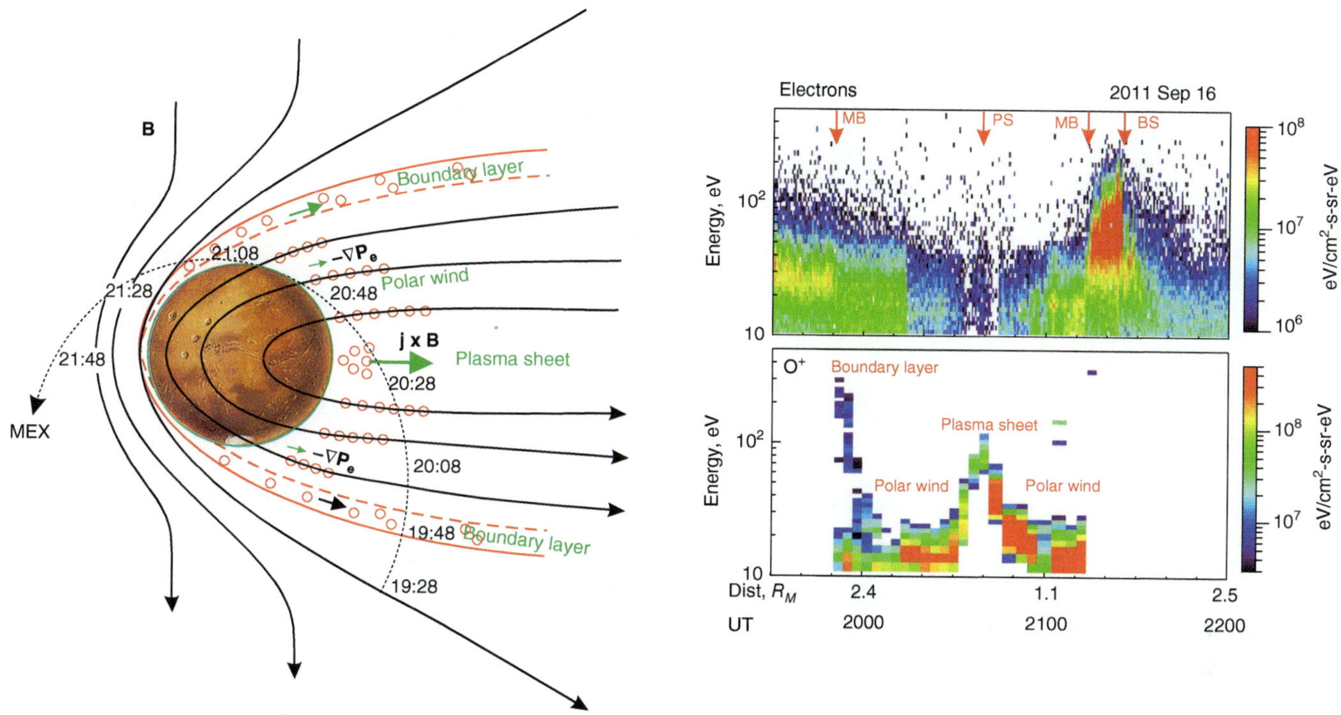

Figure 3.2 Energy-time spectrograms of electron and ion (**O**+) fluxes measured by ASPERA-3 on MEX crossing the tail. The main plasma domains, plasma sheet (PS), polar wind, boundary layer, and the characteristic boundaries bow shock (PS), magnetospheric boundary (MB) are marked. Left panel shows a sketch of the Martian magnetotail with the main plasma domains.

monitored from the measurements of suprathermal solar electrons [*Dubinin et al.*, 1994]. Halo solar electrons streaming along the interplanetary magnetic field change the direction of propagation while the field lines drape around Mars (Figure 3.3a). We observe (Figure 3.3b) that the timing of the peak of energy gained by ions coincides with a change of sign in the V_x component of the electron speed measured in the energy range of 100–500 eV, which indicates a crossing of the current sheet.

A map of the tailward fluxes of oxygen ions in the Y^*Z^* plane shown in Figure 3.4 depicts a distribution of plasma in the Martian tail. The main plasma regions, namely, the boundary layer, plasma sheet, and lobes, are clearly exposed. It is seen that although the origin of magnetotails of magnetized (Earth) and unmagnetized (Mars, Venus) planets is different a lot of similarities between their structures exist.

Since the IMF continuously varies with time, there is a certain time lag between the current sheet position determined from the instant value of the IMF clock angle and the observed position of the plasma sheet. This lag leads to a twisting of the magnetotail with the embedded plasma sheet in response to the IMF variations. Hybrid simulations by *Modolo et al.* [2012] have found that the configuration of the magnetic lobes in the near tail varies with a lag of about 2 min after the IMF rotation. The position of the plasma sheet also rotates with a lag, leading to a mismatch between its real orientation and the one expected from the IMF clock angle (Figure 3.5).

3.3. ION ACCELERATION

Since the $\boldsymbol{j}\times\boldsymbol{B}$ force due to the magnetic shear stresses of the draped field lines is the strongest in the center of the tail, the ion energy observed by a spacecraft crossing the tail gradually increases, reaches maximum at the center, and then again decreases. Such a bell-like shape of the energy variation can be easily explained. Indeed, variations of the B_x component of the magnetic field across the current sheet can be well described as a hyperbolic tangent of the distance to the center, $B_x \approx B_{xo}\tanh(l_\perp/\delta)$, where $l_\perp$ and δ are the distance of MEX to the center and the half-width of the plasma sheet, respectively (Figure 3.6). Then, the tangential forces $(1/\mu_o)(\boldsymbol{B}\cdot\nabla)\boldsymbol{B}$ and the energy W_i gained by ions will vary with distance from the center as a squared hyperbolic secant:

$$\rho\frac{d\boldsymbol{V}}{dt} = \rho(\boldsymbol{V}\cdot\nabla)\boldsymbol{V} = \boldsymbol{j}\times\boldsymbol{B} \approx 2\frac{(\boldsymbol{B}\cdot\nabla)\boldsymbol{B}}{\mu_o} \approx 2\frac{B_\perp(\partial B_x/\partial l_\perp)}{\mu_o} \quad (3.1)$$

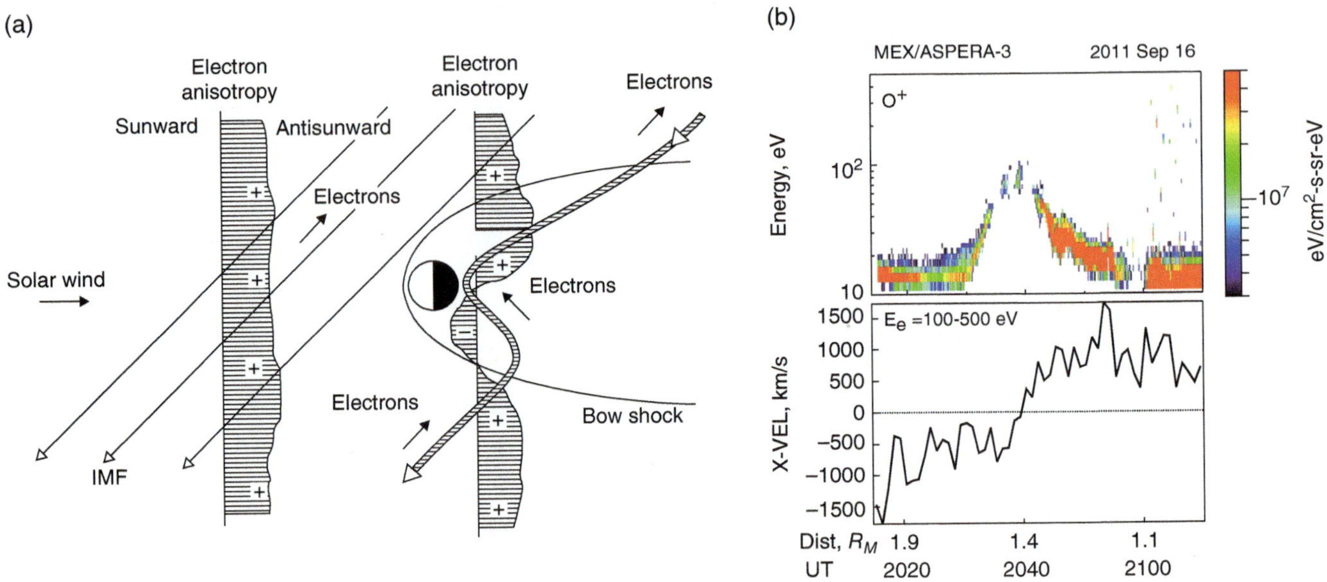

Figure 3.3 Suprathermal solar electrons can be used to monitor the position of the current sheet in the Martian tail by a reversal of the direction of their propagation (there is no magnetometer on MEX).

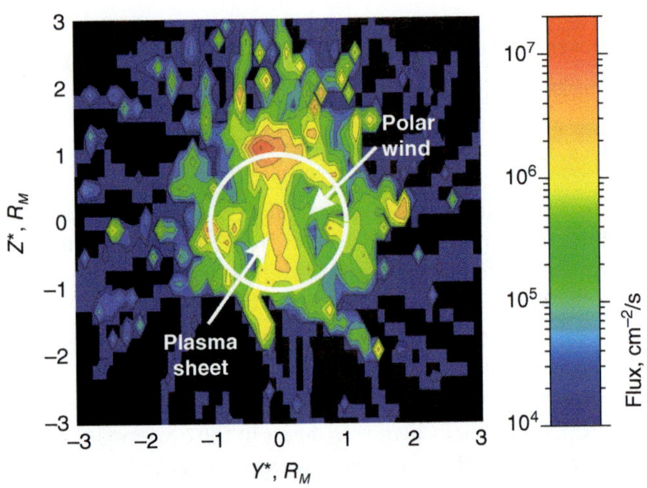

Figure 3.4 Distribution of tailward fluxes of oxygen ions in the Martian tail with clear signatures of the plasma sheet and polar wind.

and

$$W_i \approx 2\frac{B_\perp B_{xo}}{\mu_o n_i}\sec\mathrm{h}^2\left(\frac{l_\perp}{\delta}\right)\frac{L}{\delta}. \tag{3.2}$$

Here B_x and $B_\perp$ are the X component and the cross-flow component of the magnetic field in the tail, respectively, L is a characteristic scale of acceleration, and n_i is the ion number density. The additional factor 2 appears due to the sum of the normal and tangential magnetic forces. This expression describes not only the shaping of the ion spectrograms across the current sheet but also the characteristic energies of ions in the PS [see also *Dubinin et al.*, 1993, 2011, 2013a].

An interesting feature of ion acceleration on Mars is that atomic and molecular oxygen ions are accelerated up to almost the same energy. It was suggested that such an effect is related to a large ion Larmor radius as compared to the characteristic size [*Dubinin et al.*, 1993]. Since O^+ and O_2^+ ions are not magnetized, the $\boldsymbol{j} \times \boldsymbol{B}$ force accelerates electrons. The corresponding Hall electric field extracts and accelerates ions that gain the same energy/charge, which is proportional to the potential drop. Lighter ions with smaller Larmor radius are accelerated to lower energies. A similar mechanism of ion extraction and acceleration was observed in the near tail of the artificial AMPTE comet [*Haerendel et al.*, 1986; *Szego et al.*, 2000].

In the Venus tail the main ion species are O^+, H^+ (of planetary origin), and He^+ ions, and therefore their strong differentiation in energy is well displayed. Figure 3.7 gives an example of energy spectrograms of different sorts of ions across the plasma sheet (droplike picture of ion fluxes is due to a slow (~ 3 min) electrostatic elevation scan of the ion sensor). We observe a bell shaping of the spectrograms for all ion species while more heavy ions gain higher energies but lower velocities.

In some cases we observe that the ratio of energies of ion species is approximately proportional to the square root of their masses as, for example, in the case shown in Figure 3.8 [see also *Barabash et al.*, 2007c]. Presently, it is not known whether these relations correspond to a stationary flow of multi-ion species with different speeds or characterize only

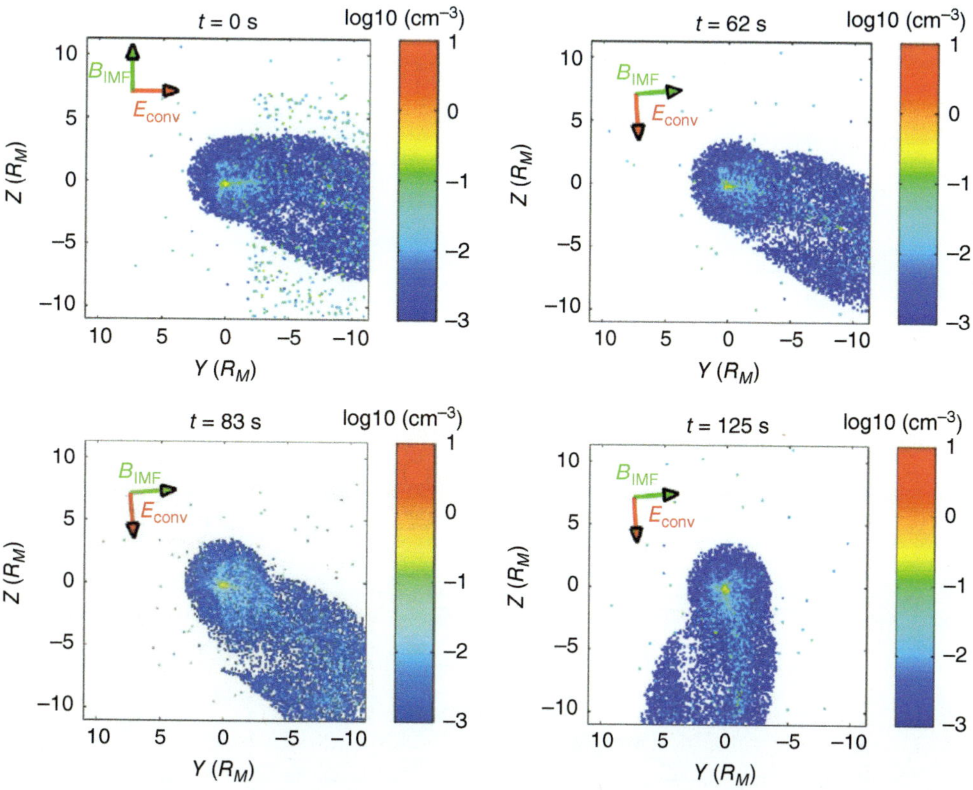

Figure 3.5 Hybrid simulations of dynamic features of solar wind interaction with Mars show a time lag (~**2** min) between the field rotations in the solar wind and orientation of the plasma sheet in the near-Mars tail.

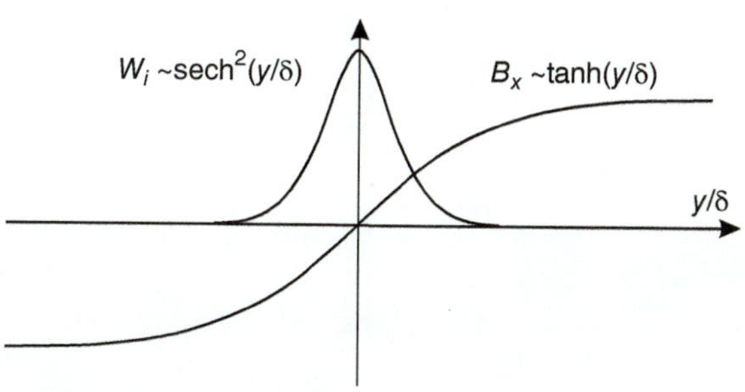

Figure 3.6 Hyperbolic tangent and squared hyperbolic secant approximately describe variations of the $\boldsymbol{B}_x$-component of the magnetic field and the magnetic tangential stresses across the current sheet.

a transition to the flow with equal velocities. Note that the steady presence of differential streaming between the main proton population and minor solar wind ions remains an intriguing problem of solar wind physics since the early observations [see, e.g., *Neugebauer*, 1981; *Marsch et al.*, 1982; *Dubinin et al.*, 2005].

For nominal solar wind conditions ion beams in the plasma sheet are usually rather cold ($T_i \leq 3-5$ eV) (Figure 3.9a), indicating an ionospheric origin of plasma filling the plasma sheet. During the disturbed periods a significant heating with appearance of suprathermal ions with much higher energies is observed, signifying an important role of other acceleration mechanisms (Figure 3.9b). For example, ion-ion instabilities can result in anomalous friction between interpenetrating magnetosheath and ionospheric plasmas and significant ion heating with generation of suprathermal tails in the ion distributions [see, e.g., *Dobe et al.*, 1999].

3.4. BURSTY FLOWS

The ion and electron measurements performed at higher time resolution have observed bursty flows in the plasma sheets of Mars and Venus which resemble a bullet like ejection of planetary plasma (Figure 3.10) [*Dubinin et al.*, 2012]. A period of bursts is about 1–2 min. Their appearance might be related to a different kind of large-amplitude waves propagating in current sheets driven, for example, by sausage- or kink-type instabilities. Such perturbations are exposed as flapping motions of the plasma sheet. It is interesting to note that flapping motions are also typical for the geotail. *Sergeev et al.* (2006) have found a close relationship between the occurrence of flapping motions and bursty plasma flows

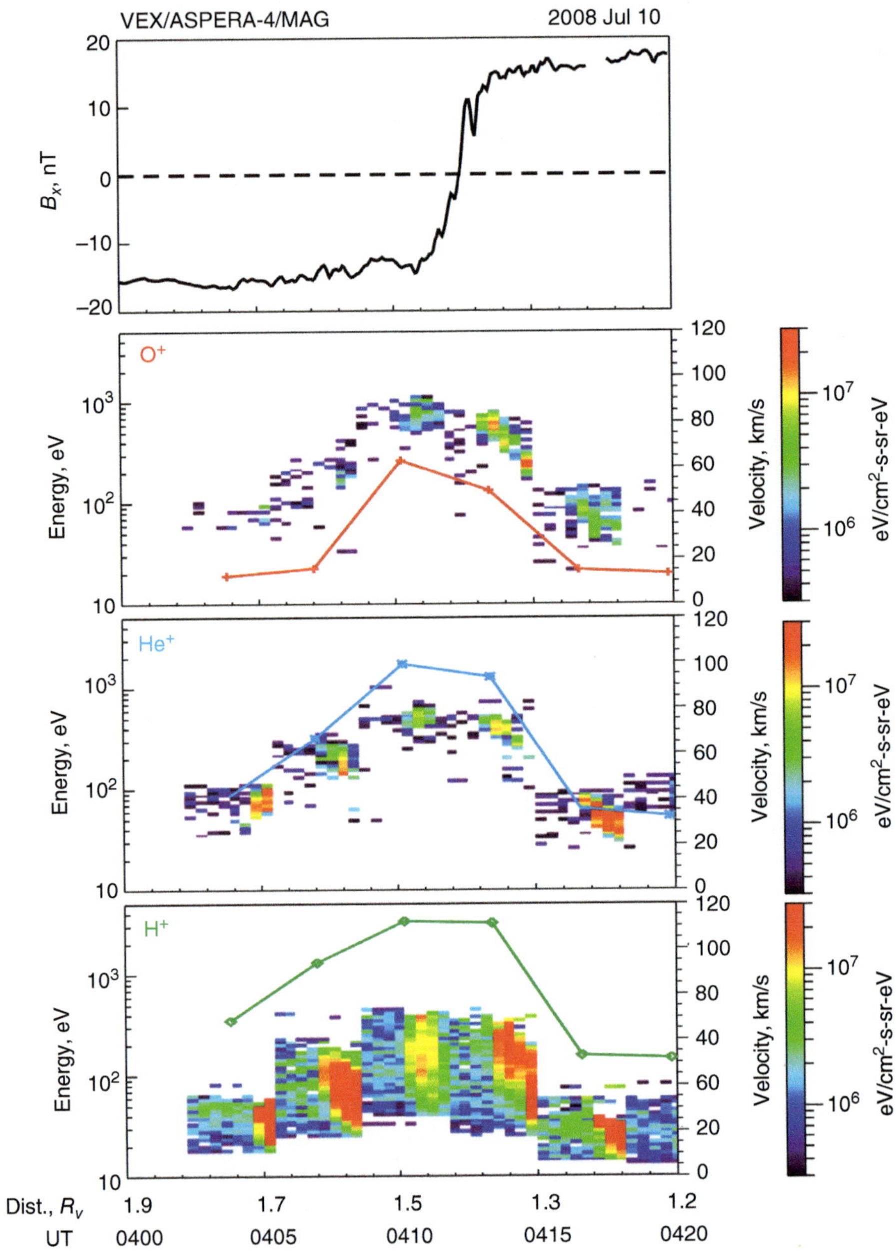

Figure 3.7 Typical variations of energy spectra and bulk speeds of different ion species measured by VEX crossing the plasma sheet on Venus.

in Earth's magnetic tail. It was suggested that both phenomena might be triggered by reconnections that launch large-amplitude waves propagating from a reconnection site.

Another source of spiky flows might be the waves generated in the foreshock or magnetosheath. Almost coherent structures in plasma flow are often observed on Mars and Venus in the magnetosheath [see, e.g., *Winningham et al.*, 2006; *Dubinin et al.*, 2011]. Such pressure pulses impacting the induced magnetosphere can modulate the slingshot force accelerating ions in the plasma sheet. One also cannot exclude a link between periodic pressure pulses and instability or reconnection processes in the plasma sheet. *Lundin et al.* [2011] have suggested that ultralow-frequency (ULF) waves (3–20 mHz) penetrating into the induced magnetosphere of Mars not only modulate the macroscopic forces but also directly accelerate ions in the tail.

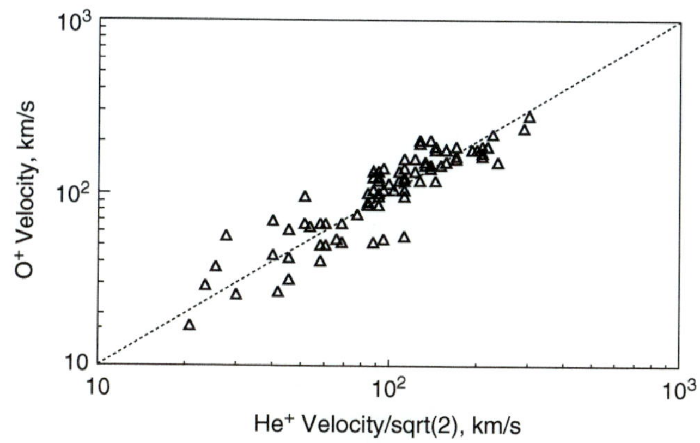

Figure 3.8 Velocities of **O**$^+$ and **He**$^+$ and ions in the central plasma sheet of Venus. Dotted curve corresponds to the mutual relation between the ion energies proportional to the square root of their mass ratio.

Large-amplitude coherent structures in the Mars and Venus tails might also be generated by a differential streaming of different ion species with respect to the net center of charge. Indeed, in the coordinate systems where these ions are at rest, they experience a different electric field and will intermittently gain and lose momentum, producing their periodic differential streaming (ion-ion waves). Such a periodic exchange of the momentum between ions was observed when the spacecraft (Phobos-2) was moving for a long time along the tail [*Dubinin et al.*, 1996].

3.5. RECONNECTION IN INDUCED TAILS

Remote observations of cometary tails, which have also an induced origin, gave us a lot of examples of sudden cometary tail disconnection that appear as a detachment of a very large plasma cloud from the coma [see, e.g., *Niedner and Brandt*, 1978]. It was suggested by

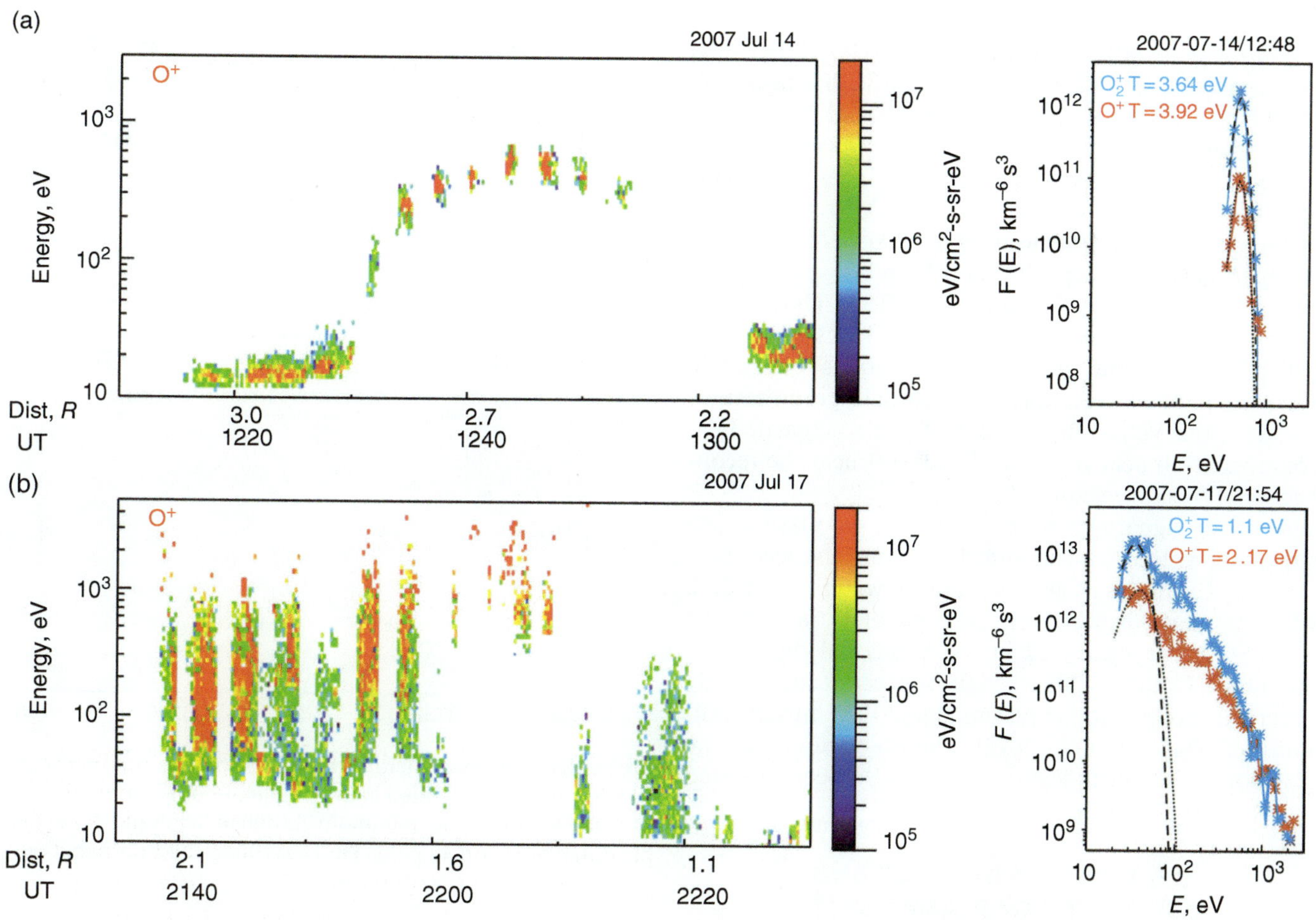

Figure 3.9 Energy-time spectrograms of oxygen ions measured by MEX crossing the plasma sheet on Mars during the "quiet" (a) and the "disturbed" (b) conditions. Examples of typical distribution functions are also shown. Temperatures of the cold ion cores are given.

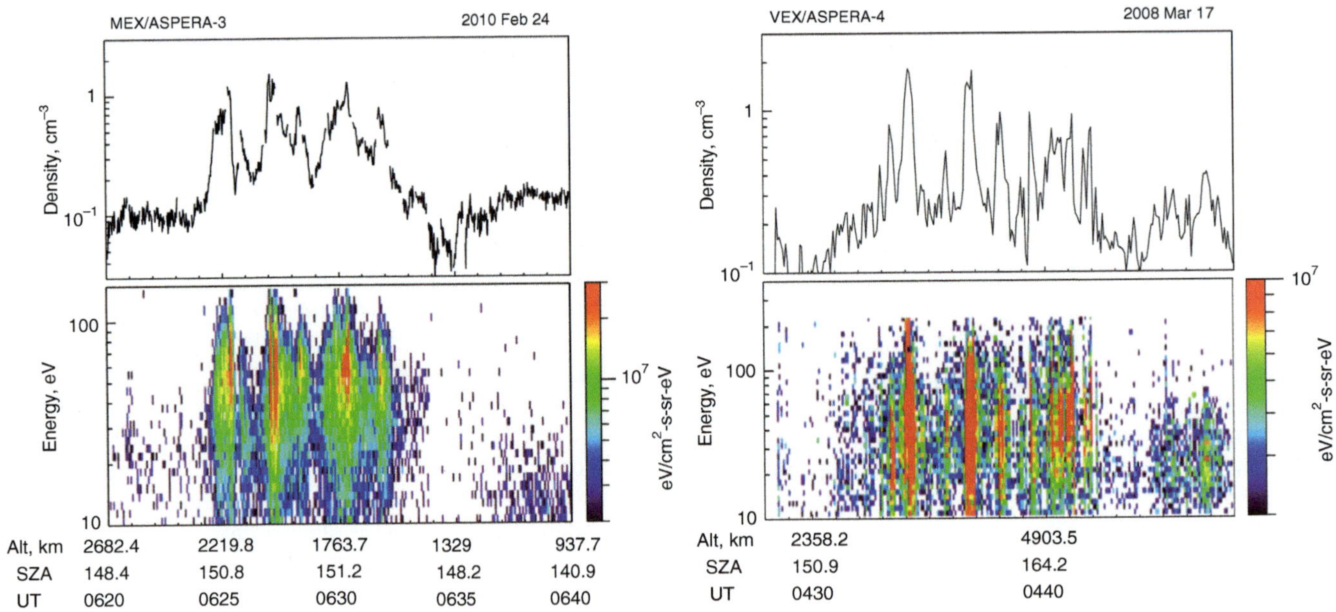

Figure 3.10 Spiky structure of the Martian and Venus plasma sheets observed in the fast mode of the electron measurements.

Podgornyi et al. [1984] and *Russell et al.* [1986] that such events may occur due to reconnection in the near tail. Similar processes can happen on Mars and Venus. The indirect evidence of the reconnection process in their induced magnetotails might be Hall magnetic fields appearing during a thinning and probably reconnection of the current sheet [*Eastwood et al.*, 2008; *Halekas et al.* 2009]. *Halekas et al.* [2009] have found 26 events with clear features of the Hall magnetic field. The average ratio of the Hall field to the main field was about 0.5, which is consistent with the observations on Earth. Figure 3.11 shows examples of such field perturbations on Mars and Venus. Bipolar out-of-plane magnetic field components appear due to Hall currents near the reconnection diffusion region.

Another signature of reconnection is plasma jets streaming in the opposite directions from the reconnection site. Examples of such jets are shown in Figure 3.12 which presents distribution functions of O^+ ions in the *XY*-VSO plane measured in the Venus tail during the event on 4, October 2008 when the spacecraft successively observed entries into the plasmoid and then into the tailward jet [*Dubinin et al.*, 2012]. The observations of the plasmoid in the near-Venus tail have also been reported by *Zhang et al.* [2012].

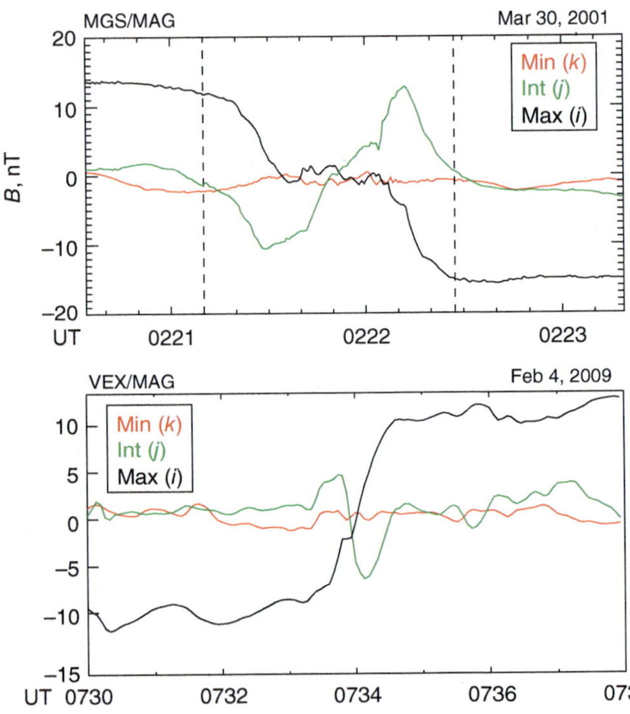

Figure 3.11 Magnetic field variations across the current sheets of Mars and Venus plotted in the principal-component axes ***i***, ***j***, ***k*** obtained from the minimum variance analysis show the appearance of the bi polar Hall magnetic field perturbations associated with the reconnection.

3.6. PRESSURE BALANCE AND ASYMMETRY OF PLASMA SHEET

Low temperatures of ion beams in the plasma sheet of Mars (see, e.g., Figure 3.9a) imply that the magnetic pressure in the lobes can scarcely be in balance with the thermal pressure of plasma confined in the current sheet. Indeed, the measurements in the Venus tail show an absence of pressure balance. Plasma pressure is several times less than the magnetic pressure (Figure 3.13).

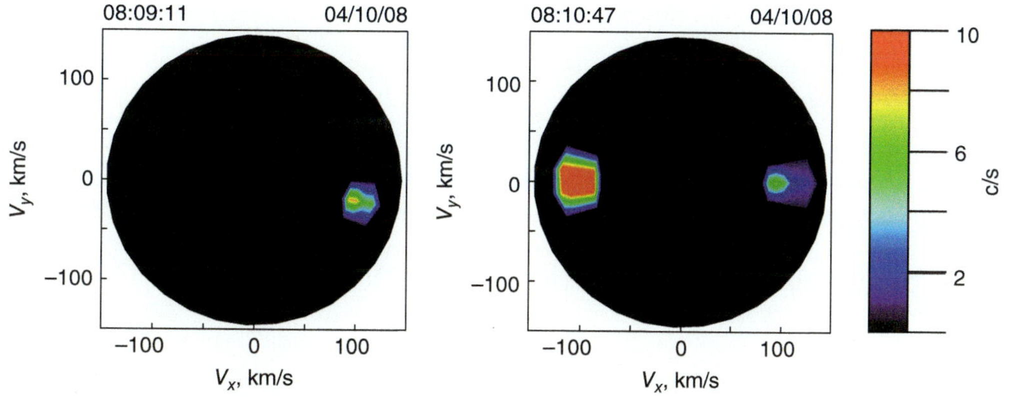

Figure 3.12 Examples of the velocity distributions of oxygen ions in the **XY – VSO** plane. Plasma jets streaming in the opposite directions indicate the existence of reconnection in the near-Venus tail.

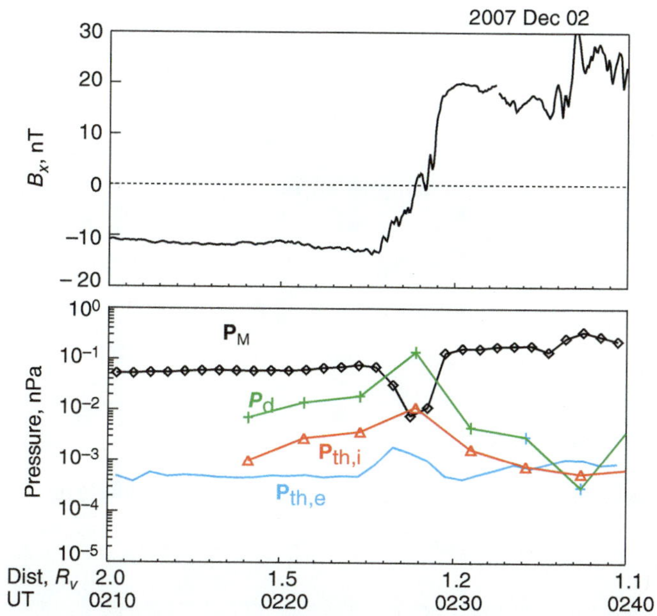

Figure 3.13 Variations of the magnetic pressure ($\mathbf{P}_M$), ion ($\mathbf{P}_{th,i}$), and electron ($\mathbf{P}_{th,e}$) thermal pressures and dynamic pressure of the moving tailward plasma ($\boldsymbol{P}_d$) across the current sheet on Venus show that the plasma sheet does not subsist in a static equilibrium.

On the other hand, the dynamic pressure of the streaming planetary plasma provides us the required pressure balance. This implies that the plasma sheet does not subsist in a static equilibrium, and the electric currents that support the observed magnetic field configuration are mainly inertial currents rather than pressure gradient currents.

Another important feature is a strong asymmetry of the Venusian plasma sheet. Such an asymmetry is observed between the hemisphere pointing in the direction of the motional electric field (E^+ hemisphere) and the hemisphere pointing in the opposite direction (E^- hemisphere). The asymmetry is exposed not only in plasma characteristics but also in an asymmetrical supply of the plasma sheet. Figure 3.14 depicts the projections of the velocity vectors of oxygen ions onto the X^*Z^* plane. Instead of an expected symmetric pattern where the magnetic field lines slipping around the poles from the dayside drag the ionospheric plasma to the tail by their tangential stresses (left panel in Figure 3.14), we observe a strong asymmetry (right panel in Figure 3.14). Indeed, in the E^+ hemisphere, the $\boldsymbol{j} \times \boldsymbol{B}$ force accelerates plasma tailward, filling the plasma sheet, while in the E^- hemisphere, the flow pattern occurs less regular, with a frequent reversal of the velocity vector. A similar picture is observed also for other ion species (He^+, H^+).

Asymmetry appears not only in the velocity distribution but also in the density. Density in the E^- hemisphere is significantly higher than in the E^+ hemisphere. Figure 3.15 compares maps of distributions of the velocities and densities of oxygen ions in the Y^*Z^* and X^*Z^* planes.

Note that the possibility that the magnetotail of Venus may be resupplied with magnetic flux tubes in an asymmetrical fashion was already suggested by *Slavin et al.* [1989] on the base of the PVO observations at $X = -8$ to $X = -12R_V$. *Slavin et al.* [1989] have shown that significant fluxes of H^+ and O^+ were detected mainly in the E^+ hemisphere near the cross-tail current sheet.

The observed strong asymmetry is controlled by the direction of the motional electric field in the upstream solar wind and probably has its origin already on the dayside. The magnetic field observations and simulations show that pile-up of the IMF on the dayside is asymmetrical [*Luhmann et al.*, 1985; *Vennerstrom et al.*, 2003, *Modolo et al.*, 2006]. The magnetic barrier that separates the magnetosheath from the ionosphere is shifted from the subsolar region to the E^+ hemisphere. This asymmetry is attributed to the action of the motional electric

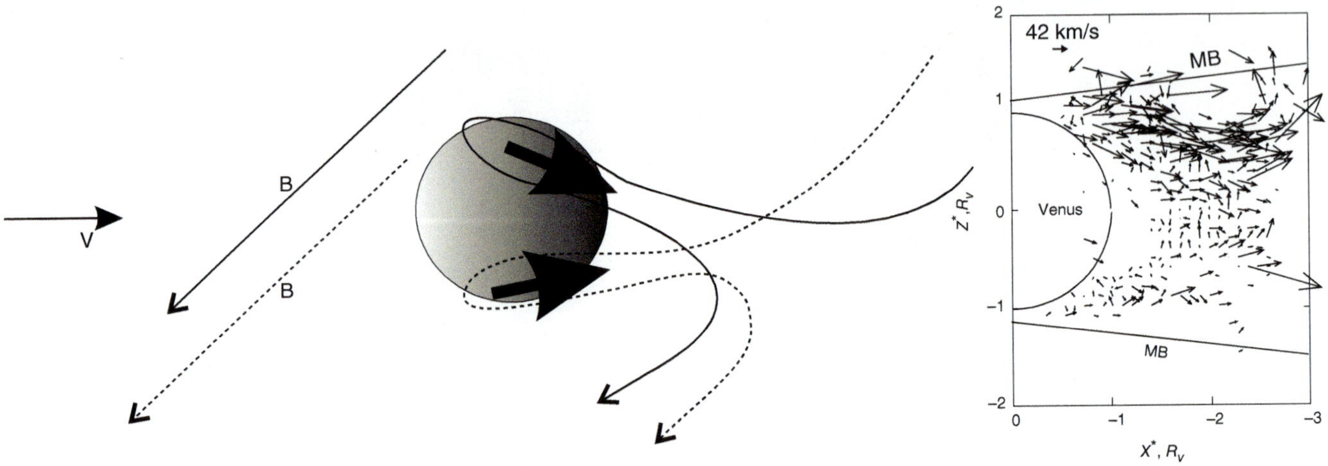

Figure 3.14 Projections of the velocity vectors of oxygen ions onto the **X*****Z*** plane in the near Venus tail. A strong asymmetry between the **E**$^{+}$ and **E**$^{-}$ hemispheres is seen. Left panel illustrates a symmetrical supply of the plasma sheet.

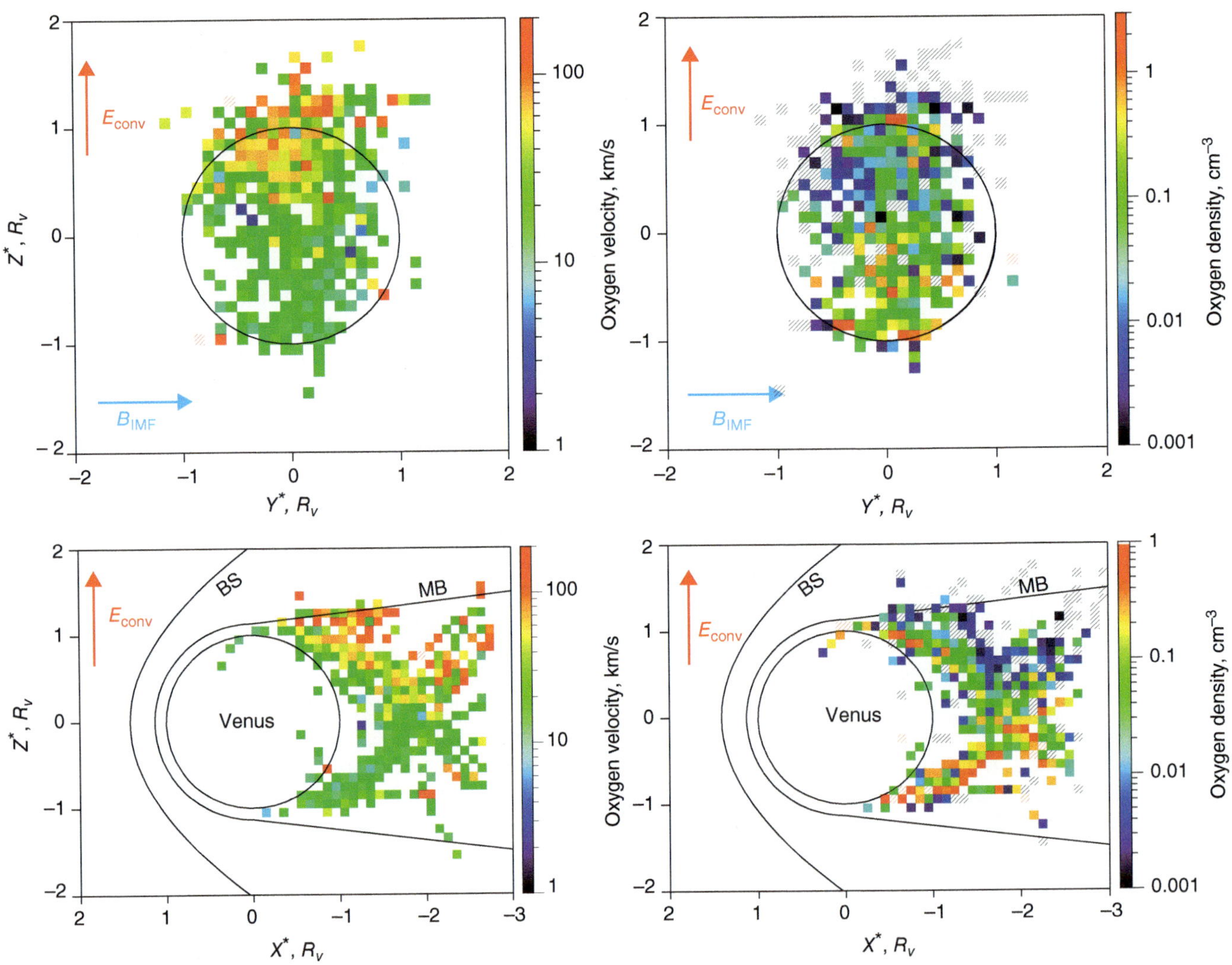

Figure 3.15 Maps of the velocity and density of oxygen ions plotted as projections onto the **Y*****Z*** and **X*****Z*** planes in the near-Venus tail. Plasma that fills the tail in the **E**$^{+}$ hemisphere is less dense as compared to that in the **E**$^{-}$ hemisphere and flows with a higher speed.

field. While oxygen ions that are not magnetized are extracted and accelerated by the electric field, the electrons drift in the crossed electric and magnetic field, feeding the Hall current, which enhances the magnetic field and, correspondingly, the tension forces in the E^+ hemisphere.

Another factor that influences the dynamics of ions in the near-Venus tail is related to a difference in the draping pattern in both hemispheres [*Zhang et al.*, 2010]. The observations and hybrid simulations show that in the hemisphere with inward directed electric field (E^- hemisphere) the magnetic field lines are wrapped more tightly around Venus than in the opposite E^+ hemisphere (Figure 3.16a). Such a reversal of the cross-flow component (B_y^*) of the magnetic field produces the inward directed field stresses that can modify a plasma transport to the tail in the E^- hemisphere. Figure 3.16b, which presents the B_y^* component in the near - Venus tail, shows the predominance of its reversal at $Z^* < 0$ (bins marked by minus sign). We also observe a frequent reversal of the V_x component of the proton velocity (bins marked by plus symbol).

3.7. ION ESCAPE THROUGH TAILS

Magnetotails of Mars and Venus are the main channels through which solar wind scavenges planetary atmospheric/ionospheric material. Such an erosion of atmospheres of nonmagnetized planets can be an important factor for evolution of their atmosphere and

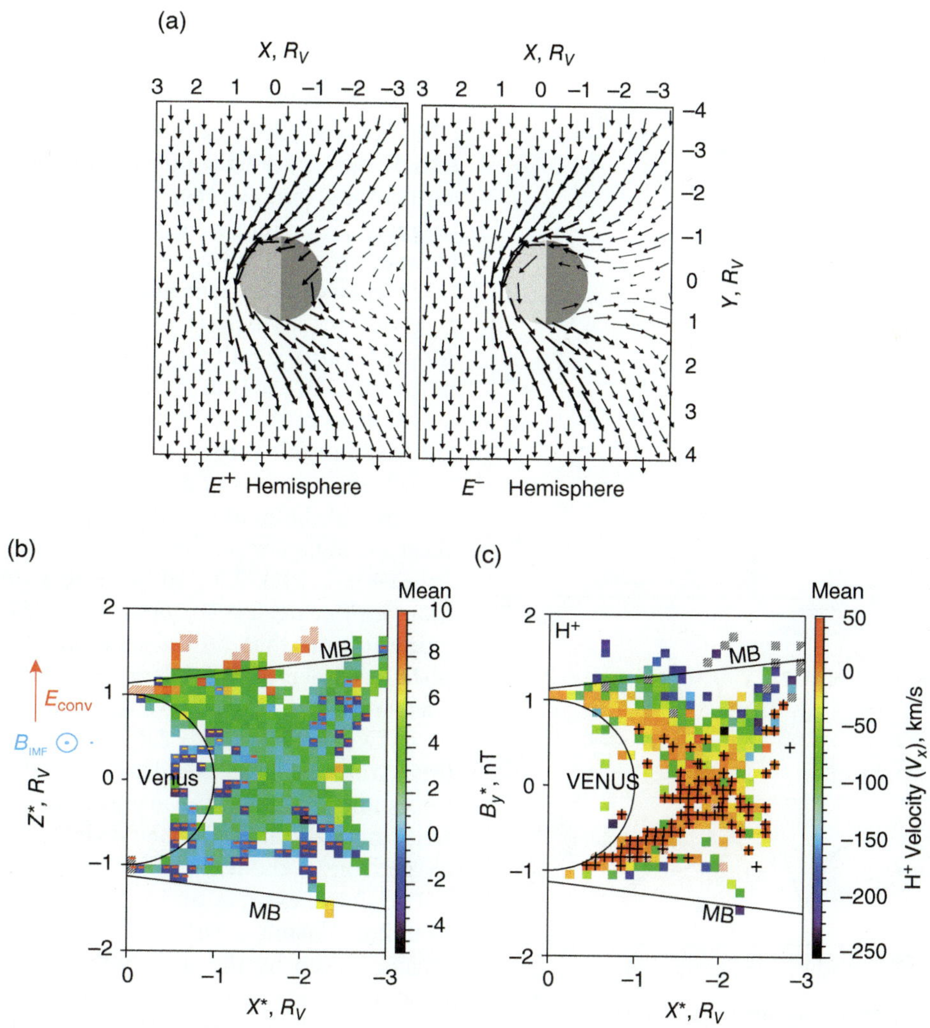

Figure 3.16 (a) Asymmetry in the draping pattern on Venus with a wrapping in the $\mathbf{E}^-$ hemisphere observed in the hybrid simulations (*Zhang et al.*, 2010) and observations (b) is probably responsible for the irregular plasma flows in this hemisphere and reversals of the $\mathbf{V}_x$ component of the ion velocities (c).

water inventory. For example, it is believed that Mars in the past possessed a much thicker atmosphere and water oceans. Nonthermal atmospheric escape forced by the direct solar wind/atmosphere interaction might be an important factor for the Mars dehydration [see, e.g., *Luhmann and Bauer*, 1992; *Luhmann and Kozyra*, 1991; *Chassefiere and Leblanc*, 2004; *Lammer et al.*, 2008; *Lundin et al.*, 2007; *Lundin, 2011; Dubinin et al.*, 2011].

It is intuitively clear that solar wind is the external driver for ion acceleration because the tension forces on the nightside take origin from a pileup of the magnetic field on the dayside that in turn is controlled by the solar wind dynamic pressure. Figure 3.17 shows how the escape flux of atomic and molecular oxygen in the center of the Martian plasma sheet varies with solar wind variations. Variations in ion escape fluxes reach a factor of ~20 even for the observed moderate solar wind variations. Assuming that

$$\frac{1}{2}n_{O^+}m_{O^+}\frac{dV_{O^+}^2}{dx}+\frac{1}{2}n_{O_2^+}m_{O_2^+}\frac{dV_{O_2^+}^2}{dx}=-\nabla_x\frac{B_{tail}^2}{\mu_o} \quad (3.3)$$

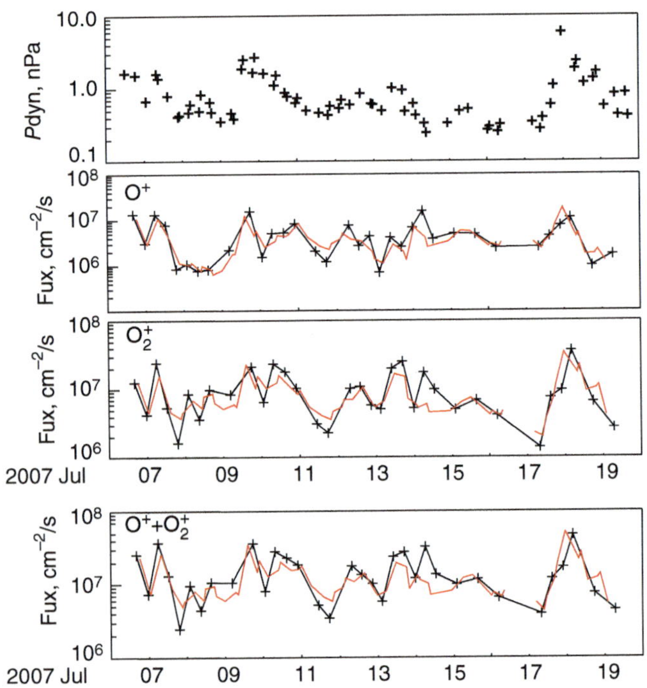

Figure 3.17 Variations in the upstream solar wind dynamic pressure cause the variations in fluxes of atomic and molecular oxygen ions in the Martian plasma sheet. Simple examination of the momentum coupling gives reasonable agreement with observations.

yields

$$n_{O^+}W_{O^+}+n_{O_2^+}W_{O_2^+}\sim\frac{B_{tail}^2}{\mu_o}=2kn_{sw}m_pV_{sw}^2, \quad (3.4)$$

where the factor of 2 appears on the right-hand side of equation 3.3 due to the combined action of the normal and tangential magnetic field stresses and we use $n_{sw}m_pV_{sw}^2=B_{mb}^2/2\mu_o$ and $k=B_{tail}^2/B_{mb}^2$. Here B_{mb}, B_{tail}, W_{O^+}, and $W_{O_2^+}$ are the magnetic field strengths in the magnetic barrier and tail and the peak ion energies in the plasma sheet, respectively. This expression can be simplified using $W_{O^+}=W_{O_2^+}$. Then, the escape flux is given by:

$$f_{O^+}=n_{O^+}V_{O^+}=2\beta f_{sw}\left(k\frac{n_{O^+}}{n_{sw}}\frac{m_p}{m_{O^+}}\right)^{1/2},$$

where f_{sw} is the solar wind flux, m_p and m_{O^+} are the masses of proton and oxygen ions, and β is the factor determined by the density ratio $\beta=\left[1+n_{O_2^+}/n_{O^+}\right]^{-1/2}$ [*Dubinin et al.*, 2011]. Comparison of the observed fluxes (black curves) and values evaluated from the above expressions (red curves) gives a reasonable agreement (Figure 3.17). Here we assumed that the factor k, which characterizes the efficiency of the momentum transfer from solar wind to ions accelerated in the plasma sheet, is about 0.1, which is reasonable if we compare the squared ratio of the magnetic field in the magnetic barrier and in the tail.

Estimates of the total escape rate of the ions energized in the Martian tail vary in the range of $2\times10^{23}-2\times10^{24}\,s^{-1}$ [*Barabash et al.* 2007b; *Dubinin et al.*, 2011]. Average total loss rates of different ion species through the Venus plasma sheet during solar minimum conditions in 2006–2007 are $7.1\times10^{24}\,s^{-1}$ (H^+), $7.9\times10^{22}\,s^{-1}$ (He^+), and $2.7\times10^{24-1}$ (O^+) [*Barabash et al.*, 2007c; *Fedorov et al.*, 2011]. It is interesting to note that *McComas et al.* [1986] have also suggested a similar $\boldsymbol{j}\times\boldsymbol{B}$ acceleration mechanism in the Venus plasma sheet. Using the average magnetic fields measured by PVO at $-12R_V\leq X\leq-8R_V$ and self-consistent currents, these authors have calculated the electromagnetic $\boldsymbol{j}\times\boldsymbol{B}$ forces and examined the average consistent plasma properties in the distant tail. Their estimate of the escape rate of oxygen ions is about $6\times10^{24}\,s^{-1}$.

The above expression for the escape flux through the Martian plasma sheet can be used to "backcast" the escape forced by the solar wind during the early solar system. Based on studies of solarlike stars at different stages of their evolution, solar wind density and velocity evolution was estimated from 100 My after solar formation up to now [see, e.g., *Lundin et al.*, 2007, and references therein]. Assuming that the density in the plasma sheet is close to the ambient density in solar wind to neutralize the charge of ejected ions [*Haerendel et al.*, 1986] and noting that the dependence on the density ratio is a weak one can easily evaluate the escape flux of atomic and molecular ions and total losses in the past using the values of solar wind flux in the early epoch (Figure 3.18). We estimate that during the epoch of the young Sun losses of oxygen from Mars could be higher than $10^{28}\,s^{-1}$, i.e., about of 3.6×10^4 tonnes per day.

The present observations in the Martian tail also show that low-energy (≤10 eV) ionospheric ions outflowing through the magnetic lobes ("polar wind") can essentially contribute to the total losses. According to *Fraenz et al.* [2010], losses through the lobes reach $10^{25}\,s^{-1}$. It is very likely that the loss of this low-energy plasma to the solar wind is one of the primary pathways for atmospheric escape not only for Mars and Venus but also for Earth [*Andre and Cully*, 2012]. Since a relationship between polar wind losses and solar wind conditions is not established, an extrapolation to the early solar epoch requires further study.

3.8. INDUCED MAGNETIC TAILS FOR FLOW ALIGNED IMF

An interesting situation can appear when the IMF is almost aligned with the solar wind velocity. In this case the induced currents driven by the $-V_{sw} \times B_{IMF}$ electric field become very small and effects of Faraday induction might be dominant [*Dubinin et al.*, 2013b]. The currents in the conductive ionosphere induced by a "switch on" of the external magnetic field produce a dipole magnetic field whose axis is antiparallel to the IMF (Figure 3.19a).

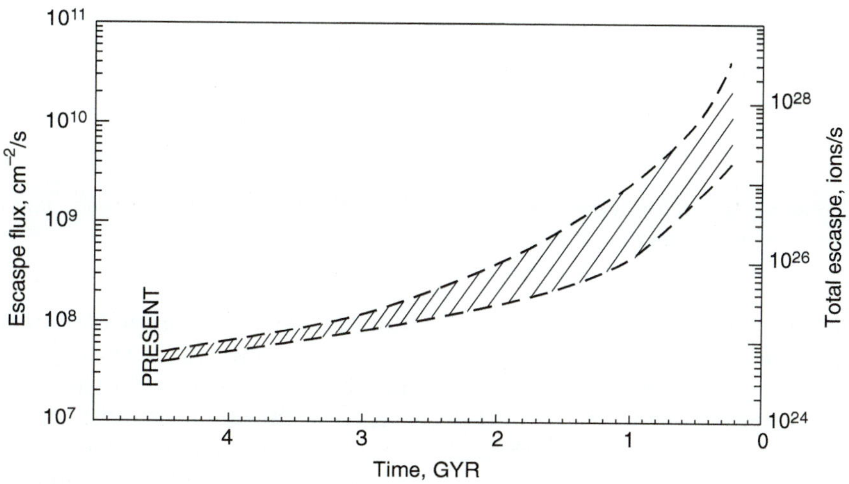

Figure 3.18 Escape rate and total losses of atomic and molecular oxygen ions (it is assumed that $\mathbf{n}_{O^+} = \mathbf{n}_{O_2^+}$) through the Martian plasma sheet as a function of time for the range of solar wind variations at different epochs.

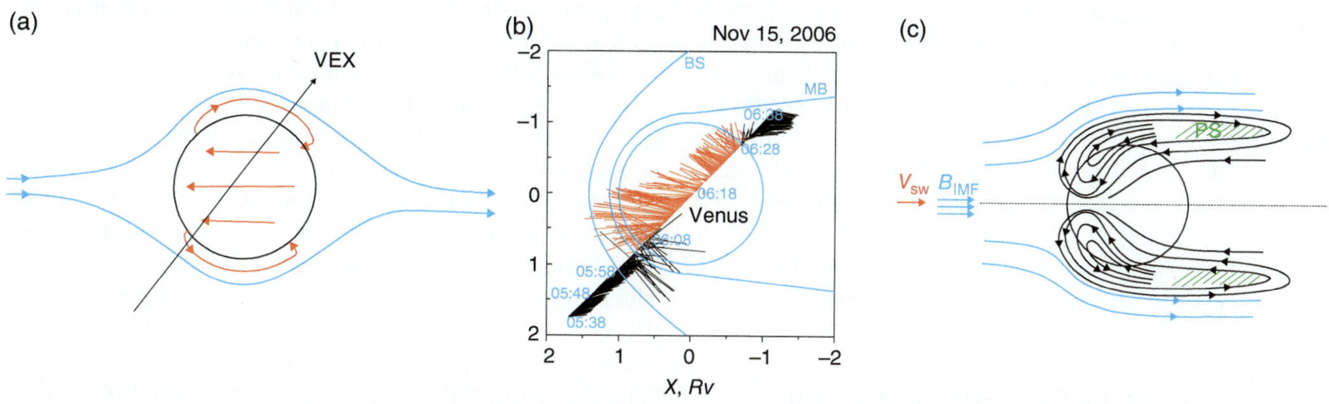

Figure 3.19 (a) Faraday eddy currents induced in the conducting Venus ionosphere can produce a dipole magnetic field whose axis is antiparallel to the IMF. For the IMF aligned with the solar wind velocity, this field might be dominant. (b) Example of the VEX observations which confirm the existence of such a type of the induced magnetosphere. (c) A cylindrical plasma sheet is a characteristic feature of the solar wind interaction with a tiny degenerated dipole with axis parallel or antiparallel to solar wind flow.

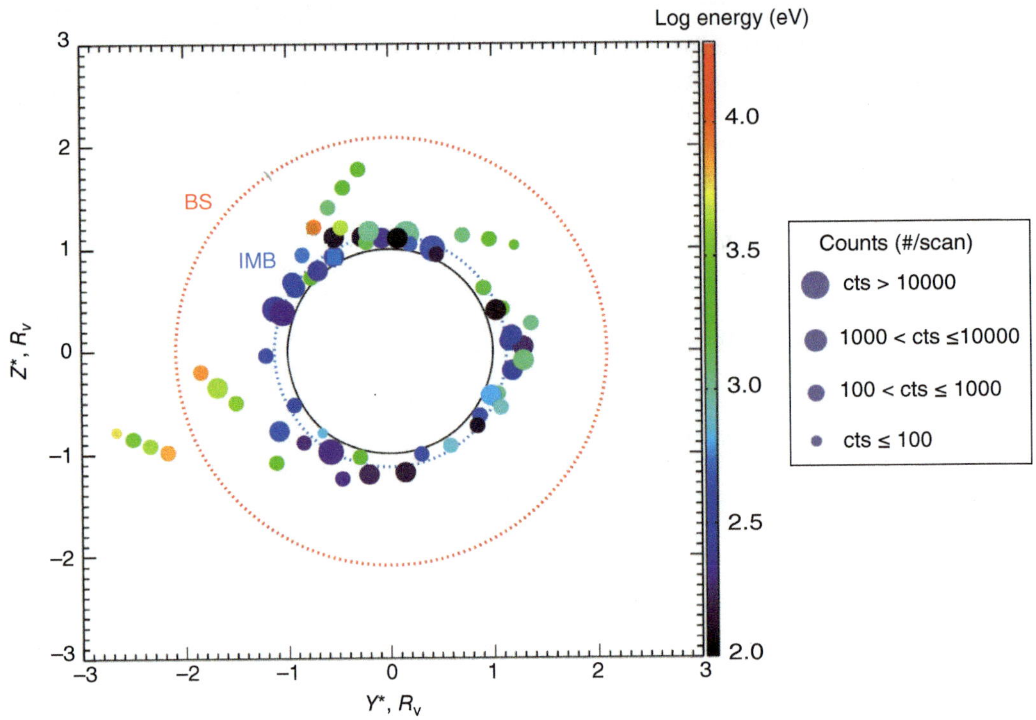

Figure 3.20 VEX observations of ions when the IMF direction is dominated by the X component show a circular spatial distribution of the ion fluxes (*Masunaga et al.*, 2011).

Early laboratory simulation experiments have confirmed the validity of this scenario [*Podgornyi et al.*, 1982]. Recent measurements on Venus bring us new evidence of such a type of induced magnetospheres. Figure 3.19b provides an example of a topology observed by VEX on Venus. The solar wind conditions were rather steady and the IMF had a dominant negative B_x component ($B_\perp \sim 1.4$ nT, $B_x \sim -7.1$ nT).

A change of sign in the B_x component inside the magnetospheric cavity is not consistent with a simple wrapping pattern and can be interpreted as the manifestation of the degenerated dipolelike magnetospheric configuration. The dipole axis for such a type of flow occurs parallel or antiparallel to solar wind velocity. The characteristic feature of solar wind interaction with a flow-aligned dipole is the existence of a circular plasma sheet (Figure 3.19c) [*Podgornyi et al.*, 1977; *Voigt et al.*, 1983]. Therefore, it is probably not surprising that the spatial distribution of fluxes of energized oxygen ions occurs rather circularly symmetrical (Figure 3.20) [*Masunaga et al.*, 2011].

3.9. EFFECT OF CRUSTAL FIELDS ON NEAR-MARS TAIL

Strong localized magnetic anomalies discovered at Mars [*Acuña et al.*, 1998] add the features typical for the Earth magnetosphere (mini-magnetopauses, mini-cusps, auroras, ion energization and escape along the "open" crustal field lines, etc.). Since the solar wind parameters and the IMF continuously vary with time, it is very difficult to document magnetospheric modifications caused by the crustal magnetic fields using only single satellite measurements. Revisiting Phobos-2 observations *Verigin et al.* [2001] have found that the width of the magnetotail at $2.8R_M$ was modulated by the location of the strongest magnetic anomalies. Global simulations of the solar wind interaction with Mars give us a better tool to monitor such modifications. Using three-dimensional magnetohydrodynamics (MHD) simulations, *Harnett and Winglee* [2005] have studied effects of anomalies on the near-Mars tail structure. It was shown that anomalies lead to large-scale asymmetries in plasma characteristics which are controlled by the location of the strongest magnetic anomalies with respect to the Sun direction and the IMF orientation. Modifications were observed at several Martian radii down the tail. A very tangled geometry of the crustal field causes the appearance of successive plasma voids bounded by regions with enhanced plasma density. In dependence on the IMF orientation quasi-closed or open minimagnetospheric configurations arise. In the latter case, IMF reconnects to the crustal magnetic field over a significant portion of the anomalous region. Flux ropes with sizes comparable to half a planetary radius initiated by reconnection were observed in

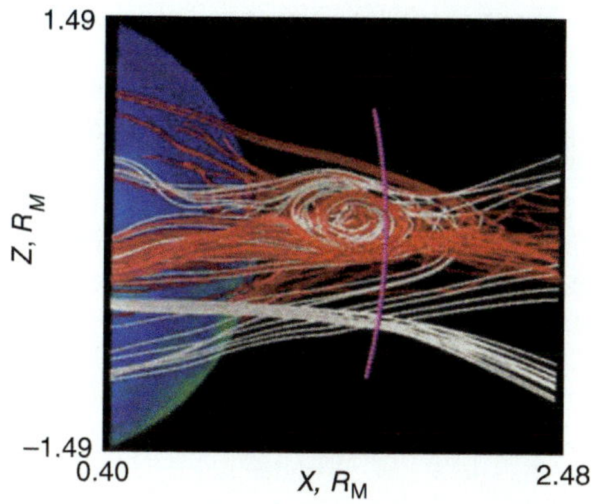

Figure 3.21 Flux rope emerging to the Martian tail in the global simulations. Pink curve shows the trajectory of the virtual spacecraft.

simulations [*Ma et al.*, 2002; *Harnett*, 2009]. Propagating after its formation down to the tail the flux rope increases in size reaching ~ $R_M/2$ in diameter and dissipates at a distance of about $3R_M$ [*Harnett*, 2009]. Figure 3.21 reproduces the magnetic field lines (different colors are only for clarity) illustrating the propagation of the flux rope in the near-Mars tail [*Harnett*, 2009]. Hundreds of large-scale flux ropes, with diameters of hundreds of kilometers, occurring near the strong crustal fields have been detected by Mars Global Surveyor at an altitude of 400 km [*Brain et al.*, 2010]. A regular ejection to the tail of large-magnetic-flux ropes can carry away significant amounts of atmosphere.

3.10. CONCLUSIONS

Although the origin and topology of the induced magnetotails at Mars and Venus are very different from those of the intrinsic magnetotail at Earth, there are a lot of similarities between their subsystems and processes therein. Since the ionospheres of Mars (most of the time) and Venus (during solar minimum periods) are mostly magnetized, the ionospheric ions extracted by the ambipolar electric field produced there an analogue of the polar wind on Earth. A plasma sheet formed by the slingshot effect of the draping IMF lines is the counterpart of Earth's plasma sheet. An analogue of the boundary layer on Earth can also be found on Mars and Venus.

Since the IMF continuously varies with time, a relative twisting of the field lines forming the induced tails is the characteristic feature of Mars and Venus magnetotails.

Large-amplitude flapping motions of plasma sheets at Mars and Venus and clear signatures of reconnection demonstrate the unique origin of such processes in both intrinsic and induced magnetotails.

The major part of plasma in the tails on Mars and Venus is of the atmospheric/ionospheric origin. Loss processes of this material through the tail forced by solar wind interaction might essentially affect the evolution of the planetary atmospheres and water inventory.

Strong crustal field anomalies on Mars add features peculiar to Earth's magnetotail that significantly complicate the magnetic field topology near Mars by the combined effects of crustal sources, IMF variations, and Mars rotation.

The strong variability inherent to induced magnetospheres calls for a continuous solar wind monitoring and multispacecraft observations of our planetary neighbors.

ACKNOWLEDGMENTS

This work was supported by grant 50QM0801 of the German Aerospace Agency (DLR) and grant MO539/17-1 of the German Science Foundation (DFG).

REFERENCES

Acuña, M., et al. (1998), Magnetic field and plasma observations at Mars: Initial results of the Mars Global Surveyor mission, *Science*, *279*, 1676–1680.

Alfven, H. (1957), On the theory of comet tails, *Tellus*, *IX*, 92.

Andre, M., and C. M. Cully (2012), Low-energy ions: A previously hidden solar system particle population, *Geophys. Res. Lett.*, *39*, L03101, doi:10.1029/2011GL050242.

Barabash, S., et al. (2006), The analyzer of space plasmas and energetic atoms (ASPERA-3) for the Mars Express Mission, *Space Sci. Rev.*, doi:10.1007/s11214-006-9124-8.

Barabash, S., et al. (2007a) , The analyzer of space plasmas and energetic atoms (ASPERA-4) for the Venus Express mission, *Planet. Space Sci.*, *55*, 1772–1792.

Barabash, S., A. Fedorov, R. Lundin, and J-A. Sauvaud (2007b), Martian atmospheric erosion rates, *Science*, *315*, 501–503.

Barabash, S., et al. (2007c), The loss of ions from Venus through the plasma wake, *Nature*, *450*, 650–653.

Bertucci, C., F. Duru, N. Edberg, et al. (2011), The induced magnetospheres of Mars, Venus and Titan, *Space Sci. Rev.*, *162*, 113–171, doi:10.1007/s11214-011-9845-1.

Brain, D. A., A. H. Baker, J. Briggs, et al. (2010), Episodic detachment of Martian crustal magnetic fields leading to bulk atmospheric plasma escape, *Geophys. Res. Lett.*, *37*, L14108, doi:10.1029/2010GL043916.

Chassefière, E., and F. Leblanc (2004), Mars atmospheric escape and evolution, interaction with the solar wind, *Planet. Space Sci.*, *52*, 1039–1058.

Coates, A., et al. (2008), Ionospheric photoelectrons at Venus: Initial observations by ASPERA-4, *Planet. Space Sci.*, *56*, 802–806, doi:10.1016/j.pss.2007.12.008.

Dobe, Z., K. Quest. V. Shapiro, K. Szego, and J. Huba (1999), Interaction of solar wind with unmagnetized planets, *Phys. Rev. Lett.*, *83*, 260–263.

Dolginov, S. S., E. Dubinin, Y. Yeroshenko, et al. (1981), About configuration of magnetic field in the Venus tail, *Cosmic Res.*, *19*, 624.

Dubinin, E., I. Podgornyi, Y. Potanin, and S. Shkolnikova (1978), Determining the magnetic moment of Venus by magnetic measurements in the tail, *Cosmic Res.*, *16*, 693–697.

Dubinin, E., P. Israelevich, I. Podgornyi, and S. Shkolnikova (1983), On the origin of magnetic fields near Mars, *Cosmic Res.*, *21*, 111–115.

Dubinin, E., R. Lundin, O. Norberg, and N. Pissarenko (1993), Ion acceleration in the Martian tail: Phobos observations, *J. Geophys. Res.*, *98*, 3991–3997.

Dubinin, E., R. Lundin, and K. Schwingenschuh (1994), Solar wind electrons as tracers of the Martian magnetotail topology, *J. Geophys. Res.*, *98*, 21,233–21,240.

Dubinin, E., K. Sauer, R. Lundin, et al. (1996), Structuring of the transition region (plasma mantle) of the martian magnetosphere, *Geophys. Res. Lett.*, *23*, 785–788.

Dubinin, E., K. Sauer, and J. F. McKenzie (2005), Differentiating ion streaming in the solar wind as an equilibrium state, *J. Geophys. Res.*, *110*, A07101, doi:10.1029/2004JA010826.

Dubinin, E., M. Fraenz, J. Woch, et al. (2006), Plasma morphology at Mars: ASPERA-3 observations, *Space Sci. Rev.*, *126*, 209–238.

Dubinin, E., M. Fraenz, A. Fedorov, et al. (2011), Ion energization and escape on Mars and Venus, *Space Sci. Rev.*, *162*, 173–211, doi:10.1007/s11214-011-9831-7.

Dubinin, E., M. Fraenz, J. Woch, et al. (2012), Bursty escape fluxes in plasma sheets of Mars and Venus, *Geophys. Res. Lett.*, *39*, L01104, doi:10.1029/2011GL049883.

Dubinin, E., M. Fraenz, T. L. Zhang, et al. (2013a), Plasma in the near Venus tail. Venus Express observations, *J. Geophys. Res.* , *118*, 1–11, doi:10.1002/2013JA019164.210.

Dubinin, E., M. Fraenz, J. Woch, et al. (2013b), Toroidal and poloidal magnetic fields at Venus: Venus Express observations, *Planet. Space Sci.*, *87*, 19–29, doi:10.1016/j.pss.2012.12.003.

Eastwood, J. P., D. A. Brain, J. S. Halekas, et al. (2008), Evidence for collisionless magnetic reconnection at Mars, *Geophys. Res. Lett.*, *35*, L02106, doi:10.1029/2007GL032289.

Fedorov, A., E. Budnik, J.-A. Sauvaud, et al., (2006), Structure of the martian wake, *Icarus*, *182*(2), 329–336.

Fedorov, A., C. Ferrier, J. A. Sauvaud, et al. (2008), Comparative analysis of Venus and Mars magnetotails, *Planet. Space Sci.*, *56*, 812–817, doi:10.1016/j.pss.2007.12.012.

Fedorov, A., S. Barabash, J.-A. Sauvaud, et al. (2011), Measurements of the ion escape rates from Venus for solar minimum, *J. Geophys. Res.*, *116*, A07220, doi:10.1029/2011JA016427.

Fraenz, M., E. Dubinin, E. Nielsen, et al. (2010), Transterminator ion flux in the Martian ionosphere, *Planet. Space Sci.*, *58*, 1442–1454, doi:10.1016/j.pss.2010.06.009.

Frahm, R. A., et al. (2006), Location of atmospheric photoelectron energy peaks within the Mars environment, *Space Sci, Rev.*, *126*, 389.

Grünwaldt, H., et al. (1997), Venus tail ray observations near Earth, *Geophys. Res. Lett.*, *24*, 1163–1166.

Haerendel, G., G. Paschmann, W. Baumjohann, and C. W. Carlson (1986), Dynamics of the AMPTE artificial comet, *Nature*, *320*, 720–723.

Halekas, J. S., J. P. Eastwood, D. A. Brain, et al. (2009), In situ observations of reconnection Hall magnetic fields at Mars: Evidence for ion diffusion region encounters, *J. Geophys. Res.*, *114*, A11204, doi:10.1029/2009JA014544.

Harnett, E. M. (2009), High-resolution multifluid simulations of flux ropes in the Martian magnetosphere, *J. Geophys. Res.*, *114*, A01208, doi:10.1029/2008JA013648.

Harnett, E. M., and R. M. Winglee (2005), Three-dimensional fluid simulations of plasma asymmetries in the Martian magnetotail caused by the magnetic anomalies, *J. Geophys. Res.*, *110*, A07226, doi:10.1029/2003JA010315.

Lammer, H., J. F. Kasting, E. Chassefière, et al. (2008), Atmospheric escape and evolution of terrestrial planets and satellites. *Space Sci. Rev.*, *139*, 399–436.

Luhmann, J. G., and S. Bauer (1992), *Venus and Mars: Atmospheres, Ionospheres and Solar Wind Interactions*, Monogr., 66, AGU, Washington, D.C.

Luhmann, J. G., and J. U. Kozyra (1991), Dayside pickup oxygen ion precipitation at Venus and Mars: Spatial distributions, energy deposition and consequences, *J. Geophys. Res.*, *96*, 5457–5467.

Luhmann, J. G., C. T. Russell, J. R. Spreiter, and S. S. Stahara (1985), Evidence for mass-loading of the Venus magnetosheath, *Adv. Space Res.*, *5*, 307.

Lundin, R. (2011), Ion acceleration and outflow from Mars and Venus: An overview, *Space Sci. Rev.*, *162*, 309–334, doi:10.1007/s11214-011-9811-y.

Lundin, R., H. Lammer, and I. Ribas (2007), Planetary magnetic fields and solar forcing: Implications for atmospheric evolution,. *Space Sci. Rev.*, *129*, 245–278.

Lundin, R. , S. Barabash, E. Dubinin, et al. (2011), Low-altitude acceleration of ionospheric ions at Mars, *Geophys. Res. Lett.*, *38*, L08108, doi:10.1029/2011GL047064.

Ma, Y., A. F. Nagy, K. C. Hansen, et al. (2002), Three-dimensional multispecies MHD studies of the solar wind interaction with Mars in the presence of crustal fields, *J. Geophys. Res.*, *107*(A10), 1282, doi:10.1029/2002JA009293.

Marsch, E., K.-H. Muhlhauser, H. Rosenbauer, et al. (1982), Solar wind Helium ions, Observations of the Helios solar probes between 0.3 and 1 AU, *J. Geophys. Res.*, 87, 35.

Masunaga, K., et al. (2011), O+ outflow channels around Mars controlled by directions of the IMF: Observations of high energy O^+ ions around terminator, *J. Geophys. Res.*, *116*, A09326, doi:10.1029/2011JA01670.

McComas, D., H. E. Spence, C. T. Russell, and M. A. Saunders (1986), The average magnetic field draping and consistent plasma properties of the Venus magnetotail, *J. Geophys. Res.*, *91*, 7939–7953.

Modolo R., G. Chanteur, E. Dubinin, and P. Matthews (2006), Simulated solar wind plasma interaction with the Martian exosphere: Influence of the EUV flux on the bow shock and the magnetic pile-up boundary, *Ann. Geophys.*, *24*, 3403.

Modolo, R., G. Chanteur, and E. Dubinin (2012), Dynamic Martian magnetosphere : Transient twist induced by a rotation of the IMF, *Geophys. Res. Lett.*, *39*, L01106, doi:10.1029/2011GL049895.

Neugebauer, M. (1981), Observations of solar wind Helium, *Fund. Cosmic Phys.*, *7*, 131.

Niedner, Jr., M. B., and J. Brandt (1978), Interplanetary gas XXIII. Plasma tail disconnection events in comets: Evidence for magnetic field reconnection at interplanetary sector boundaries? *Astrophys. J.*, *223*(2), 655–670.

Podgornyi, I., E. Dubinin, P. Israelevich, and Yu. Potanin (1977), Dynamics of plasma in laboratory models of Earth and Uranus magnetospheres, *Herold Russ. Acad. Sci., Phys.*, *41*, 1870.

Podgornyi, I., E. Dubinin, P. Israelevich, and C. P. Sonett (1982), Comparison of measurements of electromagnetic induction in the magnetosphere of Venus with laboratory simulations, *The Moon and the Planets*, *27*, 397-406.

Podgornyi, I., E. Dubinin, P. Israelevich, and S. I. Schkolnikova (1984), Dynamics of plasma in cometary tails of I-type, *Comets and Meteors*, *35*, 1–6.

Russell, C. T., and O. Vaisberg (1983), The interaction of solar wind with Venus, in *Venus*, edited by D. M. Hunten, et al., pp. 873–940, Univ. of Arizona Press, Tucson, Ariz.

Russell, C. T., M. A. Saunders, J. L. Phillips, and L. J. Fedder (1986), Near-tail reconnection as the case of cometary tail disconnections, *J. Geophys. Res.*, *91*, 1417–1423.

Russell, C. T., J. G. Luhmann, and R. J. Strangeway (2006), The solar wind interaction with Venus through the eyes of the Pioneer Venus Orbiter, *Planet. Space Sci.*, *54*, 1482–1495.

Saunders, M. A., and C. T. Russell (1986), Average dimension and magnetic structure of the distant Venus Magnetotail, *J. Geophys. Res.*, *91*(5), 5589–5604.

Sergeev, V. A., et al. (2006), Survey of large-amplitude flapping motions in the midnight current sheet, *Ann. Geophys.*, *24*, 205.

Slavin, J. A. , D. S. Intriligator, and E. J. Smith (1989), Pioneer Venus Orbiter magnetic field and plasma observations in the Venus magnetotail, *J. Geophys. Res.*, *94*(3), 2383–2398.

Smirnov, V., and P. Israelevich (1983), On the possible configuration of the Martian magnetosphere, *Space Res.*, *22*, 450–456.

Szego, K., K.-H. Glassmeier, R. Bingham, et al. (2000), Physics of mass-loaded plasmas, *Space Sci. Rev.*, *94*, 429–671.

Vennerstrom, S., N. Olsen, M. Purucker, and M. Acuna (2003), The magnetic field in the pile-up region at Mars and its variation with the solar wind, *Geophys. Res. Lett.*, *30*, 1369, doi:10.1029/2003GL016883.

Verigin, M. I., et al. (2001), Evidence of the influence of equatorial Martian crustal magnetization on the position of the planetary magnetotail boundary by Phobos 2 data, *Adv. Space Res.*, *28*, 885–889.

Voigt, G.-H., T. Hill, and A. J. Dessler (1983), The magnetosphere of Uranus, plasma sources, convection and field configuration, *Astrophys. J.*, *266*, 390.

Winningham, J. D., et al. (2006), Electron oscillations in the induced Martian magnetosphere, *Icarus*, *182*, 360.

Yeroshenko, Ye. (1979), Unipolar induction effects in the magnetic tail of Venus, *Cosmic Res.*, *17*, 77–87.

Yeroshenko, Ye., W. Riedler, K. Schwingenschuh, et al. (1990), Magnetotail of Mars: Phobos observations, *Geophys. Res. Lett.*, *17*, 885–888.

Zhang, T. L., et al. (2006), Magnetic field investigation of the Venus plasma environment: Expected new results from Venus Express, *Planet. Space Sci.*, *54*, 1336–1343.

Zhang, T. L., W. Baumjohann, J. Du, et al. (2010), Hemispheric asymmetry of the magnetic field wrapping pattern in the Venusian magnetic tail, *Geophys. Res. Lett.*, *37*, L14202, doi:10.1029/2010GL044020.

Zhang, T. L., Q. M. Lu, W. Baumjohann, et al. (2012), Magnetic reconnection in the Near Venusian Magnetotail, *Science*, *336*, 567–569, doi:10.1126/science.1217013.

4

Earth's Magnetotail

Robert L. McPherron

4.1. INTRODUCTION

Earth's magnetic tail (*magnetotail*) was discovered 50 years ago by the first spacecraft to fly downstream of Earth. The magnetotail is a complex structure created by the solar wind. Its size and orientation depend on solar wind velocity and dynamic pressure, the strength and direction of the interplanetary magnetic field (IMF), and the orientation of Earth's magnetic axis. It is a dynamic structure with internal processes responsible for most types of geomagnetic and auroral activity. An early review of magnetotail structure and dynamics is presented in *Magnetotail Physics* [*Lui*, 1987]. A more recent discussion is provided by *New Perspectives on the Earth's Magnetotail* [*Nishida et al.*, 1998]. A description of the relation of aurora to processes in the magnetotail is found in *Auroral Phenomenology and Magnetospheric Processes: Earth and Other Planets* [*Keiling et al.*, 2012]. The volume of which this chapter is part continues the discussion of Earth's magnetotail collecting presentations made at a recent AGU Chapman Conference, "Fundamental Properties and Processes of Magnetotails."

Earth's magnetotail is generated by two processes as discussed in the introduction (Chapter 1) to this volume by *Vasyliunas*, [2014]. These are the viscous interaction [*Axford and Hines*, 1961; *Piddington*, 1963; *Axford*, 1964] and magnetic reconnection [*Dungey*, 1961; *Levy*, 1964; *Levy et al.*, 1964]. The viscous interaction transports closed magnetic field lines from dayside to nightside in boundary layers on the flanks of the magnetotail. After sufficient stretching these field lines return to the dayside just inside the boundary layers. Magnetic reconnection opens dipole field lines near the subsolar point and the solar wind transports these to the nightside, forming two lobes of oppositely directed field lines. These eventually drift to the center of the tail, reconnect at an X line, and return to the dayside between the return flows of the two boundary layers.

The first detailed observations of the tail were obtained by IMP 1 [*Ness*, 1965]. This spacecraft surveyed the dawn side of the magnetopause from noon to midnight out to a distance 32 Re behind Earth. The observations showed that the magnetic field is drawn out into two oppositely directed bundles of magnetic field lines separated by a plasma sheet [*Montgomery et al.*, 1965]. A schematic interpretation of the original data is shown in Figure 4.1. The weak field region at the center of the plasma sheet is called the neutral sheet. At large distances from Earth this sheet is parallel to the solar wind and to Earth's magnetic equator. At about 11 Re distance from Earth the neutral sheet connects to the nightside magnetic equator. When the dipole tilts toward or away from the Sun, the neutral sheet is raised or lowered relative to the ecliptic plane. The tilt also causes the neutral sheet to become elliptical [*Hammond et al.*, 1994]. The tilt of the dipole orthogonal to the Earth-Sun line causes the neutral sheet and plasma sheet to rotate about this line. To remove this effect, *Ness* [1965] introduced the Geocentric Solar Magnetospheric (GSM) coordinate system which changes with time so that the plasma sheet is always parallel to the GSM equatorial plane.

The magnetotail contains several boundary layers. The first is the low-latitude boundary layer (LLBL) created by the viscous interaction [*Hones et al.*, 1972; *Eastman et al.*, 1976]. A second is the plasma sheet boundary layer (PSBL) at the top and bottom of the plasma sheet. This is a region of electron and ion beams produced near the

Department of Earth, Planetary, and Space Sciences, University of California Los Angeles, Los Angeles, California

Magnetotails in the Solar System, Geophysical Monograph 207. First Edition.
Edited by Andreas Keiling, Caitríona M. Jackman, and Peter A. Delamere.

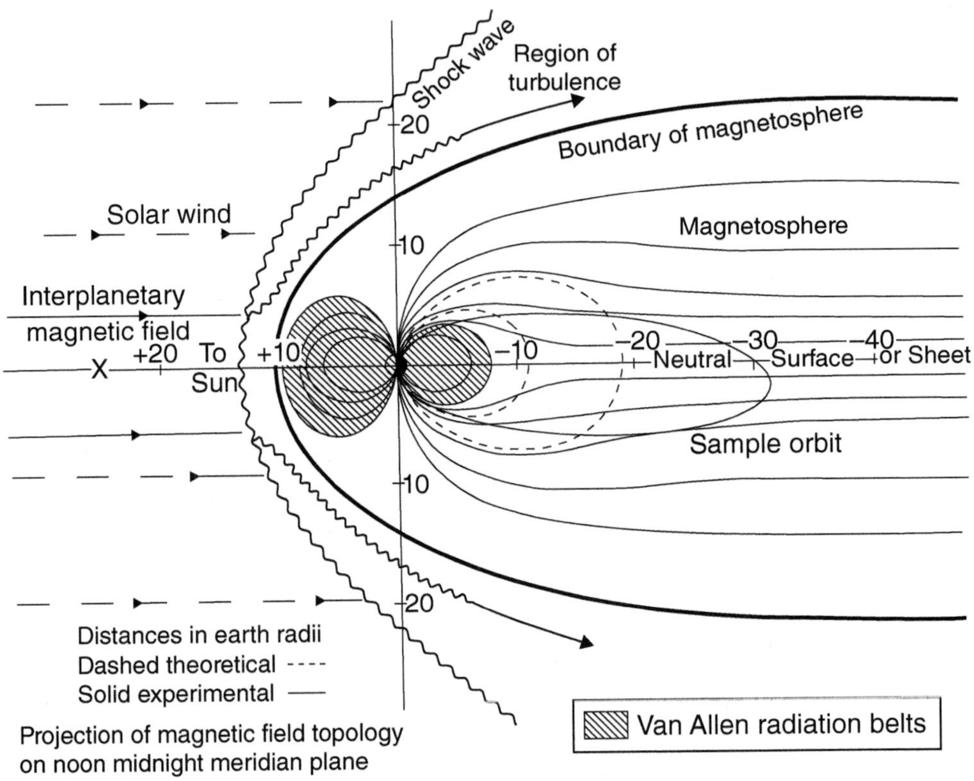

Figure 4.1 The Earth's magnetotail projected into the noon-midnight meridian based on IMP-1 magnetic field data [*Ness*, 1965] Field lines on the night side are drawn away from the Earth by the solar wind. The tail consists of two lobes of open magnetic field lines connected to the solar wind. The lobes are separated by a plasma sheet that carries the current that generates the tail field. The center of the plasma sheet is called the neutral sheet as it was first thought to have zero field, however later observations showed field lines crossing this region.

tail X line by the Speiser mechanism [*Speiser*, 1965, 1967]. The third is at the top of the north lobe and the bottom of the south lobe. These layers are called the plasma mantles and are formed by reflection of particles that enter the polar cusps [*Rosenbauer et al.*, 1975].

4.2. DYNAMIC MAGNETOTAIL

The preceding description of Earth's magnetotail might suggest the tail is a static structure flapping in the solar wind as it and Earth's dipole change direction and expanding and shrinking as solar wind dynamic pressure changes. The tail is in fact a very dynamic structure with many internal processes and rapid changes [*Sharma et al.*, 2008]. At the time the tail was discovered it was noticed that the strength of its magnetic field increases with the level of magnetic activity [*Behannon and Ness*, 1966; *Heppner et al.*, 1967]. It was quickly found that magnetic activity depends on the orientation of the IMF beginning when the IMF turns southward antiparallel to Earth's dipole field [*Fairfield and Cahill*, 1966]. These observations were interpreted as evidence that magnetic reconnection is the primary cause of magnetic activity [*Axford et al.*, 1965; *Axford*, 1967; *Axford*, 1969]. This idea led *Rostoker et al.* [1967] to suggest that the parameter controlling magnetic activity is the solar wind dawn-dusk electric field ($E_y = VB_z$) since this represents the rate at which magnetic flux is brought to the subsolar magnetopause. It was subsequently found that more complex coupling functions are better correlated with magnetic activity indices [*Newell et al.*, 2007].

As more data became available, the response of the tail to a southward turning was investigated. It was found during substorms that the strength of the tail field increases and decreases [*Caan et al.*, 1973, 1978; *Hsu and McPherron*, 2012]; thickness of the plasma sheet decreases and then increases [*Hones*, 1972; *Hones et al.*, 1984c; *McPherron et al.*, 1987]; high-speed flows occur in the plasma sheet sometimes earthward and sometimes tailward [*Hones et al.*, 1967; *Hones*, 1973; *Hones and Schindler*, 1979]; the magnetic field at synchronous orbit becomes very tail-like and then returns to a more dipolar configuration [*McPherron*, 1972b; *Kokubun and McPherron*, 1981]; and energetic particles are injected into the inner magnetosphere [*Konradi*, 1967; *Lezniak and Winckler*, 1970]. All of these changes are closely associated with auroral and magnetic disturbances and presumed to be related to their cause. These observations were soon organized into what became *the near-Earth neutral line model* (NENL) of the *magnetospheric substorm,* which we discuss below [*Hones*, 1973; *McPherron et al.*, 1973a, 1973b; *Nishida and Nagayama*, 1973; *Russell and McPherron*, 1973b]. About 15 years after the NENL model was developed new observations at the inner edge of the plasma sheet provided by the AMPTE spacecraft challenged the NENL model and the *current disruption* (CD) model was born [*Lui et al.*, 1988; *Lui*, 1991]. This model and an almost uncountable number of others were proposed as alternate explanations of the observations. Many of these are reviewed by *McPherron* [1991].

Soon after the development of the NENL model it was noted there are phenomena it could not explain. One of these is the *pseudo breakup* [*Elvey*, 1957; *Akasofu*, 1964]. The pseudo breakup starts like a typical auroral expansion but for unknown reasons dies away within a few minutes. Another is the *convection bay* [*Pytte et al.*, 1978]. The convection bay is a long interval of relatively steady magnetic activity but without any evidence of auroral expansions. Today this phenomenon is referred to as *steady magnetospheric convection* (SMC). More recently a quasi-periodic sequence of very large substorms has been identified as a separate phenomenon and named *sawtooth injection events* [*Borovsky et al.*, 1993; *Belian et al.*, 1995]. Finally there exists a phenomenon seen in prolonged activity that is called *poleward boundary intensification* (PBI) [*De La Beaujardiere et al.*, 1994]. In contrast to substorms and pseudo breakups, which start with auroral activations at the equatorward edge of the auroral oval, PBIs are auroral activations at the poleward edge.

These different phenomena represent different modes of response of the magnetosphere to driving by the solar wind [*Baumjohann*, 2002; *McPherron et al.*, 2008; *Pulkkinen et al.*, 2010]. Substorms occur in response to a fluctuating IMF B_z that allows the magnetotail to proceed through a well-defined sequence of events. SMC events are caused by moderate and steady driving and are virtually always initiated by a substorm. Sawtooth events occur during stronger driving where it appears that no steady state is possible. Pseudo breakups tend to occur in the growth phase of substorms and precede its main onset. PBIs are most frequent in the recovery phase of substorms. They frequently occur during SMC events, which might be viewed as a prolonged substorm recovery phase. PBIs also occur during sawtooth events. During prolonged solar wind driving these various response modes become intermixed and it becomes very difficult to classify activity as being one single mode.

It is important to emphasize that magnetospheric activity and its effects on the ionosphere and radiation belts are fundamentally a consequence of magnetic reconnection. As magnetic reconnection proceeds on the dayside, dipole field lines are opened and added to the tail lobes. This is an inherently unstable situation; as the magnetic pressure in the tail lobes increases, the plasma sheet becomes thinner and moves closer to Earth. Eventually reconnection must occur on the nightside to return magnetic flux to the dayside. In fact, all of the phenomena mentioned above have been shown to involve the closure of open field lines in the tail: pseudo breakup [*Hubert et al.*, 2006], substorm [*Milan et al.*, 2003], poleward boundary intensification [*De La Beaujardiere et al.*, 1994], steady magnetospheric convection [*Sergeev et al.*, 2001], and sawtooth event [*Hubert et al.*, 2008].

In the following sections we review these various modes of response with the goal of identifying major discoveries and their interpretation. We conclude by listing some outstanding questions.

4.3. MAGNETOSPHERIC SUBSTORM

The magnetospheric substorm is the name given to the collection of phenomena that occur throughout the magnetosphere at the time of an expansion of the aurora and westward electrojet near midnight [*Akasofu*, 1964, 1968; *Coroniti et al.*, 1968]. The substorm has three phases: the growth phase [*McPherron*, 1970], the expansion phase, and the recovery phase. After the global nature of the substorm was recognized, a phenomenological model was quickly developed to explain observations in the magnetotail and on the ground [*McPherron*, 1972a; *Hones*, 1973; *McPherron et al.*, 1973a, 1973b; *Nishida and Nagayama*, 1973; *Russell and McPherron*, 1973b; *Nishida and Hones*, 1974]. The basic elements of this model were the following. The IMF turns southward and reconnection begins on the dayside. Dipole field lines are opened and transported over the polar caps and forced into the tail lobes. The dayside magnetopause is eroded [*Aubry et al.*, 1970] by the removal of magnetic flux increasing the angle of attack of the magnetopause to the solar wind [*Coroniti and Kennel*, 1972]. Close to Earth this leads to both an increase in the strength of the lobe field and an increase in open magnetic flux. Far down the tail it simply increases the diameter of the tail. The increased drag on the tail boundary is compensated by the tail current moving closer to Earth increasing the force of the dipole on the current [*Siscoe*, 1966]. Plasma and the imbedded magnetic field at the inner edge of the plasma sheet flows toward the dayside reconnection region, thinning the plasma sheet and the tail current. In addition, the increased magnetic pressure of the tail lobes adds to the thinning. Eventually, after about 1 h, the tail current becomes thin enough that reconnection starts in the center of the plasma sheet at some point relatively close to Earth (the NENL). It was assumed and subsequently demonstrated [*Baker et al.*, 1984a; *Slavin et al.*, 1985; *Scholer et al.*, 1986] that an X line exists far down the tail and that field lines mapping to this point form the boundary of the plasma sheet.

Reconnection close to Earth forms a pair of X and O lines inside the plasma sheet altering the configuration of magnetic field lines. Newly closed field lines create a bubble of magnetic loops around the O line. *McPherron et al.* [1973b] speculated that reconnection would eventually involve lobe field lines and the length of the X line would increase, diverting tail current into the magnetosphere. This diverted tail current system was named the *substorm current wedge* [*McPherron*, 1972a; *McPherron et al.*, 1973a, 1973b; *Clauer and McPherron*, 1974]. At this time it was unclear what the behavior of the X and O lines would be. However, *Hones* [1977] described this development in what is now a classic picture of a bubble of plasma being pulled out of the tail by the solar wind) (see Figure 4.2). Shortly later *Hones* [1979] named this structure a *plasmoid*, defined as "a closed loop magnetic structure." When the plasmoid moves down the tail, it takes the distant X line with it as its leading edge. The recently formed near-Earth X line remains active close to Earth until it fills the plasma sheet earthward of its location. Then this X line moves down the tail, expanding the plasma sheet and reestablishing the distant X line. In the original version of the NENL model it was suggested that tail current was diverted from the region of the X line which was thought to form just tailward of 15 Re. More recent simulations and theoretical arguments show that the diversion is

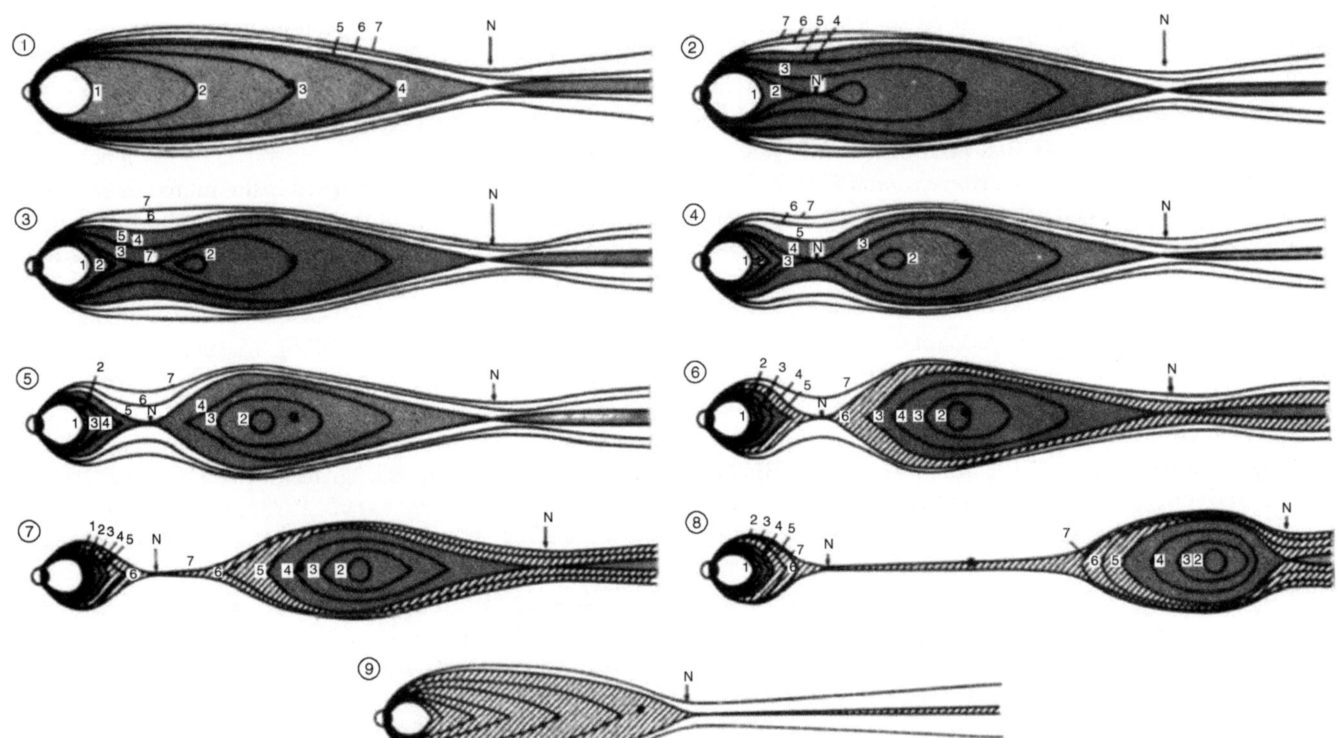

Figure 4.2 The diagram shows the sequence of changes in the magnetotail plasma sheet proposed as the explanation for a magnetospheric substorm by [*Hones*, 1977]. Panel 1 shows the plasma sheet confined by field lines connected to a distant neutral line. Panel 2 displays a near-Earth neutral line (NENL) forming at the center of the plasma sheet close to the Earth. Subsequent panels represent the plasma sheet thinning while a bubble of closed magnetic field lines (plasmoid) grows behind the NENL. In panel 5 the last closed field lines connected to the distant neutral line are cut releasing the plasmoid to be pulled from the tail by loops of solar wind field lines wrapped around it. As it departs the plasma sheet behind the NENL collapses. In panel 9 the inner plasma sheet is expanding as it is filled with plasma from the neutral line that is moving away from the Earth to replace the line previously lost to the solar wind.

actually at the earthward end of the flow channel produced by the X line [*Birn and Hesse*, 1991; *Birn and Hesse*, 1996; *Birn et al.*, 1999]. The diversion is explained as a consequence of bending of the field lines by the diverted flows and by the pressure gradients established by the flux pileup.

In 1983 the ISEE 3 spacecraft was moved into the deep tail (220 Re) where it was established that the structure of the tail at this distance was essentially the same as seen closer to Earth [*Slavin et al.*, 1983; *Tsurutani et al.*, 1984]. A new observation was the appearance of plasmoids traveling away from Earth at high speeds in agreement with predictions of the NENL model [*Baker et al.*, 1984b; *Hones et al.*, 1984a, 1984b; *Baker et al.*, 1987]. These arrived at ISEE 3 about 30 min after the onset of a major substorm expansion at Earth. Sometimes the spacecraft was completely in the tail lobe during the passage of a plasmoid and a *traveling compression region* (TCR) was observed. These are produced by the plasmoid compressing and tilting the lobe magnetic field above (or below) it as it travels down the tail [*Slavin et al.*, 1984]. During one fortuitous interval with 13 substorms seen in the lower auroral electrojet index, AL, every substorm produced either a TCR or a plasmoid, and neither were observed in the absence of substorms [*Slavin et al.*, 1992]. A comprehensive survey of TCRs in the ISEE 3 data was reported by *Slavin et al.* [1993]. Altogether 116 TCRs were observed between 20 and 220 Re behind Earth with 91% of these following an obvious substorm expansion. Time delays at various distances were consistent with the plasmoid being launched at substorm onset closer to Earth than 30 Re.

The original concept of the plasmoid was changed significantly by observations of strong B_y inside plasmoids. *Hones et al.* [1982] suggested this was due to a bent O line. An alternative interpretation was provided by *Sibeck et al.* [1984] who pointed out that strong B_y is an identifying characteristic of a magnetic *flux rope*. *Hughes and Sibeck* [1987] describe how the presence of a finite B_y in the tail naturally produces flux ropes. A numerical simulation of the Hughes and Sibeck hypothesis was performed by *Birn et al.* [1989]. An analysis of the results shows that the topology of field lines and separator surfaces becomes extremely complex as compared to those envisioned in the original plasmoid picture.

Data from the AMPTE/IRM were used to investigate plasma flows earthward of 19 Re in the tail [*Baumjohann et al.*, 1988, 1989, 1990, 1991]. A major finding of this work was that high-speed earthward plasma flows are most frequently seen in the substorm expansion phase. However, these flows are short-lived and very bursty. A superposed epoch analysis relative to substorm onset showed that the flow bursts increase the average flow speed in the central plasma sheet, they increase the temperature of the plasma, they increase the elevation angle (dipolarize) the magnetic field, and they decrease the plasma density [*Baumjohann et al.*, 1991]. This phenomenon was also studied by *Angelopoulos et al.* [1992], who found that the flow bursts were organized into brief intervals of about 10 min duration which he named *bursty bulk flows* (BBFs). Each BBF consists of several high-speed *flow bursts* of about 1 min duration. One event studied in detail occurred during a low value of the auroral electrojet index (AE) and appeared to not be associated with a change in field line topology, i.e., did not involve reconnection of lobe field lines. *Angelopoulos et al.* [1993] investigated the flows associated with the quiet state of the inner plasma sheet by first removing all intervals containing BBFs. He demonstrated that there is a small net sunward flow corresponding to a very low cross-tail potential of 12–16 keV but this flow has a very high variance, much larger than the average. This result suggests that the idea of smooth magnetospheric convection is probably wrong. Instead it proceeds through a sequence of BBFs that create turbulence in the plasma sheet that hides a weak sunward flow. *Angelopoulos et al.* [1994] investigated the statistical properties of BBFs and demonstrated they can occur at all levels of activity as measured by AE but are 4–5 times more likely to be associated with high AE values (25–30% vs. 6%). BBFs are concentrated in the midnight sector and the probability of a tailward flow with negative B_z (signature of reconnection) increases out to the limits of observation by the ISEE spacecraft (22 Re). The authors conclude that "they represent the predominant mechanism of earthward convection at distances of 15–22 RE and close to midnight."

A major difficulty noted by *Angelopoulos et al.* [1994] was that the observed median transport rates for magnetic flux and energy were too low to explain a complete substorm if the area of a flow channel was only 10 Re^2. However, if flow channels were much larger, they should be observed much more frequently. *Angelopoulos et al.* [1996] performed a detailed study of a single substorm and found that the observed rates were so large that a flow channel of only 1–2 Re^2 was sufficient provided they were continuously present in a substorm. This suggested that flow channels are quite frequent but either are not observed by a single spacecraft or the magnitude and duration of the flow and the field they carry are not properly measured. A number of subsequent studies definitely established that the width of the flow channels was 2–5 Re [*Sergeev et al.*, 1996a; *Angelopoulos et al.*, 1997; *Kauristie et al.*, 2000; *Nakamura et al.*, 2000, 2004, 2005]. This work also showed that the equatorial projection of a flow burst is that of a blunt object as previously assumed, but the noon-midnight meridian projection is that of a collapsing flux tube with somewhat larger radius of curvature that field lines to either side.

The Viking spacecraft launched in 1986 provided challenges to the NENL model. This spacecraft carried an auroral imager with higher time and space resolution than any previous instrument. We discuss some of the results below in the section on pseudo breakups. We note here that much of the auroral data seem inconsistent with the NENL model as originally proposed. In fact, in a review of this work, *Elphinstone et al.* [1996] state: "This has led to a new way of understanding the active aurora as a set of processes or modules which occur either coupled together or independent of one another to form a particular event." Basically he argued that auroral activity seldom develops in a consistent manner that can be represented by a simplistic model. Instead it requires many different processes assembled in a wide variety of sequences.

The Geotail spacecraft was launched into the distant tail in July 1992 with the objective of obtaining a better understanding of substorms. An initial survey of the data by *Nagai et al.* [1994] demonstrated that every moderate and large substorm is associated with a plasmoid. *Lepping et al.* [1995] examined many of these and showed that most are flux ropes. *Nagai et al.* [1997] performed a more detailed study showing that plasmoids are ejected from a location earthward of 30 Re within a minute or two of substorm onset. Additional studies demonstrated that virtually all substorm expansions produce a plasmoid traveling down the tail within a few minutes of onset [*Machida et al.*, 1999; *Miyashita et al.*, 1999, 2000].

A major problem for the NENL model was discovered when auroral breakup was mapped to the magnetotail by statistical magnetic field models modified to fit observations during specific substorms [*Pulkkinen et al.*, 1991a,b, 1994]. The mapping showed the initial breakup projects to a region 6–12 Re behind Earth. This is the region where turbulence called current disruption is observed [*Takahashi et al.*, 1987; *Lui et al.*, 1988] and where all proponents of current sheet instability as the cause of substorm expansion place the onset of breakup. A possible explanation of this turbulence in terms of the NENL model was proposed by *Haerendel* [1992]. He argued that the turbulence was the result of a pileup of fast flows against the dipole field. The fast flows are decelerated by the inward gradient of total pressure producing an eastward inertial current. This current connects to field-aligned currents at the edges of the flow channel which close westward in the auroral bulge creating a narrow substorm current wedge. A numerical simulation of this process by *Birn and Hesse* [1991] suggested that the current wedge is instead a consequence of vorticity caused by an azimuthal deflection of the earthward flows. *Shiokawa et al.* [1997] examined the braking hypothesis quantitatively and concluded the total diverted current was of order 0.1 MA, a factor of 10 smaller than seen in larger substorms. They suggested there must be an additional process creating the substorm current wedge.

A more likely explanation of how the current wedge is generated was provided by *Birn and Hesse* [1996] using a simulation that included a thin current sheet at the inner edge of the plasma sheet. Reconnection is much faster with a thin sheet and produces stronger electric fields concentrated in the flux pileup region than are present in simulations with thick current sheets. This simulation showed that the main source of the current wedge is not the inertial current but is instead caused by a misalignment of the pressure and flux tube volume gradients. Strong flux tube gradients pointing azimuthally away from midnight develop as a result of the dipolarization caused by localized flux pileup. Cross-tail current is deflected into field-aligned current at some distance away from the neutral sheet. The process can be visualized as a deflection of the outer part of field lines by the azimuthal flows in the pileup region. Additional details from this simulation are provided by *Birn et al.* [1999].

A detailed study of the preceding model of current wedge formation is reported by *Shiokawa et al.* [1998]. They argue that the inertial current produced by deceleration of the flow starts the current wedge and Pi 2 pulsations, then flow deflections and pressure gradients enhance and maintain the current, and finally the strong electric field from the flux pileup energizes and injects particles. *McPherron and Fairfield* [1998] noted that this modification of the NENL model provides a possible explanation for how an X line at $X = -25$ Re can create turbulence and a current wedge at 10 Re. Additional support for the concept of flux pileup was provided by a superposed epoch analysis of Geotail data. *Baumjohann et al.* [1999] demonstrated that dipolarization of the premidnight field at 11–16 Re begins immediately at substorm onset and reaches a maximum in 15 min. The dipolarization is progressively delayed at greater distances such that beyond 26 Re it is just beginning 45 min after onset. He introduced the term *dipolarization front* to describe the tailward motion of the boundary of the dipolarized field.

The consequences of fast earthward flow were investigated by *Fairfield et al.* [1999] using Geotail observations 10–15 Re behind Earth. They found that the fast flow bursts were associated with auroral brightenings seen by the Polar spacecraft, auroral Kilometric radiation (AKR) onsets, geosynchronous particle injections, and ground magnetic activity. Also the onset times of various phenomena support the hypothesis that localized reconnection 20 Re behind Earth is the cause of the fast flows. In a companion study *Ieda et al.* [2001] used Geotail data between 21 and 29 Re behind Earth to examine plasmoid formation and ejection. They found that "plasmoids are invariably associated with auroral brightenings" and, furthermore, timing suggests the plasmoids are born a few Re earthward of the spacecraft a few minutes before being observed. They also argue that the auroral brightenings are an indirect effect of reconnection associated with flux pileup and do not originate at the reconnection site. *Nakamura et al.* [2001] also conclude that all flow bursts are closely associated with some form of auroral activation.

There is another serious problem with this modified NENL model as discussed by *Ohtani* [2004]. The auroral arc that breaks up at the beginning of the expansion phase is usually equatorward of other arcs [*Akasofu*, 1964]. There have been no reports of disruption of these arcs or any evidence of an auroral streamer that might be related to the earthward flow. The absence of such disruption led *Lyons et al.* [2002a] to state that "significant dynamical changes... do not occur prior to substorm onset tailward of the onset field line region." This is the primary observation that has led many to propose that the auroral breakup is a process independent of reconnection such as ballooning [*Ohtani and Tamao*, 1993; *Hurricane*, 1997; *Pu et al.*, 1999; *Miura*, 2001].

Contrary to the view that there is no auroral activity poleward of the breakup arc, *Kadokura et al.* [2002] describe a phenomenon they call NPSBL (near plasma sheet boundary layer) aurora that precedes by a few minutes the breakup of the growth phase arc. A similar observation was reported by *Kepko et al.* [2009]. Six minutes before an auroral expansion a dim patch of aurora in the 630.00 nm band was seen moving equatorward from a position well south of the open/closed boundary. Nearly 2 min before the expansion onset fast earthward flows were seen by THEMIS spacecraft.

The two divergent views of the cause of the substorm expansion were summarized schematically by *Ohtani*, [2004] and are illustrated in Figure 4.3. Proponents of the inside-out view argue that the expansion begins with instability of the inner edge of the plasma sheet and only later involves magnetic reconnection. Those who favor the outside-in view propose reconnection begins first and that the expansion phase is triggered by the pileup of plasma and magnetic flux from the reconnection site. The continuing controversy about the processes responsible for the onset of the substorm expansion led Vassilis Angelopoulos and co-investigators to propose a five spacecraft mission whose "primary goal of THEMIS, which drove the mission design, is to elucidate which magnetotail process is responsible for substorm onset"

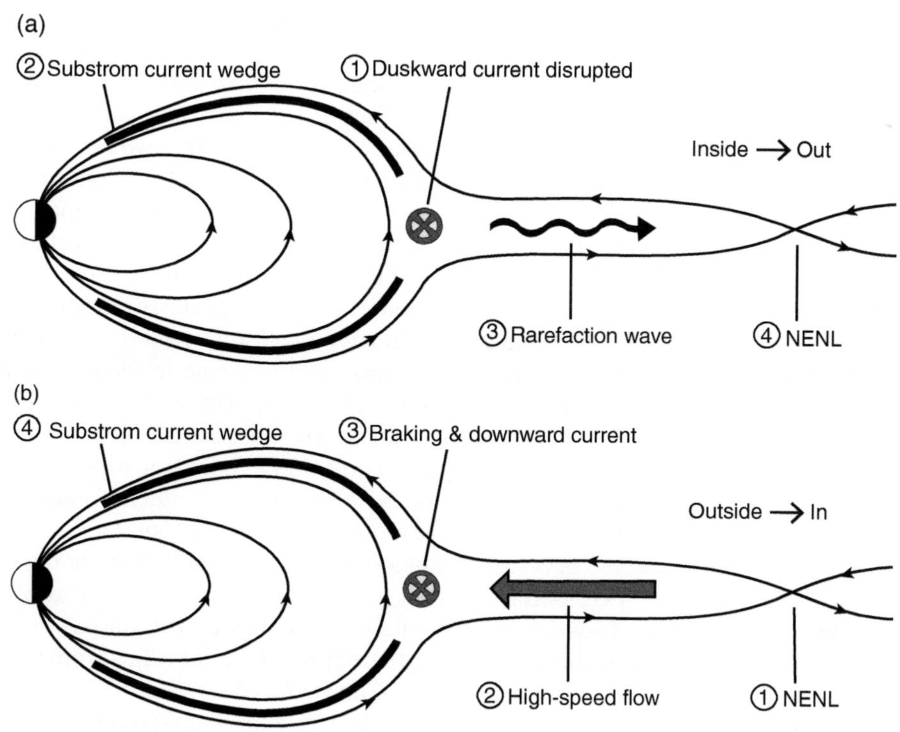

Figure 4.3 The diagram is a schematic illustration contrasting two conflicting views of the cause of the expansion phase of the magnetospheric substorm [*Ohtani*, 2004]. Several models introduced as alternatives to the near-Earth neutral line model (see Figure 4.2) propose that instability of the inner edge of the tail current or the pressure gradient at the outer edge of the dipole magnetic field expands from inside out into the tail (panel a). The alternative NENL model proposes reconnection in the outer plasma sheet causes inward flows that pileup at the inner edge of the plasma sheet (panel b).

[*Angelopoulos*, 2008]. Prior to its launch in February 2007 this author was asked to estimate the probability of success [*McPherron and Hsu*, 2006]. Using a probability argument, we estimated that it was likely than only one perfect substorm would be observed: "If pessimism is justified we can be certain that data obtained during only one substorm will not convince the proponents of either of the two substorm models .... it seems more likely that THEMIS will provide an unprecedented collection of new data that rather than solve all the problems of substorm physics it is more likely that THEMIS will reveal additional details concerning the complexity of substorms."

In the first major THEMIS publication related to substorm observations, *Angelopoulos et al.* [2008a] concluded that their "results demonstrate that substorms are likely initiated by tail reconnection." *Sergeev et al.* [2008] also conclude that reconnection is the cause of three events observed with a collection of spacecraft including Cluster. These events appear to be caused by X lines in the region 10–12 Re behind Earth. In another event observed with THEMIS *Runov et al.* [2008] examined a substorm onset preceded 6 min earlier by a pseudo breakup. The pseudo breakup was associated with signatures of reconnection while the following expansion onset displayed most of the signatures of current disruption. In the interval between the two events field-aligned currents characteristic of a flow channel were recorded. The interpretation was that mid tail reconnection initiated an earthward flow that subsequently caused the expansion.

In contrast to the preceding view, *Donovan et al.* [2008] "assert that the expansion phase began earthward of the four satellites and evolved radially outwards." *Keiling et al.* [2008] found that the observations "support a wave phenomenon (such as a ballooning-type mode) for the onset and development of the substorm/intensification." *Lui et al.* [2008] used THEMIS data for a small substorm, concluding that it could not have been caused by reconnection. In a subsequent publication *Lui* [2009] disputed the evidence presented by *Angelopoulos et al.* [2008b] claiming: "We provide evidence that (i) near-Earth current disruption, occurring before the conventional tail reconnection signatures, triggered the onset; (ii) the observed auroral intensification and tail reconnection are not causally linked; and (iii) the onset they identified is a continuation of earlier substorm activities." *Angelopoulos et al.* [2009] replied using better magnetic indices and satellite data and reiterated the conclusion that reconnection precedes the onset of the auroral expansion. *Lui* [2011] continued this argument, stating that the evidence supports a near-Earth current disruption and was not caused by magnetic flux pileup.

More recently *Lyons et al.* [2010] have developed a new explanation for the onset of the substorm expansion. In their view the expansion is caused by instability of the inner edge of the plasma sheet, but it is triggered by the arrival of a flow burst from the distant X line. The path of this flow burst is such that it does not interact with preexisting arcs poleward of the breakup arc. We discuss this idea in greater detail below in the section on poleward boundary intensifications.

Clearly our pessimism was justified as the controversy concerning the cause of the substorm expansion has not yet been resolved. Apparently the problem of substorm expansion onset still requires more work and a better understanding of the relation of magnetotail processes to the aurora.

4.4. STEADY MAGNETOSPHERIC CONVECTION

At the beginning of the space age Jim Dungey suggested that the aurora and geomagnetic activity might be explained by the process of magnetic reconnection [*Dungey*, 1961]. He visualized an IMF pointing southward antiparallel to Earth's field on the dayside merging with Earth's field at a neutral point (X-line). The solar wind carries the two halves of the newly opened field line to the nightside adding them to the north and south lobes of the magnetic tail. The field lines drift toward the neutral sheet and reconnect at a second X line becoming a closed field line moving sunward and an interplanetary field line moving away from the Sun. The feet of these field lines in the polar regions trace out a two cell system with foci at the dawn and dusk meridian. Provided the reconnection rates at the two neutral points are the same, the system will be in equilibrium and there will be SMC throughout the magnetosphere and polar ionospheres.

This is not what is observed most of the time. The IMF is continuously fluctuating and there are time delays in the system that require changes in the magnetic configuration. If the reconnection rate on the nightside is zero, magnetic flux accumulates in the lobes and the plasma sheet thins. The NENL substorm model is based on the idea that the distant X line is too far downstream to respond before the near-Earth region becomes sufficiently unstable for a second X line to form near Earth. When reconnection reaches the boundary of the plasma sheet it reconnects the last closed field lines that link Earth to the distant X line [*Caan et al.*, 1973]. This allows the bundle of magnetic flux produced by reconnection (plasmoid) to be pulled away from Earth, removing the old X line from the tail and leaving the new X line close to the Earth. The proximity of this X line to the dayside allows rapid return of newly closed magnetic flux to the dayside. In their study of substorms *Caan et al.* [1973] noted that at times there exists continuous geomagnetic activity without obvious substorm phases. *Hones et al.* [1976] also noted such intervals in which there were flows in the plasma sheet without accompanying evidence of substorm expansions. He speculated that this might represent a steady-state response to continuous strong southward IMF. A detailed study of one such event with multiple satellites was performed by *Pytte et al.* [1978]. Because of a well-defined and strong two-cell convection system in the polar regions the authors named this phenomenon "*convection bays*". Convection bays are characterized by continuous southward IMF; active aurora without obvious poleward expansions; strong negative bays in the morning sector; no or weak Pi 2 pulsations; no midlatitude positive bays; prolonged tail-like field at synchronous orbit with effects of strong field-aligned currents; no magnetopause erosion, no change in flaring; and no change in lobe field strength. *Kokubun and McPherron* [1981] showed that in such intervals the synchronous field remains very tail-like with strong perturbations from field-aligned currents. *Sergeev and Lennartsson* [1988] examined the behavior of the tail during several SMC events. They found a relatively weak and steady AL index (200–300 nT) over 5–10 h; moderate to weak lobe field strength; thick plasma sheet with strong B_z; a convection pattern that on average is stable but locally variable; strong field-aligned beams in the plasma sheet boundary layer; and weak bulk flows that on average have an azimuthal component as large as the earthward component.

A joint Russian-Finnish project to study steady magnetospheric convection was begun in 1990. Project results derived from five events have been published in a number of papers and are reviewed by *Sergeev et al.*, [1996b]. The criteria by which these events were selected included stable continuously southward IMF for more than 4–6 h; AE > 200 nT; no signatures of substorm expansion; and no current disruptions or plasmoids in the tail. New facts that emerged from this project include the following:

- A *double auroral oval* forms during SMC.
- There is equatorward flow in the midnight sector through a wide convection throat.
- There are frequent magnetic transients of duration ~10 min that repeat every 10–40 min.
- These are associated with short-duration ionospheric flows at the polar edge of the oval.
- They are associated with north-south auroral streamers.
- They are correlated with earthward flow bursts in the tail.
- The field at synchronous orbit is unchanged during the events.

These preceding results are based on only five events selected under the constraints that the IMF was steadily southward for 4–6 h and that there were no signatures of substorm expansions. *O'Brien et al.* [2002] argued that the IMF conditions that cause SMC were not known and so should not be used in selection. They instead used the requirements that the 1 min AE index is continuously disturbed with a value greater than 200 nT and that AL does not drop more rapidly than 25 nT/min. Two thousand segments of duration greater than the typical 90 min duration of a substorm recovery phase were identified in the interval 1978–1988. These were classified as SMC intervals and used to determine conditions in the solar wind responsible for SMC. The cumulative probability distribution for solar wind speed indicated that SMCs occur for lower speeds than typical solar wind and that longer SMCs correspond to lower median speeds. The GSM B_z distribution shows that SMCs are typically produced by an IMF B_z of −2 to −2.5 nT. Longer SMCs tend to have slightly smaller fluctuations than seen in all B_z data. Additional properties of this set of 2365 SMC events were identified by *McPherron et al.* [2005]. SMCs are quite common (~240 per year) provided shorter intervals are accepted as SMCs. A SMC typically starts with the beginning of the recovery phase of the preceding substorm. The upper auroral electrojet index, AU, becomes constant at this time at about 150 nT while the lower auroral electrojet index, AL, recovers for about 90 min to a constant value of ~ −200 nT. The fluctuations in AU and AL are much smaller than typical of all disturbed data.

A new approach to the study of SMC using satellite auroral images was introduced by *Dejong and Clauer* [2005]. Their underlying assumption was that SMCs represent balanced reconnection so that there are no changes in the polar caps or tail lobes. The authors identified 22 intervals of duration longer than 3 h during which the polar cap area was constant while AE > 200 nT. They found that most of these began with a substorm but a few had no identifiable preceding substorm in the AE indices. Many of these did not end with a substorm, contrary to the results reported by *Sergeev et al.* [1996b]. A few SMC intervals appear to include a pseudo breakup. *Dejong et al.* [2007] performed a statistical comparison of isolated substorms, SMC, and sawtooth events using absolute polar cap flux and change relative to event maximum flux. Isolated substorms and SMCs occur at moderate levels of flux (0.6 GWb) while sawtooth events at their peak have about 1 GWb. In relative terms SMCs display no change while both isolated substorms and sawtooth events show similar 25% reduction in area during 50 min after onset. A new name for the SMC phenomenon, *balanced reconnection interval* (BRI), was suggested by *Dejong et al.* [2008]. The statistical comparison of three modes of activity was extended by *Dejong et al.* [2009] to include many more events. These authors introduce a new measure of the state of different variables called "steadiness," defined as the ratio of the standard deviation to the mean in a moving window. They find that both SMC and sawtooth events are driven by an IMF B_z or E_y that are steadier than typical of the solar wind while isolated substorms are indistinguishable from background. Surprisingly they find that sawtooth events are associated with less steady solar wind density.

Another comparison of these three response modes is made by *Partamies et al.* [2009a]. They show that sawtooth events are seen around the equinoxes but not the solstices. SMC events are most probable at summer solstice. Substorms are more probable near winter solstice. These results are compared to three magnetic indices, K_p, AE, and D_{st}. Both K_p and D_{st} show peak activity at the equinoxes and minima at the solstices while AE shows a peak at summer solstice. *McPherron et al.* [2005] also found that the maximum of SMC occurrence is in the summer and the occurrence rate decreases after solar maximum. They attributed this behavior to changes in the strength of AU, which depends strongly on ionospheric conductivity. *McWilliams et al.* [2008] used a longer interval to show that SMC occurrence has a peak in summer every solar maximum at the time F10.7 peaks. By assuming that the SMC distribution should not depend on season, the authors developed a new selection criterion that varies between AE > 200 nT in summer and 90 nT in winter. It is not known if this is a valid assumption. It is clear that this dependence of AU on the season affects the behavior of AE and the apparent occurrence rate of SMC. *Partamies et al.* [2009a] examined the coupling efficiency for the various response modes, finding that the ratio AL/E_y is largest for SMC, less strong for isolated substorms, and weakest for sawtooth events. However, the parameter that best organizes the type of activity is solar wind speed with storm-time activity at 560 km/s, sawtooth at 480 km/s, substorms at 440 km/s, and SMC at 380 km/s [*Partamies et al.*, 2009b]. A similar result was reported by *Pulkkinen et al.* [2010].

An interesting observation reported by *Kissinger et al.* [2010] is that the probability of observing SMC events is highest in the 12 h after passage of the stream interface at the center of a corotating interaction region (CIR). The lowest probability is 1.0–1.5 days before the interface. A more detailed study of this association was carried out by *Kissinger et al.* [2011]. The solar cycle was divided into rising and declining phases and stream interfaces in each phase were classified according to the geoeffectiveness of the streams before and after the interface. Based on the Russell-McPherron (RM) effect [*Russell and McPherron*, 1973a] there are four possible classes of interface: II, EE, IE, EI where I is ineffective and E is effective. An effective stream is one in which the IMF B_y component will have a negative projection on the GSM z axis and so contribute to reconnection. In both phases of the solar cycle CIRs for which the high-speed stream is geoeffective (E), SMC occurrence peaks near the interface. The effect is stronger in the declining phase and strongest for interfaces of class IE. When the high-speed stream is ineffective, there is no occurrence peak in either phase of the solar cycle.

The explanation for the preceding behavior is unclear. The solar wind coupling efficiency AL/E_s is weaker in the rising phase by 30% than it is in the declining phase (115/160), which implies the same driver produces weaker activity. The rectified electric field is stronger in a high-speed than a low-speed stream and will be higher still if the RM effect biases the driver toward larger values. These facts combined with the definition of an SMC that depends on the strengths of the electrojet indices suggests it is likely that fewer SMCs are identified in slow-speed streams during the rising phase of the solar cycle because of weak indices.

To obtain balanced reconnection over long intervals there must be frequent fast flows in the plasma sheet returning newly closed field lines to the dayside. *Kissinger et al.* [2012a] investigated this possibility using nine years of Geotail magnetic field and plasma data combined with three years of THEMIS data. Intervals when the spacecraft were in the tail were separated into six classes: quiet, pre-SMC, SMC, substorm growth, expansion, and recovery. Many earthward flow bursts were identified for each of these classes ranging from 1000 to more than 5000 bursts. The normalized occurrence rates show that the probability of observing a flow burst is highest during SMCs. In all classes the flow probability decreases radially inward becoming nearly undetectable inside 10 Re. The flows are strongly peaked at about 23 h local time (LT) but can be seen from 20 to 04 LT. The average flow pattern projected in the GSM X-Y is somewhat erratic during substorms but is more organized for SMC. During SMC there is a broad channel of earthward flow around the noon-midnight meridian and then azimuthal deflection around Earth at about 10 Re. Everywhere in the tail outside the broad channel, flows are systematically deflected toward the flanks. The authors created equatorial maps of plasma density, temperature, total pressure, and flux transport. The maps for SMC show the plasma density near the center of the tail is low but the temperature is high so that the total pressure is high. During SMC the radial distribution of total pressure has a plateau of high pressure extending to 15 Re with a sharp gradient beyond this distant. It appears that the substorm that precedes the SMC establishes a pressure distribution that deflects subsequent flows.

A study of fast flows seen by the THEMIS spacecraft inside 12 Re was reported by *McPherron et al.* [2011]. They found that fast flows (V > 150 km/s) almost never penetrate as deep as synchronous orbit and are either reflected near midnight or deflected around Earth beyond 8 Re at other local times. Flows seen between 8 and 12 Re have precisely the same local time distribution as the distribution of expansion onsets seen by the IMAGE spacecraft [*Frey et al.*, 2004]. In addition these flow dipolarize the field in this region and are associated with the sudden onset of negative bays. More than half of these flow bursts are associated with a sudden reduction in solar wind, E, a few minutes before they are seen, suggesting they were triggered by northward turnings of IMF B_z.

These results for substorms were contrasted with what is seen during SMC events. During substorms there are frequent fast flows (at the level of 1%) but during SMC virtually no fast flows penetrate this inner part of the tail. As shown by *Kissinger et al.* [2012a], these are deflected around Earth at distances greater than 12 Re.

It is frequently noted that SMCs begin in the recovery phase of a preceding substorm [e.g., *Sergeev et al.*, 1996b; *McPherron et al.*, 2005]. Some have argued that this is simply a matter of definition. SMCs are intervals without expansions so they must be bracketed by substorms since substorms typically occur four to five times a day. However, *Kissinger et al.* [2012b] find that 92% of all SMCs are preceded by an obvious substorm expansion onset and recovery that transitions into SMCs and that only 1% have no preceding activity. There is also a dramatic difference between the AL and E_y signatures of isolated substorms and substorms that become SMCs. Isolated substorms are driven by a 2–3 h pulse of solar wind, E_s, while substorms before SMC are driven by step function increases in E_y that persist for as long as the following SMC. It is also seen that median solar wind speed decreases as the length of the SMC increases. At the same time IMF B_z becomes more negative such that E_y remains constant. As V, B_z, and E_y become steadier, the length of the SMC increases. The 1% of SMC with no evident precursor exhibit different behavior in AL and E_y; E_y grows more slowly with time and |AL| does not peak at the start of the SMC but instead simply reaches a plateau. *McPherron et al.* [2005] note that only about half of all SMCs end with a substorm. The rest simply end when the IMF turns northward and the activity fades below the selection threshold.

These observations suggest the original concept of a magnetotail with X lines front and back can exist. However, the situation must be created by a substorm and be maintained by moderate steady driving. We believe that the initial substorm creates a pressure distribution close to Earth that deflects flows away from midnight. Because of this distribution, the X line is forced tailward and creates the high-latitude branch of the double oval. The X line impulsively reconnects sufficient flux to balance the rate at which it is added to the lobe. Each burst is seen as a PBI.

4.5. SAWTOOTH INJECTION EVENTS

LANL spacecraft at synchronous orbit occasionally exhibit quasi-periodic sequences of ion injections. Assuming these were substorms, *Borovsky et al.* [1993] examined the waiting time distribution between onsets, finding two types of events. The first consists of random onsets satisfying an exponential distribution with 5 h e-folding time. The second is quasi-periodic with period 2–4 h. *Hsu and McPherron* [2012] find distributions for AL onset, maxima in lobe field, minima in lobe B_z, and separation of Pi 2 main onsets all have peaks between 1 and 4 h. This phenomenon has been named *sawtooth injection event* by *Belian et al.* [1995]. Ion fluxes gradually decrease over a period of about 2 h and then suddenly increase in less than an hour. When many such events occur in sequence, the waveform of ion flux has the shape of a saw blade, hence the name. It was found that these events tend to occur during high solar wind speed, strongly southward IMF B_z, and very large values of AE (as high as 5000 nT!). The association with AE suggested these might be very large substorms. *Cai and Clauer* [2009] found the average spacing of sawtooth injections is 3 h but there is considerable variability even in the same event with a range of periods from 1 to 5 h. What controls the tooth separation is not evident.

The first detailed investigation of what was later shown to be one of these events was reported by *Huang* [2002]. The event was recorded by Geotail in the north lobe at distances between 22 and 30 Re behind Earth. A sequence of 10 peaks in the lobe field was associated with a similar sequence of sudden drops in B_z to negative values. Successive peaks were separated by roughly 2.5 h. The observations were interpreted as remote sensing of a sequence of plasmoids each initially centered a little earthward of Geotail and subsequently moving tailward. The event occurred during a moderate magnetic storm driven by strongly southward IMF. This work was extended by *Huang et al.* [2003a, 2003b], who examined several other sawtooth events. All events occurred during the main phase or early recovery phase of moderate magnetic storms. Some sequences appear to have been triggered by a sudden increase in dynamic pressure but the quasi-periodicity of 2–3 h was interpreted as an intrinsic property of the magnetosphere. The sawtooth injections were found to be closely associated with AE increases, auroral brightening, synchronous field dipolarization, and tailward flows with negative B_z in the plasma sheet.

At the time of this work it had been postulated that sawtooth events were quite different from substorms. The synchronous data seemed to suggest that particle injections occurred at all local times simultaneously. *Henderson* [2004] rejected this conclusion by reexamining an event that had been previously studied as a large substorm showing that it was a tooth in a sawtooth sequence. The event was particularly interesting because the tail data implied that an X line had formed earthward of 11 Re. He speculated that "sawtooth events actually are quasi-periodic substorms that occur extremely close to the Earth during intervals of strong and sustained magnetospheric convection (e.g., during storms)."

A completely different interpretation of the periodicity of sawtooth oscillations was put forward by *Lee et al.* [2004]. The authors used time series of log dynamic pressure and of midlatitude magnetometer H traces to argue that every tooth is triggered by quasi-periodic pressure pulses in the solar wind. *Huang et al.* [2005] evaluated this proposal and rejected it on a basis of quantitative arguments. Particle flux changes caused by pressure pulses are much too small to be accounted for by the changes in pressure observed in these events [*Huang*, 2005]. Also, the size of changes in midlatitude magnetograms that were associated with each sawtooth are much larger than expected for the changes in pressure shown by *Lee et al.* [2004]. Another objection to the suggestion that sawtooth events are substorms triggered by near-Earth reconnection was raised by *Lui et al.* [2004].

In an attempt to find characteristics of sawtooth events different from substorms, *Cai et al.* [2006a] used AMIE (assimilative modeling ionospheric electrodynamics) potential field patterns to statistically study isolated substorms and individual sawtooth injections. For both types of activity a DP-1 potential pattern develops during the expansion phase. However, the pattern for a sawtooth event has larger area and has a stronger cross polar cap potential. The patterns for individual sawtooth injections were more variable than for single isolated substorms. In another study *Cai et al.* [2006b] examined the behavior of the synchronous magnetic field. Development of a tail-like field followed by dipolarization happens during the growth and expansion phases of isolated substorms. It also happens in association with the dropout of ion fluxes and the subsequent injections. The most obvious differences are that the synchronous field is more stretched before a sawtooth injection and the change in field inclination is larger than during an isolated substorm. Also the field changes cover a much wider range of local time. Sawtooth dipolarizations are seen from 16 to 04 LT, i.e., over 12 h of local time. Isolated substorms are typically seen between 20 and 02 LT with a 6 h extent. *Clauer et al.* [2006] used ground magnetograms to study the current systems associated with sawtooth events and found them no different than those in isolated substorms except for their larger extent.

The auroral disturbances associated with sawtooth injections have been documented by *Henderson et al.* [2006a, 2006b]. A sawtooth begins with a double auroral oval that gradually narrows during a growth phase. The aurora brightens on the lower branch and expands poleward. As this is happening, there are intense omega bands drifting eastward on the dawn side of the auroral bulge. Multiple auroral streamers emerge from the higher branch of the double oval. The next tooth begins as did the preceding tooth by a gradual narrowing of the double oval. All the usual signatures of substorm expansion are seen, including auroral bulge, midlatitude positive bays, Pi 2 pulsations, stretching and dipolarization of synchronous magnetic field, and dispersionless particle injections. During these events the solar wind driver is usually very steady and one might expect the magnetosphere to enter the SMC response mode. The authors speculate that the solar wind driver is too strong for the SMC mode to return enough flux to the dayside and that X lines must continually form close to Earth to return the flux.

Hubert et al. [2008] used satellite auroral images and SuperDARN radars to calculate the rate at which open flux is added to the polar cap and the rate at which it is removed. During sawtooth events the total open flux becomes very large (~1 GWb). With each tooth there is a reduction in flux to normal substorm levels (~0.5 GWb) but open flux is being added so fast that the magnetotail cannot relax and another tooth is created. This process continues until the level of driving is reduced. Similar results were obtained by *Huang et al.* [2009] and *Huang and Cai* [2009], who argue that the actual flux load creating a sawtooth depends on both the driver and the strength of the ring current.

The extreme stretching of the magnetic field at synchronous orbit during sawtooth events has been modeled by *Pulkkinen et al.* [2006] and *Kubyshkina et al.* [2008]. Their modeling shows that a strong current sheet develops near synchronous orbit, primarily in the premidnight to afternoon sector. The field stretching leads to very fast drift of ions on open drift paths. The sequence of injections is responsible for large changes in the asymmetric ring current but lesser changes in the symmetric ring. Cluster data were used by *Ohtani et al.* [2007] to demonstrate dipolarization and dispersionless injection as close to Earth as 4.6 Re during sawtooth events.

Recently work reported by *Troshichev et al.* [2011] casts doubt on the entire concept of systematic behavior during sawtooth substorms. The authors examine 25 individual sawtooth injections and conclude that there are significant deviations in the times of different events relative to auroral expansions. Particle injections can occur from 10 min before to 30 min after expansion onset. Magnetic onsets occur anywhere in the range 0–50 min after onset. Magnetic onsets can occur from 10 min before to 40 min after a particle injection. These are large differences compared to the typical 3 h separation of injection events. The authors argue that because of the high level of energy input there is continuous auroral and magnetic activity and that it is virtually impossible to identify the onset of specific features in the complex data. They also suggest that different parts of the plasma sheet may behave independently and that timing thus depends strongly on the location of observations relative to activity centers.

This interpretation seems at odds with the results reported by *Huang* [2002] and *Henderson et al.* [2006a], who describe a much more organized sequence of events. A more careful analysis of timing of different phenomena during sawtooth events using all available data is needed.

4.6. PSEUDO BREAKUPS

The pseudo breakup is a weak activation of the aurora that does not develop into a full-fledged substorm. This term was first defined in a paper written to establish the research agenda for the IGY [*Elvey*, 1957]. The author says: "In a system of arcs the pseudobreakup is defined as the disruption of one of the arcs near to the center of the auroral zone." He continues: "with the pseudobreakup sometimes forming a rayed arc or breaking into bands and various rayed structures. These pseudo-breakups appear to coincide with the minor bays that give the irregularity to the main bay in the horizontal component of the magnetic record."

The pseudo breakup was also described in the classic paper that presented the concept of the auroral substorm [*Akasofu*, 1964]. "If the substorm is very weak the brightening and some development of irregular folds are the only consequences. In most cases the brightening of an arc is followed by its rapid poleward motion. This results in a "bulge" around the midnight sector. When the substorm is weak the poleward motion lasts only for a few minutes and other arcs not may be seriously affected. Such a substorm results in the pseudo break-up." Later he states: "The pseudo break-up is usually associated with a surge which is propagated along a brightened arc, without affecting others."

This author observed pseudo break ups while flying balloons measuring substorm electron precipitation [*Parks et al.*, 1968]. In *McPherron* [1991] we commented on this work saying: "An event begins with a gradual increase in energetic electron precipitation and a Pi 2 micropulsation burst. Within a few minutes a small section of an auroral arc becomes visibly brighter and begins to develop rays and folds. These move rapidly along the arc. Often the lower edge of the arc turns red. A ground magnetometer measures a weak negative bay. The entire event lasts 5–10 min before it fades back into a quiet arc."

A study of auroral spirals by *Davis and Hallinan* [1976] found that major folds and spirals along an auroral arc are associated with upward field-aligned currents and weak negative bays under the spiral. They speculated that pseudo breakups are actually a manifestation of the development of an auroral spiral. From the time of this work until 1993 there were almost no studies of this phenomenon. One of the first detailed investigations after this period is that of *Koskinen et al.* [1993]. They found that a pseudo breakup preceded the main substorm onset by about 20 min and was associated with a dipolarization of the magnetic field at 8.6 Re near midnight, particle injection at synchronous orbit, an intensification of the westward electrojet, a Pi 2 burst, and a weak substorm current wedge. They note these are also signatures of expansion onset. The event did not appear to be particularly localized but was quite weak and did not result in a significant substorm expansion. Similar conclusions were drawn by *Ohtani et al.* [1993], who noted that the primary difference between a pseudo breakup and a substorm is that the pseudo breakup does not expand azimuthal or radially much beyond its initial location in space. *Jacquey et al.* [1994] used Geotail observations 100 Re down the tail to show that streams of energetic ions (60–220 keV) were seen minutes after pseudo breakups were observed at Earth just as are seen in full substorms. These conclusions were supported by the work of *Nakamura et al.* [1994], who also found little difference between the signatures of pseudo breakups and substorm expansions except that the former are short-lived and exhibit hardly any expansion. However, they noted that the expansion phase of substorms seems to consist of multiple events similar to pseudo breakups (called *wedgelets*) and speculated that an expansion is simply an ensemble of wedgelets in rapid succession in different locations. Together the collection of wedgelets creates a macroscopic substorm current wedge.

The Viking spacecraft launched in February 1986 carried an auroral imager to study substorm phenomena. In a series of papers Elphinstone and colleagues examined the variety of developments of the aurora during substorms [*Elphinstone et al.*, 1993, 1995a,b,c, 1996]. The authors discuss the signatures of pseudo breakups at considerable length and conclude that there is no objective way to decide whether a particular auroral activation is a pseudo breakup or a full substorm expansion because the only features that identify the pseudo breakup are its short duration and limited poleward expansion. The authors emphasize that most substorm expansions end with the poleward edge of the auroral bulge creating a second band of aurora at high latitudes separated from another band at lower latitudes (the *double auroral oval*). During the recovery phase the higher latitude band experiences quasi-periodic activations that produce *auroral streamers* that move equatorward and sometimes appear to trigger lower latitude activations. They argue that these events can also be mistaken for pseudo breakups. Today we call these activations *poleward boundary intensifications* (PBIs) and discuss them in the following section.

Elphinstone et al. [1995a] discuss the importance of azimuthally spaced auroral forms (AAFs) that appear before the substorm expansion onset. When they use the beginning of the rapid poleward motion of the auroral bulge as substorm onset, they find that the AAF can appear as much as 13 min earlier but more typically only a few minutes earlier. They argue that some factor besides the AAF is needed to initiate the auroral expansion. In their study of a large number of substorms viewed from space they conclude that there can be a wide range of developments of auroral activity. They proposed a modular approach in which a variety of different processes can occur in different sequences depending on the history of solar wind driving and past magnetospheric activity. In particular, *Elphinstone et al.* [1995c] and *Elphinstone et al.* [1995b] emphasize the importance of the formation of the double oval at the end of the first auroral expansion and how it affects subsequent activity. A detailed review of these ideas is presented by *Elphinstone et al.* [1996].

The idea that there is no distinct sequence that distinguishes a pseudo breakup from a substorm, or a substorm from steady magnetospheric convection, was proposed also by *Sergeev et al.* [1996c]. The authors characterize their work as a generalization of the modified near-Earth neutral line model articulated by *Baker et al.* [1996]. The authors argue that two competing processes produce all forms of magnetic and auroral activity: The first is a global and slow reconfiguration of the magnetotail brought about by dayside reconnection and transport of magnetic flux to the tail lobes (the growth phase). This process causes earthward displacement of the inner edge of the tail current and the development of a very thin current sheet terminating in the region where the tail connects to the dipole field. The second process is the impulsive dissipation event that occurs in a thin current sheet and corresponds to localized and transient reconnection in the thin current sheet. The slow process adds open magnetic flux to the tail while impulsive dissipation events subtract flux by closing field lines. They argue that a pseudo breakup is one of these events during a growth phase with duration and extent too small to alter the growing changes in the configuration of the tail. Also, they suggest that a pseudo breakup is a reconnection event that does not involve lobe field lines. In contrast, an expansion phase is a rapid sequence of impulsive dissipation events with larger scale that are located at various points in the thin current sheet that were not thickened by previous activations. The recovery phase is seen as the consequence of a northward turning of the IMF during the expansion phase that causes the current sheet to thicken and the neutral line to move tailward. In this model steady magnetospheric convection (SMC) events develop when the IMF remains constantly southward and a pressure and field distribution develop at the inner edge of the current sheet that precludes a thin current sheet near Earth and deflects flows around Earth.

A question that past work leaves unanswered is what limits the development of pseudo breakup so that it does not become a full expansion. In the *Sergeev et al.* [1996c] model it is the degree of development of the thin current sheet and perhaps reconnection involving only closed field lines. *Lewis et al.* [1998] suggest that it might be insufficient ionospheric conductivity in the growth phase to allow the development of a full substorm current wedge and that a pseudo breakup promotes the later expansion phase by producing ionospheric conductivity. Another possibility is that a pseudo breakup is caused by a reconnection event inside the plasma sheet involving only closed field lines. *Nagai et al.* [1998] reported Geotail observations of tailward flow with southward B_z at 15 Re when near the neutral sheet but no flow when the spacecraft was a little away from the neutral sheet but still inside the plasma sheet. *Petrukovich et al.* [1998] used an Interball-Geotail conjunction event to study a plasmoid generated somewhere between 10 and 28 Re behind Earth. Observations at the more distant Geotail revealed that the departing plasmoid was wrapped with closed field lines and so the pseudo breakup observed near the spacecraft meridian in Alaska was probably caused by reconnection that did not propagate into the lobes. If all pseudo breakups behaved in this way, it might be claimed that the difference between them and a substorm expansion is a breakthrough of reconnection into the lobe field.

The idea that pseudo breakups only involve closed field lines inside the plasma sheet seems to be contradicted by a comparison of observations with simulation results [*Pulkkinen et al.*, 1998]. The authors point out that the simulation shows reconnection of open flux but the absence of earthward flow during a pseudo breakup. *Hubert et al.* [2006] support the simulation results with optical observations of the amount of open flux in the tail lobes. During constant southward IMF B_z the flux in the lobe stops increasing and begins to decrease during a sequence of three pseudo breakups that preceded the onset of a major expansion. This should not happen unless pseudo breakups reconnect open flux. To the contrary, *Tang et al.* [2010] suggest that THEMIS observations support the idea of closed field line reconnection during a pseudo breakup. Several other authors have argued that there is no discernable difference between a pseudo breakup and an expansion except for the duration and intensity [*Aikio et al.*, 1999; *Kamide*, 2001; *Rostoker*, 2002]. Recent work divides pseudo breakups into several classes [*Kullen and Karlsson*, 2004; *Kullen et al.*, 2009, 2010]. These include quiet time pseudo breakups during northward IMF; growth phase events occurring at a northward turning after a relatively short interval of weakly southward IMF; and recovery phase events in the high-latitude part of a double oval.

The terminology pseudo breakup–expansion onset–intensification implies that a substorm has a "main onset" with some events before and some events after this time. By definition the main onset is the time the aurora starts a "substantial" poleward expansion. Past work has established that the sudden onset of an auroral zone negative bay, the start of a midlatitude positive bay, and Pi 2 pulsations are associated with the onset of auroral expansion. In the absence of auroral observations the onset of Pi 2 at the beginning of the negative bay may be used to time the expansion onset [*Hsu and McPherron*, 1996; *Hsu and McPherron*, 1998]. This time is called the main onset and was used by *Hsu and McPherron* [2012] to show that the tail lobe field transitions from loading to unloading at this time. In many cases the main onset is preceded by a pseudo breakup. When the time of this pseudo breakup is used to superpose tail data, it is found that the time of lobe transition is shifted about 20 min later. This was interpreted as evidence that the tail response to a pseudo breakup is different from that of the expansion phase.

When only ground magnetic data are available, it is difficult to separate negative bays and Pi 2 associated with pseudo breakups and the same phenomena associated with poleward boundary intensifications since both create these phenomena. The problem is compounded by the fact that there appear to be many types of Pi 2 and processes that might create these ultralow-frequency (ULF) waves [*Keiling and Takahashi*, 2011]. As mentioned above, fast flows obviously produce Pi 2, but several studies have shown that there might be two different types of flows [*Shue et al.*, 2008]. Pseudo breakups and full expansions primarily differ in their effect on the configuration of the magnetotail. There is no accepted explanation for why pseudo breakups are quenched and soon die out while very similar events develop into a full expansion.

4.7. POLEWARD BOUNDARY INTENSIFICATIONS

The phenomenon named *poleward boundary intensification* (PBI) is a brightening of aurora at the poleward edge of the auroral oval. It is generally thought that the brightening is caused by a burst of magnetic reconnection. Radar observations can be used to infer the reconnection rate across the open/closed separatrix (e.g., *De La Beaujardiere et al.*, 1994). The measured electric field due to reconnection in two substorms varied from a low of 1.5 mV/m in the growth phase to a high of 40 mV/m in the expansion phase. *De La Beaujardiere et al.* [1994] examined another example of this phenomenon during quiet times using two different radars separated in both latitude and longitude. The observations indicated that intensifications at the poleward edge occur every 10–20 min. For one event this bright arc moved in a southwest direction with speeds up to about 1000 m/s. The reconnection rate for this event was about 25 mV/m, in agreement with their previous results.

The Viking satellite produced high-time-resolution auroral images during magnetic activity [*Elphinstone et al.*, 1995a, 1995b, 1995c] that provide insight into the radar observations. Of particular importance is the formation of a double oval at the end of the expansion phase of a substorm. The images show that the high-latitude band of aurora is quasi-periodically activated every 20 min, producing auroral streamers that move equatorward. On some occasions these streamers reach the equatorward branch of the oval and appear to trigger substorm expansions. The authors map the high- and low-latitude branches of the oval to the plasma sheet boundary and the inner edge of the plasma sheet, respectively. *Fairfield et al.* [1999] used Geotail data at the inner edge of the plasma sheet and Polar spacecraft auroral images to study the relation between auroral activations and high-speed earthward flows. He found that most high-speed flows were related to auroral activations at the poleward edge of the oval and that shortly after the flow was detected there were bursts of AKR, geosynchronous particle injections, and ground magnetic activity.

The name poleward boundary intensification first appears in a paper by *Lyons et al.* [1999] that discusses their association with plasma flows in the tail. They show that if these intensifications are present in auroral images there are frequent plasma sheet flows and if they are absent there are fewer or no flows. The authors conclude that the phenomenon is distinct from substorms. *Lyons et al.* [2000] emphasize this view stating there is a dramatic difference between the two states of the plasma sheet. *Lyons* [2000] introduced the abbreviation PBI that is now used to refer to this phenomenon. In doing this, he distinguished three types of auroral activity: PBI, substorms, and aurora due to changes in dynamic pressure. *Zesta et al.* [2000] present particularly clear examples of PBI in a keogram (auroral intensity maps in latitude–universal time) illustrating how a growth phase arc breaks up and then expands poleward followed by a second expansion to even higher latitudes. After these are completed, a double oval is formed and numerous PBIs are seen moving south toward the equatorward branch of the oval. Each of these is shown to be associated with a fast flow in the plasma sheet. *Sergeev et al.* [1999] show that these flow burst can reach all the way to synchronous orbit and cause short-duration particle injections.

Lyons et al. [2002b] relate ULF waves to the various phenomena shown to be associated with PBI and bursty flows. They tentatively suggest that the 10–30 min periodicity seen so frequently in BBF, Pi 2, PBI, and other phenomena may be evidence of some global oscillation of the magnetotail. This idea is reminiscent of the thermal catastrophe model of substorms developed somewhat earlier by *Smith et al.* [1986].

The two-dimensional structure of PBI was examined by *Zesta et al.* [2002]. After identifying possible PBI in keograms, they used all-sky images to determine what type of aurora had created the keogram trace. They classified intensifications as nonpropagating or equatorward propagating. Propagating PBI were either north-south structures moving equatorward or east-west arcs moving equatorward. Non-propagating PBI were identified as a beaded auroral arc at high latitudes. The north-south structures are narrow in azimuth and centered around midnight. The east-west arcs are more often seen in the evening sector. This work was extended by *Zesta et al.* [2006] using Geotail data and magnetic field models. They find that a flow will be seen in space only if the satellite is near the projected equatorial crossing of the streamer seen in aurora images. Furthermore, the magnetic mapping shows that the streamers project as narrow channels parallel to the *X*-axis, even for the east-west arcs near dusk. They also argue that many flow channels exist simultaneously in space, all feeding plasma into the inner magnetosphere.

Lyons et al. [2010] and *Nishimura et al.* [2010a] and a number of additional papers by this group use THEMIS data to postulate a new model for substorm expansion onset. The essential ideas are the following. Reconnection occurs at the distant X line (open/closed field line boundary) producing a poleward boundary intensification. Low-entropy plasma flows across this boundary and streams earthward in a narrow flow channel. As this flow burst reaches the inner edge of the plasma sheet, it first bends eastward around the high-latitude part of the Harang discontinuity (HD) and then bends westward around the low-latitude part. It eventually impacts the growth phase arc lying within the discontinuity. Sometimes this arc brightens and then expands azimuthally and poleward. The expansion is attributed to instability of a pressure distribution established by the background convection and the arrival of the previous growth phase flow bursts [*Nishimura et al.*, 2010b; *Xing et al.*, 2010]. *Lyons et al.* [2011] and *Shi et al.* [2012] suggest that the PBI is generated at a point along the tail X line at which a flow channel in the open field line region of the polar cap hits the X line. *Nishimura et al.* [2011a] postulate that if the arrival of a streamer at the growth phase arc does not trigger an expansion, it will instead brighten the arc and make it more susceptible to instability with the arrival of the next streamer. Alternate paths to auroral expansion are given by *Nishimura et al.* [2011b]. They include processes labeled poleward boundary contact (very thin double oval with a connecting PBI) and Harang aurora deformation (a sudden deformation of the growth phase arc in the HD). Evidence of flux transport into the open/closed field line boundary prior to a PBI has been reported by *Shi et al.* [2012] using radar observations close to the nightside boundary. *Shi et al.* [2013] have examined this transport in greater detail using the motion of patches and arcs in the polar cap to track flux transport toward the nightside.

Recently the THEMIS configuration of satellites was altered placing the outer two spacecraft in orbit around the Moon. These spacecraft then provide the opportunity to detect the expected signature of a near-Earth neutral line: fast tailward flow with southward magnetic field. A reversal of the flow direction indicates movement of the reconnection site tailward of the moon. *Nishimura et al.* [2013b] have used these spacecraft to study flow bursts in different phases of the substorm. The results are very interesting (see their Table 1). During quiet times and substorm growth phase, flows are earthward and originate at a neutral line beyond the Moon. During substorm expansion phase, flows are tailward and originate at a near-Earth neutral line. In the recovery phase and steady magnetospheric convection, flows are again earthward from either the retreated new X line or the distant X line. *Nishimura et al.* [2013a] use this same configuration to study the origin of PBI and earthward flows prior to expansion onset. For several cases they find that the source of the flows is beyond the Moon. This leads the authors to suggest that distant neutral line reconnection generates the flows, which on reaching the inner edge of the plasma sheet trigger an instability that creates the expansion phase. This expansion then creates the near-Earth neutral line. They summariz this view by the statement "suggesting that NENL is not related to preonset flow and auroral activity but becomes active during the expansion phase." In a recent review *Lyons et al.* [2012] states the following: "The results suggest a unifying view that substorms, PBIs, streamers, and other disturbances may be all related to mesoscale flow channels, which come from the distant plasma sheet and couple with the more slowly varying large-scale background."

Frey [2010] took issue with the analysis of *Nishimura et al.* [2010a] claiming that flaws in the analysis lead to "grossly exaggerated results." They point out that the shortest time between events in this list is 3 min! We have calculated the waiting time distribution between auroral onset events in their list and find that a third are separated by less than 40 min. The peak of this distribution is at 10 min. Ten minutes is shorter than most expansion phases and 40 min is the typical duration of midlatitude positive bays. *Frey* [2010] compares his independent determination of expansion onset using the same database to the overlapping portion of the *Nishimura et al.* [2010a] database finding 522 substorms where they find only 106. He questions whether all the substorms in the larger list exhibit the same behavior. Subsequently *Mende et al.*

[2011] worked with Nishimura to collectively evaluate a subset of his list. They conclude that in some cases PBIs and subsequent flows can trigger an expansion provided the inner tail is in the proper state. A more important observation is that "All substorms that clearly show the Nishimura et al. [2010a] pattern are intensifications, not isolated substorms."

PBIs, flow bursts, and auroral streamers are obviously an important element of geomagnetic activity in the tail. Where these flow bursts are generated is not obvious. When reconnection breaks through the boundary of the plasma sheet, it will map via field lines to the open/closed boundary in the ionosphere. This site could actually be very close to Earth at the onset of the expansion phase, somewhat further away during the expansion and recovery phase, and at a distant location in quiet times. According to *Mende et al.* [2011], most of the PBI events are substorm intensifications, not substorm expansions. Events with a waiting time distribution peaked at 10 min clearly do not satisfy the traditional definition of substorms. If it were true that all tail reconnection events occurred at the distant X line, then plasmoids would never be observed traveling down the tail.

4.8. QUESTIONS RELATED TO DYNAMICS OF MAGNETOTAIL

The structure and dynamics of the magnetotail are primarily a consequence of magnetic reconnection. The purpose of a magnetospheric substorm, steady magnetospheric convection, and sawtooth events is to return magnetic flux to the dayside. We do not yet understand the physics of magnetic reconnection or how to represent it properly in physical models. It is believed that it requires a thin current sheet and probably an external driver to bring antiparallel fields together. Why reconnection occurs where it does or why it might begin and then suddenly end have not been explained. This is the problem in understanding the pseudo breakup. Why is reconnection an impulsive process creating short bursts of high-speed flow for only brief intervals of time (the bursty bulk flow and its flow peaks)? How does an X line expand at an existing site or does it simply jump to a new region having appropriate conditions (multiple onsets)? Do low-frequency waves in the tail play any role in creating the local geometry needed for reconnection? What process creates the growth phase arc? What makes this arc brighten, extend azimuthally, develop beads, and then suddenly expand poleward? The inside-out model of substorms argues this is evidence of instability of the pressure distribution established during the growth phase. In this case the complex aurora of the bulge would be created by the instability. However, it is unclear how the arrival of high-speed flows, reduction of open flux, and evidence of a new X line can be so tightly linked to this instability. Why is it not common for instability to occur without the formation of an X line? In the outside-in model it is suggested that magnetic flux and plasma pile up in the inner magnetosphere and that the bulge is a tailward motion of a dipolarization front. It seems quite likely that the rapid arrival, pileup, and deflection of flows will create conditions for instability and that the bulge is a projection of this instability onto the ionosphere.

If the auroral expansion is a consequence of instability and independent of reconnection, then it is difficult to understand how this region can be stable for up to 10 hours during an SMC during which no flows manage to penetrate this region. In either case what is the relation of aurora inside the bulge to processes in the tail? Presumably every equatorial counterclockwise vortex or velocity shear will receive a field-aligned current from the ionosphere and possibly generate a spot or strip of bright aurora. To create such a complex pattern of shears would require multiple reconnection sites and interacting flow channels or a very turbulent instability. It seems very likely that the poleward edge of the expanding bulge is a mapping of the near-Earth X line. At the end of the expansion it is this arc that produces the multiple PBIs that create auroral streamers moving toward lower latitudes and these are known to be caused by reconnection. Why are these PBIs quasi-periodic with a separation of 10–30 min? This same waiting time distribution is seen in BBF, Pi 2, the separation of pseudo breakups, and intensifications of the expansion (midlatitude positive bays) and so must represent some internal time constant of the magnetosphere. If each substorm has multiple reconnection sites each producing a flux rope, then how do these flux ropes interact to produce the relatively simple plasmoids reported in the distant tail?

The behavior of the near-Earth neutral line is not well understood nor is it known what factors control its development. It can form as close to Earth as 13 Re, but more typically around 22 Re. As seen in flows and auroral disturbances, it usually forms just premidnight but at times can occur anywhere on the nightside. It appears to move tailward just after the end of the expansion, but how far tailward is not known. In an SMC it is clearly beyond 30 Re and probably beyond the Moon. It is not known where it is located during sawtooth events.

The formation of the substorm current wedge that closes westward through the auroral bulge in the expansion phase is not understood. Each flow channel produces a narrow current wedge called a wedglet. But the current in these is weak and dies as the flow comes to a halt in the inner magnetosphere. The strength of the current wedge seen by midlatitude stations requires a much larger and more persistent source of field-aligned current. In the NENL model this is generally attributed to combination of the dipolarization near midnight combined with pressure gradients at the edges of the bulge. How a ballooning instability might produce the current wedge remains to be explained.

The properties of the distant X line remained to be developed. Where is it located? What is its shape? Once established by an interval of northward IMFs, does its position change? What controls these properties? At what rate does the distant X line reconnect open magnetic flux during northward IMFs and how long can the lobes remain filled with open field lines? Does this X line create narrow flow channels that survive all the way to the earthward end of the plasma sheet?

An appropriate model of magnetotail dynamics must take into account historical observations as well as recent observations. It should not be forgotten that phenomena such as loading and unloading of open magnetic flux and the creation of plasmoids traveling down the tail are well-established facts that must be included in any theory. Also there is persuasive evidence that there are at least three distinct modes of response: the substorm, steady convection, and sawtooth events. Each of these is a phenomenon that lasts for several hours or more. Any theory of tail dynamics must also account for the existence of these response modes and their dependence on the conditions in the solar wind.

ACKNOWLEDGMENTS

The author would like to thank the meeting organizers for the opportunity to prepare this chapter and particularly to A. Keiling for his suggestions on the preparation of this manuscript. We acknowledge support from the following grants: NASA NNX07AG16G, NASA NNX10AE61G, NASA NNX10AF25G, and NSF-GEM - AGS-1003854.

REFERENCES

Aikio, A., et al. (1999), Characteristics of pseudobreakups and substorms observed in the ionosphere, at the geosynchronous orbit, and in the midtail, *J. Geophys. Res.*, *104*(A6), 12263–12287, doi:10.1029/1999JA900118.

Akasofu, S.-I. (1964), The development of the auroral substorm, *Planet. Space Sci.*, *12*(4), 273–282.

Akasofu, S. I. (1968), *Polar and Magnetospheric Substorm*, Reidel Publishing Co., Dordrecht, Netherlands.

Angelopoulos, V. (2008), The THEMIS Mission, *Space Sci. Rev.*, *141*(1–4), 5–34, doi:10.1007/s11214-008-9336-1.

Angelopoulos, V., et al. (1992), Bursty bulk flows in the inner central plasma sheet, *J. Geophys. Res.*, *97*(A4), 4027–4039, doi:10.1029/91JA02701.

Angelopoulos, V., et al. (1993), Characteristics of ion flow in the quiet state of the inner plasma sheet, *Geophys. Res. Lett.*, *20*(16), 1711–1714, doi:10.1029/93GL00847.

Angelopoulos, V., et al. (1994), Statistical characteristics of bursty bulk flow events, *J. Geophys. Res.*, *99*(A11), 21,257–21,280, doi:10.1029/94JA01263.

Angelopoulos, V., et al. (1996), Multipoint analysis of a bursty bulk flow event on April 11, 1985, *J. Geophys. Res.Space Phys.*, *101*(A3), 4967–4989, doi:10.1029/95ja02722.

Angelopoulos, V., et al. (1997), Magnetotail flow bursts: Association to global magnetospheric circulation, relationship to ionospheric activity and direct evidence for localization, *Geophys. Res. Lett.*,*24*(18),2271–2274,doi:10.1029/97GL02355.

Angelopoulos, V., et al. (2008), Tail reconnection triggering substorm onset, *Science*, *321*(5891), 931–935, doi:10.1126/science.1160495.

Angelopoulos, V., et al. (2009), Response to comment on "tail reconnection triggering substorm Onset," *Science*, *324*(5933), doi:10.1126/science.1168045.

Aubry, M., et al. (1970), Inward motion of the magnetopause before a substorm, *J. Geophys. Res.*, *75*(34), 7018–7031, doi:10.1029/JA075i034p07018.

Axford, W. (1969), Magnetospheric convection, *Rev. Geophys.*, *7*(1, 2), 421–459, doi:10.1029/RG007i001p00421.

Axford, W., et al. (1965), Tail of the magnetosphere, *J. Geophys. Res.*, *70*(5), 1231–1236, doi:10.1029/JZ070i005p01231.

Axford, W.I. (1964), Viscous interaction between the solar wind and the earth's magnetosphere, *Planet. Space Sci.*, *12*, 45–53.

Axford, W.I. (1967), Magnetic storm effects associated with tail of magnetosphere, *Space Sci. Rev.*, *7*(2–3), 149–157, doi:10.1007/bf00215592.

Axford, W.I., et al. (1961), A unifying theory of high-latitude geophysical phenomena and geomagnetic storms, *Can. J. Phys.*, *39*, 1433–1464.

Baker, D., et al. (1984a), Direct observations of passages of the distant neutral line (80-140 R E) following substorm onsets: ISEE-3, *Geophys. Res. Lett.*, *11*(10), 1042–1045, doi:10.1029/GL011i010p01042.

Baker, D.N., et al. (1984b), Correlated dynamical changes in the near-Earth and distance magnetotail regions: ISEE 3, *J. Geophys. Res.*, *89*(A6), 3855–3864, doi:10.1029/JA089iA06p03855.

Baker, D.N., et al. (1987), Average plasma and magnetic field variations in the distant magnetotail associated with near-Earth substorm effects, *J. Geophys. Res.*, *92*(A1), 71–81, doi:10.1029/JA092iA01p00071.

Baker, D., et al. (1996), Neutral line model of substorms: Past results and present view, *J. Geophys. Res.*, *101*(A6), 12,975–13,010, doi:10.1029/95JA03753.

Baumjohann, W. (2002), Modes of convection in the magnetotail, *Phys. Plasmas*, *9*(9), 3665–3667, doi:10.1063/1.1499116.

Baumjohann, W., et al. (1988), Average ion moments in the plasma sheet boundary layer, *J. Geophys. Res.*, *93*(A10), 11,507–11,520, doi:10.1029/JA093iA10p11507.

Baumjohann, W., et al. (1989), Average plasma properties in the central plasma sheet, *J. Geophys. Res.*, *94*(A6), 6597–6606, doi:10.1029/JA094iA06p06597.

Baumjohann, W., et al. (1990), Characteristics of high-speed ion flows in the plasma sheet, *J. Geophys. Res.*, *95*(A4), 3801–3809, doi:10.1029/JA095iA04p03801.

Baumjohann, W., et al. (1991), Superposed epoch analysis of the substorm plasma sheet, *J. Geophys. Res.*, *96*(A7), 11,605–11,608, doi:10.1029/91JA00775.

Baumjohann, W., et al. (1999), Substorm dipolarization and recovery, *J. Geophys. Res.*, *104*(A11), 24,995–25,000, doi:10.1029/1999JA900282.

Behannon, K. W., and N. F. Ness (1966), Magnetic storms in Earth's magnetic tail, *J. Geophys. Res.*, *71*(9), 2327, doi:10.1029/JZ071i009p02327.

Belian, R. D., et al. (1995), Quasi-periodic, substorm associated, global flux variations observed at geosynchronus orbit, in *Space Plasmas: Coupling between Small and Medium Scale Processes*, edited by M. Ashour-Abdalla et al., p. 143, AGU, Washington, D.C.

Birn, J., et al. (1989), Filamentary structure of a three-dimensional plasmoid, *J. Geophys. Res.*, *94*(A1), 241–251, doi:10.1029/JA094iA01p00241.

Birn, J., and M. Hesse (1991), The substorm current wedge and field-aligned currents in MHD simulations of magnetotail reconnection, *J. Geophys. Res.*, *96*(A2), 1611–1618, doi:10.1029/90JA01762.

Birn, J., and M. Hesse (1996), Details of current disruption and diversion in simulations of magnetotail dynamics, *J. Geophys. Res.*, *101*(A7), 15,345–15,358, doi:10.1029/96JA00887.

Birn, J., et al. (1999), Flow braking and the substorm current wedge, *J. Geophys. Res.*, *104*(A9), 19,895–19,903, doi:10.1029/1999JA900173.

Borovsky, J. E., et al. (1993), The occurrence rate of magnetospheric-substorm onsets: Random and periodic substorms, *J. Geophys. Res.*, *98*(A3), 3807–3813, doi:10.1029/92JA02556.

Caan, M., et al. (1973), Solar wind and substorm-related changes in the lobes of the geomagnetic tail, *J. Geophys. Res.*, *78*(34), 8087–8096, doi:10.1029/JA078i034p08087.

Caan, M. N., et al. (1978), The statistical magnetic signature of magnetospheric substorms, *Planet. Space Sci.*, *26*(3), 269–279.

Cai, X., et al. (2006a), Statistical analysis of ionospheric potential patterns for isolated substorms and sawtooth events, *Ann. Geophys.*, *24*(7), 1977–1991.

Cai, X., et al. (2006b), A statistical study of magnetic dipolarization for sawtooth events and isolated substorms at geosynchronous orbit with GOES data, *Ann. Geophys.*, *24*(12), 3481–3490.

Cai, X., et al. (2009), Investigation of the period of sawtooth events, *J. Geophys. Res.Space Phys.*, *114*, 9, doi:10.1029/2008ja013764.

Clauer, C., and R. McPherron (1974), Mapping the local time-universal time development of magnetospheric substorms using mid-latitude magnetic observations, *J. Geophys. Res.*, *79*(19), 2811–2820, doi:10.1029/JA079i019p02811.

Clauer, C. R., et al. (2006), Characterizing the 18 April 2002 storm-time sawtooth events using ground magnetic data, *J. Geophys. Res. P. A Space Phys.*, *111*(A4), 13.

Coroniti, F. V., et al. (1968), Studies of the magnetospheric substorm: 3. Concept of the magnetospheric substorm and its relation to electron precipitation and micropulsations, *J. Geophys. Res.*, *73*(5), 1715–1722, doi:10.1029/JA073i005p01715.

Coroniti, F. V., et al. (1972), Changes in magnetospheric configuration during the substorm growth phase, *J. Geophy. Res.*, *77*(19), 3361–3370, doi:10.1029/JA077i019p03361.

Davis, T. N., and T. J. Hallinan (1976), Auroral spirals. 1. Observations, *J. Geophys. Res. Space Phys.*, *81*(22), 3953–3958, doi:10.1029/JA081i022p03953.

DeJong, A. D., and C. R. Clauer (2005), Polar UVI images to study steady magnetospheric convection events: Initial results, *Geophys. Res. Lett.*, *32*(24), 1–4, doi:10.1029/2005GL024498.

DeJong, A. D., et al. (2007), Aurora and open magnetic flux during isolated substorms, sawteeth, and SMC events, *Ann. Geophys.*, *25*(8), 1865–1876.

DeJong, A. D., et al. (2009), A statistical study of BRIs (SMCs), isolated substorms, and individual sawtooth injections, *J. Geophys. Res. Space Phys.*, *114*, doi:10.1029/2008ja013870.

DeJong, A. D., et al. (2008), Balanced reconnection intervals: Four case studies, *Ann. Geophys.*, *26*(12), 3897–3912.

De La Beaujardiere, O., et al. (1994), Quiet-time intensifications along the poleward auroral boundary near midnight, *J. Geophys. Res.*, *99*(A1), 287–298, doi:10.1029/93JA01947.

Donovan, E., et al. (2008), Simultaneous THEMIS in situ and auroral observations of a small substorm, *Geophys. Res. Lett.*, *35*(17), 5, doi:10.1029/2008gl033794.

Dungey, J. W. (1961), Interplanetary magnetic field and the auroral zones, *Phys. Res. Lett.*, *6*, 47–48.

Eastman, T. E., et al. (1976), The magnetospheric boundary layer: Site of plasma, momentum and energy transfer from the magnetosheath into the magnetosphere, *Geophys. Res. Lett.*, *3*(11), 685–688, doi:10.1029/GL003i011p00685.

Elphinstone, R., et al. (1995a), Observations in the vicinity of substorm onset: Implications for the substorm process, *J. Geophys. Res.*, *100*(A5), 7937–7969, doi:10.1029/94JA02938.

Elphinstone, R., et al. (1995b), The double oval UV auroral distribution, 2. The most poleward arc system and the dynamics of the magnetotail, *J. Geophys. Res.*, *100*(A7), 12,093–12,102, doi:10.1029/95JA00327.

Elphinstone, R., et al. (1995c), The double oval UV auroral distribution, 1. Implications for the mapping of auroral arcs, *J. Geophys. Res.*, *100*(A7), 12,075–12,092, doi:10.1029/95JA00326.

Elphinstone, R. D., et al. (1993), The auroral distribution and its mapping according to substorm phase, *J. Atmos. Terrest. Phys.*, *55*(14), 1741–1762, doi:10.1016/0021-9169(93)90142-l.

Elphinstone, R. D., et al. (1996), What is a global auroral substorm? *Rev. Geophys.*, *34*(2), 169–232, doi:10.1029/96rg00483.

Elvey, C. T. (1957), Problems of auroral morphology, *Proc. Natl. Acad. Sci.*, *43*(1), 63–75.

Fairfield, D., et al. (1999), Earthward flow bursts in the inner magnetotail and their relation to auroral brightenings, AKR intensifications, geosynchronous particle injections and magnetic activity, *J. Geophys. Res.*, *104*(A1), 355–370, doi:10.1029/98JA02661.

Fairfield, D. H., et al. (1966), Transition region magnetic field and polar magnetic disturbances, *J. Geophys. Res.*, *71*(1), 155–169.

Frey, H. U. (2010), Comment on "Substorm triggering by new plasma intrusion: THEMIS all-sky imager observations" by Y. Nishimura et al, *J. Geophys. Res. Space Phys.*, *115*, 5, doi:10.1029/2010ja016113.

Frey, H. U., et al. (2004), Substorm onset observations by IMAGE-FUV, *J. Geophys. Res.*, *109*(A10), 1–6, doi:10.1029/2004JA010607.

Haerendel, G. (1992), Disruption, ballooning or auroral avalanche — On the cause of substorms, in *Proceedings of the International Conference on Substorms (JCS-1)*, pp. 417–420, ESA Publications Div., Paris.

Hammond, C., et al. (1994), Imaging the effect of dipole tilt on magnetotail boundaries, *J. Geophys. Res.*, *99*(A4), 6079–6092, doi:10.1029/93JA01924.

Henderson, M. G. (2004), The May 2–3, 1986 CDAW-9C interval: A sawtooth event, *Geophys. Res. Lett.*, *31*(11), 5.

Henderson, M. G., et al. (2006a), Magnetospheric and auroral activity during the 18 April 2002 sawtooth event, *J. Geophys. Res.*, *111*(A1), 1–15, doi:10.1029/2005JA011111.

Henderson, M. G., et al. (2006b), Substorms during the 10–11 August 2000 sawtooth event, *J. Geophys. Res.*, *111*(A6), 1–32, doi:10.1029/2005JA011366.

Heppner, J. P., et al. (1967), OGO-A magnetic field observations, *J. Geophys. Res.*, *72*(21), 5417–5471, doi:10.1029/JZ072i021p05417.

Hones, E. W., Jr. (1972), Plasma sheet variations during substorms, *Planet. Space Sci.*, *20*(9), 1409–1431, doi:10.1016/0032-0633(72)90048-7.

Hones, E. W., Jr. (1973), Plasma flow in the plasma sheet and its relation to substorms, *Radio Sci.*, *8*(11), 979–990, doi:10.1029/RS008i011p00979.

Hones, E. W., Jr. (1977), Substorm processes in the magnetotail: Comments on "On hot tenuous plasmas, fireballs, and boundary layers in the Earth's magnetotail" by L. A. Frank, K. L. Ackerson, and R. P. Lepping, *J. Geophys. Res.*, *82*(35), 5633–5643, doi:10.1029/JA082i035p05633.

Hones, E. W., Jr. (1979), Transient phenomena in the magnetotail and their relation to substorms, *Space Sci. Rev.*, *23*(3), 393–410.

Hones, E. W., Jr., et al. (1967), Outward flow of plasma in the magnetotail following geomagnetic bays, *J. Geophys. Res.*, *72*(23), 5879–5892, doi:10.1029/JZ072i023p05879.

Hones, E. W., Jr., et al. (1972), Measurements of magnetotail plasma flow made with Vela 4B, *J. Geophys. Res.*, *77*(28), 5503–5522, doi:10.1029/JA077i028p05503.

Hones, E. W., Jr., et al. (1976), Proton flow measurements in the magnetotail plasma sheet made with Imp 6, *J. Geophys. Res.*, *81*(1), 227–234, doi:10.1029/JA081i001p00227.

Hones, E. W., Jr., and K. Schindler (1979), Magnetotail plasma flow during substorms: A survey with IMP 6 and IMP 8 satellites, *J. Geophys. Res.*, *84*(A12), 7155–7169, doi:10.1029/JA084iA12p07155.

Hones, E. W. J., et al. (1982), On the three-dimensional magnetic structure of the plasmoid created in the magnetotail at substorm onset, *Geophys. Res. Lett.*, *9*(3), 203–206, doi:10.1029/GL009i003p00203.

Hones, E. W., Jr., et al. (1984a), Structure of the magnetotail at 220 R E and its response to geomagnetic activity, *Geophys. Res. Lett.*, *11*(1), 5–77, doi:10.1029/GL011i001p00005.

Hones, E. W., Jr., et al. (1984b), Detailed examination of a plasmoid in the distant magnetotail with ISEE 3, *Geophys. Res. Lett.*, *11*(10), 1046–1049, doi:10.1029/GL011i010p01046.

Hones, E. W., Jr., et al. (1984c), Associations of geomagnetic activity with plasma sheet thinning and expansion: A statistical study, *J. Geophys. Res.*, *89*(A7), 5471–5478, doi:10.1029/JA089iA07p05471.

Hsu, T.-S., and R. L. McPherron (1996), Occurrency frequency of substorm field and plasma signatures observed near-earth by ISEE-1/2, in *International Conference on Substorms-3 (ICS-3)*, edited by E. J. Rolfe, and B. Kaldeich, pp. 333–340, ESA, Versailles, France.

Hsu, T.-S., and R. L. McPherron (1998), The main onset of a magnetospheric substorm, in *International Conference on Substorms (ICS-4)*, edited by S. Kokubun, and Y. Kamide, pp. 79–82, Terra Scientific Publishing Company and Kluwer Academics, Tokyo and Dordrecht.

Hsu, T.-S., and R. L. McPherron (2012), A statistical analysis of substorm associated tail activity, *Adv. Space Res.*, *50*(10), 1317–1343, doi:10.1016/j.asr.2012.06.034.

Huang, C. S. (2002), Evidence of periodic (2–3 hour) near-tail magnetic reconnection and plasmoid formation: Geotail observations, *Geophys. Res. Lett.*, *29*(24), 4, doi:10.1029/2002gl016162.

Huang, C.-S. (2005), Variations of polar cap index in response to solar wind changes and magnetospheric substorms, *J. Geophys. Res.*, *110*(A1), 17, doi:10.1029/2004JA010616.

Huang, C.-S., et al. (2003a), Periodic magnetospheric substorms: Multiple space-based and ground-based instrumental observations, *J. Geophys. Res.*, *108*(A11), SMP 16-1–SMP 16–17, doi:10.1029/2003JA009992.

Huang, C.-S., et al. (2003b), Periodic magnetospheric substorms and their relationship with solar wind variations, *J. Geophys. Res.*, *108*(A6), SMP 19-1–SMP 19–12, doi:10.1029/2002JA009704.

Huang, C.-S., et al. (2005), Are sawtooth oscillations of energetic plasma particle fluxes caused by periodic substorms or driven by solar wind pressure enhancements? *J. Geophys. Res.*, *110*(A7), 1–19, doi:10.1029/2005JA011018.

Huang, C.-S., and X. Cai (2009), Magnetotail total pressure and lobe magnetic field at onsets of sawtooth events and their relation to the solar wind, *J. Geophys. Res.*, *114*(A4), 1–13, doi:10.1029/2008JA013807.

Huang, C. S., et al. (2009), Magnetic flux in the magnetotail and polar cap during sawteeth, isolated substorms, and steady magnetospheric convection events, *J. Geophys. Res. Space Phys.*, *114*, doi:10.1029/2009ja014232.

Hubert, B., et al. (2006), Dayside and nightside reconnection rates inferred from IMAGE FUV and Super Dual Auroral Radar Network data, *J. Geophys. Res.*, *111*(A3), 1–16, doi:10.1029/2005JA011140.

Hubert, B., et al. (2008), Open magnetic flux and magnetic flux closure during sawtooth events, *Geophys. Res. Lett.*, *35*(23), 1–6, doi:10.1029/2008GL036374.

Hughes, W. J., and D. G. Sibeck (1987), On the 3-dimensional structure of plasmoids, *Geophys. Res. Lett.*, *14*(6), 636–639, doi:10.1029/GL014i006p00636.

Hurricane, O. A. (1997), MHD ballooning stability of a sheared plasma sheet, *J. Geophys. Res. Space Phys*, *102*(A9), 19,903–19,911, doi:10.1029/97ja01595.

Ieda, A., et al. (2001), Plasmoid ejection and auroral brightenings, *J. Geophys. Res.*, *106*(A3), 3845–3857, doi:10.1029/1999JA000451.

Jacquey, C., et al. (1994), Tailward energetic ion streams observed at ~100 RE by GEOTAIL-EPIC associated with geomagnetic activity intensification, *Geophys. Res. Lett.*, *21*(25), 3015–3018, doi:10.1029/94GL01606.

Kadokura, A., et al. (2002), Detailed analysis of a substorm event on 6 and 7 June 1989—1. Growth phase evolution of nightside auroral activities and ionospheric convection toward expansion phase onset, *J. Geophys. Res. Space Phys.*, *107*(A12), doi:10.1029/2001ja009127.

Kamide, Y. (2001), Some "missing" elements of constraint in substorm initiation modeling, *J. Atmos. Solar Terrest. Phys.*, *63*(7), 635–642, doi:10.1016/S1364-6826(00)00187-5

Kauristie, K., et al. (2000), Ionospheric current signatures of transient plasma sheet flows, *J. Geophys. Res. Space Phys.*, *105*(A5), 10,677–10,690, doi:10.1029/1999ja900487.

Keiling, A., et al. (2008), Multiple intensifications inside the auroral bulge and their association with plasma sheet activities, *J. Geophys. Res. Space Phys.*, *113*(A12), 20, doi:10.1029/2008ja013383.

Keiling, A., and K. Takahashi (2011), Review of Pi2 Models, *Space Sci. Revi.*, *161*(1–4), 63–148, doi:10.1007/s11214-011-9818-4.

Keiling, A., et al. (2012), *Auroral Phenomenology and Magnetospheric Processes: Earth and Other Planets*, Geophysical Monogr. Ser., 197, AGU, Washington, D.C.

Kepko, L., et al. (2009), Equatorward moving auroral signatures of a flow burst observed prior to auroral onset, *Geophys. Res. Lett.*, *36*, 6, doi:10.1029/2009gl041476.

Kissinger, J., et al. (2010), An investigation of the association between steady magnetospheric convection and CIR stream interfaces, *Geophys. Res. Lett.*, *37*, 5, doi:10.1029/2009gl041541.

Kissinger, J., et al. (2011), Steady magnetospheric convection and stream interfaces: Relationship over a solar cycle, *J. Geophys. Res. Space Phys.*, *116*, doi:10.1029/2010ja015763.

Kissinger, J., et al. (2012a), Diversion of plasma due to high pressure in the inner magnetosphere during steady magnetospheric convection, *J. Geophys. Res. Space Phys.*, *117*, 15, doi:10.1029/2012ja017579.

Kissinger, J., et al. (2012b), Necessity of substorm expansions in the initiation of steady magnetospheric convection, *Geophys. Res. Lett.*, *39*, doi:10.1029/2012gl052599.

Kokubun, S., and R. McPherron (1981), Substorm signatures at synchronous altitude, *J. Geophys. Res.*, *86*(A13), 11,265–11,277, doi:10.1029/JA086iA13p11265.

Konradi, A. (1967), Proton events in magnetosphere associated with magnetic bays, *J. Geophys. Res.*, *72*(15), 3829–3841, doi:10.1029/JZ072i015p03829.

Koskinen, H., et al. (1993), Pseudobreakup and substorm growth phase in the ionosphere and magnetosphere, *J. Geophys. Res.*, *98*(A4), 5801–5813, doi:10.1029/92JA02482.

Kubyshkina, M., et al. (2008), Magnetospheric currents during sawtooth events: Event-oriented magnetic field model analysis, *J. Geophys. Res.*, *113*(A8), 1–13, doi:10.1029/2007JA012983.

Kullen, A., and T. Karlsson (2004), On the relation between solar wind, pseudobreakups, and substorms, *J. Geophys. Res.*, *109*(A12), 20.

Kullen, A., et al. (2009), Geomagnetic signatures of auroral substorms preceded by pseudobreakups, *J. Geophys. Res. Space Phys.*, *114*, 14, doi:10.1029/2008ja013712.

Kullen, A., et al. (2010), Occurrence and properties of substorms associated with pseudobreakups, *J. Geophys. Res. Space Phys.*, *115*, 17, doi:10.1029/2010ja015866.

Lee, D.-Y., et al. (2004), Sawtooth oscillations directly driven by solar wind dynamic pressure enhancements, *J. Geophys. Res.*, *109*(A4), 1–16, doi:10.1029/2003JA010246.

Lepping, R. P., et al. (1995), The WIND magnetic field investigation, *Space Sci. Rev.*, *71*(1–4), 207–229.

Levy, R. H. (1964), Interaction of a streaming plasma with the magnetic field of a 2-dimensional dipole, *Phys. Fluids*, *7*(6), 822–825, doi:10.1063/1.1711292.

Levy, R. H., et al. (1964), Aerodynamic aspects of the magnetospheric flow, *AIA J.*, *2*(12), 2065–2076, doi:10.2514/6.1964-87.

Lewis, R. V., et al. (1998), The behavior of the electric field within the substorm current wedge, *J. Geophys. Res. Space Phys.*, *103*(A1), 179–190, doi:10.1029/97ja01987.

Lezniak, T. W., and J. R. Winckler (1970), Experimental study of magnetospheric motions and the acceleration of energetic electrons during substorms, *J. Geophys. Res.*, *75*(34), 7075–7098, doi:10.1029/JA075i034p07075.

Lui, A. (1991), A synthesis of magnetospheric substorm models, *J. Geophys. Res.*, *96*(A2), 1849–1856, doi:10.1029/90JA02430.

Lui, A. T. Y. (1987), *Magnetotail Physics*, John Hopkins Univ. Press, Baltimore and London.

Lui, A. T. Y. (2009), Comment on "Tail reconnection triggering substorm onset," *Science*, *324*(5933), 3, doi:10.1126/science.1167726.

Lui, A. T. Y. (2011), Reduction of the cross-tail current during near-Earth dipolarization with multisatellite observations, *J. Geophys. Res. Space Phys.*, *116*, 11, doi:10.1029/2011ja017107.

Lui, A. T. Y., et al. (1988), A case study of magnetotail current sheet disruption and diversion, *Geophys. Res. Lett.*, *15*(7), 721–724.

Lui, A. T. Y., et al. (2004), Magnetotail behavior during storm time "sawtooth injections", *J. Geophys. Res.*, *109*(A10), 1–8, doi:10.1029/2004JA010543.

Lui, A. T. Y., et al. (2008), Determination of the substorm initiation region from a major conjunction interval of THEMIS satellites, *J. Geophys. Res. Space Phys.*, *113*, 19, doi: 10.1029/2008ja013424.

Lyons, L. R. (2000), Geomagnetic disturbances: Characteristics of, distinction between types, and relations to interplanetary conditions, *J. Atmos. Solar Terrest. Phys.*, *62*(12), 1087–1114, doi:10.1016/s1364-6826(00)00097-3.

Lyons, L., et al. (1999), Association between geotail plasma flows and auroral poleward boundary intensifications observed by CANOPUS photometers, *J. Geophys. Res.*, *104*(A3), 4485–4500, doi:10.1029/1998JA900140.

Lyons, L. R., et al. (2000), Structured currents associated with tail bursty flows during turbulent plasma sheet conditions, in *Magnetospheric Current Systems*, edited by S. Ohtani et al., pp. 267–274, AGU, Washington, D.C.

Lyons, L. R., et al. (2002a), Relation of substorm breakup arc to other growth-phase auroral arcs, *J. Geophys. Res.*, *107*(A11), SMP 26-1–SMP 26–10, doi:10.1029/2002JA009317.

Lyons, L. R., et al. (2002b), Auroral poleward boundary intensifications and tail bursty flows: A manifestation of a large-scale ULF oscillation? *J. Geophys. Res. Space Phys.*, *107*(A11), 15, doi:10.1029/2001ja000242.

Lyons, L. R., et al. (2010), Substorm triggering by new plasma intrusion: Incoherent-scatter radar observations, *J. Geophys. Res. Space Phys.*, *115*, 13, doi:10.1029/2009ja015168.

Lyons, L. R., et al. (2011), Possible connection of polar cap flows to pre- and post-substorm onset PBIs and streamers, *J. Geophys. Res. Space Phys.*, *116*, 14, doi:10.1029/2011ja016850.

Lyons, L. R., et al. (2012), Auroral disturbances as a manifestation of interplay between large-scale and mesoscale structure of magnetosphere-ionosphere electrodynamical coupling, in *Auroral Phenomenology and Magnetospheric Processes: Earth and Other Planets*, edited by A. Keiling et al., pp. 193–204, AGU, Washington, D.C.

Machida, S., et al. (1999), GEOTAIL observations of flow velocity and north-south magnetic field variations in the near and mid-distant tail associated with substorm onsets, *Geophys. Res. Lett.*, *26*(6), 635–638, doi:10.1029/1999gl900030.

McPherron, R., et al. (1973a), Phenomenological model for substorms, *J. Geophys. Res.*, *78*(16), 3131–3149, doi:10.1029/JA078i016p03131.

McPherron, R., et al. (1973b), Substorms in space: The correlation between ground and satellite observations of the magnetic field, *Radio Sci.*, *8*(11), 1059–1076, doi:10.1029/RS008i011p01059.

McPherron, R. L. (1970), Growth phase of magnetospheric substorms, *J. Geophys. Res.*, *75*(28), 5592–5599.

McPherron, R. L. (1972a), Substorm related changes in the geomagnetic tail, *Planet. Space Sci.*, *20*(9), 1521–1539, doi:10.1016/0032-0633(72)90054-2.

McPherron, R. L. (1972b), Substorm related changes in the geomagnetic tail: The growth phase, *Planet. Space Sci.*, *20*(9), 1521–1539.

McPherron, R. L. (1991), Physical processes producing magnetospheric substorms and magnetic storms, in *Geomagnetism*, vol. *4*, edited by J. Jacobs, pp. 593–739, Academic, London.

McPherron, R. L., et al. (1987), Is near-earth current sheet thinning the cause of auroral substorm expansion onset? (extended abstract), in *Quantitative Modeling of Magnetosphere-Ionosphere Coupling Processes*, edited by Y. K. a. R. A. Wolf, pp. 252–257, Kyoto Sangyo Univ., Kyoto, Japan.

McPherron, R. L., and D. H. Fairfield (1998), Session 1 Summary: What are the major expansion phase actions as seen at various regions? in *International Conference on Substorms (ICS-4)*, edited by S. Kokubun and Y. Kamide, pp. 29–33, Terra Scientific Publishing Company and Kluwer Academics, Tokyo and Dordrecht.

McPherron, R. L., et al. (2005), Solar wind drivers for steady magnetospheric convection, in *Multiscale Coupling of Sun-Earth Processes*, edited by A. T. Y. Lui et al., pp. 113–124, Elsevier, Amsterdam.

McPherron, R. L., and T. S. Hsu (2006), Challenges of multi-spacecraft missions to end the substorm controversy, in *International Conference on Substorms*, vol. *8*, edited by E. Donovan, pp. 1–4, Dept. of Phys., Univ. of Alberta, Banff, Alberta, Canada.

McPherron, R. L., et al. (2008), Response of the Earth's magnetosphere to changes in the solar wind, *J. Atmos. Solar Terrest. Phys.*, *70*(2–4), 303–315, doi:10.1016/j.jastp.2007.08.040.

McPherron, R. L., et al. (2011), Characteristics of plasma flows at the inner edge of the plasma sheet, *J. Geophys. Res. Space Phys.*, *116*, doi:10.1029/2010ja015923.

McWilliams, K. A., et al. (2008), Steady magnetospheric convection selection criteria: Implications of global SuperDARN convection measurements, *Geophys. Res. Lett.*, *35*(9), doi:10.1029/2008gl033671.

Mende, S. B., et al. (2011), Substorm triggering by poleward boundary intensification and related equatorward propagation, *J. Geophys. Res. Space Phys.*, *116*, 17, doi:10.1029/2010ja015733.

Milan, S. E., et al. (2003), Variations in the polar cap area during two substorm cycles, *Ann. Geophys.*, *21*(5), 1121–1140.

Miura, A. (2001), Ballooning instability as a mechanism of the near-Earth onset of substorms, *Space Sci. Rev.*, *95*(1–2), 387–98, doi:10.1023/a:1005249915285.

Miyashita, Y., et al. (1999), GEOTAIL observations of total pressure and electric field variations in the near and mid-distant tail associated with substorm onsets, *Geophys. Res. Lett.*, *26*(6), 639–642, doi:10.1029/1999gl900031.

Miyashita, Y., et al. (2000), A statistical study of variations in the near and middistant magnetotail associated with substorm onsets: GEOTAIL observations, *J. Geophys. Res.*, *105*(A7), 15913–15930, doi:10.1029/1999JA000392.

Montgomery, M., et al. (1965), Spatial distribution energy spectra and time variations if energetic electrons (E 50kev) at 17.7 Earth radii, *Phys. Rev. Lett.*, *14*(7), 209–213, doi:10.1103/PhysRevLett.14.209.

Nagai, T., et al. (1994), Initial geotail survey if magnetic substorm signatures in the magnetotail, *Geophys. Res. Lett.*, *21*(25), 2991–2994, doi:10.1029/94gl01420.

Nagai, T., et al. (1997), Substorms, tail flows and plasmoids, *Adv. Space Res.*, *20*(4–5), 961–971, doi:10.1016/s0273-1177(97)00504-8.

Nagai, T., et al. (1998), Geotail observations of a fast tailward flow at X GSM = − 15 R E, *J. Geophys. Res.*, *103*(A10), 23,543–23,550, doi:10.1029/98JA02246.

Nakamura, R., et al. (1994), Particle and field signatures during pseudobreakup and major expansion onset, *J. Geophys. Res. Space Phys.*, *99*(A1), 207–221, doi:10.1029/93JA02207.

Nakamura, R., et al. (2000), Fast flow bursts and auroral activations, in *Proceedings of the Fifth International Conference on Substorms*, edited by A. Wilson, pp. 319–322, Esa Publications Div., Noordwijk, Netherlands.

Nakamura, R., et al. (2001), Flow bursts and auroral activations: Onset timing and foot point location, *J. Geophys. Res.-Space Phys.*, *106*(A6), 10,777–10,789, doi:10.1029/2000ja000249.

Nakamura, R., et al. (2004), Spatial scale of high-speed flows in the plasma sheet observed by Cluster, *Geophys. Res. Lett.*, *31*(9), doi:10.1029/2004gl019558.

Nakamura, R., et al. (2005), Multi-point observation of the high-speed flows in the plasma sheet, *Adv. Space Res.*, *36*(8), 1444–1447, doi:10.1016/j.asr.2005.05.101.

Ness, N.F. (1965), The Earth's magnetic tail, *J. Geophys. Res.*, *70*(13), 2989–3005, doi:10.1029/JZ070i013p02989.

Newell, P.T., et al. (2007), A nearly universal solar wind-magnetosphere coupling function inferred from 10 magnetospheric state variables, *J. Geophys. Res.*, *112*(A1), 1–16, doi:10.1029/2006JA012015.

Nishida, A., et al. (1973), Synoptic survey for the neutral line in the magnetotail during the substorm expansion phase, *J. Geophys. Res.*, *78*(19), 3782–3798, doi:10.1029/JA078i019p03782.

Nishida, A., and E. W. J. Hones (1974), Association of plasma sheet thinning with neutral line formation in the magnetotail, *J. Geophys. Res.*, *79*(4), 535–547, doi:10.1029/JA079i004p00535.

Nishida, A., et al. (1998), New Perspectives on the Earth'S Magnetotail, *Geophys. Monogr. Ser.*, *105*, AGU, Washington, D. C.

Nishimura, Y., et al. (2010a), Substorm triggering by new plasma intrusion: THEMIS all-sky imager observations, *J. Geophys. Res. Pt. A Space Phys.*, *115*(A7), A07222, doi:10.1029/2009ja015166.

Nishimura, Y., et al. (2010b), Preonset time sequence of auroral substorms: Coordinated observations by all-sky imagers, satellites, and radars, *J. Geophys. Res. Space Phys.*, *115*, 12, doi:10.1029/2010ja015832.

Nishimura, Y., et al. (2011a), Relations between multiple auroral streamers, pre-onset thin arc formation, and substorm auroral onset, *J. Geophys. Res. Space Phys.*, *116*, 10, doi: 10.1029/2011ja016768.

Nishimura, Y., et al. (2011b), Categorization of the time sequence of events leading to substorm onset based on THEMIS all-sky imager observations, in *Dynamic Magnetosphere*, vol. *3*, pp. 133-142, Springer-Verlag, Berlin.

Nishimura, Y., et al. (2013a), Identifying the magnetotail source region leading to preonset poleward boundary intensifications, *J. Geophys. Res. Space Phys.*, *118*(7), 4335–4340, doi: 10.1002/jgra.50407.

Nishimura, Y., et al. (2013b), Tail reconnection region versus auroral activity inferred from conjugate ARTEMIS plasma sheet flow and auroral observations, *J. Geophys. Res. Space Phys.*, *118*(9), 5758–5766, doi:10.1002/jgra.50549.

O'Brien, T.P., et al. (2002), Steady magnetospheric convection: Statistical signatures in the solar wind and AE, *Geophys. Res. Lett.*, *29*(7), doi:10.1029/2001GL014641.

Ohtani, S. (2004), Flow bursts in the plasma sheet and auroral substorm onset: Observational constraints on connection between midtail and near-Earth substorm processes, *Space Sci. Rev.*, *113*(1–2), 77–96.

Ohtani, S., et al. (1993), A multisatellite study of a pseudo-substorm onset in the near-earth magnetotail, *J. Geophys. Res. Space Phys.*, *98*(A11), 19,355–19,367, doi:10.1029/93ja 01421.

Ohtani, S., and T. Tamao (1993), Does the ballooning instability trigger substorms in the near-earth magnetotail, *J. Geophys. Res. Space Phys.*, *98*(A11), 19,369–19,379, doi:10. 1029/93ja01746.

Ohtani, S., et al. (2007), Cluster observations in the inner magnetosphere during the 18 April 2002 sawtooth event: Dipolarization and injection at r = 4.6 R E, *J. Geophys. Res.*, *112*(A8), 1–12, doi:10.1029/2007JA012357.

Parks, G. K., et al. (1968), Studies of magnetospheric substorms, I. Characteristics of modulated energetic electron precipitation occurring during auroral substorms, *J. Geophys. Res.*, *73*(5), 1685–1696.

Partamies, N., et al. (2009a), Different magnetospheric modes: Solar wind driving and coupling efficiency, *Ann. Geophys.*, *27*(11), 4281–4291.

Partamies, N., et al. (2009b), Statistical survey on sawtooth events, SMCs and isolated substorms, *Adv. Space Res.*, *44*(3), 376–384, doi:10.1016/j.asr.2009.03.013.

Petrukovich, A., et al. (1998), Two spacecraft observations of a reconnection pulse during an auroral breakup, *J. Geophys. Res.*, *103*(A1), 47–59, doi:10.1029/97JA02296.

Piddington, J. H. (1963), Theories of the geomagnetic storm main phase, *Planet. Space Sci.*, *11*(11), 1277–1288, doi:10.1016/0032-0633(63)90232-0.

Pu, Z., et al. (1999), Ballooning instability in the presence of a plasma flow: A synthesis of tail reconnection and current disruption models for the initiation of substorms, *J. Geophys. Res.*, *104*(A5), 10,235–10,248, doi:10.1029/ 1998JA900104.

Pulkkinen, T. I., et al. (1991a), Modeling the growth-phase of a substorm using the Tsyganenko model and multi-spacecraft observations—CDAW-9, *Geophys. Res. Lett.*, *18*(11), 1963–1966, doi:10.1029/91gl02002.

Pulkkinen, T. I., et al. (1991b), Mapping of auroral arcs during substorm growth phase, *J. Geophys. Res.*, *96*(A12), 21,087–21094, doi:10.1029/91ja01960.

Pulkkinen, T. I., et al. (1994), Thin current sheets in the magnetotail during substorms: CDAW 6 revisited, *J. Geophys. Res.*, *99*(A4), 5793–5803, doi:10.1029/93JA03234.

Pulkkinen, T. I., et al. (1998), Pseudobreakup and substorm onset: Observations and MHD simulations compared, *J. Geophys. Res. Space Physics*, *103*(A7), 14,847–14,854, doi:10.1029/97ja03244.

Pulkkinen, T. I., et al. (2006), Magnetospheric current systems during stormtime sawtooth events, *J. Geophys. Res.*, *111*(A11), 1–13.

Pulkkinen, T. I., et al. (2010), Magnetospheric modes and solar wind energy coupling efficiency, *J. Geophys. Res.*, *115*(A3), 1–9, doi:10.1029/2009JA014737.

Pytte, T., et al. (1978), Multiple-satellite studies of magnetospheric substorms: Distinction between polar magnetic substorms and convection-driven negative bays, *J. Geophys. Res.*, *83*(A2), 663–679, doi:10.1029/JA083iA02p00663.

Rosenbauer, H., et al. (1975), Heos-2 plasma observations in distant polar magnetosphere–plasma mantle, *J. Geophys. Res. Space Phys.*, *80*(19), 2723–2737, doi:10.1029/JA080i019p02723.

Rostoker, G. (2002), Identification of substorm expansive phase onsets, *J. Geophys. Res. Lett.*, *107*(A7), doi:10.1029/2001JA 003504.

Rostoker, G., et al. (1967), Relationship between changes in the interplanetary magnetic field and variations in the magnetic field at the Earth's surface, *J. Geophys. Res.*, *72*(23), 5853–5863, doi:10.1029/JZ072i023p05853.

Runov, A., et al. (2008), Multipoint in situ and ground-based observations during auroral intensifications, *J. Geophys. Res. Space Phys.*, *113*(A00C07), 16, doi:10.1029/2008ja013493.

Russell, C., and R. McPherron (1973a), Semiannual variation of geomagnetic activity, *J. Geophys. Res.*, *78*(1), 92–108, doi:10.1029/JA078i001p00092.

Russell, C. T., and R. McPherron (1973b), The magnetotail and substorms, *Space Sci. Rev.*, *15*(2–3), 205–266.

Scholer, M., et al. (1986), ISEE 3 observations during a plasma sheet encounter at 140 RE: Evidence for enhancement of reconnection at the distant neutral line, *J. Geophys. Res.*, *91*(A2), 1451–1458.

Sergeev, V. A., and W. Lennartsson (1988), Plasma sheet at X-20 RE during steady magnetospheric convection, *Planet. Space Sci.*, *36*(4), 353–370, doi:10.1016/0032-0633(88)90124-9.

Sergeev, V., et al. (1996a), Detection of localized, plasma-depleted flux tubes or bubbles in the midtail plasma sheet, *J. Geophys. Res.*, *101*(A5), 10817–10826, doi:10.1029/96JA00460.

Sergeev, V. A., et al. (1996b), Steady magnetospheric convection: A review of recent results, *Space Sci. Rev.*, *75*(3–4), 551–604.

Sergeev, V. A., et al. (1996c), Coupled-mode scenario for the magnetospheric dynamics, *J. Geophys. Res. Space Phys.*, *101*(A6), 13,047–13,065, doi:10.1029/95ja03192.

Sergeev, V., et al. (1999), Development of auroral streamers in association with localized impulsive injections to the inner magnetotail, *Geophys. Res. Lett.*, *26*(3), 417–420, doi:10.1029/1998GL900311.

Sergeev, V. A., et al. (2001), Substorm and convection bay compared: Auroral and magnetotail dynamics during convection bay, *J. Geophys. Res. Space Phys.*, *106*(A9), 18,843–18,855, doi:10.1029/2000ja900087.

Sergeev, V., et al. (2008), Study of near-Earth reconnection events with Cluster and Double Star, *J. Geophys. Res.*, *113*(A7), 1–19, doi:10.1029/2007JA012902.

Sharma, A. S., et al. (2008), Transient and localized processes in the magnetotail: A review, *Ann. Geophys.*, *26*(4), 955–1006.

Shi, Q. Q., et al. (2013), THEMIS observations of ULF wave excitation in the nightside plasma sheet during sudden impulse events, *J. Geophys. Res. Space Phys.*, *118*(1), 284–298, doi:10.1029/2012ja017984.

Shi, Y., et al. (2012), Multipoint observations of substorm pre-onset flows and time sequence in the ionosphere and magnetosphere, *J. Geophys. Res. Space Phys.*, *117*, 18, doi:10.1029/2011ja017185.

Shiokawa, K., et al. (1997), Braking of high-speed flows in the near-Earth tail, *Geophys. Res. Lett.*, *24*(10), 1179–1182, doi:10.1029/97gl01062.

Shiokawa, K., et al. (1998), High-speed ion flow, substorm current wedge, and multiple Pi 2 pulsations, *J. Geophys. Res.*, *103*(A3), 4491–4507, doi:10.1029/97JA01680.

Shue, J. H., et al. (2008), Two classes of earthward fast flows in the plasma sheet, *J. Geophys. Res. Pt. A Space Phys.*, *113*(A2), A02205, doi:10.1029/2007ja012456.

Sibeck, D. G., et al. (1984), Magnetotail flux ropes, *Geophys. Res. Lett.*, *11*(10), 1090–1093, doi:10.1029/GL011i010p01090.

Siscoe, G. L. (1966), A unified treatment of magnetospheric dynamics with applications to magnetic storms, *Planet. Space Sci.*, *14*, 947–967.

Slavin, J. A., et al. (1983), Average configuration of the distant (<220 Re) magnetotail: Initial ISEE-3 magnetic field results, *Geophys. Res. Lett.*, *10*(10), 973–976, doi:10.1029/GL010i010p00973.

Slavin, J. A., et al. (1984), Substorm associated traveling compression regions in the distant tail: ISEE-3 geotail observations, *Geophys. Res. Lett.*, *11*(7), 657–60, doi:10.1029/GL011i007p00657.

Slavin, J., et al. (1985), An ISEE 3 study of average and substorm conditions in the distant magnetotail, *J. Geophys. Res.*, *90*(A11), 10,875–10,895, doi:10.1029/JA090iA11p10875.

Slavin, J. A., et al. (1992), ISEE 3 plasmoid and TCR observations during an extended interval of substorm activity, *Geophys. Res. Lett.*, *19*(8), 825–828, doi:10.1029/92gl00394.

Slavin, J. A., et al. (1993), ISEE 3 observations of traveling compression regions in the Earth's magnetotail, *J. Geophys. Res.*, *98*(A9), 15,425–15,4466, doi:10.1029/93ja01467.

Smith, R., et al. (1986), Thermal catastrophe in the plasma sheet boundary layer, *Geophys. Res. Lett.*, *13*(13), 1380–1383, doi:10.1029/GL013i013p01380.

Speiser, T. W. (1965), Particle trajectories in model current sheets: 1. Analytical solutions, *J. Geophys. Res.*, *70*(17), 4219–4226, doi:10.1029/JZ070i017p04219.

Speiser, T. W. (1967), Particle trajectories in model current sheets .2. Applications to auroras using a geomagnetic tail model, *J. Geophys. Res.*, *72*(15), 3919–3932, doi:10.1029/JZ072i015p03919.

Takahashi, K., et al. (1987), Disruption of the magnetotail current sheet observed by AMPTE/CCE, *Geophys. Res. Lett.*, *14*(10), 1019–1022.

Tang, C. L., et al. (2010), Precursor activation and substorm expansion associated with observations of a dipolarization front by Time History of Events and Macroscale Interactions during Substorms (THEMIS), *J. Geophys. Res. Pt. A Space Phys.*, *115*(A7), A07215, doi:10.1029/2009ja014879.

Troshichev, O., et al. (2011), Saw-tooth substorms: Inconsistency of repetitive bay-like magnetic disturbances with behavior of aurora, *Adv. Space Res.*, *47*(4), 702–709, doi:10.1016/j.asr.2010.09.026.

Tsurutani, B. T., et al. (1984), Magnetic structure of the distant geotail from −60 to −220 Re: ISEE-3, *Geophys. Res. Lett.*, *11*(1), 1–4, doi:10.1029/GL011i001p00001.

Vasyliunas, V. M. (2014), The magnetotail: An unsolved fundamental problem of magnetospheric physics, in *Magnetotails in the Solar System*, edited by A. Keiling et al. Washington D. C.

Xing, X., et al. (2010), Substorm onset by new plasma intrusion: THEMIS spacecraft observations, *J. Geophys. Res. Space Phys.*, *115*, 13, doi:10.1029/2010ja015528.

Zesta, E., et al. (2000), The auroral signature of earthward flow bursts observed in the magnetotail, *Geophys. Res. Lett.*, *27*(20), 3241–3244, doi:10.1029/2000GL000027.

Zesta, E., et al. (2002), Two-dimensional structure of auroral poleward boundary intensifications, *J. Geophys. Res.*, *107*(A11), SIA 6-1–SIA 6-20, doi:10.1029/2001JA000260.

Zesta, E., et al. (2006), Auroral poleward boundary intensifications (PBIs): Their two-dimensional structure and associated dynamics in the plasma sheet, *J. Geophys. Res.*, *111*(A5), 1–14, doi:10.1029/2004JA010640.

5
Jupiter's Magnetotail

Norbert Krupp[1], Elena Kronberg[1], and Aikaterini Radioti[2]

5.1. INTRODUCTION

Burke and Franklin [1955] inferred from synchrotron radiation measurements that Jupiter, the largest planet in our solar system, must have an intrinsic magnetic field and, therefore a magnetosphere. Jupiter's magnetosphere is a plasma laboratory of superlatives with the strongest radiation belts, the harshest radiation environment, the largest intrinsic magnetic field, and the most powerful aurorae. Jupiter has the largest number of satellites (currently 63), including the four Galilean moons Io, Europa, Ganymede, and Callisto. It is the largest entity in our solar system. Jupiter's magnetopause extends from 63 to 92 Jovian radii R_J (1 R_J = 71, 492 km) [*Joy et al.*, 2002] on the dayside and up to ~200 R_J from dawn to dusk. Even more impressive is Jupiter's magnetotail, extending out to at least up to the orbit of Saturn [*Scarf*, 1979; *Sittler et al.*, 1987; *Lepping* et al., 1983; *Kurth* et al., 1982], more than 5 astronomical units (AU; 1 AU = 149.6 × 10^6 km) from the planet.

The spacecraft exploration of Jupiter and its environment has a history of approximately 40 years. There have been seven flyby missions (Pioneer 10 and 11, P10 and P11; Voyager 1 and 2, VG1 and VG2; Ulysses, ULY; Cassini, CAS; New Horizons, NH) and one orbiter (Galileo, GLL) between 1973 and 2007. All the spacecraft trajectories inside the Jovian magnetosphere projected into the equatorial plane are shown in Figure 5.1. P10, P11, VG1, VG2, as well as ULY flew by the planet and exited the magnetosphere either on the dayside or through the dawn and dusk flanks. GLL orbited the planet and measured through the entire magnetotail out to 150R_J. CAS only briefly skimmed the dusk magnetosphere at about 200R_J for a few hours. NH traversed Jupiter's magnetotail out to~2500R_J [e.g., *Hill et al.*, 2009] and VG2 explored the magnetotail out to ~7000R_J to 8000R_J on its transit to Saturn [*Scarf*, 1979]. In addition to spacecraft observations, there have been many efforts using magnetohydrodynamic (MHD) simulations over the past decades [*Ogino et al.*, 1998; *Walker et al.*, 2001; *Fukazawa et al.*, 2005, 2006, 2010; *Chané et al.*, 2013]. Details of the global configuration of the Jovian magnetosphere have been described in several books and review articles [*Dessler*, 1983; *Bagenal et al.*, 2004; *Gehrels*, 1976; *Bagenal*, 1992, 2013; *Khurana et al.*, 2004; *Krupp et al.*, 2004; *Russell et al.*, 2000; *Russell*, 2001, 2004; *Mauk et al.*, 2004].

Traditionally, the Jovian magnetosphere has been subdivided into inner (<10R_J), middle (10 R_J to 40R_J) and outer magnetosphere (>40R_J) [*Khurana et al.*, 2004]. *Ness et al.* [1979] nicely distinguished between the near tail (out to 140R_J) and the distant tail (out to 9000R_J). The dominant plasma sources of the Jovian magnetosphere are the moons Io (>1000 kg/s of sulfur and oxygen ions) and Europa (>100 kg/s protons and oxygen ions) in the inner magnetosphere and the power source is the fast planetary rotation rather than the solar wind, as pointed out by *Brice and Ioannidis* [1970]. Due to the centrifugal force of the fast rotating planet, the mass-loaded magnetic field lines frozen into the plasma are continuously stretched near the equator and deviate significantly from a dipole configuration [*Mauk and Krimigis*, 1987]. Oppositely directed field lines come close together and a stable configuration can only be reached through a

[1] *Max-Planck-Institut für Sonnensystemforschung, Göttingen, Germany*

[2] *Laboratoire de Physique Atmosphérique et Planétaire, Institut d'Astrophysique et de Géophysique, Université de Liège, Liège, Belgium*

Magnetotails in the Solar System, Geophysical Monograph 207. First Edition.
Edited by Andreas Keiling, Caitríona M. Jackman, and Peter A. Delamere.

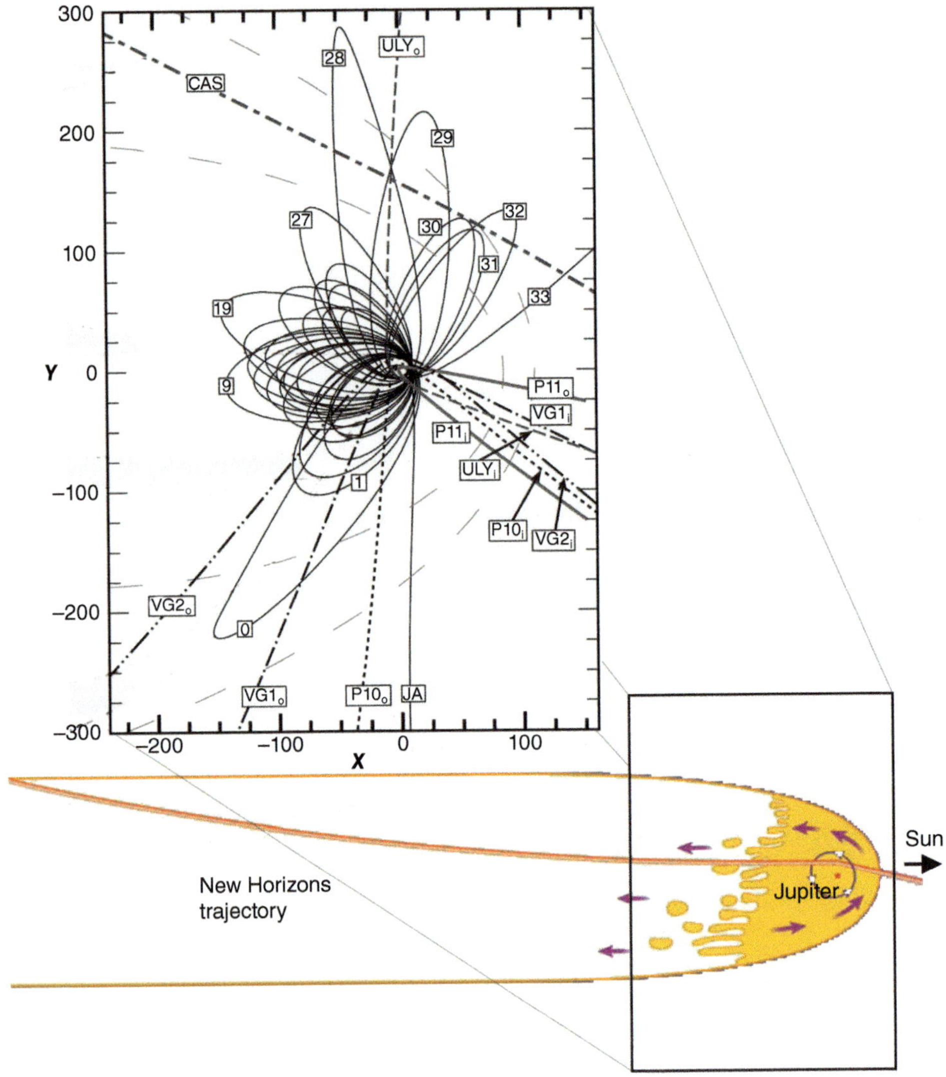

Figure 5.1 Top: Trajectories of flyby missions P10, P11, VG1, VG2, ULY, CAS and the orbiting mission GLL (0–34). Adapted from *Krupp et al.* [2004]. Bottom: Trajectory of the NH spacecraft in the Jovian magnetosphere. Adapted from *Krupp* [2007].

building-up current sheet and an associated plasma sheet. An equatorially confined magnetodisc is formed that is wobbling up and down with respect to the equator due to the 9.6° tilt between Jupiter's magnetic dipole axis and the planetary rotation axis. This wobbling pancake-like magnetodisc is filled with hot plasma consisting of sulfur and oxygen from the moon Io. A theory describing the overall plasma transport in Jupiter's magnetosphere is the "Vasyliunas cycle" [*Vasyliūnas*, 1983] illustrated in Figure 5.2: Mass-loaded flux tubes in the outer magnetosphere on the dayside are confined by the solar wind dynamic pressure, stretch out as they rotate into the dusk sector, and evolve into tailward propagating plasmoids. The closed, empty flux tubes then return to the dayside via dawn where they again become massloaded and start a new cycle [*Kivelson and Southwood*, 2005; *Badman and Cowley*, 2007]. The closed magnetic field lines are connected to the ionosphere of the planet, but the further away from the planet, the less the plasma is able to rigidly corotate with the planet. The plasma sub-corotates with the magnetic field lines in the disc, coupled to the ionosphere via field-aligned currents. The related field-aligned current system from that region is accelerating electrons into the ionosphere, creating the main auroral oval. The location of this "corotation breakdown" is dependent on local time (~40R_J in the predawn sector and~20R_J to 25R_J near dusk [*McNutt et al.*, 1981]). The dawn-dusk asymmetry is a consequence of rotational stress in the magnetodisc leading to heating and increased plasma anisotropy [*Kivelson and Southwood*, 2005].

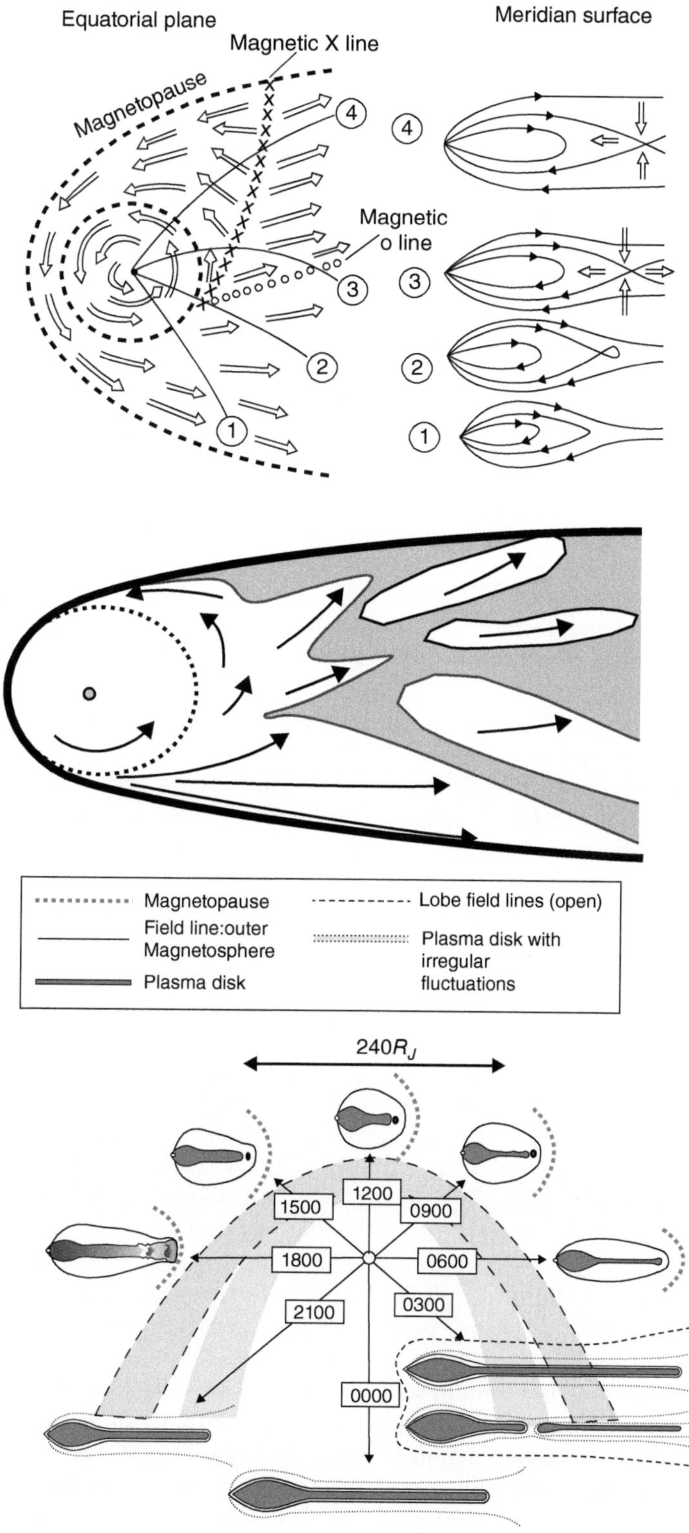

Figure 5.2 Upper: Schematics of the local time variation of the Jovian plasma. Upper right: Vasyliūnas cycle [*Vasyliūnas*, 1983] illustrated in the equatorial plane and meridional cuts. Upper left: An alternate interpretation of magnetotail flow. Bubbles of plasma that have broken off through an explosive ballooning instability are shown continuing to move across the tail at close to their velocity of release. The outer edge of the plasma sheet retracts after a release and then starts again to expand. Reconnected lobe flux tubes are shown flowing slowly sunward; lower: equatorial cut of the magnetosphere combined with meridional cuts taken at different local times indicated by arrows [from *Kivelson and Southwood*, 2005, Figures 5.4 and 5.7].

The magnetotail downstream of a planet is a direct consequence of momentum transfer between the magnetosphere and the solar wind. As pointed out by *Khurana et al.* [2004], the viscous interaction between the magnetosphere and the solar wind and the direct entry of the solar wind plasma into the magnetosphere can contribute to both the structure and dynamics of a magnetotail, depending on whether the system is open or closed. At Jupiter, we find a magnetodisc in the equatorial plane (half thickness of $2R_J$ in the dawn hemisphere and about $7.6R_J$ at dusk) with closed magnetic field lines and two lobe regions above and below the magnetodisc where open fields are found. Plasma in the magnetotail sub-corotates with the planet at least out to $\sim 150R_J$ [*Krupp et al.*, 2001] where GLL measurements were recorded. As pointed out by *Kivelson and Southwood* [2005], the nightside magnetodisc can become unstable during the thinning of the plasma sheet as material flows down the tail (see Figure 5.2).

The understanding of magnetodiscs in general and large-scale magnetotail configurations has an astrophysical aspect. Most probably all the "hot Jupiters" among the exoplanets outside our solar system will have intrinsic magnetic fields and therefore will have magnetospheres. If they rotate fast and if they have internal plasma sources, they may have a similar configuration as Jupiter and Saturn.

5.2. PARTICLE PARAMETERS OF JUPITER'S MAGNETOTAIL

The thermal properties in the Jovian magnetotail have been derived by *Frank et al.* [2002]. Three primary ion populations were detected: cool hydrogen ions with temperatures of 10 eV, hot hydrogen ions with temperatures of approximately 10 keV, and a third population of heavy ions such as O^{2+}, O^{+}, S^{2+}, and S^{3+} with temperatures in the range of 500 eV. The bulk flows of the thermal plasmas in the magnetotail exhibit substantial components in the corotational and radially outward directions, but the bulk speeds of 50, 200 km/s are significantly less than those for rigid corotation. Plasma moments derived from one GLL orbit are shown in Figure 5.3. *Krimigis et al.* [1981] pointed out that the particle pressure in the middle and outer Jovian magnetosphere exceeds the magnetic pressure. The major component of the particle pressure is coming from energetic particles with energies above 20–30 keV. The plasma beta reaches the value of 1 at a distance of $15R_J$. Beyond that distance beta values of 10–100 and greater are measured [*Mauk et al.*, 2004]. This means that hot plasma dominates the entire Jovian magnetotail. As shown in Figure 5.4, the particles sub-corotate with occasional radial inward and outward flow bursts. the bursts occur quasi-periodically in the predawn sector with a repetition rate of a few days.

5.3. ENERGETIC EVENTS AND MAGNETIC RECONNECTION

Jupiter's magnetotail is highly dynamic and changes in the overall particles and field parameters occur on time scales of minutes up to several days. As pointed out in the previous section, the plasma essentially sub-corotates with the planet, lagging behind rigid corotation. However, the corotating motion is disrupted by so-called energetic events or reconnection signatures. The first evidence of such a reconnection event was pointed out by *Krimigis et al.* [1980] and later by *Nishida* [1983] where an increased number of particles moving in the antisunward direction were identified in the VG2 data. During that period VG2 was on its outbound trajectory in the predawn sector of the Jovian magnetosphere at 03:00 local time (LT) at a distance of $\sim 150R_J$. During the event, the magnetic field changed its direction from southward to northward, indicative of reconnection.

Seventeen years later, GLL went into orbit around Jupiter and spent weeks in the same region of the magnetosphere. The particle and field instruments onboard GLL clearly rediscovered these energetic events [*Krupp et al.*, 1998; *Woch et al.*, 1998, 1999; *Russell et al.*, 1998] and found them to occur quasi-periodically every 2–3 days in the predawn sector of the magnetosphere. Correlated changes in plasma wave and radio emissions were also observed [*Louarn et al.*, 1998, 2000]. Enhancements in the fluxes of various auroral radio emissions have been reported in correlation with new radio sources in the Io torus and large fluctuations in the magnetodisc density. Figure 5.5 shows examples of particles, magnetometer, and radio emission signatures of some of those energetic events. *Woch et al.* [1998, 1999] and later *Kronberg et al.* [2005] interpreted the quasi-periodic bursts as a sudden release of energy into the Jovian magnetotail in a two-step scenario: (i) a signature of massloading over about 2 days and (ii) a mass release most probably in the form of plasmoids down the magnetotail followed by about one day of recovery phase, as sketched in Figure 5.6. This two-step process of mass loading and release occurs periodically every 2–4 days [*Kronberg et al.*, 2009]. If a spacecraft happens to be in the region downtail from the reconnection site, the onboard instruments will measure the radially outward moving plasmoid. However, if the spacecraft is somewhere else in the magnetotail, the signature of the reconnection event but no plasmoid will be measured in the magnetosphere, i.e., in periodic changes of the energy spectral slope or in the form of accelerated field-aligned particles creating additional enhancements in auroral emissions [*Louarn et al.*, 2001; *Woch et al.*, 2002, 2004;

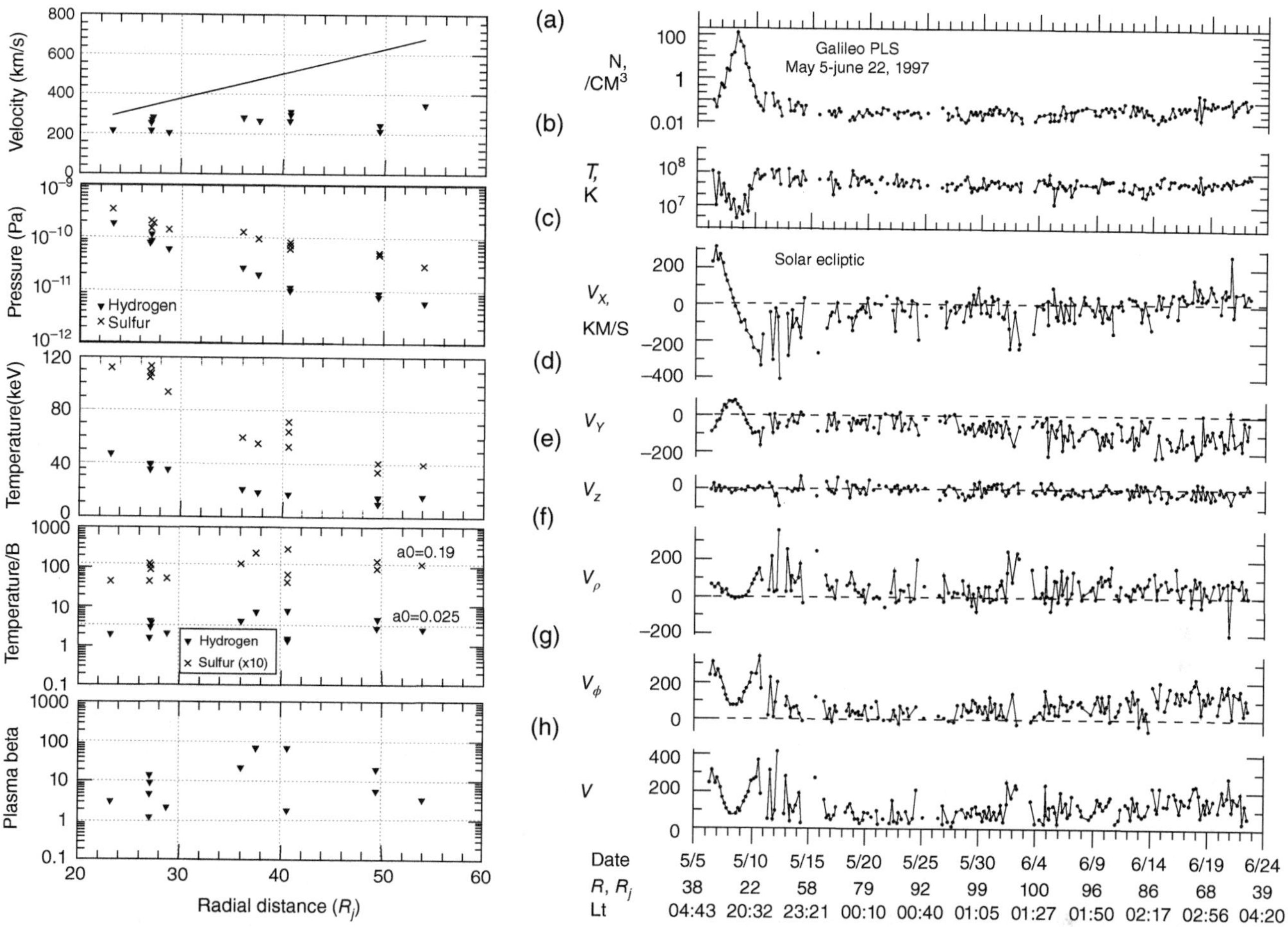

Figure 5.3 Left: Ion parameters derived from energetic ion measurements as a function of radial distance from the planet in the predawn magnetosphere. From top to bottom: velocity, pressure, temperature, ratio of temperature relative to the magnetic field magnitude B, plasma beta [from *Kane et al.*, 1999]. Right: Thermal plasma parameters measured during the period 5 May to 22 June, 1997: (a) ion densities, (b) ion temperatures, (c, e) ion bulk flows in solar-ecliptic coordinates, (f) ion bulk flow component in the radial direction, (g) ion bulk flow component in the corotational direction, and (h) bulk flow speed [From *Frank et al.* 2002].

Kronberg et al., 2005, 2008b, 2009]. *Louarn et al.* [2000] concluded that the periodicity of radio emissions vary from orbit to orbit but that they are more frequent the denser the magnetodisc is as a consequence of plasma mass loading and progressive evacuation. The authors showed that the energetic events are related to an instability in the outer parts of the Io torus or close by in which plasma is injected into the more distant magnetodisc. The relative composition of energetic ions has been studied by *Radioti et al.* [2007], who showed changes in the abundance ratios of sulfur/oxygen ions at 39 keV/nuc during energetic events, as shown in Figure 5.7. It was shown that, during the Jovian substorm like events, the heavier ions are more efficiently energized than the lighter ions. The authors proposed ion acceleration by small-scale magnetic field variations along the reconfiguration events. *Artemyev et al.* [2013] describes another possibility to accelerate ions in the Jovian magnetotail due to the interaction of ions with dipolarization fronts of nonlinear waves in transient reconnection regions embedded into the magnetotail horizontal current sheet.

Kronberg et al. [2012] investigated energetic electron and ion field-aligned beams in GLL data associated with some of the energetic events analyzed by *Kronberg et al.* [2005, 2008a, 2008b] and *Louarn et al.* [2001]. They concluded that most of the field-aligned acceleration can be associated to reconnection. *Russell* [2001], *Ge et al.* [2007, 2010], and later (extended by) *Vogt et al.* [2010] analyzed magnetic field data taken onboard GLL and classified the tail reconnection signatures into four types (dipolarizations, strong northward excursions, tailward moving plasmoids, and planetward moving plasmoids), out of which the most probable location of a tail reconnection line was inferred. Figure 5.8 summarizes the studies on

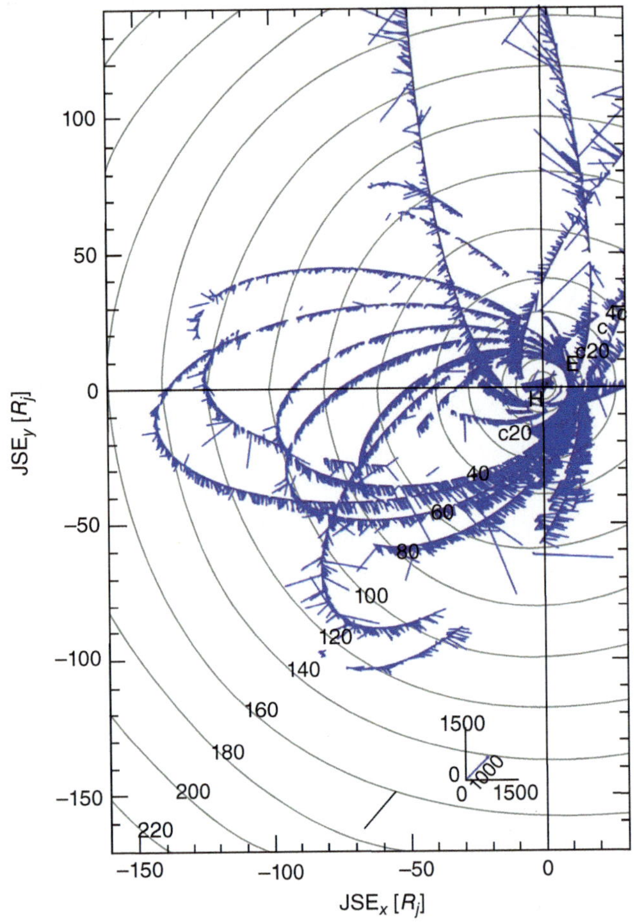

Figure 5.4 Global flow patterns derived from energetic ion measurements of the energetic particles detector (EPD) onboard GLL. The values are calculated via directional particle anisotropies described in detail by *Krupp et al.* [2001]. The plot is an updated version of Figure 5.1 in the work of *Krupp et al.* [2001] and includes data from all GLL orbits. The data are projected into Jupiter's equatorial plane x-y in the Jupiter Solar Ecliptic coordinate system (JSE).

the location of reconnection X-lines in Jupiter's magnetotail based on particle and field observations. It shows a very nice agreement of earlier studies by *Woch et al.* [2002] where the spatial distribution of mainly inward flow bursts vs. mainly outward directed flow bursts in energetic ion data were compared to magnetometer analyses by *Vogt et al.* [2010]. The location of the X-line is approximately between $60R_J$ and $80R_J$ in the predawn sector and at $120R_J$ to $150R_J$ near midnight. This is in relatively good agreement of early theoretical work by *Zimbardo* [1991, 1993] on the basis of collisionless reconnection and the growth of a tearing instability in Jupiter's magnetotail. It was established that the latter theoretical reconnection conditions become satisfied at the location of the X line [*Kronberg et al.*, 2007].

After the GLL era, it was known that plasmoids were released quasi-periodically down the Jovian magnetotail, but it was unknown how long these plasmoids remain in the magnetotail. The trajectory of the NH spacecraft (see Figure 5.1) through the Jovian system provided data beyond the maximum distance covered by GLL. Particle measurements onboard NH revealed that the plasma distributions evolved from a hot plasma disk configuration out to ~$100R_J$ into a highly variable plasma configuration with sharp intensity changes, smooth variations, and periodic plasmoid releases [*McComas et al.*, 2007] shown in Figure 5.9. This observation seems to be consistent with the plasmoids identified closer to the planet in the GLL data described above. *Hill et al.* [2009] showed that recurrent bursts of energetic protons and iogenic material were detected every 4–5 days, even out to ~$2500R_J$. Using the dispersive characteristic of these particle events, the authors concluded that the source of those particles was in the magnetotail at ~$150{\pm}40R_J$. This result was consistent with the suggested location of the reconnection -line from the GLL studies by *Woch et al.* [2002] and *Vogt et al.* [2010]. In addition, *Hill et al.* [2009] suggested that narrow spatial structures, or filaments, with a diameter of ~$5R_J$ exist in the $400R_J$ wide and more than $9000R_J$ long magnetotail. The ion composition ratios measured in the distant tail also match the GLL results. *McNutt et al.* [2007] and *Haggerty et al.* [2009] analyzed the ion composition ratio S/O to be 0.8, which nicely confirms the GLL values reported during the energetic events reported by *Radioti et al.* [2007] shown in Figure 5.7.

Magnetotail dynamics and especially the energetic events can also be studied remotely by analyzing the Jovian auroral emissions in detail. Field lines connecting the auroral zone and the magnetotail are found poleward of the main oval emission. *Prangé et al.* [2001] defined an auroral activity index and found auroral emission variability of 5–10 days (besides other time periods of a few hours and 6 weeks). This is slightly larger than the 2–3 day periodicity seen in the energetic particle data onboard GLL and matched the observations of low-energy plasma measurements from the PLS instrument onboard GLL [*Vasyliūnas et al.*, 1997].

Auroral observations have shown the occasional appearance of transient polar arcs and spots [*Grodent et al.*, 2003, 2004] located poleward of the main emission, which given their observed location and properties were proposed to be triggered by reconnection processes in the Jovian magnetotail. A more detailed analysis based on daily UV auroral observations by *Radioti et al.* [2008] revealed the presence of "polar dawn spots" consisting of transient auroral emissions in the polar dawn region (Figure 5.10), with a characteristic recurrence period of 2–3 days. Because of their periodic recurrence and observed location, the polar dawn spots were interpreted as auroral signatures of

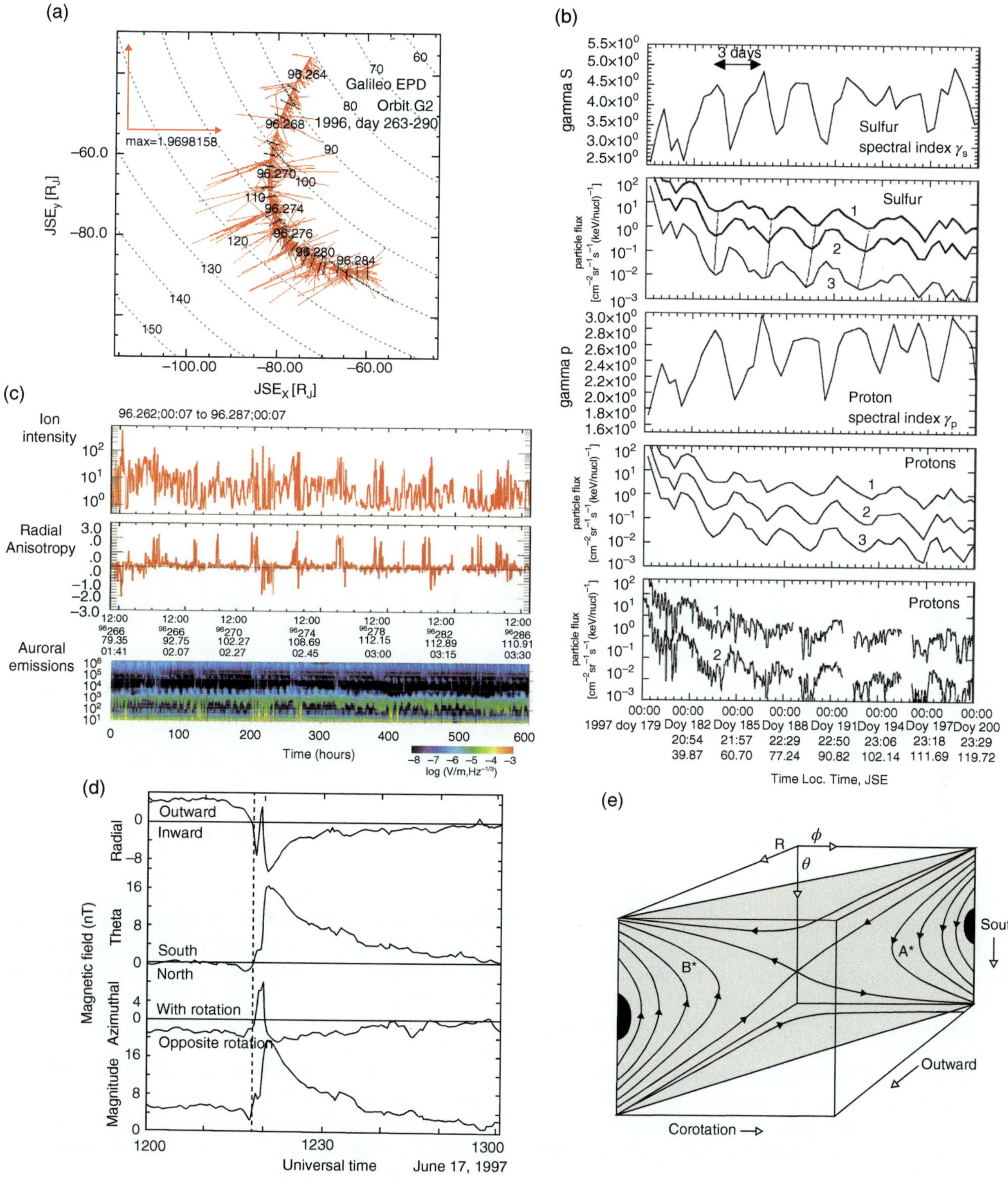

Figure 5.5 Energetic events as measured by GLL in the predawn Jovian magnetotail. (a) Energetic ion anisotropies in the equatorial plane of Jupiter. (b) Response of particle parameters as a consequence of the energetic events. (c) particle intensity, radial particle anisotropy, and plasma wave emissions during the quasi-periodic energetic events. (d) Example of a reconnection signature in magnetometer data at $60R_J$. (e) Interpretation of the magnetic configuration resulting from the reconnection event. adapted from *Krupp et al.* [1998]; *Woch et al.* [1998]; *Louarn et al.* [1998]; and *Russell* [2004].

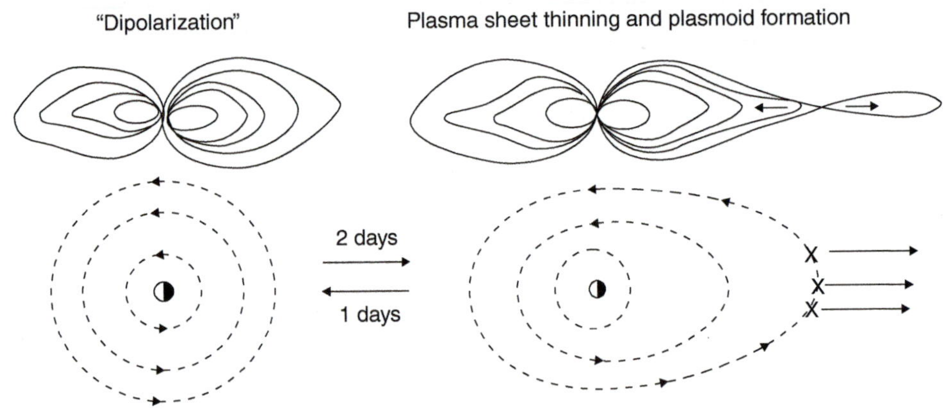

Figure 5.6 Two-step scenario of energetic events in the Jovian magnetotail. Left: Quiet state with more dipolarlike magnetic fieldlines. Right: Disturbed state during mass loading, thin plasma sheet, and plasmoid release. [From *Woch et al.*, 1998, Figure 5.4].

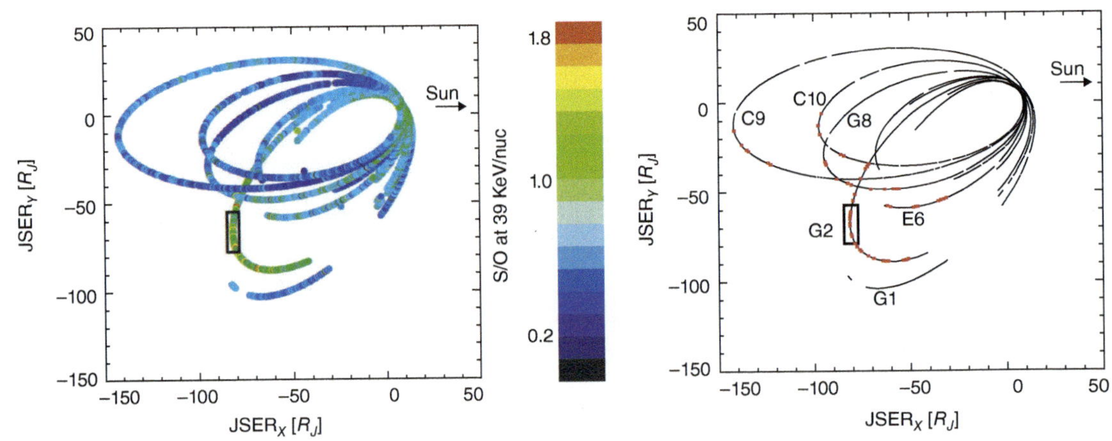

Figure 5.7 Left: Color-coded relative ion abundance ratio of sulfur/oxygen ions at 39 keV/nuc along the first 10 GLL orbits. Right: Reconfiguration events marked along the first 10 GLL orbits. [From *Radioti et al.*, 2007, Figure 5.1].

internally driven magnetic reconnection in the Jovian magnetotail (Vasyliunas cycle). Particularly, they were associated with the inward moving flow bursts released during magnetotail reconnection in Jupiter's tail [*Radioti et al.*, 2010, 2011]. The association of the polar dawn auroral spots with tail reconnection was further confirmed by *Ge et al.* [2010]. Apart from auroral spots, long-lived quasi-sun-aligned polar auroral filaments are reported [*Nichols et al.*, 2009] to span the highly variable region poleward of the main emission. They presumably map to the tail region, implying a relation to the tail dynamics.

5.3.1. Solar Wind Influence

As explained in the previous sections, Jupiter is rotationally driven and the solar wind does not play the major role as at Earth. However, some of the data have been interpreted by Earth-like solar wind interaction scenarios. Based on the conclusion that the solar wind has a major influence on the dynamics in Earth's magnetosphere, early studies by *Brice and Ioannidis* [1970] assumed that similar effects might also occur at Jupiter. Since then, the solar wind interaction at Jupiter has been hotly debated [e.g., *McComas and Bagenal*, 2007, 2008; *Cowley et al.*, 2008]. The review by *Delamere and Bagenal* [2010] nicely summarizes the current knowledge. If an Earth-like "Dungey-cycle" is assumed [*Dungey*, 1961], magnetic flux opened through dayside reconnection is transported across the polar region and is closed in the tail again with a return flow in the dawn sector of the magnetosphere [*Cowley et al.*, 2003]. *Delamere and Bagenal* [2010], on the other hand, proposed that magnetic flux is opened and closed intermittently in small-scale structures in turbulent interaction regions on the flanks of the magnetosphere. In that model the interaction with the solar wind is more a viscous interaction between the high-beta heavy-ion

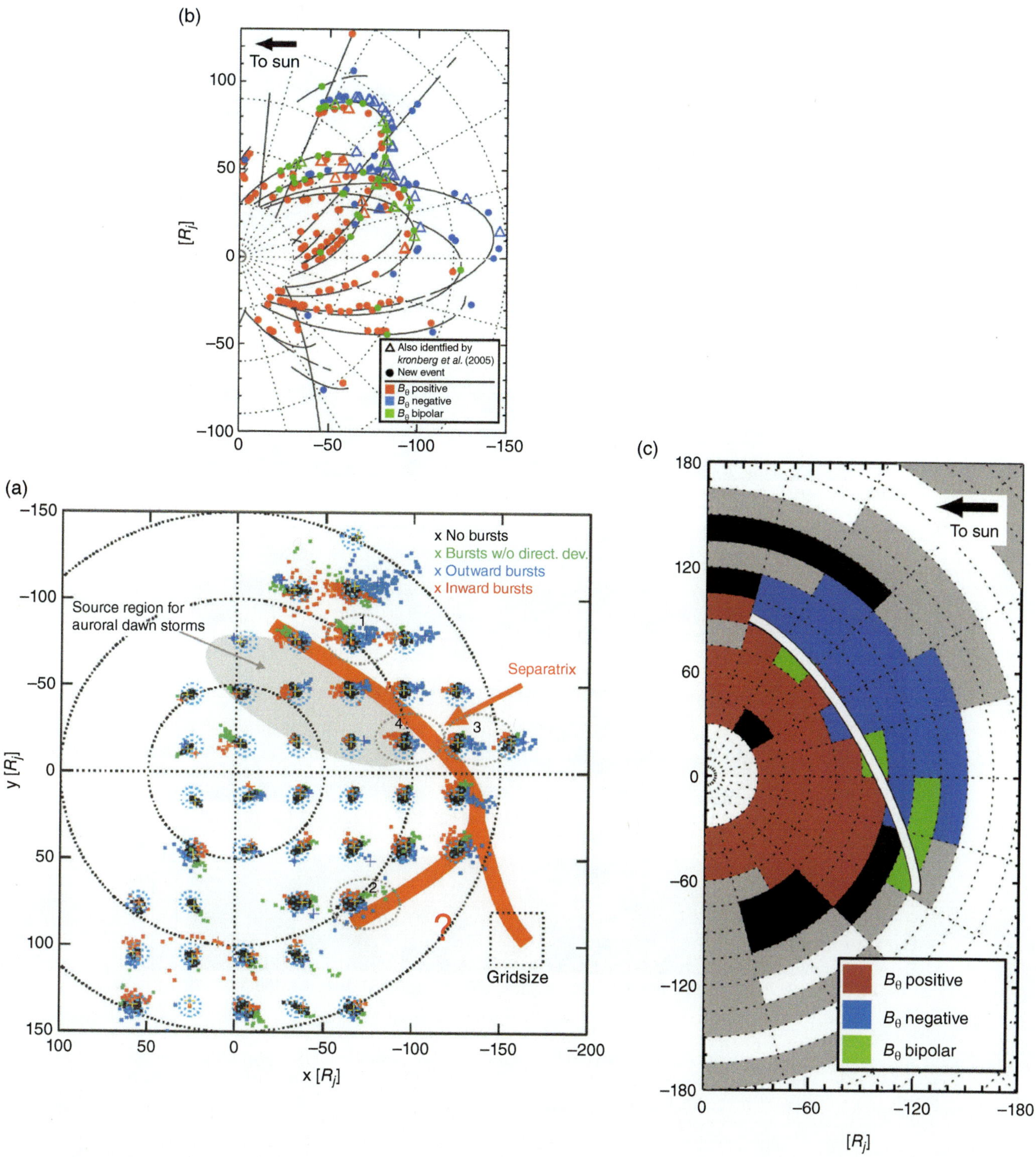

Figure 5.8 (a) Distribution of particle flow bursts, separatrix between mainly inward and mainly outward bursts as thick red line [From *Woch et al.*, 2002, Figure 5.3], (b) Location of energetic event signatures in the north-south component of the magnetic field (colored boxes) compared to events in energetic particle data (triangles) [From *Vogt et al.*, 2010, Figure 5.8. (c) Location of a reconnection separatrix (white line) in the Jovian magnetotail as inferred from 249 events in Galileo magnetometer data [From *Vogt et al.*, 2010, Figure 5.11].

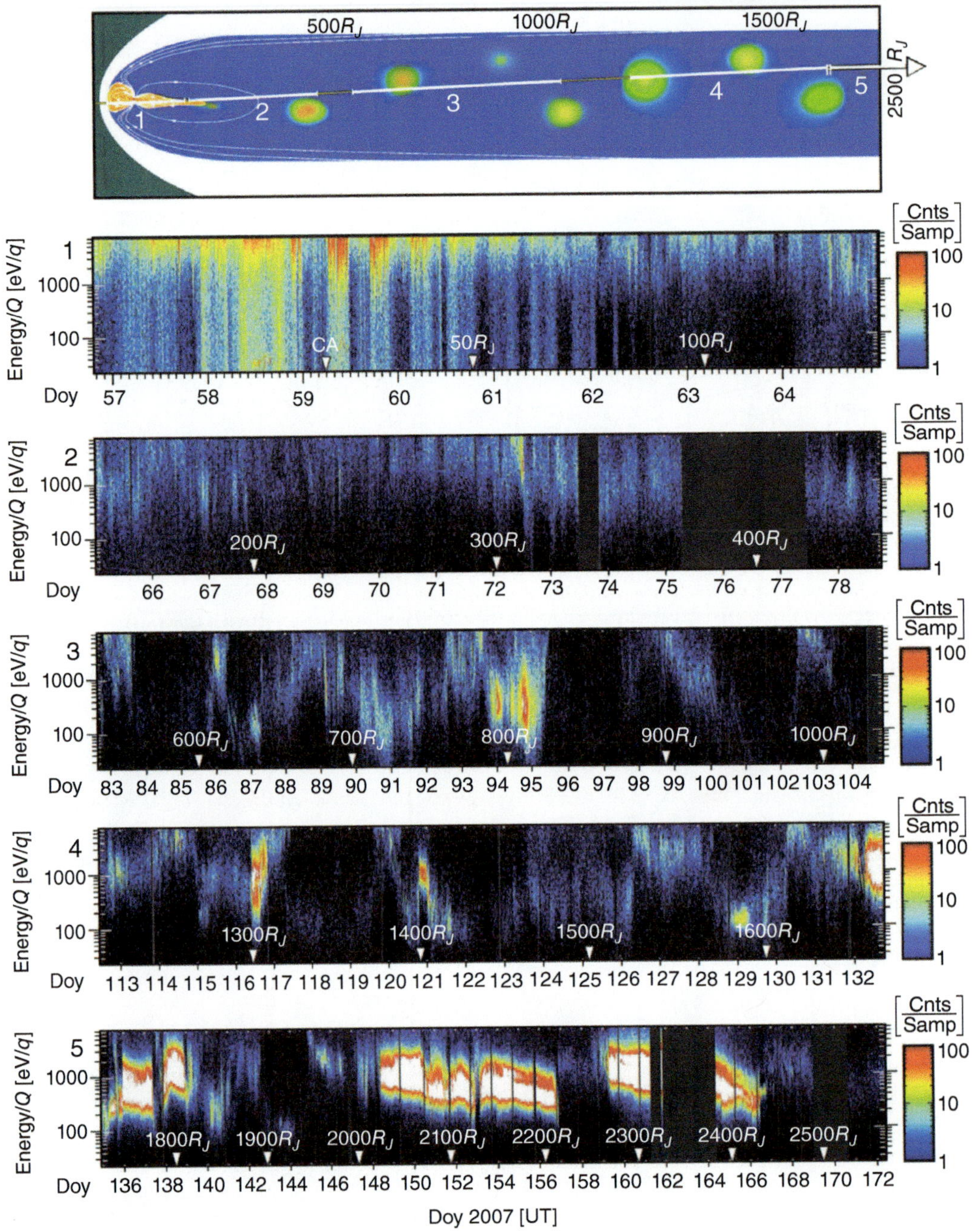

Figure 5.9 Top panel: Trajectory of NH through the Jovian magnetotail. Other panels: Color-coded energy/charge spectrograms of ions measured by the SWAP instrument onboard the NH spacecraft as a function of distance from Jupiter. [From *McComas et al.*, 2007, Figure 5.1].

plasma of the magnetosphere and the high Alfvén Mach number in the solar wind. Other evidence of solar wind interaction at Jupiter is based on radio wave data analyses. *Ye et al.* [2012] analyzed data from the radio and plasma wave science investigation (RPWS) onboard CAS during the Jupiter flyby in 2000/2001 and between 2000 and 2004 before the spacecraft inserted into the Kronian magnetosphere. They found that the wave emissions came from the entire Jovian magnetotail from a line source rather than from a point source and were modulated by the SIII longitude period of Jupiter (SIII is a longitude system defining the Jovian rotation period based on the orientation of the dipole axis on Jan 1, 1965). Using the lower cutoff frequency of the emission, which is related to the plasma frequency in the Jovian magnetosheath, solar wind parameters can be derived. Simulations also support the notion that the solar wind can influence the global configuration and the dynamics of the Jovian

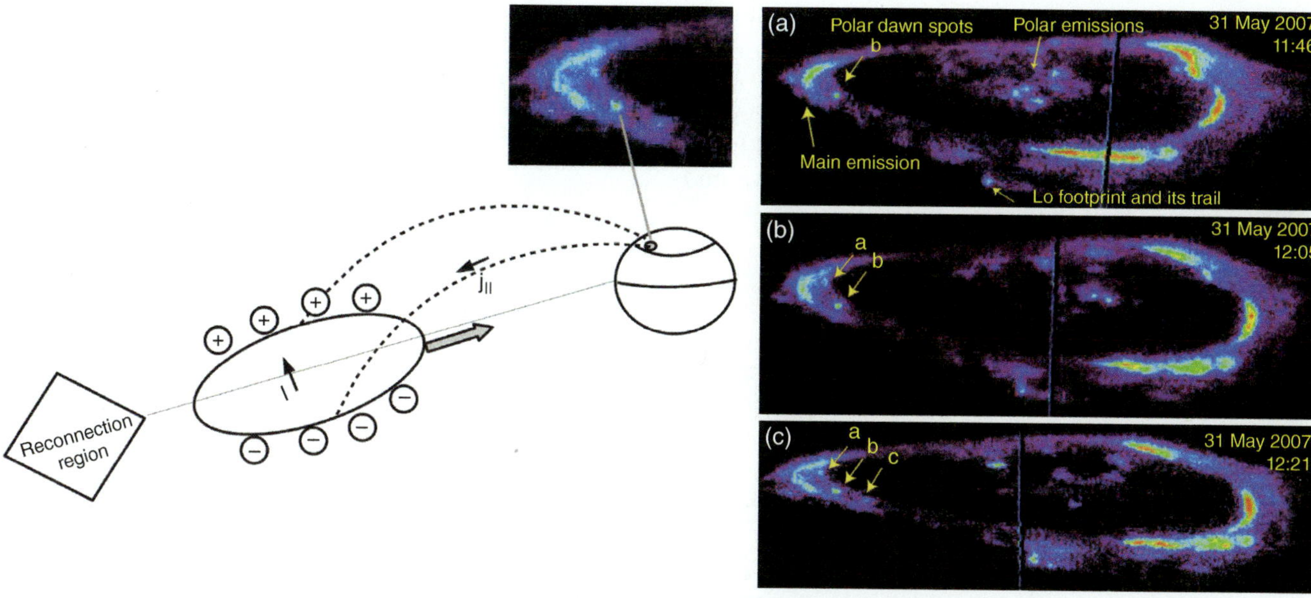

Figure 5.10 Left: Relationship between auroral spots in the ionosphere and the flowing bubble released during magnetic reconnection. The auroral spots are related to the upward field-aligned current (directed from the ionosphere to the equator) at the dawn side of the flow burst. Right: Sequence of three FUV auroral emissions at the north pole of Jupiter measured by HST/ACS on 31 May 2007 at (a) 11:46, (b) 12:05, and (c) 12:21 UT. The arrows indicate the main oval (main emission), the Io footprint and its trail, the polar emissions, and the polar dawn spots. [From *Radioti et al.*, 2010, Figures 5.1 and 5.4].

magnetosphere. *Fukazawa et al.* [2010] performed an MHD simulation of the dynamics in the distant Jovian magnetotail and found that periodic plasmoid ejections from a reconnection line in Jupiter's plasma sheet down the magnetotail were favored during low solar wind dynamic pressure and northward interplanetary magnetic field (IMF).

5.4. SUMMARY

In this chapter, the current knowledge about particle and field observations from Jupiter's outer magnetodisc and magnetotail is summarized. It is believed that Jupiter's magnetotail extends at least out to the orbit of Saturn. The data analysis efforts from GLL and NH suggest that protons and heavy ions from Io are transported and lost down the tail in the form of plasmoids and, perhaps, small-scale drizzle along the magnetotail flanks. The plasma in the Jovian tail is hot and the particle pressure dominated by energetic heavy ions is 10–100 times higher that the magnetic pressure.

There is evidence of a reconnection site located between $60R_J$ and $150R_J$, the distance being dependent on local time. The observed "energetic events" measured in energetic particles, plasma and radio waves, magnetic field north-south component, and auroral emissions occur throughout the magnetotail. They are most likely related to reconnection in the tail and appear quasi-periodically every few days, especially in the predawn sector of the magnetosphere.

Observations in the distant tail, several thousand R_J away from the planet, confirm the release of plasmoids (every few days) originating near $150R_J$.

5.5. FUTURE EXPLORATION

The investigation of the Jovian magnetosphere and especially its magnetotail will continue in the future. The NASA mission JUNO, launched in 2011, will arrive at Jupiter in 2016 and will return data from the high-latitude magnetosphere inside of $35R_J$. At times, the spacecraft will be connected along magnetic field lines that map back deep into the magnetotail. Additionally, JUNO will provide us with unprecedented views of the nightside aurora where the ionospheric signatures of tail reconnection are expected to be located. These measurements will help us understand the acceleration mechanisms in the Jovian system and the importance of reconnection.

In addition, the European Space Agency (ESA) is currently preparing the JUICE mission (Jupiter Icy Moon Explorer), which will arrive in the Jovian system in 2030 [*Grasset et al.*, 2013]. The main goal of the mission is to study the icy moons Ganymede, Europa, and Callisto as well as the Jovian magnetosphere and the atmosphere of the planet. The orbits of the JUICE spacecraft will partially cover the Jovian magnetotail during the initial insertion and later orbits out to ~$40R_J$ to $50R_J$, which will fill in some gaps left by GLL. The JUICE magnetospheric instrumentation will be able to globally monitor the processes ongoing in the Jovian magnetotail by using energetic neutral atom (ENA) imaging. This technique has been used very successfully in the Kronian and Earth magnetosphere with the INCA instrument on CAS and onboard of the IMAGE mission, respectively.

ACKNOWLEDGMENTS

This work is supported by the Max Planck Society and by the German Space Agency DLR under contract 50 OH 1101. A. Radioti is funded by the Belgian Fund for Scientific Research (FNRS).

REFERENCES

Artemyev A. V., S. Kasahara, A. Y., Ukhorskiy, and M. Fujimoto (2013), Acceleration of ions in the Jupiter magnetotail: Particle resonant interaction with dipolarization fronts, *Planet. Space Sci.*, *82*, 134–148.

Badman, S. V., and S. W. H. Cowley (2007), Significance of Dungey-cycle flows in Jupiter's and Saturn's magnetospheres, and their identification on closed equatorial field lines, *Ann. Geophys.*, *25*, 941–951.

Bagenal, F. (1992), Giant planet magnetospheres, *Annu. Rev. Planet. Sci.*, *20*, 289–328.

Bagenal, F. (2013), Planetary Magnetospheres, in *Planets, Stars and Stellar Systems: Solar and Stellar Planetary Systems*, edited by T. D. Oswalt , L. M. French, and P. Kalas, vol. 3, p. 251.

Bagenal, F., T. Dowling, and W. McKinnon, (Eds.) (2004), *Jupiter, the Planet, Satellites and Magnetosphere*. Cambridge Univ. Press, New York.

Brice, N. M., and G. A. Ioannidis (1970), The magnetospheres of Jupiter and Earth, *Icarus 13*, 173, doi:10.1016/0019-1035(70)90048-5.

Burke, B. F., and K. L. Franklin (1955), Observations of a variable radio source associated with the planet Jupiter, *J. Geophys. Res. 60*, 213–217.

Chané, E., J. Saur, and S. Poedts (2013), Modeling Jupiter's magnetosphere: Influence of the internal sources, *J. Geophys. Res. (Space Phys.)*, *118*, 2157–2172.

Cowley, S. W. H., E. J. Bunce, T. S. Stallard, and S. Miller (2003), Jupiter's polar ionospheric flows: Theoretical interpretation, *Geophys. Res. Lett.*, *30*, 1220.

Cowley, S. W. H., A. J. Deason, and E. J. Bunce (2008), Axisymmetric models of auroral current systems in Jupiter's magnetosphere with predictions for the Juno mission, *Annal. Geophys.*, *26*, 4051–4074.

Delamere, P. A., and F. Bagenal (2010), Solar wind interaction with Jupiter's magnetosphere, *J. Geophys. Res. (Space Phys.)*, *115*, 10,201.

Dessler, A. J. (Ed.) (1983), *Physics of the Jovian Magnetosphere*, Cambridge Univ. Press, New York.

Dungey, J. W. (1961), Interplanetary magnetic field and the auroral zones, *Phys. Rev. Lett.*, *6*, 47–48.

Frank, L. A., W. R. Paterson, and K. K. Khurana (2002), Observations of thermal plasmas in Jupiter's magnetotail, *J. Geophys. Res. (Space Phys.)*, *107*, 1003, doi:10.1029/2001JA000077.

Fukazawa, K., T. Ogino, and R. J. Walker (2005), Dynamics of the Jovian magnetosphere for northward interplanetary magnetic field (IMF), *Geophys. Res. Lett.*, *32*, L03202, doi:10.1029/2004GL021392.

Fukazawa, K., T. Ogino, and R. J. Walker (2006), Configuration and dynamics of the Jovian magnetosphere, *J. Geophys. Res., (Space Phys.)*, *111*, 10,207.

Fukazawa, K., T. Ogino, and R. J. Walker (2010), A simulation study of dynamics in the distant Jovian magnetotail, *J. Geophys. Res. (Space Phys.)*, *115*, 9219.

Ge, Y. S., L. K. Jian, and C. T. Russell (2007), Growth phase of Jovian substorms, *Geophys. Res. Lett.*, *34*, 23,106.

Ge, Y. S., C. T. Russell, and K. K. Khurana (2010), Reconnection sites in Jupiter's magnetotail and relation to Jovian auroras, *Planet. Space Sci.*, *58*, 1455–1469.

Gehrels, T. (Ed.) (1976), *Jupiter*, Univ. of Arizona Press, Tucson, Ariz.

Grasset, O., et al. (2013), JUpiter ICy moons Explorer (JUICE): An ESA mission to orbit Ganymede and to characterise the Jupiter system, *Planet. Space Sci.*, *78*, 1–21.

Grodent, D., J. T. Clarke, J. H. Waite, S. W. H. Cowley, J.-C. Gerard, and J. Kim (2003), Jupiter's polar auroral emissions, *J. Geophys. Res*, *108*, 1366.

Grodent, D., J.-C. Gerard, J. T. Clarke, G. R. Gladstone, and J. H. Waite (2004), A possible auroral signature of a magnetotail reconnection process on Jupiter, *J. Geophys. Res.*, *109*, 5201.

Haggerty, D. K., M. E. Hill, R. L. McNutt, and C. Paranicas, (2009), Composition of energetic particles in the Jovian magnetotail, *J. Geophys. Res. (Space Phys.)*, *114*, 2208.

Hill, M. E., D. K. Haggerty, R. L. McNutt, and C. P. Paranicas (2009), Energetic particle evidence for magnetic filaments in Jupiter's magnetotail, *J. Geophys. Res. (Space Phys.)*, *114*, 11,201.

Joy, S. P., M. G. Kivelson, R. J. Walker, K. Khurana, and C. T. Russell (2002), Probabilistic models of the Jovian magnetopause and bowshock locations, *J. Geophys. Res.*, *107*, SMP 17–1, doi:10.1029/2001JA009146.

Kane, M., D. J. Williams, B. H. Mauk, R. W. McEntire, and E. C. Roelof (1999), Galileo energetic particles detector measurements of hot ions in the neutral sheet region of Jupiter's magnetodisk, *Geophys. Res. Lett.*, *26*, 5–8.

Khurana, K. K., M. G. Kivelson, V. M. Vasyliunas, N. Krupp, J. Woch, A. Lagg, B. H. Mauk, and W. S. Kurth (2004), The configuration of Jupiter's magnetosphere, in *Jupiter: The Planet, Satellites and Magnetosphere*, edited by F. Bagenal, T. E. Dowling, and W. B. McKinnon, pp. 593–616.

Kivelson, M. G., and D. J. Southwood (2005), Dynamical consequences of two modes of centrifugal instability in Jupiter's outer magnetosphere, *J. Geophys. Res.*, *110*, 12,209.

Krimigis, S. M., T. P. Armstrong, W. I. Axford, C. O. Bostrom, C. Y. Fan, G. Gloeckler, L. J. Lanzerotti, D. C. Hamilton, and R. D. Zwickl (1980), Energetic ≈ 100-keV tailward-directed ion beam outside the Jovian plasma boundary, *Geophys. Res. Lett.*, *7*, 13–16.

Krimigis, S. M., J. F. Carbary, E. P. Keath, C. O. Bostrom, W. I. Axford, G. Gloeckler, L. J. Lanzerotti, and T. P. Armstrong (1981), Characteristics of hot plasma in the Jovian magnetosphere: Results from the Voyager spacecraft, *J. Geophys. Res.*, *86*, 8227–8257, doi:10.1029/JA086iA10p08227.

Kronberg, E. A., J., Woch, N. Krupp, A. Lagg, K. K. Khurana, and K.-H. Glassmeier (2005), Mass release at Jupiter: Substorm-like processes in the Jovian magnetotail, *J. Geophys. Res.*, *110*, 3211.

Kronberg, E. A., K.-H. Glassmeier, J. Woch, N. Krupp, A. Lagg, and M. K. Dougherty (2007), A possible intrinsic mechanism for the quasi-periodic dynamics of the Jovian magnetosphere, *J. Geophys. Res. 112*, 5203.

Kronberg, E. A., J. Woch, N. Krupp, and A. Lagg (2008a), Mass release process in the Jovian magnetosphere: Statistics on particle burst parameters, *J. Geophys. Res. (Space Phys.)*, *113*, 10,202.

Kronberg, E. A., J. Woch, N. Krupp, A. Lagg, P. W. Daly, and A. Korth (2008b), Comparison of periodic substorms at Jupiter and Earth, *J. Geophys. Res. (Space Phys.)*, *113*, 4212.

Kronberg, E. A., J. Woch, N. Krupp, and A. Lagg (2009), A summary of observational records on periodicities above the rotational period in the Jovian magnetosphere, *Ann. Geophys. 27*, 2565–2573.

Kronberg, E. A., S. Kasahara, N. Krupp, and J. Woch (2012), Field-aligned beams and reconnection in the Jovian magnetotail, *Icarus*, *217*, 55–65.

Krupp, N. (2007), New surprises in the largest magnetosphere of our solar system. *Science*, *318*, 216.

Krupp, N., J. Woch, A. Lagg, B. Wilken, S. Livi, and D. J. Williams (1998), Energetic particle bursts in the predawn Jovian magnetotail, *Geophys. Res. Lett. 25*, 1249–1253.

Krupp, N., A. Lagg, S. Livi, B. Wilken, J. Woch, E. C. Roelof, and D. J. Williams (2001), Global flows of energetic ions in Jupiter's equatorial plane: First-order approximation, *J. Geophys. Res. 106*, 26 017–26,032.

Krupp, N., et al. (2004), Dynamics of the Jovian magnetosphere, in *Jupiter: The Planet, Satellites and Magnetosphere*, edited by F. Bagenal, T. E. Dowling, and W. B. McKinnon, pp. 617–638.

Kurth, W. S., D. A. Gurnett, J. D. Sullivan, H. S. Bridge, F. L. Scarf, and E. C. Sittler, Jr. (1982), Observations of Jupiter's distant magnetotail and wake, *J. Geophys. Res.*, *87*, 10,373–10,383.

Lepping, R. P., M. D. Desch, E. C. Sittler, Jr., K. W. Behannon, L. W. Klein, J. D. Sullivan, and W. S. Kurth (1983), Structure and other properties of Jupiter's distant magnetotail, *J. Geophys. Res. 88*, 8801–8815.

Louarn, P., A. Roux, S. Perraut, W. Kurth, and D. Gurnett (1998), A study of the large-scale dynamics of the Jovian magnetosphere using the Galileo Plasma Wave Experiment, *Geophys. Res. Lett.*, *25*, 2905–2908.

Louarn, P., A. Roux, S. Perraut, W. Kurth, and D., Gurnett (2000), A study of the Jovian "energetic magnetospheric events" observed by Galileo: Role in the radial plasma transport, *J. Geophys. Res.*, *105*(A6), 13,073–13,088.

Louarn, P., B. H. Mauk, M. G. Kivelson, W. S. Kurth, A. Roux, C. Zimmer, D. A. Gurnett, and D. J. Williams (2001), A multi-instrument study of a Jovian magnetospheric disturbance, *J. Geophys. Res.*, *106*, 29,883–29,898.

Mauk, B. H., and S. M. Krimigis (1987), Radial force balance within Jupiter's dayside magnetosphere, *J. Geophys. Res.*, *92*, 9931–9941.

Mauk, B. H., D. G. Mitchell, R. W. McEntire, C. P. Paranicas, E. C. Roelof, D. J. Williams, S. M. Krimigis, and A. Lagg (2004), Energetic ion characteristics and neutral gas interactions in Jupiter's magnetosphere, *J. Geophys. Res. 109*, 9.

McComas, D. J., and F. Bagenal (2008), Reply to comment by S. W. H. Cowley et al. on "Jupiter: A fundamentally different magnetospheric interaction with the solar wind," *Geophys. Res. Lett.*, *35*, 10, 103.

McComas, D. J., and F. Bagenal (2007), Jupiter: A fundamentally different magnetospheric interaction with the solar wind, *Geophys. Res. Lett.*, *34*, 20,106.

McComas, D. J., F. Allegrini, F. Bagenal, F. Crary, R. W. Ebert, H. Elliott, A. Stern, and P. Valek (2007), Diverse plasma populations and structures in Jupiter's magnetotail, *Science*, *318*, 217.

McNutt, Jr., R. L., J. W. Belcher, and H. S. Bridge (1981), Positive ion observations in the middle magnetosphere of Jupiter, *J. Geophys. Res.*, *86*, 8319–8342.

McNutt, R. L., et al. (2007), Energetic particles in the Jovian magnetotail, *Science*, *318*, 220.

Ness, N. F., M. H. Acuna, R. P. Lepping, K. W. Behannon, L. F. Burlaga, and F. M. Neubauer (1979), Jupiter's magnetic tail, *Nature* , *280*, 799–802.

Nichols, J. D., J. T. Clarke, J. C. Gerard, and D. Grodent (2009), Observations of Jovian polar auroral filaments, *Geophys. Res. Lett. 36*, 8101.

Nishida, A. (1983), Reconnection in the Jovian magnetosphere, *Geophys. Res. Lett. 10*(6), 451–454.

Ogino, T., R. J. Walker, and M. G. Kivelson (1998), A global magnetohydrodynamic simulation of the Jovian magnetosphere, *J. Geophys. Res.*, *103*, 225–235.

Prangé, R., G. Chagnon, M. G. Kivelson, T. A. Livengood, and W. Kurth (2001), Temporal monitoring of Jupiter's auroral activity with IUE during the Galileo mission: Implications for magnetospheric processes, *Planet. Space Sci.*, *49*, 405–415.

Radioti, A., J. Woch, E. A. Kronberg, N. Krupp, A. Lagg, K.-H. Glassmeier, and M. K. Dougherty (2007), Energetic ion composition during reconfiguration events in the Jovian magnetotail, *J. Geophys. Res. (Space Phys.)*, *112*, 6221.

Radioti, A., D. Grodent, J.-C. Gérard, B. Bonfond, and J. T. Clarke (2008), Auroral polar dawn spots: Signatures of internally driven reconnection processes at Jupiter's magnetotail, *Geophys. Res. Lett.*, *35*, 3104.

Radioti, A., D. Grodent, J.-C. Gérard, and B. Bonfond (2010), Auroral signatures of flow bursts released during magnetotail reconnection at Jupiter, *J. Geophy. Res. (Space Phys.) 115*, 7214, doi:10.1029/2007GL032460.

Radioti, A., D. Grodent, J.-C. GéRard, M. F. Vogt, M. Lystrup, and B. Bonfond (2011), Nightside reconnection at Jupiter: Auroral and magnetic field observations from 26 July 1998, *J. Geophys. Res. (Space Phys.)*, *116*, 3221.

Russell, C. T. (2001), The dynamics of planetary magnetospheres, *Planet. Space Sci.*, *49*, 1005–1030.

Russell, C. T. (2004), Outer planet magnetospheres: A tutorial, *Adv. Space Res.*, *33*, 2004–2020.

Russell, C. T., K. K. Khurana, D. E. Huddleston, and M. G. Kivelson (1998), Localized reconnection in the near Jovian magnetotail, *Science*, *280*, 1061.

Russell, C. T., M. G. Kivelson, K. K. Khurana, and D. E. Huddleston (2000), Circulation and dynamics in the Jovian magnetosphere, *Adv. Space Res.*, *26*, 1671–1676.

Scarf, F. L. (1979), Possible traversals of Jupiter's distant magnetic tail by Voyager and by Saturn, *J. Geophys. Res.*, *84*, 4422–4424.

Sittler, E. C., R. P. Lepping, B. H. Mauk, and S. M. Krimigis (1987), Detection of a hot plasma component within the core regions of Jupiter's distant magnetotail, *J. Geophys. Res.*, *92*, 9943–9948.

Vasyliūnas, V. M., L. A. Frank, K. L. Ackerson, and W. R. Paterson (1997), Geometry of the plasma sheet in the midnight-to-dawn sector of the Jovian magnetosphere: Plasma observations with the Galileo spacecraft: *Geophys. Res. Lett.*, *24*, 869–872.

Vasyliunas, V. M. (1983), Plasma distribution and flow, in *Physics of the Jovian Magnetosphere*, edited by A. J. Dessler, pp. 395–453, Cambridge Univ. Press, New York.

Vogt, M. F., M. G. Kivelson, K. K. Khurana, S. P. Joy, and R. J. Walker (2010), Reconnection and flows in the Jovian magnetotail as inferred from magnetometer observations, *J. Geophys. Res. (Space Phys.) 115*, 6219.

Walker, R. J., T. Ogino, and M. G. Kivelson (2001), Magnetohydrodynamic simulations of the effects of the solar wind on the Jovian magnetosphere, *Planet. Space Sci.*, *49*, 237–245.

Woch, J., et al. (1999), Plasma sheet dynamics in the Jovian magnetotail: Signatures for substorm-like processes? *Geophys. Res. Lett.*, *26*, 2137–2140.

Woch, J., N. Krupp, and A. Lagg (2002), Particle bursts in the Jovian magnetosphere: Evidence for a near-Jupiter neutral line, *Geophys. Res. Lett.*, *29* (7), doi:10.1029/2001GL014080.

Woch, J., N. Krupp, A. Lagg, and A. Tomás (2004), The structure and dynamics of the Jovian energetic particle distribution, *Adv. Space Res.*, *33*, 2030–2038, doi:10.1016/j.asr.2003.04.050.

Woch, J., N. Krupp, A. Lagg, B. Wilken, S. Livi, and D. J. Williams (1998), Quasi-periodic modulations of the Jovian magnetotail, *Geophys. Res. Lett.*, *25*, 1253–1256.

Ye, S.-Y., D. A. Gurnett, J. D. Menietti, W. S. Kurth, G. Fischer, P. Schippers, and G. B. Hospodarsky (2012), Cassini observation of Jovian anomalous continuum radiation, *J. Geophys. Res. (Space Phys.)*, *117*, 4211.

Zimbardo, G. (1991), Collisionless reconnection in Jupiter's magnetotail, *Geophys. Res. Lett.*, *18* (4), 741–744.

Zimbardo, G. (1993), Observable implications of tearing-mode instability in Jupiter's nightside magnetosphere, *Planet. Space Sci.*, *41*, 357–361.

6

Saturn's Magnetotail

Caitríona M. Jackman

6.1. INTRODUCTION

The gas giant planet Saturn has a strong internal magnetic field and possesses a large magnetosphere. Saturn's magnetosphere has been studied by many authors, and the reader is referred to reviews by *Bagenal* [1992, 2009], *Russell* [2001], *Kivelson* [2007], and *Jackman* et al. [2014a] for comparison of the Kronian system with other planets in our solar system. Saturn has often been described as intermediate between Earth and Jupiter in terms of its magnetospheric dynamics, and some of these features will be addressed throughout this review. Saturn has a well-developed magnetotail, the region which stretches vast distances on the nightside of the planet, shaped by the solar wind and interplanetary magnetic field (IMF) flowing past the magnetosphere. The magnetotail is a very important region because it is the site of plasma and energy buildup and explosive release. In situ exploration of the magnetotail reveals it to be a very dynamic environment, while remote sensing of the aurora and planetary radio emissions can also shed light on magnetotail dynamics.

The first spacecraft to visit Saturn's magnetosphere were Pioneer 11 (1979), Voyager 1 (1980), and Voyager 2 (1981). Figure 6.1 shows the trajectories of these spacecraft as they flew past. It is clear from this figure that these spacecraft barely skimmed the nightside of the planet. Indeed, only Voyager 1 sampled the tail in any detail, glimpsing the plasma sheet and northern lobe but not the current sheet itself (due to the high latitude of the pass). Voyager data were used to infer the presence of a nightside magnetodisk current sheet of thickness ~$4R_S$ ($1R_S = 60{,}268$ km = 1 Saturn radius) [e.g., *Connerney et al.*, 1981]. The diameter of Saturn's magnetotail was also estimated to be ~$80R_S$, with an average field strength of ~3 nT at $25R_S$ downtail [*Behannon et al.*, 1981; *Ness et al.*, 1981]. These results served to inspire us to return to Saturn with a dedicated orbiter, and thus the Cassini spacecraft was planned and ultimately launched in 1997. It reached Saturn in 2004 and is scheduled to orbit until 2017 (as currently planned at time of writing), giving unprecedented coverage of many magnetospheric regions spanning more than a solar cycle.

The arrival of the Cassini spacecraft and its subsequent exploration of the Kronian system has revolutionized our ideas of Saturn's magnetospheric dynamics. The trajectory coverage of the Cassini spacecraft from 2004 to 2013 is also shown in Figure 6.1, and this illustrates how comprehensive the coverage has been compared to the previous Pioneer 11 and Voyager encounters. The deepest coverage of the tail occurred in 2006, when the spacecraft traveled as far as $68R_S$ from the planet post-midnight. Overall there is a strong asymmetry in the nightside coverage, with many more orbits on the dawn flank than at dusk. The Cassini spacecraft has a wonderfully sophisticated suite of instruments, and throughout this review we will refer to data from the magnetometer (MAG) [*Dougherty et al.*, 2004], the Cassini plasma spectrometer (CAPS) [*Young et al.*, 2004], the magnetospheric imaging instrument (MIMI) [*Krimigis et al.*, 2004], and the radio and plasma wave investigation (RPWS) [*Gurnett et al.*, 2004]. Table 6.1 shows several characteristics of Saturn's magnetosphere as measured remotely and in situ by spacecraft. In this review, section 6.2 will describe large-scale tail structure,

Department of Physics and Astronomy, University of Southampton, Southampton, United Kingdom

Magnetotails in the Solar System, Geophysical Monograph 207. First Edition.
Edited by Andreas Keiling, Caitríona M. Jackman, and Peter A. Delamere.

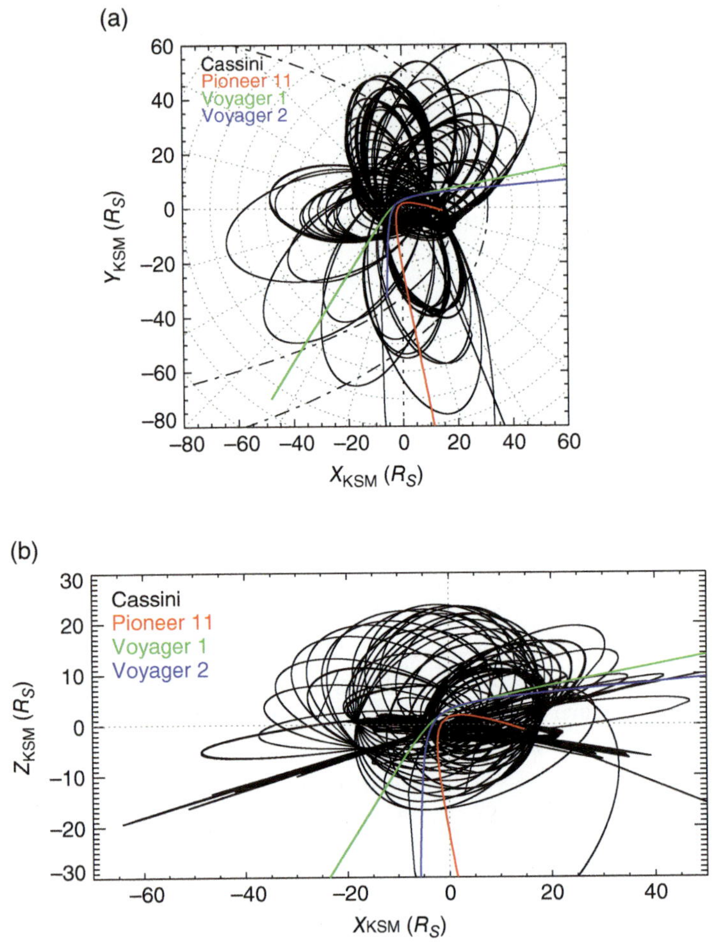

Figure 6.1 Trajectories of the Pioneer 11, Voyager 1, Voyager 2, and Cassini (2004–2013) spacecraft as they encountered Saturn's magnetosphere. The plots are in the kronocentric solar magnetospheric (KSM) coordinate system. KSM is the Kronian analogue of the terrestrial GSM system, where the *X* axis coincides with the direction to the Sun, the *XZ* plane contains the planetary dipole axis, and the *Y* component is azimuthal, positive toward dusk. Dot-dashed lines show the position of the magnetopause, as determined from the *Kanani et al.* [2010] model, for solar wind dynamic pressures of 0.1 and 0.01 nPa. (1a) The *X-Y* plane. (1b) The *X-Z* plane. The Sun is to the right in both plots.

section 6.3 dynamics, and section 6.4 remote sensing, and section 6.5 will present a summary and outline open questions for the future.

6.2. LARGE-SCALE TAIL STRUCTURE

The canonical picture of a planetary magnetotail has a current sheet at the very center, colocated with a plasma sheet with a larger-scale height, and lobes at increasingly higher latitudes before the magnetopause is reached. Intermediate layers (such as the plasma sheet boundary layer, plasma mantle, or magnetopause cushion region) may exist, but for the purposes of this review we focus on the three key regions of Saturn's magnetotail (current sheet, plasma sheet, and lobes) and describe their average properties as sampled in situ by spacecraft.

6.2.1. Saturn's Current Sheet

Saturn's current sheet lies at the center of the magnetotail, and traversals of the current sheet center may be identified by changes in the sign of the radial component of the magnetic field. It is a region of high-beta plasma (where beta is the ratio of thermal to magnetic pressure), and a statistical study using data from Cassini's CAPS electron spectrometer (ELS, at $E < 28$ keV) showed that electron flux, number density, and partial electron beta maximize at current sheet crossings [*Arridge et al.*, 2009]. The electron temperature was found to vary by no more than a factor of two between the current sheet and the magnetotail lobes. Similar to Jupiter's current sheet, the current sheet in the tail region of Saturn's magnetosphere is an extension of the magnetospheric current sheet with an inner edge near $6R_S$ [*Connerney et al.*, 1983; *Arridge et al.*, 2008a]. At large distances in the magnetotail (at least beyond the ~$68R_S$ that Cassini has explored), it is not clear whether the current sheet continues as an azimuthal disc confined inside a cylindrical volume or whether the current closes on the magnetopause in a manner similar to the terrestrial magnetotail. Saturn's current sheet geometry has been investigated by several authors and found to exhibit two interesting features, warping and flapping, which are now described in more detail.

Table 6.1 List of key planetary and magnetospheric parameters for Saturn.

Average distance from the Sun (AU, where 1 AU = 149,597,871 km)	9.54
Equatorial planetary radius, R_P (km)	60,268
Surface field strength (nT)	21,084
Dipole moment (A m^2)	4.6×10^{25}
Planetary field direction (at equator)	South
Dipole tilt (deg)	<1
Siderial rotation rate (hh:mm)	~10:48
Average solar wind D_P (nPa)	0.01 [*Jackman and Arridge*, 2011a]
Average IMF spiral angle (deg)	83 [*Jackman et al.*, 2008a]
Asymptotic tail radius (R_P)	60 [*Arridge et al.*, 2006]
Tail lobe flux content (GWb)	~15–50 [*Badman et al.*, 2005]
Average MP standoff distance (R_P)	22 and 27 [*Achilleos et al.*, 2008]

6.2.1.1. Current Sheet Warping. First, the current sheet is found to be warped out of the equatorial plane both in the tail and on the dayside [*Arridge et al.*, 2008a, 2011], forming a bowl or basin shape. *Arridge et al.* [2008a] modeled this and included a parameter called the "hinging distance," the characteristic distance beyond the planet where the current sheet no longer lies in the magnetic dipole equator. This distance is typically similar to the subsolar magnetopause standoff distance and is modelled at ~$25R_S$ but in reality is highly sensitive to solar wind conditions. A link between the hinging distance of Earth's plasma sheet and the solar wind dynamic pressure was made by *Tsyganenko and Fairfield* [2004]. At Saturn, *Arridge et al.* [2011] have suggested that increases in solar wind dynamic pressure or decreases in plasma sheet mass content can result in changes in the hinging distance that displace the plasma sheet further from the equator (and vice versa for opposite conditions). Figure 6.2 illustrates this hinging schematically in a meridional view (left panel) and a three-dimensional representation of the distortion (right panel). This hinging is due to the action of the solar wind and is a function of the planet's small dipole tilt and large (26.7°) obliquity. The motion of the current sheet has strong implications, for example, for the local environment of the moon Titan, as shown in the right-hand panel of Figure 6.2. Titan orbits at a radial distance of $20R_S$ and may alternately find itself in a current sheet-like, plasma-sheet-like, or lobe-like field and plasma regime depending on the degree of current sheet hinging [*Simon et al.*, 2010].

The Cassini spacecraft first arrived at Saturn during southern hemisphere summer, but the mission timeline has crossed equinox (August 2009) and beyond into

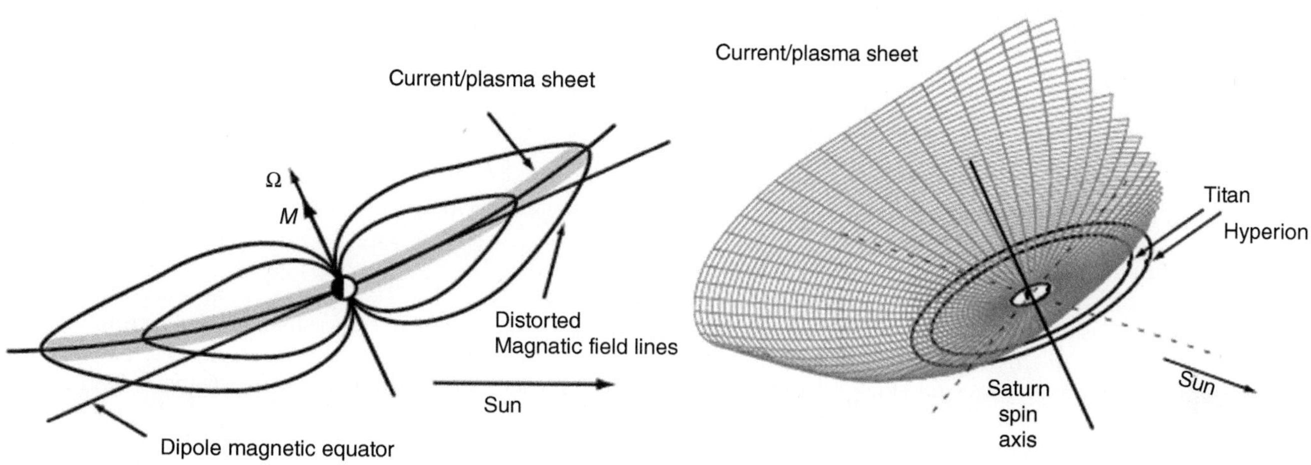

Figure 6.2 Schematic illustrating the hinging of Saturn's current sheet out of the equatorial plane (left) and a three-dimensional noon-midnight meridian view (right). The Sun is to the right in both plots. [From *Arridge et al.*, 2008a].

southern hemisphere winter, allowing observers the opportunity to see Saturn's current sheet across the seasons. Additionally, the long duration of the Cassini mission means that by the end the spacecraft will have been sampling Saturn's magnetosphere for almost a full solar cycle, thus exploring magnetospheric dynamics under a wide range of solar wind conditions, making the study of the dynamic current sheet a fascinating topic.

6.2.1.2. Current Sheet Flapping. The second noteworthy feature of Saturn's current sheet is its tendency to oscillate vertically with a period of ~10 h, in phase with other periodic signals in Saturn's magnetosphere [e.g., *Arridge et al.*, 2008b, 2011, *Khurana et al.*, 2009; *Simon et al.*, 2010; *Volwerk et al.*, 2013]. Such flapping is puzzling given the almost exact alignment of Saturn's magnetic and spin axes. Figure 6.3 shows four days worth of data from the Cassini magnetometer and plasma instruments, illustrating the regular motion of the current sheet, resulting in alternating encounters with the center of the current sheet, the plasma sheet, and the magnetotail lobes.

Many efforts have been made to model the motion of the current sheet. For example, *Arridge et al.* [2011] found the sheet flaps in phase with B_R and B_θ and in quadrature with the B_ϕ component in the cam region of Saturn's magnetosphere. They also found that the phase of the flapping has a strong local time asymmetry. *Provan et al.* [2012] examined the phase of magnetic oscillations in Saturn's tail and found that the oscillations exhibit dual periodicities with different influences in the northern and southern hemispheres for reasons which are still under debate. They calculated that the current sheet can flap north-south with an amplitude of $3R_S$ when the oscillations are in phase and that the current sheet thickness can vary by factors of ~2 when in antiphase. A full

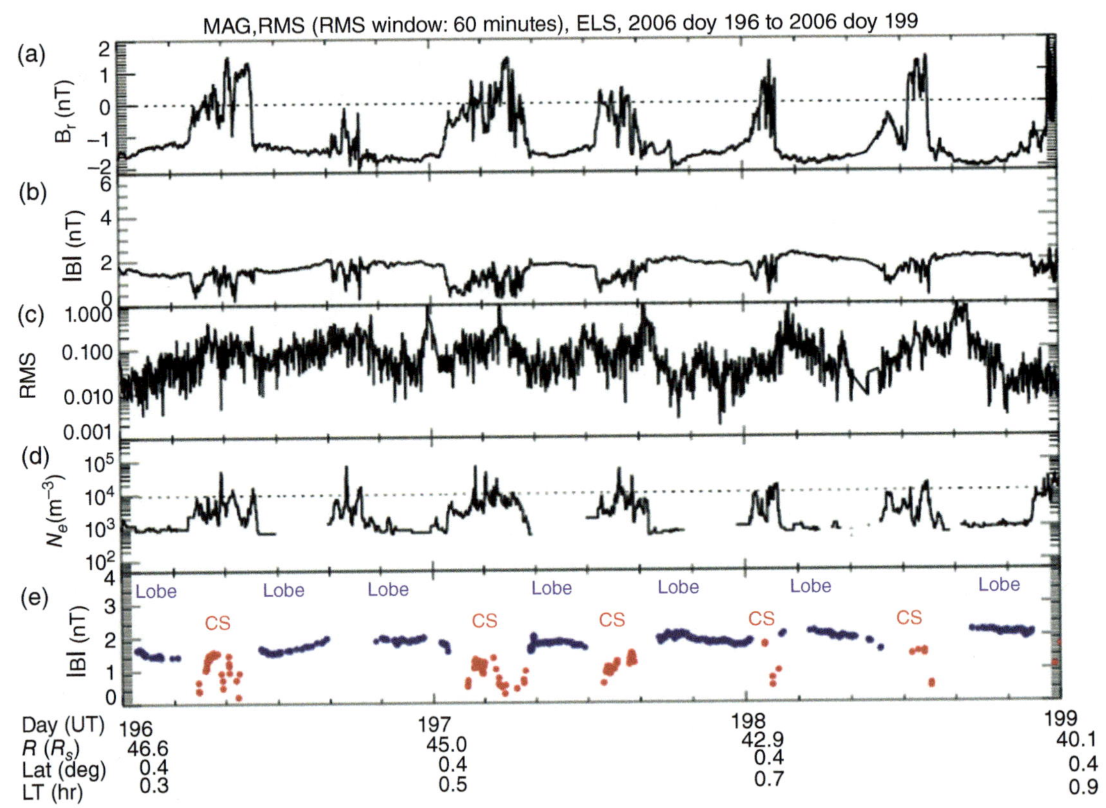

Figure 6.3 Plasma and magnetic field data from 2006 days 196–199 showing regular encounters with the current sheet and lobes alternately. The top panel shows the radial component of the magnetic field, the second panel shows the total magnetic field strength, and the middle panel shows the relative RMS of the magnetic field, calculated over a 60-min averaging window. The fourth panel shows the measured electron density from the Cassini CAPS electron spectrometer (with the model density for the central plasma sheet from *Arridge et al.* [2011] shown as a dashed line). The bottom panel also shows the total field strength (like panel 2), but this time only for regions classified as lobe (blue) and current sheet (red). [From *Jackman and Arridge*, 2011b].

understanding of Saturn's current sheet oscillations and their relationship to other periodicities in Saturn's magnetosphere is certainly a key goal for the remainder of the Cassini mission.

6.2.2. Plasma Sheet

Encompassing the central current sheet is the plasma sheet, the region of closed magnetic field lines and high plasma beta, where energetic particles are concentrated into a disklike layer [e.g., *Krupp et al.*, 2005]. *Arridge et al.* [2009] observed electrons in the plasma sheet in three states: a quiet plasma sheet characterized by electrons with energies of 0.1 keV, a disturbed plasma sheet with energies ~1 keV, and a state in which bimodal electron distributions containing both of the above components are observed as well as bimodal distributions with a cold ~0.01 keV component. *Arridge et al.* [2009] also defined a central and outer plasma sheet for Saturn with densities of 0.001–0.1 and 0.0001–0.01 cm^{-3} and temperatures of 0.06–0.2 keV and 0.1–0.2 keV, respectively. Data from the Cassini CAPS ion mass spectrometer (IMS) showed that ion flows in the plasma sheet are typically 150–200 km/s and predominantly in the corotational direction out to $30R_S$ (where rigid corotation is ~300 km/s), with some outward flow beyond this distance. Ion temperatures were found to be ~1 keV for heavy W^+ ions and ~0.1 keV for protons. Ion densities fall from 1 cm^{-3} near $20R_S$ to 0.1 cm^{-3} near $45R_S$. The composition is variable, with the observed number density ratio n_{W+}/n_p varying between 1 and >10, but typically 2–3 [*McAndrews et al.*, 2009].

Sergis et al. [2011] took advantage of several vertical cuts through the plasma sheet by the Cassini spacecraft both pre- and postequinox to study the scale height, vertical displacement, tilt angle, and hinging distance. One such example (preequinox) is shown in Figure 6.4, which presents energetic particle data from MIMI, electron data from CAPS ELS, and magnetic field data from Cassini MAG, plotted as a function of vertical (z) distance from the equatorial plane such that time increases from right to left. The changing character of field and plasma from lobes to plasma sheet to current sheet is clear in data from all instruments. In this particular example, Cassini moved from the northern to the southern lobe, passing through the plasma and current sheets on its way and observing a vertical offset of ~$1.8R_S$ from the equatorial plane. The grey shaded interval is a Titan flyby, and this illustrates further the importance of understanding the field and plasma regime near the moon (as pointed out by *Simon et al.* [2010]). Overall, the results of this work showed that the plasma sheet can be thicker in the south than in the north, with a strongly energy dependent scale height (~50% thicker in energetic particle range compared to cold electron plasma). Its position is offset from the equatorial plane with a hinge distance of ~$16R_S$, and clear changes in tilt

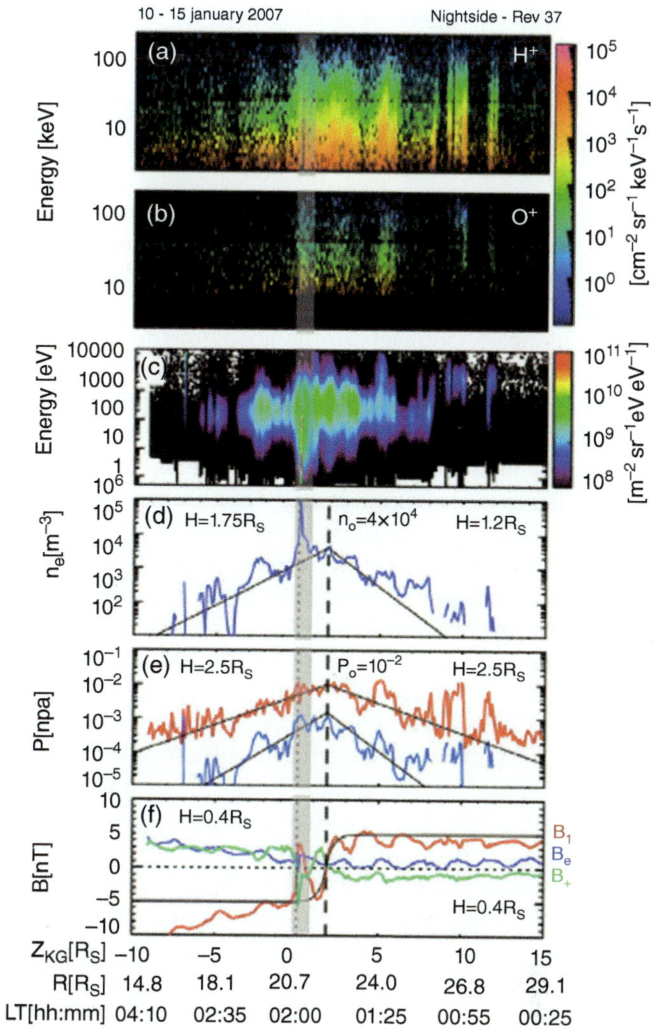

Figure 6.4 Vertical cut through Saturn's plasma sheet: (a) energetic ($E>3$ keV) H^+ intensity spectrogram; (b) energetic ($E>9$ keV) O^+ intensity spectrogram; (c) low-energy ($E<28$ keV) electron differential energy flux spectrogram; (d) electron number density with an exponential function fit $n=n_0 \exp(-z/H)$; (e) energetic ($E>3$ keV) ion and low-energy electron pressure in red and blue, respectively, together with exponential decay functions $P=P_0 \exp(-z/H)$; and (f) magnetic field vector components (in spherical polar coordinates) together with an exponential sheet fit (black line) to the radial component B_r (red). All panels are arranged in increasing z. The vertical short dashed line at $z=0$ denotes the equatorial plane, while the dashed line corresponds to the center of the sheet. The vertical shaded stripes correspond to Titan encounters of Cassini. [From *Sergis et al.*, 2011.]

were observed as the current sheet became progressively more aligned to the solar wind direction as equinox approached.

6.2.3. Lobes

Moving to higher latitudes still, we encounter the magnetotail lobes. The lobes are regions of negligible plasma density and quiet magnetic field and can house the "open" magnetic flux stored in the tail, where one end of a field line magnetically connects to the northern (or southern) polar cap, with the other end of the field line extending far downstream and generally believed to connect into the solar wind. We do note, however, that in rapidly rotating magnetospheres such as Jupiter and Saturn, low plasma density measured at higher latitudes can be a result of centrifugal confinement of plasma close to the equatorial plane [e.g., *Hill and Michel*, 1976; *Mauk et al.*, 1985], and thus low plasma density is not necessarily evidence of an open field.

Jackman and Arridge [2011b] performed a statistical study of Saturn's magnetotail lobes, identifying them using criteria based on electron density, magnetic field strength, and level of magnetic field fluctuation. They applied these criteria to the Cassini dataset from the deepest tail orbits in 2006 and derived a functional fit to the falloff of Saturn's lobe field strength with radial distance from the planet as $B_{LOBE} = 251 \times R^{-1.20}$. They compared this to previous fits for Earth, $B = 1659.2 \times R^{-1.46} + 7.47$ [*Fairfield and Jones*, 1996], and Jupiter, $B = 2900 \times \rho^{-1.37}$ [*Kivelson and Khurana*, 2002]. Figure 6.5 shows these gradients plotted against each other in the top panel and scaled to a common distance in the bottom panel, taking into account varying dipolar field strengths and ratios of subsolar magnetopause standoff distances to downtail distances. The near magnetotails of Earth, Jupiter, and Saturn show a similar falloff in the region with spacecraft coverage (solid lines). Significant differences begin to emerge beyond a common scaled distance of ~$30R_E$, with the Jovian and Kronian traces falling off much more rapidly than the terrestrial trace. However, we note that this is based purely on extrapolation of measured values, and thus confirmation of the validity of this behavior will only be possible with future exploration of the deeper tails of Jupiter and Saturn by spacecraft equipped with magnetometers. *Coroniti and Kennel* [1972] noted that Earth's distant tail ceases flaring beyond an asymptotic distance where the external static pressure of the solar wind is balanced by the internal magnetic pressure of the magnetotail lobes. For the terrestrial case, this asymptotic distance and tail radius are approximately $140R_E$ and $30R_E$, respectively. *Arridge et al.* [2006] adapted this calculation to Saturn and found the corresponding numbers to be $220R_S$ and $60R_S$. This leads us to the topic of the deep Kronian tail and in particular the basic property of its length.

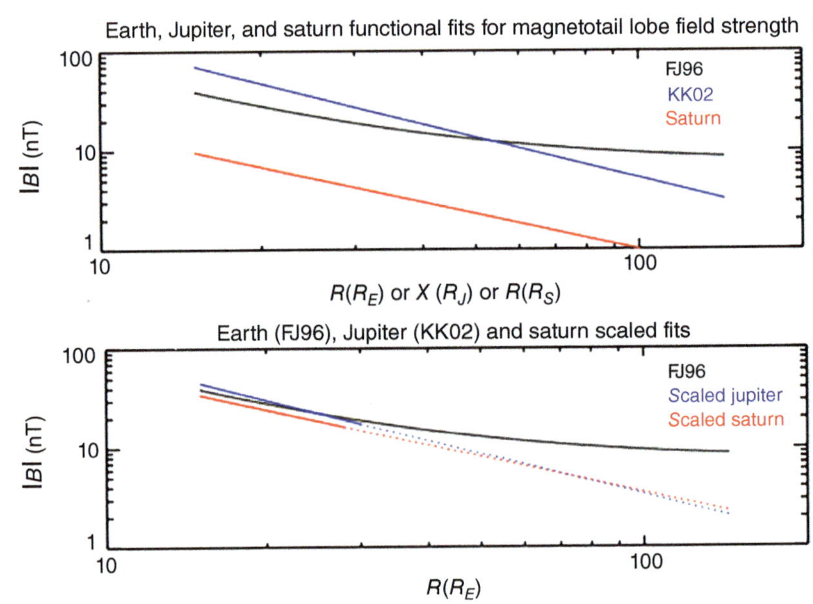

Figure 6.5 Plots of the falloff of the lobe magnetic field strength at Earth (black), Jupiter (blue), and Saturn (red): (top) unscaled relationships and (bottom) all values scaled to Earth. The solid lines in the bottom panel are valid for comparison as they represent areas sampled by spacecraft, while the behavior represented by the dashed lines may not be physically representative of the deeper magnetotails of Jupiter and Saturn. [From *Jackman and Arridge*, 2011b.]

6.2.4. Saturn's Tail Length

The length of a planetary magnetotail is an important property in terms of our understanding of the scale of influence that a planet can exert on its local environment. As detailed above, the Cassini spacecraft has only explored as far as $68R_S$ in Saturn's tail, and we infer that the tail is likely much longer than this, based on analogy with other magnetospheres and by extrapolation of our current data. Studies of coherent plasma and magnetic structures downstream of various planets have indicated that a magnetotail can stretch to very long distances downstream, extending beyond the region of closed field lines and trapped plasma. For example, it is thought that Jupiter's tail can at times extend to distances beyond ~$9000R_J$ ($1R_J$ = 71,492 km; *Lepping et al.* [1983]) and even as far as the orbit of Saturn based on observations taken during the Voyager flyby [e.g., *Scarf et al.*, 1982; *Kurth et al.*, 1982], while the New Horizons spacecraft has sampled the coherent Jovian tail in situ to distances from $1600R_J$ [*McNutt et al.*, 2007] to 2500 R_J [*McComas et al.*, 2007].

For the case of Earth, *Dungey* [1965] estimated the tail length based on the picture where field lines are opened at the dayside through reconnection and dragged over the poles to form the tail before ultimately sinking in toward the center plane of the tail and reconnecting. He calculated that if field lines remain open for 4 h during their convection cycle and are stretched at a solar wind speed of 450 km/s, the tail can extend to a length of ~$1000R_E$. This picture has been explored in detail for the terrestrial tail in the light of more recent spacecraft measurements of polar cap magnetic flux content as well as dayside and nightside reconnection rates [e.g., *Milan*, 2004]. *Milan et al.* [2005] examined the case at Saturn where the polar cap refresh time (i.e., the time for magnetic flux to be cycled through dayside to nightside reconnection) is expected to be many days [e.g., *Jackman et al.*, 2004; *Badman et al.*, 2005], whereas the planetary rotation period (~10–11 h) is much shorter than the refresh time. In this case the magnetotail lobes are thought to become twisted by a rotational torque applied to the ionospheric ends of the open field lines [e.g., *Isbell et al.*, 1984]. *Milan et al.* [2005] calculated that each turn of a twisted lobe field line can be ~$900R_S$ in length. They argued that Saturn is a last-in-first-out system, in which recently opened field lines are the first to be closed by reconnection in the tail, due to the imposed twist. This implies that Saturn's magnetotail lobes contain a core of old open field lines that can be stretched to incredible lengths

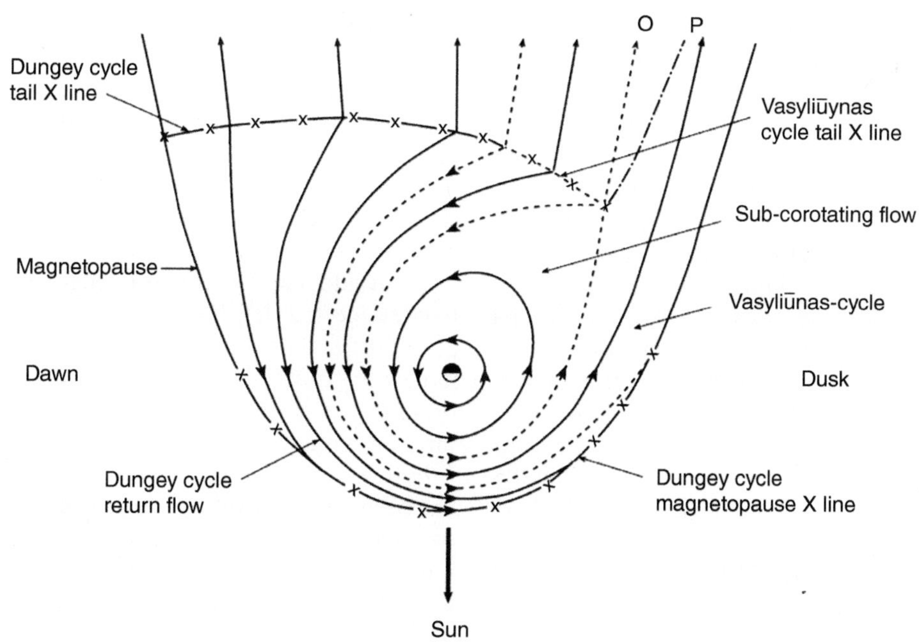

Figure 6.6 Schematic of plasma flows in Saturn's magnetosphere. The figure looks down onto the equatorial plane, with dawn to the left, dusk to the right, and Sun to the bottom. The solid lines represent plasma flow streamlines. Short-dashed lines show boundaries between flow regimes. The location of suggested reconnection lines are shown by solid and dashed lines marked with crosses. The curve indicated by the O marks the path of the plasmoid O line in the Vasyliunas cycle flow (also a streamline), while P marks the outer limit of the plasmoid field lines, which eventually asymptotes to the dusk tail magnetopause. The Sun is to the bottom in this plot. [From *Cowley et al.*, 2004.]

(further than 15,000R_S). In situ observations that indicate the closure of open field lines via tail reconnection will be discussed in section 6.3, where the most recent debates regarding the role of dayside reconnection and Dungey cycle in driving Saturn's magnetospheric dynamics are also briefly outlined.

6.3. MAGNETOSPHERIC DYNAMICS

6.3.1. Plasma Flows

There are two key cycles of plasma flow thought to operate in Saturn's magnetosphere: the Dungey cycle and the Vasyliūnas cycle, as illustrated schematically in Figure 6.6. In the cycle originally proposed for Earth by *Dungey* [1961], magnetic reconnection at the dayside magnetopause opens magnetic flux, and the solar wind interaction carries these open magnetic field lines from dayside to nightside, where they are stretched out to form the tail lobes. As they are stretched out downtail, open flux tubes sink in toward the center plane of the tail, where they reconnect again, closing the flux that was opened on the dayside. The "Dungey cycle time scale" refers to the length of time from the opening of the field lines at the dayside to the closing of the field lines on the nightside (this is the time scale used to estimate the tail length above). The return of empty flux tubes back toward the planet and the release of plasmoids may be consequences of the Dungey cycle. In addition to the solar wind-driven Dungey cycle, the internally driven Vasyliūnas cycle is a key driver of Saturn's magnetospheric dynamics. Given the size of the planet and its rapid rotation rate, centrifugal effects become important, and energy transferred to internally generated plasma by this fast rotation drives the "planetary wind" or Vasyliūnas cycle, as first suggested for Jupiter [*Hill et al.*, 1974; *Michel and Sturrock*; 1974; *Vasyliūnas*, 1983]. The plasma is accelerated by magnetic stresses from the ionosphere, gains energy, and moves outward from the planet. Centrifugal forces cause the field lines to stretch to the point where a thin current sheet forms, across which the closed field lines reconnect. This reconnection simultaneously shortens the flux tube and (like the Dungey cycle) releases plasma down the tail in the form of a plasmoid.

Cowley et al. [2005] considered the relative roles that these cycles may play in magnetospheric dynamics at Saturn, with a strong emphasis on the auroral counterpart. It was suggested that following reconnection in the tail and associated "tail collapse" hot plasma may be injected from the nightside into the outer magnetosphere, first in the midnight and dawn sector and second at increasing local times via noon and dusk (We note that the "outer" magnetosphere has been defined by *Arridge et al.* [2012] as the region beyond 15R_S in the equatorial plane). *Cowley et al.* [2005] specifically suggested that precipitation of hot plasma from Dungey-type tail reconnection should occur on a layer of streamlines passing from the mapped tail to the mapped magnetopause reconnection regions via dawn. *Badman and Cowley* [2007] made more detailed theoretical predictions as to the nature of "return" flows following magnetotail reconnection. They suggest that following flux closure on the nightside via the Dungey cycle, plasma flows are expected to form few-planetary-radius-wide layers inside the dawn and morning magnetopause. In the absence of cross-field plasma diffusion, these layers will be characterized by the presence of hot light ions originating from either the planetary ionosphere or the solar wind. In contrast to the Dungey cycle flows, the inner Vasyliūnas cycle and middle magnetosphere transport will be dominated by hot (few-keV) heavy ions originating from plasma sources (such as the rings and Enceladus).

While this chapter focuses primarily on Saturn's tail, it is important to briefly mention the work concerning the in situ observations (or lack thereof) of reconnection at Saturn's dayside magnetopause, driving the Dungey cycle flows and hence impacting the tail. Initial predictions based on an empirical formula adapted from Earth suggested an average dayside reconnection voltage of 50 kV [*Jackman et al.*, 2004]. Subsequently *McAndrews et al.* [2008] reported evidence of electron and ion heating and locally open magnetic field indicating reconnection at Saturn's dayside magnetopause with a voltage of 48 kV. However, many authors have suggested that the rate of reconnection may be suppressed by the high Mach number and high plasma beta regime at Saturn [e.g., *Scurry and Russell*, 1991; *Masters et al.*, 2012] although this is refuted by some [e.g., *Grocott et al.*, 2009]. Others have reported a lack of flux transfer events as evidence that dayside reconnection is negligible [*Lai et al.*, 2012] or have suggested that the Kelvin-Helmholtz instability can account for the required plasma mixing at the boundary [*Delamere et al.*, 2013]. A more detailed discussion of these topics is outside the scope of this review, and the interested reader is referred to Chapters 1 and 13.

Following these and other theoretical predictions, it has been a high priority to examine the nature of the flow patterns in the in situ plasma data at Saturn. The first hints of plasma behavior came from the Pioneer 11 flyby, when *Frank et al.* [1980] found the plasma flow fell below the corotation speed beyond ~10R_S and suggested that quasi-periodic outward flows of plasma could occur beyond ~6.5R_S, associated with reconfiguration of the magnetosphere. Subsequently *Richardson* [1998] used Voyager data to demonstrate that the flow velocities in Saturn's outer magnetosphere displayed significant and variable nonazimuthal flow, with a tendency toward outward flow. Such observations were from mere glimpses of the tail, and the arrival of Cassini 2004 afforded many more opportunities to investigate the nature of plasma flow on the nightside of Saturn.

McAndrews et al. [2009] presented a bulk ion flow map from the nightside, equatorial region of Saturn's magnetosphere as derived from the Cassini CAPS IMS data. Figure 6.7 shows this map, with colored arrows denoting the direction and magnitude of measured flows. The map clearly illustrates that corotation flows dominate over radial flows, which might suggest that the flux tubes sampled within the radial range shown (inside of $50R_S$) are still closed and attached to the planet, with the exception of a small number of plasmoid intervals which are labeled. This implies that conditions for flux tube breaking (of the relatively dense populations detectable by the IMS) have not been exceeded.

Thomsen et al. [2010] complemented the *McAndrews et al.* [2009] study by analyzing the CAPS IMS data from the first 4.5 years of Cassini's Saturn orbit. Their derived flow velocities revealed that the plasma is dominantly in the corotational direction inside of $20R_S$. Beyond this radial distance at dusk they showed that flows are roughly aligned with the magnetopause, while in the premidnight-to-dawn sector the flows typically have a tailward and dawnward direction. Thus overall they also found no evidence for return circulation from the nightside to the dayside beyond ~$20R_S$. The suggestion was made that the return flows perhaps take the form of lower-density flux tubes which are not detected by the instrument.

While neither the *McAndrews et al.* [2009] nor the *Thomsen et al.* [2010] papers revealed significant populations of depleted, returning flux tubes, *McAndrews et al.* [2009] noted that electron observations suggest that such a process must take place in order to produce the low-density plasma populations observed in Saturn's inner magnetosphere (again, we use the definition of *Arridge et al.* [2012] for the "inner" magnetosphere as that region inside $6R_S$ in the equatorial plane). Indeed,

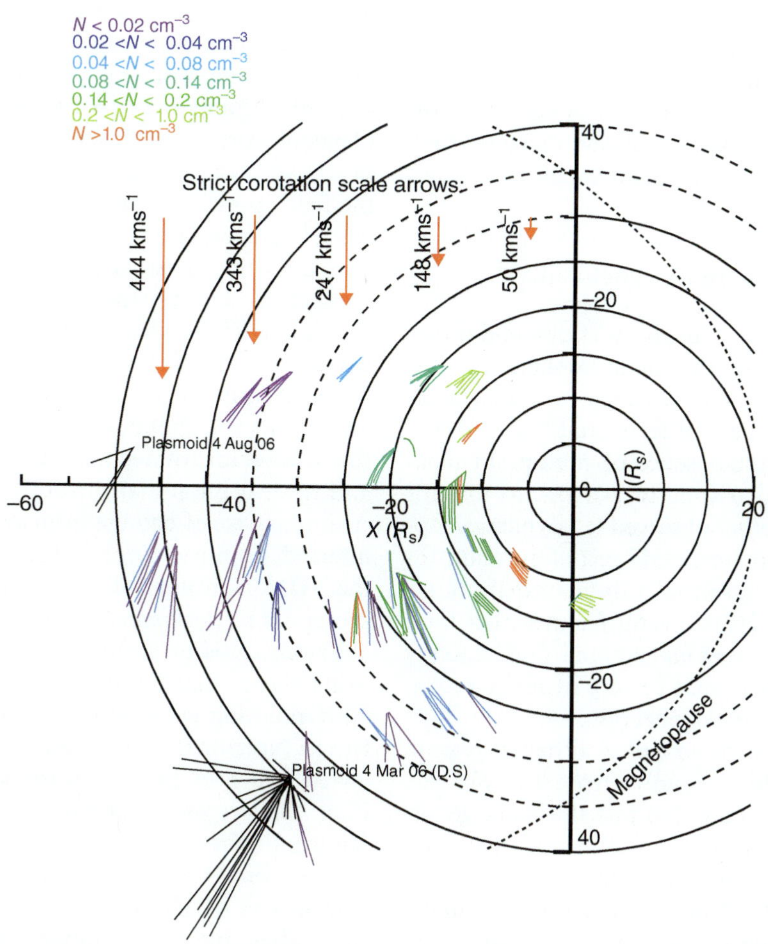

Figure 6.7 Ion flows in the equatorial plane, where arrow length represents velocity and color represents total ion density. Scale arrows for strict corotation at various radial distances are shown. The flows during two plasmoid encounters on 4 March 2006 and 4 August 2006 are shown. The flow speed on 4 March 2006 has been scaled down by 50% relative to the rest of the flows for clarity. The Sun is to the right in this plot. [From *McAndrews et al.*, 2009.]

such a process was reported subsequently by *Masters et al.* [2011] via a serendipitous encounter with return flow ~17.5R_S from the planet in the postdawn magnetosphere. The Cassini CAPS instrument measured plasma flow which was composed in part of water group ions (internally produced), with temperatures of order 1 keV, and flow velocities ~200 km/s, which is 120% of the corotation velocity in that region. *Masters et al.* [2011] suggest that return flows at Saturn are time variable and they were unable to make any conclusions regarding Dungey cycle flow.

Most recently, *Thomsen et al.* [2013] surveyed the CAPS IMS measurements from Cassini's dusk orbits in 2010. Unlike previous studies conducted over all local times, this study did not find significant evidence of outflow in the dusk sector. They did observe inflow which they attribute to dipolarization of the field following reconnection in the tail. The lack of outflow is suggested to indicate that tail reconnection on the dusk flank has not yet proceeded to involve lobe field lines, and thus any plasma that has become disconnected via Vasyliūnas-type reconnection may simply continue to move in the azimuthal direction after release.

In the next section we discuss the topic of reconnection in detail, in the knowledge that we still have much to learn from the large-scale plasma flow measurements in Saturn's magnetosphere.

6.3.2. Reconnection in Saturn's Magnetotail

In addition to the quasi-steady plasma flows and transport outlined above, Saturn's magnetosphere is also observed to undergo dramatic, explosive energy release events, the Kronian equivalent of terrestrial "substorms." For the case of Saturn, the absence of multipoint measurements or upstream solar wind measurements make it difficult to determine whether observed tail dynamics represent a systemwide magnetospheric reconfiguration (in the stricter sense of the terrestrial substorm definition). Nonetheless, here we describe several features that have been observed in situ in Saturn's magnetotail, and remotely in the auroral zones, which point to very dynamic reconfiguration of the tail during reconnection.

Reconnection begins in the so-called diffusion region, where the magnetic field decouples from the plasma, magnetic field is reconfigured, and fast flows are generated. The search for the starting point for reconnection, the "diffusion region," is ongoing at the time of writing, and such a discovery will ultimately represent an important advance in the understanding of the microphysics of reconnection at Saturn. In the absence of a direct encounter with the reconnecting region, the existence of the reconnection process can be inferred by the observation of its products. In the case of a magnetotail, the primary reconnection product is a plasmoid, a magnetic island which, once disconnected from the planet, is free to move downtail, releasing mass from the magnetosphere [*Hones*, 1976, 1977]. As plasmoids move downtail, magnetic forces within them can act to reduce the dimensions in the radial direction, allowing the structures to thicken in the north-south direction [e.g., *Slavin et al.*, 1989, 1993]. Thus plasmoids may produce "bulges" in the tail. Lobe field lines in the vicinity of plasmoids then drape around the bulges as they move downtail, resulting in the formation of signatures known as "traveling compression regions" (TCRs) [*Slavin et al.*, 1984].

Plasmoids and TCR signatures have been identified in data from the Cassini magnetometer by first looking for changes in the north-south component of the magnetic field. *Jackman et al.* [2007] published the first observations of such events in Saturn's tail. *Hill et al.* [2008] subsequently published the Cassini plasma data to accompany two of these three events, and in Figure 6.8 we show one example. The magnetometer data in the top panel are in kronocentric radial theta phi (KRTP) coordinates, where the radial component (B_r) is positive outward from Saturn, the theta component (B_θ) is positive southward, and the azimuthal component (B_ϕ) is positive in the direction of planetary corotation. The primary magnetic field signature of interest is the strong southward-to-northward turning of the field (positive-to-negative B_θ) centered near 2300 UT when the spacecraft was 44R_S downtail, which is interpreted as the tailward passage of a plasmoid over the spacecraft. The position of Cassini close to the magnetotail current sheet meant that it was possible to derive reliable plasma parameters from the ion fluxes. These indicate that the plasmoid was dominated by water group ions and that during the approach of the plasmoid the flow vector rotated away from the azimuthal direction and toward the radial direction, implying tailward flow. A similar case of change from azimuthal to tailward flow, inferred from a change in the energetic ion fluxes from the MIMI ion neutral camera (INCA) instrument, was presented by *Jackman et al.* [2008b] for a different event.

The analysis of further plasmoid and TCR signatures using plasma data has been hampered somewhat by spacecraft pointing constraints and by low electron and ion fluxes. Nonetheless, a large body of work has been published that uses data from the Cassini magnetometer to infer the average properties of reconnection events at Saturn. Figure 6.9 shows the results of a superposed epoch analysis on 34 tailward moving plasmoids identified by *Jackman et al.* [2011], where the zero epoch time is the time when the B_θ component goes through zero. The striking feature in the second panel is the southward-to-northward turning of the field, which is interpreted as the passage of a tailward moving plasmoid over the spacecraft. The magnitude of this field deflection is ~1 nT, but

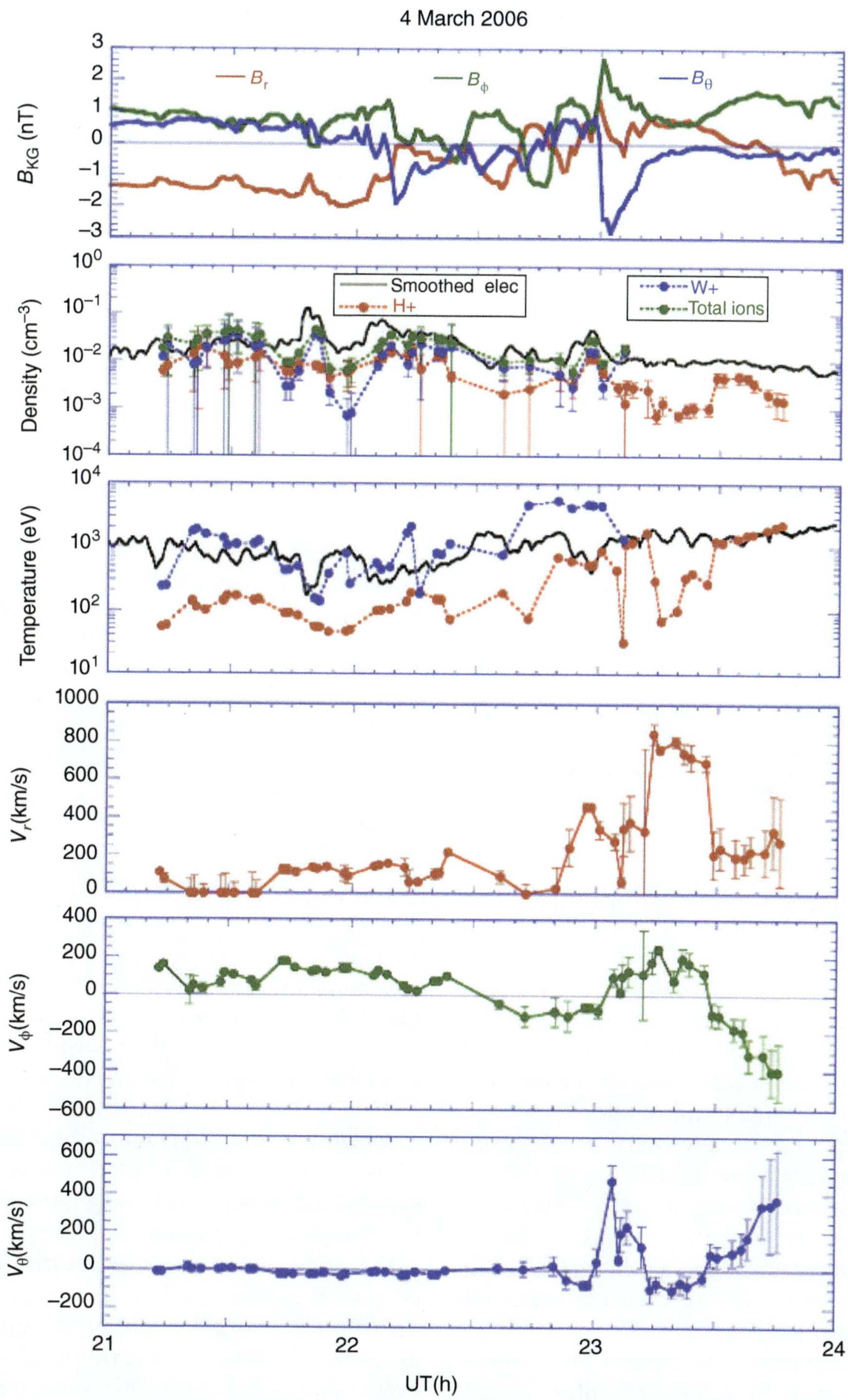

Figure 6.8 Example of plasmoid originally observed in the Cassini magnetometer data by *Jackman et al.* [2007]. Top panel shows magnetic field data, with lower panels showing number densities and temperatures of electrons, water group ions (W^+), and protons (H^+). The bottom three panels show the ion velocity components. Rigid corotation would imply $v\phi = 430$ km/s, off the top of the $v\phi$ scale. [From *Hill et al.*, 2008.]

it is highly asymmetric with a local maximum of ~0.3 nT and local minimum of ~−0.7 nT. We must consider this field signature in the light of the trajectory of the Cassini spacecraft during 2006 when these events were sampled. At this time the spacecraft was primarily orbiting in the equatorial plane and thus sampling the southern lobe beneath the hinged current sheet (discussed in section 6.2.1). Thus the spacecraft may only see the bottom part of any

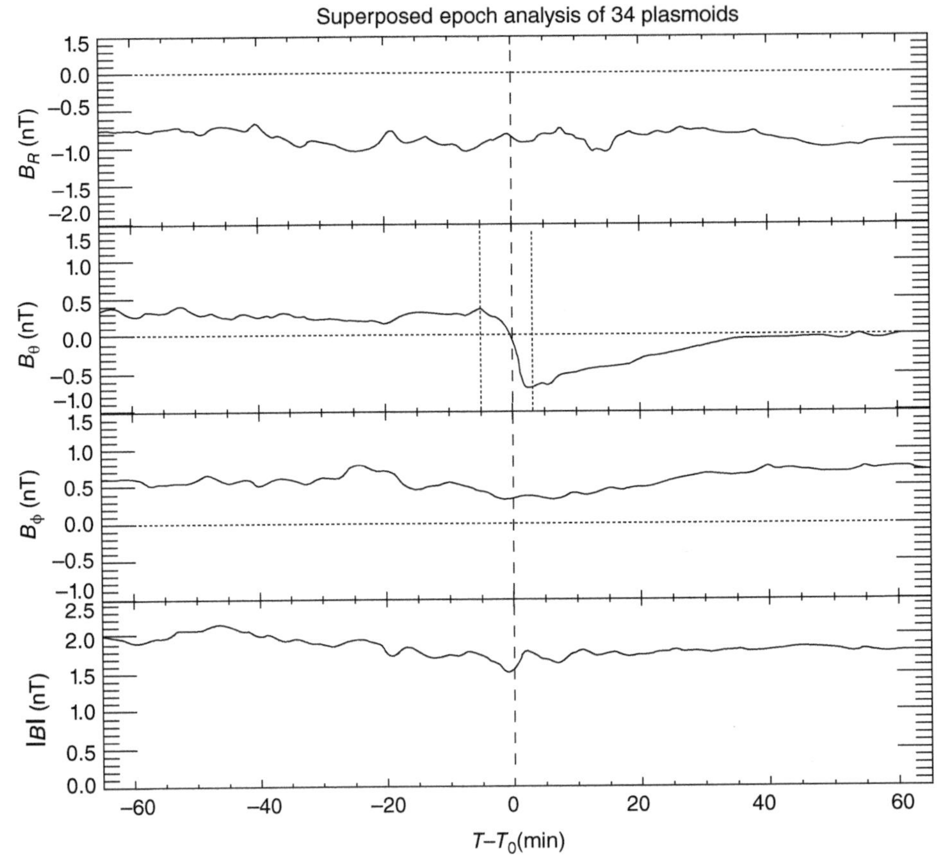

Figure 6.9 Superposed epoch analysis of 34 tailward moving plasmoids as sampled by the Cassini magnetometer. The zero epoch time is where B_θ goes through zero. [From *Jackman et al.*, 2011.]

plasmoids formed near the current sheet center (which would lead to a smaller field change than if the spacecraft passed through the middle of the structures). The asymmetry in the south-north field deflection can come about if the spacecraft passes through the plasmoids at an angle. Another striking feature of the southward-to-northward turning is that following the initial field deflection the field remains northward for an extended interval. This feature has been interpreted as the Kronian equivalent of the terrestrial "postplasmoid plasma sheet" (PPPS) [*Richardson et al.*, 1987], which represents an interval where previously open flux is being closed by reconnection.

The expected field signatures associated with reconnection on closed and open field lines are illustrated in Figure 6.10. In the top panel, the so-called initial, or steady, state, the B_θ component is small and positive. This is indeed what is typically observed in the "quiet" tail at Saturn [*Jackman and Arridge*, 2011b]. The second panel represents the case where reconnection is occurring on closed field lines via the Vasyliūnas cycle. In this case, the signature of a tailward moving plasmoid is a bipolar southward-to-northward change in B_θ. By contrast, the third panel illustrates the case where reconnection is ongoing on closed and open field lines. In this case, open magnetic flux that was stored in the lobes can be closed. The field signature then displays the characteristic extended interval of the northward field known as the PPPS.

Jackman et al. [2011] interpreted the extended interval of the northward field after plasmoid passage in the superposed epoch analysis (Figure 6.9) as representing the Kronian equivalent of a PPPS. They thus inferred that reconnection events in Saturn's tail can involve the closure of significant flux. By making some assumptions about the azimuthal extent of plasmoids at Saturn, they estimated the extent of the flux closure at ~3.01 GWb per event. If one compares this value with an average total tail flux content of ~35 GWb, based on auroral images at Saturn [*Badman et al.*, 2005, 2014], one can estimate that it would take ~12 such plasmoid-PPPS events to cycle all of the flux in Saturn's tail. Thus *Jackman et al.* [2011] concluded that when the Dungey cycle is active at Saturn, reconnection can be a significant driver of magnetospheric dynamics.

However, subsequent work has indicated that reconnection cannot account for the loss of all the material

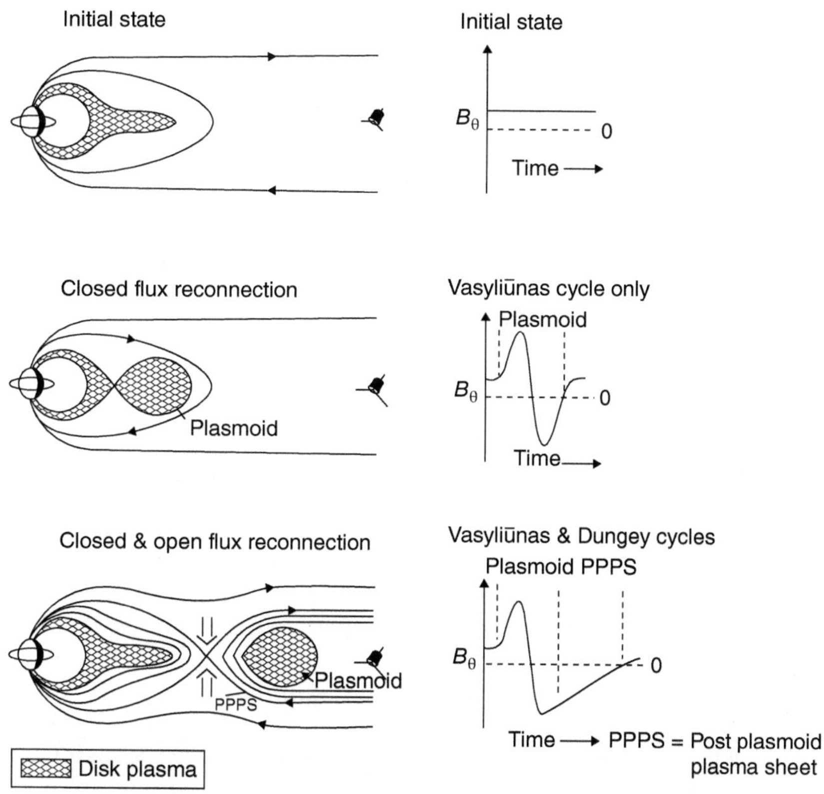

Figure 6.10 Schematic of three possible states of the magnetotail and the associated B_θ signatures as measured by a spacecraft situated down the tail. The Sun is to the left in these plots. [From *Jackman et al.*, 2011.]

that is added to Saturn's magnetosphere and hence that some other processes must be at work in addition to large-scale plasmoid release to remove plasma from Saturn's nightside. *Bagenal and Delamere* [2011] estimated the average mass of plasmoids at Saturn and compared this to the suggested mass loading rates at Saturn of 8–250 kg/s, primarily from the moon Enceladus [e.g., *Fleshman et al.*, 2010; *Jurac and Richardson*, 2005; *Pontius and Hill*, 2009; *Chen et al.*, 2010]. They assumed a plasmoid of volume $(10R_S)^3$ with a density of $0.01\,cm^{-3}$ of 18 amu ions. From this, they calculate that plasmoids would need to be ejected at a rate of 200 per day to remove just 100 kg/s. Such calculations come with larger errors, and thus a major target of future work is to refine these size and mass estimates through improved observation [e.g., *Jackman et al.*, 2014b]. Should it be confirmed that the number of observed plasmoids cannot account for the required mass loss from Saturn's magnetosphere, the next step may involve the search for other loss mechanisms, such as the smaller-scale diffusive-type processes proposed to operate on the dusk flank.

Another unanswered question regarding the nature of reconnection in Saturn's tail concerns the location of the reconnection site. *Jackman et al.* [2011] conducted a large survey of the magnetometer data from Cassini and produced a catalogue of 34 tailward moving plasmoids. Very little work has been published to date on the planetward moving products of reconnection (e.g., dipolarizations). *Bunce et al.* [2005], *Russell et al.* [2008], and *Jackman et al.* [2013] all reported northward-to-southward turnings of the field that they attribute to planetward moving dipolarizations of the field following reconnection events. However, it is clear that the overwhelming majority of large B_θ deflections observed in situ in the tail represent tailward moving plasmoids. Thus it has been impossible to derive a statistical separatrix between north-to-south and south-to-north events, as had been done, for example, for the case of Jupiter by *Vogt et al.* [2010]. Work is ongoing to extend the catalogue and to understand more about the spread of planetward moving and tailward moving events [*Jackman et al.*, 2014b]. *Mitchell et al.* [2005] (and subsequent expansions by *Hill et al.* [2008] and *Mitchell et al.* [2009]) approached this problem through analysis of data from the Cassini magnetospheric imaging instrument, which can remotely sense magnetospheric processes. They showed strong energetic neutral atom (ENA) emissions coming from the $20R_S$ to $30R_S$ region downtail, suggesting that this is close to the site of explosive energy release via reconnection. Attempts have also been made to model the reconnection process in

Saturn's tail. *Jia et al.* [2012], for example, suggested the average X-line position to be between ~25R_S and 40R_S, based on a magnetohydrodynamic global simulation of Saturn's magnetosphere. They noted that this location is highly sensitive to external solar wind conditions and other factors and may be highly mobile. The recurrence rate of the events themselves also remains somewhat of a mystery and will be the subject of intense future work.

6.4. REMOTE SENSING OF TAIL DYNAMICS

In addition to in situ exploration of Saturn's tail, planetary radio emissions and auroral images can be used as excellent remote diagnostics of tail dynamics. In the following sections we discuss Saturn's radio emissions and auroral images in more detail.

6.4.1. Radio Emissions

Saturn's radio emissions, in particular Saturn kilometric radiation (SKR), have been studied since the Voyager era [e.g., *Kaiser et al.*, 1980] and are an excellent proxy for magnetospheric conditions, often illuminating the response of the magnetosphere to the solar wind. Strong correlations have been found between SKR intensity and solar wind ram pressure [*Desch*, 1982] and solar wind momentum and kinetic energy [*Desch and Rucker*, 1983; *Jackman et al.*, 2005; *Taubenschuss et al.*, 2006; *Badman et al.*, 2008, *Rucker et al.*, 2008]. The period of the SKR emission (thought to be linked in some way to planetary rotation, but a subject of intense debate) is also sensitive to upstream solar wind conditions [*Zarka et al.*, 2007], varying systematically by ~1% with a characteristic time scale of ~20–30 days. However, it is the response of the SKR to dynamics in Saturn's magnetotail that make it an intriguing terrestrial analogue and also a potentially useful remote proxy of tail dynamics.

One of the first links between SKR and Saturn's magnetotail dynamics was made by *Bunce et al.* [2005], who showed that an intense burst of SKR was linked to an episode of solar wind compression-induced tail reconnection. Subsequently, *Jackman et al.* [2009] explored the radio data for nine tail reconnection events at Saturn. This comparison revealed a very similar frequency response to that observed with the auroral kilometric radiation (AKR) at terrestrial substorm onset [e.g., *Morioka et al.*, 2007]. In situ reconnection signatures were found to be associated with intensification and continuous extension to low frequencies of the SKR. The extension to lower frequencies is interpreted physically as a growth/expansion of the radio source region to higher altitudes along auroral field lines in response to precipitation of energetic particles into the auroral zone and the formation of a large potential drop, driven by the energetic reconnection process. In addition, *Louarn et al.* [2007] described distinctive signatures whereby the SKR exhibits a sudden intensification, followed by the detection of periodic narrowband radiation. They suggested that such a combination of radio signatures is linked to plasma evacuation from Saturn's plasma disk. Finally, enhancements in SKR coincident with ENA emission have been reported to be linked to several plasma acceleration events in the midnight-to-dawn quadrant at radial distance of 15R_S to 20R_S, related to tail reconnection [*Mitchell et al.*, 2009]. We do not cover ENA emission in detail in this chapter and refer the reader to Chapter 19.

Thus far, radio emissions have been shown to have the potential to be an extremely valuable tool in remotely sensing magnetospheric dynamics. More study is required to establish the exact nature of the correlation between the solar wind, tail dynamics, and SKR and narrowband emission such that radio emissions can be used as a robust remote proxy for magnetotail conditions in the absence of in situ spacecraft in the tail.

6.4.2. Aurora

Saturn's auroral emissions are spectacular and dynamic, exhibiting rapid changes in intensity and morphology. Here we discuss two main classes of auroral response to magnetospheric (and in particular magnetotail) dynamics: large-scale changes in the morphology of the main auroral oval and small-scale changes in the form of discrete auroral features adjacent to the main emission.

Large-scale changes in the aurora manifest themselves as changes in the size and intensity of the main auroral oval. The main auroral ring at Saturn is suggested to be produced by the magnetosphere–solar wind interaction, through the shear in rotational flow across the open-closed field line boundary [e.g., *Cowley and Bunce*, 2003; *Bunce et al.*, 2008]. Opening and closing of magnetic flux via dayside and magnetotail reconnection result in changes in the size of the polar cap, the region which contains the open flux. Figure 6.11 shows a schematic of the magnetospheric and ionospheric response to opening and closing of magnetic flux. There is a clear relationship between the size of the main auroral oval and the flux content of the magnetosphere. *Badman et al.* [2005, 2014] mapped these changes using images from the Hubble Space Telescope (HST), where they marked the inner boundary of the auroral emission and calculated the size of the dark polar cap region inside that, thus calculating the flux content of the tail lobes whose field lines map to this region. They found the typical flux content to be ~10–50 GWb, with an average of 35 GWb. The typical change between consecutive observations was found to be ~5 GWb (based on the January 2004 HST campaign, which consisted of 68 images during 13 "visits" over 23 days

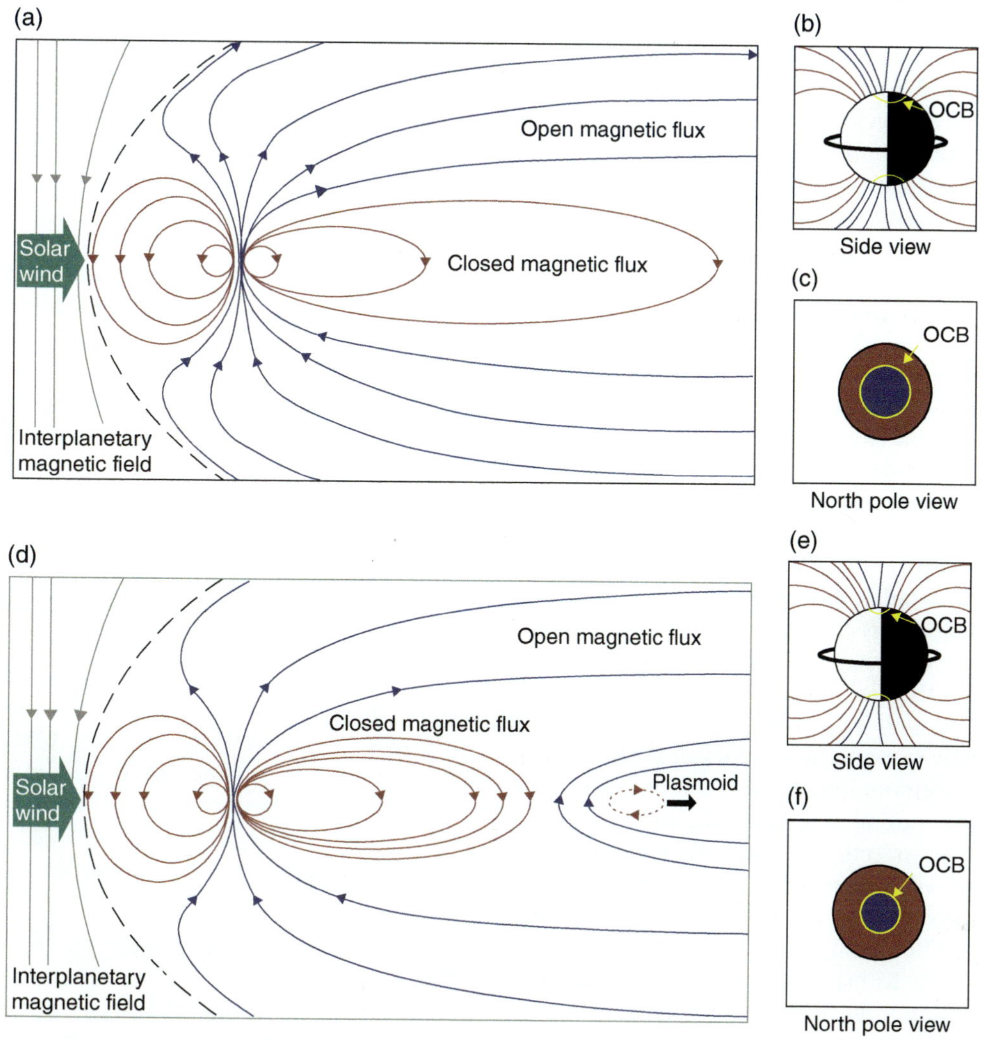

Figure 6.11 Schematic of Saturn's open magnetosphere (top panels) during "steady state" and (bottom panels) following a large tail reconnection event that closes some of the open flux that had previously been stored in the tail lobes: (a) view of the magnetosphere; (b) location of the open-closed field line boundary (OCB) in the ionosphere; (c) polar view of the OCB. Panels (d), (e), and (f) are in the same format as (a), (b), and (c), but for the case following reconnection. The Sun is to the left in these plots. [From *Badman et al.*, 2014.]

and consecutive observations at least every ~2 days). While it is impossible from this method to separate the contributions of flux opening at the dayside and concurrent flux closure at the nightside, this number is of the same order as the ~3 GWb estimated by *Jackman et al.* [2011] to be closed via a large tail reconnection event with a significant PPPS.

In addition to large-scale changes in the size of Saturn's main auroral oval, magnetotail dynamics can also result in smaller-scale auroral changes. Small discrete spots of auroral emission lying postmidnight and just poleward of the main oval emission were reported by *Jackman et al.* [2013]. These spots, akin to the "polar dawn spots" reported for Jupiter by *Grodent et al.* [2004] and *Radioti et al.* [2008], were postulated to be linked to dipolarizations of the field in the tail following reconnection, which in turn results in a diversion of the cross-tail current, and the generation of field-aligned currents which flow toward and away from the polar region. It was estimated that, following a strong dipolarization at Saturn, such currents are strong enough to produce detectable, discrete spots of auroral emission. These auroral features were also suggested to be the precursor to more intense activity associated with recurrent energization via particle injections from the tail following reconnection [*Mitchell et al.*, 2009].

A final small-scale auroral emission is that linked with injected plasma from the magnetotail and takes the form of a bright region of auroral emission that grows and elongates with time. *Radioti et al.* [2013] used simultaneous observations of UV aurora (from the Cassini UVIS instrument) and ENA emissions together with simulations of magnetospheric behavior to describe the evolution of an injected population, possibly related to tail reconnection as well as its auroral counterpart. They infer that pitch angle diffusion and electron scattering may not be the only mechanisms producing this injection-related auroral emission and that field-aligned currents driven by pressure gradients along the boundaries of the injected hot plasma may also play an important role.

6.5. DISCUSSION

In this chapter we reviewed some of the key advances in the physics of Saturn's magnetotail since the first spacecraft reached the planet in the late 1970s. The large-scale structure of the tail has been probed, and many of the properties of the current sheet, plasma sheet, and lobes have been uncovered through multi-instrument studies. The competition between solar-wind-driven and internally driven flows has been explored and identified as an important topic in studies of Saturn's magnetosphere. Reconnection has been shown to be a very dynamic process in Saturn's tail, reconfiguring the field, energizing the plasma, and changing the magnetic flux content of the tail. The effects of tail dynamics, in particular, energetic reconnection events, are not confined to the local area. Remote sensing of radio emissions and images of Saturn's aurora in a multitude of wavelengths show the far-reaching effects of tail reconfigurations.

The Pioneer and Voyager spacecraft gave a mere glimpse of the Kronian system and left many tantalizing unanswered questions. The Cassini mission thus far has been a huge success, and in terms of magnetotail physics, our exploration has resulted in many new insights. However, we also now have a large new set of unanswered questions, some of which can be tackled during the remainder of the Cassini tour and some of which undoubtedly will require further in situ exploration by future spacecraft, in conjunction with remote sensing, and more advanced modeling. Some open questions below are as follow:

• What is the relative role of solar-wind-driven (Dungey cycle) versus internally driven (Vasyliūnas cycle) dynamics at Saturn?

• What is the recurrence rate of reconnection at Saturn? Models and observations disagree on the frequency of plasmoid ejection. Should we expect a link between the planetary rotation period and the frequency of energy release events on the nightside, and if so, how is this relationship communicated from the planet to the tail?

• What is the nature of Saturn's deep tail? How do plasmoids evolve with increasing distance? How is the planet able to communicate periodicities to large distances? Are there multiple X lines (as one might expect given analogy with the terrestrial tail)?

• What is the nature of mass balance at Saturn? Given the apparent imbalance between inferred mass loading and observed mass loss via plasmoids, how is this extra mass lost and from where?

• What is the nature of plasma flow at Saturn? The limited observations thus far indicate some evidence of outward/inward/azimuthal flow, but the picture is incomplete. How can magnetometer observations of plasmoid loss and field dipolarization be reconciled with plasma flow observations, and how strong of an influence does corotation have in the tail?

• What is the global effect of magnetotail reconnection? How can we improve our understanding of radio emissions and auroral images such that they can be used as a remote proxy of tail dynamics?

We now look to the future for more pioneering exploration of Saturn's complex and dynamic magnetosphere in anticipation of new answers and of many new questions.

ACKNOWLEDGMENTS

CMJ's work at UCL was funded through a Leverhulme Trust Early Career Fellowship and a Royal Astronomical Society Fellowship. CMJ's work at Southampton was funded through a Royal Astronomical Society Fellowship. CMJ acknowledges Nick Achilleos for assistance and reviewers for helpful suggestions. CMJ acknowledges lively discussion at the AGU Chapman meeting on "Fundamental Properties and Processes of Planetary Magnetotails" in Reykjavik in March 2013 for inspiration for many of the topics covered in this review.

REFERENCES

Achilleos, N., C. S. Arridge, C. Bertucci, C. M. Jackman, M. K. Dougherty, K. K. Khurana, and C. T. Russell (2008), Large-scale dynamics of Saturn's magnetopause: Observations by Cassini, *J. Geophys. Res.*, *113*, A11209, doi:10.1029/2008JA013265.

Arridge, C. S., N. Achilleos, M. K. Dougherty, K. K. Khurana, and C. T. Russell (2006), Modeling the size and shape of Saturn's magnetopause with variable dynamic pressure, *J. Geophys. Res.*, *111*, A11227, doi:10.1029/2005JA011574.

Arridge, C. S., K. K. Khurana, C. T. Russell, D. J. Southwood, N. Achilleos, M. K. Dougherty, A. J. Coates, and H. K. Leinweber (2008a), Warping of Saturn's magnetospheric and magnetotail current sheets, *J. Geophys. Res.*, *113*, A08217, doi:10.1029/2007JA012963.

Arridge, C. S., N. André, N. Achilleos, K. K. Khurana, C. L. Bertucci, L. K. Gilbert, G. R. Lewis, A. J. Coates, and M. K. Dougherty (2008b), Thermal electron periodicities at 20 RS in Saturn's magnetosphere, *Geophys. Res. Lett.*, *35*, L15107, doi:10.1029/2008GL034132.

Arridge, C. S., et al. (2009), Plasma electrons in Saturn's magnetotail: Structure, distribution and energisation, *Planet. Space Sci.*, *57*(14–15), 2032–2047, doi:10.1016/j.pss.2009.09.007.

Arridge, C. S., et al. (2011), Periodic motion of Saturn's nightside plasma sheet, *J. Geophys. Res.*, *116*, A11205, doi: 10.1029/2011JA016827.

Arridge, C. S., et al. (2012), Mapping magnetospheric equatorial regions at Saturn from Cassini Prime mission equatorial observations, *Space Sci. Rev. (2011)*, *164*, 1–83, doi:10.1007/s11214-011-9850-4

Badman, S. V., E. J. Bunce, J. T. Clarke, S. W. H. Cowley, J.-C. Gérard, D. Grodent, and S. E. Milan (2005), Open flux estimates in Saturn's magnetosphere during the January 2004 Cassini-HST campaign, and implications for reconnection rates, *J. Geophys. Res.*, *110*, A11216, doi:10.1029/2005JA011240.

Badman, S. V., and S.W.H. Cowley (2007), Significance of Dungey-cycle flows in Jupiter's and Saturn's magnetospheres, and their identification on closed equatorial field lines. *Ann. Geophys.*, *25*, 941–951

Badman, S. V., S. W. H. Cowley, L. Lamy, B. Cecconi, and P. Zarka (2008), Relationship between solar wind corotating interaction regions and the phasing and intensity of Saturn kilometric radiation bursts, *Ann. Geophys.*, *26*, 3641–3651.

Badman, S. V., C. M. Jackman, J. D. Nichols, J. T. Clarke, and J.-C. Gérard (2014), Open flux in Saturn's magnetosphere, *Icarus*, 231, 137–145, http://dx.doi.org/10.1016/j.icarus.2013.12.004.

Bagenal, F. (1992), Giant planet magnetospheres, *Annu. Rev. Earth Planet. Sci.*, *20*, 289–328, doi:10.1146/annurev.ea.20.050192.001445.

Bagenal, F. (2009), Comparative planetary environments, in *Heliophysics: Plasma Physics of the Local Cosmos*, edited by C. J. Schrijver and G. L. Siscoe, pp. 360–398. Cambridge Univ. Press, Cambridge, United Kingdom.

Bagenal, F., and P. A. Delamere (2011), Flow of mass and energy in the magnetospheres of Jupiter and Saturn, *J. Geophys. Res.*, *116*, A05209, doi:10.1029/2010JA016294.

Behannon, K. W., J. E. P. Connerney, and N. F. Ness (1981), Saturn's magnetic tail: Structure and dynamics, *Nature*, *292*, 753.

Bunce, E. J., S. W. H. Cowley, D. M. Wright, A. J. Coates, M. K. Dougherty, N. Krupp, W. S. Kurth, and A. M. Rymer (2005), In situ observations of a solar wind compression-induced hot plasma injection in Saturn's tail, *Geophys. Res. Lett.*, *32*, L20S04, doi:10.1029/2005GL022888.

Bunce, E. J., et al. (2008), Origin of Saturn's aurora: Simultaneous observations by Cassini and the Hubble Space Telescope, *J. Geophys. Res.*, *113*, A09209, doi:10.1029/2008JA013257.

Chen, Y., T. W. Hill, A. M. Rymer, and R. J. Wilson (2010), Rate of radial transport of plasma in Saturn's inner magnetosphere, *J. Geophys. Res.*, *115*, A10211, doi:10.1029/2010JA015412.

Connerney, J. E. P., M. H. Acuna, and N. F. Ness (1981), Saturn's ring current and inner magnetosphere, *Nature*, *292*, 724–726.

Connerney, J.E.P., M.H. Acuña, and N.F. Ness (1983), Currents in Saturn's magnetosphere. *J. Geophys. Res*. *88*(A11), 8779–8789, doi:10.1029/JA088iA11p08779.

Coroniti, F. V., and C. F. Kennel (1972), Changes in magnetospheric configuration during the substorm growth phase, *J. Geophys. Res.*, *77*(19), 3361–3370.

Cowley, S. W. H., and E. J. Bunce (2003), Corotation-driven magnetosphere-ionosphere coupling currents in Saturn's magnetosphere and their relation to the auroras, *Ann. Geophys.*, *21*, 1691–1707.

Cowley, S. W. H., E. J. Bunce, and J. M. O'Rourke (2004), A simple quantitative model of plasma flows and currents in Saturn's polar ionosphere, *J. Geophys. Res.*, *109*, A05212, doi:10.1029/2003JA010375.

Cowley, S. W. H., S. V. Badman, E. J. Bunce, J. T. Clarke, J.-C. Gérard, D. Grodent, C. M. Jackman, S. E. Milan, and T. K. Yeoman (2005), Reconnection in a rotation-dominated magnetosphere and its relation to Saturn's auroral dynamics, *J. Geophys. Res.*, *110*, A02201, doi:10.1029/2004JA010796.

Delamere, P. A., R. J. Wilson, S. Eriksson, and F. Bagenal (2013), Magnetic signatures of Kelvin-Helmholtz vortices on Saturn's magnetopause: Global survey, *J. Geophys. Res.*, *118*, 393, 404, doi:10.1029/2012JA018197.

Desch, M. D. (1982), Evidence for solar wind control of Saturn radio emission, *J. Geophys. Res.*, *87*, 4549–4554.

Desch, M. D., and H. O. Rucker (1983), The relationship between Saturn kilometric radiation and the solar wind, *J. Geophys. Res.*, *88*, 8999–9006.

Dougherty, M. K., et al. (2004), The Cassini magnetic field investigation, *Space Sci. Rev.*, *114*, 331.

Dungey, J. W. (1961), Interplanetary magnetic field and the auroral zones, *Phys. Rev. Lett.*, *6*, 47–48, doi:10.1103/PhysRevLett.6.47.

Dungey, J. W. (1965), The length of the magnetospheric tail, *J. Geophys. Res.*, *70*, 1753.

Fairfield, D. H., and J. Jones (1996), Variability of the tail lobe field strength, *J. Geophys. Res.*, *101*, 7785–7791, doi:10.1029/95JA03713.

Fleshman, B. L., P. A. Delamere, and F. Bagenal (2010), The source of Saturn's extended neutral cloud, Abstract SM11C-1768, presented at 2010 Fall Meet., AGU, San Francisco, Calif., 13–17 December

Frank, L. A., B. G. Burek, K. L. Ackerson, J. H. Wolfe, and J. D. Mihalov (1980), Plasmas in Saturn's magnetosphere, *J. Geophys. Res.*, *85*(A11), 5695–5708, doi:10.1029/JA085iA11p05695.

Grocott, A., S. V. Badman, S. W. H. Cowley, S. E. Milan, J. D. Nichols, and T. K. Yeoman (2009), Magnetosonic Mach number dependence of the efficiency of reconnection between planetary and interplanetary magnetic fields, *J. Geophys. Res.*, *114*, A07219, doi:10.1029/2009JA014330.

Grodent, D., J.-C. Gérard, J. T. Clarke, G. R. Gladstone, and J. H. Waite, Jr. (2004), A possible auroral signature of a magnetotail reconnection process on Jupiter, *J. Geophys. Res.*, *109*, A05201, doi:10.1029/2003JA010341.

Gurnett, D. A., et al. (2004), The Cassini radio and plasma wave investigation, *Space Sci. Rev.*, *114*(1–4), 395–463.

Hill, T. W., and F. C. Michel (1976), Heavy ions from the Galilean satellites and the centrifugal distortion of the Jovian magnetosphere, *J. Geophys. Res.*, *81*, 4561.

Hill, T. W., A. J. Dessler, and F. C. Michel (1974), Configuration of the Jovian magnetosphere, *Geophys. Res. Lett.*, *1*, 3–6. doi:10.1029/GL001i001p00003.

Hill, T. W., et al. (2008), Plasmoids in Saturn's magnetotail, *J. Geophys. Res.*, *113*, A01214, doi:10.1029/2007JA012626.

Hones, E. W., Jr. (1976), The magnetotail: Its generation and dissipation, in *Physics of Solar Planetary Environments*, edited by D. J. Williams, pp. 559–571, AGU, Washington, D. C.

Hones, E. W., Jr. (1977), Substorm processes in the magnetotail: Comments on "On hot tenuous plasma, fireballs, and boundary layers in the Earth's magnetotail" by L. A. Frank et al., *J. Geophys. Res.*, *82*, 5633.

Isbell, J., A. J. Dessler, and J. H. Waite, Jr. (1984), Magnetospheric energization by interaction between planetary spin and the solar wind, *J. Geophys. Res.*, *89*, 10,716–10,722.

Jackman, C. M., and C. S. Arridge (2011a), Solar cycle effects on the dynamics of Jupiter's and Saturn's magnetospheres, *Solar Phys.*, *274*, 481–502, doi:10.1007/s11207-011-9748-z.

Jackman, C. M., and C. S. Arridge (2011b), Statistical properties of the magnetic field in the Kronian magnetotail lobes and current sheet, *J. Geophys. Res.*, *116*, A05224, doi:10.1029/2010JA015973.

Jackman, C. M., N. Achilleos, E. J. Bunce, S. W. H. Cowley, M. K. Dougherty, G. H. Jones, S. E. Milan, and E. J. Smith (2004), Interplanetary magnetic field at ~9 AU during the declining phase of the solar cycle and its implications for Saturn's magnetospheric dynamics, *J. Geophys. Res.*, *109*, 11,203, doi:10.1029/2004JA010614.

Jackman, C. M., N. Achilleos, E. J. Bunce, B. Cecconi, J. T. Clarke, S. W. H. Cowley, W. S. Kurth, and P. Zarka (2005), Interplanetary conditions and magnetospheric dynamics during the Cassini orbit insertion fly-through of Saturn's magnetosphere, *J. Geophys. Res.*, *110*, A10212, doi:10.1029/2005JA011054.

Jackman, C. M., C. T. Russell, D. J. Southwood, C. S. Arridge, N. Achilleos, and M. K. Dougherty (2007), Strong rapid dipolarizations in Saturn's magnetotail: In situ evidence of reconnection, *Geophys. Res. Lett.*, *34*(11), Art. No. L11203.

Jackman, C. M., R. J. Forsyth, and M. K. Dougherty (2008a), The overall configuration of the interplanetary magnetic field upstream of Saturn as revealed by Cassini observations, *J. Geophys. Res.*, *113*, A08114, doi:10.1029/2008JA013083.

Jackman, C. M., et al. (2008b), A multi-instrument view of tail reconnection at Saturn, *J. Geophys. Res.*, *113*, A11213, doi:10.1029/2008JA013592.

Jackman, C. M., L. Lamy, M. P. Freeman, P. Zarka, B. Cecconi, W. S. Kurth, S. W. H. Cowley, and M. K. Dougherty (2009), On the character and distribution of lower-frequency radio emissions at Saturn, and their relationship to substorm-like events, *J. Geophys. Res.*, *114*, A08211, doi:10.1029/2008JA013997.

Jackman, C. M., J. A. Slavin, and S. W. H. Cowley (2011), Cassini observations of plasmoid structure and dynamics: Implications for the role of magnetic reconnection in magnetospheric circulation at Saturn, *J. Geophys. Res.*, *116*, A10212, doi:10.1029/2011JA016682.

Jackman, C. M., N. Achilleos, S. W. H. Cowley, E. J. Bunce, A. Radioti, D. Grodent, S. V. Badman, M. K. Dougherty, and W. Pryor (2013), Auroral counterpart of magnetic field dipolarizations in Saturn's tail, *Planet. Space Sci.*, *82–83*, 34–42, doi:10.1016/j.pss.2013.03.010

Jackman, C. M., et al. (2014a), Large-scale structure and dynamics of the magnetotails of Mercury, Earth, Jupiter and Saturn, *Space Sci. Rev.*, 182(1), 85–154, doi:10.1007/s11214-014-0060-8.

Jackman, C. M., et al. (2014b), Saturn's dynamic magnetotail: A comprehensive magnetic field and plasma survey of plasmoids and travelling compression regions, and their role in global magnetospheric dynamics, *J. Geophys. Res. Space Physics*, *119*, 5465–5494, doi:10.1002/2013JA019388.

Jia, X., K. C. Hansen, T. I. Gombosi, M. G. Kivelson, G. Tóth, D. L. DeZeeuw, and A. J. Ridley (2012), Magnetospheric configuration and dynamics of Saturn's magnetosphere: A global MHD simulation, *J. Geophys. Res.*, *117*, A05225, doi:10.1029/2012JA017575.

Jurac, S., and J. D. Richardson (2005), A self-consistent model of plasma and neutrals at Saturn: Neutral cloud morphology, *J. Geophys. Res.*, *110*, A09220, doi:10.1029/2004JA010635.

Kaiser, M. L., M. D. Desch, J. W. Warwick, and J. B. Pearce (1980), Voyager detection of nonthermal radio emission from Saturn, *Science*, *209*, 1238–1240.

Kanani, S. J., et al. (2010), A new form of Saturn's magnetopause using a dynamic pressure balance model, based on in situ, multi-instrument Cassini measurements, *J. Geophys. Res.*, *115*, A06207, doi:10.1029/2009JA014262.

Khurana, K. K., D. G. Mitchell, C. S. Arridge, M. K. Dougherty, C. T. Russell, C. Paranicas, N. Krupp, and A. J. Coates (2009), Sources of rotational signals in Saturn's magnetosphere, *J. Geophys. Res.*, *114*, A02211, doi:10.1029/2008JA013312.

Kivelson, M. G., (2007), Planetary magnetospheres, in *Solar-Terrestrial Environment*, edited by Y. Kamide and C.-L. Chian, pp. 470–492, Springer Verlag, New York.

Kivelson, M. G., and K. K. Khurana (2002), Properties of the magnetic field in the Jovian magnetotail, *J. Geophys. Res.*, *107*(A8), 1196, doi:10.1029/2001JA000249.

Krimigis, S. M. et al. (2004), Magnetosphere imaging instrument (MIMI) on the Cassini mission to Saturn/Titan, *Space Sci. Rev.*, *114*(1–4), 233–329.

Krupp, N., et al. (2005), The Saturnian plasma sheet as revealed by energetic particle measurements, *Geophys. Res. Lett.*, *32*, L20S03, doi:10.1029/2005GL022829.

Kurth, W. S., J. D. Sullivan, D. A. Gurnett, F. L. Scarf, H. S. Bridge, and E. C. Sittler, Jr. (1982), Observations of Jupiter's distant magnetotail and wake, *J. Geophys. Res.*, *87*(A12), 10,373–10,383, doi:10.1029/JA087iA12p10373.

Lai, H. R., H. Y. Wei, C. T. Russell, C. S. Arridge, and M. K. Dougherty (2012), Reconnection at the magnetopause of Saturn: Perspective from FTE occurrence and magnetosphere size, *J. Geophys. Res.*, *117*, A05222, doi:10.1029/2011JA017263.

Lepping, R. P., M. D. Desch, L. W. Klein, E. C. Sittler, Jr., J. D. Sullivan, W. S. Kurth, and K. W. Behannon (1983), Structure and other properties of Jupiter's distant magnetotail, *J. Geophys. Res.*, *88*(A11), 8801–8815, doi:10.1029/JA088iA11p08801.

Louarn, P., et al. (2007), Observation of similar radio signatures at Saturn and Jupiter: Implications for the magnetospheric dynamics, *Geophys. Res. Lett.*, *34*, L20113, doi:10.1029/2007GL030368.

Masters, A., M. F. Thomsen, S. V. Badman, C. S. Arridge, D. T. Young, A. J. Coates, and M. K. Dougherty (2011), Supercorotating return flow from reconnection in Saturn's magnetotail, *Geophys. Res. Lett.*, *38*, L03103, doi:10.1029/2010GL046149.

Masters, A., J. P. Eastwood, M. Swisdak, M. F. Thomsen, C. T. Russell, N. Sergis, F. J. Crary, M. K. Dougherty, A. J. Coates, and S. M. Krimigis (2012), The importance of plasma β conditions for magnetic reconnection at Saturn's magnetopause, *Geophys. Res. Lett.*, *39*, L08103, doi:10.1029/2012GL051372.

Mauk, B. H., S. M. Krimigis, and R. P. Lepping (1985), Particle and field stress balance within a planetary magnetosphere, *J. Geophys. Res.*, *90*, 8253–8264.

McAndrews, H. J., C. J. Owen, M. Thomsen, B. Lavraud, A. Coates, M. K. Dougherty, and D. T. Young (2008), Evidence for reconnection at Saturn's magnetopause, *J. Geophys. Res.*, *113*, A04210, doi:10.1029/2007JA012581.

McAndrews, H. J. et al. (2009), Plasma in Saturn's nightside magnetosphere and the implications for global circulation, *Planet. Space Sci.*, *57* (14–15), 1714–1722, doi:10.1016/j.pss.2009.03.003.

McComas, D. J., F. Allegrini, F. Bagenal, F. Crary, R. W. Ebert, H. Elliott, A. Stern, and P. Valek (2007), Diverse plasma populations and structures in Jupiter's magnetotail, *Science*, *318*, doi:10.1126/science.1147393.

McNutt, R. L. Jr., et al. (2007), Energetic particles in the Jovian magnetotail, *Science*, *318*, 220.

Michel, F. C., and P. A. Sturrock (1974), Centrifugal instability of the Jovian magnetosphere and its interaction with the solar wind, *Planet. Space Sci.*, *22*, 1501.

Milan, S. E. (2004), A simple model of the flux content of the distant magnetotail, *J. Geophys. Res.*, *109*, A07210, doi:10.1029/2004JA010397.

Milan, S. E., E. J. Bunce, S. W. H. Cowley, and C. M. Jackman (2005), Implications of rapid planetary rotation for the Dungey magnetotail of Saturn, *J. Geophys. Res.*, *110*, A03209, doi:10.1029/2004JA010716.

Mitchell, D. G., et al. (2009), Recurrent energization of plasma in the midnight to-dawn quadrant of Saturn's magnetosphere, and its relationship to auroral UV and radio emissions, *Planet. Space Sci.*, *57*, 1732–1742, doi:10.1016/j.pss.2009.04.002.

Morioka, A., Y. Miyoshi, F. Tsuchiya, H. Misawa, T. Sakanoi, K. Yumoto, R. R. Anderson, J. D. Menietti, and E. F. Donovan (2007), Dual structure of auroral acceleration regions at substorm onsets as derived from AKR spectra, *J. Geophys. Res.*, *112*, A06245, doi:10.1029/2006JA012186.

Ness, N. F., M. H. Acuña, R. P. Lepping, J. E. P. Connerney, K. W. Behannon, L. F. Burlaga, and F. M. Neubauer (1981), Magnetic field studies by Voyager 1: Preliminary results at Saturn, *Science*, *212*, 211.

Pontius, D., and T. W. Hill (2009), Inertial corotation lag and mass loading in Saturn's magnetosphere, *Geophys. Res. Lett.*, *36*, L23103, doi:10.1029/2009GL041030.

Provan, G., D. J. Andrews, C. S. Arridge, A. J. Coates, S. W. H. Cowley, G. Cox, M. K. Dougherty, and C. M. Jackman (2012), Dual periodicities in planetary-period magnetic field oscillations in Saturn's tail, *J. Geophys. Res.*, *117*, A01209, doi:10.1029/2011JA017104.

Radioti, A., D. Grodent, J.-C. Gérard, B. Bonfond, and J. T. Clarke (2008), Auroral polar dawn spots: Signatures of internally driven reconnection processes at Jupiter's magnetotail, *Geophys. Res. Lett.*, *35*, L03104, doi:10.1029/2007GL032460.

Radioti, A., E. Roussos, D. Grodent, J.-C. Gérard, N. Krupp, D. G. Mitchell, J. Gustin, B. Bonfond, and W. Pryor (2013), Signatures of magnetospheric injections in Saturn's aurora, *J. Geophys. Res.*, *118*, 1922–1933, doi:10.1002/jgra.50161.

Richardson, I. G., S. W. H. Cowley, E. W. Hones, Jr., and S. J. Bame (1987), Plasmoid-associated energetic ion bursts in the deep geomagnetic tail: Properties of plasmoids and the post-plasmoid plasma sheet, *J. Geophys. Res.*, *92*, 9997–10,013, doi:10.1029/JA092iA09p09997.

Richardson, J. D. (1998), Thermal plasma and neutral gas in Saturn's magnetosphere, *Rev. Geophys.*, *36*(4), 501–524, doi:10.1029/98RG01691.

Rucker, H. O., et al. (2008), Saturn kilometric radiation as a monitor for the solar wind?, *Adv. Space Res.*, *42*, 40–47.

Russell, C. T. (2001), The dynamics of planetary magnetospheres, *Planet. Space Sci.*, *49*, doi:10.1016/S0032-0633(01)00017-4.

Russell, C. T., C. M. Jackman, H. Y. Wei, C. Bertucci, and M. K. Dougherty (2008), Titan's influence on Saturnian substorm occurrence, *Geophys. Res. Lett.*, *35*, L12105, doi:10.1029/2008GL034080.

Scarf, F. L., D. A. Gurnett, W. S. Kurth, and R. L. Poynter, (1982), Voyager 2 plasma wave observations at Saturn, *Science*, *215*, 587.

Scurry, L., and C. T. Russell (1991), Proxy studies of energy transfer to the magnetosphere, *J. Geophys. Res.*, *96*(A6), 9541–9548, doi:10.1029/91JA00569.

Sergis, N., C. S. Arridge, S. M. Krimigis, D. G. Mitchell, A. M. Rymer, D. C. Hamilton, N. Krupp, M. K. Dougherty, and A. J. Coates (2011), Dynamics and seasonal variations in Saturn's magnetospheric plasma sheet, as measured by Cassini, *J. Geophys. Res.*, *116*, A04203, doi:10.1029/2010JA016180.

Simon, S., et al. (2010), Dynamics of Saturn's magnetodisk near Titan's orbit: Comparison of Cassini magnetometer observations from real and virtual Titan flybys, *Planet. Space Sci.*, *58*, 1625–1635.

Slavin, J. A., E. J. Smith, B. T. Tsurutani, D. G. Sibeck, H. J. Singer, D. N. Baker, J. T. Gosling, E. W. Hones, Jr., and F. L. Scarf (1984), Substorm-associated traveling compression regions in the distant tail: ISEE 3 geotail observations, *Geophys. Res. Lett.*, *11*, 657.

Slavin, J. A., et al. (1989), CDAW-8 observations of plasmoid signatures in the geomagnetic tail: An assessment, *J. Geophys. Res.*, *94*, 15,153.

Slavin, J. A., M. F. Smith, E. L. Mazur, D. N. Baker, E. W. Hones, Jr., T. Iyemori, and E. W. Greenstadt (1993), ISEE 3 observations of traveling compression regions in the Earth's magnetotail, *J. Geophys. Res.*, *98*(A9), 15,425–15,446, doi:10.1029/93JA01467.

Taubenschuss, U., H. O. Rucker, W. S. Kurth, B. Cecconi, P. Zarka, M. K. Dougherty, and J. T. Steinberg (2006), Linear prediction studies for the solar wind and Saturn kilometric radiation, *Ann. Geophys.*, *24*, 3139–3150.

Thomsen, M. F., D. B. Reisenfeld, D. M. Delapp, R. L. Tokar, D. T. Young, F. J. Crary, E. C. Sittler, M. A. McGraw, and J. D. Williams (2010), Survey of ion plasma parameters in Saturn's magnetosphere, *J. Geophys. Res.*, *115*, A10220, doi:10.1029/2010JA015267.

Thomsen, M. F., R. J. Wilson, R. L. Tokar, D. B. Reisenfeld, and C. M. Jackman (2013), Cassini/CAPS observations of duskside tail dynamics at Saturn, *J. Geophys. Res.*, *118*, 5767–5781, doi:10.1002/jgra.50552.

Tsyganenko, N. A., and D. H. Fairfield (2004), Global shape of the magnetotail current sheet as derived from Geotail and Polar data, *J. Geophys. Res.*, *109*, A03218, doi:10.1029/2003JA010062.

Vasyliunas, V. M. (1983), Plasma distribution and flow, in *Physics of the Jovian Magnetosphere*, edited by A.J. Dessler, 395 pp., Cambridge Univ. Press, New York.

Vogt, M. F., M. G. Kivelson, K. K. Khurana, S. P. Joy, and R. J. Walker (2010), Reconnection and flows in the Jovian magnetotail as inferred from magnetometer observations, *J. Geophys. Res.*, *115*, A06219, doi:10.1029/2009JA015098.

Volwerk, M., et al. (2013), Comparative magnetotail flapping: an overview of selected events at Earth, Jupiter and Saturn, *Ann. Geophys.*, *31*, 817–833, doi:10.5194/angeo-31-817-2013.

Young, D. T., et al. (2004), Cassini plasma spectrometer investigation, *Space Sci. Rev.* *114*(1–4), 1–112.

Zarka, P., L. Lamy, B. Cecconi, R. Prangé, and H. O. Rucker (2007), Modulation of Saturn's radio clock by solar wind speed, *Nature*, *450*, 265–267.

7
Magnetotails of Uranus and Neptune

C. S. Arridge

7.1. INTRODUCTION

Uranus and Neptune were explored in situ with flybys of the Voyager 2 spacecraft in 1986 and 1989, respectively, with the closest approach to Uranus on 24 January 1986 and to Neptune on 25 August 1989. These brief snapshots of the Uranian and Neptunian magnetospheres revealed complex internal magnetic fields, with significantly offset and tilted configurations, and highly variable signatures as Voyager 2 observed a number of diurnal cycles in each magnetosphere. The Voyager 2 Uranus flyby results can be found in *Journal of Geophysical Research: Space Physics*, volume 92, issue A13, pages 14,873–15,375 (December 1987), and specifically in papers by *Behannon et al.* (1987), *Mauk et al.* [1987], *McNutt et al.* [1987], *Voigt et al.* [1987], and *Kane et al.* [1991] and for the Neptune flyby in *Journal of Geophysical Research: Space Physics*, volume 96, issue S01, pages 18,903–19,268 (October 1991), and specifically in papers by *Belcher et al.* [1989], *Ness et al.* [1989], *Mauk et al.* [1991], and *Schulz et al.* [1995]. We also direct the reader to the book volumes written after the Voyager 2 encounters with Uranus [*Matthews et al.* 1991] and Neptune [*Cruikshank*, 1995] and the last review of the magnetotail configuration at Uranus and Neptune by *Lepping* [1994].

More than 20 years have now elapsed since these flybys during which time studies of other magnetotails in the solar system have significantly progressed our understanding of magnetotail physics. What new knowledge can we now apply to the data gathered by Voyager 2? Studies of the solar system magnetospheres have demonstrated, for example, a new understanding of reconnection physics and its dependence on the plasma beta, the presence of waves in the magnetotails, and how internal magnetospheric plasma sources can be variable. Furthermore, what lessons might the magnetotails of Uranus and Neptune have for our studies of other bodies in the solar system?

In this review we first present an overview of the magnetospheres of Uranus and Neptune, focusing on the inner and outer boundary conditions (provided by the internal field and ionosphere at the inner boundary and the solar wind at the outer boundary), diurnal and seasonal influences, and plasma sources and transport. The configuration of the Uranian and Neptunian magnetotails are described and then we discuss observations of dynamics that were observed by Voyager 2 and inferences of dynamics from recent UV observations of auroral emissions from Uranus. We conclude by considering what new understanding we can apply to Uranus and Neptune and consider what lessons Uranus and Neptune have for magnetotail physics.

7.2. MAGNETOSPHERES OF URANUS AND NEPTUNE

7.2.1. Scales and Boundary Conditions

Table 7.1 shows the scales and basic parameters for Uranus and Neptune and their magnetospheres. Both planets are around four times larger than Earth (4.0 and 3.9, respectively) and the distance to the subsolar magnetopause is around 20 times the planetary radius, producing magnetospheres that are around a factor of six larger in linear dimension than Earth's magnetosphere. In terms of rotation rate, both Uranus and Neptune are intermediate between Earth and the rapid rotators and Jupiter and Saturn,

Mullard Space Science Laboratory, Department of Space and Climate Physics, University College London, Surrey, United Kingdom, and Now at Department of Physics, Lancaster University, Lancaster, United Kingdom

Magnetotails in the Solar System, Geophysical Monograph 207. First Edition.
Edited by Andreas Keiling, Caitríona M. Jackman, and Peter A. Delamere.

Table 7.1 Basic scales and parameters for the magnetospheres of Uranus and Neptune

	Uranus	Neptune
Equatorial radius (km)	25559	24764
Spin period[a] (h)	17.2	16.1
Dipole equatorial,[b] $\|\mathbf{B}\|$ (nT)	22800	14200
Dipole tilt[b]	−58.6°	46.9°
Offset[b] (x,y,z), R_P	(−0.157,0.025,−0.314)	(0.047,0.483,0.002)
Obliquity	97.9°	29.6°
Magnetopause subsolar distance,[a] R_P	18	23–26
Plasma source[a]	Atmosphere	Triton
Plasma source rate[a] (kg/s)	0.02	0.2
Major ion species[a]	H^+	H^+, N^+
Plasma lifetime[a] (days)	1–30	1

[a]From *Bagenal* (1992).
[b]From *Connerney* (1993).

although the quantitative effect of centrifugal stress and centrifugally driven interchange is not sufficiently understood to know if Uranus and Neptune are closer to rapid rotators in this regard. The obliquity of Neptune is similar to Saturn, but Uranus has a very large obliquity such that its angular momentum vector points south of its orbital plane, making Uranus a retrograde rotator. This large obliquity produces the largest seasonal variability of any planet in the solar system. At least at the encounters with Voyager 2 both of the ice giant magnetospheres were found to have dominant internal plasma sources: the ionosphere and hydrogen corona at Uranus and Triton at Neptune, with the solar wind representing a secondary source of plasma.

The internal fields of both Uranus and Neptune are highly complex with significant asymmetries. Planetary magnetic fields are usually described by a set of spherical harmonic coefficients [e.g., *Connerney*, 1993; *Holme and Bloxham*, 1996; *Herbert*, 2009] but are often reduced to tilted offset dipoles for simplicity. The resulting fields are highly tilted (−58.6° and 46.9° for Uranus and Neptune, respectively) and offset from the center of the planet: southward by ~$0.3R_U$ at Uranus and in the equator ~$0.5R_N$ at Neptune, although in reality these dipoles are also offset in other directions. The ionosphere of Uranus is unique in that, due to the large obliquity of the planet, the poles spend 42 years (half a Uranus year) in darkness, rather than around 5 hours at Saturn, for example. Estimates of the ionospheric Pedersen conductivity at Uranus vary between 0.4 and 14 S [e.g., *McNutt et al.*, 1987; *Selesnick and McNutt*, 1987; and references therein]. Using corotation breakdown near Triton, an estimate for Neptune's ionospheric Pedersen conductivity was found to be 0.02 S [*Richardson et al.*, 1991].

Figure 7.1 shows solar wind properties at Uranus and Neptune derived from Voyager 2 plasma and magnetic field data up to 2 AU upstream of each planet. These were calculated using the same assumptions as *Jackman and Arridge* [2011]. Some care should be used in interpreting differences between the upstream conditions at Uranus and Neptune since the data for Uranus were collected at solar minimum at the end of solar cycle 21, whereas the data at Neptune approached solar maximum in the rising phase of solar cycle 22. Also, pickup ions (not measured at these heliocentric distances) contribute to the plasma β and Mach number and may also produce a difference between the data at Uranus and Neptune. Median and the 10th and 90th percentile solar wind parameters are shown in Table 7.2. The solar wind density is around a factor of two to four smaller at Neptune, but the solar wind speed is approximately the same, resulting in a solar wind dynamic pressure that is a factor of four smaller at Neptune. Peak pressures of 0.042 nPa (0.012 nPa) at Uranus (Neptune) result in a Chapman-Ferraro magnetopause subsolar distance of $18R_U$ ($19R_N$) whereas using median pressures gives $25R_U$ ($24R_N$). The peak solar wind plasma beta is smaller at Neptune, but the interplanetary magnetic field (IMF) strength is larger, presumably due to near-solar maximum conditions when the data were collected so we do not suspect that the beta is significantly different between Uranus and Neptune. A similar argument can be applied to the slightly smaller Alfvén and fast-mode Mach numbers. As can be seen from the data in Figure 7.2 and Table 7.2, the Uranian bow shock is typically a very high Mach number and high β shock [e.g., *Bagenal et al.*, 1987] and more recent evidence has been presented showing that the shock structure is nonstationary with evidence of reformation [*Tiu et al.*, 2011]. The different upstream conditions are naturally expected to lead to different magnetosheath particle distributions.

We also investigated correlations between solar wind parameters, for example, in selecting a particular value of the solar wind dynamic pressure as an input to a simulation, is there a value for the field strength that is more appropriate? Correlations were determined using Spearman's rank correlation coefficient and a one-tailed Student's *t*-test was used to determine if these correlations

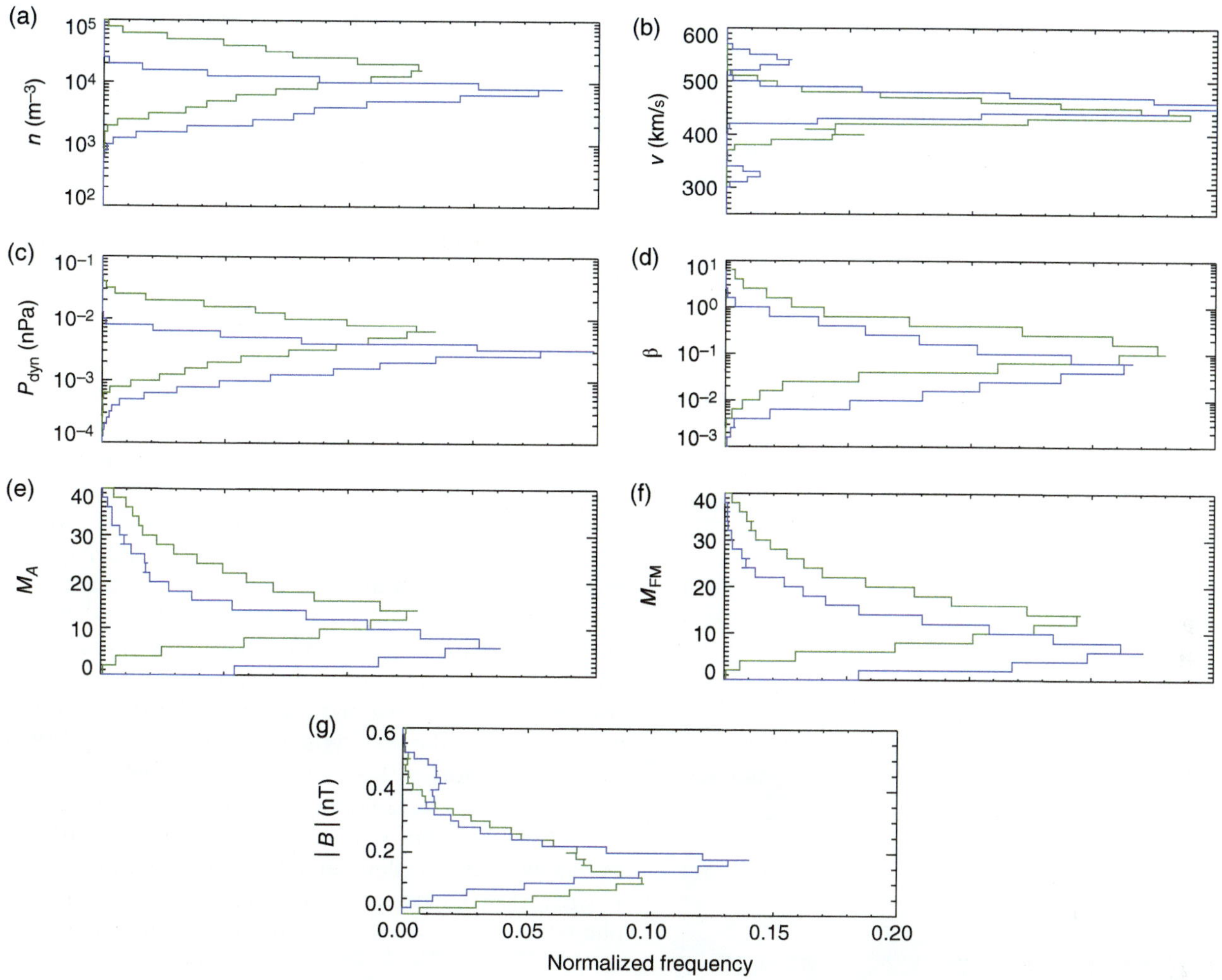

Figure 7.1 Solar wind properties as derived from Voyager 2 data upstream (2 AU upstream of each planet) of Uranus (green) and Neptune (blue): (a) solar wind number density (H and He^{++}); (b) proton speed; (c) dynamic pressure; (d) plasma beta; (e) Alfvén Mach number; (f) fast-mode Mach number; (g) field strength. These frequencies are normalised to the total number of points.

Table 7.2 Solar wind statistics near Uranus and Neptune based on Voyager 2 magnetic field and plasma data showing median and 10th and 90th percentile values (in parentheses) of various solar wind parameters

	Uranus	Neptune
Number density (m^{-3})	16500 (5620, 41600)	6780 (2720, 12300)
Proton speed (km/s)	448 (413, 478)	461 (439, 488)
Dynamic pressure (pPa)	6.25 (2.13, 15.4)	2.71 (1.10, 5.01)
Field strength (nT)	0.169 (0.0730, 0.309)	0.180 (0.110, 0.360)
Plasma β	0.176 (0.0513, 0.760)	0.0673 (0.0148, 0.334)
Alfvén Mach number	16.4 (9.34, 33.3)	9.57 (3.92, 21.4)
Fast-mode Mach number	15.3 (8.93, 27.0)	9.17 (3.90, 18.7)

Note: Values are all shown to three significant figures.

were significantly different from zero at a level of 1%. Correlation coefficients below 0.4 were associated with no correlation. At Uranus, a moderate positive correlation was found between the field strength and the dynamic pressure (r_s = 0.468). A moderate negative correlation was found between the field strength and the plasma β (r_s = −0.448). Strong negative correlations were found between the field strength and the Alfvén and fast-mode Mach numbers

(r_s = −0.750 and r_s = −0.764, respectively). At Neptune, strong negative correlations were found between the field strength and the plasma β (r_s = −0.719) and the Alfvén (r_s = −0.877) and fast-mode (r_s = −0.878) Mach numbers. Strong positive correlations were found between the solar wind pressure and the plasma β (r_s = 0.700), Alfvén (r_s = 0.769), and fast-mode (r_s = −0.763) Mach numbers.

7.2.2. Diurnal and Seasonal Magnetospheric Configurations

The large dipole tilt at both Uranus and Neptune produces significant diurnal variations in the magnetosphere. Seasonal variations produced by nonzero obliquities are also important and the very large obliquity of Uranus produces extreme seasonal variation. Figure 7.2 shows the angle between the dipole axis and the Sun, called the angle of attack for Earth, Uranus, and Neptune based on a similar figure by *Lepping* [1994]. Each panel starts at equinox where the spin axis of the planet is perpendicular to the Sun direction. The angle between the spin axis and the Sun direction is given by the black curve. The gray regions show the range of angles of attack produced by the diurnal rotation of the dipole around the spin axis. For Earth and Neptune the gray regions and black curves coincide, but they are opposite for Uranus because the dipole is orientated opposite to the spin axis. Figure 7.3 shows source-surface models for the magnetosphere for a range of angles of attack from pole-on to perpendicular configurations.

For Neptune we see the same sort of behavior as Earth except the large dipole tilt produces a much larger envelope of angles of attack. This results in an Earth-like magnetospheric configuration at some phases of planetary rotation but "pole-on" configurations at others where the dipole axis is parallel or antiparallel to the solar wind flow. For Uranus at solstice the spin axis points nearly away or nearly toward the Sun with the dipole vector rotating around the spin axis, producing a nearly Earth-like configuration, albeit rotating around the solar wind flow vector every 17.2 hours. At equinox we have a pole-on configuration but switching which pole is pointing at the Sun every half a planetary rotation, with an Earth-like configuration in between. It is not clear how the magnetosphere reorganizes itself when in this configuration.

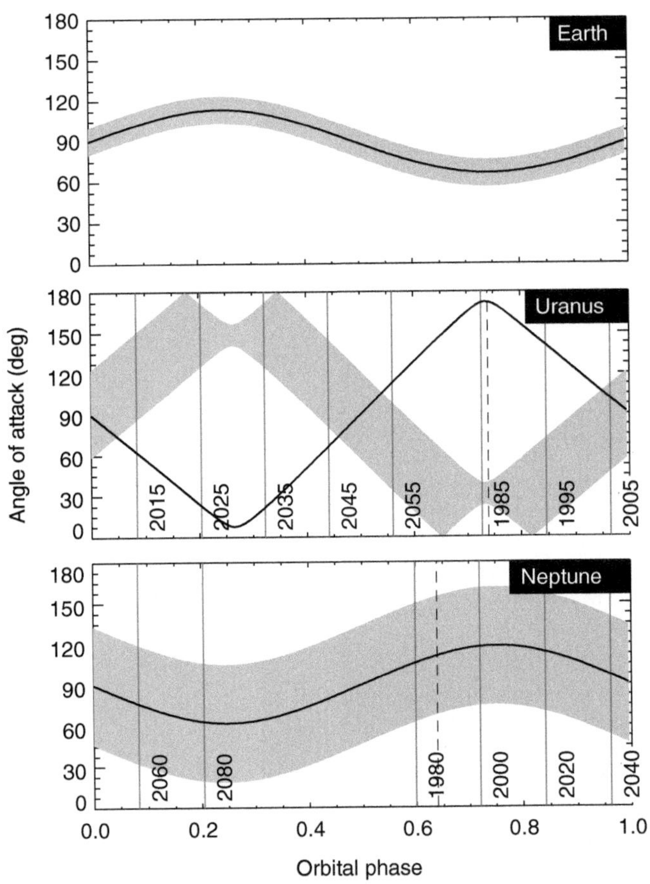

Figure 7.2 Angle of attack (angle between a vector from the planet to the Sun and the dipole axis) for Earth, Uranus, and Neptune as a function of the fractional time along the planet's orbit around the Sun. The solid black curves show the angle between the Sun vector and the spin axis of the planet. The gray shaded regions show the range of angles of the angle of attack due to the tilt of the dipole and planetary rotation. The black curve and the shaded region for Uranus are opposite since Uranus is a retrograde rotator. The vertical dashed lines indicate the encounters of Voyager 2 with Uranus and Neptune and the vertical solid gray lines highlight the expected configuration for future years. Adapted from *Lepping* [1994].

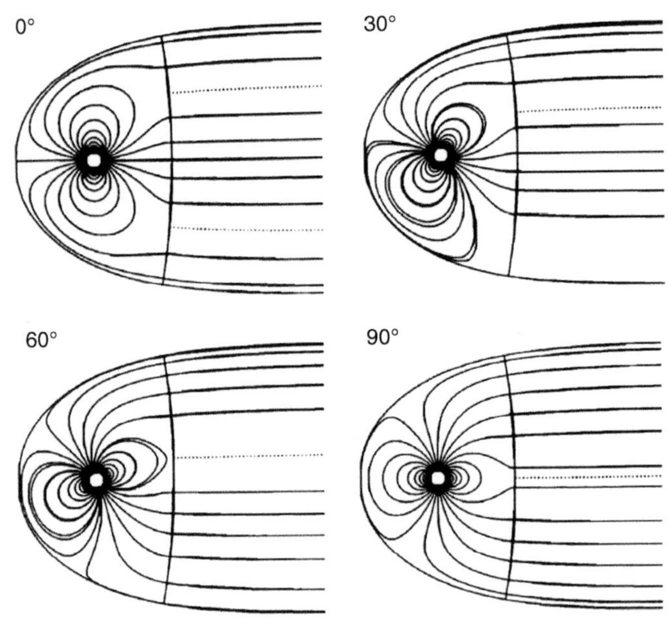

Figure 7.3 Configurations of the magnetosphere as a function of the angle of attack. Adapted from *Schulz and McNab* [1996, Figures 3a–c].

7.2.3. Plasma Sources and Transport

The presence of icy moons in the Uranian magnetosphere produces a neutral tori of H_2O, OH, O, and O_2 due to sputtering from micrometeoroid, photon, and charged particle impacts on their surfaces. The neutral tori at Titania and Oberon extend well outside the magnetosphere and so most of the neutrals never enter the inner magnetosphere, but those due to Miranda, Ariel, and Umbriel lie well within the magnetosphere and are able to contribute to internal mass loading. The total neutral number density due to the tori at Miranda, Ariel, and Umbriel peaks at around $3\times10^5\,m^{-3}$ [*Cheng*, 1987]. However, the ionization of this torus by photoionization and electron impact ionization and subsequent rapid convection of plasma out of the system (at least during the Voyager 2 encounter) produce a very low heavy ion density of $\sim130\,m^{-3}$, below the limit of detection by the Voyager plasma instrumentation [*Selesnick and McNutt*, 1987; *Cheng*, 1987]. However, the proton number density was found to be around $10^6\,m^{-3}$ and adequately explained by an ionospheric source, driven by ionospheric photoelectron escape, and ionization of the Uranus hydrogen corona [*Selesnick and McNutt*, 1987; *Cheng*, 1987].

However, at Neptune escape of H and N from a collisional atmosphere at Triton and subsequent spreading of this cloud via charge exchange reactions have been shown to be a viable plasma source for the Neptunian magnetosphere and produce a significant heavy ion (N^+) component in the magnetosphere [*Cheng*, 1990; *Richardson*, 1993; *Decker and Cheng*, 1994]. The escape rate of H from Triton is much larger than N, but H is collisionally ejected from the system and precipitates into Neptune before it can be ionized, thus producing a larger ratio of N^+ to H^+ than is reflected by the neutral source rate [*Decker and Cheng*, 1994]. It is not clear if Triton represents a variable plasma source.

At both Uranus and Neptune the solar wind represents a secondary source of plasma, but this may be highly seasonally dependent or solar wind plasma may enter the magnetosphere at certain phases of rotation where dayside reconnection is more favorable.

At both Uranus and Neptune the particular orientations of the angle of attack imply that plasma is rapidly convected out of both magnetospheres, despite both of these systems having important corotational electric fields [*Vasyliunas*, 1986; *Selesnick and Richardson*, 1986; *Selesnick*, 1990].

7.3. MAGNETOTAIL CONFIGURATION AT URANUS AND NEPTUNE

Macek et al. [1992] modeled the distant magnetotails of the outer planets by conserving fluxes of momentum density, magnetic field, plasma density, and energy density at the tail boundaries to determine the shapes of the magnetotail and the dependence of plasma density and magnetic field strength along the tail. For the distant ($\pm1000R_U$) magnetotail of Uranus they found a significantly flattened tail with a Z dimension of $\sim60R_U$ and in $170R_U$ in Y with an average radius of $\sim100R_U$. The tail field was calculated to be 0.3 nT with a plasma density of $1000\,m^{-3}$. Similar to Uranus, for Neptune *Macek et al.* [1992] found a flattened tail at $1000R_N$ with a Z radius of $50R_N$ and a Y radius of $150R_N$ and an average radius of $80R_N$. The tail field was less than 0.1 nT and the tail plasma density around $200\,m^{-3}$.

7.3.1. Solstitial Configuration from Voyager 2 Observations and Modeling

At solstice the spin axis of Uranus is oriented toward or away from the Sun and so the magnetospheric tilt angle is roughly constant with the dipole simply rotating around the solar wind direction (see Figure 7.4). The rotation of the dipole coils up the tail into a helix of opposite polarities. From magnetohydrodynamic (MHD) simulations, *Tóth et al.* (2004) found that the tail twisted up at a roughly constant rate of $0.4°/R_U$, which when combined with the rotation rate of Uranus represents a propagation speed of 370 km/s ($52R_U$ hour^{-1}). This corresponds to a wavelength along the tail of $900R_U$. This twist is illustrated in Figure 7.5. The propagation of the twist, i.e., the rotation of the dipole, is communicated as an Alfvén wave that propagates along the tail but competes with friction against the solar wind. The Alfvén speed in the tail is between 450 and 600 km/s throughout most of the tail so the propagation speed of the twist is reduced due to this friction with the solar wind. In the absence of friction the lobes would be twisted up at a rate of $0.28°/R_U$.

Unless the tail lobes coil up as a perfectly rigid structure, the field lines in the tail lobes are also twisted into a helical structure due to merging with the IMF or loading of the field lines with plasma from the solar wind or internal sources or as a result of the finite wave speed down the tail [*Behannon et al.*, 1987, and references therein]. This helical structure within each tail lobe is described by the pitch angle of the field lines, and this was measured as $5.5°\pm3.0°$ from Voyager 2 magnetometer data [*Behannon et al.*, 1987]. One can also express this pitch angle as $\tan^{-1}(2\pi R/Tv)$, where R is the radius of one of the tail lobes, T is the period of the Alfvén wave, and v is the local Alfvén speed. From this expression one finds that the wavelength of the helical twist in the tail lobes is $2740R_U$ [*Behannon et al.*, 1987], much larger than the helical twisting of the tail lobes.

Figure 7.6 shows a terminator plane cut of the magnetotail from *Behannon et al.* [1987] with the trajectory of

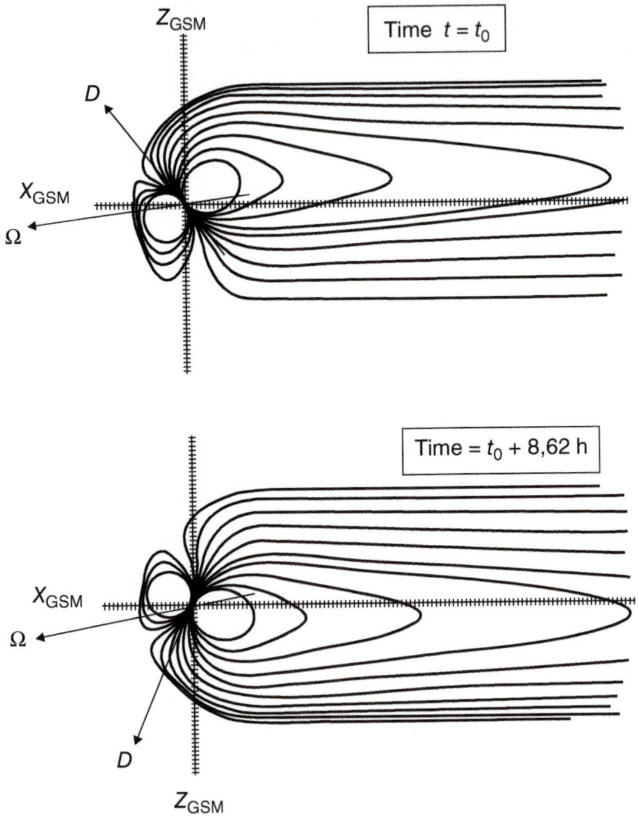

Figure 7.4 Configuration of the Uranian magnetosphere as seen from a fixed observer's point of view, separated half a rotation apart, as derived from an MHD equilibrium model [from *Voigt et al.*, 1987].

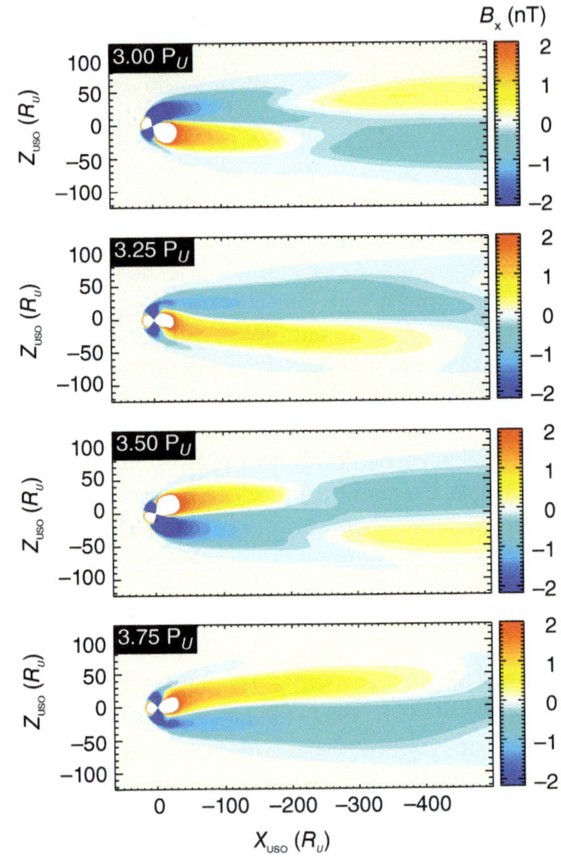

Figure 7.5 Results from an MHD model of the Uranus magnetosphere showing the coiling up of the tail as Uranus rotates near solstice. The color scale shows the B_x component (in Uranus solar orbital (USO) coordinates, analogous to geocentric solar ecliptic coordinates) of the magnetic field in the Y_{USO}=0 plane. The four panels cover one Uranus rotation (with a period of P_U). After *Tóth et al.* [2004].

Voyager 2; in this projection the spacecraft trajectory is a spiral since the rotation of the planet has been removed. The dashed lines demark regions where the field measured by Voyager 2 had 50, 75, and 90% of the lobe field value and these curves have been smoothly connected together from the different traversals of the plasma sheet. The shaded region thus depicts the plasma sheet. The thick line shows the center of the current sheet. From this we can see that the plasma and current sheets are asymmetrical with a curved shape due to the tilted nature of the Uranian dipole [*Voigt et al.*, 1987] and a thicker plasma sheet on the side of the tail where the first and third crossings occurred. It is important to recognize that this is not a dawn-dusk asymmetry. The plasma sheet shape was also modeled using a linear MHD equilibrium approach by *Voigt et al.* [1987] and was found to agree with these observations. Estimates of the current density in the magnetotail was found to range between 2.4 and 3.5×10^{-11} A/m^2, several orders of magnitude smaller than in Earth's tail current sheet [*Behannon et al.*, 1987], consistent with modeling that showed that the current sheet was not as stretched as in Earth's tail, also resulting in a thicker plasma sheet [*Voigt et al.*, 1987]. Pressure balance between the lobes and current sheet center can be adequately explained by the pressure from <6 keV H^+ and electrons.

The shaded arcs were anomalous plasma sheet encounters adjacent to the third crossing and were initially interpreted as dynamical by *Behannon et al.* [1987] (discussed further in section 7.4.1.) However, *Hammond et al.* [1990] used an MHD simulation adapted from Earth to demonstrate that the plasma sheet encounter prior to the third crossing could be attributed to a "pincer-shaped" plasma sheet that is a natural feature of the Uranian plasma sheet at certain tilt angles. However, no similar structure was found in the *Tóth et al.* [2004] MHD simulations, suggesting they have a dynamical origin.

7.3.2. Equinoctial Configuration from Modeling

A pole-on configuration magnetosphere is illustrated by Figure 7.7. In this configuration the tail plasma sheet is cylindrical with currents closing entirely in the magnetotail.

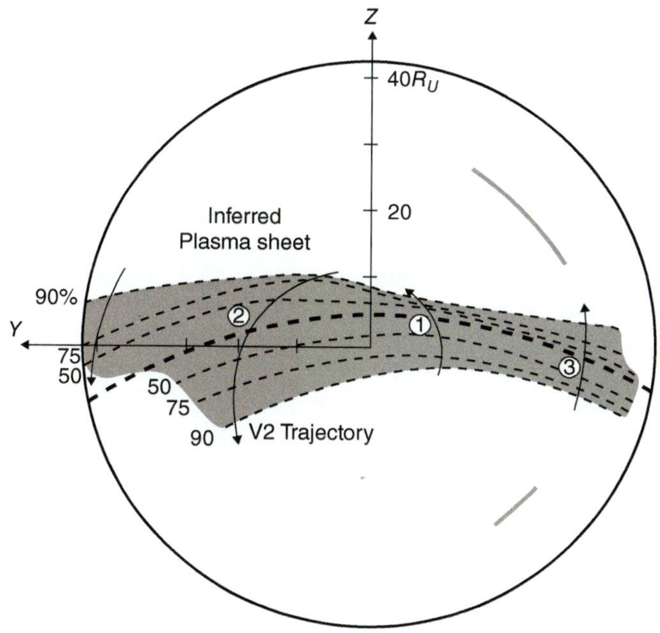

Figure 7.6 Uranian magnetotail plasma sheet in a stationary frame such that the trajectory of Voyager 2 (bold curves) is a spiral shape. The dashed line shows points at which the measured field is 50, 75, and 90% of the measured lobe field strength; thus the shaded region marks out the plasma sheet. The thick dashed line is the current sheet. These lines have been smoothly connected with observations taken along the Voyager 2 trajectory. The circled numbers indicate observed current sheet crossings. The gray portions of the trajectory in the northern and southern lobes indicate transient entries into the plasma sheet. [from *Behannon et al.*, 1987].

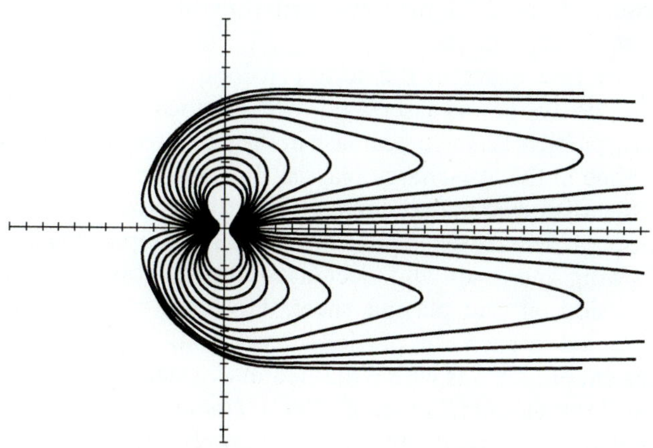

Figure 7.7 Symmetric pole-on magnetospheric configuration from the MHD equilibrium model of *Voigt* [1986] as relevant for Uranus at equinox. The axis of symmetry in this model is the tail axis. [From *Voigt*, 1986].

Away from equinox the tail current sheet is more asymmetrical in crosssection and for some angles of attack the tail current partially closes on the magnetopause. Figure 7.8, shows cross sections of the magnetotail for a range of angles of attack. These are drawn from the perspective of looking at Uranus from the magnetotail. In a pole-on configuration the field goes toward Uranus in the middle of the tail and out on the edges with the tail current sheet forming a cylinder. As the angle of attack increases, the in-sector moves toward the magnetopause, deforming as it does so. At an attack angle of 25° the tail current sheet partially closes on the magnetopause. At angles greater than this the current sheet is a curved surface until it becomes a planar current sheet at an attack angle of 90°. Near equinox the angle of attack varies by around 60° (see Figure 7.2), oscillating between near-pole-on and near-perpendicular configurations, changing between the two roughly every quarter of a planetary rotation. The solstitial configuration encountered by Voyager 2 was between the 30° and 60° cases in Figure 7.8. The configuration of

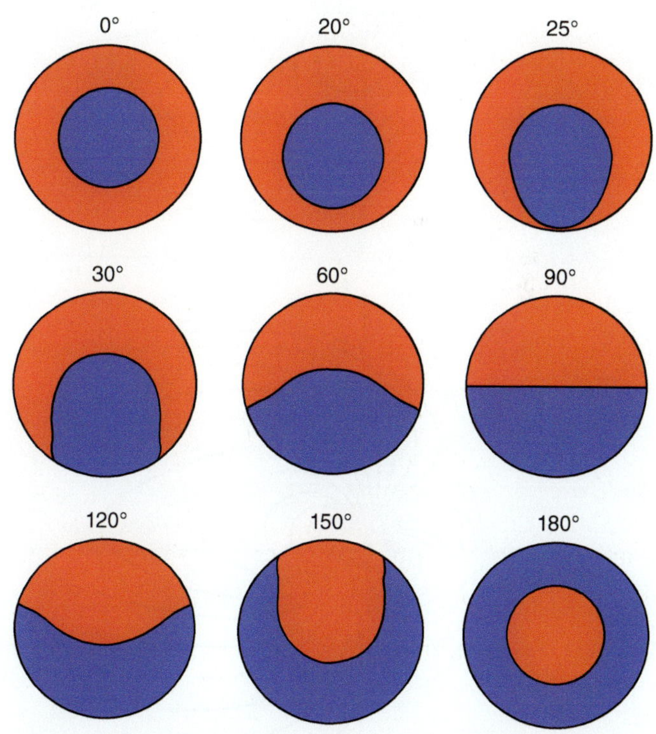

Figure 7.8 Diagram showing the polarity of the field in the magnetotail as a function of the angle of attack of the dipole. These diagrams are drawn from the perspective of an observer in the tail looking toward a planet. Red indicates field lines coming toward the observer, blue away. The tail current sheet is located at the boundary between the blue and red regions with the tail magnetopause as the circular boundary. Adapted from *Schulz and McNab* [1996], Figures 3a–c and similar to Figure 5 in *Lepping* [1994].

the Uranus magnetosphere during equinox is similar to that expected during periods of geomagnetic reversal in the terrestrial paleomagnetosphere.

7.3.3. Magnetotail Configuration at Neptune

During the Voyager 2 flyby the tail periodically oscillated between the two configurations depicted in Figure 7.9, thus representing an extreme time dependence. The tail cross sections depicted in Figure 7.8 are qualitatively applicable here also. During Voyager 2 the angle of attack ranged between 70° and 156°, thus rapidly varying between a curved but near-planar magnetotail current sheet and a highly curved current sheet, similar to the configurations for attack angles of 20°–25° but with opposite polarity. See also Figure 2 in *Voigt and Ness* [1990]. From modelling, *Voigt and Ness* [1990] found that the plasma beta in the tail was somewhere intermediate between a vacuum case and Earth's case and so, similar to Uranus, Neptune's tail is not as stretched as Earth. From modeling arguments they also suggest that Neptune's magnetotail might transit through a range of quasi-equilibrium states during a planetary rotation. *Vasko et al.* [2012] have studied the consequences of such extreme time dependence for particle acceleration and found that the inductive electric field, associated with rapid reconfiguration, can result in particle acceleration and loss.

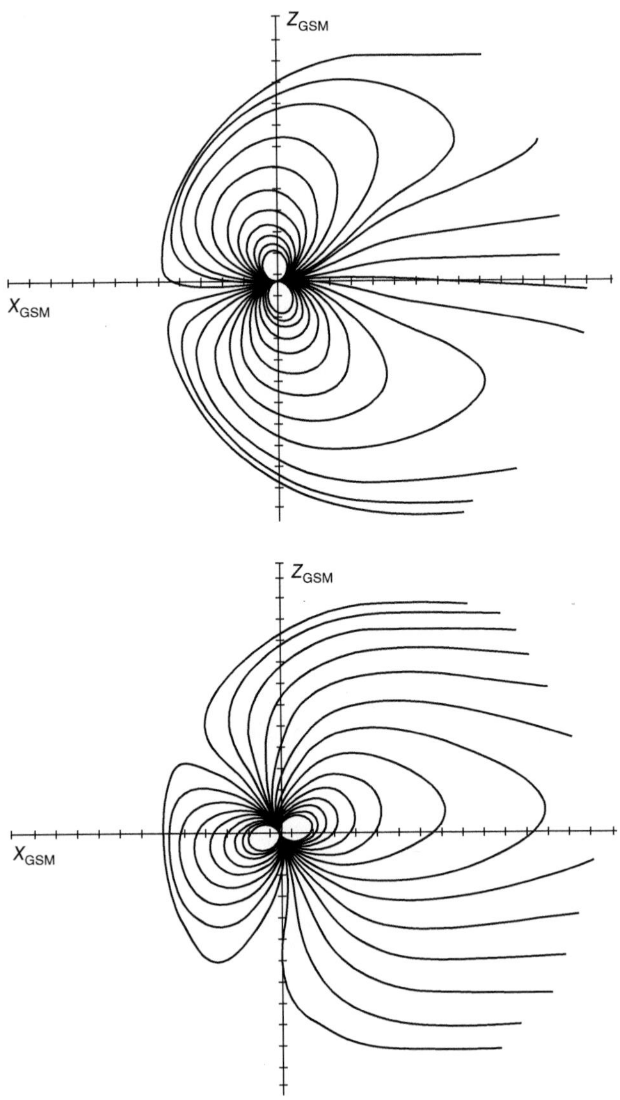

Figure 7.9 Field lines in Neptune's magnetosphere in the noon-midnight meridional plane showing the two configurations (assumed to be in equilibrium) expected during the Voyager 2 encounter with Neptune. This is from an MHD equilibrium model of Neptune's magnetosphere. [From *Voigt and Ness*, 1990].

7.4. MAGNETOTAIL DYNAMICS

7.4.1. Uranus

The position of Voyager 2, in the stationary frame of the magnetosphere, during the Uranus flyby is presented in Figure 7.10. Three clear immersions in the plasma sheet and accompanied by current sheet crossings were found in the magnetic field and energetic particle data [*Behannon et al.*, 1987; *Kane et al.*, 1991] and coincided with the expected position of the plasma/current sheet. These are marked by "N" in Figure 7.10. In addition, two anomalous current sheet crossings were found above and below the expected location of the sheet and surrounded the third and final full traversal of the tail plasma/current sheet. These are observed several 10s of R_U from the nominal location of the plasma sheet and so Voyager may have entered a region where the spatial structure of the plasma/current sheet was highly distorted or Voyager observed some filamentary structure of the tail [*Kane et al.*, 1991].

The first event in the southern lobe, before the third sheet crossing, was associated with enhancements in energetic particle fluxes, a decrease in the field strength, but no change in the low-energy plasma. For a simple encounter with a distorted planar current sheet the gradients in energetic particle fluxes should have a fairly simple structure showing increases and decreases during approach and recession of the plasma sheet. However, this was not observed in energetic particles, and it was suggested that this encounter was with a filamentary or detached part of the plasma sheet [*Kane et al.*, 1991]. *Hammond et al.* [1990] used an MHD model to provide an alternative interpretation for this anomalous encounter in terms of distortions in the current sheet produced by the rotation of Uranus (see also section 7.3.1). However, these features were not found in later MHD simulations [*Tóth et al.*, 2004].

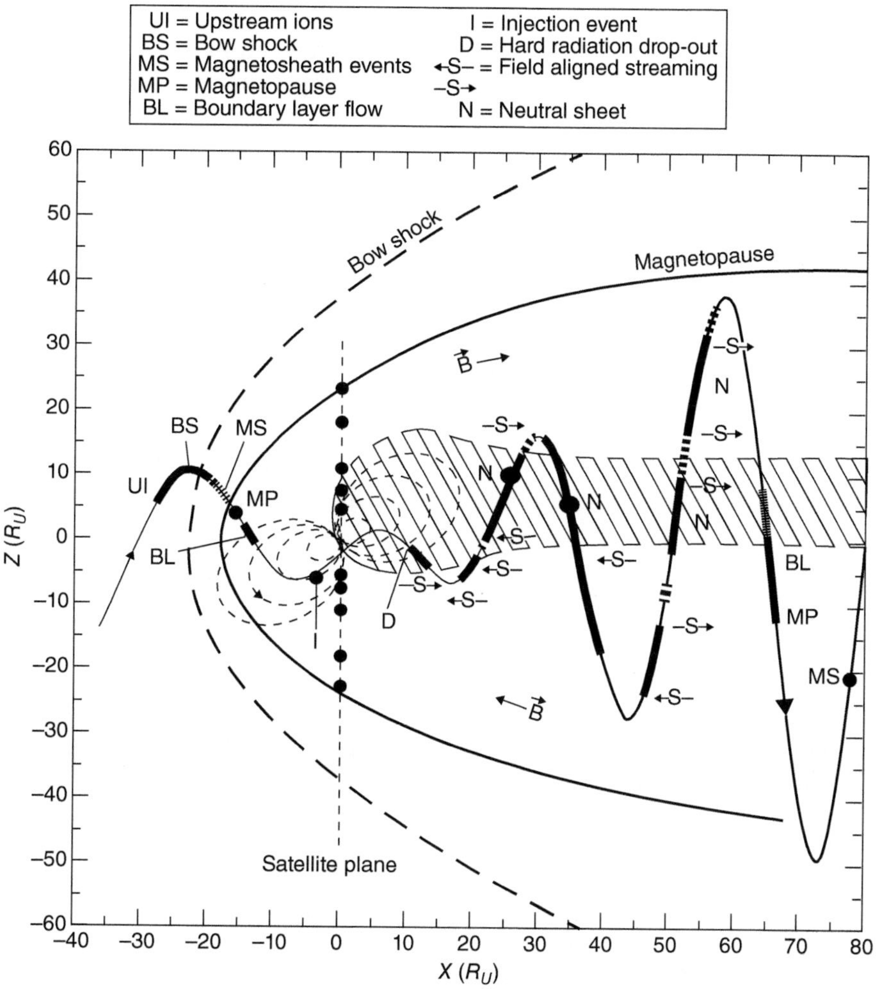

Figure 7.10 Trajectory of Voyager 2 through the Uranian magnetosphere in a rotating solar magnetospheric coordinate system summarizing the energetic particle results from Voyager 2 [from *Mauk et al.*, 1987].

In addition, this interpretation does not agree with the observed gradients in the energetic particles.

The second event after the third sheet crossing occurred in the northern lobe and was accompanied with a flattening of the B_x component of the magnetic field and dramatic increases in energetic particle fluxes. The particle activity was observed to increase by almost an order of magnitude in less than 1 s and the spectral index changed indicating a hardening of the spectrum. The energetic particle anisotropies found during this event suggested the plasma sheet was highly turbulent. Field-aligned energetic particle signatures near the three full current sheet crossings were found to be energy dispersed, as would be expected from particle acceleration at some remote site and then the more energetic particles arrive at the observer first [*Kane et al.*, 1991]. This was not observed during the second event after the third current sheet crossing. This suggests that Voyager 2 was located near the acceleration region for the field-aligned streaming found in the magnetotail closer to Uranus, or Voyager 2 was entering a new acceleration region. Therefore, this was interpreted as evidence of "substorm" activity in the Uranian tail [*Kane et al.*, 1991]. Some combination of flapping or decoupling of the plasma sheet therefore produced this anomalous partial plasma/current sheet encounter were a region of particle acceleration was found.

In summary, beyond around $45R_U$ the evidence from the energetic particle data suggests that the Uranian tail is not a simple rotating contiguous planar current sheet, but the tail is filled with filamentary or partially detached structures [*Kane et al.*, 1991], possibly analogous to the structure found in the distant Jovian tail by New Horizons [*McComas et al.*, 2007].

The aurora of Uranus was first detected in International Ultraviolet Explorer observations of H Lyman-α [*Clarke et al.*, 1986] and subsequently imaged by Voyager 2 [e.g., *Broadfoot et al.*, 1986; *Herbert*, 2009]. It was redetected in Hubble Space Telescope images taken during November 2011 and also identified in earlier observations from 1998 [*Lamy et al.*, 2012]. The November 2011 campaign was specifically designed to image the aurora during the passage of a solar wind compression region at Uranus. During the November 2011 period the attack angle was between 74° and 137° with the spin axis almost perpendicular to the Sun and as such the system alternated between an almost pole-on and near-perpendicular conditions. Auroral emissions were found on two occasions producing two distinct auroral spots with similar auroral power to the Voyager 2 observations. Unlike Earth and Jupiter, whose auroral emissions match the duration of interplanetary shocks, the emissions at Uranus were more localized and transient, variable on a time scale of minutes [*Lamy et al.*, 2012]. It is possible that such emissions could be produced from impulsive plasma injections through the dayside magnetopause when the orientation of the Uranian dipole and the solar wind are conducive to dayside reconnect, expected to occur once per planetary rotation. Interestingly, evidence for periodic reconnection with a ~17 h period during the Voyager 2 encounter at solstice was presented by *Richardson et al.* [1988].

Cowley (2013) argued that the wrapping of open flux across the dayside magnetopause could suppress dayside reconnection preventing a large auroral response to the arrival of an interplanetary shock, thus consistent with the auroral observations by *Lamy et al.* [2012]. Figure 7.11 shows a series of schematics of the Uranian magnetosphere drawn for the conditions at the November 2011 observations. Magnetic flux is opened at the dayside magnetopause in Figure 7.11a and is transported into the

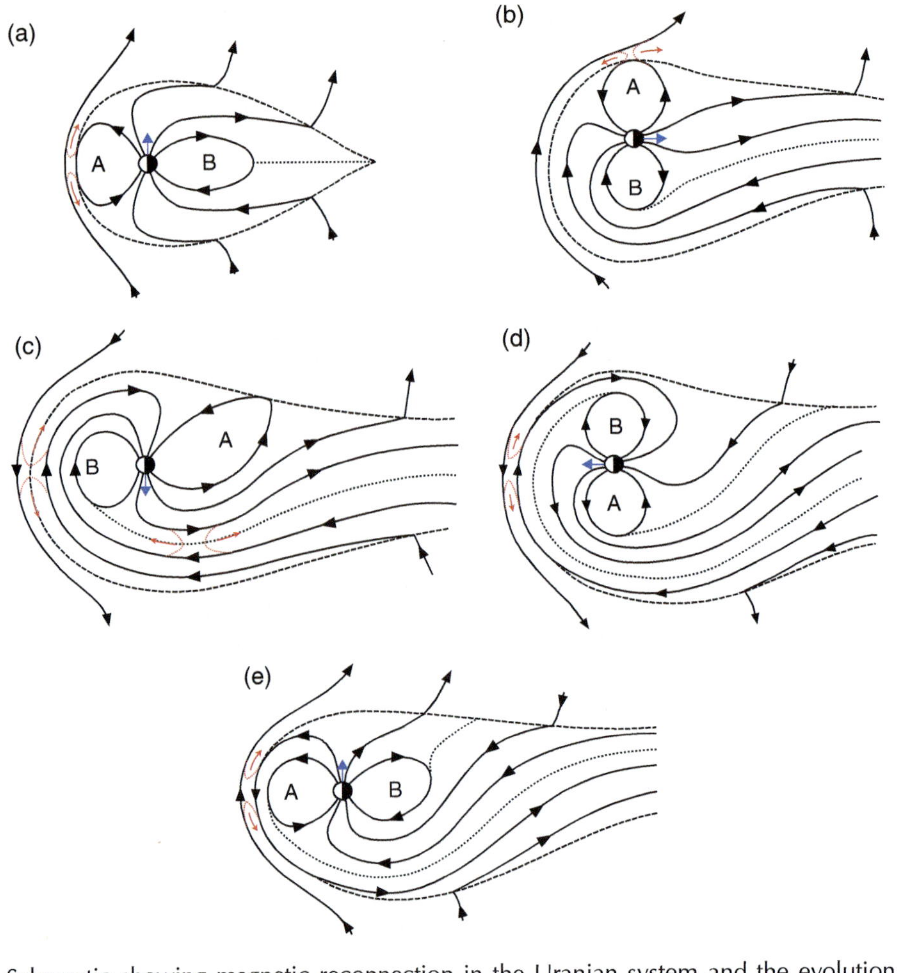

Figure 7.11 Schematic showing magnetic reconnection in the Uranian system and the evolution of open flux tubes over a rotation of Uranus [*Cowley*, 2013]. Each diagram shows a plane containing Uranus, with the Sun on the left and the spin axis pointing into the page, and is approximately perpendicular to the orbital plane of Uranus. The dipole axis is denoted by the blue arrow and rotates in the plane of the figure. Solid curves with arrows indicate magnetic field lines, the dashed line is the magnetopause, and the dotted line indicates the boundaries between open flux regimes that have different connectivities. The red dashed lines and short arrows indicate the field and flows associated with reconnecton. The lettered regions are used to mark different sides of the planet.

magnetotail. As the dipole rotates into the nightside, the magnetic flux in region A is less exposed at the dayside magnetopause and so the reconnection rate drops. Furthermore, open flux from the southern lobe is wrapped over the closed flux at the dayside, thus "protecting" closed flux from being opened, although the addition of an IMF B_Y component may reduce this shielding effect. As Uranus rotates, the open flux is compressed (Figure 7.11c), allowing the open flux to be closed. The flux that is closed in region B was originally opened in region A, thus requiring an inner magnetospheric convection system to return the flux to region A. In Figure 7.11d the open lobe flux might reconnect with the solar wind. Although this will not change the amount of open flux in the system, it will result in a rearrangement of the existing open flux, producing a more complex tail structure consisting of three lobes. These considerations led *Cowley* [2013] to suggest the presence of a structured and weak tail at Uranus during equinox and might lead to the weak auroral response observed by *Lamy et al.* [2012].

7.4.2. Neptune

Similar to Uranus, field-aligned ion streaming was observed in Neptune's magnetosphere; however, only streaming toward Neptune was observed. Two intervals where planetward streaming of plasma sheet ions was observed were found in the tail [*Mauk et al.*, 1991]. The first example was found at high magnetic latitudes (marked PLST in Figure 7.12, around $15R_N$ from

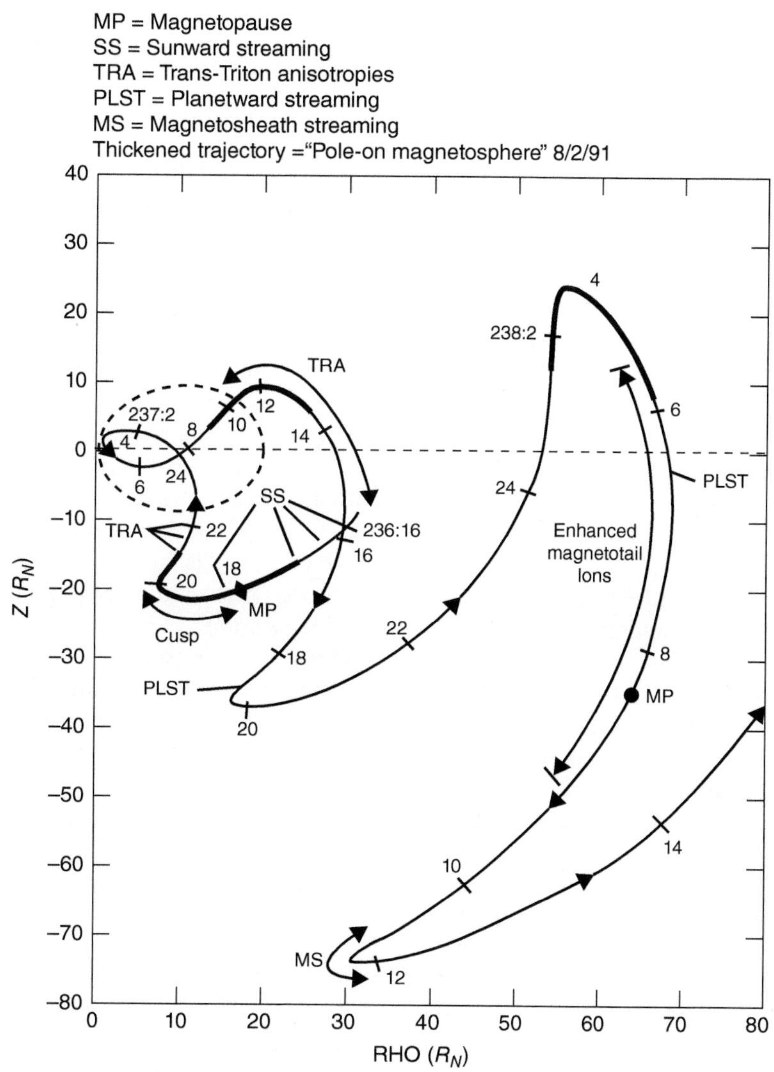

Figure 7.12 Summary of the energetic particle observations from Voyager 2 along the spacecraft trajectory in cylindrical magnetic coordinates such that ρ is the distance from the magnetic axis and *z* is the distance from the magnetic equatorial plane [from *Mauk et al.*, 1991].

Neptune's magnetic dipole axis) and another near the magnetic equator (again marked PLST in Figure 7.12, at $70R_N$). The observation of such planetward streaming at very high magnetic latitudes where the plasma sheet would not be expected suggests detached or filamentary regions of the plasma sheet, similar to those suggested for the Uranus tail observations. The second event was found near the tail magnetopause and also after an interval of time when Neptune was emerging from a pole-on configuration (bold segments in Figure 7.12). Without more advanced time-dependent models it is not clear what configuration the plasma sheet should be in; however, at the symmetrical position before the onset of the pole-on configuration (near 0000 UT on day 238, 1989) a similar energetic particle signature was not observed, suggesting some dynamical process (*Mauk et al.*, 1991). The presence of such confusing observations probably demonstrates the inadequacy of equilibrium models in explaining the structure of the tail in such a rapidly varying system.

In summary, Voyager 2 found a quiet magnetotail during the Neptune encounter. The presence of a generally quiet tail at Neptune is supported by the lack of substorm-generated asymmetries in the inner magnetosphere [*Mauk et al.*, 1991]. The finding of a small auroral response at Uranus during equinox [*Lamy et al.*, 2012], when Uranus was in a similar configuration to Neptune during the Voyager 2 encounter, and the theoretical work of *Cowley* [2013] suggest that tail dynamics are suppressed at Neptune when in such a similar highly inclined configuration.

7.5. DISCUSSION

7.5.1. Summary

The magnetotails of Uranus and Neptune are inherently three-dimensional and time-dependent systems. The Voyager 2 observations of Uranus at solstice could be readily interpreted in terms of a rotating Earth-like configuration. However, Neptune and Uranus at equinox present significant challenges, even at an elementary level, to understanding the time-dependent configuration of the tail. The application of quasi-equilibrium models to understand their structure appears questionable at best. Both tails were found to be less stretched than at Earth, probably due to lower plasma pressure and amounts of open flux relative to Earth. Evidence for filamentary/detached plasma sheet structures were found in both the Uranian and Neptunian magnetotails. It is not clear if this is a natural consequence of rapid rotation and the inherent asymmetries in these systems. Evidence for "substorm"-like activity was found at Uranus during solstice conditions, although indirect evidence for a quiet tail at Uranus during equinox has been found from auroral observations and theoretical considerations. Neptune's magnetotail and magnetosphere appeared quiet when observed by Voyager 2 in a similar configuration to the Uranus equinoctial configuration.

7.5.2. Applying New Understanding

Since the epoch of the Voyager 2 flybys there has been considerable research in how flux loading and transport work in giant planet magnetospheres with rapid rotation and diagnosing such dynamics through auroral observations [e.g., *Cowley et al.*, 2005; *Jackman*, 2011]. Such new understanding might be applied in order to understand the Voyager 2 observations at Uranus and Neptune and to make new predictions [e.g., *Cowley*, 2013].

The magnetotails of Earth, Saturn, and Jupiter have been found to be the sites of oscillations (flapping) in the current sheet associated with reconnection and other disturbances (see *Volwerk et al.* [2013] and references therein for a recent review). *Popoudin et al.* [2012] have studied the growth of kink mode waves in a cylindrical current sheet, as expected at Neptune and Uranus (near equinox). For the case of Uranus they estimated growth time scales for such waves of the order of 1 hour with periods of 9 minutes. Such waves might be excited not only by dynamical processes but also by the rapid reconfiguration of the magnetosphere. The further application of flapping wave modes may shed light on the partial plasma sheet encounters observed by Voyager 2 at Uranus and help to distinguish between displacements/flapping of the plasma sheet and filamentation. Such fine structure in the plasma sheet at Earth has been probed in more detail with multipoint measurements [e.g., *Lennartsson et al.*, 2007] that show that indeed the high-latitude plasma sheet at Earth is highly and frequently filamented and therefore not necessarily related to rapid rotation/reconfiguration.

There is a considerable body of evidence for time-dependent mass loading due to impulsive volcanism at Io and resulting time-dependent magnetospheric dynamics at Jupiter [e.g., *Brown and Bouchez*, 1997; *Thomas et al.*, 2004]. Although our understanding of these processes is at an early stage, such understanding might be applied to predict such impulsive mass loss from Triton and understand subsequent magnetospheric and magnetotail dynamics.

7.5.3. Lessons for Other Magnetospheres

What lessons might the study of the magnetotails of Uranus and Neptune have for the other magnetotails in the solar system? The extreme time dependence and resulting time dependence of magnetospheric configurations at Uranus and Neptune cannot be ignored as they might be at other magnetospheres, for example, in assuming aligned rotators. The development and application of new time-dependent models are clearly a priority. Such extreme time dependence of the whole magnetosphere might also be an internal trigger for tail reconnection.

The effect of large solar wind plasma beta and Mach numbers on dayside reconnection rates has recently received considerable attention for the magnetosphere of Saturn [*Scurry and Russell*, 1991; *Grocott et al.*, 2009; *Masters et al.*, 2012] as well as the influence of viscous processes at Jupiter and Saturn [e.g., *Desroche et al.*, 2012, 2013]. Such properties of the solar wind are even more extreme at Uranus and Neptune and combined with the shielding effects discussed by *Cowley* [2013] might make Uranus and Neptune uniquely weakly driven by the solar wind. Further attention of these aspects at Uranus/Neptune thus may shed light on these processes at the gas giants. Uranus and Neptune also provide an example of weak solar wind drive but apparently also weak internal plasma sources and slower planetary rotation with weaker ionospheres. This provides an interesting comparison with the magnetospheres of Saturn and Jupiter where the role of internal plasma and energy sources is much clearer.

The transmission of waves down the tail communicating the rotation of the dipole is affected by a combination of field line bend-back, wave delay, and MHD outflow [e.g., *Northrop et al.*, 1974; *Khurana and Schwarzl*, 2005]. This raises the question of what exactly is the wave speed and how does this depend on the configuration of the magnetotail where the plasma density, field strength, and thus Alfvén wave speed vary between the lobes and plasma sheet. This is also a current problem in understanding the communication of periodic signals in Saturn's magnetosphere. Study of this problem at Uranus and Neptune may shed light on these issues at Saturn.

7.5.4. Open Questions

The further study of the magnetotails of Uranus and Neptune clearly require additional data with instrumentation specifically designed to study these environments. For example, the particle fluxes often fell below the Voyager 2 instrument noise levels at Uranus and Neptune. In the absence of new in situ data from Uranus or Neptune for the next few decades progress might be made by examining the following open questions, which might be addressed by modeling (MHD, semi-empirical, quasi-equilibrium), theoretical considerations, and reanalysis of the Voyager 2 data in the light of the discussions in sections 7.5.2. and 7.5.3:

• How can we address the inherent time dependence and make sensible predictions?

• How can we apply new modeling techniques to address this time dependence and make predictions for new spacecraft missions?

• What exactly is the role of the solar wind in driving dynamics in the magnetotails of Uranus and Neptune?

• How much external driving might be masked by drastic diurnal variations?

ACKNOWLEDGMENTS

CSA was supported in this work by a Royal Society University Research Fellowship. CSA acknowledges Norman F. Ness (Bartol Research Institute, University of Delaware) and John Richardson (MIT) for the use of Voyager 2 solar wind data, which was obtained from the NASA National Space Science Data Center (NSSDC). CSA would like to thank the referees for their useful comments on the manuscript.

REFERENCES

Bagenal, F. (1992), Giant planet magnetospheres, *Annu. Rev. Earth Planet. Sci.*, *20*, 289–328.

Bagenal, F., J. W. Belcher, E. C. Sittler, Jr., and R. P. Lepping (1987), The uranian bow shock: Voyager 2 inbound observations of a high Mach number shock, *J. Geophys. Res.*, *92*(A8), 8603–8612.

Behannon, K. W., R. P. Lepping, E. C. Sittler, Jr., N. F. Ness, B. H. Mauk, S. M. Krimigis and R. L. McNutt, Jr. (1987), The magnetotail of Uranus, *J. Geophys. Res.*, *92*(A13), 15,354–15,366.

Belcher, J. W., et al. (1989), *Science*, *246*(4936), 1478–1483.

Broadfoot, A. et al. (1986), Ultraviolet spectrometer observations of Uranus, *Science*, *233*(4759), 74–79, doi:10.1126/science.233.4759.74.

Brown, M. E., and A. H. Bouchez (1997), The response of Jupiter's magnetosphere to an outburst on Io, *Science*, *278*, 268–271.

Cheng, A. F. (1987), Proton and oxygen plasmas as Uranus, *J. Geophys. Res.*, *92*(A13), 15,309–15,314.

Cheng, A. F. (1990), Triton torus and Neptune aurora, *Geophys. Res. Lett.*, *17*(10), 1669–1672.

Clarke, J., et al. (1986) Continued observations of the H Ly α emission from Uranus, *J. Geophys. Res.*, *(91)*A8, 8771–8781.

Connerney, J. E. P. (1993), Magnetic fields of the outer planets, *J. Geophys. Res.*, *98*(E10), 18,695–18,679.

Cowley, S. W. H. (2013), Response of Uranus' auroras to solar wind compressions at equinox, *J. Geophys. Res. Space Phys.*, *118*, 2897–2902, doi:10.1002/jgra.50323.

Cowley, S. W. H., S. V. Badman, E. J. Bunce, J. T. Clarke, J.-C. Gérard, D. Grodent, C. M. Jackman, S. E. Milan, and T. K. Yeoman (2005), Reconnection in a rotation-dominated magnetosphere and its relation to Saturn's auroral dynamics, *J. Geophys. Res.*, *110*, A02201, doi:10.1029/2004JA010796.

Cruikshank, D. P. (1995), *Neptune and Triton*, Univ. of Arizona Press, Tucson, Ariz.

Decker, R. B., and A. F. Cheng (1994), A model of Triton's role in Neptune's magnetosphere, *J. Geophys. Res.*, *99*(E9), 19,027–19,045.

Desroche, M., F. Bagenal, P. A. Delamere, and N. Erkaev (2012), Conditions at the expanded jovian magnetopause and implications for the solar wind interaction, *J. Geophys. Res.*, *117*, A07202, doi:10.1029/2012JA017621.

Desroche, M., F. Bagenal, P. A. Delamere, and N. Erkaev (2013), Conditions at the magnetopause of Saturn and implications for the solar wind interaction, *J. Geophys. Res. Space Phys.*, *118*(6), 3087–3095, doi:10.1002/jgra.50294.

Grocott, A., S. V. Badman, S. W. H. Cowley, S. E. Milan, J. D. Nichols, and T. K. Yeoman (2009), Magnetosonic Mach number dependence of the efficiency of reconnection between planetary and interplanetary magnetic fields, *J. Geophys. Res.*, *114*, A07219, doi:10.1029/2009JA014330.

Hammond, C. M., R. J. Walker, and M. G. Kivelson (1990), A pincer-shaped plasma sheet at Uranus, *J. Geophys. Res*, *95*(A9), 14,987–14,994.

Herbert, F. (2009), Aurora and magnetic field of Uranus, *J. Geophys. Res.*, *114*, A11206, doi:10.1029/2009JA014394.

Holme, R., and J. Bloxham (1996), The magnetic fields of Uranus and Neptune: Methods and models, *J. Geophys. Res.*, *101*(E1), 2177–2200.

Jackman, C. M. (2011), Saturn's radio emissions and their relation to magnetospheric dynamics, in *Proceedings of the Seventh International Workshop on Planetary, Solar and Heliospheric Radio Emissions (PRE VII)*, G. Fischer, W. S. Kurth, and P. Louarn. pp. 1–12, Austr. Acad. of Sci. Press.

Jackman, C. M, and C. S. Arridge (2011), Solar cycle effects on the dynamics of Jupiter's and Saturn's magnetospheres, *Solar Phys.*, *274*(1–2), 481–502, doi:10.1007/s11207-011-9748-z.

Kane, M., B. H. Mauk, E. P. Keath, and S. M. Krimigis (1991), Structure and dynamics of the uranian magnetotail: Results from hot plasma and magnetic field observations, *J. Geophys. Res.*, *96*(A7), 11,485–11,499.

Khurana, K. K., and H. K. Schwarzl (2005), Global structure of Jupiter's magnetospheric current sheet, *J. Geophys. Res.*, *110*, A07227, doi:10.1029/2004JA010757.

Lamy, L., R. et al. (2012), Earth-based detection of Uranus' aurorae, *Geophys. Res. Lett.*, *39*, L07105, doi:10.1029/2012GL051312.

Lennartsson, O. W., L. Kistler, and H. Rème (2007), Plasma sheet fine structure at high latitude, *Geophys. Res. Lett.*, *34*, L18103, doi:10.1029/2007GL030753.

Lepping, R. P. (1994), Comparisons of the field configurations of the magnetotails of Uranus and Neptune, *Planet. Space Sci.*, *42*(10), 847–857.

Macek, W. M., W. S. Kurth, R. P. Lepping, and D. G. Sibeck (1992), Distant magnetotails of the outer magnetic planets, *Adv. Space Res.*, *12*(8), 847–855.

Masters, A., J. P. Eastwood, M. Swisdak, M. F. Thomsen, C. T. Russell, N. Sergis, F. J. Crary, M. K. Dougherty, A. J. Coates, and S. M. Krimigis (2012), The importance of plasma ß conditions for magnetic reconnection at Saturn's magnetopause, *Geophys. Res. Lett.*, *39*, L08103, doi:10.1029/2012GL051372.

Matthews, M. S., J. T. Bergstralh, and E. D. Miner (1991), *Uranus*, Univ. of Ariz. Press, Tucson, Ariz.

Mauk, B. H., E. P. Keath, M. Kane, S. M. Krimigis, A. F. Cheng, M. H. Acuña, T. P. Armstrong, and N. F. Ness (1991), The magnetosphere of Neptune: Hot plasmas and energetic particles, *J. Geophys. Res.*, *96* (Suppl.), 19,061–19,084.

Mauk, B. H., S. M. Krimigis, E. P. Keath, A. F. Cheng, T. P. Armstrong, L. J. Lanzerotti, G. Gloeckler, and D. C. Hamilton (1987), The hot plasma and radiation environment of the uranian magnetosphere, *J. Geophys. Res.*, *92*(A13), 15,283–15,308.

McComas, D. J., F. Allegrini, F. Bagenal, F. Crary, R. W. Ebert, H. Elliott, A. Stern, and P. Valek (2007), Diverse plasma populations and structures in Jupiter's magnetotail, *Science*, *318*(5848), 217, doi:10.1126/science.1147393.

McNutt, R. L., Jr., R. S. Selesnick, and J. D. Richardson (1987), Low-energy plasma observations in the magnetosphere of Uranus, *J. Geophys. Res.*, *92*(A5), 4399–4410.

Ness, N. F., M. H. Acuña, L. F. Burlaga, J. E. P. Connerney, R. P. Lepping, and F. M. Neubauer (1989), *Science*, *246*(4936), 1473–1478.

Northrop, TG., C. K. Goertz, and M. F. Thomsen (1974), The magnetosphere of Jupiter as observed with Pioneer 10: 2. Nonrigid rotation of the magnetodisc, *J. Geophys. Res.*, *79*(25), 3579–3582.

Popoudin, S. Y., A. V. Artemyev, and K. V. Malova (2012), Kink instability of a cylindrical current sheet, *Cosmic Res.*, *50*(4), 282–292, doi:10.1134/S0010952512040053.

Richardson, J. D. (1993), A quantitative model of plasma in Neptune's magnetosphere, *Geophys. Res. Lett.*, *20*(14), 1467–1470.

Richardson, J. D., J. W. Belcher, R. S. Selesnick, M. Zhang, G. L. Siscoe, and A. Eviatar (1988), Evidence for periodic reconnection at Uranus? *Geophys. Res. Lett.*, *15*(8), 733–736.

Richardson, J. D., J. W. Belcher, A. Szabo, and R. L. McNutt, Jr. (1991), The plasma environment of Neptune, In *Neptune and Triton*, edited by D. P. Cruickshank, Univ. of Ariz. Press, Tucson, Ariz.

Schulz, M., and M. C. McNab (1996), Source-surface modeling of planetary magnetospheres, *J. Geophys. Res.*, *101*(A3), 5095–5118.

Schulz, M., M. C. McNab, R. P. Lepping, and G.-H. Voigt (1995), Magnetospheric configuration of Neptune, *Neptune and Triton*, edited by D. P. Cruickshank, Univ. of Ariz. Press, Tucson, Ariz.

Scurry, L., and C. T. Russell (1991), Proxy studies of energy transfer to the magnetosphere, *J. Geophys. Res.*, *96*, 9541–9548.

Selesnick, R. S. (1990), Plasma convection in Neptune's magnetosphere, *Geophys. Res. Lett.*, *17*(10), 1681–1684.

Selesnick, R. S., and R. L. McNutt, Jr. (1987), Voyager 2 plasma ion observations in the magnetosphere of Uranus, *J. Geophys. Res.*, *92*(A13), 15,249–15,262.

Selesnick, R. S., and J. D. Richardson (1986), Plasmasphere formation in arbitrarily oriented magnetospheres, *Geophys. Res. Lett.*, *13*(7), 624–627

Thomas, N., F. Bagenal, T. W. Hill, and J. K. Wilson (2004), The Io neutral clouds and plasma torus, in *Jupiter: The Planet, Satellites and Magnetosphere*, edited by F. Bagenal et al., Cambridge Univ. Press, Cambridge, England.

Tiu, D., I. H. Cairns, X. Yuan, and P. A. Robinson (2011), Evidence for reformation of the uranian bow shock: Hybrid simulation and comparisons with Voyager data, *J. Geophys Res.*, *116*, A04228, doi:10.1029/2010JA016057.

Tóth, G., D. Kovács, K. C. Hansen, and T. I. Gombosi (2004) Three-dimensional MHD simulations of the magnetosphere of Uranus, *J. Geophys. Res.*, *109*, A11210, doi:10.1029/2004JA010406.

Vasko, I. Y., H. V. Malova, A. V. Artemyev, and L. M. Zelenyi (2012), Charged particle acceleration by induction electric field in Neptune magnetotail, *Planet. Space Sci.*, *73*, pp. 168–177, doi:10.1016/j.pss.2012.09.010.

Vasyliunas, V. M. (1986), The convection-dominated magnetosphere of Uranus, *Geophys. Res. Lett.*, *13*(7), pp. 621–623.

Voigt, G.-H. (1986), Field line twist and field-aligned currents in an axially symmetric equilibrium magnetosphere, *J. Geophys. Res.*, *91*(A10), 10995–11002.

Voigt, G.-H., and N. F. Ness (1990), The magnetosphere of Neptune: Its response to daily rotation, *Geophys. Res. Lett.*, *17*(10), 1705–1708.

Voigt, G.-H., K. W. Behannon, and N. F. Ness (1987), Magnetic field and current structures in the magnetosphere of Uranus, *J. Geophys. Res.*, *92*(A13), 15337–15346.

Volwerk, M., et al. (2013), Comparative magnetotail flapping: an overview of selected events at Earth, Jupiter and Saturn, *Ann. Geophys.*, *31*(5), 817–833, doi:10.5194/angeo-31-817-2013.

8

Satellite Magnetotails

Xianzhe Jia

8.1. INTRODUCTION

The interaction of the solar wind with planets gives rise to planetary magnetospheres with an extended region stretching in the antisunward direction, called the "magnetotail" (see Chapter 1). Like planets, planetary moons can also be considered to possess magnetotails. Here, for moons, "magnetotail" refers to the wake region behind a moon formed as a result of its interaction with the ambient flowing plasma within the magnetosphere of its parent planet or the solar wind. Long-term studies of planetary magnetospheres have established that the properties and structures of planetary magnetotails differ greatly at different planets (as reviewed in Chapters 2–7) depending on various factors, such as the external solar wind conditions, the internal magnetic field and the planetary rotation rate, the properties of the atmosphere/ionosphere, and internal plasma sources. Similarly, how the particles and fields are distributed in satellite magnetotails is determined by a number of factors, including the ambient plasma conditions and the intrinsic properties of the moons, such as their magnetic character (magnetized vs. unmagnetized), and their abilities to supply neutrals and plasma (which in turn depends on whether or not they possess an atmosphere/ionosphere).

The two gas giants in our solar system, Jupiter and Saturn, both have a large number (over 60) of satellites. Most of these satellites are small bodies (with radii <200 km) and irregular in shape, and their physical properties are not well known due to lack of in situ spacecraft observations. The major satellites (whose radii exceed ~200 km), on the other hand, have been visited by a number of spacecraft with one or more close encounters. Observations acquired in the vicinities of the major satellites in the two systems have revealed that they exhibit great diversity in their magnetic properties and particle environments (see reviews by *Kivelson et al.* [2004], *Jia et al.* [2010a], *Mauk et al.* [2009], and *Krupp et al.* [2010] and references therein). As a result, their interactions with surrounding medium differ in many ways, resulting in wake regions with different structures.

In this chapter, we use the major satellites of Jupiter and Saturn as examples to illustrate the variety of ways in which planetary moons interact with their external plasma environments as inferred from in situ and remote observations as well as physics-based numerical simulations, with a focus on the structures and properties of their resulting magnetotails (or wake regions). Satellites that will be discussed in this chapter are listed in Table 8.1, which provides an overview of their general properties that are particularly pertinent to the topic of this chapter, such as the sources of magnetic field and neutral gas. The table also lists the typical conditions of the ambient plasma flows characterized in terms of the various Mach numbers, which, as will be discussed later, are useful for understanding important aspects of the plasma interaction and wake structure at the moons. It should be noted that although some of Saturn's outer moons, such as Titan and Iapetus, may spend a fraction of their orbital period outside of Saturn's magnetosphere, most of the moons that will be discussed here always remain inside of their parent magnetospheres. At the orbits of these moons, the ambient magnetospheric plasma approximately corotates with the planet at bulk flow speeds much larger than the Keplerian speeds of the moons' orbital motion. Therefore, the corotating plasma continually overtakes the moons from their trailing sides, resulting in wakes or tails on the leading sides.

Department of Atmospheric, Oceanic and Space Sciences, University of Michigan, Ann Arbor, Michigan

Magnetotails in the Solar System, Geophysical Monograph 207. First Edition.
Edited by Andreas Keiling, Caitríona M. Jackman, and Peter A. Delamere.

Table 8.1 General properties of the major satellites of Jupiter and Saturn.

Satellite	Radius (km)	Orbital Dist.	Neutral Source[a]	Source(s) of Magnetic Field[b]	Ambient Flow Conditions[c]			
					M_A	M_s	M_{ms}	β
			Jupiter					
Io	1821.6	5.91	Yes(SO_2)	Plasma, Ind	0.3	2.0	0.3	0.3
Europa	1560.8	9.40	Yes(O_2)	Ind, Plasma	0.5	0.9	0.4	0.3
Ganymede	2631.2	14.97	Yes(O_2)	Intrinsic, Plasma, Ind	0.7	0.5	0.4	2.4
Callisto	2410.3	26.33	Yes (O_2/CO_2?)	Ind, Plasma	2.8	0.4	0.4	64
			Saturn					
Enceladus	252.1	3.95	Yes(H_2O)	Plasma (Ind?)	0.1	1.4	0.1	0.005
Tethys	533.0	4.89	No	Inert	0.2	1.3	0.2	0.03
Dione	561.7	6.26	Yes(O_2)	Inert	0.4	1.1	0.2	0.14
Rhea	764.3	8.74	Yes (O_2/CO_2)	Inert	0.6	1.3	0.6	0.26
Titan	2575.5	20.27	Yes(N_2)	Plasma(Ind?)	~1	0.6	0.6	~11
Iapetus[d]	735.6	59.09	No	?	10	8	10	0.5

[a]This column indicates whether or not a moon possesses a substantial source of neutrals and only major neutral species are listed in the parentheses.
[b]This column describes the origin of magnetic fields in the vicinity of the moons. "Intrinsic" refers to intrinsic magnetic field, "Ind" refers to internally induced field through electromagnetic induction, "Plasma" refers to plasma-induced magnetic field, and "Inert" refers to moons that are unmagnetized and do not possess substantial sources of neutrals. The question mark means that situation is unclear.
[c]The dimensionless numbers used to characterize the ambient flow conditions refer to: $M_A = V_{flow}/V_A$, the Alfvénic Mach number: $M_s = V_{flow}/C_s$, the sonic Mach number: $M_{ms} = V_{flow}/\sqrt{V_A^2 + C_s^2}$, the magnetosonic Mach number (where V_{flow}, V_A, and C_s represent the flow speed, Alfvén speed and sonic speed of the ambient plasma, respectively), and β, the ratio of thermal to magnetic pressure. For Jupiter's moons, are listed the average inferred values near Jupiter's central plasma sheet obtained from *Kivelson et al.* [2004]. The average flow conditions of Saturn's magnetospheric plasma near the Saturnian moons come from *Neubauer et al.* [2006] and *Khurana et al.* [2008] .
[d]Because Iapetus spends much of its orbital time in the solar wind due to its large orbital distance, the flow conditions listed here correspond to the typical solar wind conditions near Saturn's orbit.

The chapter is organized as follows. The large-scale interaction and wake structure associated with inert moons are described in section 8.2. Conducting/mass-loading moons are discussed in section 8.3, and the magnetotail structure in the case of a strongly magnetized moon, Ganymede, is covered in section 8.4. Finally, section 8.5 provides a summary and concluding remarks.

8.2. INERT MOONS

For an inert moon, which refers to a moon that is nonconducting and possesses neither an appreciable global magnetic field nor a substantial source of neutral gas, its interaction with the external plasma environment can be characterized essentially as that of plasma flowing past an absorbing body, a situation analogous to that of Earth's Moon (see Chapter 9). Plasma flowing onto an inert moon directly interacts with the moon's surface and, to a large extent, gets absorbed on the ram side. The interaction then leads to a plasma void immediately behind the moon, forming the so-called "plasma wake" (Figure 8.1a). In the central wake, plasma density generally drops to very low levels compared to the background, and as a result of maintaining the balance of total pressure (plasma pressure plus magnetic pressure), the magnetic field lines are drawn into the plasma depletion region from the outside, resulting in an enhancement in the field strength. The wake is being gradually refilled with distance from the moon as the plasma moves toward the center of the wake along the magnetic field lines. The global interaction geometry is largely determined by the propagation of various magnetohydrodynamics (MHD) wave modes (slow, fast, and Alfvén) triggered near the surface of the absorbing body. The three MHD wave modes produce magnetic field and plasma perturbations with different characteristics. In the fast mode, which can propagate in any direction, magnetic field and plasma density vary in phase. The slow mode, which propagates mainly in the field-aligned direction, causes magnetic field and plasma density to vary in antiphase. The Alfvén mode, which also propagates mainly along the magnetic

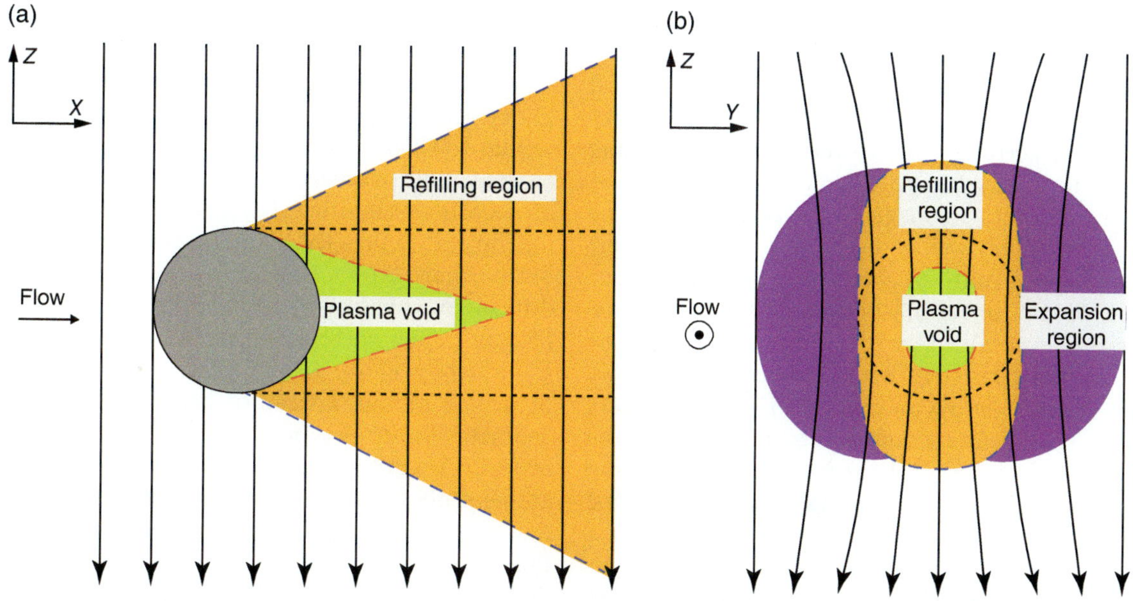

Figure 8.1 Schematic representation of the geometry of plasma interaction with an inert moon for a case when the incident flow is perpendicular to the ambient magnetic field, a situation appropriate for the Saturnian inert moons of interest. The illustration is shown in a moon-centered Cartesian coordinate system commonly used in studying plasma-moon interactions, where $+\hat{X}$ is along the incident flow direction, $+\hat{Y}$ points toward the parent planet, and $+\hat{Z}$ completes the right-handed system pointing northward. (a) View in an *XZ* plane that contains the directions of the unperturbed flow and the background magnetic field. (b) View in a *YZ* plane at $X > 1R_{moon}$ (R_{moon} is the moon's radius) in the wake. Various regions of the wake are highlighted with different colors. The red dashed lines mark the boundaries of the plasma void near the center of the wake while the blue dashed lines show the wavefronts of the rarefaction wave. The dotted lines delineate the boundaries of the geometric wake.

field direction, introduces perturbations in the transverse components of the magnetic field and thus does not change the magnetic field strength and plasma density. As a result of the nonisotropic propagation of different wave modes, the resulting wake structure is three dimensional in nature with properties differing in the directions parallel and perpendicular to the magnetic field (except when the incident flow is parallel to the background field, for which case the wake is symmetric around the flow direction). In the direction perpendicular to the ambient flow and magnetic field (Figure 8.1b), adjacent to the field enhancement region near the center of the wake there exists a pair of "expansion regions" formed as a result of the propagation of rarefaction fast-mode waves into the external flow. Within the expansion region, magnetic field strength and plasma density vary in phase, both decreasing compared to the background. Details of the wake structure, including the extent and strength of the flow and magnetic field perturbations, depend on the ambient flow and field conditions, such as the angle between the incident flow and the background field (the wake refills more quickly when the external magnetic field is oblique to the incident flow) and the ratios between the flow speed and the various characteristic MHD wave speeds (given by the Mach numbers), which, as will be discussed below, are particularly important for understanding the differences between a lunar-type interaction and the interaction with the Saturnian inert moons discussed below.

Several of Saturn's icy moons may be assigned to the category of inert interaction. These include Tethys, Dione, and Rhea, all of which are located in the inner part of Saturn's magnetosphere with mean orbital distances of $4.9R_S$, $6.3R_S$, and $8.7R_S$ (R_S = 60, 268 km is Saturn's radius), respectively. In addition, Saturn's outermost regular satellite, Iapetus, may possibly belong to this category, but the magnetic field observations obtained during the only close flyby from Cassini, which occurred when the moon was in the solar wind, did not allow for definitive assessment. Direct in situ magnetic field measurements acquired by the Cassini spacecraft in the close vicinity of these icy moons show no evidence of any appreciable internal magnetic field for any of the moons [*Khurana et al.*, 2008; *Simon et al.*, 2011]. Signatures of a thin, sputtering-induced oxygen exosphere at Dione [*Tokar et al.*, 2012] and Rhea [*Teolis et al.*, 2010] have been detected through Cassini measurements. The weak exospheres interact with their surrounding plasmas by producing pickup ions [*Tokar et al.*, 2012], causing absorptions of energetic charged particles [*Roussos et al.*, 2012], and, to some extent, perturbing the upstream plasma and fields [*Simon et al.*, 2011], but they do not appear to produce significant perturbations to the large-scale plasma interactions. To a large degree, the global interactions of these moons with Saturn's magnetospheric plasma can be understood reasonably well in the context of plasma inert moon interaction as outlined above.

Magnetic field observations obtained near the moons, especially those obtained in the wake regions, show signatures largely consistent with those expected for an inert moon interaction. As an example, Figure 8.2 shows Cassini magnetic field measurements acquired on the downstream flyby of Tethys [*Khurana et al.*, 2008] during which the spacecraft traversed Tethys' wake at a distance of ~2 Tethys radii south of the moon. The data show a weak enhancement in the magnetic field strength near closest approach with field depressions on both sides (corresponding to the "expansion region" discussed above). The field enhancement in the central wake is consistent with absorption of the incident flow by the moon on the upstream side and plasma depletion in the wake. In addition, there is bipolar signature (from negative to positive) in the transverse component of the magnetic field (B_y) near closest approach, consistent with magnetic field lines being sucked into the wake (see Figure 8.1b) to maintain the balance of total pressure. Magnetic field data from wake flybys of Rhea [*Khurana et al.*, 2008] and Dione [*Simon et al.*, 2011] show features qualitatively similar to those seen at Tethys.

The Cassini observation of a plasma depletion region at locations significantly below the geometric wake during the Tethys flyby points to an important aspect of the inert interactions at the Saturnian moons that differs considerably from that at Earth's Moon. The plasma depletion and field enhancement region extend significantly along the magnetic field direction at the Saturnian moons, instead of being confined close to the geometric wake as typically seen in the lunar-type interaction. This difference arises due to the properties of the ambient plasma flow. The net flux of sputtering particles that impinges on a moon's surface during the time a flux tube takes to flow past the moon's diameter depends on the plasma density as well as the ratio of the sound speed (which determines the flow speed along the flux tube) to the flow speed (which establishes the duration of the transit across the moon), or the inverse of the sonic Mach number M_s. For Earth's Moon, which spends much of its orbital time in the supersonic solar wind, that ratio typically is $\ll 1$ (or $M_s \gg 1$). For the Saturnian moons discussed here (Table 8.1), the ambient plasma flow

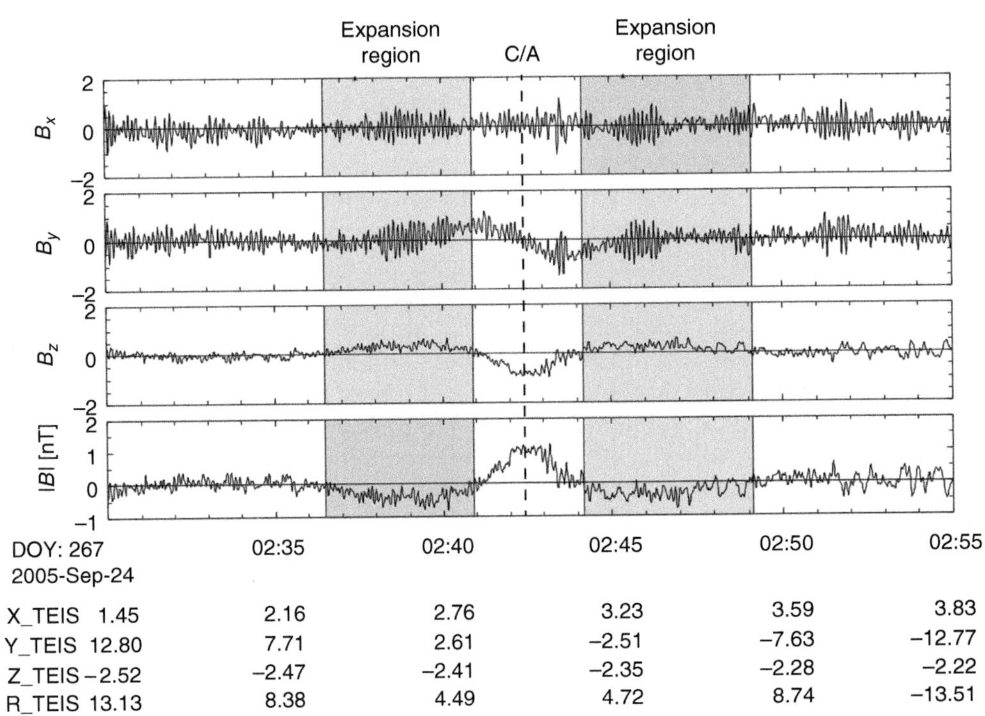

Figure 8.2 Detrended magnetic field data obtained by Cassini during a close encounter with Tethys in the downstream region. The data are presented in a Tethys-centered coordinate system defined in Figure 8.1. The vertical dashed line marks the time of closest approach. Adapted from *Khurana et al.* [2008].

typically is transonic ($M_s \sim 1$). As a result, the absorption of particles streaming along a flux tube in contact with the moon leads to a more extended plasma depletion region (in which magnetic field strength increases) around the moon in a transonic interaction than in a supersonic interaction.

Global simulations, especially hybrid simulations, have been applied to study the plasma interactions with the Saturnian inert moons [*Roussos et al.*, 2008; *Simon et al.*, 2009, 2011], providing useful global context for interpreting in situ observations. In a hybrid model, electrons are treated as a fluid while ions are represented as individual macroparticles, thereby allowing for resolving ion kinetic effects (e.g., finite gyroradius effects and non-Maxwellian particle distributions) that are important for understanding the refilling process and the structure of the wake. By treating the moons as pure plasma-absorbing bodies, the hybrid simulations of the Saturnian moons are able to reproduce reasonably well the observed magnetic signatures, thereby confirming that the principal interaction mode at these moons can be characterized, to a good approximation, as inert interaction. The hybrid simulations have also provided important physical insights into the 3D wake structure at the moons. In particular, they show that for transonic upstream flows the plasma depletion region at the Saturnian moons extends over large distances along the magnetic field direction as suggested by the in situ observations.

8.3. CONDUCTING/MASS-LOADING MOONS

A number of moons in the two gas giant systems have been found to possess global conducting regions (ionosphere or subsurface conducting layer) and/or significant sources of neutrals that can appreciably perturb the ambient plasma and fields. In particular, all four Galilean satellites of Jupiter, Io, Europa, Ganymede (which also has a strong intrinsic magnetic field and will be discussed separately in section 8.4), and Callisto, have tenuous atmospheres (see a review by *McGrath et al.* [2004]). Saturn's largest moon, Titan, has a dense atmosphere and a well-developed ionosphere [e.g., *Sittler et al.*, 2010]. Another Saturnian moon, Enceladus, albeit small in size (radius is ~252 km), has been found during the Cassini mission to possess active water plumes near its south pole. These plumes outgas neutral and dust particles into the Saturn system [e.g., *Dougherty et al.*, 2006; *Waite et al.*, 2006; *Hansen et al.*, 2006], providing a major source of plasma to Saturn's magnetosphere.

Common to the moons mentioned above is the existence of a significant region of electrical conductivity, which may arise from an ionosphere (Pedersen and Hall conductivities) or result from the so-called mass-loading process in which neutrals emanating from the moon are ionized through processes like photoionization, electron impact ionization, or charge exchange, and the freshly ionized particles are subsequently picked up by the ambient flow generating what is referred to as a "pickup current" [*Goertz*, 1980]. The contribution of mass loading to the plasma currents in the interaction can be equivalently described as a Pedersen conductivity [e.g., *Neubauer*, 1998]. For most of the moons of interest here, which are embedded within their parent magnetospheres (with the exception of Titan, which may sometimes be exposed to Saturn's magnetosheath flow or the solar wind), the ambient plasma flows at the moons' orbits are typically sub-Alfvénic ($M_A < 1$) with possible exceptions at Callisto and Titan (see Table 8.1).

Much of the large-scale interaction of a conducting/mass-loading moon with a sub-Alfvénic flow ($M_A < 1$) generally can be understood in terms of the "Alfvén wing" model (Figure 8.3), which has been initially applied to describe Io's interaction with the plasma torus [*Neubauer*, 1980; *Southwood et al.*, 1980] and later expanded to study the plasma interactions at other planetary satellites in the Jovian and Saturnian systems (see reviews by *Kivelson et al.* [2004] and *Jia et al.* [2010a]). As a magnetized plasma flows toward a conducting obstacle, it is slowed down and diverted around the moon because the incoming plasma and field are unable to penetrate into the conductor. The deceleration of the flow produces perturbations carried by various MHD wave modes, among which compressional perturbations can propagate effectively across the magnetic field, thereby producing disturbances (e.g., magnetic field enhancement) upstream of the moon if the flow is submagnetosonic ($M_{ms} < 1$), as is the case for most of the moons discussed here. The slowing of the flow near the obstacle also generates Alfvén waves propagating away from the obstacle along magnetic field lines at the Alfvén speed (V_A) and consequently slowing the flow at large distances above and below the moon. As a result, those field lines are bent (Figure 8.3a) and draped around the moon (Figure 8.3b). These Alfvén waves carry field-aligned currents, which close through perpendicular currents in the conducting region of the obstacle, such as the Pedersen currents in the ionosphere or the pickup currents through the mass-loading region. In the rest frame of the moon, the perturbations associated with the field-aligned currents cause the field lines to bend back within the region called the "Alfvén wing," whose boundaries are tilted relative to the background field at an angle $\theta = \tan^{-1}\left(V_{flow}/V_A\right)$, where V_{flow} is the speed of the unperturbed plasma flow in

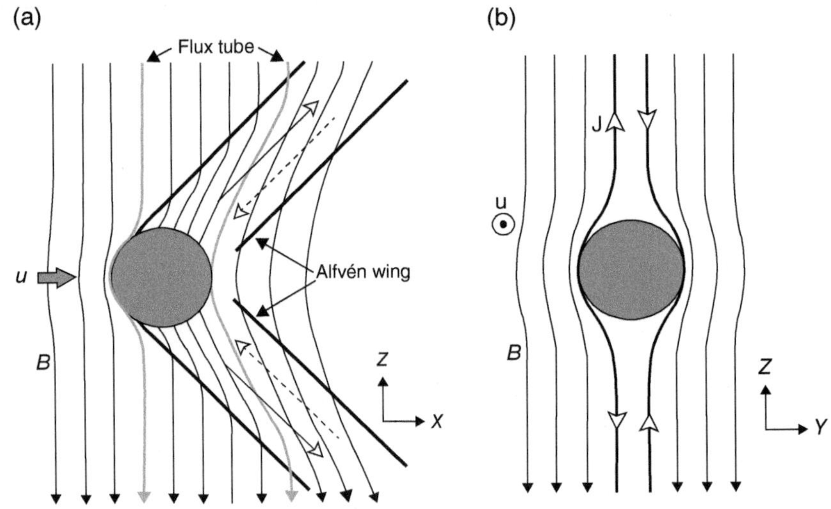

Figure 8.3 Schematic of the Alfvén wing structure arising from the sub-Alfvénic plasma interaction with a conducting moon or a moon with a significant mass-loading source (Adapted from *Kivelson et al.* [2004].) The coordinate system used here is the same as that defined in Figure 8.1. (a) Plane (*XZ* plane) that contains the directions of the unperturbed flow ($+\hat{X}$) and the background magnetic field ($-\hat{Z}$). (b) Cross section of the interaction region in a plane (*YZ* plane) normal to the plasma flow direction. The Alfvén wing currents flowing along the field lines are represented by arrows. Solid and dashed lines represent the currents flowing on the $-\hat{Y}$ and $+\hat{Y}$ side, respectively.

the rest frame of the moon [*Neubauer*, 1980; *Southwood et al.*, 1980].

As a consequence of the interaction described above, the structure of the wake region associated with a conducting/mass-loading moon exhibits notable differences from that of an inert moon, as discussed in section 8.2. For a conducting/mass-loading moon, the magnetic field strength is generally depressed in the central wake (as opposed to being enhanced in the case of an inert moon) where the slowed flows due to the interaction with the conducting body (including ionosphere) or the mass-loading process are reaccelerated to the incoming flow speed by the curvature force associated with those kinked field lines (Figure 8.3a). As viewed in a plane perpendicular to the incident flow direction, magnetic field lines are draped around a conducting obstacle and bulge out in the wake region (Figure 8.3b), rather than being sucked into the wake in the case of an inert moon (Figure 8.1b).

Many aspects of the observations in the wakes of conducting/mass-loading moons can be well understood in terms of the model described above. As an example, here we provide some brief highlights on the observed wake structure of Jupiter's moon Io, which represents a prototype of the plasma interaction with a conducting/mass-loading moon that has been extensively studied (see the review by *Saur et al.* [2004] and references therein). We focus on Galileo's first close encounter with Io (I0 flyby) during which the spacecraft flew through the wake near the moon's equator (Figure 8.4). In the central wake, the magnetometer detected a strong depression of $\sim 40\%$ in the magnetic field strength [*Kivelson et al.*, 1996a]. The field depression region contains cold and dense plasma with nearly stagnant flows [*Frank et al.*, 1996], a population believed to originate from Io's downstream ionosphere. Adjacent to the cold plasma wake on the two sides are regions of hot and tenuous plasma with fast flows. These are pickup ions that are produced on flux tubes convecting from the upstream to the downstream along the flanks, where the flow speeds up and consequently the temperature of pickup ions is high. Field-aligned bidirectional electron beams with energies from hundreds of eV to several keV have been observed in Io's wake [*Frank and Paterson*, 1999; *Williams et al.*, 1996]. Although their generation mechanism is not well understood, these energetic electrons provide an important ionization source that may contribute to the formation of Io's downstream ionosphere and cold plasma wake observed by Galileo [*Saur et al.*, 2002; *Dols et al.*, 2012].

In addition to sources such as mass loading or a conducting ionosphere that can generate significant perturbations to the inflow plasma, another important source of perturbation arises if the moon has an internally induced magnetic field. When a conductor is embedded in an environment with a time-changing magnetic field, it responds to the driving field by generating electric currents that flow near the surface of the conductor. Those

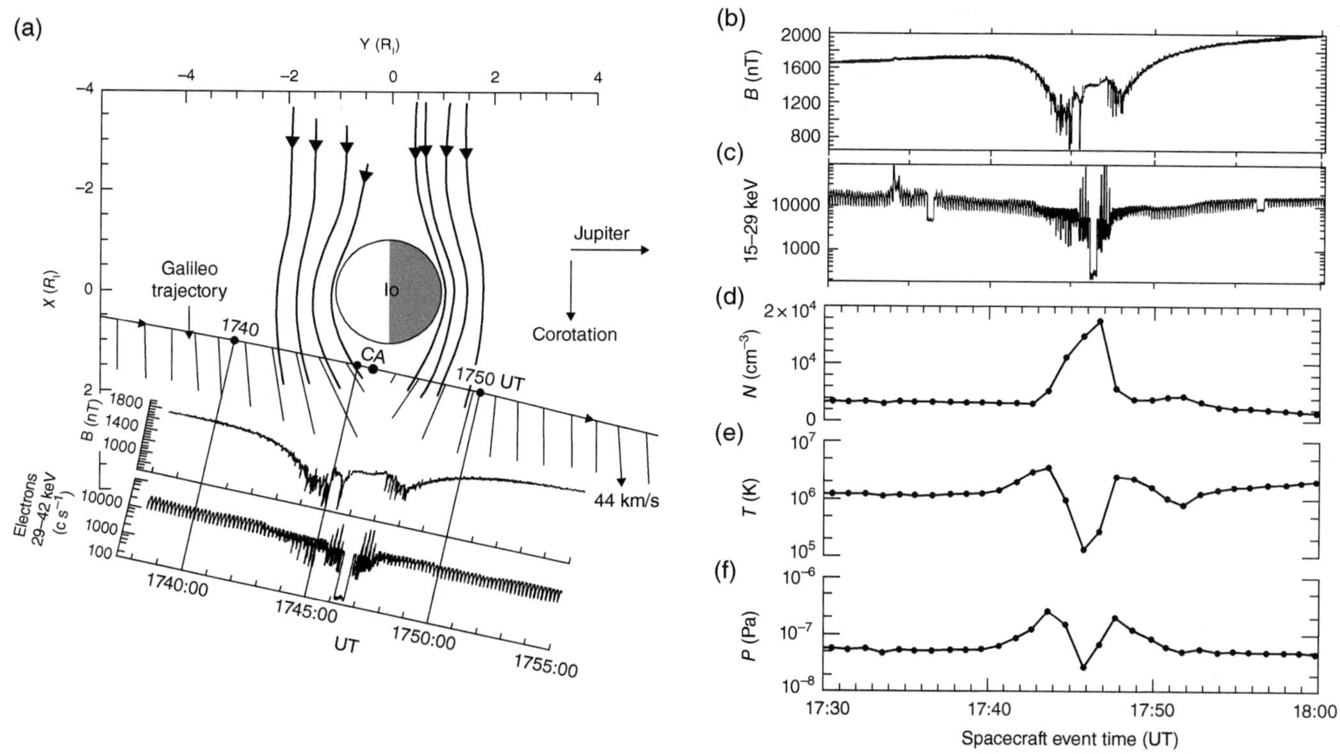

Figure 8.4 Overview of Galileo particles and field measurements in Io's wake during the I0 flyby. (Adapted from *Williams et al.* [1999].) (a) Galileo trajectory along with plasma flow vectors measured by the Galileo PLS [*Frank et al.*, 1996]. Also plotted are the magnetic field strength measured by the Galileo magnetometer [*Kivelson et al.*, 1996a] and the 29–42 keV electron intensities measured by the Galileo EPD [*Williams et al.*, 1999]. (b–f) Time series of measured magnetic field strength, 15–29 keV electron intensities, plasma density, plasma temperature, and plasma thermal pressure, respectively.

currents induce an external magnetic field perturbation that acts to prevent the driving field from penetrating into the conductor. At Saturn, because of the high degree of axial symmetry of the planet's internal field, the magnetospheric field at the orbits of Saturn's moons does not contain a significant time-varying component that can drive strong magnetic induction. At Jupiter, however, due to the $\sim 10°$ tilt of its magnetic axis from the rotational axis, the Jovian dipole equator sweeps over the Galilean satellites at their synodic periods (the rotation period of Jupiter seen in the rest frame of the moon), resulting in periodically changing background magnetic field at the orbits of the moons that provides a strong driving signal for magnetic induction. It is now known based on in situ magnetic field observations that each of the four Galilean moons harbors a global subsurface liquid layer generating an induced magnetic field in response to Jupiter's time-varying magnetospheric field. For the three icy moons Europa, Callisto, and Ganymede, the liquid layers are composed primarily of water [*Khurana et al.*, 1998; *Kivelson et al.*, 2000, 2002], whereas for Io the subsurface conducting layer is made of partially molten rocks [*Khurana et al.*, 2011]. The presence of an induced magnetic field alters the global interaction geometry in that it tends to displace and skew the Alfvén wings and breaks the symmetry of the wake [*Neubauer*, 1999]. Such an effect may be most readily detected at Europa, where the induction field has strength comparable to that arising from the external plasma interaction whereas at the other Galilean moons the effect of the induction field may be masked by either strong plasma interactions (for example, at Io) or strong intrinsic field (for example, at Ganymede). Using Galileo magnetic field observations from flybys that passed through Europa's Alfvén wing, *Volwerk et al.* [2007] confirmed the prediction of shrinkage and displacement of the Alfvén wing structure due to the induced field. Furthermore, the induced field also appears to modify the wake geometry by shifting the region of enhanced plasma density off the equatorial plane, and this effect may account for the plasma density structure observed in Europa's wake [*Paterson et al.*, 1999], as suggested by global MHD simulations of Europa's plasma interaction [*Schilling et al.*, 2008].

The recent discovery of a significant source of dust particles at Saturn's moon Enceladus [e.g., *Farrell et al.*,

2009; *Morooka et al.*, 2011] presents interesting new perspectives on plasma interaction with mass-loading moons. Upon becoming electrically charged (either positively or negatively), a dust particle can interact with the surrounding plasma environment through the electric and magnetic fields. Analysis of in situ data and numerical simulations suggest that charged dust plays a key role in the Enceladus interaction with Saturn's magnetospheric plasma. It not only contributes significantly to the slowing of the flow in the vicinity of the moon [*Omidi et al.*, 2012] but also greatly modifies the field structure and plasma flow pattern within the Alfvén wings [*Simon et al.*, 2011; *Kriegel et al.*, 2011; *Jia et al.*, 2012]. In particular, the absorption of electrons by the heavy dust grains within the Enceladus' plume makes the water group ions the major current carriers. Under such circumstances, the resultant Hall current flows in the opposite direction to that in a system where only ions and electrons are present, leading to the so-called anti-Hall effect [*Simon et al.*, 2011]. It is likely that charged dust also modifies Enceladus' wake structure and associated tail current system, and this effect needs to be assessed in future investigations of Enceladus' interaction with Saturn's magnetosphere.

Although the Alfvén wing model is often invoked to describe sub-Alfvénic plasma interactions, it provides a valid description of the large-scale interaction for cases where the incident flow is trans- or super-Alfvénic ($M_A \geq 1$) but sub-magnetosonic ($M_{ms} < 1$), as for the ambient conditions near Saturn's moon Titan. Under such conditions, the global shape of the interaction region can be regarded as a continuum that follows the "Alfvén wing" geometry arising from a sub-Alfvénic interaction. The resultant wake region exhibits the form in which the draped magnetic field lines pinch together on the downstream side forming a magnetotail with a current sheet separating oppositely oriented magnetic-field lines above and below. Cassini observations from more than 80 close flybys of Titan, which covered various regions of the Titan system under a wide range of ambient field and plasma conditions, have revealed many new aspects of the Titan system, including complex plasma and field structures of the induced magnetotail. Discussions on Titan's interaction with Saturn's magnetosphere, especially in light of recent Cassini observations, can be found in a recent review by *Sittler et al.* [2010].

The electromagnetic interaction of a planetary satellite (with either a significant mass-loading source or a strong intrinsic field) with its parent magnetosphere produces disturbances not only in the close vicinity of the moon but also in the ionosphere of its parent planet. One manifestation of this remote coupling is the auroral footprints associated with planetary moons [e.g., *Bonfond*, 2012], which have been observed for Jupiter's Galilean satellites, Io, Europa, and Ganymede [e.g., *Connerney et al.*, 1993; *Clarke et al.*, 2002], and Saturn's moon Enceladus [*Pryor et al.*, 2011]. These auroral emissions are excited by precipitating electrons that are accelerated by the Alfvén waves produced due to the plasma interactions with the moons. Although some of the Alfvén wave energy is reflected at boundaries of strong plasma density gradient or converted into other means of particle energy, a fraction of the wave energy can escape the relatively dense plasma regions around the moons (possibly facilitated by turbulent wave filamentation [*Hess et al.*, 2011]) and reach the high-latitude regions of their parent planets, where parallel electric fields develop and accelerate electrons to produce auroral emissions. While the footprints associated with Ganymede and Enceladus have so far been observed as individual spots, the auroral footprints of Io and Europa are seen to be accompanied by extended downstream tails [*Clarke et al.*, 2002; *Grodent et al.*, 2006]. When magnetically mapped from Jupiter's ionosphere to the moons' orbits, these long emission trails correspond to wake regions with scales much larger than the moons themselves. A long wake is expected for a moon with a significant mass-loading source. As pointed out by *Southwood and Dunlop* [1984], if there is substantial ion pickup near the moon (such as Io), the reacceleration of the mass-loaded flux tubes downstream of the moon would occur over a distance that is inversely proportional to α^2, where α denotes the fractional reduction of velocity due to mass loading. Depending on the strength of mass loading, the length of the wake associated with a mass-loading moon can be much larger than the distance over which pickup ions are produced. The field-aligned currents associated with the reacceleration of the mass loaded flux tubes in the extended wake region that link the moons to Jupiter's ionosphere may, therefore, account for the long tails downstream of Io's and Europa's footprints [e.g., *Hill and Vasyliūnas*, 2002; *Delamere et al.*, 2003].

8.4. A STRONGLY MAGNETIZED MOON: GANYMEDE

For a strongly magnetized moon, the interaction of its internal magnetic field with the flowing plasma predominantly controls the global structure of the interaction region. Jupiter's moon Ganymede provides the only known example in the solar system that possesses a strong intrinsic field, which was discovered during the Galileo mission [*Kivelson et al.*, 1996b]. Ganymede's internal magnetic field is sufficiently strong to overpower Jupiter's magnetospheric field and stand off the Jovian plasma above its surface, forming a mini-magnetosphere with a magnetotail resembling those of magnetized planets.

In situ measurements from Galileo during its close flybys of the moon, which passed through different regions of the moon's magnetosphere (Figure 8.5) under different external conditions, have shown that Ganymede's magnetosphere shares similarities with planetary magnetospheres in terms of common properties familiar from the long-term studies of planetary magnetospheres. For instance, Ganymede's magnetospheric interaction not only generates a rich set of wave phenomena like those seen at Earth and Jupiter [*Gurnett et al.*, 1996; *Volwerk et al.*, 2013] but also produces its own aurora that appears at the boundary between open and closed field lines [*Feldman et al.*, 2000; *McGrath et al.*, 2013] and ionospheric outflows [*Frank et al.*, 1997a; *Vasyliūnas and Eviatar*, 2000] like those observed in Earth's ionosphere.

Galileo observations have also revealed that, despite many similarities, Ganymede's magnetosphere differs from those of planets due to the nature of the external flow and field conditions as well as the properties of the moon itself. At Ganymede's orbit, the ambient Jovian plasma flow is typically sub-Alfvénic ($M_A < 1$) and sub-sonic ($M_s < 1$). Under such circumstances, Jupiter's corotating plasma directly impinges on and interacts with Ganymede's magnetosphere without being modified by a bow shock, which is a common feature of the planetary magnetospheres. As shown in Figure 8.5, the sub-Alfvénic interaction at Ganymede gives rise to a cylinder-shaped magnetospheric configuration differing dramatically from the bullet shape of planetary magnetospheres produced by interaction with the super-magnetosonic solar wind. The magnetosphere contains a large polar cap consisting of open field lines that link to Jupiter and a small closed field line region near the equator extending above the moon's surface at a distance of $\sim 1\, R_G$ ($R_G = 2634$ km is Ganymede's radius) on the upstream side. Besides the familiar structures such as the magnetopause and tail current sheet, Ganymede's magnetosphere generates a pair of nearly vertically extended Alfvén wings that mediate the interaction of Ganymede with the local plasma and Jupiter's ionosphere through field-aligned currents. The field-aligned currents in the Alfvén wings close near Ganymede not only through the moon and its ionosphere (like that in the case of a conducting moon discussed in section 8.3) but also through the magnetopause and tail current sheet [*Jia et al.*, 2008].

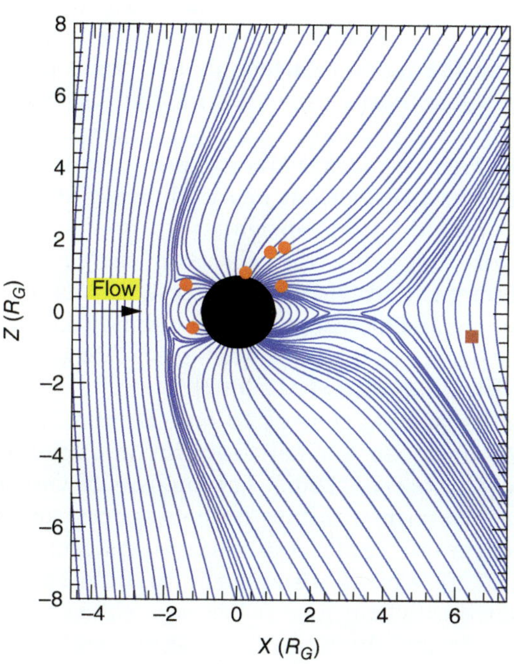

Figure 8.5 Trajectories of the Galileo spacecraft during its flybys of Ganymede's magnetosphere. The symbols show the intersections of the trajectories with the *XZ* plane (in GphiO coordinate system, a Ganymede-centered coordinate system similar to that defined in Figure 8.1) of Ganymede's magnetosphere. Sample field lines extracted from the global MHD simulation of Jia et al. [2009] are shown to provide a global context of Ganymede's magnetosphere. The flybys within Ganymede's magnetosphere are shown as dots whereas the distant tail flyby (G12A) is shown as a square. Another distant flyby from Galileo (G9A) that occurred further downstream (at $X \sim 31\, \boldsymbol{R}_G$) is not shown.

At all times, the Jovian magnetic field remains in a direction nearly opposite to Ganymede's internal field near the equator such that magnetic reconnection should occur. Reconnection is expected to be the dominant process that couples the mini-magnetosphere with Jupiter's giant magnetosphere, enabling the plasma and energy entry into the system. Plasma measurements in Ganymede's magnetosphere [*Frank et al.*, 1997b] indicate that although some of the incident Jovian flow diverts around the magnetosphere and is accelerated along the flanks, much of the ambient plasma is able to gain access into the magnetosphere through reconnection, and plasma that enters the Alfvén wings as a result of reconnection is convected across the polar caps toward the downstream region. Since the convection flow speed over the polar caps is directly dependent on the cross-polar cap potential imposed by the reconnection, it provides a measure of the reconnection efficiency at the upstream magnetopause. Flow measurements along with magnetic field observations inside Ganymede's polar cap [*Frank et al.*, 1997b; *Kivelson et al.*, 1997; *Williams et al.*, 1997a; *Vasyliūnas and Eviatar*, 2000] together indicate a high reconnection efficiency at the magnetopause, probably a result of the low Alfvénic Mach number and low β plasma conditions of the ambient flow near Ganymede. In contrast to the highly fluctuating solar wind with unpredictable variations in plasma and magnetic field conditions, Jupiter's magnetospheric plasma at Ganymede's orbit imposes external field and plasma conditions that vary slowly (of the order of a few hours) with respect to the time required for Jovian plasma to flow across Ganymede's magnetosphere (of the order of minutes). Therefore, Ganymede's magnetosphere provides a unique opportunity to investigate whether reconnection is intermittent or steady in a relatively stable external environment. Analysis of Galileo particles and field measurements of the crossings of Ganymede's upstream magnetopause together with global magnetosphere simulations suggest that even under steady external conditions, reconnection is predominantly intermittent rather than steady, leading to the formation of flux transfer events (FTEs) on the magnetopause [*Jia et al.*, 2010b]. It is found that the bursty reconnection on Ganymede's magnetopause tends to occur on relatively short time scales of the order of tens of seconds, which may be related to the low Alfvénic Mach number of the ambient plasma flow.

The reconnected flux tubes generated at the upstream magnetopause must reconnect again in the downstream region after having convected over the polar caps in order to conserve magnetic flux. For this reason, reconnection is expected to actively occur in Ganymede's magnetotail that returns part of the flow back toward the moon and the upstream magnetosphere and ejects the rest down the tail. The exact nature of the tail reconnection, such as the location and spatial extent of the neutral line and steady vs. intermittent, is largely unknown, due to lack of direct sampling of the low-latitude tail region (see Figure 8.5). Nonetheless, it can be expected that reconnection plays a crucial role in driving dynamics in Ganymede's magnetotail, resulting in magnetic reconfigurations and particle acceleration and energization. Galileo made two flybys of Ganymede's distant wake at downstream distances of ~$6R_G$ and ~$31R_G$, respectively. During both flybys, magnetic perturbations (e.g., depressions in field strength) along with signatures of heated plasma were detected, which may be products of tail reconnection (K. Khurana, personal communication, 2013). Close to the moon, energetic ions and electrons with pitch angle distributions consistent with trapping have been detected by Galileo on closed field lines in the upstream magnetosphere [*Williams et al.*, 1997b; *Williams*, 2001]. It has been suggested that these trapped energetic particles may have resulted from injections through tail reconnection followed by gradient/curvature drifts through the magnetotail reaching the upstream magnetosphere [*Williams et al.*, 1997b; *Eviatar et al.*, 2000].

Some important aspects of the large-scale plasma convection and transport of energetic particles in Ganymede's magnetosphere arise due to the intrinsic properties of the moon. Because Ganymede is tidally locked to Jupiter and hence there is no corotation electric field in the rest frame of plasma, the circulation of the plasma flow and the return of the magnetic flux from the tail to upstream through this tiny magnetosphere, whose space is largely filled by the moon, are expected to differ markedly from Earth's case. Depending on the ionospheric conductance, the return bulk flows within the range of the moon's diameter do not necessarily divert around the moon and instead they may flow into the moon's ionosphere and be absorbed [*Eviatar et al.*, 2000; *Jia et al.*, 2009]. If the ionosphere is highly conducting, return flows carrying magnetic flux will divert fully around the moon returning to the upstream side. If the ionosphere is weakly conducting, which is likely to be the case for Ganymede, the return flow will primarily be absorbed as it encounters the ionosphere, while the magnetic field will continue to diffuse through the moon's ionosphere and interior. For energetic particles that undergo energy-dependent drift motions (e.g., gradient and curvature drifts) in addition to the $\boldsymbol{E} \times \boldsymbol{B}$ drift imposed by the cross-tail electric field, their drift paths are sensitive to the magnetic field geometry and may be impacted by kinetic effects such as losses to the moon's surface due to finite gyroradius effects. Quantitative analyses of particle motion in Ganymede's magnetosphere that incorporate these effects, such as particle tracing using realistic electric and magnetic fields from physics-based global models, should provide useful insights into processes responsible for generating the observed energetic particle distributions and also help assess effects of space weathering arising from the interaction of the magnetospheric particles with the moon's surface [e.g., *Paranicas et al.*, 1999; *Khurana et al.*, 2007].

Like its satellite siblings Io and Europa, Ganymede also produces an auroral footprint in Jupiter's ionosphere [*Clarke et al.*, 2002]. The intensity of Ganymede's auroral footprint exhibits both long-term (~5 h) and short-term (10–40 min and ~100 s) variations [*Grodent et al.*, 2009]. While the long-term variation appears well correlated with Ganymede's location relative to Jupiter's central plasma sheet, the generation mechanisms for the short-term variations are less clear. However, the variation on the 100 s time scale may be related to the intermittent magnetopause reconnection that has a similar characteristic repetition time of tens of seconds [*Jia et al.*, 2010b]. The size of Ganymede's footprint, when mapped from Jupiter to Ganymede's orbit, corresponds roughly to the size of the moon's magnetosphere [*Grodent et al.*, 2009]. While recent Hubble Space Telescope (HST) observations detected a second spot in Ganymede's auroral footprint [*Bonfond et al.*, 2013], there has been no reported observation of a downstream tail trailing Ganymede's footprint like those of Io's and

Europa's. However, perturbations in Jupiter's magnetosphere have been measured near Ganymede's L-shell far downstream of the moon. On its inbound pass of Jupiter's magnetosphere, the Pioneer 11 spacecraft crossed the L-shell of Ganymede (south of the moon) at a downstream distance of ~$310R_G$ and observed transient magnetic perturbations characteristic of field-aligned currents [*Kivelson and Winge*, 1976], which may be an indication that disturbances produced by Ganymede's interaction with the Jovian magnetosphere also propagate over large distances downstream, forming an extended wake like those of Io and Europa.

8.5. SUMMARY

Planetary satellites interact with their surroundings in many different ways depending on their magnetic characters, their roles in supplying neutrals and plasma, and the conditions of the ambient plasma and field by which they are engulfed. The various interaction modes result in wake regions with different structures and properties. Using the major satellites of Jupiter and Saturn as examples, we have discussed the general features of the plasma-moon interactions focusing on the structures of the resultant wakes.

Saturn's icy moons, Tethys, Dione, Rhea, and Iapetus, to a good approximation, can be characterized as inert moons where neither appreciable internal magnetic fields nor significant sources of neutrals are present, and they basically act as absorbing obstacles to the external plasma. Absorption of the incident plasma by the upstream hemisphere leads to an extended region downstream of these moons in which plasma density is reduced and magnetic field strength is enhanced. The inert interactions at the Saturnian moons share many features with that at Earth's Moon but differ in ways that reflect the nature of their ambient plasma conditions.

A number of satellites in the two gas giant systems have been found to possess regions of significant electrical conductivity, which may arise from an ionosphere (such as at Io and Titan), a subsurface liquid layer (such as at Europa and Callisto), or mass loading (such as at Enceladus). Their interactions with the ambient plasma not only produce significant field and plasma perturbations in the moons' close vicinities over volumes large compared to their sizes but may also generate disturbances that can propagate far away from them leaving imprints in the magnetospheres/ionospheres of their parent planets, such as auroral footprints. While many aspects of the plasma interactions with conducting/mass-loading moons can be understood in the context of the Alfvén wing model, recent advances made in observations and modeling of the various moons in the two systems, such as the Cassini discovery of a significant source of dusty plasma at Enceladus, have provided us with new perspectives on plasma-moon interactions.

Jupiter's moon Ganymede presents a unique example in the solar system in that its plasma interaction is predominantly controlled by the moon's strong intrinsic magnetic field that creates a mini-magnetosphere with a magnetotail resembling those of planetary magnetospheres. Because of the large shear angle between the external Jovian field and Ganymede's internal field and the low Alfvénic Mach number and low plasma β conditions of the external flow, magnetic reconnection is the major process that couples the Jovian plasma with Ganymede's magnetosphere, dictates the large-scale plasma convection, and drives magnetospheric dynamics. As suggested by in situ measurements and global simulations, magnetopause reconnection is intermittent even in a relatively stable external environment like that of Ganymede. However, properties and structure of Ganymede's magnetotail are poorly documented by observations, and there remain many outstanding questions (such as the nature of tail reconnection, particle energization and acceleration, aurora generation, etc.), the answers to which will shed valuable light on our understanding of physical processes operating in planetary magnetotails. In that regard, observations from future planetary missions to Ganymede, such as the recently announced Jupiter Icy Moon Explorer (JUICE) mission led by the European Space Agency [*Grasset et al.*, 2013], are eagerly anticipated.

ACKNOWLEDGMENTS

The author thanks the organizers of the AGU Chapman conference on "Fundamental Properties and Processes of Magnetotails" for the opportunity to talk about the subject of this paper. The author is grateful to Margaret Kivelson and Krishan Khurana for helpful discussions on this topic. The research at the University of Michigan was supported by the NASA Outer Planets Research Program through grant NNX12AM74G and the NASA Cassini Data Analysis Program through grant NNX12AK34G, and by the NASA Cassini mission under contract 1409449 with JPL.

REFERENCES

Bonfond, B. (2012), When moons create aurora: The satellite footprints on giant planets, in *Auroral Phenomenology and Magnetospheric Processes: Earth and Other Planets, Geophys. Monogr. Ser.*, *197*, 133–140, doi:10.1029/2011GM001169.

Bonfond, B., S. Hess, F. Bagenal, J.-C. Gérard, D. Grodent, A. Radioti, J. Gustin, and J. T. Clarke (2013), The multiple spots of the Ganymede auroral footprint, *Geophys. Res. Lett.*, *40*, 4977–4981, doi:10.1002/grl.50989.

Clarke, J. T., et al. (2002), Ultraviolet emissions from the magnetic footprints of Io, Ganymede and Europa on Jupiter, *Nature*, *415*, 997–1000.

Connerney, J. E. P., R. Baron, T. Satoh, and T. Owen (1993), Images of excited H_3^+ at the foot of the Io flux tube in Jupiter's atmosphere, *Science*, *262*, 1035–1038, doi:10.1126/science.262.5136.1035.

Delamere, P. A., F. Bagenal, R. Ergun, and Y.-J. Su (2003), Momentum transfer between the Io plasma wake and Jupiter's ionosphere, *J. Geophys. Res.*, *108*, 1241, doi:10.1029/2002JA009530.

Dols, V., P. A. Delamere, F. Bagenal, W. S. Kurth, and W. R. Paterson (2012), Asymmetry of Io's outer atmosphere: Constraints from five Galileo flybys, *J. Geophys. Res.*, *117*, E10010, doi:10.1029/2012JE004076.

Dougherty, M. K., K. K. Khurana, F. M. Neubauer, C. T. Russell, J. Saur, J. S. Leisner, and M. E. Burton (2006), Identification of a dynamic atmosphere at Enceladus with the Cassini magnetometer, *Science*, *311*, 1406–1409, doi:10.1126/science.1120985.

Eviatar, A., D. J. Williams, C. Paranicas, R. W. McEntire, B. H. Mauk, and M. G. Kivelson (2000), Trapped energetic electrons in the magnetosphere of Ganymede, *J. Geophys. Res.*, *105*(A3), 5547–5553.

Farrell, W. M., W. S. Kurth, D. A. Gurnett, R. E. Johnson, M. L. Kaiser, J.-E. Wahlund, and J. H. Waite (2009), Electron density dropout near Enceladus in the context of water-vapor and water-ice, *Geophys. Res. Lett.*, *36*, L10203, doi:10.1029/2008GL037108.

Feldman, P. D., M. A. McGrath, D. F. Strobel, H. W. Moos, K. D. Retherford, and B. C. Wolven (2000), HST/STIS ultraviolet imaging of polar aurora on Ganymede, *Astrophys. J.*, *535*, 1085–1090.

Frank, L. A., and W. R. Paterson (1999), Intense electron beams observed at Io with the Galileo spacecraft, *J. Geophys. Res.*, *104*, 28,657–28,670, doi:10.1029/1999JA900402.

Frank, L. A., W. R. Paterson, K. L. Ackerson, V. M. Vasyliūnas, F. V. Coroniti, and S. J. Bolton (1996), Plasma observations at Io with the Galileo spacecraft, *Science*, *274*, 394–395, doi:10.1126/science.274.5286.394.

Frank, L. A., W. R. Paterson, K. L. Ackerson, and S. J. Bolton (1997a), Outflow of hydrogen ions from Ganymede, *Geophys. Res. Lett.*, *24*(17), 2151–2154.

Frank, L. A., W. R. Paterson, K. L. Ackerson, and S. J. Bolton (1997b), Low-energy electrons measurements at Ganymede with the Galileo spacecraft: Probes of the magnetic topology, *Geophys. Res. Lett.*, *24*(17), 2159–2162.

Goertz, C. K. (1980), Io's interaction with the plasma torus, *J. Geophys. Res.*, *85*, 2949–2956, doi:10.1029/JA085iA06p02949.

Grasset, O., et al. (2013), JUpiter ICy moons Explorer (JUICE): An ESA mission to orbit Ganymede and to characterise the Jupiter system, *Planet. Space Sci.*, *78*, 1–21, doi:10.1016/j.pss.2012.12.002.

Grodent, D., J.-C. Gérard, J. Gustin, B. H. Mauk, J. E. P. Connerney, and J. T. Clarke (2006), Europa's FUV auroral tail on Jupiter, *Geophys. Res. Lett.*, *33*, 6201, doi:10.1029/2005GL025487.

Grodent, D., B. Bonfond, A. Radioti, J. Gérard, X. Jia, J. D. Nichols, and J. T. Clarke (2009), Auroral footprint of Ganymede, *J. Geophys. Res.*, *114*, 7212, doi:10.1029/2009JA014289.

Gurnett, D. A., W. S. Kurth, A. Roux, S. J. Bolton, and C. F. Kennel (1996), Evidence for a magnetosphere at Ganymede from plasma-wave observations by the Galileo spacecraft, *Nature*, *384*, 535–537.

Hansen, C. J., L. Esposito, A. I. F. Stewart, J. Colwell, A. Hendrix, W. Pryor, D. Shemansky, and R. West (2006), Enceladus' water vapor plume, *Science*, *311*, 1422–1425, doi:10.1126/science.1121254.

Hess, S. L. G., P. A. Delamere, V. Dols, and L. C. Ray (2011), Comparative study of the power transferred from satellite-magnetosphere interactions to auroral emissions, *J. Geophys. Res.*, *116*, A01202, doi:10.1029/2010JA015807.

Hill, T. W., and V. M. Vasyliūnas (2002), Jovian auroral signature of Io's corotational wake, *J. Geophys. Res.*, *107*, 1464, doi:10.1029/2002JA009514.

Jia, X., R. J. Walker, M. G. Kivelson, K. K. Khurana, and J. A. Linker (2008), Three-dimensional MHD simulations of Ganymede's magnetosphere, *J. Geophys. Res.*, *113*, 6212, doi:10.1029/2007JA012748.

Jia, X., R. J. Walker, M. G. Kivelson, K. K. Khurana, and J. A. Linker (2009), Properties of Ganymede's magnetosphere inferred from improved three-dimensional MHD simulations, *J. Geophys. Res.*, *114*, 9209, doi:10.1029/2009JA014375.

Jia, X., M. G. Kivelson, K. K. Khurana, and R. J. Walker (2010a), Magnetic fields of the satellites of Jupiter and Saturn, *Space Sci. Rev.*, *152*, 271–305, doi:10.1007/s11214-009-9507-8.

Jia, X., R. J. Walker, M. G. Kivelson, K. K. Khurana, and J. A. Linker (2010b), Dynamics of Ganymede's magnetopause: Intermittent reconnection under steady external conditions, *J. Geophys. Res.*, *115*, A12202, doi:10.1029/2010JA015771.

Jia, Y.-D., Y. J. Ma, C. T. Russell, H. R. Lai, G. Toth, and T. I. Gombosi (2012), Perpendicular flow deviation in a magnetized counter-streaming plasma, *Icarus*, *218*, 895–905, doi:10.1016/j.icarus.2012.01.017.

Khurana, K. K., M. G. Kivelson, D. J. Stevenson, G. Schubert, C. T. Russell, R. J. Walker, and C. Polanskey (1998), Induced magnetic fields as evidence for subsurface oceans in Europa and Callisto, *Nature*, *395*, 777–780, doi:10.1038/27394.

Khurana, K. K., R. T. Pappalardo, N. Murphy, and T. Denk (2007), The origin of Ganymede's polar caps, *Icarus*, *191*, 193–202, doi:10.1016/j.icarus.2007.04.022.

Khurana, K. K., C. T. Russell, and M. K. Dougherty (2008), Magnetic portraits of Tethys and Rhea, *Icarus*, *193*, 465–474.

Khurana, K. K., X. Jia, M. G. Kivelson, F. Nimmo, G. Schubert, and C. T. Russell (2011), Evidence of a global magma ocean in Io's interior, *Science*, *332*, 1186–1189, doi:10.1126/science.1201425.

Kivelson, M., K. Khurana, C. Russell, M. Volwerk, R. Walker, and C. Zimmer (2000), Galileo magnetometer measurements: A stronger case for a subsurface ocean at Europa, *Science*, *289*(5483), 1340–1343.

Kivelson, M. G., and C. R. Winge (1976), Field-aligned currents in the Jovian magnetosphere: Pioneer 10 and 11, *J. Geophys. Res.*, *81*, 5853–5858, doi:10.1029/JA081i034p05853.

Kivelson, M. G., K. K. Khurana, R. J. Walker, J. Warnecke, C. T. Russell, J. A. Linker, D. J. Southwood, and C. Polanskey (1996a), Io's interaction with the plasma torus: Galileo magnetometer report, *Science*, *274*, 396–398, doi:10.1126/science.274.5286.396.

Kivelson, M. G., K. K. Khurana, C. T. Russell, R. J. Walker, J. Warnecke, F. V. Coroniti, C. Polanskey, D. J. Southwood, and G. Schubert (1996b), Discovery of Ganymede's magnetic field by the Galileo spacecraft, *Nature*, *384*, 537–541.

Kivelson, M. G., K. K. Khurana, F. V. Coroniti, S. Joy, C. T. Russell, R. J. Walker, J. Warnecke, L. Bennett, and C. Polanskey (1997), The magnetic field and magnetosphere of Ganymede, *Geophys. Res. Lett.*, *24*, 2155.

Kivelson, M. G., K. K. Khurana, and M. Volwerk (2002), The permanent and inductive magnetic moments of Ganymede, *Icarus*, *157*, 507– 522.

Kivelson, M. G., F. Bagenal, W. S. Kurth, F. M. Neubauger, C. Paranicas, and J. Saur (2004), Magnetospheric interactions with satellites, in *Jupiter: The Planet, Satellites and Magnetosphere*, edited by F. Bagenal et al., pp. 513–536, Cambridge Univ. Press, Cambridge, U. K.

Kriegel, H., S. Simon, U. Motschmann, J. Saur, F. M. Neubauer, A. M. Persoon, M. K. Dougherty, and D. A. Gurnett (2011), Influence of negatively charged plume grains on the structure of Enceladus' Alfvén wings: Hybrid simulations versus Cassini Magnetometer data, *J. Geophys. Res.*, *116*, A10223, doi:10.1029/2011JA016842.

Krupp, N., K. K. Khurana, L. Iess, V. Lainey, T. A. Cassidy, M. Burger, C. Sotin, and F. Neubauer (2010), Environments in the outer solar system, *Space Sci. Rev.*, *153*, 11–59, doi:10.1007/s11214-010-9653-z.

Mauk, B. H., et al. (2009), Fundamental plasma processes in Saturn's magnetosphere, in *Saturn from Cassini-Huygens*, edited by M. K. Dougherty, L. W. Esposito, and S. M. Krimigis, p. 281, Springer, Dordrecht Heidelberg London New York.

McGrath, M. A., E. Lellouch, D. F. Strobel, P. D. Feldman, and R. E. Johnson (2004), Satellite atmospheres, in *Jupiter: The Planet, Satellites and Magnetosphere*, edited by F. Bagenal et al., pp. 457–483, Cambridge Univ. Press, Cambridge, U. K.

McGrath, M. A., X. Jia, K. Retherford, P. D. Feldman, D. F. Strobel, and J. Saur (2013), Aurora on Ganymede, *J. Geophys. Res.*, *118*, 2043–2054, doi:10.1002/jgra.50122.

McPherron, R. (this volume), Earth's magnetotail, in *Magnetotails in the solar system, Geophys. Monogr. Ser.*

Morooka, M. W., J.-E. Wahlund, A. I. Eriksson, W. M. Farrell, D. A. Gurnett, W. S. Kurth, A. M. Persoon, M. Shafiq, M. André, and M. K. G. Holmberg (2011), Dusty plasma in the vicinity of Enceladus, *J. Geophys. Res.*, *116*, A12221, doi:10.1029/2011JA017038.

Neubauer, F. M. (1980), Non-linear standing Alfvén wave current system at Io: Theory, *J. Geophys. Res.*, *85*, 1171–1178.

Neubauer, F. M. (1998), The sub-Alfvénic interaction of the Galilean satellites with the Jovian magnetospheres, *J. Geophys. Res.*, *103*(E9), 19,843–19,866.

Neubauer, F. M. (1999), Alfvén wings and electromagnetic induction in the interiors: Europa and Callisto, *J. Geophys. Res.*, *104*, 28,671–28,684, doi:10.1029/1999JA900217.

Neubauer, F. M., et al. (2006), Titan's near magnetotail from magnetic field and electron plasma observations and modeling: Cassini flybys TA, TB, and T3, *J. Geophys. Res.*, *111*, 10,220, doi:10.1029/2006JA011676.

Omidi, N., R. L. Tokar, T. Averkamp, D. A. Gurnett, W. S. Kurth, and Z. Wang (2012), Flow stagnation at Enceladus: The effects of neutral gas and charged dust, *J. Geophys. Res.*, *117*, A06230, doi:10.1029/2011JA017488.

Paranicas, C., W. R. Paterson, A. F. Cheng, B. H. Mauk, R. W. McEntire, L. A. Frank, and D. J. Williams (1999), Energetic particle observations near Ganymede, *J. Geophys. Res.*, *104*(A8), 17,459–17,469.

Paterson, W. R., L. A. Frank, and K. L. Ackerson (1999), Galileo plasma observations at Europa: Ion energy spectra and moments, *J. Geophys. Res.*, *104*, 22,779–22,792, doi:10.1029/1999JA900191.

Pryor, W. R., et al. (2011), The auroral footprint of Enceladus on Saturn, *Nature*, *472*, 331–333, doi:10.1038/nature09928.

Roussos, E., J. Muller, S. Simon, A. Boswetter, U. Motschmann, N. Krupp, M. Franz, J. Woch, K. K. Khurana, and M. K. Dougherty (2008), Plasma and fields in the wake of Rhea: 3-D hybrid simulation and comparison with Cassini data, *Ann. Geophys.*, *26*, 619–637.

Roussos, E., et al. (2012), Energetic electron observations of Rhea's magnetospheric interaction, *Icarus*, *221*, 116–134, doi:10.1016/j.icarus.2012.07.006.

Saur, J., F. M. Neubauer, D. F. Strobel, and M. E. Summers (2002), Interpretation of Galileo's Io plasma and field observations: I0, I24, and I27 flybys and close polar passes, *J. Geophys. Res.*, *107*, 1422, doi:10.1029/2001JA005067.

Saur, J., F. M. Neubauer, J. E. P. Connerney, P. Zarka, and M. G. Kivelson (2004), Plasma interaction of Io with its plasma torus, in *Jupiter: The Planet, Satellites and Magnetosphere*, edited by F. Bagenal et al., pp. 537–560, Cambridge Univ. Press, Cambridge, U. K.

Schilling, N., F. M. Neubauer, and J. Saur (2008), Influence of the internally induced magnetic field on the plasma interaction of Europa, *J. Geophys. Res.*, *113*(A3), doi:10.1029/2007JA012842.

Simon, S., J. Saur, F. M. Neubauer, U. Motschmann, and M. K. Dougherty (2009), Plasma wake of Tethys: Hybrid simulations versus Cassini MAG data, *Geophys. Res. Lett.*, *36*, L04108, doi:10.1029/2008GL036943.

Simon, S., J. Saur, F. M. Neubauer, A. Wennmacher, and M. K. Dougherty (2011), Magnetic signatures of a tenuous atmosphere at Dione, *Geophys. Res. Lett.*, *38*, L15102, doi:10.1029/2011GL048454.

Sittler, E. C., R. E. Hartle, C. Bertucci, A. Coates, T. Cravens, I. Dandouras, and D. Shemansky (2010), Energy Deposition processes in Titan's upper atmosphere and its induced magnetosphere, in *Titan from Cassini-Huygens*, edited by R. H. Brown et al., pp. 393–453, Springer Dordrecht Heidelberg London New York.

Southwood, D. J., and M. W. Dunlop (1984), Mass pickup in sub-Alfvénic plasma flow: A case study for Io, *Planet. Space Sci.*, *32*, 1079–1086, doi:10.1016/0032-0633(84)90133-8.

Southwood, D. J., M. G. Kivelson, R. J. Walker, and J. A. Slavin (1980), Io and its plasma environment, *J. Geophys. Res.*, *85*, 5959–5968.

Teolis, B. D., et al. (2010), Cassini finds an oxygen-carbon dioxide atmosphere at Saturn's icy moon Rhea, *Science*, *330*, 1813–1815, doi:10.1126/science.1198366.

Tokar, R. L., R. E. Johnson, M. F. Thomsen, E. C. Sittler, A. J. Coates, R. J. Wilson, F. J. Crary, D. T. Young, and G. H. Jones (2012), Detection of exospheric O_2^+ at Saturn's moon Dione, *Geophys. Res. Lett.*, *39*, L03105, doi:10.1029/2011GL050452.

Vasyliūnas, V. M., and A. Eviatar (2000), Outflow of ions from Ganymede: A reinterpretation, *Geophys. Res. Lett.*, *27*(9), 1347–1349.

Volwerk, M., K. Khurana, and M. Kivelson (2007), Europa's Alfvén wing: shrinkage and displacement influenced by an induced magnetic field, *Ann. Geophys.*, *25*, 905–914, doi:10.5194/angeo-25-905-2007.

Volwerk, M., X. Jia, C. Paranicas, W. S. Kurth, M. G. Kivelson, and K. K. Khurana (2013), ULF waves in Ganymede's upstream magnetosphere, *Ann. Geophys.*, *31*, 45–59, doi:10.5194/angeo-31-45-2013.

Waite, J. H., et al. (2006), Cassini ion and neutral mass spectrometer: Enceladus plume composition and structure, *Science*, *311*, 1419–1422, doi:10.1126/science.1121290.

Williams, D. J. (2001), Ganymede's ionic radiation belts, *Geophys. Res. Lett.*, *28*(19), 3793–3796.

Williams, D. J., B. H. Mauk, R. E. McEntire, E. C. Roelof, T. P. Armstrong, B. Wilken, J. G. Roederer, S. M. Krimigis, T. A. Fritz, and L. J. Lanzerotti (1996), Electron beams and ion composition measured at lo and in its torus, *Science*, *274*, 401–403, doi:10.1126/science.274.5286.401.

Williams, D. J., B. H. Mauk, R. W. McEntire, E. C. Roelof, T. P. Armstrong, B. Wilken, S. M. K. J. G. Roederer, T. A. Fritz, L. J. Lanzerotti, and N. Murphy (1997a), Energetic particle signatures at Ganymede: Implications for Ganymede's magnetic field, *Geophys. Res. Lett.*, *24*(17), 2163–2166.

Williams, D. J., B. Mauk, and R. W. McEntire (1997b), Trapped electrons in Ganymede's magnetic field, *Geophys. Res. Lett.*, *24*(23), 2953–2956.

Williams, D. J., R. M. Thorne, and B. Mauk (1999), Energetic electron beams and trapped electrons at Io, *J. Geophys. Res.*, *104*, 14,739–14,754, doi:10.1029/1999JA900115.

9

Moon's Plasma Wake

J. S. Halekas[1,4], D. A. Brain[2,4] and M. Holmström[3]

9.1. INTRODUCTION

9.1.1. Lunar Wake: Solar System Context

An uncountable number of objects interact with flowing plasma of stellar and/or planetary origin in our solar system and beyond. These objects range in scale from submicrometer dust to giant planets and have varying degrees of intrinsic magnetization, internal conductivity, and atmospheric density. The Moon, as an airless and unmagnetized (or nearly so) body, represents an end-member case in its interactions with flowing plasma in the solar wind and terrestrial foreshock and magnetosphere. Whereas a significant atmosphere and/or intrinsic magnetic field can slow and deflect the incoming plasma upstream from the obstacle and initiate currents that produce draped magnetic field geometries, plasma impacts the surface of the Moon with only minimal upstream modification or inductive effects, resulting in a downstream plasma void but largely unchanged plasma flow and magnetic field outside of the wake.

We can usefully consider the interaction of nonmagnetic bodies with flowing plasma in terms of obstacle scale size, conductivity, and mass loading, as shown in Figure 9.1. Magnetized objects, not shown here, have their own unique interactions that depend on the scale of the magnetic field [*Omidi et al.*, 2002]. The Moon, though it has localized crustal magnetic sources that can affect the plasma, does not have a global magnetic field [*Dyal et al.*, 1974].

Nonmagnetic bodies with a significant atmosphere perturb incoming plasma by mass loading (or momentum loading) and slowing the flow and forming induced ionospheric current systems that oppose the magnetic field carried in the plasma. The resulting interaction depends on the magnetosonic Mach number of the flow. For bodies embedded in a sub-magnetosonic flow, such as Io, Europa, and other giant planet moons, the presence of a conductive obstacle drives currents that carry an outward propagating magnetic perturbation laterally at the Alfvén speed, resulting in a moderately draped magnetic field geometry known as an Alfvén wing. Bodies embedded in a super-magnetosonic flow, such as Mars, Venus, and comets, have similar current systems, but the faster flow also results in the formation of an upstream shock that bounds the interaction region. Mass-loading and viscous effects further slow plasma near the object and help pull the draped fields into an extended two-lobed magnetotail [*Baumjohann et al.*, 2010].

In contrast, a nonconductive body such as the Moon, having no significant atmosphere to produce an ionosphere or mass load the incident plasma, has almost no global effect on the incoming flow (kinetic processes driven by reflected solar wind electrons and escaping photoelectrons [*Halekas et al.*, 2012], as well as solar wind reflection and magnetic pileup over strong crustal magnetic fields [*Lue et al.*, 2011; *Saito et al.*, 2011] can affect incoming plasma in localized regions). Nonetheless, the interaction has similarities to those discussed above. The removal of plasma at the dayside surface results in a downstream void bounded by a rarefaction wave that propagates out into the plasma, similar in some ways to

[1] *Department of Physics and Astronomy, University of Iowa, Iowa City, Iowa*

[2] *Laboratory for Atmospheric and Space Physics, University of Colorado, Boulder, Colorado*

[3] *Swedish Institute of Space Physics, Kiruna, Sweden*

[4] *NASA's Lunar Science Institute, NASA Ames Research Center, Moffett Field, California*

Magnetotails in the Solar System, Geophysical Monograph 207. First Edition.
Edited by Andreas Keiling, Caitríona M. Jackman, and Peter A. Delamere.

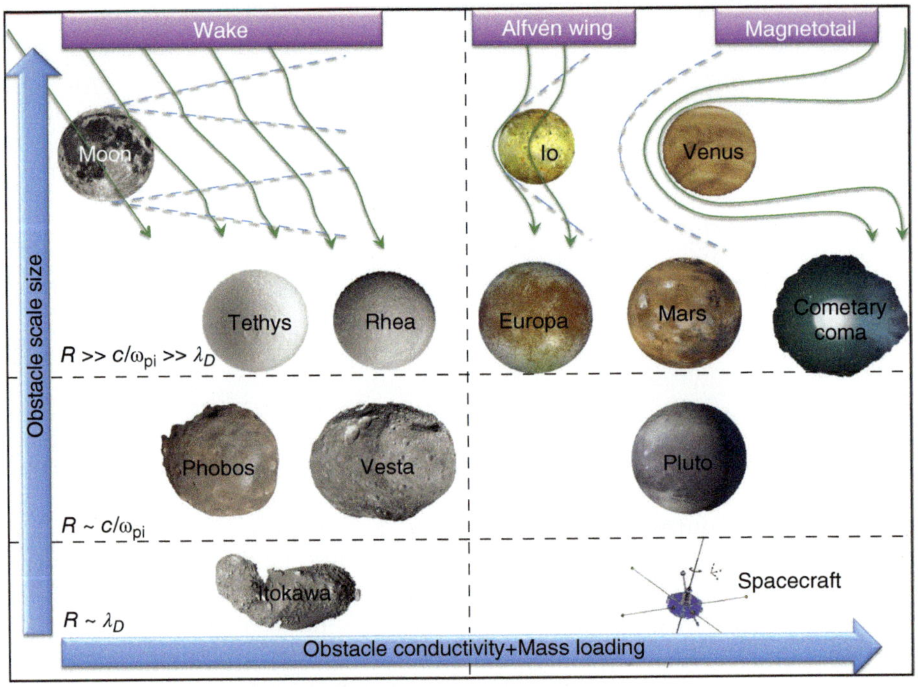

Figure 9.1 Moon-plasma interaction in context with other nonmagnetized solar system bodies. Nonconductive objects only slightly perturb the magnetic field, while obstacles with more significant conductivity (intrinsic or ionospheric) and/or mass-loading slow and deflect the incoming flow and drive currents that significantly alter the magnetic field, with the geometry of the resulting interaction depending on plasma Mach number. For smaller objects, kinetic effects become even more important, especially for objects near the Debye scale, for which the surface sheath increasingly dominates the interaction.

an Alfvén wing but associated with a rather different current system.

Wakelike interactions occur in both super-magnetosonic and sub-magnetosonic plasma flows. For the super-magnetosonic case, including the Moon in the solar wind, the lateral dimensions of the interaction region have nearly the same scale, corresponding to the only moderately anisotropic fast-mode propagation of the rarefaction wave [*Zhang et al.*, 2012; *Vernisse et al.*, 2013]. For the sub-magnetosonic case, exemplified by Tethys and Rhea, but also valid for the Moon in some regions of the terrestrial magnetosphere, the extent and flaring angle of the interaction region in the plane containing the magnetic field increases dramatically, while the interaction perpendicular to the field has a limited extent [*Roussos et al.*, 2007; *Khurana et al.*, 2007; *Simon et al.*, 2009; *Jia et al.*, 2010], primarily because the fast mode does not contribute to the sub-magnetosonic interaction [*Neubauer*, 1998].

The character of the interaction also changes as the obstacle size decreases. When the scale approaches the Debye length, the charged surface and nonneutral plasma sheath become as or more important than wake effects, as exemplified by numerous observations of spacecraft charging [*Whipple*, 1981] and simulations of small obstacle wakes [*Kimura and Nakagawa*, 2008; *Nakagawa and Kimura*, 2011; *Zimmerman et al.*, 2011]. In addition, at these scales, induced current systems cannot significantly perturb the plasma flow. The intermediate regime, at the scale of the ion inertial length, remains poorly explored for both conductive and nonconductive obstacles in space, but future in situ plasma measurements at bodies like Pluto, Vesta, and Phobos should help elucidate the unique aspects of this parameter regime.

9.1.2. History of Lunar Wake Research

The first observations of the lunar wake, early in the space age, came from Explorer 35 and the Apollo subsatellites, as reviewed in detail by *Schubert and Lichtenstein* [1974] and *Ness* [1972]. Early studies focused on the more easily observed magnetic field signature [*Colburn et al.*, 1967; *Ness et al.*, 1967, 1968], with charged particle observations sufficient to show the presence of a downstream plasma cavity but not to fully characterize distribution functions [*Lyon et al.*, 1967; *Siscoe et al.*, 1969]. As a result, theoretical efforts focused on a magnetohydrodynamic (MHD) treatment of the wake from the view of current systems [*Colburn et al.*, 1967] and from a continuum picture of wake refilling [*Michel*, 1968;

Spreiter et al., 1970]. Significant effort also focused on sporadic magnetic enhancements outside of the rarefaction, termed "limb shocks" or "limb compressions," since shown to result from interactions with crustal magnetic fields [*Sonett and Mihalov*, 1972; *Russell and Lichtenstein*, 1975; *Halekas et al.*, 2006].

Nearly 30 years after these observations, modern spacecraft provided a new window on the lunar wake. The Wind spacecraft made a number of lunar flybys, some through the wake at distances as close as ~6 lunar radii downstream, enabling new measurements of magnetic fields [*Owen et al.*, 1996] as well as more detailed measurements of particle distributions [*Bosqued et al.*, 1996; *Ogilvie et al.*, 1996; *Clack et al.* 2004] and plasma waves both in and upstream from the wake [*Farrell et al.*, 1996; *Kellogg et al.*, 1996, *Bale et al.*, 1997]. Geotail also made remote observations of upstream plasma waves on field lines threading the lunar wake [*Nakagawa et al.*, 2003], and a lunar flyby by the Nozomi spacecraft provided evidence of counterstreaming electrons on field lines threading the wake [*Futaana et al.*, 2001]. These observations, along with the rediscovery and application of relevant theory for unmagnetized plasma expansion into vacuum [*Gurevich et al.*, 1969; *Crow et al.*, 1975; *Denavit*, 1979; *Samir et al.*, 1983], focused theoretical activity on the importance of kinetic processes associated with plasma refilling along magnetic field lines and inspired new simulation efforts [*Farrell et al.*, 1998; *Birch and Chapman*, 2001a, 2001b, 2002; *Kallio*, 2005; *Travnicek et al.*, 2005].

The next view of the lunar wake came from Lunar Prospector (LP), which had limited plasma instrumentation but by virtue of its low-altitude polar orbit provided a thorough statistical picture of the low-altitude lunar wake and the electron dynamics therein. LP results for the first time highlighted the importance of the solar wind suprathermal electron population in driving the refilling of the wake [*Halekas et al.*, 2005].

In the past five years, we have again seen a new surge in our understanding of the wake, driven by observations from an international suite of lunar spacecraft, including the Japanese mission Kaguya, the Chinese mission Chang'E, and the Indian mission Chandrayaan. These missions have brought highly capable new plasma instrumentation to bear on the low-altitude wake and have demonstrated the importance of finite gyroradius effects [*Nishino et al.*, 2009a, 2009b; *Wang et al.*, 2010; *Holmstrom et al.*, 2010], particularly those associated with protons that reflect from the surface [*Saito et al.*, 2008] and crustal magnetic fields [*Lue et al.*, 2011], as well as providing new insight on plasma wave activity driven by the plasma processes in and around the wake [*Hashimoto et al.*, 2010; *Nishino et al.*, 2010].

Most recently, the ARTEMIS mission [*Angelopoulos*, 2011] has repurposed two comprehensively instrumented probes from the THEMIS mission, placing them in elliptical orbits around the Moon, providing two-point coverage of the wake and its surroundings. Though not yet fully explored, this extensive data set has already provided new insight on kinetic plasma processes [*Halekas et al.*, 2011b; *Tao et al.*, 2012] and the global structure of the wake [*Zhang et al.*, 2012, 2014]. At the same time, advances in simulation capabilities have led to an evolved view of the 3D structure of the lunar wake [*Fatemi et al.*, 2012, 2013; *Holmstrom et al.*, 2012, *Wiehle et al.*, 2011; *Wang et al.*, 2011; *Israelivich and Ofman*, 2012; *Umeda*, 2012; *Nakagawa*, 2013; *Vernisse et al.*, 2013], as discussed further in section 9.4.

9.1.3. Lunar Wake: Plasma Physics Context

Plasma processes that operate in the lunar wake have relevance for a number of other disciplines. Gurevich and colleagues [*Gurevich et al.*, 1969; *Gurevich and Pitaevsky*, 1975] first developed much of the basic theory applicable to the refilling of the wake parallel to the magnetic field, showing that the expansion of plasma into a vacuum produces not only a rarefaction region but also electric fields that accelerate ions into the void. Researchers have since applied a similar theory to topics as diverse as polar wind [*Singh and Schunk*, 1982], spacecraft-plasma interactions [*Stone et al.*, 1988], laser plasmas [*Felber and Decoste*, 1978], and Mach probes [*Hutchinson*, 2008a, 2008b]. Meanwhile, the refilling of the wake in the direction perpendicular to the magnetic field has clear analogies to the well-studied AMPTE barium release [*Bernhardt et al.*, 1987] and to plasma confinement. The lunar wake forms a natural laboratory for studying plasma instabilities, as discussed further in section 9.2.8.

9.2. STRUCTURE AND DYNAMICS OF LUNAR WAKE

9.2.1. Global Structure of Lunar Wake

The wake varies significantly in response to changes in the flowing plasma impinging on the upstream lunar surface; however, some basics of the interaction remain constant. Most (>70%) solar wind ions impacting the dayside surface do not reflect (except over strongly magnetic crustal regions [*Lue et al.*, 2011]), and a large majority of those that do reflect first lose their charge and become neutral [*McComas et al.*, 2009; *Wieser et al.*, 2009]. Given the long ionization lifetime of hydrogen, the reflected neutral population does not perturb the incoming solar wind in any significant way. The solar wind therefore flows unimpeded to the dayside surface, with the influence of the Moon on the upstream flow limited, though measurable [*Halekas et al.*, 2012]. As a result, the lunar wake forms downstream as a nearly perfect interface between the solar wind and a cavity devoid of plasma. In the high-velocity flow of the solar wind and the terrestrial magnetosheath, the cavity extends to large distances downstream (at lunar orbit, the bulk flow generally remains super-magnetosonic even in the sheath, given the highly tangential surface of the terrestrial bow shock at the far downstream location of the Moon).

Figure 9.2 shows the expected global structure of the wake in the solar wind, as predicted by hybrid simulations [*Holmstrom et al.*, 2012] and in large part confirmed by statistical studies [*Colburn et al.*, 1971; *Whang and Ness*, 1970; *Halekas et al.*, 2005; *Zhang et al.*, 2012, 2014]. The nearly complete plasma void extending well downstream from the Moon provides the defining feature of the wake for all field orientations. The wake has moderately enhanced magnetic fields in the central wake and reduced fields in boundary regions where the field has a significant component perpendicular to the boundary normal. These effects result from the diamagnetic current $J=(B\times\nabla p)/B^2$ from the pressure gradient at the wake boundary. The outer boundary of the rarefaction region and the recompression seen at the perpendicular flank of the central void correspond to additional (nondiamagnetic) current systems associated with standing magnetosonic wavefronts [*Fatemi et al.*, 2013]. Other current systems may also exist, in part to help close those mentioned here [*Vernisse et al.*, 2013].

A great deal of effort has been expended in debating whether we should consider the lunar wake a fluid or kinetic structure. In fact, to form a complete picture, we must reconcile both views. For unmagnetized plasma flow the refilling would have cylindrical symmetry, but instead the interplanetary magnetic field (IMF) controls the parallel infilling of the lunar wake [*Clack et al.*, 2004]. On field lines that thread the wake cavity, plasma expansion occurs primarily parallel to the field, in a

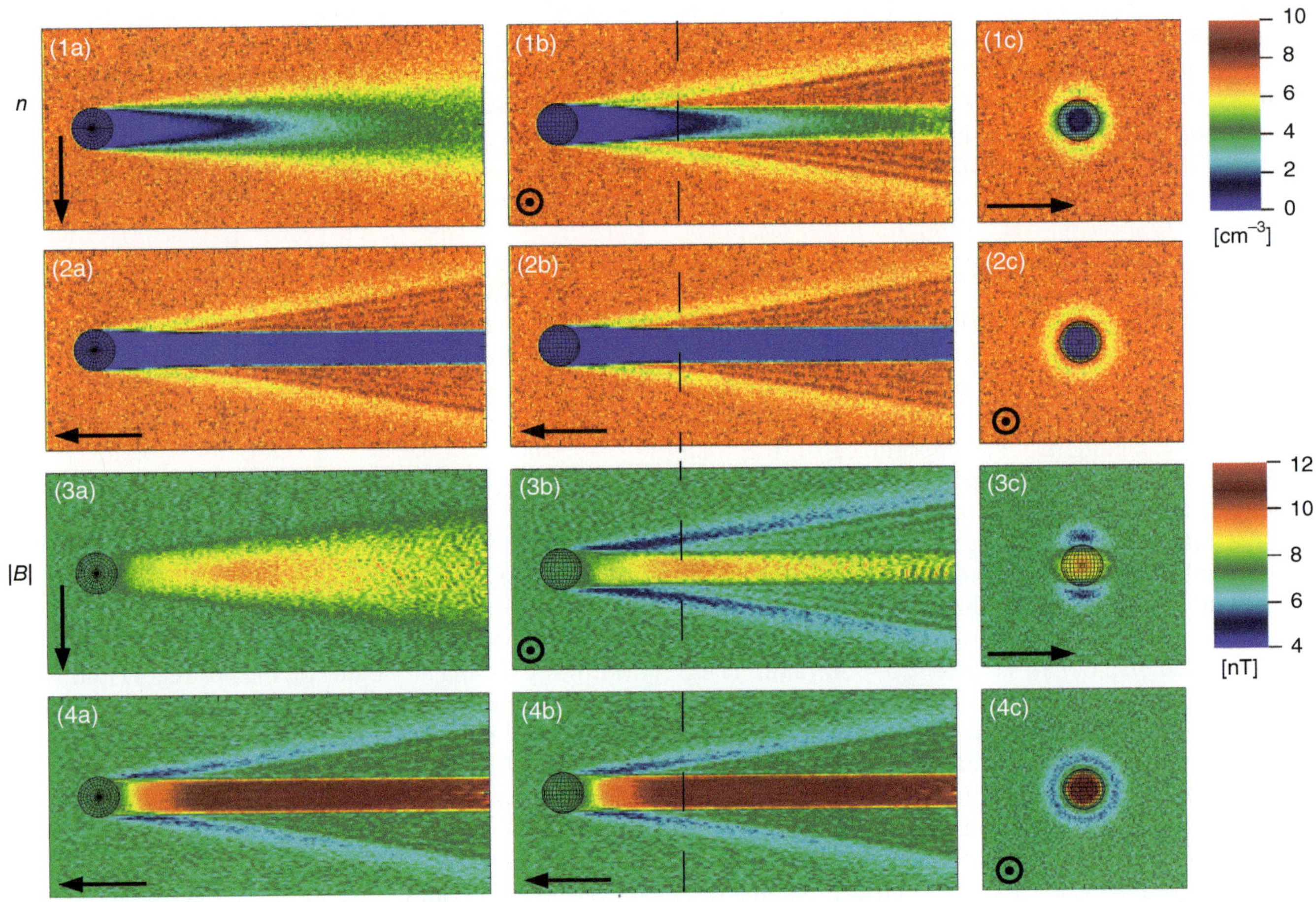

Figure 9.2 Global structure of the wake for two different magnetic field orientations as seen in the plasma density (top panels) and magnetic field magnitude (bottom panels), from a 3D hybrid simulation. The solar wind flows from the left, and the arrows show the direction of the IMF. The wake cuts lie 10,000 km downstream from the center of the Moon, and the simulation results shown extend 32,500 km downstream, 4000 km upstream, and 8500 km laterally in each direction. See *Holmstrom et al.* [2012] for more details.

manner that a fluid picture cannot fully capture, with a characteristic velocity related to the ion sound speed (leading some to argue for the dominance of ion sonic processes [*Farrell et al.*, 1998]). Rarefaction and recompression, most simply treated as fluid processes, play a role in the perpendicular dimension, especially on field lines that do not thread the central cavity. Observations of the outer wake boundary largely confirm the importance of magnetosonic (fluid) effects [*Sonett and Mihalov*, 1972; *Whang and Ness*, 1970; *Zhang et al.*, 2012], though sporadic external perturbations caused by crustal magnetic fields near the limb have caused a great deal of confusion [*Schubert and Lichtenstein*, 1974]. In the end, no real discrepancy exists as long as we acknowledge the importance of multiple physical processes in different regions of the wake. In fact, the interplay between processes that refill the wake along field lines and those that occur perpendicular to the field should provide a rich topic for future study. The refilling of plasma along the magnetic field complicates the wake structure and dynamics significantly (except in the singular case with parallel flow and field), since these processes supply plasma to regions that would otherwise never refill [see *Michel*, 1968]. In turn, perpendicular compression and rarefaction must likewise affect parallel refilling processes by altering the magnetic field geometry and mirroring particles.

Comparisons of wake observations to simulations and theory face the complication that the solar wind rarely remains steady for the time that a spacecraft takes to cross the wake. Figure 9.2 demonstrates that the expected structure of the wake varies significantly for different magnetic field orientations. The wake also changes in response to variations in other plasma parameters, particularly plasma beta, which correlates with the magnitude of the diamagnetic response [*Ogilvie and Ness*, 1969; *Zhang et. al.*, 2014]. Therefore, a spacecraft typically encounters a variety of wake geometries during a single orbit, with reconfigurations occurring on a time scale commensurate with the convection time (several seconds for typical flow speeds). Figure 9.3 shows an example of this variability, utilizing ARTEMIS MAG [*Auster et al.*, 2008] and ESA [*McFadden et al.*, 2008] data from an orbit that effectively crossed multiple regions of the wake (as a result of field rotations during the crossing). The probe enters at the parallel flank, observing parallel refilling, but with no diamagnetic reduction in the magnetic field. In the central wake we find a plasma cavity together with an increase in the field magnitude. The probe then exits the wake on the perpendicular flank, observing a combination of parallel and perpendicular refilling processes and passing through a diamagnetic current system and rarefaction region.

9.2.2. Measurement Issues in Lunar Wake

Measurement issues also complicate the comparison of data to simulation and theory. Even for instruments with high sensitivity, the low density of the central wake proves hard to accurately measure, especially for ions, which typically have low thermal velocities. Worse yet, a combination of natural radioactivity and penetrating cosmic rays produces a background on the order of several counts per second in microchannel plate detectors, comparable to the signal from real particles for very low densities. As a result, in the central wake, background counts can dominate the measured ion density. Background counts also corrupt other ion moments, since they have an even distribution over all energies and angles and act to artificially decrease the velocity moment and increase the temperature/pressure moments. Corrected moments typically show a constant or increasing velocity in the wake and a nearly constant ion temperature, in agreement with basic theory [*Samir et al.*, 1983].

Another issue affects ion instruments (like ARTEMIS ESA in some modes) that do not uniformly sample the angular distribution. Because of the extremely cold ion distributions that can result from parallel cooling in the wake [*Clack et al.*, 2004, *Gurevich et al.*, 1969], any angular undersampling can result in significant ion density underestimates for some orientations of the instrument.

In addition, spacecraft in shadow can charge to negative values on the order of the electron temperature (itself elevated in the central wake), potentially making part of the electron spectrum unmeasurable. Thanks to a special bias mode for the electric field probes, the ARTEMIS probes typically do not charge to large values; however, some artificial discontinuities still result from changes in the spacecraft charge resulting from bias mode changes and shadow entry and exit.

A final complication affects missions that rely on sun sensors to track spacecraft orientation. In shadow, the spin period changes due to thermal effects, resulting in a slow drift in the apparent phase of all measured vector quantities. Only careful thermal modeling of the spacecraft allowed the ARTEMIS mission to correct for this effect [*Georgescu et al.*, 2011].

9.2.3. Physical Processes at Perpendicular Flank

Many early studies treated the wake using gas-dynamic or MHD formulations [*Michel*, 1968; *Wolf*, 1968; *Spreiter et al.*, 1970]. We would expect these theories to have the most success in wake regions with the magnetic field quasi-perpendicular to the boundary normal, especially on flux tubes that do not thread the plasma void. Indeed, one can understand much of the plasma dynamics in the

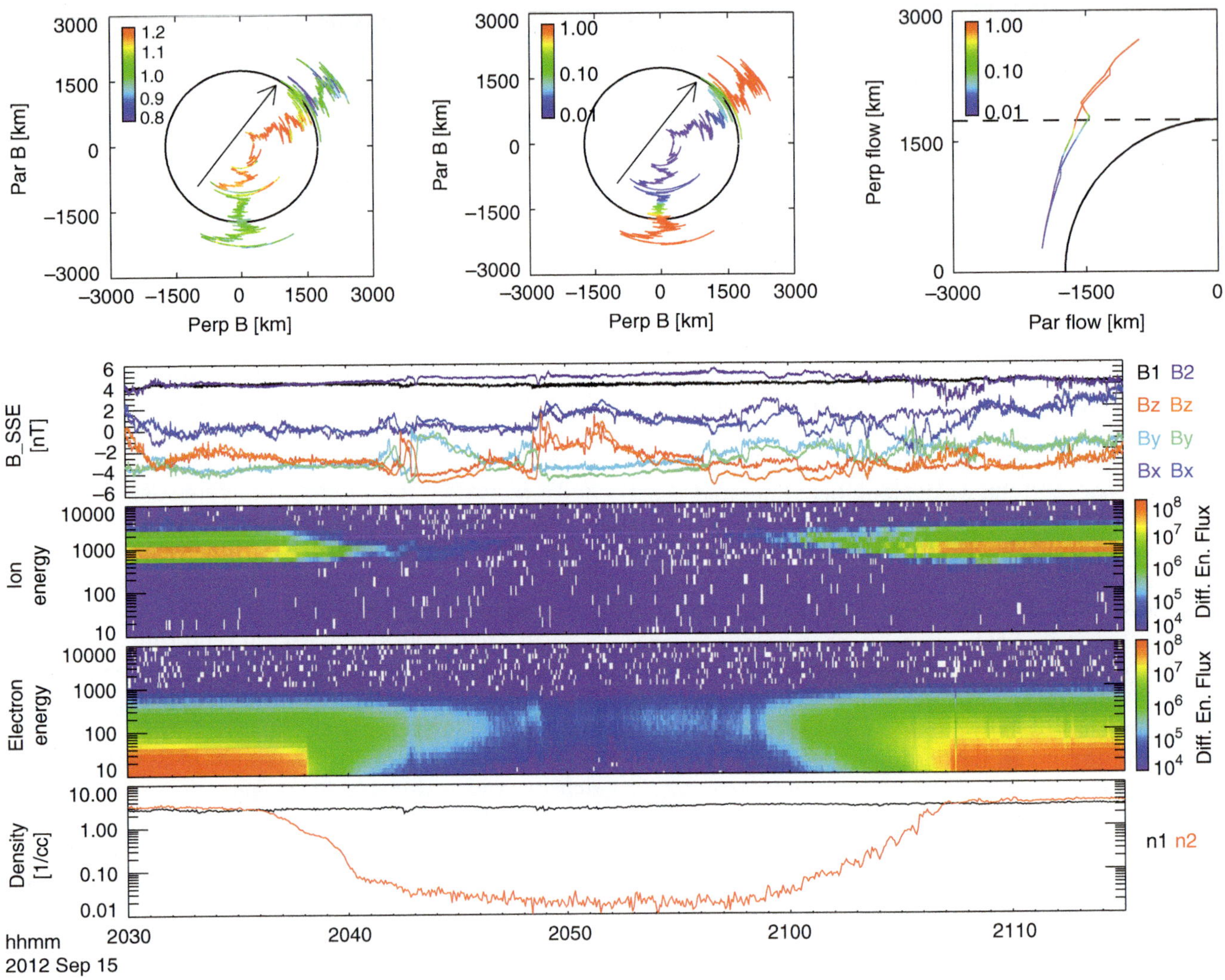

Figure 9.3 Orbit through the wake for the ARTEMIS P2 probe showing the trajectory in coordinates referenced to the magnetic field and flow velocity in the top three panels (arrows show the direction of the P2 trajectory in these coordinates), with time series of upstream (B1) and downstream (B2) magnetic fields in SSE coordinates, ion and electron energy spectra in units of differential energy flux [eV/(cm^2 s sr eV)], and a comparison of the upstream (n1) and downstream ion density (n2). Colors in the upper left panel correspond to the magnetic field ratio B2/B1 and in the upper right two panels to the density ratio n2/n1.

outer perpendicular flank of the wake in terms of fast magnetosonic wave propagation [*Zhang et al.*, 2012, 2014; *Vernisse et al.*, 2013]. The plasma-vacuum discontinuity launches a rarefaction wave that propagates outward into the plasma, resulting in a bounding depression in the field magnitude and the density. At the same time, plasma-filled flux tubes expand into the cavity, compressing plasma-free flux tubes (which pass through the Moon largely unaltered) in the central wake. Though we can derive much of this picture from considerations of pressure balance, the near-Moon wake does not satisfy static pressure or force balance (even with the addition of other terms such as magnetic tension and electric field). This should be intuitively obvious, given that the wake cavity does in fact refill. Farther downstream, as the cavity collapses, pressure balance more nearly holds, and infilling plasma piles up against the compressed magnetic field in the central wake to form a recompression region. Fluid theory suggests that the leading edge of this recompression can steepen to form a slow-mode shock [*Michel*, 1968; *Wolf*, 1968], and 1D hybrid simulations predict the same effect for large electron temperatures [*Israelivich and Ofman*, 2012]; however, no confirming observations exist.

In most respects, the observed magnetic signature matches that expected from diamagnetic currents driven by the pressure gradient at the wake boundary [*Owen et al.*, 1996; *Colburn et al.*, 1967]; however, observations show a more gradual increase in the cavity magnetic field inward from the flanks than predicted from diamagnetic currents alone [*Halekas et al.*, 2005]. As pointed out by *Michel* [1968], $\nabla \cdot \boldsymbol{B} = 0$ requires a gradual magnetic field amplification (unless, perhaps, significant field compression occurs within the Moon). We can resolve the seeming discrepancy by including polarization currents associated with the rarefaction and recompression magnetosonic wavefronts [*Fatemi et al.*, 2013; *Vernisse et al.*, 2013]; with these accounted for, LP data and hybrid simulations closely agree [*Fatemi et al.*, 2012]. Polarization current systems (which act to bound the interaction) exist because, though quasi-static in the Moon frame, the wake interaction has rapidly evolving time-dependent effects in the convecting plasma frame.

Many aspects of the perpendicular refilling process have seen only sparse experimental confirmation. In large part, this results from an unlucky coincidence of orbital geometry and solar wind physics. The rarefaction and recompression regions only fully form thousands of kilometers downstream for most plasma conditions; as a result, low-altitude orbiters cannot observe them. Furthermore, the prevailing IMF lies in the ecliptic such that equatorial orbits primarily cover the plane containing the magnetic field. As a result, until ARTEMIS, few observations of the rarefaction and recompression regions existed. Even ARTEMIS encounters this region too seldom to allow us to make robust statistical conclusions [*Zhang et al.*, 2014].

In Figure 9.4, we show an ARTEMIS observation of the recompression region from an orbit in the outer terrestrial magnetosheath. During this period, the sheath flow had a low magnetosonic Mach number (thanks to the combination of low flow speed and high temperature), allowing ARTEMIS P1 to sample an effective distance far downstream. At the entry flank, farther downwake, we find a very clear rarefaction and recompression region, with correlated density and field magnitude consistent with fast-mode wave propagation, before entering a central cavity with reduced density and compressed fields. At the exit flank, closer to the Moon, we see less obvious rarefaction and recompression signatures, consistent with the convection time scale required for these to fully form.

At both the entry and exit edges of the central cavity, anticorrelated features in the field and density suggest the existence of compressive slow-mode oscillations, boundary layer instabilities, or perhaps even mirror-mode or interchange instabilities. At the entry side, these oscillations appear correlated with significant mixing of plasma entering the wake along field lines, seen in the ion spectra as separate high- and low-energy components (the energy gain and loss result from electric fields parallel to the tilted magnetic fields, as discussed further in section 9.2.4.). Sporadic magnetic connection to the surface also complicates the ion spectra, eliminating one of the components during connected intervals.

Throughout the central wake, we observe measurable levels of plasma, indicating that parallel processes affect perpendicular refilling, at least to some degree. In addition, the interface between the recompression region and the cavity appears less sharp than predicted theoretically and in most simulations. In part, this may result from the significant ion gyroradius during this time period, which acts to smooth out sharp gradients in plasma pressure (see section 9.2.6. for more discussion of finite gyroradius effects). However, it could result in part from pressure-gradient or shear-driven instabilities that act at the discontinuity.

9.2.4. Physical Processes at Parallel Flank

Wake encounters by the Wind spacecraft in 1994 drew attention to the physics associated with the refilling of the wake along magnetic field lines. Wind observed parallel ion acceleration into the wake and an exponential density decrease from the boundary [*Ogilvie et al.*, 1996]. Though these data provided experimental confirmation, the basic theory dates earlier [*Gurevich et al.*, 1969], and scientists had anticipated that it should apply to the wake [*Samir et al.*, 1983].

In the absence of electric fields, ions refilling the wake would obey a Boltzmann transport equation [*Michel*, 1968; *Gurevich and Pitaevsky*, 1975], resulting in the following density distribution (with n the density, s the distance from the wake boundary, and t the expansion time):

$$n = \frac{n_0}{2}\left[1 - \mathrm{erf}\left(\frac{s}{t}\sqrt{\frac{M_i}{2kT_i}}\right)\right]. \tag{9.1}$$

One can derive equation 9.1 by considering travel time effects for initially Maxwellian ions at the wake boundary and integrating the portion of the distribution that can reach a given point at a given time. This formula and others of its kind only apply to the 1D expansion of a stationary plasma, but one can map them to the wake by putting $t = -x/V_{sw}$, the solar wind plasma convection time past the lunar limb. Note that the combination

$$\frac{s}{t}\sqrt{\frac{M_i}{2kT}} = \tau \quad \text{or} \quad \frac{s}{t}\sqrt{\frac{M_i}{kT}} = \xi = \sqrt{2}\tau$$

crops up often. This dimensionless quantity indicates a process that has no characteristic scale length, conveniently tractable in terms of "self-similar" solutions. The

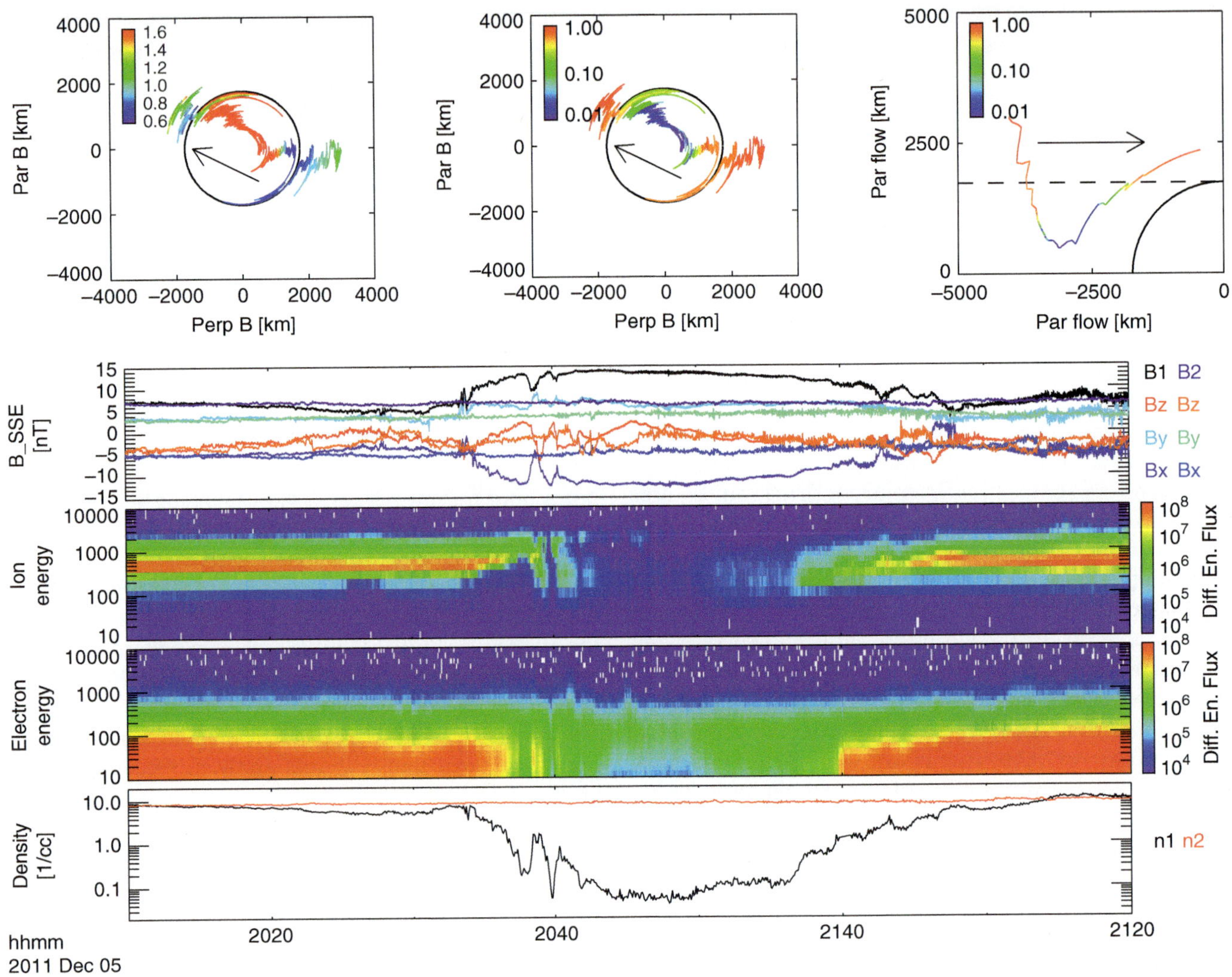

Figure 9.4 Orbit through the wake for the ARTEMIS P1 probe showing the trajectory in wake coordinates in the top three panels (same format as Figure 9.3), with time series of upstream (B2) and downstream (B1) magnetic fields in SSE coordinates, ion and electron energy spectra in units of differential energy flux [eV/(cm² s sr eV)], and a comparison of the upstream (n2) and downstream ion density (n1).

self-similar approximation holds in the outer wake but breaks down at the edge of the expansion front where the characteristic length scale of the expansion reaches the Debye scale [*Gurevich and Pitaevsky*, 1975; *Samir et al.*, 1983] in the rarefied plasma of the near-Moon central wake.

Though early efforts to understand the wake utilized equation 9.1, it applies only to a noncharged noncollisional gas (or the unusual case with $T_e << T_i$). In the opposite limit of cold ions and warm electrons, the following widely used equations instead apply (with the T in the variable ξ now T_e) [*Samir et al.*, 1983; *Denavit*, 1979; *Ogilvie et al.*, 1996]:

$$n = n_0 \exp\left[-\left(1+\xi\right)\right], \tag{9.2}$$

$$V_i = \sqrt{\frac{T_e}{M_i}}\left(1+\xi\right) = C_S\left(1+\xi\right). \tag{9.3}$$

One can derive equations 9.2 and 9.3 by solving the 1D ion continuity and momentum equations with an electric field term assuming Maxwellian electrons that follow the Boltzmann relation

$$n_e = n_0 \exp\left(\frac{e\phi}{kT_e}\right)$$

to relate electrostatic potential and electron density. One finds that ions accelerate into the wake, with an increasing velocity related to the ion acoustic speed. We can understand this result by considering the disparate thermal speeds of electrons and ions. Thermal electrons would expand into the vacuum more quickly than ions absent electric fields. Instead, a parallel electric field develops in order to maintain quasi-neutrality and acts to accelerate ions into the cavity, with a higher velocity than derivable from a fluid treatment. This solution holds until the edge of the expansion front, where quasi-neutrality fails [*Crow et al.*, 1975; *Farrell et al.*, 2008].

In space plasmas, one generally does not find purely Maxwellian electron distributions; however, we can derive similar equations for arbitrary electron distributions, including commonly utilized Kappa distributions [*Manfredi et al.*, 1983; *Sack and Schamel*, 1987]. *Halekas et al.* [2005] showed that these Kappa-based solutions more closely matched LP electron data. In particular, the modified theory correctly captures velocity filtration effects for electrons. The parallel potential along the field line ensures that only a truncated "velocity-filtered" portion of the electron distribution reaches any point in the wake. For Maxwellian electrons, the truncated distribution has the same temperature (but lower density). For Kappa electrons, the truncated distribution has a higher effective temperature, in agreement with observations [*Ogilvie et al.*, 1996; *Halekas et al.*, 2005; *Zhang et al.*, 2014]. Some have attributed the electron temperature increase to travel time effects. Travel time effects are indeed very important for the slower ions, but much less so for the electrons, since their velocity significantly exceeds the solar wind speed for most of the distribution ($V \sim 600$ km/s for a 1 eV electron, as compared to a typical solar wind electron temperature of 10–20 eV and a typical central wake electron temperature of ~100 eV). Therefore, one cannot obtain the temperature increase in the wake without incorporating velocity filtration by electrostatic potentials.

Early theoretical work also treated the more general case with finite temperatures for ions and electrons [*Gurevich et al.*, 1969]; however, few have applied this to the lunar wake, most likely because one cannot analytically solve the resulting equations. Finite ion temperature mainly affects the cold-ion solution by "blurring" the density profile, mainly near the wake boundary. The full theory also correctly predicts parallel cooling of the ion distribution, with an exponent twice that of the density [*Gurevich and Pitaevsky*, 1975], consistent with observations and with adiabatic derivations [*Michel*, 1968; *Clack et al.*, 2004]). Like the cold-ion theory, one can easily extend the warm-ion theory to deal with non-Maxwellian electrons.

The theories described above apply most directly to the case with the magnetic field perpendicular to the flow. One can treat other field orientations (except the singular parallel case) by transforming reference frames [*Ogilvie et al*, 1996; *Hutchinson*, 2012; *Halekas et al.*, 2011b]. In tilted field geometries one typically observes an ion velocity increase on one side of the wake and an ion velocity decrease on the other due to vector addition of the refilling velocity along the magnetic field with the solar wind velocity. The parallel theory cannot account for the effects of perpendicular rarefaction and compression or the resulting change in path length along the field or magnetic mirroring of incident particles. The parallel theories are highly general and apply to a variety of objects but will tend to become less applicable when perpendicular effects become more important, for instance, when the ion gyroradius reaches a scale comparable to or larger than the object scale size.

Figure 9.5 shows a comparison of the theories described above to ARTEMIS observations. The crossing takes place primarily in the plane of the field in a moderately tilted magnetic field orientation. The ion distribution shows the characteristic energy gain and loss and a nearly exponential density decrease into the wake, in agreement with previous observations in similar geometries [*Ogilvie et al.*, 1996; *Halekas et al.*, 2011b]. We calculate expected ion densities in the wake using the five different theories described above, updating the wake coordinates at each point to take into account field rotations for example. As expected, the purely gas-dynamic theory proves unsuccessful in capturing the refilling process, with much larger densities observed in the central wake, highlighting the importance of parallel electric fields. All four formulations incorporating warm electrons match the observed density reasonably well (those with Kappa electron distributions much more closely match the electron temperature, not shown). The two cases that incorporate finite ion temperature do a better job of matching the shape of the observed density profile at the edge of the refilling region, as expected. Overall, theory more accurately matches data at the entry flank, closer to the Moon, most likely because perpendicular effects have had less time to affect the parallel entry. None of the theories perfectly capture the density variations at the rapid field rotations at ~20:42–20:45, but they do qualitatively show the same effects (which effectively correspond to brief transitions into different regions of the wake).

We note the presence of strong compressional fluctuations from ~19:56–20:04 visible in both the density and magnetic field and correlated with an additional high–energy population in the spectra, composed of ions reflected from crustal magnetic fields (see section 9.2.5). The parallel refilling theories obviously cannot capture these features.

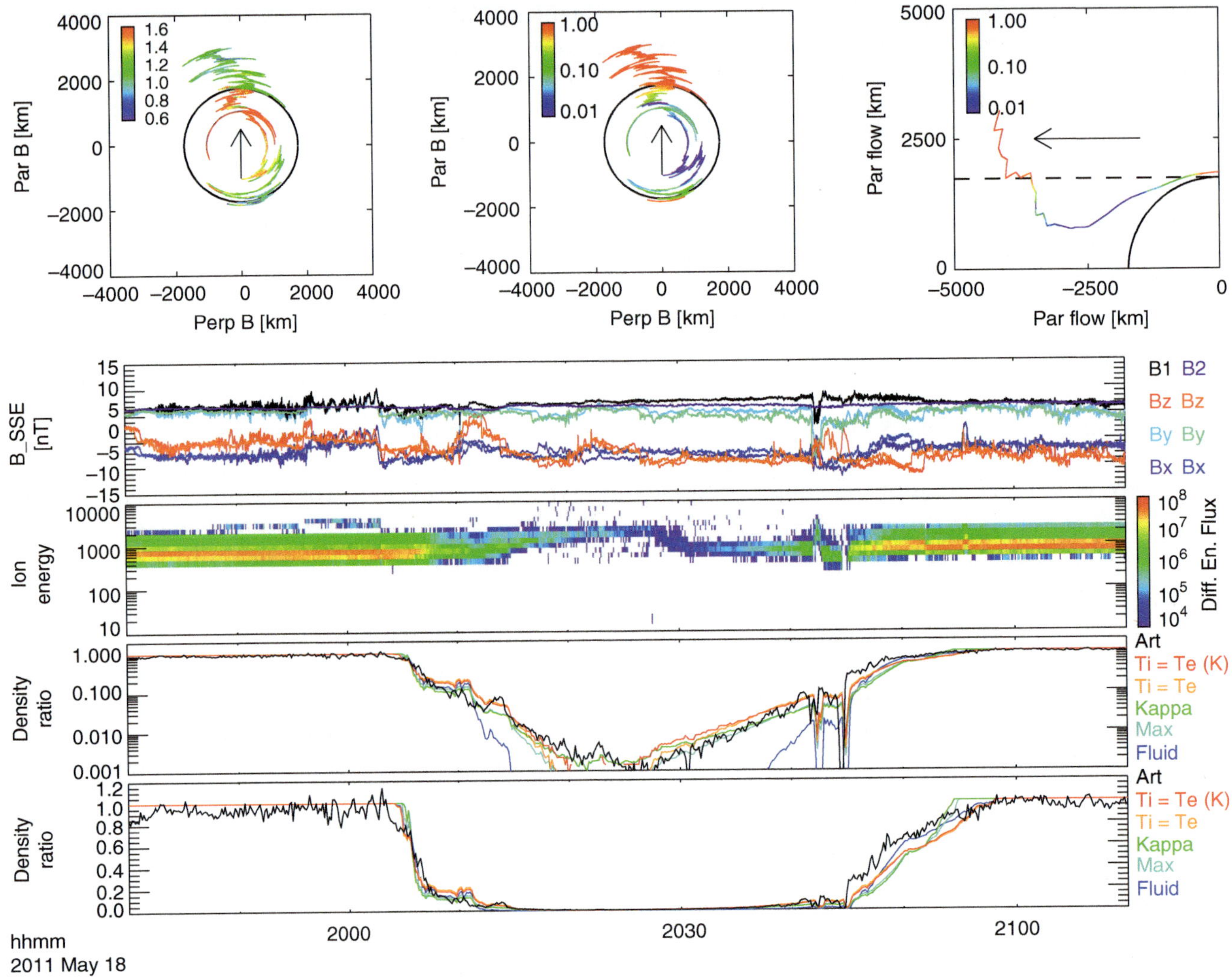

Figure 9.5 Orbit through the wake for the ARTEMIS P1 probe showing the trajectory in wake coordinates (same format as Figure 9.3) and time series of upstream (B2) and downstream (B1) magnetic fields in SSE coordinates, background-subtracted ion energy spectra in units of differential energy flux [eV/(cm^2 s sr eV)], and a comparison of the measured wake/upstream density ratio (background subtracted) to predictions from five different models (purely fluid, cold-ion and warm-ion ($T_i = T_e$) approximations for Maxwellian and Kappa electron distributions) for plasma expansion into a vacuum on both linear and logarithmic scales.

9.2.5. Perturbations Due to Lunar Surface, SubSurface, and Crustal Magnetic Fields

The theories of the wake described above only consider an obstacle with no extent in the flow direction, no magnetic fields, and no interior conductivity. In fact, the spherical shape of the Moon perturbs the structure of the near-Moon wake. However, *Hutchinson* [2008a, 2008b, 2012, 2013] has shown that the parallel electrostatic solutions remain valid for obstacles of arbitrary shape (and in the presence of perpendicular drifts) given straightforward modifications. The presence of the lunar surface eliminates some trajectories that would otherwise reach the central wake; this results in lower density and higher ion velocity than predicted by the unmodified theory. This extension to the theory might explain Chandrayaan observations of high ion refilling velocities at low inductive effects, which have been used to probe the lunar subsurface conductivity [*Dyal et al.*, 1974] have much more significant manifestations for icy moons [*Neubauer*, 1998].

Significant modifications to the basic wake structure result from crustal magnetic fields. The physics of their interaction with the solar wind remains poorly understood, but observations show that they drive compressional enhancements, observed outside of the rarefaction and recompression regions of the main wake, and variously termed penumbral increases, limb shocks, limb compressions, and lunar external magnetic enhancements [*Ness et al.*, 1968; *Colburn et al.*, 1971; *Sonett and Mihalov*, 1972; *Russell and Lichtenstein*, 1975; *Halekas et al.*, 2006]. These features proved a great source of confusion for some time, but greater understanding of their statistical occurrence clarified their correlation to crustal magnetic fields [*Sonett and Mihalov*, 1972; *Russell and Lichtenstein*, 1975; *Halekas et al.*, 2006]. It still remains uncertain to what degree these external magnetosonic disturbances perturb the wake. Their presence alters the plasma flow [*Siscoe et al.*, 1969], so they must modify the wake to some degree. However, it remains unknown whether this perturbation mainly affects the outer regions of the wake or whether it significantly changes the refilling dynamics in the central wake.

Figure 9.6 shows an example of a wake crossing with significant crustal magnetic field effects. At the entry flank, ARTEMIS observes a large compressional feature (apparent in both magnetic field and density), along with smaller leading oscillations, outside of the wake rarefaction region and diamagnetic current system. Its appearance well downstream from the limb indicates something other than a purely local interaction with crustal magnetic fields. At the exit flank, we find a much smaller compression outside of the diamagnetic current region, despite the much closer distance. At the entry flank, the strongly magnetic southern far side lies at the limb, while near the exit flank only weak crustal sources exist to perturb the flow, explaining the dichotomy. Outside of the compressional features, near the limb at the exit flank, transverse oscillations in the magnetic field appear, associated with protons reflected from more distant magnetic fields.

9.2.6. Finite Gyroradius Effects

We have briefly discussed the importance of finite ion temperature on the parallel refilling process (section 9.2.4). However, recent observations from Kaguya have also highlighted the importance of finite ion temperature (and gyroradius) for perpendicular refilling processes. The parallel electric field that develops along magnetic field lines implies a charge distribution that also produces inward electric fields at the perpendicular flank (other Ohm's law terms such as ∇P_e and $\boldsymbol{J} \times \boldsymbol{B}$ also contribute). Ion gyromotion at the perpendicular flanks, in the presence of these inward fields, results in an effective $\boldsymbol{E} \times \boldsymbol{B}$ drift parallel or antiparallel to the initial solar wind velocity, with a resulting energy gain/loss at the two flanks, and enhanced access to the central wake, termed type I entry [*Nishino et al.*, 2009a]. This pairwise energy gain/loss occurs mainly at low altitudes and has a completely different nature from the pairwise energy gain/loss associated with the parallel ion entry along tilted magnetic fields. Similar signatures have also been observed at Rhea [*Wilson et al.*, 2010; *Roussos et al.*, 2007], but it remains unclear whether these require the same electric field effects. The Rhea effects could result purely from gyromotion of an ion population with a large temperature and a large gyroradius compared to the obstacle scale and the resulting gyrophase and energy-dependent access to the two flanks.

A small fraction of the solar wind reflects in charged form from the lunar surface [*Saito et al.*, 2008], with a much larger fraction reflecting from regions with crustal magnetic fields [*Lue et al.*, 2011]. These reflected ions, with a significant velocity relative to the solar wind, feel the convection electric field and follow cycloidal trajectories much like pickup ions, but with their initial velocity resulting in a prolate cycloid rather than a perfect cycloid. Reflected protons can reach energies of up to 9 times that of the bulk solar wind flow (see Figure 9.5 for an example of this population, just before wake entry) and for some combinations of flow velocity and magnetic field can penetrate the lunar wake, a process termed type II entry [*Nishino et al.*, 2009b]. In the wake, the charge density associated with these ions can dominate, resulting in electric fields that pull electrons into the wake to restore quasi-neutrality [*Nishino et al.*, 2010]. These electric fields, opposite to those associated with the normal refilling processes, change the structure of the wake in regions where these ions penetrate. A combination of type I and type II entry, wherein reflected ions gyrate into the wake close to the limb, also occurs [*Wang et al.*, 2010].

9.2.7. Wake in Different Plasma Regimes

As described in section 9.2.1, the wake almost never remains quasi-stationary for extended periods of time. Instead, it reconfigures on time scales commensurate with solar wind variability. As yet, it remains unknown what happens to the instantaneous wake structure as a discontinuity passes; however, observations show that the entire wake rapidly reaches a new equilibrium configuration on the convection time scale, as expected.

The wake also varies significantly in response to changes in the ambient plasma. The Moon only spends about two weeks of every month in the undisturbed solar wind. The rest of the time, it encounters plasma affected to varying degrees by the presence of Earth and its

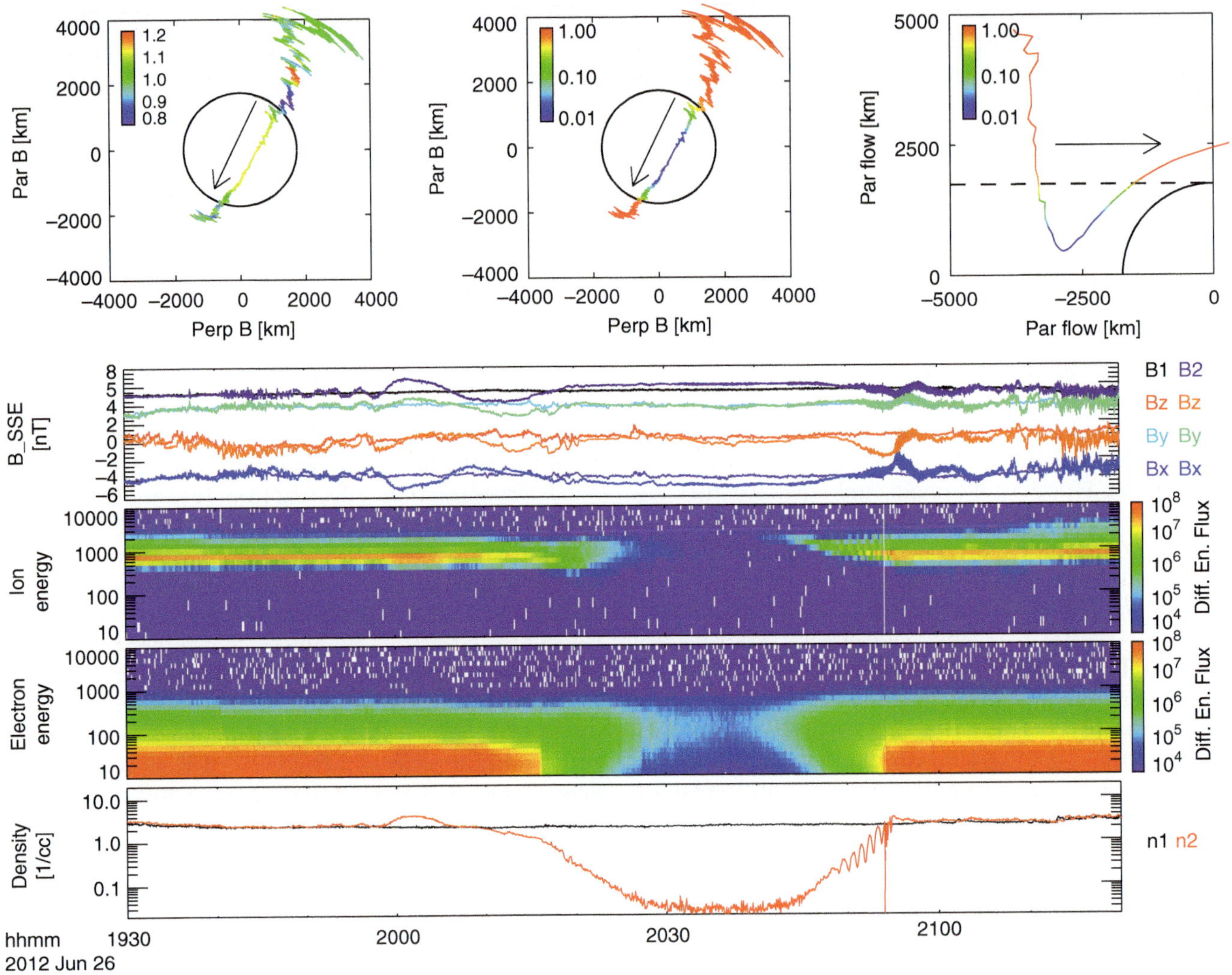

Figure 9.6 Orbit through the wake for the ARTEMIS P2 probe showing the trajectory in wake coordinates in the top three panels (same format as Figure 9.3), with time series of upstream (B1) and downstream (B2) magnetic fields in SSE coordinates, ion and electron energy spectra in units of differential energy flux [eV/(cm^2 s sr eV)], and a comparison of the upstream (n1) and downstream ion density (n2). Unphysical oscillations in the density at the exit of the wake result from a combination of undersampling of the cold ion distribution (cooled in the parallel direction) and the spin-phase drift described in section 9.2.2.

magnetic field. In the central terrestrial magnetosphere, in the tail lobes and plasma sheet, a wakelike structure may still develop when sufficient flow exists, but this regime has as yet seen little attention, other than research showing that smaller and less reproducible magnetic perturbations exist [*Schubert and Lichtenstein*, 1974]. For sub-magnetosonic flows, one could expect a wake more like the laterally extended structures observed for outer planet moons; however, the case with significant parallel flow along the lobe fields (rather than perpendicular convection) could prove unique. Additionally, ions of lunar origin can dominate over tenuous lobe plasma [*Harada et al.*, 2013], possibly significantly perturbing any wakelike structure.

In the terrestrial magnetosheath at lunar orbit, the Moon typically still encounters a super-magnetosonic flow, and the wake structure remains similar to that in the solar wind. However, in the frequently observed cases where the plasma has a higher beta, more significant magnetic field compression can result. The flow also typically has a lower Mach number, resulting in more rapid refilling and therefore a shorter wake (see Figure 9.4).

At the boundary between the magnetosheath and the pure solar wind (the bow shock) interesting effects also occur. The presence of the wake and its very tenuous plasma can in fact significantly perturb the structure of the bow shock. This occurs in part because the Mach number of the plasma changes in the wake, resulting in both a deformation and a change in the character of the shock in the wake [*Nishino et al.*, 2011].

Even entirely outside of the magnetosphere, the presence of Earth can affect the wake. In the ion foreshock region, though in the solar wind, the Moon still lies exposed to ions reflected from the terrestrial bow shock, which can travel upstream from the shock, reaching significant energies well above that of the solar wind. These ions can penetrate the wake from downstream, providing a charge density in the central wake that can exceed that resulting from normal refilling processes. In analogy to the case of type II entry, one expects this additional plasma population to significantly perturb the wake and alter the normal refilling dynamics.

9.2.8. Plasma Instabilities of Lunar Wake

Numerous plasma instabilities occur in and around the wake. Fluid instabilities driven by pressure gradients or velocity shear have seen perhaps the least study but may prove important. As shown by *Borisov and Mall* [2000], the perpendicular flank of the wake cavity satisfies flute instability criteria, possibly allowing interchange between plasma-filled flux tubes and compressed plasma-free flux tubes, as postulated to occur at Rhea [*Roussos et al.*, 2012]. Even if flute instabilities cannot grow, shorter-wavelength drift-wave instabilities could result from the pressure gradient at the boundary, as postulated to occur for expanding plasmas such as the AMPTE barium cloud [*Bernhardt et al.*, 1987; *Winske*, 1988]. Velocity shear instabilities may also operate at the perpendicular flank, at the interface between flux tubes in the cavity (subject to parallel refilling processes) and those that do not penetrate the cavity. No one has quantified whether sufficient shear exists to overcome stabilizing forces and enable these instabilities.

Both perpendicular and parallel refilling processes naturally result in highly anisotropic distributions for both electrons and ions. As described in section 9.2.4, the parallel refilling process results in heating of the electrons (due to velocity filtration of the non-Maxwellian solar wind distributions) [*Halekas et al.*, 2005] and the development of a loss cone distribution (due to combined electrostatic and magnetic reflection of infilling electrons) as well as exponential parallel cooling of the ion distribution [*Gurevich and Pitaevsky*, 1975; *Clack et al.*, 2004]. The electron loss cone should drive whistler turbulence, while the ion parallel cooling could drive ion cyclotron waves or mirror mode instabilities. The low beta in the wake may significantly moderate or even completely damp the development of the ion anisotropy instabilities. However, even if parallel cooling alone cannot produce temperature anisotropies sufficient to drive significant instability growth, type II ion entry provides an additional very large effective enhancement of the perpendicular ion temperature capable of generating plasma waves.

The wake particularly excels as a laboratory for electrostatic instabilities. The parallel electric fields developed in the refilling process naturally produce a variety of distributions unstable to beam-beam or bump-on-tail instabilities. Ions accelerated into the wake eventually meet in the central region, where they can drive a beam-beam electrostatic instability [*Farrell et al.*, 1997, 1998]. Similarly, a small portion of the tail of the electron distribution, though decelerated during its entry to the central wake, can penetrate the central void and regain energy as it exits the opposite side of the wake, forming a distribution with a significant positive slope and driving electrostatic waves (or phase-space holes) [*Birch and Chapman*, 2001b], as observed by ARTEMIS [*Halekas et al.*, 2011b; *Tao et al.*, 2012]. The velocity-filtered electrons exiting the wake or the initially reflected portion of the entering population may prove responsible for producing the whistler waves observed on magnetic field lines that thread the wake [*Farrell et al.*, 1996; *Nakagawa et al.*, 2003] and possibly also the ion acoustic waves observed on the same field lines [*Bale et al.*, 1997]. They also may generate the electrostatic solitary waves observed by Kaguya near the wake boundary [*Hashimoto et al.*, 2010].

Hutchinson [2012] has shown that a convective electron bump-on-tail instability can also develop from the combination of the flow downstream and acceleration/deceleration by the wake potential. The detailed growth of this instability, which relies on electrons that cross the symmetry axis of the wake, depends critically on the potential structure in the center of the wake. Here, where plasma from the two sides of the wake interacts, analytical models of the parallel refilling break down, leaving the electrostatic structure somewhat of a question mark.

The type II entry of reflected protons that gyrate across magnetic field lines into the wake generates localized regions of parallel electric fields that point in the opposite direction and pull counterstreaming electrons into the wake, driving significant broadband electrostatic activity [*Nishino et al.*, 2010]. Foreshock ions could in principle have similar effects. Type II ions themselves have significant velocity relative to the solar wind and refilling wake plasma and could drive instabilities both inside and outside the wake. However, the growth rate appears problematic, given the small spatial extent of the type II ion structures (resulting in a short time for wave growth in the solar wind frame). These ions do drive compressional fluctuations, as seen in Figures 9.5 and 9.6, despite their limited spatial extent.

Finally, the presence of the lunar surface plays a role in driving instabilities in the wake. The shadowed surface typically charges negative and emits secondary electrons that accelerate outward through the sheath [*Halekas et al.*, 2011a]. These secondary beams should radiate as well as interact with infilling plasma, driving still more electrostatic instabilities.

9.3. SIMULATIONS OF LUNAR WAKE

9.3.1. Overview of Wake Models

We now describe some models used to study the plasma wake of the Moon (also applicable to other solid objects without a significant atmosphere or internal magnetic fields). The wake problem is closely related to that of expansion of plasma into vacuum, often modeled with an initial condition of a half space with no plasma adjoining a half space containing plasma. We discussed analytical 1D models of this plasma expansion process in section 9.2.4. As described previously, one can apply such models to the lunar wake by placing the plasma-vacuum interface at the lunar terminator at the start of the simulation and letting the simulation domain convect with the solar wind plasma flow, with each specific time corresponding to a downstream distance. An improved 1D model utilizes a central cavity initially empty of plasma, surrounded by uniform plasma [e.g., *Birch and Chapman*, 2002]. This initial condition allows for the interaction of plasma populations expanding into vacuum from opposite sides of the cavity, more relevant to studies of the real lunar wake, especially far downstream from the Moon.

For numerical modeling of the lunar wake, and more generally of plasma expansion into vacuum, one must choose both the dimensionality of the model and the physics included in the model. Wake models can have from one to three spatial dimensions. Higher spatial dimensions enable better treatment of the interaction of multiple refilling particle populations and allow for a more comprehensive set of physical processes, at the cost of computational complexity. In terms of physics, we can make a distinction between two classes of models that have been applied to the lunar wake, namely fluid and kinetic simulations. Fluid models, including both gas-dynamic and MHD models, assume that we can describe the velocity space distribution of the plasma with a limited set of parameters, e.g., a Maxwellian distribution with a certain density, bulk velocity, and temperature. In addition to single-fluid MHD models, multifluid MHD models exist that allow for several plasma populations with different moments. Kinetic models, on the other hand, make no assumptions regarding the velocity space distribution, instead representing plasma populations with discrete particles or groups of particles and tracking electric and magnetic fields on a grid. As an alternative, one can use a Vlasov method and solve for the phase space density on a velocity space grid (*Umeda et al.*, [2011] utilized this approach for the 2D problem of flow onto an infinite cylinder). *Ledvina et al.* [2008] give an in-depth description of general methods for modeling collisionless plasmas.

Most early models of the lunar wake utilized MHD or gas-dynamic treatments [*Spreiter et al.*, 1970; *Michel*, 1968]. *Cui and Lei* [2008] have published a more recent 2D MHD model of the wake. The fluid approach has several limitations for the wake. First, the process of plasma expansion into vacuum itself changes the velocity distributions, typically resulting in highly non-Maxwellian distributions as faster particles stream ahead into the wake. Second, refilling particle populations can meet and interact in the central wake, resulting in counterstreaming distributions intractable in a fluid approach. Finally, important finite gyroradius effects result from the absorption and reflection of particles by the surface of the object and the resulting gyro-phase dependence of the distribution in the wake.

The particle velocity distributions that develop in the wake can be more accurately represented by kinetic methods, such as particle-in-cell (PIC) models. Examples include the 1D electrostatic PIC models of *Farrell et al.* [1998] and *Guio and Pecseli* [2004] and the electromagnetic PIC models used by *Birch and Chapman* [2001a]. One disadvantage of 1D models comes from the requirement for divergence-free magnetic fields. The dimensionality of the model requires that the field only vary along the coordinate perpendicular to the initial plasma-vacuum interface, and the zero-divergence condition therefore requires a constant perpendicular field. This leads to some inaccuracy, since the field component across the wake boundary in actuality can increase in the wake. To overcome this, one can increase the dimensionality of the model; however, kinetic methods rapidly become computationally expensive due to the need to resolve the velocity dimensions. A 3D simulation with three velocity components results in a 6D problem that proves very difficult to solve with reasonable resolution. Even 2D models have enough computational complexity that one often must introduce unphysical simplifications such as a reduced ion–electron mass ratio or object scale size.

The computational complexity of kinetic models makes it very difficult to perform 3D simulations of the lunar wake with a realistic obstacle size and ion–electron mass ratio. As an intermediate between fluid and fully kinetic implementations, one can utilize a hybrid approach (as shown in Figures 9.2 and 9.7), treating ions

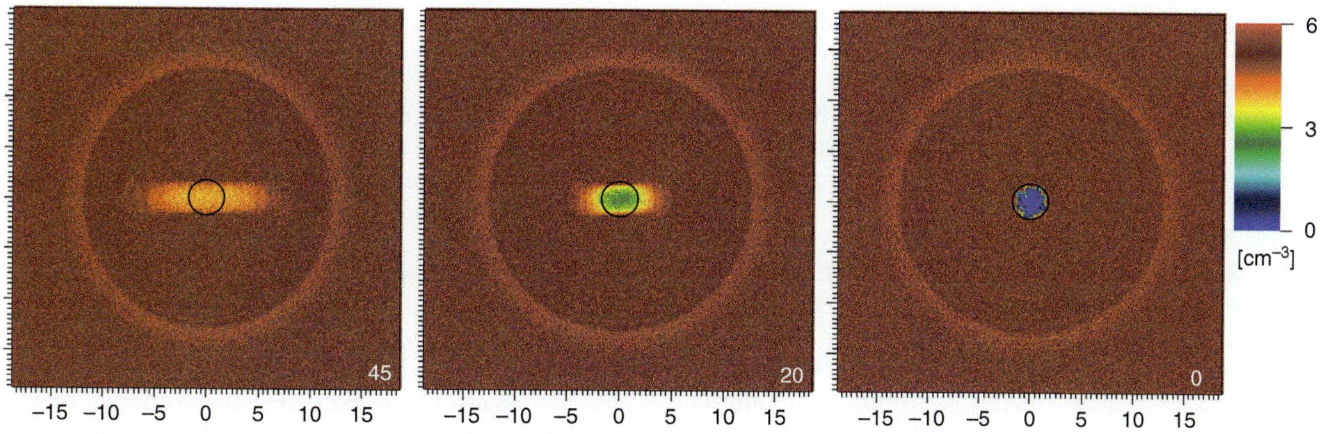

Figure 9.7 High-resolution 2D hybrid simulation of the lunar wake in a convecting frame showing proton number density in the plane perpendicular to the flow for IMF orientations of 45°, 20°, and 0° away from the solar wind direction (in the horizontal plane). The black circle shows the projection of the Moon along the flow axis and the simulation results shown extend 20,000 km laterally from the center of the Moon (axes in units of 1000 km). For a solar wind velocity of 500 km/s the simulation time of 200 s corresponds to a distance of 100,000 km ≈ 58 lunar radii downstream from the center of the Moon. The simulation uses a solar wind number density of 5 cm^{-3}, an ion temperature of 5×10^4 K, electron temperature of 10^4 K, and IMF magnitude of 5 nT. For these plasma parameters the ion plasma beta is 0.35.

kinetically and electrons as a fluid (typically massless). The assumption of zero electron mass makes the electric field a dependent variable, leaving only the magnetic field and the ion positions and velocities as unknowns. This also removes the electron temporal and spatial scales from the problem, leaving the ion temporal and spatial scales to govern the solution. The Moon's scale, only ~1–2 orders of magnitude greater than either the solar wind proton thermal gyroradius or the ion inertial length, allows the application of a hybrid model, enabling global 3D simulations prohibitive in a fully kinetic approach. *Israelivich and Ofman* [2012] used a 1D convecting hybrid model, and *Travnicek et al.* [2005] investigated the lunar wake using a 2D hybrid model, allowing a large simulation domain and high resolution, with an emphasis on wave phenomena. *Kallio* [2005)] and *Wang et al.* [2011] utilized 3D hybrid models to study the general morphology of the wake. Comparisons with LP observations [*Fatemi et al.*, 2012], Wind observations [*Holmstrom et al.*, 2012], and ARTEMIS observations [*Wiehle et al.*, 2011] show that hybrid models reproduce many observed wake features. While earlier works have focused on magnetic fields in the wake, *Fatemi et al.* [2013] and *Vernisse et al.* [2013] have also studied the current systems in and around the wake. *Lipatov et al.* [2012] considered the effects of pickup ions on the lunar environment.

Finally, test particle models, which do not self-consistently compute the electromagnetic fields but rather follow the trajectories of particles in prescribed fields [*Holmstrom et al.*, 2010], allow the reconstruction of velocity space distributions with high resolution. For example, *Fatemi et al.* [2012] utilized a backward Liouville method to study how surface absorption affects velocity distributions in the lunar wake (see also section 9.2.5).

For a 2D model, one can utilize either a convecting simulation domain oriented perpendicular to the flow (with a circular initial vacuum region) [*Birch and Chapman*, 2002] or an infinite cylindrical absorbing obstacle in a stationary simulation domain containing the plane of the flow [*Kimura and Nakagawa*, 2008]. Figure 9.7 shows an example of the first case, a 2D convecting-frame simulation, at a location very far downstream (difficult to reach in a stationary simulation). We see an external density rarefaction that expands from the Moon at the fast magnetosonic speed. We also find that the wake crosssection becomes flattened in the plane of the IMF (except for the parallel-IMF case), due to the fact that ions refill the cavity along the magnetic field. The width of the wake perpendicular to the plane of the IMF is somewhat smaller than the diameter of the Moon, because of perpendicular infilling and recompression. For the singular case with IMF perfectly parallel to the flow, a (deformed) vacuum region still exists. The distortions in the boundary of the cavity at this downstream location could perhaps result from a flute instability, as discussed in section 9.2.8. Surprisingly, the wake remains quite observable at this downstream location; in practice, even small solar wind fluctuations could make it difficult to observe at this distance, though *Clack et al.* [2004] did see the wake at $25R_L$.

9.3.2. Numerical Issues for Wake Simulations

We now describe some of the particular numerical issues that arise for global models of the lunar wake. When considering the Moon as an insulating spherical obstacle to the solar wind, one important decision relates to the boundary conditions at the surface of the Moon. For fluid models, *Xie et al.* [2012] considers absorption of the fluid on the dayside but flow around a sphere on the night side. Most particle (hybrid) models remove the solar wind ions that hit the lunar surface. For the magnetic field, most models consider free propagation through the Moon, so that the magnetic field convects undisturbed through the Moon (corresponding to a perfectly resistive obstacle). For hybrid models, the near vacuum close to the night side of the Moon represents the most difficult aspect of the wake to treat numerically. The low-density cells there result in large electric fields, since the electric field computation in a hybrid model involves division by local charge density. Furthermore, the Alfvén velocity becomes very large in the wake, placing difficult requirements on the simulation time step. To deal with these singularities, one can limit the magnitude of the electric field by increasing the charge density in low-density cells. *Kallio* [2005], *Trávnicek et al.* [2005], and *Wiehle et al.* [2011] have all utilized a minimum charge density. As an alternative, one can introduce unphysical particles. *Holmstrom et al.* [2012] reduced the weight of particles hitting the surface of the Moon at each subsequent time step, rather than deleting them. Reducing the density of these particles to the minimum allowable by simulation constraints greatly improves the agreement with observations [*Fatemi et al.*, 2012]. *Wang et al.* [2011] launched ions from the night side of the Moon with solar wind velocity. Instead of increasing the charge density in low-density regions, one can instead solve a Laplace equation in vacuum regions. *Holmstrom* [2013] considered this approach as the limit of a large, but finite, resistivity in vacuum regions, leading to the solution of a magnetic diffusion equation. Using this scheme, one can solve for the magnetic field in the interior of the Moon using the same diffusion equation. *Wang et al.* [2011], *Wiehle et al.* [2011], *Vernisse et al.* [2013], and *Holmstrom* [2013] have used similar schemes for modeling a moon with spatially varying internal resistivity.

9.4. FRONTIERS IN STUDY OF LUNAR WAKE

Though our understanding of the lunar wake and its structure and dynamics has increased tremendously over the past 50 years of study, some important aspects remain incompletely understood. We now know about a number of different refilling processes, including fluid (frozen-in) convection perpendicular to the magnetic field, parallel refilling along the magnetic field, and kinetic infilling perpendicular to the field (by gyrating particles, including reflected protons). However, we do not yet fully understand how these processes interact with each other, especially in regions where multiple refilling modes operate. The global effects of reflected solar wind protons on the wake have not yet been studied self-consistently, but only with test particle models [*Holmstrom et al.*, 2010]. A 3D model fully incorporating this reflected population, validated by detailed observational studies of the regions of the wake where multiple refilling processes occur, would go far toward solidifying our understanding of the wake.

Study of the wake's reaction to time-dependent solar wind conditions (from both an observational and a numerical approach), particularly the instantaneous behavior as the wake reconfigures in response to a change in the incoming plasma, represents another important frontier in our understanding of the wake. The effect of the lunar internal conductivity on the wake may prove an important aspect of this problem. Though *Wang et al.* [2011] have considered a steady case with uniform internal lunar resistivity, no one has yet addressed the effects of internal conductivity on the wake's response to time-varying solar wind conditions, from either an observational or a simulation standpoint.

Finally, the consistent inclusion of crustal magnetic fields represents a critical future model development. To date, no numerical study has investigated the global effects of crustal fields on the wake, though *Kallio* [2012] has performed localized studies. The strength of the magnetic fields, reaching hundreds or even thousands of nanotesla close to the surface, makes global simulations computationally expensive, since this increase over the magnetic field in the solar wind requires a correspondingly small time step to resolve the increased ion gyrofrequency. However, the global effects of these magnetic sources represent perhaps the most outstanding problem in lunar plasma physics and should prove worthy of the effort.

ACKNOWLEDGMENTS

We thank NASA's Lunar Science Institute for support and acknowledge ISSI for hosting a workshop that inspired portions of this review. We acknowledge NASA contract NAS5-02099 and V. Angelopoulos for use of data from ARTEMIS and specifically C. W. Carlson and J. P. McFadden for use of ESA data and K. H. Glassmeier, U. Auster, and W. Baumjohann for use of FGM data.

REFERENCES

Angelopoulos, V. (2011), The ARTEMIS mission, *Space Sci Rev*, doi:10.1007/s11214-010-9687-2.

Auster, H. U., et al. (2008), The THEMIS fluxgate magnetometer, *Space Sci. Rev.*, *141*, 235–264.

Bale, S. D., C. J. Owen, J.-L. Bougeret, K. Goetz, P. J. Kellogg, R. P. Lin, R. Manning, and S. J. Monson (1997), Evidence of currents and unstable particle distributions in an extended region around the lunar wake, *Geophys. Res. Lett.*, *24*, 1427–1430.

Baumjohann, W., M. Blanc., A. Fedorov, and K.-H. Glassmeier (2010), Current systems in planetary magnetospheres and ionospheres, *Space Sci. Rev.*, *152*, 99–134.

Bernhardt, P. A., R. A. Roussel-Dupre, M. B. Pongratz, G. Haerendel, A. Valenzuela, D. A. Gurnett, and R. R. Anderson (1987), Observations and theory of the AMPTE magnetotail Barium releases, *J. Geophys. Res.*, *92*, 5777–5794.

Birch, P. C., and S. C. Chapman (2001a), Particle-in-cell simulations of the lunar wake with high phase space resolution, *Geophys. Res. Lett.*, *28*, 219.

Birch, P. C., and S. C. Chapman (2001b), Detailed structure and dynamics in particle-in-cell simulations of the lunar wake, *Phys. Plasmas*, *8*, 4551–4559.

Birch, P. C., and S. C. Chapman (2002), Two dimensional particle-in-cell simulations of the lunar wake, *Phys. Plasmas*, *9*, 1785–1789.

Borisov, N., and U. Mall (2000), Plasma distribution and electric fields behind the Moon, *Phys. Lett. A*, *265*, 369–376.

Bosqued, J. M., et al. (1996), Moon-solar wind interaction: First results from the WIND/3DP experiment, *Geophys. Res. Lett.*, *23*, 1259–1262.

Clack, D., J. C. Kasper, A. J. Lazarus, J. T. Steinberg, and W. M. Farrell (2004), Wind observations of extreme ion temperature anisotropies in the lunar wake, *Geophys. Res. Lett.*, *31*, L06812, doi:10.1029/2003GL018298.

Colburn, D. S., R. G. Currie, J. D. Mihalov, and C. P. Sonett (1967), Diamagnetic solar-wind cavity discovered behind moon, *Science*, *158*, 1040.

Colburn, D. S., J. D. Mihalov, and C. P. Sonett (1971), Magnetic observations of the lunar cavity, *J. Geophys. Res.*, *76*, 2940–2957.

Crow, J. E., P. L. Auer, and J. E. Allen (1975), The expansion of plasma into a vacuum, *J. Plasma Phys.*, *14*, 65–76.

Cui, W., and L. Lei (2008), 2D MHD simulation of the lunar wake, *Chin. J. Space. Sci.*, *28*, 189–193.

Denavit, J. (1979), Collisionless plasma expansion into a vacuum, *Phys. Fluids*, *22*, 1384.

Dyal, P., C. W. Parkin, and W. D. Daily (1974), Magnetism and the interior of the Moon, *Rev. Geophys. Space Phys.*, *12*, 568–591.

Farrell, W. M., R. J. Fitzenreiter, C. J. Owen, J. B. Byrnes, R. P. Lepping, K. W. Ogilvie, and F. Neubauer (1996), Upstream ULF waves and energetic electrons assocated with the lunar wake: Detection of precursor activity, *Geophys. Res. Lett.*, *23*, 1271–1274.

Farrell, W. M., M. L. Kaiser, and J. T. Steinberg (1997), Electrostatic instability in the central lunar wake: A process for replenishing the plasma void?, *Geophys. Res. Lett.*, *24*, 1135–1138.

Farrell, W. M., M. L. Kaiser, J. T. Steinberg, and S. D. Bale (1998), A simple simulation of a plasma void: Applications to Wind observations of the lunar wake, *J. Geophys. Res.*, *103*, 23,653–23,660.

Farrell, W. M., T. J. Stubbs, J. S. Halekas, G. T. Delory, M. R. Collier, R. R. Vondrak, and R. P. Lin (2008), Loss of solar wind plasma neutrality near the lunar terminator and polar regions, *Geophys. Res. Lett.*, *35*, L015105, doi:10.1029/2007GL032653.

Fatemi, S., M. Holmstrom, and Y. Futaana (2012), The effects of lunar surface plasma absorption and solar wind temperature anisotropies on the solar wind proton velocity space distributions in the low-altitude lunar plasma wake, *J. Geophys. Res.*, *117*, A10105, doi:10.1029/2011JA017353.

Fatemi, S., M. Holmstrom, Y. Futaana, S. Barabash, and C. Lue (2013), The lunar wake current systems, *Geophys. Res. Lett.*, *40*, 17–21.

Felber, F. S., and R. Decoste (1978), Fast expansions of laser-heated plasmas, *Phys. Fluids*, *21*, 520–522.

Futaana, Y., S. Machida, Y. Saito, A. Matsuoka, and H. Hayakawa (2001), Counterstreaming electrons in the near vicinity of the Moon observed by plasma instruments on board NOZOMI, *J. Geophys. Res.*, *106*, 18,729–18,740.

Futaana, Y., et al. (2010), Protons in the near lunar wake observed by the Sub-keV Atom Reflection Analyzer on board Chandrayaan-1, *J. Geophys. Res.*, *115*, A10248, doi:10.1029/2010JA015264.

Georgescu, E., F. Plaschke, U. Auster, K.-H. Fornacon, and H. U. Frey (2011), Modelling of spacecraft spin period during eclipse, *Ann. Geophys*, *29*, 875–882.

Guio, P., and H. L. Pecseli (2004), Phase space structures generated by an absorbing obstacle in a streaming plasma, *Geophys. Res. Lett.*, *31*, L03806.

Gurevich, A. V., and L. P. Pitaevsky (1975), Non-linear dynamics of a rarefied ionized gas, *Prog. Aerospace Sci.*, *16*, 227–272.

Gurevich, A. V., L. P. Pitaevskii, and V. V. Smirnova (1969), Ionospheric aerodynamics, *Space Sci. Rev.*, *9*, 805–871.

Halekas, J. S., S. D. Bale, D. L. Mitchell, and R. P. Lin (2005), Magnetic fields and electrons in the lunar plasma wake, *J. Geophys. Res.*, *110*, A07222, doi:10.1029/2004JA010991.

Halekas, J. S., D. A. Brain, D. L. Mitchell, R. P. Lin, and L. Harrison (2006), On the occurrence of magnetic enhancements caused by solar wind interaction with lunar crustal fields, *Geophys. Res. Lett.*, *33*, L01201, doi:10.1029/2006GL025931.

Halekas, J. S., Y. Saito, G. T. Delory, and W. M. Farrell (2011a), New views of the lunar plasma environment, *Planet. Space Sci.*, *59*, 1681–1694, doi:10.1016/j.pss.2010.08.011.

Halekas, J. S., et al. (2011b), First results from ARTEMIS, a new two-spacecraft lunar mission: Counter-streaming plasma populations in the lunar wake, *Space Sci. Rev.*, doi:10.1007/s11214-010-9738-8.

Halekas, J. S., et al. (2012), Lunar precursor effects in the solar wind and terrestrial magnetosphere, *J. Geophys. Res*, *117*, A05101, doi:10.1029/2011JA017289.

Harada, Y., S. Machida, J. S. Halekas, A. R. Poppe, and J. P. McFadden (2013), ARTEMIS observations of lunar dayside plasma in the terrestrial magnetotail lobe, *J. Geophys. Res.*, *118*, 3042–3054, doi:10.1002/jgra.50296., doi:10.1002/jgra.50296.

Hashimoto, K., et al. (2010), Electrostatic solitary waves associated with magnetic anomalies and wake boundary of the Moon observed by KAGUYA, *Geophys. Res. Lett.*, *37*, L19204, doi:10.1029/2010GL044529.

Holmstrom, M. (2013), Handling vacuum regions in a hybrid plasma solver, Numerical Modeling of Space Plasma Flows (ASTRONUM-2012), ASP Conference Series, arXiv:1301.0272

Holmstrom, M., M. Weiser, S. Barabash, Y. Futaana, and A. Bhardwaj (2010), Dynamics of solar wind protons reflected by the Moon, *J. Geophys. Res.*, A06206, doi:10.1029/2009JA014843.

Holmstrom, M., S. Fatemi, Y. Futaana, and H. Nilsson (2012), The interaction between the Moon and the solar wind, *Earth Planets Space*, *64*, 237–245.

Hutchinson, I. H. (2008a), Ion collection by oblique surfaces of an object in a transversely flowing strongly magnetized plasma, *Phys. Rev. Lett.*, *101*, 035,004, doi:10.1103/PhysRevLett.101.035004.

Hutchinson, I. H. (2008b), Oblique ion collection in the drift approximation: How magnetized Mach probes really work, *Phys. Plasmas 15*(12), 123,503, doi:10.1063/1.3028314.

Hutchinson, I. H. (2012), Electron velocity distribution instability in magnetized plasma wakes and artificial electron mass, *J. Geophys. Res.*, *117*, A03101, doi:10.1029/2011JA017119.

Hutchinson, I. H. (2013), Near-lunar proton velocity distribution explained by electrostatic acceleration, *J. Geophys. Res.*, *118*, 1825–1827 doi:10.1002/jgra.50277.

Israelivich, P., and L. Ofman (2012), Hybrid simulation of the shock wave trailing the Moon, *J. Geophys. Res.*, *117*, A08223, doi:10.1029/2011JA017358.

Jia, Y.-D., C. T. Russell, K. K. Khurana, G. Toth, J. S. Leisner, and T. I. Gombosi (2010), Interaction of Saturn's magnetosphere and its moons: 1. Interaction between corotating plasma and standard obstacles, *J. Geophys. Res.*, *115*, A04214, doi:10.1029/2009JA014630.

Kallio, E. (2005), Formation of the lunar wake in quasi-neutral hybrid model, *Geophys. Res. Lett.*, *32*, L06107, doi:10.1029/2004GL021989.

Kallio, E., et al. (2012), Kinetic simulations of finite gyroradius effects in the lunar plasma environment on global, meso, and microscales, *Planet. Space Sci.*, *74*, 146–155.

Kellogg, P. J., K. Goetz, S. J. Monson, J.-L. Bougeret, R. Manning, and M. L. Kaiser (1996), Observations of plasma waves during a traversal of the Moon's wake, *Geophys. Res. Lett.*, *23*, 1267–1270.

Khurana, K. K., C. T. Russell, and M. K. Dougherty (2007), Magnetic portraits of Tethys and Rhea, *Icarus*, doi.10.1016/j.icarus.2007.08.005.

Kimura, S., and T. Nakagawa (2008), Electromagnetic full particle simulation of the electric field structure around the moon and the lunar wake, *Earth Planets Space*, *60*, 591–599.

Ledvina, S. A., Y.-J. Ma, and E. Kallio (2008), Modeling and simulating flowing plasmas and related phenomena, *Space Sci. Rev.*, *139*, 143–189.

Lipatov, A. S., J. F. Cooper, E. C. Sittler, Jr., and R. E. Hartle (2012), Effects of Na^+ and He^+ pickup ions on the lunar-like plasma environment: 3D hybrid modeling, *Adv. Space Res.*, *50*(12), 1583–1591.

Lue, C., et al. (2011), Strong influence of lunar crustal fields on the solar wind flow, *Geophys. Res. Lett.*, *38*, L03202, doi:10.1029/2010GL046215.

Lyon, E. F., H. S. Bridge, and J. H. Binsack (1967), Explorer 35 plasma measurements in the vicinity of the moon, *J. Geophys. Res.*, *72*, 6113–6117.

Manfredi, G., S. Mola, and M. R. Feix (1983), Rescaling methods and plasma expansions into vacuum, *Phys. Fluids B*, *5*, 388–401.

McComas, D. J., et al. (2009), Lunar backscatter and neutralization of the solar wind: First observations of neutral atoms from the Moon, *Geophys. Res. Lett.*, *36*, L12104, doi:101029/2009GL038794.

McFadden, J. P., C. W. Carlson, D. Larson, M. Ludlam, R. Abiad, B. Elliott, P. Turin, M. Marckwordt, and V. Angelopoulos (2008), The THEMIS ESA plasma instrument and in-flight calibration, *Space Sci. Rev.*, *141*, 277–302.

Michel, F. C. (1968), Magnetic field structure behind the Moon, *J. Geophys. Res.*, *73*, 1533–1541.

Nakagawa, T. (2013), Ion entry into the wake behind a nonmagnetized obstacle in the solar wind: Two-dimensional particle-in-cell simulations, *J. Geophys. Res.*, *118*, 1849–1860, doi:10.1002/jgra.50129.

Nakagawa, T., and S. Kimura (2011), Role of the solar wind magnetic field in the interaction of a non-magnetized body with the solar wind: An electromagnetic 2-D particle-in-cell simulation, *Earth Planets Space*, *67*, 477–486.

Nakagawa, T., Y. Takahashi, and M. Iizima (2003), GEOTAIL observation of upstream ULF waves associated with the lunar wake, *Earth Planets Space*, *55*, 569–580.

Ness, N. F. (1972), Interaction of the solar wind with the Moon, in *Solar Terrestrial Physics/1970*, Part II, edited by Dyer, pp. 159–205, D. Reidel Publ. Co., Dordrecht.

Ness, N. F., K. W. Behannon, C. S. Searce, and S. C. Cantarano (1967), Early results from the magnetic field instrument on lunar Explorer 35, *J. Geophys. Res.*, *72*, 5769–5778.

Ness, N. F., K. W. Behannon, H. E. Taylor, and Y. C. Whang (1968), Perturbations of the interplanetary magnetic field by the lunar wake, *J. Geophys. Res.*, *73*, 3421–3440.

Neubauer, F. M. (1998), The sub-Alfvénic interaction of the Galilean satellites with the Jovian magnetosphere, *J. Geophys. Res.*, *103*, 19,843–19,866.

Nishino, M. N., et al. (2009a), Pairwise energy gain-loss feature of solar wind protons in the near-Moon wake, *Geophys. Res. Lett.*, *36*, L12108, doi:10.1029/2009GL039049.

Nishino, M. N., et al. (2009b), Solar-wind proton access deep into the near-Moon wake, *Geophys. Res. Lett.*, *36*, L16103, doi:10.1029/2009GL039444.

Nishino, M. N., et al. (2010), Effect of the solar wind proton entry into the deepest lunar wake, *Geophys. Res. Lett.*, *37*, L12106, doi:10.1029/2010GL042607.

Nishino, M. N., et al. (2011), Anomalous deformation of the Earth's bow shock in the lunar wake: Joint measurement by Chang'E-1 and SELENE, *Planet. Space Sci.*, *59*, 378–386.

Ogilvie, K. W., and N. F. Ness (1969), Dependence of the lunar wake on solar wind plasma characteristics, *J. Geophys. Res.*, *74*, 4123.

Ogilvie, K. W., J. T. Steinberg, R. J. Fitzenreiter, C. J. Owen, A. J. Lazarus, W. M. Farrell, and R. B. Torbert (1996), Observations of the lunar plasma wake from the WIND spacecraft on December 27, 1994, *Geophys. Res. Lett.*, *10*, 1255–1258.

Omidi, N., X. Blanco-Cano, C. T. Russell, H. Karimabadi, and M. Acuña (2002), Hybrid simulations of solar wind interaction with magnetized asteroids: General characteristics, *J. Geophys. Res.*, *107*, 1487, doi:10.1029/2002JA009441.

Owen, C. J., R. P. Lepping, K. W. Ogilvie, J. A. Slavin, W. M. Farrell, and J. B. Byrnes (1996), The lunar wake at 6.8 R_L: WIND magnetic field observations, *Geophys. Res. Lett.*, *10*, 1263–1266.

Roussos, E., J. Muller, S. Simon, A. Boesswetter, U. Motschmann, N. Krupp, M. Franz, J. Woch, K. K. Khurana, and M. Dougherty (2007), Plasma and fields in the wake of Rhea: 3d hybrid simulation and comparison with Cassini data, *Ann. Geophys.*, *26*, 619–637.

Roussos, E., et al. (2012), Energetic electron observations of Rhea's magnetospheric interaction, *Icarus*, *221*, 116–134.

Russell, C. T., and B. R. Lichtenstein (1975), On the source of lunar limb compression, *J. Geophys. Res.*, *80*, 4700.

Sack, Ch., and H. Schamel (1987), Plasma expansion into vacuum: A hydrodynamic approach, *Phys. Rep.*, *156*, 311–395.

Saito, Y., et al. (2008), Solar wind proton reflection at the lunar surface: Low energy ion measurements by MAP-PACE onboard SELENE (KAGUYA), *Geophys. Res. Lett.*, *35*, L24205, doi:10.1029/2008GL036077.

Saito, Y., M. N. Nishino, M. Fujimoto, T. Yamamoto, S. Yokota, H. Tsunakawa, H. Shibuya, M. Matsushima, H. Shimizu, and F. Takahashi (2011), Simultaneous observation of the electron acceleration and ion deceleration over lunar magnetic anomalies, *Earth Planets Space*, *64*, 83–92.

Samir, U., K. H. Wright, Jr., and N. H. Stone (1983), The expansion of a plasma into a vacuum: Basic phenomena and processes and applications to space plasma physics, *Rev. Geophys.*, *21*, 1631–1646.

Schubert, G., and B. R. Lichtenstein (1974), Observations of Moon-plasma interactions by orbital and surface experiments, *Rev. Geophys.*, *12*, 592–626.

Siscoe, G. L., E. F. Lyon, J. H. Binsach, and H. S. Bridge (1969), Experimental evidence for a detached lunar compression wave, *J. Geophys. Res.*, *74*, 59–69.

Simon, S., J. Saur, F. M. Neubauer, U. Motschmann, and M. K. Dougherty (2009), Plasma wake of Tethys: Hybrid simulations versus Cassini MAG data, *Geophys. Res. Lett.*, *36*, L04108, doi:10.1029/2008GL036943.

Singh, N., and R. W. Schunk (1982), Numerical calculations relevant to the initial expansion of the polar wind, *J. Geophys. Res.*, *87*, 9154–9170.

Sonett, C. P., and J. D. Mihalov (1972), Lunar fossil magnetism and perturbations of the solar wind, *J. Geophys. Res.*, *77*, 588–603.

Spreiter, J. R., M. C. Marsch, and A. L. Summers (1970), Hydromagnetic aspects of solar wind flow past the Moon, *Cosmic Electrodynam*, *1*, 5.

Stone, N. H., K. H. Wright, Jr., U. Samir, and K. S. Hwang (1988), On the expansion of ionospheric plasma into the near-wake of the Space Shuttle orbiter, *Geophys. Res. Lett.*, *15*, 1169–1172.

Tao, J. B., et al. (2012), Kinetic instabilities in the lunar wake: ARTEMIS observations, *J. Geophys. Res.*, *117*, A03106, doi:10.1029/2011JA017364.

Trávnicek, P., P. Hellinger, D. Schriver, and S. D. Bale (2005), Structure of the lunar wake: Two-dimensional global hybrid simulations, *Geophys, Res. Lett.*, *32*, L06102, doi:10.1029/2004GL022243.

Umeda, T. (2012), Effect of ion cyclotron motion on the structure of wakes: A Vlasov simulation, *Earth Planets Space*, *64*, 231–231, doi:10.5047/eps.2011.05.035.

Umeda, T., T. Kimura, K. Togano, K. Fukazawa, Y. Matsumoto, T. Miyoshi, N. Terada, T. K. M. Nakamura, and T. Ogino (2011), Vlasov simulation of the interaction between the solar wind and a dielectric body, *Phys. Plasmas*, *18*, 012908.

Vernisse, Y., H. Kriegel, S. Wiehle, U. Motschmann, and K. H. Glassmeir (2013), Stellar winds and planetary bodies simulations: Lunar type interaction in super-Alfvénic and sub-Alfvénic flows, *Planet. Space Sci.*, *84*, 37–47, doi:10.1016/j.pss.2013.04.004.

Wang, X.-D., et al. (2010), Accelerationg of scattered solar wind protons at the poalr terminator of the Moon: Results from Chang'E-1/SWIDS, *Geophys. Res. Lett.*, *37*, L07203, doi:10.1029/2010GL042891.

Wang, Y.-C., J. Muller, W.-H. Ip, and U. Motschmann (2011), A 3D hybrid simulation study of the electromagnetic field distributions in the lunar wake, *Icarus*, *216*, 415-425.

Whang, Y. C., and N. F. Ness (1970), Observations and interpretation of the lunar Mach cone, *J. Geophys. Res.*, *75*, 6002–6010.

Whipple, E. C. (1981), Potentials of surfaces in space, *Rep. Prog. Phys.*, *44*, 1197–1250.

Wiehle, S., F. et al. (2011), First lunar wake passage of ARTEMIS: Discrimination of wake effects and solar wind fluctuations by 3D hybrid simulations, *Planet. Space Sci.*, *59*, 661–71.

Wieser, M., et al. (2009), Extremely high reflection of solar wind protons as neutral hydrogen atoms from regolith in space, *Planet. Space. Sci.*, doi:10.1016/j.pss.2009.09.012.

Wilson, R. J., R. L. Tokar, W. S. Kurth, and A. M. Persoon (2010), Properties of the thermal ion plasma near Rhea as measured by the Cassini plasma spectrometer, *J. Geophys. Res.*, *115*, doi:10.1029/2009JA014679.

Winske, D. (1988), Short-wavelength modes on expanding plasma clouds, *J. Geophys. Res.*, *93*, 2539–2552.

Wolf, R. A. (1968), Solar-wind flow behind the Moon, *J. Geophys. Res.*, *73*, 4281–4289.

Xie L. H., L. Li, Y. T. Zhang, and D. L. De Zeeuw (2012), Three-dimensional MHD simulation of the lunar wake, *Sci. china Earth Sci.*, doi:10.1007/s11430-012-4383-6.

Zhang, H., et al. (2012), Outward expansion of the lunar wake: ARTEMIS observations, *Geophys. Res. Lett.*, *39*, L18104, doi:10.1029/2012GL052839.

Zhang, H., K. K. Khurana, M. G. Kivelson, V. Angelopoulos, W. X. Wan, L. B. Liu, Q.-G. Zong, Z. Y. Pu, Q. Q. Shi, and W. L. Liu (2014), Three-dimensional lunar wake reconstructed from ARTEMIS data, *J. Geophys. Res.*, *119*, 5220–5243, doi:10.1002/2014JA020111.

Zimmerman, M. I., W. M. Farrell, T. J. Stubbs, J. S. Halekas, and T. L. Jackson (2011), Solar wind access to lunar polar craters: Feedback between surface charging and plasma expansion, *Geophys. Res. Lett.*, *38*, L19202, doi:10.1029/2011GL048880.

10

Physics of Cometary Magnetospheres

Tamas I. Gombosi

10.1. INTRODUCTION

Our present understanding of cometary atmospheres is based on Whipple's "dirty iceball" idea in which the nucleus consists of a mixture of frozen volatiles and nonvolatile dust [*Whipple*, 1950]. Figure 10.1 shows images of all cometary nuclei photographed by spacecraft so far.

As cometary nuclei approach the Sun, water vapor and other volatile gases sublimate from the surface layers generating a rapidly expanding cloud of dust and gas. The freshly sublimated gas leaves the surface at slightly less than the local speed of sound, it undergoes a sonic transition near the surface, and it propagates as a "cometary wind" into interplanetary space. The sublimating gases are capable of blowing away specks of dust from the surface [cf. *Mendis et al.*, 1985, *Gombosi et al.*, 1986, *Tenishev et al.*, 2008, *Zakharov et al.*, 2009].

It was recognized a long time ago that the expanding cometary exosphere represents an extensive, "soft" obstacle for the supersonic and super-Alfvénic solar wind flow [*Biermann et al.*, 1967]. The resulting interaction is very different from that with other solar system bodies with gravitationally bound dense atmospheres and/or significant intrinsic magnetic fields. Neutral atoms and molecules of cometary origin become ionized (because of photoionization, charge transfer, or electron impact ionization) with characteristic ionization scale lengths of $10^5 - 10^7$ km. The ionization process introduces a new, practically stationary particle into the high-speed magnetized flow of the solar wind. Spacecraft instrumentation at comets detected large-amplitude low-frequency magnetic field fluctuations. These fluctuations grow from the relatively low solar wind turbulence level (at large cometocentric distances) to very large amplitudes near the comet. The enhanced fluctuation level is due to instabilities associated with "mass loading," due to the interaction of the solar wind with the newly ionized cometary material.

In their pioneering work *Biermann et al.* [1967] predicted that the deceleration of the solar wind flow by mass loading leads to the gradual deceleration of the solar wind flow at large cometocentric distances and eventually to the formation of a weak shock ($M \approx 2$) where the flow is impulsively decelerated to subsonic velocities. This was a revolutionary idea, since it predicted that the presence of the comet can be "felt" far upstream from the nucleus in the supersonic solar wind flow. Later *Galeev et al.* [1985] recognized that implanted cometary ions carry most of the "thermal" pressure (i.e., the second moment of the distribution function) and that charge exchange cooling of the implanted plasma population can play a very important role in the dynamics of the contaminated solar wind flow. Their model predicted a weak and highly structured shock with a viscous subshock, a continuously decelerated and cooled plasma flow behind the shock, and finally a stagnation region. In situ measurements later confirmed the gross features predicted by this model [cf. *Gringauz et al.*, 1986; *Johnstone et al.*, 1986; *Neubauer et al.*, 1986].

Downstream of the bow shock the plasma population is a varying mixture of shocked plasma (solar wind contaminated with upstream pickup particles) and cometary plasma ionized in the subsonic region. This region is called the cometosheath. The cometosheath is characterized by a rapidly increasing rate of ion pickup (as the plasma moves toward the comet), resulting in continuous deceleration (and eventual stagnation) of the

Department of Atmospheric, Oceanic and Space Sciences, University of Michigan, Ann Arbor, Michigan

Magnetotails in the Solar System, Geophysical Monograph 207. First Edition.
Edited by Andreas Keiling, Caitríona M. Jackman, and Peter A. Delamere.

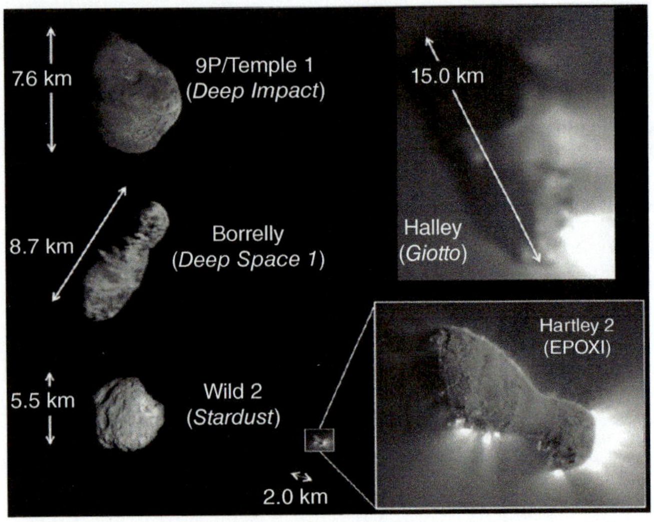

Figure 10.1 Cometary nuclei visited by spacecraft prior to the Rosetta mission (Courtesy of Michael A'Hearn, EPOXI team/ University of Maryland/JPL/NASA).

plasma flow, accompanied by increasing plasma density and magnetic field magnitude [*Balsiger et al.*, 1986; *Neubauer et al.*, 1986]. The inner, nearly stagnating region of the cometosheath is primarily photochemically controlled and the plasma density varies with the inverse of the cometocentric distance [cf. *Balsiger* et al., 1986].

In the innermost region of the cometary environment the plasma is of cometary origin and it is strongly coupled to the outflowing neutral atmosphere. This part of the coma is called the "cometary ionosphere" and it is magnetic field free, since the heavily mass loaded solar wind plasma cannot penetrate into this region [cf. *Neubauer et al.*, 1986; *Balsiger et al.*, 1986]. A schematic view of the dayside cometary atmosphere is shown in Figure 10.2.

Our understanding of the main physical processes controlling the "ionopause" (the surface separating the magnetized cometosheath plasma from the magnetic field-free inner cometary ionosphere) was significantly modified by Giotto's encounter with comet Halley. A diamagnetic cavity boundary is formed at a location where the inward pointing (toward the comet) total magnetic force (the sum of the magnetic pressure gradient and magnetic curvature forces) is balanced by the outward pointing ion-neutral dragforce. Before the cometary encounters, an inner shock was predicted inside the "ionopause" to decelerate the supersonic outflow of the cometary ions and divert them toward the tail. In reality, the drag by the rapidly expanding neutral gas forces the plasma to maintain supersonic velocity up to the immediate vicinity of the diamagnetic cavity boundary, where it undergoes a shock transition [*Cravens*, 1989; *Goldstein et al.*, 1989]. The shocked ionospheric plasma piles up and is rapidly removed by recombination.

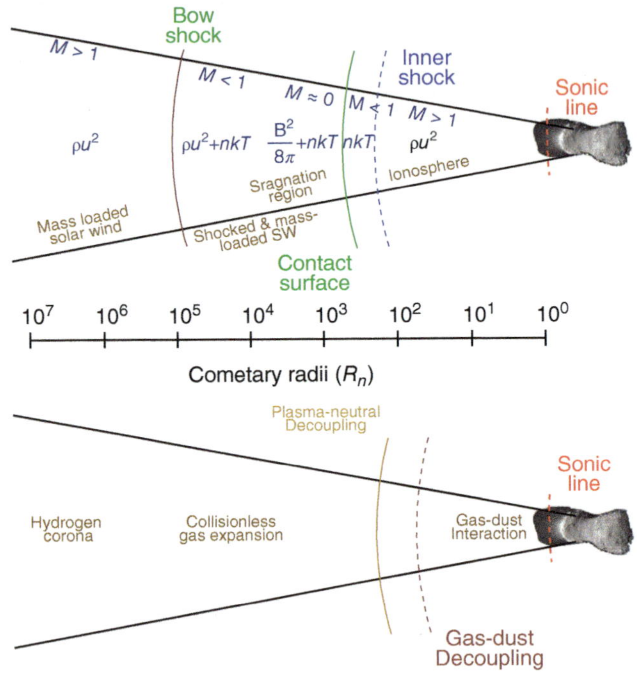

Figure 10.2 Schematic representation of the plasma (upper panel) and neutral gas (lower panel) environment of an active comet. After *Combosi et al.* [1986].

Figure 10.3 Comet Hale-Bopp (1997) showing two distinct tails: a dust tail (white) and an ion tail (blue). (From: http:// www.tivas.org.uk/solsys/tas_solsys_comet.html.)

Comets exhibit two distinct types of tails. A nice example is comet Hale-Bopp, shown in Figure 10.3.

For active comets the straight, narrow plasma tails (also called type I tails) are 10^7–10^8 km long and, within a few degrees, always point away from the Sun. This observation inspired *Biermann* [1951] to postulate the existence of a continuous "solar corpuscular radiation" (solar wind). Assuming that the antisunward acceleration of small irregularities in type I comet tails were due to collisional coupling between a radially outward plasma flow from the Sun and newly ionized cometary particles, *Biermann* [1951] inferred a solar wind density and velocity of $n_{sw} \approx 1000$ cm^{-3} and $u_{sw} \approx 1000$ km/s that represents a particle flux ~500 times larger than was later observed.

A few years later *Alfvén* [1957] pointed out that the plasma densities inferred by *Biermann* [1951] were inconsistent with observed coronal densities (assuming that the plasma density decreases with the square of the heliocentric distance as the solar corpuscular radiation moves outward). *Alfvén* [1957] offered an alternative explanation for the formation of cometary plasma tails that is depicted in Figure 10.4. The main assumption in this model was that the solar corpuscular radiation was carrying a "frozen-in" magnetic field that "hangs up" in the high-density inner coma where the solar particles strongly interact with the cometary atmosphere and consequently the solar plasma flow considerably slows down. This interaction results in a "folding" of the magnetic field around the cometary coma that creates the long plasma tail. Disturbances along the folded magnetic field lines propagate as magnetohydrodynamic (MHD) waves and can reach velocities of 100 km/s even if the solar plasma has a density of $\sim 5\,\text{cm}^{-3}$.

The second type of tails are broad and curved dust tails (also called type II tails) that usually lag behind the Sun-comet line (opposite to the direction of cometary orbital motion). Since the gravity of the cometary nucleus is negligible, the motion of the dust particles is controlled by an interplay between solar gravity and solar radiation pressure. Assuming that dust grains have a more or less constant mass density ρ_d, the anti-sunward radiation pressure F_{rad} is inversely proportional tothe square of the heliocentric distance, d_h and proportional to the cross section of the dust grain, πa^2 (a is the equivalent radius of a spherical dust grain), $F_{rad} \propto a^2/d_h^2$. The sunward pointing gravitational force is proportional to the particle mass ($4\pi\rho_d a^3/3$) and inversely proportional to the heliocentric distance: $F_{grav} \propto a^3/d_h^2$. Since the two forces point in the opposite direction and have the same heliocentric dependence but exhibit different dependence on the grain size ($F_{rad} \propto a^2$ and $F_{grav} \propto a^3$), the resulting effect is a complex "dust mass spectrometer" where each particle size is moving under the influence of its own "reduced solar gravity." This effect results in the broad and curved dust tail.

10.2. THE COMA

10.2.1. Gas Production

The atmospheres of comets, commonly referred to as comas, are different in a number of important ways from planetary atmospheres. The most important distinguishing characteristics of comas are (1) the lack of any significant gravitational force, (2) the relatively fast radial outflow velocities, and (3) the time-dependent nature of their physical properties. A direct consequence of these features is the expanding nature of cometary atmospheres.

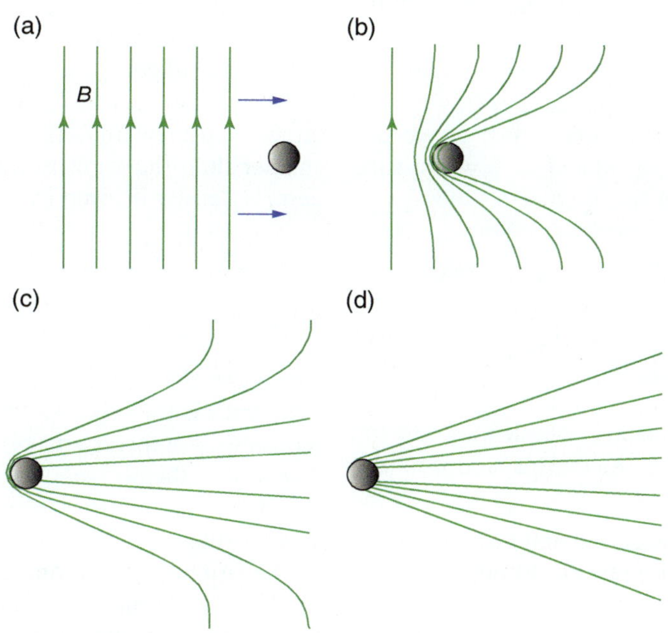

Figure 10.4 Alfvén's scenario of the formation of comet tails. A plasma beam with a frozen-in magnetic field approaches the head of a comet (panel a); the field is deformed (panels b and c); the final state is reached when the beam has passed (panel d) [Alfvén, 1957].

The source of cometary activity is the sublimation of cometary volatiles. There are many sophisticated sublimation models (for a nice review we refer to *Davidsson* [2008]), but the basic concept is well illustrated by the classic *Delsemme and Swings* [1952] theory, which considered a cometary nucleus covered by a homogeneous surface of sublimating water ice. They assumed that the surface did not contain macroscopic irregularities; i.e., surface irregularities were much smaller than the mean free path of vaporized particles. As the pressure of the cometary atmosphere is much smaller at the nucleus surface than the critical pressure of the phase transition triple point, the liquid phase is unstable and sublimation of frozen volatiles is responsible for gas production. Assuming that the sublimated gas was in equilibrium with the surface, *Delsemme and Swings* [1952] applied the Clausius-Clapeyron equation to determine the steady-state saturated gas pressure:

$$p_s(T_s) = p_r \exp\left[\frac{L(T_s)}{kN_A}\left(\frac{1}{T_r} - \frac{1}{T_s}\right)\right], \tag{10.1}$$

where T_s is the surface temperature, p_s is the vapor pressure, p_r is the saturated vapor pressure at a reference temperature T_r, $L(T_s)$ is the latent heat of vaporization, k is the Boltzmann constant, and N_A is Avogadro's number. For water ice the latent heat is fairly insensitive to temperature and it can be approximated by a value of $L_w = 4.80 \times 10^4$ J/mol [*Delsemme and Miller*, 1971]. The sublimation vapor pressure of boiling water is $p_r = 10^5$ Pa at $T_r = 373$ K.

Delsemme and Swings [1952] also assumed that the sublimated molecules behave as a perfect gas ($p_s = n_s kT_s$), thus defining the gas number density n_s just above the surface. In addition, they considered a low-pressure kinetic model and calculated the flux of scattered gas particles moving away from the nucleus:

$$\dot{z}(T_s) = 1/4\, n_s u_{\text{th}} = n_s(T_s)\sqrt{\frac{kT_s}{2\pi m_c}}. \tag{10.2}$$

Here m_c is the mass of a cometary volatile molecule (in our case $m_c = 18$ amu) and u_{th} is the thermal speed of the gas ($u_{\text{th}} = \sqrt{8kT_s / \pi m_c}$).

In the simplest scenario one can assume that the incident solar energy flux is balanced by blackbody radiation and sublimation:

$$\frac{J_{\text{sol}}}{d_h^2} = \dot{z}(T_s)\frac{L_w}{N_A} + \sigma_{\text{SB}} T_s^4, \tag{10.3}$$

where $J_{\text{sol}} = 1365$ W is the solar radiation energy flux at 1 AU and σ_{SB} is the Stefan-Boltzmann constant. We note that in equation 10.3 the heliocentric distance is measured in units of AU. Figure 10.5 shows the sublimation temperature (upper panel) and gas production rate (lower panel) obtained by solving equations (10.1)–(10.3) for an iceball with radius of 1 km. It was also assumed that only about 5% of the surface area is active at any given time [cf. *Combi et al.*, 2012]. The "effective" sublimating surface can be significantly increased by grain sublimation in the coma. This process is able to increase the apparent sublimating area by a large factor [*Fougere et al.*, 2013].

Inspection of Figure 10.5 reveals that water production "turns on" around 3.5 AU. It reaches about 10^{28} S^{-1} near 1 AU and a spectacular 10^{30}S^{-1} for a sun-grazing comet. The collisional mean free path near the surface of active comets is less than a meter, therefore the gas rapidly engulfs the nucleus, even though only a small fraction of the surface area is active. In a first approximation spherically symmetric gas coma models provide reasonable results for most comets [cf. *Zakharov et al.*, 2009; *Combi et al.*, 2012].

10.2.2. Ion-Neutral Coupling

Here we present a very simple, spherically symmetric description of the innermost region of active comet atmospheres that captures most of the basic physics without going into too much detail. This description follows the derivation of *Cravens* [1989] and it is a nice tool for a tutorial. However, those who are interested in more details are referred to appropriate review papers [cf. *Strobel*, 2002; *Rubin et al.*, 2011].

The neutral composition in the inner coma is assumed to be pure water, and for cometocentric distances much larger than a few times the radius of the nucleus but much smaller than the ionization scale length ($\lambda \approx 10^6$ km), the neutral density is given by

$$n_n = \frac{Q}{4\pi u_n r^2}, \tag{10.4}$$

where Q is the total gas production rate of the comet and $u_n \approx 1$ km/s is the radial outward velocity of neutral water molecules. Expression (4) neglects the exponential decrease of the neutral density on scales of the ionization length (λ), but here we are describing a region much smaller than λ.

The primary photoionization products are H_2O^+ and OH^+, but the dominant ion in the collisional inner comas is H_3O^+ that is rapidly formed by the reaction

$$H_2O^+ + H_2O \rightarrow H_3O^+ + OH \tag{10.5}$$

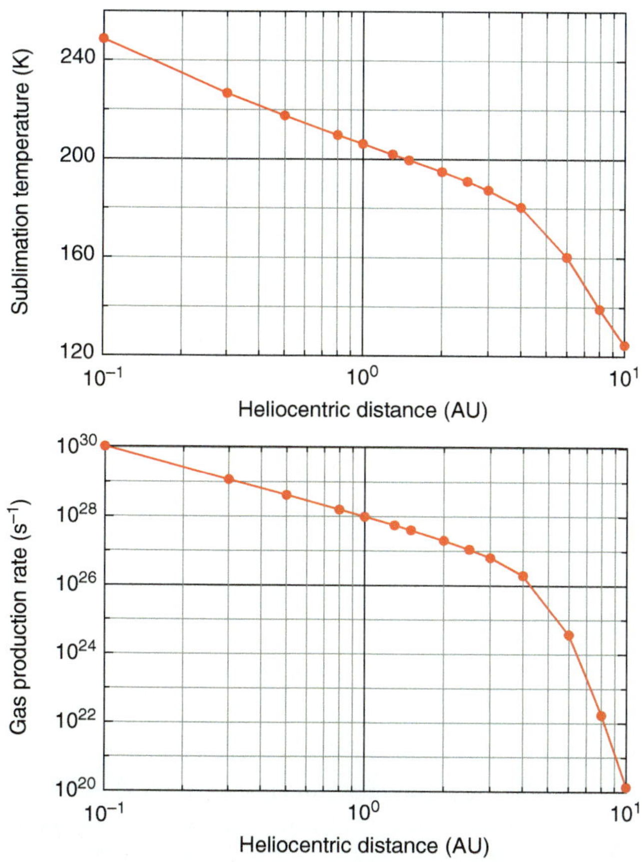

Figure 10.5 Sublimation temperature (upper panel) and gas production rate (lower panel) of a "*Delsemme and Swings* [1952] type" comet with 1 km radius, assuming that only 5% of subsolar area is active.

[cf. *Mendis et al.*, 1985]. The production rate of cometary ions is dominated by photoionization and can be approximated as

$$S_i = \frac{I_0}{d_h^2} n_n, \tag{10.6}$$

where $I_0 \approx 10^{-6}\text{s}^{-1}$ is the ionization frequency at 1 AU [*Cravens*, 1989]. The H_3O^+ ions are destroyed by dissociative recombination that produces a neutral water molecule and a hydrogen atom. The reaction rate coefficient of this dissociative recombination is dependent on the electron temperature: $\alpha = 1.21\times10^{-5}/\sqrt{T_e}\,\text{cm}^3/\text{s}$ resulting in an ion loss rate $L_i = \alpha n_i^2$ (assuming that the ion and electron densities are equal) [cf. *Mendis et al.*, 1985]. This expression for the recombination rate is valid below approximately 800 K [cf. *Schunk and Nagy*, 2000; *Rubin et al.*, 2014].

The continuity and momentum equations for the radially expanding unmagnetized ionosphere near the nucleus can be written as

$$\frac{\partial n_i}{\partial t} + \frac{1}{r^2}\frac{\partial}{\partial r}\left(r^2 u_i n_i\right) = \frac{I_0}{d_h^2} n_n - \alpha n_i^2 \tag{10.7}$$

and

$$m_i n_i \frac{\partial u_i}{\partial t} + m_i n_i u_i \frac{\partial u_i}{\partial r} + \frac{\partial p_i}{\partial r} = \left(m_i n_i \nu_{in} + \frac{I_0}{d_h^2} m_n n_n \right)(u_n - u_i), \tag{10.8}$$

where n_i and p_i are the density and pressure of H_3O^+ ions, $m_n = 18$ amu, $m_i = 19$ amu, and $\nu_{in} = k_{in} n_n$ is the ion-neutral charge transfer collision frequency ($k_{in} = 1.1 \times 10^{-9}\,\text{cm}^3/\text{s}$) [*Cravens and Körösmezey*, 1986]. The right-hand side of equation 10.8 describes the ion-neutral momentum coupling due to charge exchange (first term) and the production of new ions from neutrals (second term).

Quite a few physical insights can be gained by inspecting equations 10.7 and 10.8. Let us first examine the ion-neutral coupling term. For an active comet the ion-neutral coupling time scale can be approximated by

$$\tau_{in} \approx \frac{4\pi u_n r^2}{k_{in} Q}. \tag{10.9}$$

As long as the coupling time is shorter than the characteristic transport time, $\tau_{transp} = r/u_n$, the ions and neutral velocities remain closely coupled and in a good approximation, $u_i \approx u_n$. The radial distance where $\tau_{in} = \tau_{transp}$ is the distance where the ions decouple from the neutrals in the absence of any other process (note that so far we completely neglected the interaction of the cometary coma with the solar wind):

$$R_{in} \approx \frac{k_{in} Q}{4\pi u_n^2} = 875\left(\frac{Q}{10^{28}}\right)\ (\text{km}). \tag{10.10}$$

For a Halley class comet ($Q \approx 7 \times 10^{29}$) we get $R_{in} \approx 6 \times 10^4$ km, while for a typical active comet ($Q \approx 10^{28}$) we get $R_{in} \approx 900$ km. It should be mentioned again that the solar wind interaction might push other plasma boundaries inside this region, so the strongly coupled ionospheric plasma might not extend to this distance.

It is obvious from the above discussion that in the cometary ionosphere the ions are closely coupled to the neutrals and the ion velocity is very close to the neutral velocity, $u_i \approx u_n$. Since the neutral velocity is more or less constant, we conclude that in the ionosphere the ion velocity is also constant. This makes it possible to evaluate the radial profile of the ion density using the continuity equation 10.7.

Substituting the constant neutral velocity for the ion velocity and considering steady-state conditions the continuity equation becomes:

$$\frac{\partial}{\partial r}\left(r^2 n_i\right) = \frac{I_0}{d_h^2}\frac{Q}{4\pi u_n^2} - \frac{\alpha}{u_n} r^2 n_i^2. \tag{10.11}$$

It is obvious that the solution to equation 10.11 is $n_i \propto r^{-1}$:

$$n_i = \left[\sqrt{1+\frac{Q}{Q_0}} - 1\right]\left(\frac{u_n}{2\alpha}\right)\frac{1}{r}, \tag{10.12}$$

where

$$Q_0 = \frac{\pi d_h^2 u_n^3}{\alpha I_0} \approx 2.6\times10^{26} d_h^2 \sqrt{T_e}. \tag{10.13}$$

For active comets, $Q \gg Q_0$, and therefore

$$n_i \approx \sqrt{\frac{Q}{Q_0}}\left(\frac{u_n}{2\alpha}\right)\frac{1}{r}. \tag{10.14}$$

When deriving equation 10.12 we assumed that the radial dependence of the recombination rate coefficient can be neglected. For a typical active comet ($R_n \approx 1$ km, $Q \approx 10^{28}\,\mathrm{s}^{-1}$) we get about $n_n \approx 10^{12}\,\mathrm{cm}^{-3}$ and $n_i \approx 10^7\,\mathrm{cm}^{-3}$ at the surface.

A word of caution is in order at this point. In the immediate vicinity of the nucleus the fluid approximation is not valid and a thin kinetic "Knudsen layer" separates the nucleus from the collisional coma. Expressions (10.12) and (10.14) were obtained in the fluid approximation, so they are not strictly valid inside the Knudsen layer.

10.2.3. Inner Shock and Contact Surface

The radially outward flowing cometary ionosphere is separated from the heavily contaminated and nearly stagnating outside plasma by a tangential discontinuity, the contact surface (sometimes also called diamagnetic cavity boundary). For active comets this contact surface is located deep inside the region where ion-neutral collisions force the cometary ionosphere to expand with the velocity of the neutral gas (see section 9.2.2). Outside the contact surface the magnetic field "piles up" due to the very small plasma flow velocity.

There are two physical constraints that control the location of the contact surface: the balance between the ambient solar wind pressure and the outward ion-neutral drag and the recombination rate coefficient that is a function of electron temperature.

10.2.3.1. Electron Temperature. Let us start with the electron temperature. Deep in the collisional coma inelastic collisions between the thermal electrons and the neutral water molecules result in rotational and vibrational excitation of the molecule; the rate for this interaction is especially large because of the polar nature of the water molecule. The electron-neutral collision frequency can be approximated by [*Banks and Kockarts*, 1973]

$$\nu_{en} = 2.615\times10^{-5}\frac{n_n}{\sqrt{T_e}}, \tag{10.15}$$

where the collision frequency is given in Hz and n_n is given in units of cm^{-3}. The characteristic time scale of electron-neutral collisions is

$$\tau_{en} = 4.81\times10^{-3}\sqrt{T_e}\frac{10^{28}}{Q}\frac{r^2}{R_n^2}(\mathrm{s}). \tag{10.16}$$

Electrons and neutrals remain closely coupled as long as τ_{en} is smaller than the dynamical time scale of the expanding ionospheric plasma, $\tau_{\mathrm{transp}} = r/u_n$. This condition yields the electron-neutral decoupling distance, where the electron temperature exhibits a sharp, nearly two orders of magnitude increase [*Körösmezey et al.*, 1987]. The ratio of the electron-neutral decoupling distance to the nucleus radius is now

$$\frac{R_{en}}{R_n} = \frac{208}{\sqrt{T_e}}\frac{Q}{10^{28}}\frac{R_n}{u_n} \tag{10.17}$$

Typical electron temperatures in the collisional coma are ≈ 100 K. For a Halley class comet ($Q \approx 7\times10^{29}$, $R_n = 7$ km) we get $R_{en} \approx 7\times10^4$ km, while for a typical active comet ($Q \approx 10^{28}$, $R_n = 1$ km) the decoupling radius is $R_{en} \approx 20$ km.

In considering the thermal coupling and the electron temperature, more relevant than the elastic electron-neutral collisions is the electron-neutral water cooling process, which is dominated in the inner coma by rotational cooling [cf. *Körösmezey et al.*, 1987; *Schunk and Nagy*, 2000]. Recently, Cravens (private communication, 2013) derived an an analytic expression that fits the sum of the rotational, vibrational, and electronic cooling rates for the temperature interval $kT_n \leq kT_e \leq 0.8$ eV ($T_e \approx 9300$ K) with about 5% accuracy. The functional form of the electron cooling rate L_e is based on the tabulated results of *Gan and Cravens* [1990] and *Cravens et al.* [2011]:

$$L_e = A_1\left[1-\exp\left(-\frac{k(T_e-T_n)}{0.033\,eV}\right)\right] + A_2\left[0.415-\exp\left(-\frac{kT_e-0.10\,eV}{0.10\,eV}\right)\right], \quad (10.18)$$

where $A_1 = 4\times10^{-9}$ eV cm³/s, $A_2 = 0$ for $kT_e \le 0.188$ eV and $A_2 = 6.5\times10^{-9}$ eV cm³/s for $kT_e > 0.188$ eV.

The electron cooling time scale on water can now be expressed as

$$\tau_{ew} = \frac{kT_e}{n_n L} = \frac{kT_e}{L}\frac{4\pi u_n r^2}{Q}. \quad (10.19)$$

In the collision dominated region $T_e \approx T_n$ and we define the characteristic cooling time for an electron temperature that is only 1 K above the neutral temperature. This yields a numerical value for the quantity kT_e/L:

$$\tau_{ew} = 2.1\times10^{-3} r^2\left(\frac{u_n}{1\text{km/s}}\right)\left(\frac{10^{28}\,\text{s}^{-1}}{Q}\right), \quad (10.20)$$

where r is measured in kilometers.

The electron temperature will decouple from the ion temperature near the point where the electron cooling time scale and the plasma dynamical time scale, $\tau_{\text{transp}} = r/u_n$, are about equal:

$$R_{ew} = 480\left(\frac{1\text{km}^2/\text{s}^2}{u_n^2}\right)\left(\frac{Q}{10^{28}\,\text{s}^{-1}}\right), \quad (10.21)$$

where R_{ew} is given in kilometers. For a Halley class comet ($Q \approx 7\times10^{29}$) we get $R_{ew} \approx 3.4\times10^4$ km, and for a weak comet ($Q \approx 10^{28}$) the water cooling decoupling distance is about 500 km.

10.2.3.2. Inner Shock. The ion-acoustic speed of the collisional ionospheric plasma is ≈0.35 km/s, so the expanding plasma flow inside the contact surface is supersonic. However, this radially expanding supersonic plasma flow cannot continue beyond the contact surface that separates the cometary ionosphere from the contaminated solar wind plasma. Since a supersonic flow cannot "turn," it has to be terminated by an "inner shock" before it reaches the contact surface. On the dayside the shocked, subsonic flow nearly stagnates before it is gradually diverted toward the tail.

The ion density between the inner shock and the contact surface, however, is not determined by the local photochemical equilibrium, because plasma is continuously supplied from the expanding ionosphere [*Cravens*, 1989]. This plasma is lost by recombination. The balance between supply and loss can be expressed as

$$\alpha n_{\text{rec}}^2 \Delta = n_i u_n, \quad (10.22)$$

where Δ is the thickness of the recombination layer between the shock and the contact surface and $n_{\text{rec}} \approx 4n_i$ is the ion density in the recombination layer (this assumes a strong hydrodynamic shock). This equation can be solved for Δ:

$$\Delta \approx \frac{1}{8}\sqrt{\frac{Q_0}{Q}} R_{\text{cs}}, \quad (10.23)$$

where R_{cs} is the cometocentric distance of the contact surface.

For active comets the thickness of the recombination layer is very small compared to R_{cs}. For instance, in the case of comet Halley, Δ was so small that special data analysis was needed to identify the associated density peak. Figure 10.6 shows the comparison of a global MHD simulation with Giotto observations for the H_3O^+ hydronium ion with mass/charge of 19 amu/e [*Rubin et al.*, 2009]. H_3O^+ is the dominant ion within 30,000 km of comet 1P/Halley's nucleus. Outside 40,000 km the experimental data can be well fitted by a higher production rate curve, indicating strong time dependence of the cometary gas production.

Giotto's ion mass spectrometer (IMS) performed continuous measurements, scanning in 1/4 s through $m/e = 12-57$ amu/e in 64 steps with 1/256 s accumulation time each. During a spin period of 4 s therefore the IMS acquired 16 mass spectra including a whole mass scan each. Inside 40,000 km the data are averaged over the whole period and outside this distance over five spin periods. During a fraction of a spin period the Giotto IMS measured a distinct rise in count rate at a cometocentric distance of roughly 4550 km. *Goldstein et al.* [1989] and *Cravens* [1989] identified the sharp spike (about 47 km in width) as coming from the recombination layer between the inner shock and the contact surface.

Inspection of Figure 10.6 also shows that the plasma density in the inner cometary atmosphere decreases as r^{-1}, as predicted by equation 10.14. The narrow-cavity boundary region (with the contact surface and inner shock) was at around 4550 km. The electron-neutral decoupling took place around 2×10^4 km. Beyond this boundary the plasma density decreases as r^{-2}, because recombination is no longer important. Qualitatively similar simulation results were obtained earlier by *Gombosi et al.* [1996], *Häberli et al.* [1997], and *Benna and Mahaffy* [2007].

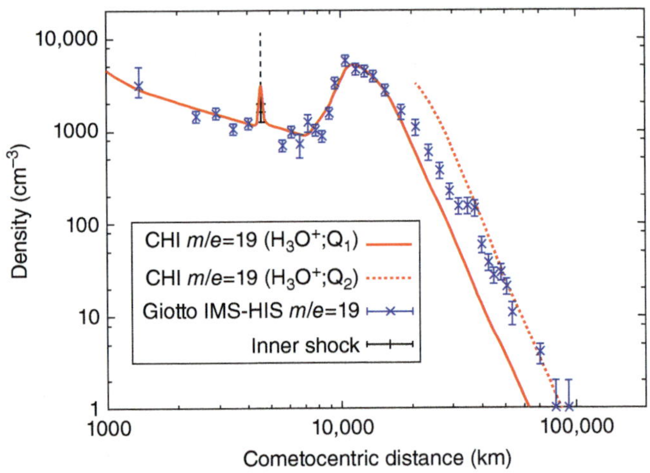

Figure 10.6 Mass/charge = 19 amu/e density profile including the H_3O^+ ion along the Giotto trajectory compared to the Giotto IMS data from the HIS sensor (points). The position of the inner shock is also given, indicated by the mean (measured over one 4 s spin period of Giotto) as well as the maximum density (dashed vertical line; density measured during a fraction of a spin period). The observations are compared to simulation results obtained with two different production rates, Q_a and Q_2 [*Rubin et al.*, 2009]

Lookingforwardtocomet Churyumov-Gerasi – menko, it needs to be pointed out that in the case of weak comets $Delta \approx 0.1$ and therefore finite gyroradius effects will come into play. In this case, at least for determining some of the structure, a hybrid simulation would be helpful [e.g., *Puhl-Quinn and Cravens,* 1995].

10.2.3.3. Contact Surface. At the contact surface the outside $j \times B$ force is balanced on the inside by the ion-neutral drag force [*Cravens*, 1986; *Ip and Axford*, 1987]:

$$\nabla \frac{B^2}{2\mu_0} - \left(\frac{\mathbf{B}}{\mu_0} \cdot \nabla\right)\mathbf{B} = \left(m_i n_i \nu_{in}\right)\left(\mathbf{u}_n - \mathbf{u}_i\right), \quad (10.24)$$

where μ_0 is the magnetic permeability of free space and for the sake of simplicity we neglected the ionization drag. The first term on the left-hand side is the magnetic pressure gradient while the second term describes the magnetic tension force. In the immediate vicinity of the contact surface the pressure gradient force dominates over the magnetic tension force, because this is the region where the "piled-up" magnetic field rapidly goes to zero inside the contact surface, thus forming the diamagnetic cavity. The maximum value of the magnetic field in the pile-up region is reached at a come to centric distance of R_{max}, and its value can be estimated by equating the magnetic pressure with the dynamic pressure of the ambient solar wind that is driving the entire interaction:

$$m_p n_{sw} u_{sw}^2 = \frac{B_{max}^2}{2\mu_0}, \quad (10.25)$$

where m_p is the proton mass.

Substituting equation 10.14 into equation 10.24 and neglecting the magnetic tension term yield the following:

$$\frac{\partial}{\partial r}\left(\frac{B^2}{2\mu_0}\right) = \frac{m_i k_{in}}{4\pi d_h}\sqrt{\frac{I_0}{4\pi\alpha u_n}}\frac{Q^{3/2}}{r^3}. \quad (10.26)$$

Equation 10.26 can be integrated from the contact surface, R_{cs}, where the magnetic field vanishes to R_{max}, where the magnetic field reaches its maximum value (see equation 10.25):

$$m_p n_{sw} u_{sw}^2 = A\left(\frac{Q}{10^{28}}\right)^{3/2}\left(\frac{R_n^2}{R_{cs}^2} - \frac{R_n^2}{R_{max}^2}\right), \quad (10.27)$$

where

$$A = 10^{42}\frac{m_i k_{in}}{4\pi d_h}\sqrt{\frac{I_0}{4\pi\alpha u_n}}\frac{1}{R_n^2}. \quad (10.28)$$

Assuming that $R_{max}^2 \gg R_{cs}^2$, the radius of the contact surface becomes

$$R_{cs} = \sqrt{\frac{m_i}{m_p}\frac{k_{in}}{4\pi d_h n_{sw} u_{sw}^2}\sqrt{\frac{I_0}{4\pi\alpha u_n}}Q^{3/4}}. \quad (10.29)$$

Substituting the numerical values and using typical solar wind parameters ($n_{sw} = 5\,\mathrm{cm}^{-3}$ at 1 AU and $u_{sw} = 400$ km/s), we get

$$R_{cs} = 7.11\times10^{-20} T_e^{1/4}\sqrt{d_h}Q^{3/4} \quad \text{(km)}. \quad (10.30)$$

Assuming $T_e \approx 10^2$ K in the collisional coma for a Halley class comet, we get $R_{cs} \approx 5500$ km and for a typical active comet $R_{cs} \approx 225$ km. We note that for a Halley class comet $R_{cs} \ll R_{en}$, while for a typical active-type comet $R_{cs} \gg R_{en}$.

Recently *Rubin et al.* [2012] performed large-scale MHD simulations for a relatively weak comet (representing comet 67P/Churyumov-Gerasimenko). Figure 10.7 shows the diamagnetic cavity. The color code represents the magnitude of plasma velocity, while the streamlines show the plasma trajectories. One can see that in the ionosphere the ions are supersonic and moving radially

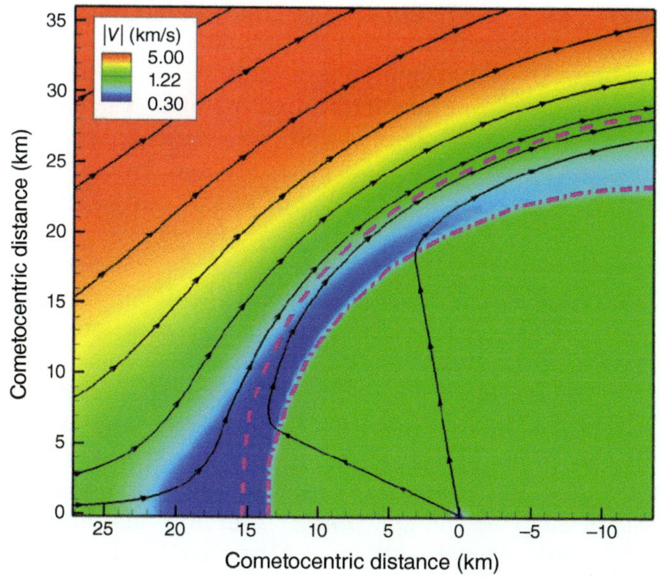

Figure 10.7 Plasma bulk flow velocity (and streamlines) for comet Churyumov-Gerasimenko with the Sun on the left-hand side and the comet centered at the origin. While inside the contact surface (dashed line) the plasma flow is dominated by collisions with the abundant neutrals flowing out radially, outside the inner shock (dash-dot line) the plasma flow is bent tailward and its flow velocity becomes parallel to the shock increases [*Rubin et al.*, 2012].

outward with the neutral velocity. The ions are practically stagnating in the recombination region. Outside the contact surface the contaminated solar wind flows around the contact surface.

10.3. MASS LOADING

A well-developed cometary atmosphere extends to distances several orders of magnitude larger than the size of the nucleus. It is the mass loading of the solar wind with newly created cometary ions, from this extended exosphere, that is responsible for the interaction with the solar wind and the formation of cometary magnetotails. Mass loading occurs when a high-speed magnetized plasma moves through a background of neutral particles that is continuously ionized. Photoionization and electron impact ionization result in the addition of plasma to the plasma flow, while charge exchange replaces fast ions with almost stationary ones. "Ion pickup" (or ion implantation) is the process of accommodation (but not thermalization) of a single newborn ion to the plasma flow. The combined effect of the various ionization processes is usually net mass addition. Conservation of momentum and energy requires that the plasma flow be decelerated as newly born charged particles are "picked up." The process of continuous ion pickup and its feedback to the plasma flow are called "mass loading." A detailed review of mass loading in space plasmas can be found in the work of *Szegö et al.* [2000], while *Coates and Jones* [2009] published an excellent summary of the cometary mass-loading process.

Freshly born ions are accelerated by the motional electric field of the high-speed solar wind flow. The ion trajectory is cycloidal, resulting from the superposition of gyration and $E \times B$ drift. The corresponding velocity space distribution is a ring beam distribution, where the gyration speed of the ring is $u \sin \alpha$ (where u is the bulk plasma speed and α is the angle between the solar wind velocity and magnetic field vectors) and the beam velocity (along the magnetic field line) is $u \cos \alpha$.

The ring beam distribution has large velocity space gradients, and it is unstable to the generation of low-frequency transverse waves. The ions both gyrate about the magnetic field and $E \times B$ drift, such that the ion trajectories are cycloidal. The ring beam distribution is highly unstable against the growth of ultra low frequency (or ULF) electromagnetic waves (i.e., frequencies near the heavy ion gyrofrequency, or ≈ 10 mHz). The wave spectrum tends to have a strong monochromatic component forless active comets or at large cometocentric distances (i.e., comet Giacobini-Zinner far upstream) but tends to be turbulent like with power spread over a large range of wave numbers for regions closer to the comet [*Sagdeev et al.*, 1986; *Lee*, 2013; *Tsurutani*, 1991; *Mazelle et al.*, 1997].

The combination of ambient and self-generated magnetic field turbulence pitch angle scatters the newborn ions from the pickup ring. In a first approximation the pickup particles interact with the low-frequency waves without significantly changing their energy in the average wave frame. As a result of this process the pitch angles of the pickup-ring particles are scattered on the spherical velocity space shell of radius u around the local solar wind velocity [cf. *Neugebauer et al.*, 1989]. This process isillustrated in Figure 10.8. The top panel shows the pickup distribution in velocity space, while the bottom panel shows Giotto observations [*Neugebauer et al.*, 1989]. In a first approximation the distribution can be represented by a velocity space shell of radius u_{sw} centered at the solar wind velocity. In this approximation each pickup particle contributes $\frac{1}{3} m_i u_{sw}^2$ to the plasma pressure.

It is interesting to note that the rate of change of the phase-space distribution function due to ion pickup (due to photoionization) can be well approximated by

$$\frac{\delta F}{\delta t} = \frac{I_0}{d_h^2} \frac{n_0}{4\pi \tau_i v^2} \delta\left(u_{sw} + v\right), \tag{10.31}$$

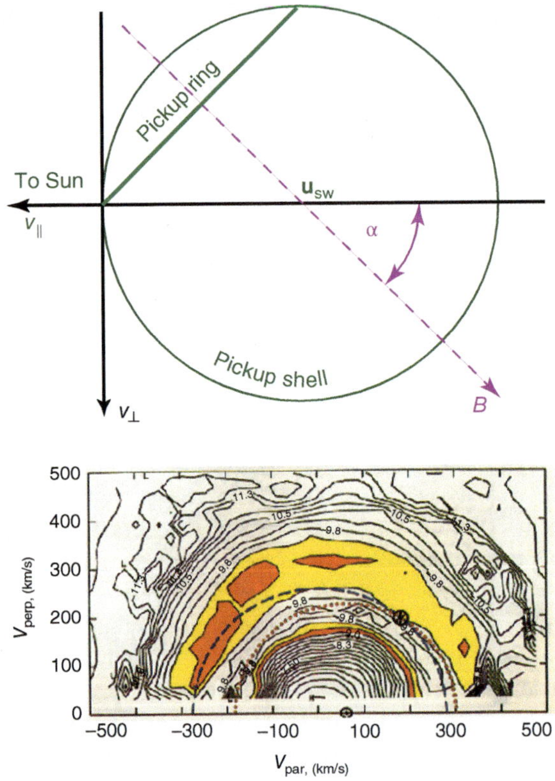

Figure 10.8 Illustration of the pickup velocity space distribution (top panel) and observed velocity space distribution at comet Halley (bottom panel) [*Neugebauer et al.*, 1989].

where τ_i is the ionization time scale and v is the speed of the particles. We note that all odd moments of the pickup distribution function vanish; therefore the pickup population can be *exactly* described by the equations of ideal MHD (since the heat flux is identically zero and therefore the set of moment equations naturally closes with the pressure equation). This is a simple counterexample to the popular (but incorrect) statement that ideal MHD implicitly assumes Maxwellian (therefore equilibrium) velocity distribution. It does not. The correct statement is that any distribution function with identically zero heat flux terms satisfy the equations of ideal MHD.

Implanted ions were detected at comets Giacobini-Zinner, Halley, Grigg-Skjellerup, and Borelli as large fluxes of energetic particles [*Coates et al.*,1993; *Hynds et al.*, 1986; *McKenna-Lawlor et al.*, 1986; *Somogyi et al.*, 1986; *Young et al.*, 2004]. A significant part of the detected energetic ion population was observed at energies considerably larger than the pickup energy, indicating the presence of some kind of acceleration process acting on implanted ions. The primary acceleration mechanism turned out to be slow velocity diffusion of lower energy implanted ions [*Coates et al.*, 1989; *Neuge-bauer et al.*, 1989; *Gombosi et al.*, 1991].

10.4. MATHEMATICAL DESCRIPTION

There are three approaches to simulate cometary magnetospheres: magnetohydrodynamics (including extended MHD) and hybrid descriptions. Hybrid simulations consider fluid electrons and kinetic ions and they solve the standard set of kinetic equations [cf. *Szegő et al.*, 2000; *Bagdonat and Motschmann*, 2002; *Lipatov et al.*, 2002; *Katoh et al.*, 2003; *Bagdonat et al.*, 2004; *Delamare*, 2006; *Hansen et al.*, 2007; *Gortsas et al.*, 2009, 2010; *Wiehle et al.*, 2011]. Magnetofluid simulations, on the other hand, incorporate quite a bit of cometary physics into the governing equations. In this section we briefly summarize the governing multifluid extended MHD (XMHD) equations that describe the mass-loaded cometary atmosphere. We will also briefly describe the appropriate source terms, but for details we refer the reader to the detailed paper of *Rubin et al.*, [2014].

The multifluid XMHD equations for ionic species s can be written in the following form:

$$\frac{\partial \rho_s}{\partial t} + \nabla \cdot (\rho_s u_s) = \frac{\delta \rho_s}{\delta t}, \tag{10.32}$$

$$\begin{aligned} \rho_s \frac{\partial \mathbf{u}_s}{\partial t} + \rho_s (\mathbf{u}_s \cdot \nabla) \mathbf{u}_s + \nabla \cdot P_s - \rho_s g \\ - \rho_s \frac{q_s}{m_s} [E + \mathbf{u}_s \times B] = \rho_s \frac{\delta u_s}{\delta t}, \end{aligned} \tag{10.33}$$

$$\begin{aligned} \frac{\partial P_s}{\partial t} + (\mathbf{u}_s \cdot \nabla) P_s + P_s (\nabla \cdot \mathbf{u}_s) + (P_s \cdot \nabla) u_s \\ + P_s \cdot (\nabla \mathbf{u}_s)^{\mathrm{T}} - \frac{q_s}{m_s} \left[P_s \times B + (P_s \times B)^{\mathrm{T}} \right] \\ + \nabla \cdot Q_s = \frac{\delta P_s}{\delta t}, \end{aligned} \tag{10.34}$$

where ρ, $\mathbf{u}$, P, and Q represent mass density, bulk velocity, pressure tensor, and heat flow tensor, respectively. The electric and magnetic field vectors are denoted by $\mathbf{E}$ and $\mathbf{B}$, while q is the particle charge and m is the particle mass. The $\rho_s \mathbf{g}$ term in the momentum equation represents the gravity force, which can often be neglected in the case of a small object such as a comet. In addition, $\delta\rho/\delta t$ is the mass source rate (due to ionization, recombination, and charge exchange), $\delta u/\delta t$ is the time rate of change of the bulk velocity, and $\delta P_s/\delta t$ is the rate of change of the pressure tensor.

We consider the scenario when the pressure tensor is gyrotropic (all off-diagonal terms are zero) but anisotropic (the pressure can be different in the direction of the magnetic field and perpendicular to it) and the heat flow tensor describes the magnetic field aligned flow of parallel and perpendicular random energies. In this case the P and Q tensors can be written in the form

$$P_{ij} = p_\perp \delta_{ij} + \left(p_\parallel - p_\perp\right) b_i b_j, \tag{10.35}$$

$$\begin{aligned} Q_{ijk} &= h_\perp \left(\delta_{ij} b_k + \delta_{ik} b_j + \delta_{jk} b_i\right) \\ &+ \left(h_\parallel - 3h_\perp\right) b_i b_j b_k, \end{aligned} \tag{10.36}$$

where **b** is the unit vector along the magnetic field, $p_\parallel$ and $p_\perp$ are the parallel and perpendicular pressure components, and $h_\parallel$ and $h_\perp$ are the-field aligned flows of parallel and perpendicular random energies. We note that the scalar pressure and isotropic heat flux can be easily expressed with the help of $p_\parallel$, $p_\perp$, $h_\parallel$, and $h_\perp$:

$$p = 1/3 P_{ii} = 1/3\left(p_\parallel + 2p_\perp\right), \tag{10.37}$$

$$\mathbf{h} = 1/2 Q_{iij} = 1/2\left(h_\parallel + 2h_\perp\right)\mathbf{b}. \tag{10.38}$$

With these definitions of the pressure and heat flow tensors the ion momentum and pressure equations become

$$\begin{aligned} &\rho_s \frac{\partial \mathbf{u}_s}{\partial t} + \rho_s (\mathbf{u}_s \cdot \nabla)\mathbf{u}_s + \nabla p_{s_\perp} - \rho_s \frac{q_s}{m_s}[\mathbf{E} + \mathbf{u}_s \times \mathbf{B}] \\ &= -\mathbf{B}\nabla_\parallel \left(\frac{p_{s_\parallel} - p_{s_\perp}}{B}\right) - \left(p_{s_\parallel} - p_{s_\perp}\right)\nabla_\parallel \mathbf{b}, \\ &+ \rho_s \mathbf{g} + \rho_s \frac{\delta \mathbf{u}_s}{\delta t} \end{aligned} \tag{10.39}$$

$$\begin{aligned} &\frac{\partial p_{s_\parallel}}{\partial t} + (\mathbf{u}_s \cdot \nabla) p_{s_\parallel} + p_{s_\parallel}(\nabla \cdot \mathbf{u}_s) + 2p_{s_\parallel} \mathbf{b} \cdot \nabla_\parallel \mathbf{u}_s \\ &+ \nabla_\parallel h_{s_\parallel} = \frac{h_{s_\parallel} - 3h_{s_\perp}}{B} \nabla_\parallel B + \frac{\delta p_{s_\parallel}}{\delta t}, \end{aligned} \tag{10.40}$$

$$\begin{aligned} &\frac{\partial p_{s_\perp}}{\partial t} + (\mathbf{u}_s \cdot \nabla) p_{s_\perp} + 2p_{s_\perp}(\nabla \cdot \mathbf{u}_s) - p_{s_\perp} \mathbf{b} \cdot \nabla_\parallel \mathbf{u}_s \\ &+ \nabla_\parallel h_{s_\perp} = \frac{5h_{s_\perp}}{2B} \nabla_\parallel B + \frac{\delta p_{s_\perp}}{\delta t}. \end{aligned} \tag{10.41}$$

Equations 10.32, 10.39, 10.40, and 10.41 represent the fluid equations for each ion species. The electron continuity equation is replaced by the condition of quasi-neutrality:

$$n_e = \sum_{s=\text{ions}} Z_s n_s, \tag{10.42}$$

where Z_s is the charge state of ion species s. The electron velocity can be expressed in terms of the average velocity of positive ions and the Hall term:

$$\mathbf{u}_e = \mathbf{u}_+ - \frac{\mathbf{j}}{e n_e}, \tag{10.43}$$

where **j** is the current density and

$$\mathbf{u}_+ = \sum_{s=\text{ions}} \frac{Z_s n_s}{n_e} \mathbf{u}_s. \tag{10.44}$$

We note that in general $u_+ \neq u_i$ because the charge-averaged velocity is different than the mass-averaged one. The electron pressure equations can be written as

$$\begin{aligned} &\frac{\partial p_{e_\parallel}}{\partial t} + (\mathbf{u}_e \cdot \nabla) p_{e_\parallel} + p_{e_\parallel}(\nabla \cdot \mathbf{u}_e) + 2p_{e_\parallel} \mathbf{b} \cdot \nabla_\parallel \mathbf{u}_e \\ &+ \nabla_\parallel h_{e_\parallel} = \frac{h_{e_\parallel} - 3h_{e_\perp}}{B} \nabla_\parallel B + \frac{\delta p_{e_\parallel}}{\delta t}, \end{aligned} \tag{10.45}$$

$$\begin{aligned} &\frac{\partial p_{e_\perp}}{\partial t} + (\mathbf{u}_e \cdot \nabla) p_{e_\perp} + 2p_{e_\perp}(\nabla \cdot \mathbf{u}_e) - p_{e_\perp} \mathbf{b} \cdot \nabla_\parallel \mathbf{u}_e \\ &+ \nabla_\parallel h_{e_\perp} = \frac{5h_{e_\perp}}{2B} \nabla_\parallel B + \frac{\delta p_{e_\perp}}{\delta t}. \end{aligned} \tag{10.46}$$

The heat flow terms are still not defined. These terms are usually taken from some other approximations, like collisional or collisionless heat conduction [cf. *Gombosi and Rasmussen,* 1991; *Hollweg,* 1976].

The transport equations are supplemented with Faraday's law, Ampère's law, and Ohm's law to form a closed system of differential equations:

$$\frac{\partial \mathbf{B}}{\partial t} = -\nabla \times \mathbf{E}, \tag{10.47}$$

$$\mu_0 \mathbf{j} = \nabla \times \mathbf{B}, \tag{10.48}$$

$$\begin{aligned} \mathbf{E} + \mathbf{u}_e \times \mathbf{B} &= -\frac{1}{e n_e} \nabla p_{e_\perp} + \eta_e \mathbf{j} \\ &- \frac{B}{e n_e} \nabla_\parallel \left(\frac{p_{e_\parallel} - p_{e_\perp}}{B} \mathbf{b}\right). \end{aligned} \tag{10.49}$$

The three terms on the right-hand side of equation 10.49 describe the ambipolar electric field (due to the mass difference between electrons and ions), the dissipation due to electrical resistivity, and the effect of adiabatic focusing.

The appropriate source term for comets are described in detail in a recent paper by *Rubin et al.*, [2014].

10.5. BOW SHOCK AND COMETOSHEATH

Conservation of momentum and energy require that the solar wind be decelerated as newly born charged particles are "picked up" by the plasma flow. Continuous deceleration of the solar wind flow by mass loading is possible only up to a certain point at which the mean molecular weight of the plasma particles reaches a critical value. At this point a weak shock forms and impulsively decelerates the flow to subsonic velocities [cf. *Biermann et al.*, 1967; *Galeev et al.*, 1985; *Koenders et al.*, 2013]. Bow shock crossings were identified in the data from each of the Halley flyby spacecraft at approximately the expected locations. The shock jumps were clearly defined in many of the observations from the plasma probes and magnetometers on Giotto, VEGA, and Suisei. For the other comets it is generally recognized that the pickup ions generated so much mass loading and turbulence that the shock crossing was extremely thick. The cometary "shock wave" is quite different than the "classical" planetary and interplanetary shocks, because the deceleration and dissipation is due to mass loading and wave-particle interaction that take place over a very large region with the "shock" being only the downstream boundary of an extended distributed process [cf. *Staines et al.*, 1991; *Neugebauer et al.*, 1990; *Coates et al.*, 1997; *Richter et al.*, 2011].

The cometosheath is located between the cometary shock and the magnetic field free region in the innermost coma. The plasma population in the cometosheath is a changing mixture of ambient solar wind and particles picked up upstream and downstream of the shock. The distinction between cometary particles picked up outside and inside the shock is important because of the large difference in their random energy. The random energy of ions in a pickup shell is typically 20 keV for water group ions picked up upstream of the shock. Cometary ions born behind the shock are picked up at smaller velocities, and consequently, the random energy of their pickup shell is significantly smaller than that of ions born upstream of the shock. Overall, ions retain (in their energy) a memory of where they were born, and the plasma frame energy of pickup ions decreases with decreasing cometocentric distance. The observed distribution functions are therefore quite complicated [cf. *Mukai et al.*, 1986; *Neugebauer et al.*, 1990].

One of the debated issues is whether or not energetic electrons are a permanent feature of the cometosheath. The electron spectrometer on Giotto observed large fluxes of energetic (0.8 – 3.6 keV) electrons in the so-called mystery region between about 8.5×10^3 and 5.5×10^3 km [*Réme*, 1991]. At a cometocentric distance of about 5.5×10^3 km these fluxes abruptly disappeared, simultaneously with a sudden decrease of the total ion density and velocity and an increase of the ion temperature. At the same time the magnetic field changed direction and became much smoother. *Réme* [1991] interpreted this change as a permanent feature of the cometosheath and found similar events in the Vega and Suisei data sets. A different view was presented by *Gringauz and Verigin* [1991], who did not see evidence of the presence of energetic electrons in the cometosheath and interpret the Giotto energetic electron event as a transient feature generated by a passing interplanetary disturbance.

Another very interesting feature in the cometosheath is the *cometopause*, discovered by the Vega plasma instrument [*Gringauz et al.*, 1986]. At around 1.65×10^5 km the PLASMAG instrument observed a sharp transition from a primarily solar wind proton dominated plasma population to a mainly cometary water group ion plasma. This transition was also accompanied by a moderate increase in the low-frequency plasma wave intensity, while there were no obvious changes in the magnetic field. The unexpected feature of the cometopause was not its existence but its sharpness. *Gombosi* [1987] suggested that an "avalanche" of charge exchange collisions in the decelerating plasma flow can rapidly deplete the solar wind proton population and replace it with slower water group ions.

10.6. COMETARY MAGNETOTAILS

Cometary magnetotails are formed by the draping of the interplanetary magnetic field (IMF) around the diamagnetic cavity boundary [*Alfvén*, 1957]. Cometary ions that are ionized outside the diamagnetic cavity get embedded into the decelerated solar wind plasma and move downstream into the tail. As the contaminated plasma flow moves downtail, it is gradually accelerated to the ambient solar wind speed while the draped magnetic field lines return to their original configuration. Near the center of the tail the two magnetic lobes are separated by a current sheet and the embedded cometary ions are concentrated near this neutral sheet. These cometary ions can be observed through resonance fluorescence processes [cf. *Feldman et al.*, 2005; *Lisse et al.*, 2005; *Ip*, 2005]. A nice summary of cometary magnetotail observations can be found in the work of *Jones et al.* [2010].

Since cometary magnetotails are essentially draped IMF lines, they are very sensitive to changes in the solar wind. When a plasma tail interacts with dynamical structures in the solar wind (such as sector boundaries, corotating interaction regions, interplanetary shocks, coronal mass ejections, and other transient phenomena), the tail itself can get distorted or even disconnected from the comet [cf. *Brandt et al.*, 1999]. Figure 10.9 shows two images of the tail of comet Kohoutek [*Lundin and Barabash*, 2004]) revealing narrow jets (or beams), wavy (Kelvin-Helmholtz unstable) structures, and disconnection events.

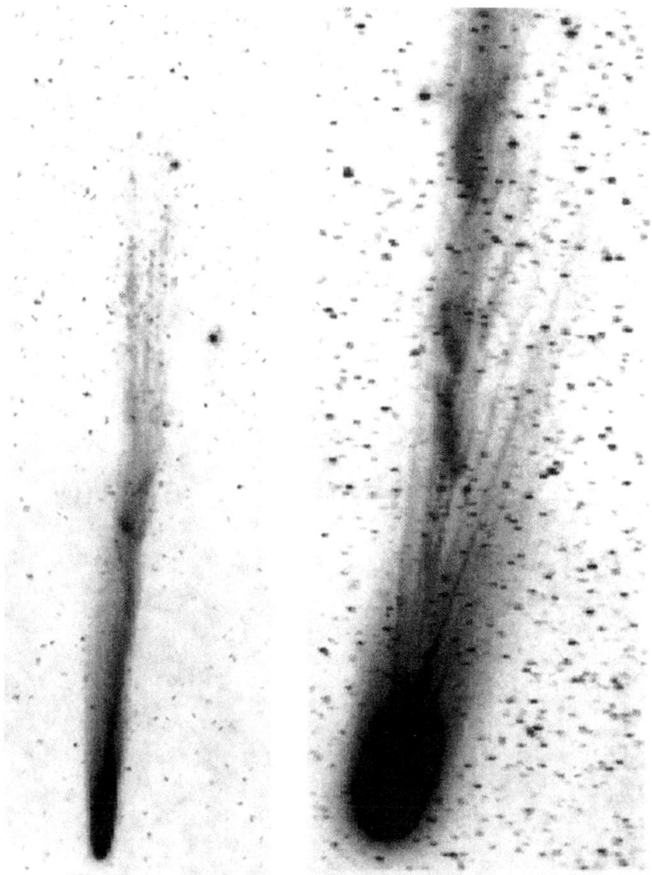

Figure 10.9 Dynamical structures in the long tail of comet Kohoutek. Narrow beams/jets, Kelvin-Helmholtz waves, and disconnection events can be seen. [From *Lundin and Barabash*, 2004].

Jia et al., [2009] simulated the tail disconnection event on 20 April 2007 on comet 2P/Encke, caused by a coronal mass ejection (CME) at a heliocentric distance of 0.34 AU. The MHD simulation reproduced the interaction process and demonstrated how the CME triggered a tail disconnection.The CME disturbed the comet with a combination of a 180° sudden rotation of the IMF, followed by a 90° gradual rotation. The simulation results are shown in Figure 10.10.

At 16:10 UT, the front of the flux rope (marked by the dashed vertical white lines) is at -10^5 km, in front of the comet bow shock. The IMF is in the x-z plane so the tail appears thinner in this plane, and wider in the x-y plane. At 16:30 UT, marked by the white dashed lines, the interface of reversed field lines has formed a cone centering at the comet head. In the x-y plane, the axial component in the flux rope starts to increase behind the front, as marked by the black field lines on the left of the white dashed line. The field on the right of the white line is the old IMF pointing northward. The comet tail appears stretched in both slices. At 17:40 UT, the tip of the interface between the reversed field lines has propagated to 400,000 km. Magnetic reconnection in the tail has generated a density pileup region, which appears as the trailing head of the disconnected tail. On the left of the density pileup, a new tail starts to grow. This new tail becomes wider in the x-z plane and thinner in the x-y plane as a consequence of field rotation in the IMF. The field on the left of the white lines has both y and z components. The old field lines pointing northward are confined in the tail on the right of the white lines. At 18:30 UT, the axis of the flux rope is significantly bent, as marked by the red dotted line in the left panel. At the top boundary, the axis of the flux rope has passed through the shown region. Along the x axis, the center of the flux rope has only propagated to the inner coma, while the front of the flux rope has also passed through the shown region. The IMF along this red line is primarily in the y direction (z axis component), while those on the right of the red line are rotating from the y direction to the z direction as they leave the red line. The red line in the right panel marks the interface of the field lines with the same orientation. This stretched red line suggests that the highly draped field lines significantly affect the force balance in the interaction process and thus control the evolution speed of this process.

The first in situ measurements of cometary magnetotails was carried out by the International Cometary Explorer (ICE) mission, which crossed the tail of comet Giacobini-Zinner at a distance of about 8000 km [*Bame et al.*, 1986; *Hynds et al.*, 1986; *Ipavich et al.*, 1986; *Ogilvie et al.*, 1986; *Scarf et al.*, 1986; *Smith et al.*, 1986]. ICE started its life as the third International Sun-Earth Explorer (ICEE-3) in 1978. In the early 1980s it was recognized that using the available fuel and five gravitational encounters with the Moon the spacecraft could intercept comet Giacobini-Zinner. Since ICE was originally designed as a space physics mission, it did not carry any cameras, but it was well instrumented to study the plasma environment of the comet. The lack of remote sensing instruments made it possible to fly ICE through the comet tail, since there was no possibility to observe the illuminated side of the comet.

During the encounter the ICE spacecraft observed very high levels of hydromagnetic turbulence generated by the ion pickup process [*Scarf et al.*, 1986] and a large increase in water group ion density [*Hynds et al.*, 1986; *Ipavich et al.*, 1986; *Ogilvie et al.*, 1986]. Figure 10.11 shows that the water group number density increased by two orders of magnitude as the spacecraft reached closest approach [*Gloeckler et al.*, 1986].

The Ulysses spacecraft had two distant unplanned comet tail crossings. On 1 May 1996 it crossed the tail of comet Hyakutake at a distance of $\sim 5 \times 10^8$ km (3.73 AU). The solar wind ion composition spectrometer (SWICS)

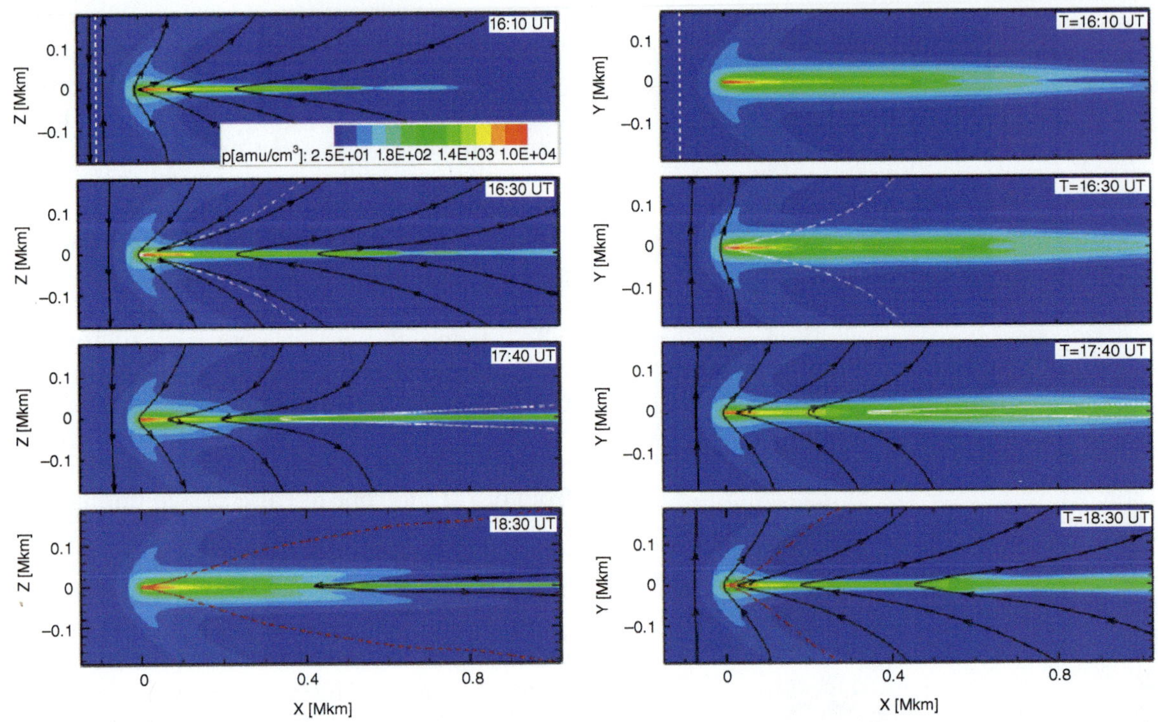

Figure 10.10 Time evolution of the disconnection event in two-dimensional slices. Color contours represent plasma density showing the bow shock and tail. The black lines represent magnetic field lines. White dotted lines show the current sheets, while red dashed lines show the location of the fux rope axis. Left panels show the meridional plane, while panels on the right show the ecliptic plane. [From *Jia et al.,* 2009].

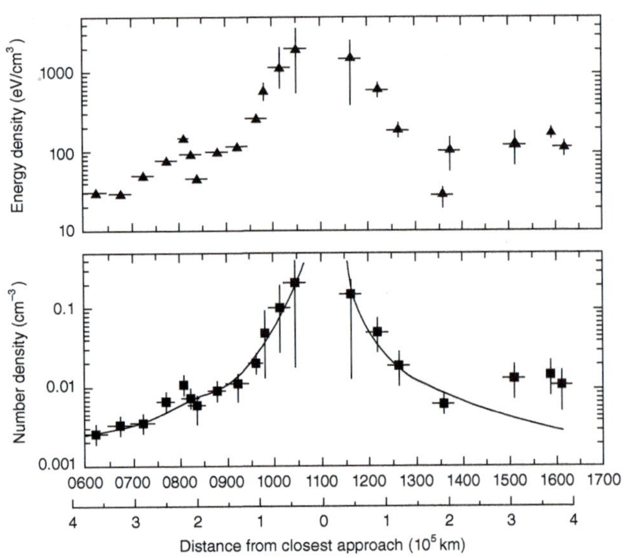

Figure 10.11 Variation of the energy density (upper panel) and number density (lower panel) of cometary water-group pick up ions with time and distance from closest approach from the nucleus of Giacobini-Zinner. [from *Gloeckler et al.*, 1986].

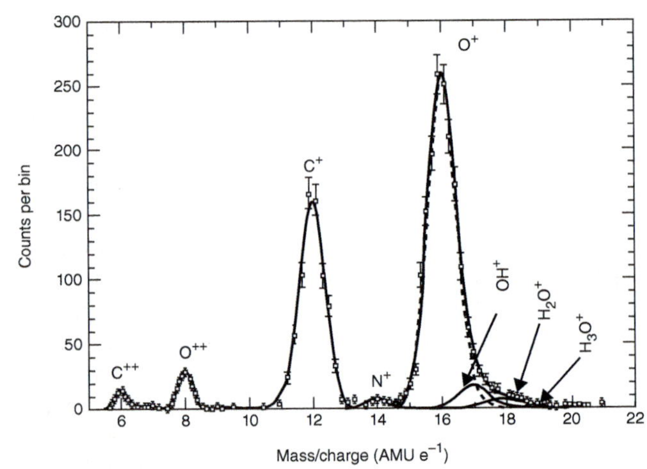

Figure 10.12 Mass-per-charge spectrum obtained during the crossing of the distant tail of comet Hyakutake. Resolved ion species include (from left to right) C^{++}, N^{++}, O^{++}, C^{+}, N^{+}, and O^{+}. [From *Gloeckler et al.*, 2000].

identified a number of species that are most likely of cometary origin [*Gloeckler et al.*, 2000]. Figure 10.12 shows the mass-per-charge spectrum of these ions. We note the dominance of water group ions (O^+, OH^+, H_2O^+, and H_3O^+). Particularly telling is the presence of H_3O^+ that is collisionally created in water vapor dominated regions by the $H_2O + H_2O^+ \rightarrow H_3O^+ + OH$ reaction.

Ulysses had a second unplanned comet tail crossing in February 2007, when it crossed the ion tail of comet McNaught [*Neugebauer et al.*, 2007]. Inspection of

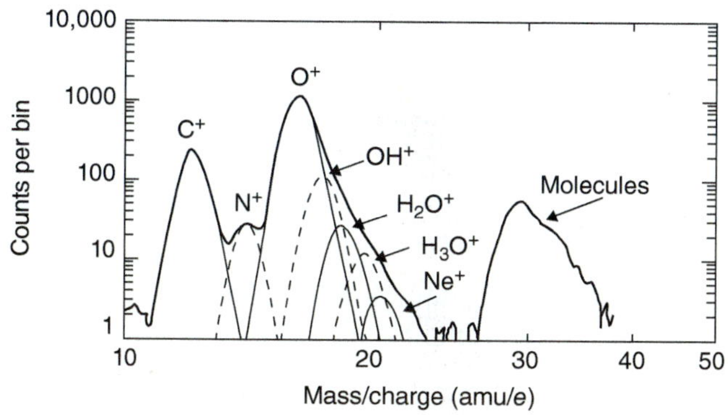

Figure 10.13 Mass-per-charge spectrum obtained during the crossing of the distant tail of comet McNaught. [From *Neugebauer et al.*, 2007].

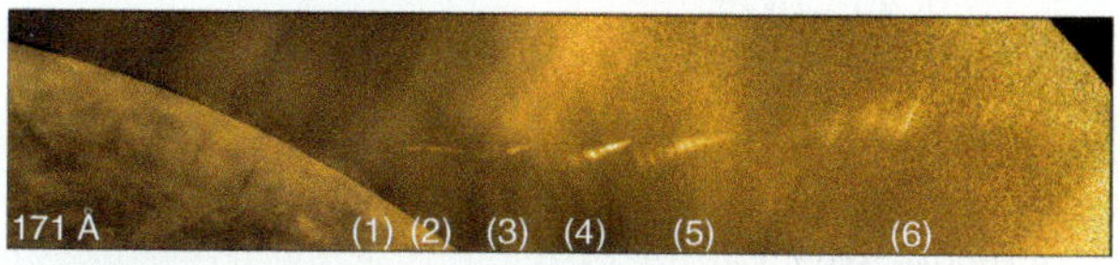

Figure 10.14 171 ÅAIA images of the tail of comet Lovejoy after perihelion. The images were taken at six times: (1) 00:41:36 UT, (2) 00:43:00, (3) 00:44:12, (4) 00:46:12, (5) 00:48:12, (6) 00:56:00. One can see the dramatic changes in the tail of the comet. [From *McCauley et al.* 2013].

Figure 10.13 reveals that the mass-per-charge spectrum of heavy ions was quite similar to the one observed during the Hyakutake tail crossing, confirming the dominance of water group ions.

Recent interest focuses on the tail dynamics of sungrazing comets. The most significant sungrazer since the launch of the Solar Dynamics Observatory (SDO) is comet Lovejoy. Detailed, high-resolution observations of the perihelion passage of this comet were carried out by the SDO/AIA instrument [*Mccauley et al.*, 2013].

Comet Lovejoy reached perihelion on 16 December 2011 as it passed approximately 140,000 km above the photosphere. Since the comet survived perihelion, it is thought that the size of the nucleus must have been about 500 m. During the coronal passage, it is believed that a significant fraction of the comet's mass was burned off.

It was a surprise that comet Lovejoy survived perihelion. Figure 10.14 shows a series of images taken after it passed $0.2R_{\odot}$ above the photosphere. The series of photographs show that the tail becomes unstable and eventually detaches from the nucleus. Sungrazing comets are becoming important not only because they provide new insights into the physics of comets but also because they help us to understand the thermal structure of the low solar corona. In effect, these comets provide a tool to probe regions that are not accessible to in situ observations.

Simulating the interaction of sungrzaing comets with the low solar corona is very challenging, because one needs to carry out a realistic low corona simulation before the rapidly moving comet (typical speeds are ~ 500 km/s) can be inserted into this background. The first attempts for such simulations are just being made [*Downs et al.*, 2013], but significant additional progress is required before we can carry out more realistic simulations.

10.7. MODEL-DATA COMPARISON

Until the arrival of the Rosetta spacecraft to comet Churyumov-Gerasimenko [*Glassmeier et al.*, 2007], comet Halley remains the best studied active comet, using observational, theoretical, and simulation methods. The Giotto spacecraft flew within $\sim$600 km of the nucleus and it provided a wealth of data not only about the global plasma environment (density, velocity, temperature, and magnetic field) but also about the composition of the ionized and neutral gases. Recently *Rubin et al.* [2014] carried out a detailed comparison between a multifluid MHD simulation and Giotto observation. In this section we review the highlights of this study.

The simulation of *Rubin et al.* [2014] solves separate equations for solar wind protons, cometary light and heavy ions, and electrons. The cometary light ions are mainly H^+, but they are separately treated from the solar wind. The cometary heavy ions are dominated by water group ions, with relatively small contribution from carbon and nitrogen bearing molecules and other species. Mass loading, charge exchange, dissociative ion-electron recombination, and collisional interactions between the fluids are taken into account. The computational domain spans over several million kilometers and the close vicinity of the comet is resolved to the details of the magnetic cavity. Simulation results are compared with the data of the JPA [*Johnstone et al.*, 1986] and IMS [*Balsiger et al.*, 1986] instruments.

Figure 10.15 shows the comparison of the simulated and measured ion mass densities along the Giotto trajectory. The solid lines show the model results while the measurements from the JPA instrument [*Formisano et al.*, 1990] and the IMS [*Altwegg et al.*, 1993] are denoted by the dots. The lack of solar wind protons in the vicinity of the comet becomes clearly visible inside approximately 100,000 km from the comet along the Giotto trajectory. The proton population originating from the Sun is deflected around the cavity and also slowed down due to the abundant elastic collisions with the neutral coma, which leads to an enhancement in density before it is then lost due to the charge-exchange with the neutral gas as discussed earlier. This leads to a slight enhancement at roughly 10,000 km which is not symmetric around the cavity.

The model reproduces the ion pile-up for both the cometary light and heavy plasma species at around 10,000–20,000 km from the nucleus although less prominently

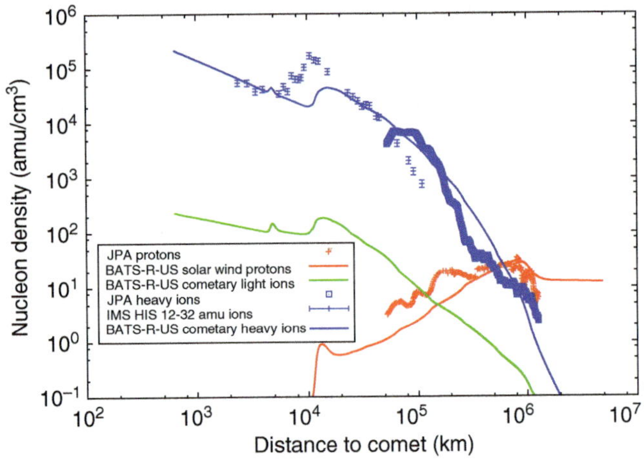

Figure 10.15 Nucleon density (in units of amu/cm³) along Giotto's inbound trajectory. The *x* axis denotes the distance to the comet (km) with the closest approach at roughly 600 km. The solid lines show the simulated densities of the solar wind protons, the cometary light ions, and the cometary heavy ions. The corresponding measurements are added, i.e. the observations of the heavy ions by the JPA [*Formisano et al.*, 1990] and the IMS [*Altwegg et al.*, 1993] instruments. [From *Rubin et al.*, 2014].

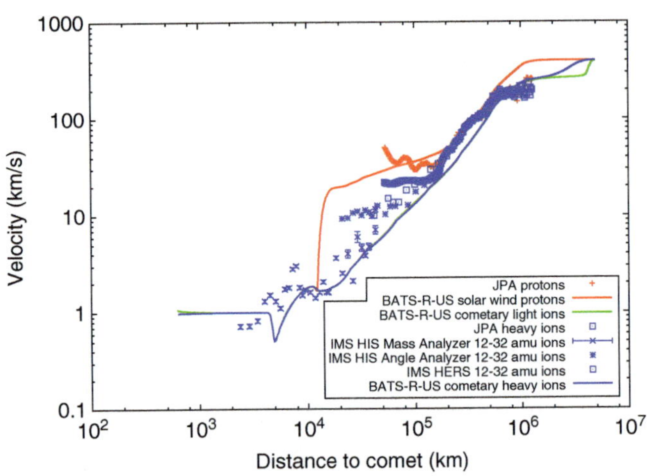

Figure 10.16 Ion velocities (km/s) along Giottos inbound trajectory. The *x*-axis denotes the distance to the comet (km) with the closest approach at roughly 600 km. The solid lines show the modeled bulk speeds of the solar wind protons, the cometary light ions, and the cometary heavy ions. The corresponding measurements are added, i.e., the heavy ion velocity by the JPA [*Formisano et al.*, 1990] and the IMS [*Altwegg et al.*, 1993] instruments. [From *Rubin et al.*, 2014].

than measured. Generally, the model indicates that the pile-up region of the light cometary ions and the heavy ones do not have to be exactly colocated given the slightly different temperature dependence of ion-electron recombination rates and the more extended gas distribution of

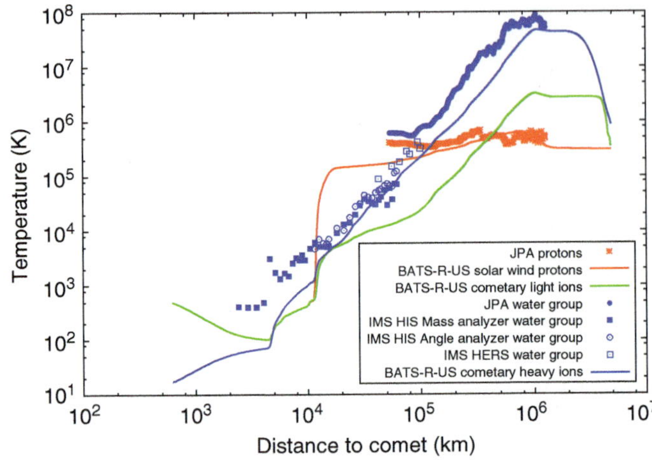

Figure 10.17 Ion temperatures (K) along Giotto's inbound path. The *x*-axis denotes the distance to the comet (km). The solid lines show the modeled temperatures of the solar wind protons, the cometary light ions, and the cometary heavy ions. The corresponding measurements are also given: the observations of the protons and the heavy ions by the JPA [*Formisano et al.*, 1990] and the IMS [*Altwegg et al.*, 1993] instruments. [From *Rubin et al.*, 2014].

the cometary daughter species. Closer to the nucleus the model shows the formation of the inner shock. Also the radial distribution of the water group ions is steeper than for the cometary light ions, again the result of the faster decrease of the abundance of the cometary parent species with distance from the nucleus.

Figure 10.16 shows the simulated ion bulk velocities compared to the measurements [*Formisano et al.*, 1990; *Altwegg et al.*, 1993]. Close to the comet (inside ~10,000 km from the nucleus) the velocity vectors and bulk speeds of the cometary ions are tightly coupled through ion–ion, ion–neutral, and ion–electron interactions while the solar wind protons are already attenuated in this region.

In Figure 10.17 the simulated temperatures of the individual ion species are compared to those derived from the IMS and JPA instruments. It can be seen that the temperature difference between the solar wind and the pickup populations is quite well reproduced by the MHD model.

Overall, the agreement between observations and the detailed simulation results of *Rubin et al.* [2014] is quite remarkable.

10.8. ROSETTA

The goal of this chapter was to summarize our pre-Rosetta mission understanding of cometary magnetospheres. This understanding is based on ground-based observations and a small number of short-duration flyby missions. It is highly likely that some of the ideas we outlined here will be modified or even rejected after we

Figure 10.18 Artist's illustration of the Rosetta orbiter comet with lander on the surface of comet Churyumov-Gerasimenko. (Courtesy of NASA JPL).

gain a detailed, long-duration insight into the way comet Churyumov-Gerasimenko behaves as it becomes active, approaches its perihelion, and then gradually fades.

In early 2014 Rosetta will wave up from hibernation and start preparations for its comet encounter. In November 2014 it will deliver a lander unit, called PHILAE. After the lander delivery the Rosetta orbiter will start its 13 months long "escort" phase. During the prime mission Churyumov-Gerasimenko will pass through perihelion in August 2015. The spacecraft is well instrumented for both in situ and remote sensing observations [cf. *Glassmeier et al.*, 2007]. Figure 10.18 showsan artist's illustration of the Rosetta orbiter and PHILAE lander at comet Churyumov-Gerasimenko.

Rosetta's target comet is moderately active near perihelion. *Snodgrass et al.* [2013] estimated comet Churyumov-Gerasimenko's preperihelion water production rate curve as

$$Q(H_2O) = \frac{2.3\times 10^{28}}{d_h^{5.9}} \quad \text{molecules / s}, \qquad (10.50)$$

implying that Rosetta will encounter production rates of $Q \sim 6\times 10^{24}$ molecules/s at $d_h = 4$ AU, $\sim 3\times 10^{25}$ at 3 AU, and 4×10^{26} at 2 AU. At perihelion ($d_h = 1.3$ AU) one can expect a water production rate of 5×10^{27} molecules/s. This value is about two orders of magnitude smaller than the production rate of comet Halley was during the Giotto encounter.

The expectation is that the nucleus will be practically inactive during the initial phase of the Rosetta mission when the heliocentric distance will be around $d_h \approx 2.5$ AU. This is the phase when the lander can carry out its main mission, the in situ examination of the nucleus. As the comet approaches the Sun, it will start to develop higher and higher activity levels. During this phase the cometary magnetosphere will gradually develop. Initially the magnetosphere will be nearly collisionless (requiring kinetic description) due to the very low density of the cometary neutral ionosphere and ionosphere.

ACKNOWLEDGMENTS

This work has been supported by the U.S. Rosetta Project (JPL subcontract 1266313 under NASA grant NMO710889) and by NSF grant AST-0707283. Valuable discussions with Drs. Michael Combi, Ying-Dong Jia, and Martin Rubin are gratefully acknowledged. The author would also like to thank two referees for their valuable comments that helped to improve the original manuscript.

REFERENCES

Alfvén, H. (1957), On the theory of comet tails, *Tellus*, *9*, 92–96.

Altwegg, K., H. Balsiger, J. Geiss, R. Goldstein, W. Ip, A. Meier, M. Neugebauer, H. Rosenbauer, and E. Shelley (1993), The ion population between 1300 km and 230 000 km in the coma of comet P/Halley, *Astron. Astrophys.*, *279*, 260–266.

Bagdonat, T., and U. Motschmann (2002), From a weak to a strong comet: 3D global hybrid simulation studies, *Earth MoonandPlanets*,*90*,305–321,doi:10.1023/A:1021578232282.

Bagdonat, T., U. Motschmann, K. Glassmeier, and E. Khurt (2004), Plasma environment of Comet Churyumov-Gerasimenko 3D hybrid code simulations, in *New Rosetta Targets: Observations, Simulations and Instrument Performances, Astrophysics and Space Science Library*, vol. 311, edited by L. Colangeli et al. pp. 153–166, Giunta Reg Della Campania; Osservator Astronom Capodimonte; Univ. Degli Studi Napoli, Naples, Italy.

Balsiger, H., et al. (1986), Ion composition and dynamics at comet Halley, *Nature*, *321*, 330–334.

Bame, S. J., et al. (1986), Comet Giacobini-Zinner: Plasma description, *Science*, *232*, 356–361, doi:10.1126/science.232.4748.356.

Banks, P. M., and G. Kockarts (1973), *Aeronomy*, Academic Press, New York.

Benna, M., and P. Mahaffy (2007), Multi-fluid model of comet 1P/Halley, *Planet. Space Sci.*, *55*, 1031–1043, doi:10.1016/j.pss.2006.11.019.

Biermann, L. (1951), Kometenschweife und Solare Korpuskularstrahlung, *Zeitschrift für Astrophysik*, *29*, 274–286.

Biermann, L., B. Brosowski, and H. U. Schmidt (1967), The interaction of the solar wind with a comet, *Sol. Phys.*, *1*, 254–284.

Brandt, J. C., F. M. Caputo, J. T. Hoeksema, M. B. Niedner, Y. Yi, and M. Snow (1999), Disconnection events (DEs) in Halley's comet 1985–1986:: The correlation with crossings of the heliospheric current sheet (HCS), *Icarus*, *137*, 69–83, doi:10.1006/icar.1998.6030.

Coates, A., and G. Jones (2009), Plasma environment of jupiter family comets, *Planet. Space Sci.*, *57*, 1175–1191, doi:10.1016/j.pss.2009.04.009.

Coates, A., A. Johnstone, B. Wilken, K. Jockers, and K. Glassmeier (1989), Velocity space diffusion of pickup ions from the water group at comet Halley, *J. Geophys. Res.*, *94*, 9983–9993, doi:10.1029/JA094iA08p09983.

Coates, A. J., A. D. Johnstone, B. Wilken, and F. Neubauer (1993), Velocity space diffusion and nongyrotropy of pickup water group ions at comet Grigg-Skjellerup, *J. Geophys. Res.*, *98*, 20,985, doi:10.1029/93JA02535.

Coates, A. J., C. Mazelle, and F. M. Neubauer (1997), Bow shock analysis at comets Halley and Grigg-Skjellerup, *J. Geophys. Res.*, *102*, 7105–7113, doi:10.1029/96JA04002.

Combi, M. R., V. M. Tenishev, M. Rubin, N. Fougere, and T. I. Gombosi (2012), Narrow dust jets in a diffuse gas coma: A natural product of small active regions on comets, *Astrophys. J.*, *749*(1), 29, doi:10.1088/0004-637X/749/1/29.

Cravens, T. E. (1986), The physics of the cometary contact surface, in *Proc. 20th ESLAB Symposiumo on the Exploration of Halley's Comet*, vol. *ESA SP-250*, pp. 241–246, Eur. Space Agency, Heidelberg, Germany.

Cravens, T. E. (1989), A magnetohydrodynamical model of the inner coma of comet Halley, *J. Geophys. Res.*, *94*, 15,025–15,040, doi:10.1029/JA094iA11p15025.

Cravens, T. E., and A. Körösmezey (1986), Vibrational and rotational cooling of electrons by water-vapor, *Planet. Space Sci.*, *34*, 961–970, doi:10.1016/0032-0633(86)90005-X.

Cravens, T. E., N. Ozak, M. S. Richard, M. E. Campbell, I. P. Robertson, M. Perry, and A. M. Rymer (2011), Electron energetics in the Enceladus torus, *J. Geophys. Res.*, *116*, A09205, doi:10.1029/2011JA016498.

Davidsson, B. J. R. (2008), Comet Knudsen layers, *Space Sci. Rev.*, *138*, 207–223, doi:10.1007/s11214-008-9305-8.

Delamere, P. A. (2006), Hybrid code simulations of the solar wind interaction with Comet 19P/Borrelly, *J. Geophys. Res.*, *111*, doi:10.1029/2006JA011859.

Delsemme, A., and D. Miller (1971), Physico-chemical phenomena in comets, III. Continuum of comet Burnham (1960 II), *Planet. Space Sci.*, *19*, 1229–1257, doi:10.1016/0032-0633(71)90180-2.

Delsemme, A. H., and P. Swings (1952), Hydrates de gaz dans les noyaux cométaires et les grains interstellaires, *Ann. Astrophys.*, *15*, 1.

Downs, C., J. A. Linker, Z. Mikic, P. Riley, C. J. Schrijver, and P. Saint-Hilaire (2013), Probing the solar magnetic field with a sun-grazing comet, *Science*, *340*(6137), 1196–1199, doi:10.1126/science.1236550.

Feldman, P. D., A. L. Cochran, and M. R. Combi (2005), Spectroscopic investigations of fragment species in the coma, in *Comets II*, edited by M. C. Festou et al. pp. 425–447, University of Ariz. Press, Tucson, Ariz.

Formisano, V., E. Amata, M. B. Cattaneo, P. Torrente, A. Johnstone, A. Coates, and B. Wilken (1990), Plasma-flow inside comet P/Halley, *Astron. Astrophys.*, *238*, 401–412.

Fougere, N., M. Combi, M. Rubin, and V. Tenishev (2013), Modeling the heterogeneous ice and gas coma of comet 103P/Hartley 2, *Icarus*, *225*, 688–702, doi:10.1016/j.icarus.2013.04.031.

Galeev, A. A., T. E. Cravens, and T. I. Gombosi (1985), Solar wind stagnation near comets, *Astrophys. J.*, *289*, 807–819.

Gan, L., and T. E. Cravens (1990), Electron energetics in the inner coma of comet halley, *J. Geophys. Res.*, *95*, 6285–6303, doi:10.1029/JA095iA05p06285.

Glassmeier, K.-H., H. Boehnhardt, D. Koschny, E. Kührt, and I. Richter (2007), The Rosetta mission: Flying towards the origin of the solar system, Space Sci. *Rev.*, *128*, 1–21, doi:10.1007/s11214-006-9140-8.

Gloeckler, G., D. Hovestadt, F. M. Ipavich, M. Scholer, B. Klecker, and A. B. Galvin (1986), Cometary pick-up ions observed near Giacobini-Zinner, *Geophys. Res. Lett.*, *13*, 251–254, doi:10.1029/GL013i003p00251.

Gloeckler, G., J. Geiss, N. A. Schwadron, L. A. Fisk, T. H. Zurbuchen, F. M. Ipavich, R. von Steiger, H. Balsiger, and B. Wilken (2000), Interception of comet Hyakutake's ion tail at a distance of 500 million kilometres, *Nature*, *404*, 576–578.

Goldstein, B. E., K. Altwegg, H. Balsiger, S. A. Fuselier, W.-H. Ip, A. Meier, M. Neugebauer, H. Rosenbauer, and R. Schwenn (1989), Observations of a shock and a recombination layer at the contact surface of comet Halley, *J. Geophys. Res.*, *94*, 17,251–17,257, doi:10.1029/JA094iA12p17251.

Gombosi, T. I. (1987), Charge exchange avalanche at the cometopause, *Geophys. Res. Lett.*, *14*, 1174–1177, doi:10.1029/GL014i011p01174.

Gombosi, T. I., and C. E. Rasmussen (1991), Transport of gyration dominated space plasmas of thermal origin I.: Generalized transport equations, *J. Geophys. Res.*, *96*, 7759.

Gombosi, T. I., A. F. Nagy, and T. E. Cravens (1986), Dust and neutral gas modeling of the inner atmospheres of comets, *Rev. Geophys.*, *24*(3), 667–700.

Gombosi, T. I., M. Neugebauer, A. Johnstone, A. Coates, and D. Huddleston (1991), Comparison of observed and calculated implanted ion distributions outside comet Halley's bow shock, *J. Geophys. Res.*, *96*, 467.

Gombosi, T. I., D. L. De Zeeuw, R. M. Häberli, and K. G. Powell (1996), Three-dimensional multiscale MHD model of cometary plasma environments, *J. Geophys. Res.*, *101*(A7), 15,233–15,253.

Gortsas, N., U. Motschmann, E. Kuehrt, J. Knollenberg, S. Simon, and A. Boesswetter (2009), Mapping of coma anisotropies to plasma structures of weak comets: A 3-D hybrid simulation study, *Ann. Geophys.*, *27*, 1555–1572.

Gortsas, N., U. Motschmann, E. Kuehrt, K. Glassmeier, K. C. Hansen, J. Mueller, and A. Schmidt (2010), Global plasma-parameter simulation of Comet 67P/Churyumov-Gerasimenko approaching the Sun, *Astron. Astrophys.*, *520*, doi: 10.1051/0004-6361/201014761.

Gringauz, K., et al. (1986), First in situ plasma and neutral gas measurements at comet Halley, *Nature*, *321*, 282–285, doi:10.1038/321282a0.

Gringauz, K. I., and M. I. Verigin (1991), Permanent and non-stationary plasma phenomena in comet Halley's head, in *Cometary Plasma Processes, AGU Geophys. Monograph Ser.*, vol. *61*, edited by A. D. Johnstone, pp. 107–116, AGU, Washington, D.C.

Häberli, R. M., M. R. Combi, T. I. Gombosi, D. L. De Zeeuw, and K. G. Powell (1997), Quantitative analysis of H_2O^+ coma images using a multiscale MHD model with detailed ion chemistry, *Icarus*, *130*, 373–386.

Hansen, K. C., T. Bagdonat, U. Motschmann, C. Alexander, M. R. Combi, T. E. Cravens, T. I. Gombosi, Y. Jia, and I. P. Robertson (2007), The plasma environment of comet 67P/Churyumov-Gerasimenko throughout the Rosetta main mission, *Space Sci. Rev.*, *128*, 133–166, doi:{10.1007/s11214-006-9142-6}.

Hollweg, J. V. (1976), Collisionless electron heat conduction in the solar wind, *J. Geophys. Res.*, *81*, 1649–1658, doi:10.1029/JA081i010p01649.

Hynds, R., S. Cowley, T. Sanderson, K. Wenzel, and J. Vanrooijen (1986), Observations of energetic ions from comet Giacobini-Zinner, *Science*, *232*, 361–365, doi:10.1126/science.232.4748.361.

Ip, W. (2005), Global solar wind interaction and ionospheric dynamics, in *Comets II*, edited by M. C. Festou et al. pp. 605–629, Univ. of Arizona Press, Tucson, Ariz.

Ip, W.-H., and W. I. Axford (1987), The formation of a magnetic-field-free cavity at comet Halley, *Nature*, *325*, 418–419, doi:10.1038/325418a0.

Ipavich, F. M., A. B. Galvin, G. Gloeckler, D. Hovestadt, B. Klecker, and M. Scholer (1986), Comet Giacobini-Zinner: In situ observations of energetic heavy ions, *Science*, *232*, 366–369, doi:10.1126/science.232.4748.366.

Jia, Y. D., C. T. Russell, L. K. Jian, W. B. Manchester, O. Cohen, A. Vourlidas, K. C. Hansen, M. R. Combi, and T. I. Gombosi (2009), Study of the 2007 April 20 CME-comet interaction event with an MHD model, *Astrophys. J. Lett.*, *696*, L56, doi:10.1088/0004-637X/696/1/L56.

Johnstone, A., et al. (1986), Ion flow at comet Halley, *Nature*, *321*, 344–347, doi:10.1038/321344a0.

Jones, G. H., R. J. Forsyth, and A. J. Coates (2010), The structure of comets' induced magnetotails: Remote and in situ observations, in *Pickup Ions Throughout the Heliosphere and Beyond, AIP Conf. Proc.*, vol. *1302*, edited by J. A. le Roux et al. pp. 225–230, Am. Instit. Phys., doi:10.1063/1.3529974.

Katoh, Y., H. Oya, M. Iizima, and T. Ono (2003), Numerical study on the spatial extent of interaction region surrounding comet nucleus: Ion pick-up process, *Earth Planets and Space*, *55*(11), 705–711.

Koenders, C., K.-H. Glassmeier, I. Richter, U. Motschmann, and M. Rubin (2013), Revisiting cometary bow shock positions, *Planet. Space Sci.*, *87*, 85–95, doi:10.1016/j.pss.2013.08.009.

Körösmezey, A., T. Cravens, T. Gombosi, A. Nagy, D. Mendis, K. Szegö, B. Gribov, R. Sagdeev, V. Shapiro, and V. Shevchenko (1987), A new model of Cometary ionospheres, *J. Geophys. Res.*, *92*, 7331–7340, doi:10.1029/JA092iA07p07331.

Lee, M. A. (2013), Ultra-low frequency waves at comets, in *Plasma Waves and Instabilities at Comets and in Magnetospheres, AGU Geophys. Monogr. Ser.*, vol. *53*, edited by B. T. Tsurutani and H. Oya, pp. 13–29, AGU, doi:10.1029/GM053p0013.

Lipatov, A. S., U. Motschmann, and T. Bagdonat (2002), 3D hybrid simulations of the interaction of the solar wind with a weak comet, *Planet. Space Sci.*, *50*, 403–411, doi: 10.1016/S0032-0633(02)00004-1.

Lisse, C. M., T. E. Cravens, and K. Dennerl (2005), X-ray and extreme ultraviolet emission from comets, in *Comets II*, edited by M. C. Festou et al. pp. 631–643, Univ. of Ariz. Press, Tucson, Ariz.

Lundin, R., and S. Barabash (2004), The wakes and magnetotails of Mars and Venus, *Adv. Space Res.*, *33*(11), 1945–1955, doi:10.1016/j.asr.2003.07.054.

Mazelle, C., J. Cao, G. Belmont, F. Neubauer, and A. Coates (1997), Compressive character of low frequency waves driven by newborn ions at comet Grigg-Skjellerup, *Adv. Space Res.*, *20*(2), 267–270, doi:10.1016/S0273-1177(97)00544-9.

McCauley, P. I., S. H. Saar, J. C. Raymond, Y.-K. Ko, and P. Saint-Hilaire (2013), Extreme-ultraviolet and X-ray observations of comet Lovejoy (C/2011 W3) in the lower corona, *Astrophys. J.*, *768*, 161, doi:10.1088/0004-637X/768/2/161.

McKenna-Lawlor, S., E. Kirsch, D. Osullivan, A. Thompson, and K. Wenzel (1986), Energetic ions in the environment of comet Halley, *Nature*, *321*, 347–349, doi:10.1038/321347a0.

Mendis, D. A., H. L. F. Houpis, and M. L. Marconi (1985), The physics of comets, *Fund. Cosmic Phys.*, *10*, 1.

Mukai, T., W. Miyake, T. Terasawa, M. Kitayama, and K. Hirao (1986), Plasma observation by Suisei of solar-wind interaction with comet Halley, *Nature*, *321*, 299–303, doi:10.1038/321299a0.

Neubauer, F. M., et al. (1986), First results from the giotto magnetometer experiment at comet Halley, *Nature*, *321*, 352–355, doi:10.1038/321352a0.

Neugebauer, M., A. Lazarus, H. Balsiger, S. Fuselier, F. Neubauer, and H. Rosenbauer (1989), The velocity distributions of cometary protons picked up by the solar-wind, *J. Geophys. Res.*, *94*, 5227–5239, doi:10.1029/JA094iA05p05227.

Neugebauer, M., A. J. Coates, and F. M. Neubauer (1990), Comparison of picked-up protons and water group ions upstream of comet Halley's bow shock, *J. Geophys. Res.*, *95*, 18,745–18,753, doi:10.1029/JA095iA11p18745.

Neugebauer, M., et al. (2007), Encounter of the Ulysses spacecraft with the ion tail of comet McNaught, *Astrophys. J.*, *667*, 1262–1266, doi:10.1086/521019.

Ogilvie, K. W., M. A. Coplan, P. Bochsler, and J. Geiss (1986), Ion composition results during the International Cometary Explorer encounter with Giacobini-Zinner, *Science*, *232*, 374–377, doi:10.1126/science.232.4748.374.

Puhl-Quinn, P., and T. E. Cravens (1995), One-dimensional hybrid simulations of the diamagnetic cavity boundary region of comet halley, *J. Geophys. Res.*, *100*, 21,631–21,644, doi:10.1029/95JA01820.

Rème, H. (1991), Cometary plasma observations between the shock and the contact surface, in *Cometary Plasma Processes, AGU Geophys. Monogr. Ser.*, vol. *61*, edited by A. D. Johnstone, pp. 87–105, AGU, doi:10.1029/GM061p0087.

Richter, I., C. Koenders, K. Glassmeier, B. Tsurutani, and R. Goldstein (2011), Deep Space 1 at comet 19P/Borrelly: Magnetic field and plasma observations, *Planet. Space Sci.*, *59*, 691 – 698, doi:10.1016/j.pss.2011.02.001.

Rubin, M., K. C. Hansen, T. I. Gombosi, M. R. Combi, K. Altwegg, and H. Balsiger (2009), Ion composition and chemistry in the coma of Comet 1P/Halley-A comparison between Giotto's Ion Mass Spectrometer and our ion-chemical network, *Icarus*, *199*(2), 505–519, doi:10.1016/j.icarus.2008.10.009.

Rubin, M., V. M. Tenishev, M. R. Combi, K. C. Hansen, T. I. Gombosi, K. Altwegg, and H. Balsiger (2011), Monte Carlo modeling of neutral gas and dust in the coma of Comet 1P/ Halley, *Icarus*, *213*, 655–677, doi:10.1016/j.icarus.2011.04.006.

Rubin, M., K. C. Hansen, M. R. Combi, L. K. S. Daldorff, T. I. Gombosi, and V. M. Tenishev (2012), Kelvin-Helmholtz instabilities at the magnetic cavity boundary of comet 67P/ Churyumov-Gerasimenko, *J. Geophys. Res.*, *117*, A06,227, doi:10.1029/2011JA017300.

Rubin, M., M. R. Combi, L. K. S. Daldorff, T. I. Gombosi, K. C. Hansen, Y. Shou, V. M. Tenishev, G. Tóth, B. van der Holst, and K. Altwegg (2014), Comet 1P/Halley multifluid MHD model for the Giotto fly-by, *Astrophys. J.*, *781*, 86, doi:10.1088/0004-637X/781/2/86.

Sagdeev, R. Z., V. D. Shapiro, V. I. Shevchenko, and K. Szego (1986), MHD turbulence in the solar wind-comet interaction region, *Geophys. Res. Lett.*, *13*, 85–88, doi:10.1029/ GL013i002p00085.

Scarf, F. L., F. V. Coroniti, C. F. Kennel, D. A. Gurnett, W. Ip, and E. J. Smith (1986), Plasma wave observations at comet Giacobini-Zinner, *Science*, *232*, 377–381, doi:10.1126/ science.232.4748.377.

Schunk, R. W., and A. F. Nagy (2000), *Ionospheres: Physics, Plasma Physics, and Chemistry*, Cambridge Univ. Press, New York.

Smith, E. J., B. T. Tsurutani, J. A. Slavin, D. E. Jones, G. L. Siscoe, and D. A. Mendis (1986), International Cometary Explorer encounter with Giacobini-Zinner: Magnetic field observations, *Science*, *232*, 382–385, doi:10.1126/science.232.4748.382.

Snodgrass, C., Tubiana, C., Bramich, D. M., Meech, K., Boehnhardt, H., and Barrera, L. (2013), Beginning of activity in 67P/Churyumov-Gerasimenko and predictions for 2014–2015, *Astron. Astrophys.*, *557*, A33, doi:10.1051/0004-6361/201322020.

Somogyi, A., et al. (1986), 1st observations of energetic particles near comet Halley, *Nature*, *321*(6067), 285–288, doi:10.1038/321285a0.

Staines, K., A. Balogh, S. Cowley, R. Hynds, T. Yates, I. Richardson, T. Sanderson, K.-P. Wenzel, D. McComas, and B. Tsurutani (1991), Cometary water-group ions in the region surrounding comet Giacobini-Zinner: Distribution functions and bulk parameter estimates, *Planet. Space Sci.*, *39*, 479–506, doi:10.1016/0032-0633(91)90008-X.

Strobel, D. F. (2002), Aeronomic systems on planets, moons, and comets, in *Atmospheres in the Solar System: Comparative Aeronomy, Geophys. Monogr. Ser.*, vol. *130*, edited by M. Mendillo et al., pp. 7–22, AGU, Washington, D.C., doi:10.1029/130GM02.

Szegő, K., et al. (2000), Physics of mass loaded plasmas, Space Sci. *Rev.*, *94*, 429–671, doi:10.1023/A:1026568530975.

Tenishev, V., M. Combi, and B. Davidsson (2008), A global kinetic model for cometary comae: The evolution of the coma of the Rosetta target comet Churyumov-Gerasimenko throughout the mission, *Astrophys. J.*, *685*(1), 659–677.

Tsurutani, B. T. (1991), Comets: A laboratory for plasma waves and instabilities, in *Cometary Plasma Processes, AGU Geophys. Monogr. Ser.*, vol. *61*, edited by A. D. Johnstone, pp. 189–209, AGU, doi:10.1029/GM061p0189.

Whipple, F. L. (1950), The acceleration of Comet Encke, *Astrophys. J.*, *111*, 375.

Wiehle, S., U. Motschmann, N. Gortsas, K. Glassmeier, J. Mueller, and C. Koenders (2011), Simulation of cometary jets in interaction with the solar wind, *Adv. Space Res.*, *48*(6), 1108–1113, doi: 10.1016/j.asr.2011.05.024.

Young, D., et al. (2004), Solar wind interactions with comet 19P/ Borrelly, *Icarus*, *167*, 80–88, doi:10.1016/j.icarus.2003.09.011.

Zakharov, V. V., A. V. Rodionov, G. A. Lukianov, and J. F. Crifo (2009), Monte-Carlo and multifluid modelling of the circumnuclear dust coma II. Aspherical-homogeneous, and spherical-inhomogeneous nuclei, *Icarus*, *201*, 358–380, doi:10.1016/j.icarus.2008.12.022.

11

Heliotail

David J. McComas

11.1. INTRODUCTION

The solar wind expands outward from the Sun, inflating a bubble, our heliosphere, in the local interstellar medium (LISM), just as other stars and their stellar winds across the galaxy and beyond interact with their local interstellar environments, producing a broad menagerie of astrospheres (see Figure 11.1).

As the solar wind moves radially outward, it incorporates cold interstellar neutral atoms that become ionized and "picked up" by the solar wind's motional electric field: These pickup ion (PUIs) constitute ~15–20% of the solar wind flow by the time it reaches the termination shock at ~100 AU toward the heliosphere's nose (the distance should increase and may be roughly twice that far on the back side). Beyond the termination shock in the inner heliosheath (between the termination shock and heliopause), the flow must turn and ultimately move back in the tailward direction, forming some sort of heliotail (see Figure 11.2).

In addition to forming PUIs, some of the cold interstellar neutral atoms charge exchange with heliospheric ions, producing energetic neutral atoms or ENAs. ENAs are not affected by electromagnetic forces and thus continue traveling in the direction they were moving at the instant of charge exchange. Most of these ENAs radiate away from the heliosphere, but a fraction propagate inward and occasionally one can be detected by a spacecraft in the inner heliosphere, the IBEX.

The Interstellar Boundary Explorer (IBEX) [*McComas et al.*, 2009a] is the first mission to remotely observe the heliosphere's global interaction with the local interstellar medium. IBEX measures ENAs over the range of ~0.1–6 keV in 14 energy bins and, by using the spacecraft spin and spin axis repointing toward the Sun, produces energy-resolved all-sky maps from the inside looking out, every six months. In the first set of sky maps [*McComas et al.*, 2009b; *Fuselier et al.*, 2009; *Funsten et al.*, 2009a; *Schwadron et al.*, 2009], the IBEX team discovered a completely unpredicted "ribbon" of enhanced ENA emissions nearly encircling the heliosphere. The ribbon appears as a narrow, circular feature superposed on a "globally distributed flux" of smoothly varying ENA emissions from all directions in the sky. The direction of the interstellar magnetic field outside the heliosphere appears to align with the ribbon center, ordering the ribbon and showing that the interstellar magnetic forces play a key role in the heliopshere's interstellar interaction. See *McComas et al.* [2011a, 2013a] for recent summaries and reviews of the IBEX observations and possible source mechanisms for the ribbon.

The Sun and interstellar medium move with respect to each other, so the cold (few eV) neutral atoms from the interstellar medium are observed to flow into the heliosphere in the rest frame of the Sun. In addition to measuring ENAs from the outer heliosphere, IBEX also directly detects these low-energy interstellar neutrals that make it to the spacecraft without being ionized by the solar wind or solar photons. These observations recently indicated that the interstellar flow relative to the heliosphere (and thus the speed of the Sun through the interstellar medium), is ~23 km/s, which is slower and from a slightly different direction than previously thought [*Bzowski et al.*, 2012; *Möbius et al.*, 2012; *McComas et al.*, 2012a]. These observations, along with the growing preponderance of evidence from IBEX [e.g., *Schwadron et al.*,

Southwest Research Institute, San Antonio, Texas and Department of Physics and Astronomy, University of Texas at San Antonio, San Antonio, Texas

Magnetotails in the Solar System, Geophysical Monograph 207. First Edition.
Edited by Andreas Keiling, Caitríona M. Jackman, and Peter A. Delamere.

Figure 11.1 Astrospheres around three stars. Differences in the stellar winds and local interstellar environments can produce vastly different interactions; remarkably, Mira's molecular hydrogen "astrotail" (bottom) stretches back ~13 light years (the distance the star covers in ~30,000 years). Courtesy of NASA/ESA/JPL-Caltech/GSFC/SwRI.

2009, 2011] as well as the Voyagers [e.g., *Stone et al.*, 2008] and modeling results [e.g., *Opher et al.*, 2006; *Washimi et al.*, 2007; *Ratkiewicz and Grygorczuk*, 2008] that the interstellar magnetic field must be relatively strong (probably ~3 μG or more), led *McComas et al.* [2012a] to conclude that there is almost certainly no bow shock ahead of the heliosphere in the interstellar medium as had been previously believed (that is, a fast magnetosonic bow shock; a spatially limited slow-mode shock might still be possible [*Zieger et al.*, 2013]).

Using magnetohydrodynamic (MHD) simulations, *Zank et al.* [2013] demonstrated that Ly-α observations of the hydrogen wall are consistent with there being no bow shock for the likely external field of ~3 μG. Similarly, models and simulations have tried to explore the structure of the plasma in the downwind region [e.g., *Izmodenov et al.*, 2009; *Zank et al.*, 2009, and references therein] and even examined the possibility of variations over the solar cycle [e.g. *Sternal et al.*, 2008]; however, until 2013, no study had actually examined the heliotail observationally.

11.2. OBSERVATIONS OF THE HELIOTAIL

Very recently, *McComas et al.* [2013b] examined the heliotail for the first time. That study began as an effort to better quantify a smaller structure noted by *Schwadron et al.* [2011] which was clearly identifiable in the first-year sky maps of the globally distributed flux. That structure, dubbed the "offset heliotail," had appeared to be centered approximately between the downwind direction and the direction of the external magnetic field on the tailward side (we refer to this direction as "down field").

McComas et al. [2013b] used a much larger data set, incorporating the full first three years of IBEX observations (2009–2011) and including Compton-Getting (CG) corrections. This data set was validated and released to the community by *McComas et al.* [2012b] and included orbit-by-orbit corrections for variations in the survival probability of ENAs reaching 1 AU from the outer heliosphere as well as a correction for the time-variable cosmic ray background. The much larger and better corrected and validated data set was key to the first exploration of

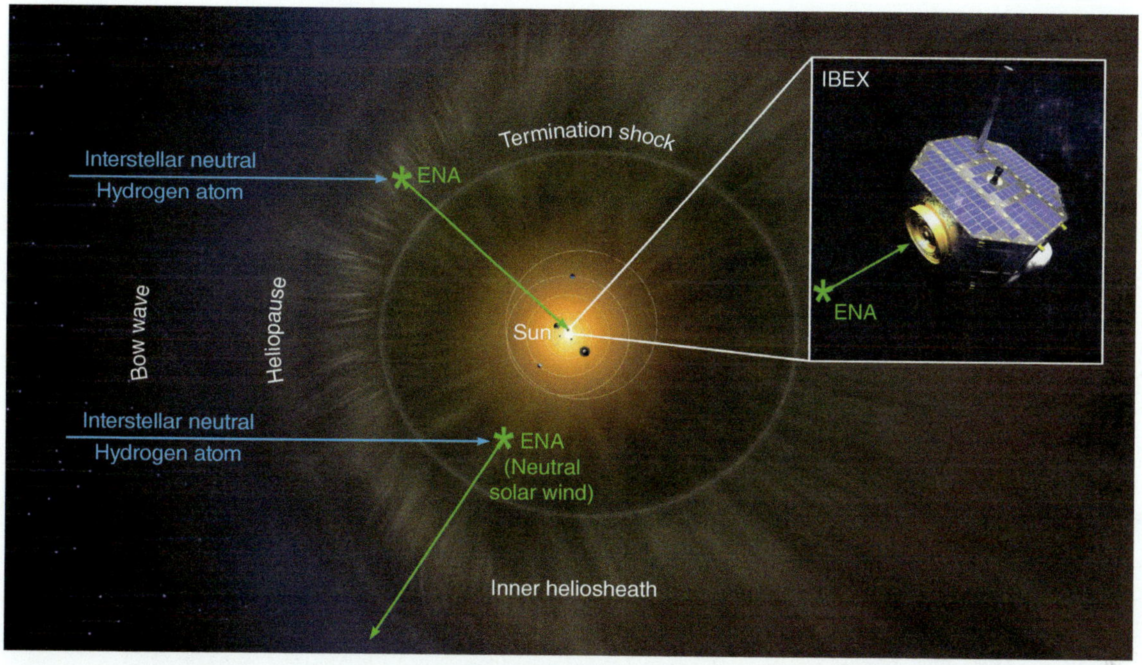

Figure 11.2 Our heliosphere. The region of space dominated by the Sun. Expanding solar wind inflates a bubble in the LISM. The outer boundary, the heliopause, separates solar from interstellar plasmas, while the termination shock slows and begins to divert the solar wind and incorporated PUIs toward the tail. The inner heliosheath lies between the termination shock and the heliopause. ENAs formed by charge exchange between cold interstellar neutrals and fast ions in this region can be directly observed by IBEX. From *McComas et al.* [2013b].

the heliotail, as the downwind direction is more sparsely sampled by IBEX owing to its inertially fixed orbit and often brighter foreground emissions of ENAs from Earth's magnetosphere which obscure that portion of the sky. We note that IBEX's orbit was modified in 2011 and its new long-term stable lunar synchronous orbit [*McComas et al.*, 2011b] slowly moves the location of the bright magnetospheric emissions, which will substantially improve the statistical accuracy of observations in the downwind direction over IBEX's extended mission.

With this much more complete and better corrected and validated set of observations, *McComas et al.* [2013b] were able to show that the small "offset heliotail" was simply one of two side "lobes" in a much larger heliotail structure spanning the entire downwind direction of the heliosphere. Figure 11.3 (top) shows a Mollweide projection centered directly on the downwind direction which color codes the spectral slope of the ENA flux from ~1 to 6 keV. In the tail, steeper spectral slopes at low/midlatitudes indicate ENAs produced by distributions of slower, lower-energy ions compared to those produced at higher latitudes by faster, hotter ion distributions.

In Figure 11.3, the port lobe is the previously identified "offset heliotail," while the starboard lobe is a new feature that appears very similar to but slightly smaller than the port lobe (note that to avoid confusion for various viewing geometries we use the nautical terms to identify the sides of the heliosphere as it moves through the interstellar medium). The broad tail structure is roughly centered on the downwind direction and is narrower in latitude near the middle, giving a two-lobed appearance. Furthermore, this two-lobed structure is slightly tilted (~10°) in the sense of the magnetic field orientation in the surrounding LISM.

In order to better quantify the two-lobed structure in the heliotail, *McComas et al.* [2013b] examined the centroid of each lobe as a function of energy. Over the top three energy bands of IBEX-Hi (central energies of 1.7, 2.7, and 4.3 keV [*Funsten et al.*, 2009b]), the center of both lobes moves toward the downwind direction with increasingly higher energies. For the port and starboard lobes, the centroids move a net angle of (+24°, +6°) and (−24°, −18°), respectively (ecliptic longitude, latitude), both toward the down-tail direction with increasing energy. This apparent motion may be due to the fact that higher-energy ENAs travel faster, have suffered fewer losses, and thus probably come from further down tail on average than lower-energy ENAs. In any case, however, the consistency of this behavior for both lobes strongly suggests that they are physically similar and related structures of the heliotail.

The summary of the heliotail observation from *McComas et al.* [2013b] is that "the heliotail includes a broad low to mid-latitude structure, which has an excess

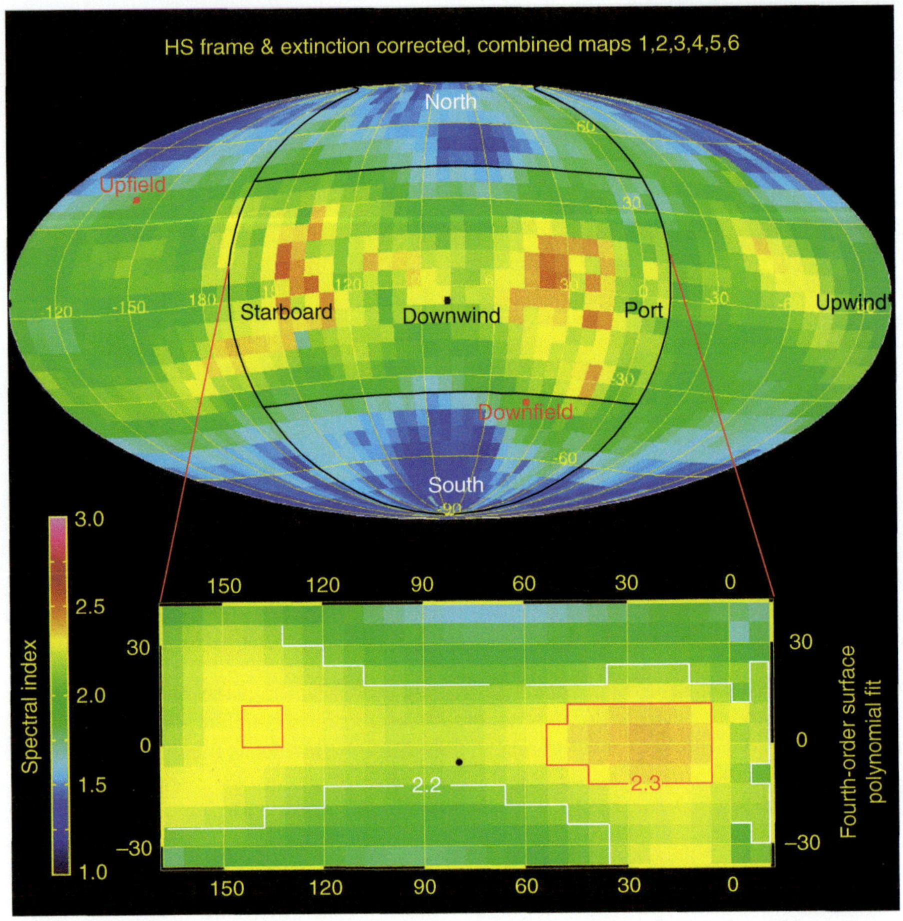

Figure 11.3 Mollweide projection of ENA spectral slopes centered on the downwind direction (top) and fourth-order polynomial smoothing for the region indicated (bottom). The heliotail comprises four regions covering the tailward side of the heliosphere (terminator indicated by black circle): port and starboard lobes at low-/midlatitudes of high spectral slope and north and south regions of lower spectral slope. From *McComas et al.* [2013b].

of lower energy ENAs (<1 keV) and a deficit at higher energies (>2 keV), producing a relatively steep power law energy spectrum ($\gamma = 2$–3) compared to the rest of the sky. This structure is large, spanning nearly 180° in longitude, but appears latitudinally thinner in the middle and at the edges in the IBEX ENA observations, and thus, we describe it as having two lobes. The lobes are roughly centered on the downwind direction, and the port lobe appears somewhat larger than the starboard one. Overall, the two lobe structure is tilted in the sense of the external magnetic fields, but at only a fraction of the angle of that field. Emissions from both lobes are centered closer to the downwind direction with increasing energies across the ENA energy bands centered at 1.7, 2.7, and 4.3 keV."

11.3. DISCUSSION

The spatial arrangement of the ENA spectral index from the heliotail clearly reflects and is consistent with the latitudinal ordering of the solar wind over much of the solar cycle around solar minimum, with slow solar wind at low/midlatitudes and fast wind at high latitudes [e.g., *McComas et al.*, 1998, *McComas et al.*, 2008]. Figure 11.4 provides a simple schematic diagram of the heliotail configuration. The observed steeper power law spectral distributions (Figure 11.4c) at low/mid-latitudes is associated with flows of the slower low/midlatitude solar wind around solar minimum (colored green in Figures 11.4a and 11.4b). In contrast, faster high-latitude solar wind around solar minimum (colored blue) is consistent with producing hotter ions beyond the termination shock and the less steep spectral slope (from more high-energy ions) of the observed ENAs at high latitudes.

The tail structure in Figure 11.3a is shown fading out with distance as charge exchange depletes the heliospheric ions (and produces the ENAs actually observed by IBEX). Because the fast wind travels further in a given time and has a longer "cooling distance" [*Schwadron et al.*, 2011] than the slow wind, it should extend further back than the slow wind, as indicated schematically in

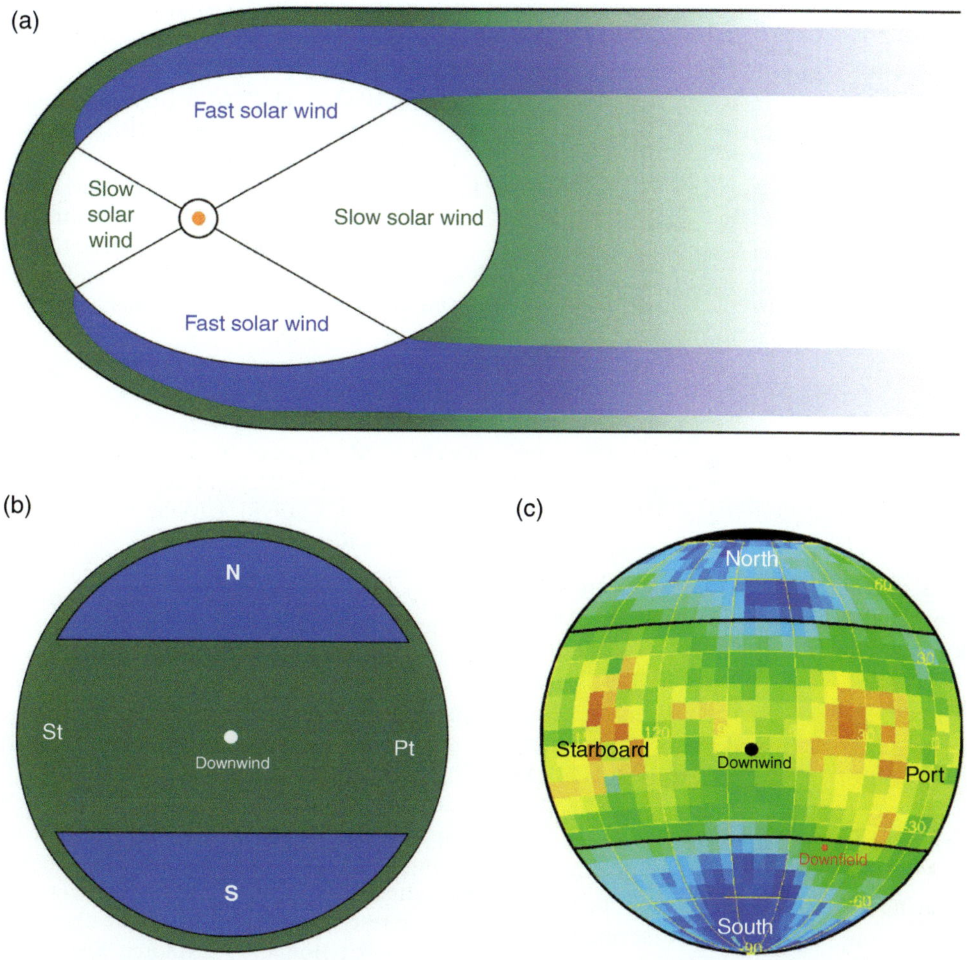

Figure 11.4 Schematic diagram showing (a) a meridional cut through the heliotail, (b) a cut as viewed looking down the tail, and (c) the IBEX spectral index observations from Figure 11.3 for comparison. Fast and slow solar wind regions are indicated by blue and green, respectively. From *McComas et al.* [2013b].

Figure 11.4a. The charge exchange process effectively "evaporates" the original heliospheric ions on scales of hundreds to maybe a thousand AU. Transverse pressure balance should squeeze the tail with increasing distance; however, multiple charge exchanges should also produce a longer coupled ion/neutral "wake" region behind the heliosphere (neither effect shown) extending much further back as ENAs also continue to reionize and ions reneutralize over the scales of thousands of AU.

Another key aspect of the heliotail observations from IBEX is the twisting of the slow solar wind plasma sheet, seen as a tilt between the lobes. The tilting is at a smaller angle but toward the direction of the external magnetic field. This shows that even over the relatively short line-of-sight observing distances for the ENAs (hundreds of AU), the external interstellar magnetic tension force is strong enough to substantially squeeze and begin rotating the heliotail, as shown schematically in Figure 11.5. It is interesting to note that the interplanetary magnetic field similarly squeezes and rotates the deep Earth's magnetotail (e.g., see Figure 4 and related discussion by *Sibeck et al.* [1985]).

So, how does the heliotail really form? It seems likely that the flow of inner heliosheath plasma simply diverts back around the sides and poles in the space between the termination shock and heliopause, as sketched in Figure 11.4. Because of the squeezing of the external field, this could produce a tail that is broader in longitude than it is in latitude, as observed. However, another interesting idea is based on the recent understanding that with no bow shock [*McComas et al.*, 2012a] the interaction and bow wave can extend many hundreds of AU ahead of the heliosphere, producing a significantly broader hydrogen wall [*Zank et al.*, 2013]. This "subsonic" sort of interaction could potentially allow "Alfven wings" [*Drell et al.*, 1965; *Neubauer*, 1980] to form around the heliosphere [*Kivelson and Jia*, 2012]. It is interesting to consider if the heliotail's two-lobed structure, with a steep

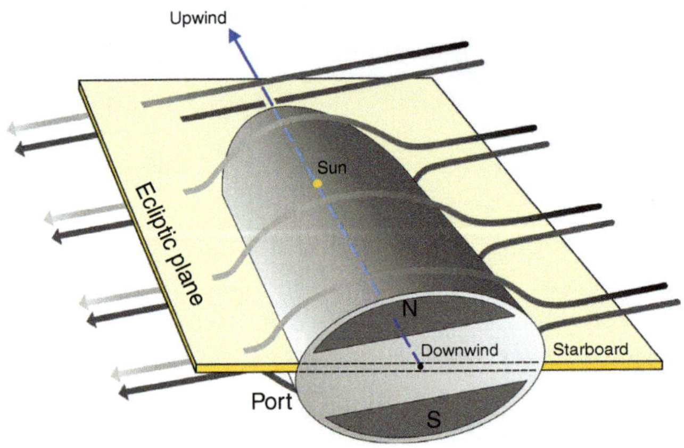

Figure 11.5 Schematic diagram of squeezing and twisting of the heliotail caused by magnetic tension forces from the external interstellar magnetic field. From *McComas et al.* [2013b].

spectral slope that is characteristic of a slow solar wind source, might represent the residual structure of Alfven wings folding back around the heliosphere into the heliotail.

Thus, IBEX's first three years of observations have enabled the first direct study of our heliosphere's heliotail. Theorists and modelers are urged to try to quantitatively reproduce the observed distributions of ENAs from the heliotail and account for all of the various features of the heliotail structure and interaction. As pointed out by *McComas et al.* [2013b], the explanation of the ecliptic latitude ordering of structure in the heliotail leads to a very specific prediction, that no (or at least much less) order will be observed when the less ordered solar wind from the current solar maximum [*McComas et al.*, 2013c] finally reaches the heliotail and produces ENAs that transit back into the near-Earth environment where we can observe them. We look forward to such future observations from the continuing IBEX extended mission and hopefully from the much higher sensitivity and resolution ENA measurements promised by a follow-on heliospheric imaging mission, such as the Interstellar Mapping Probe (IMAP) suggested by *McComas et al.* [2011a].

ACKNOWLEDGMENTS

This work was only possible because of efforts of all of the outstanding men and women who have made the IBEX mission such a wonderful success and was carried out as a part of the IBEX project with support from NASA's Explorer Program.

REFERENCES

Bzowski, M., et al. (2012), Neutral interstellar helium parameters based on IBEX-Lo observations and test particle calculations, *ApJS*, *198*, 12.

Drell, S. D., et al. (1965), Drag and propulsion of large satellites in the ionosphere: An Alfven propulsion engine in space, *J. Geophys. Res.*, *70*(13), 3131–3145.

Funsten, H. O., et al. (2009a), Structures and spectral variations of the outer heliosphere in IBEX energetic neutral atom maps, *Science*, *326*, 964.

Funsten, H. O., et al. (2009b), The Interstellar Boundary Explorer High Energy (IBEX-Hi) neutral atom imager, *Space Sci. Rev.*, *146*, 75–103, doi:10.1007/s11214-009-9504-y.

Fuselier, S. A., et al. (2009), Width and variation of the ENA flux ribbon observed by the Interstellar Boundary Explorer, *Science*, *326*, 962.

Izmodenov, V. V., et al. (2009), Kinetic-gasdynamic modeling of the heliospheric interface, *Space Sci. Rev.*, *146*, 329–351, doi:10.1007/s11214-009-9528-3.

Kivelson, M. G., and X. Jia (2012), 2012 Fall AGU Meet., Abstract SH14B-07, San Francisco, Calif., 3–7 December 2012.

McComas, D. J. (1998), Ulysses' return to the slow solar wind, *Geophys. Res. Lett.*, *25*, 1.

McComas, D. J., et al. (2008), Weaker solar wind from the polar coronal holes and the whole Sun, *Geophys. Res. Lett.*, *35*(18), L18103.

McComas, D. J., et al. (2009a), IBEX-Interstellar Boundary Explorer, *Space Sci. Rev.*, *146*, 11–33.

McComas, D. J., et al. (2009b), Global observations of the interstellar interaction from the Interstellar Boundary Explorer, *Science*, *326*, 959.

McComas, D. J., et al. (2011a), IBEX observations of heliospheric energetic neutral atoms: Current understanding and future directions, *Geophys. Res. Lett.*, *38*, L18101.

McComas, D. J., et al. (2011b), A new class of long-term stable lunar resonance orbits: Space weather applications and the Interstellar Boundary Explorer, *Space Weather*, *9*, S11002, doi:10.1029/2011SW000704.

McComas, D. J., et al. (2012a), The heliosphere's interstellar interaction: No bow shock, *Science*, doi:10.1126/science.1221054.

McComas, D. J., et al. (2012b), The first three years of IBEX observations and our evolving heliosphere, *Astrophys. J. Suppl.*, *203*, 1.

McComas, D. J., W. S. Lewis, and N. A. Schwadron (2013a) IBEX's enigmatic ribbon in the sky and its many possible sources, *Rev. Geophys.*, *52*, 118–155, doi:10.1002/2013RG000438.

McComas, D. J., et al. (2013b), The heliotail revealed by the Interstellar Boundary Explorer, *Astrophys. J.*, *771*, 77, doi:10.1088/0004-637X/771/2/77.

McComas, D. J., et al. (2013c), Weakest solar wind of the space age and the current "mini" solar maximum, *Astrophys. J.*, *779*, 2.

Möbius, E., et al. (2012), Interstellar gas flow parameters derived from Interstellar Boundary Explorer-Lo observations in 2009 and 2010: Analytical analysis, *Astrophys. J. Suppl.*, *198*, 11.

Neubauer, F. M. (1980), Nonlinear standing Alfven wave current system at Io: Theory, *J. Geophys. Res.*, *85*, A3, 1171–1178.

Opher, M., et al. (2006), The effects of a local interstellar magnetic field on Voyager 1 and 2 observations, *Astrophys. J.*, *640*, L71–L74.

Ratkiewicz, R., and J. Grygorczuk (2008), Orientation of the local interstellar magnetic field inferred from Voyagers' positions, *Geophys. Res. Lett.*, *35*, L23105.

Schwadron, N. A., et al. (2009), Comparison of Interstellar Boundary Explorer observations with 3D global heliospheric models, *Science*, *326*, 966.

Schwadron, N. A., et al. (2011), Separation of the Interstellar Boundary Explorer ribbon from globally distributed energetic neutral atom flux, *Astrophys. J.*, *731*, 56.

Sibeck, D. G., et al. (1985), The distant magnetotail's response to a strong interplanetary magnetic field By: Twisting, flattening, and field line bending, *J. Geophys. Res.*, *90*, A5, 4011–4019.

Stone, E. C., et al. (2008), An asymmetric solar wind termination shock, *Nature*, *454*, 71.

Sternal, O., H. Fichtner, and K. Scherer (2008), Calculation of the energetic neutral atom flux from a 3D time-dependent model heliosphere, *Astron. Astrophys.*, *477*, 365–371, doi:10.1051/0004-6361:20077715.

Washimi, H., et al. (2007), A forecast of the heliospheric termination-shock position by three-dimensional MHD simulations, *Astrophys. J.*, *670*(2), L139–L142.

Zank, G. P., et al. (2009), Physics of the solar wind-local interstellar medium interaction: Role of magnetic fields, *Space Sci. Rev.*, *146*, 295–327, doi 10.1007/s11214-009-9497-6.

Zank, G. P., et al. (2013), Heliospheric structure: The bow wave and the hydrogen wall, *Astrophys. J.*, *763*, 20.

Zieger, B., et al., (2013), A slow bow shock ahead of the heliosphere, *Geophys. Res. Lett.*, *40*, 2923–2928, doi:10.1002/grl.50576.

Section III: Specialized Topics

12

Formation of Magnetotails: Fast and Slow Rotators Compared

D. J. Southwood

12.1. INTRODUCTION

Magnetotails form behind planetary bodies wherever there is transfer of tangential stress between the solar wind and the planetary plasma environment. The outer atmosphere of any planet has an ionized component maintained usually by solar radiation. Where the planet is magnetized, the plasma environment extends far from the planet to form a magnetosphere. On the sunward side, the dayside, the limit is where the pressure of the planetary field in the interior is adequate to hold off and deflect the solar wind. Pressure acts normal to an interface, and if pressure were the only stress, the magnetosphere shape would be a bubble elongated in the antisolar direction.

Magnetospheres are not bubble shaped. They have extended tails reaching far behind the planet. For this to happen through interaction with the solar wind, tangential stress as well as normal stress must be important in the solar terrestrial interaction [see, e.g., *Parker*, 1958; *Piddington*, 1960; *Axford et al.*, 1965]. A theorist, Dungey, managed to jump the gun in announcing the actual experimental discovery of Earth's magnetic tail. *Dungey* [1963] used a private communication from L. J. Cahill not only to announce that 15 passes of the Explorer XII show evidence of a geomagnetic tail but also to point out that many aspects of Cahill's observations are consistent with his 1961 open magnetosphere model [*Dungey*, 1961]. The formal experimental announcements came a year or so later [*Cahill*, 1964; *Ness*, 1965].

The momentum transfer may be through viscous drag. However, for magnetized slow rotating planets, like Earth or Mercury, magnetic reconnection between solar and planetary field is the dominant process. Where there are strong internal sources of plasma, however, and strong rotation, outward motion of internal material can give rise to the magnetotail. Jupiter and Saturn are instances where rotation cannot be ignored. Indeed, for Saturn, there are models where reconnection with the solar wind is not important at all.

The chapter starts with Earth as an example of a magnetosphere where rotation is unimportant, a "slow rotator." The bulk of the chapter concerns the magnetospheres of Jupiter and Saturn. These are fast rotators. Saturn provides a particularly interesting example, at least for theorists, in that it is an aligned rotator with the rotation and magnetic symmetry axes parallel. However, not just the simplicity of the Saturn situation but also the extraordinary Cassini spacecraft data set from nine years in orbit makes it a very useful case study.

12.2. TERRESTRIAL TAIL FORMATION: SOLAR WIND RECONNECTION DOMINANT

Reconnection was first proposed for the solar wind interaction with Earth by *Dungey* [1961]. He put forward the idea of the open magnetosphere that has now effectively become the standard, as is illustrated in Figure 12.1.

In the Dungey cycle, the momentum transfer that leads to the tail formation is through magnetic viscosity. The magnetic pressure of the terrestrial field deflects the solar wind flow about the planet. However, reconnection near the nose allows a small normal component of the magnetic field to cross the magnetopause. As a result of the magnetohydrodynamic (MHD) frozen-in field theorem, there is a convection of both magnetic field and plasma as represented by the sequence of numbers in the sketch. After reconnection

The Blackett Laboratory, Imperial College, London, United Kingdom

Magnetotails in the Solar System, Geophysical Monograph 207. First Edition.
Edited by Andreas Keiling, Caitríona M. Jackman, and Peter A. Delamere.

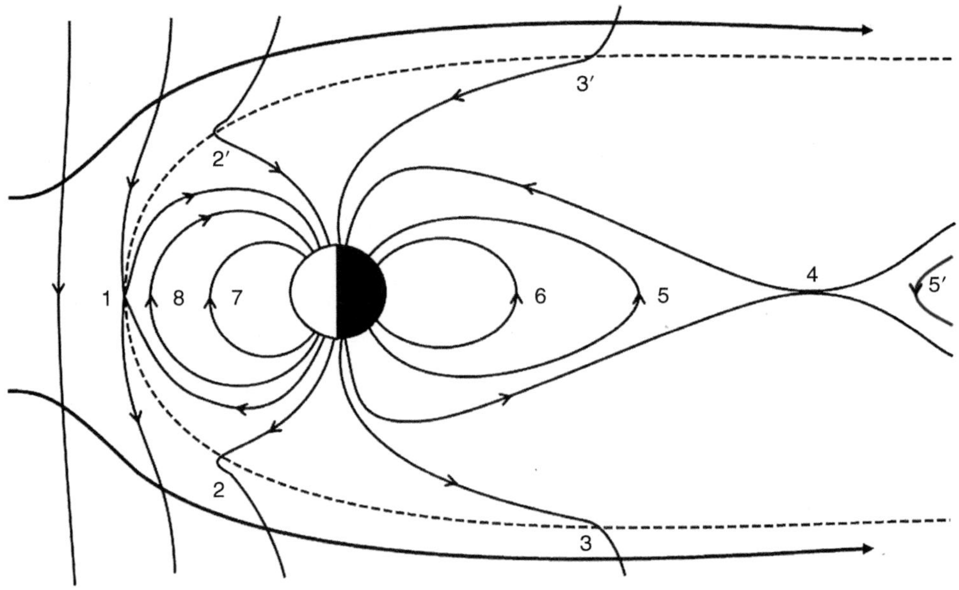

Figure 12.1 Schematic of Dungey open magnetosphere. The sketch of the noon-midnight meridian is shown for the singular, but symmetric, case where the external field is southward. The dashed curve represents the magnetopause. Two continuous curves represent the solar wind deflecting north or south around the magnetoshperic obstacle. The numbers represent the sequence of the Dungey cycles which begins with reconnection at point 1 linking the solar and terrestrial fields. The upstream bow shock has not been included in the figure.

at 1, terrestrial flux is dragged by the solar wind over the polar caps (2, 2′, 3, 3′). In this part of the flow the field is bent strongly and the associated magnetic stress must transfer momentum to magnetospheric plasma emerging from region 1 and to the collisional ionosphere at the field line feet in the polar caps. The stress not only induces an antisolar flow but at the same time results in the distortion of the magnetospheric cavity and the formation of a long (~1000R_E, Earth radii) tail. (The sketch is not to scale!) In mid tail reconnection occurs again (4); a return flow of closed field lines returns flux to the dayside (5, 6, 7, 8) to reconnect again. At the same time, the tail reconnection creates new flux (5′) that accelerates material down the tail and eventually into interplanetary space.

The open magnetosphere has proved very effective at explaining solar terrestrial coupling. In fact, the coupling is normally neither the symmetric simple case shown nor is it steady. Indeed, although in long-term average, dayside and nightside reconnection rates match, much of the time they are unbalanced as noticed first by *Aubry et al.* [1970] and later by *Freeman and Southwood* [1988] and then fully codified by *Cowley and Lockwood* [1992]. On the dayside patchy reconnection gives rise to bundles of attached flux, which are known as flux transfer events [*Russell and Elphic*, 1979]. Originally identified as tubes on which reconnection is no longer actually occurring, that is not necessarily so [see *Southwood et al.*, 1988].

In the unbalanced state where dayside and nightside reconnection rated do not match, the magnetic field in the polar cap stores and releases energy rather as an inductance does in an electrical circuit. However, in the steady state, the magnetic energy does not change and transfer of momentum and energy from the solar wind is taken up in two important magnetospheric regions, in the reconnection regions on the magnetopause and in the midtail current sheet, where, in both cases, plasma is accelerated as a by-product of the reconnection process. Lastly, energy is absorbed in the ionosphere at the feet of the magnetospheric field lines, where dissipative horizontal Pedersen currents absorb energy and exert a frictional drag on the flow induced by the solar wind.

A cross section of the magnetosphere-ionospheric part of the system is shown in Figure 12.2. In steady-state magnetohydrodynamics, field lines are electric equipotentials. Because reconnection has introduced a small normal magnetic component across the magnetopause, the solar wind voltage ΔV is imposed not only in the magnetosphere but also on the ionosphere. The effect on the postshock solar wind is that the wind is slowed in the regions where the interplanetary magnetic field is attached to Earth. The inertial currents in the wind that do the braking serve to close the current systems set up inside the terrestrial system. Only a cross section of the field-aligned current system coupling solar wind–magnetosphere–ionosphere is shown in the sketch in Figure 12.2. The system extends in the third dimension from the magnetosphere nose to the tail

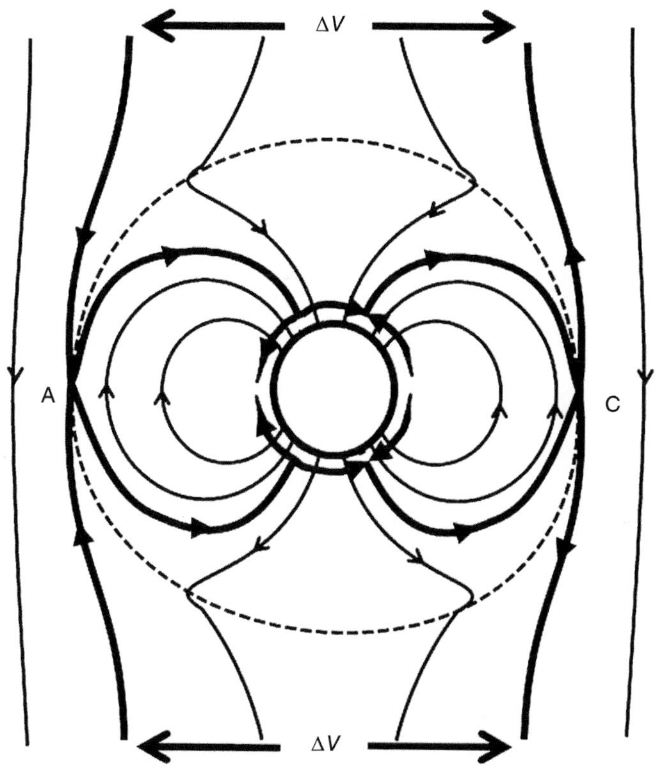

Figure 12.2 Dawn-dusk cross section of the basic terrestrial field-aligned current system. In steady-state MHD, field lines are regarded as equipotentials. The broad arrows represent the field-aligned current flow and horizontal ionospheric currents. Current flows into the ionosphere at dawn and exit at dusk to drive a dawn-dusk electric field. ΔV indicates the solar wind voltage imposed on the magnetosphere and ionosphere. Although only one plane is shown, the current system is three dimensional and extends from the magnetosphere nose to tall.

in space and from day to night in the ionosphere. A simple sketch of the ionospheric end of the global system is shown in Figure 12.3. The sketch is drawn for the northern hemisphere (field vertically down) and shows a double vortex flow system fixed in local time with flow away from the Sun over the polar cap. Field-aligned currents flow on dawn and dusk flanks of the polar cap ionosphere and connect to the magnetosphere and ultimately to the solar wind. Dissipative Pedersen currents flow horizontally in the ionosphere at right angles to the plasma flow. The Lorentz force associated with them opposes the flow. There is a return flow toward the Sun at lower latitudes.

12.3. TAIL FORMATION AND STRUCTURE IN FAST ROTATING MAGNETOSPHERE

In the idealized terrestrial magnetosphere, Earth's upper atmospheric rotational motion will be imposed on the ionosphere and thereby on the magnetosphere through frictional drag between the thermospheric neutral particles and the ions of the ionosphere. However, a quick analysis of time and length scales shows why rotation is not important. Earth rotates once every 24 h. The solar wind travelling at say 450 km/s is traveling at around $250R_E$/h, where R_E, the radius of Earth, equals 6.4×10^3 km. In one terrestrial rotation, the solar wind thus travels $\sim 6000R_E$. This much exceeds the critical length of the tail. The length of the tail was estimated by *Dungey* [1965] using a very simple argument. Knowing the speed of ionospheric motion (100s of meters per second^{-1}) and knowing the solar wind moved at 100s of kilometers per second^{-1} and that the ionospheric ends of flux tubes needed to move around $1R_E$ to move from day-side reconnection site to nightside, he concluded that the solar wind end of the connected flux tubes would move $\sim 1000R_E$ while connected. The solar wind moves that distance in ~ 4 hours during the convection cycle. In that time Earth (and the ionospheric feet of the field lines) would have rotated 60°, enough to provide a small twist in the field but no winding up.

Jupiter and Saturn make sharp contrasts with Earth. The length scales are larger by factors of 10 or 10^2 and rotation is 2.5 times faster. Rotation cannot be ignored. Moreover, in both Jupiter and Saturn's magnetosphere there are important internal sources of plasma. In such circumstances, rotation can be the most important process in the formation of the magnetotail. The standard picture of the magnetospheric flux and plasma cycle in a rotation-dominated system with internal plasma sources is the Vasyliunas cycle. Two illustrative sketches are shown in Figure 12.4. The situation envisaged is steady, and it is assumed that magnetic reconnection with the interplanetary field is unimportant. Reconnection nonetheless still plays a fundamental part.

The sketch shows two meridian cuts in the simplest case where the rotation and magnetic axes are aligned. Saturn at equinox (when the solar is incident normal to the rotation axis) has a configuration very like this. Magnetic shells at different points in the cycle are numbered in the figure. Material from an internal source is ionized deep inside the magnetosphere and so there must be a net outward mass transport. In the dipolar field regions (1) the material is transported radially by interchange motion [*Hill*, 1976; *Southwood and Kivelson*, 1989]. The net transport of plasma outward has little effect on the magnetic field. Nonetheless, the centrifugal effect of the plasma rotation about the planet will not only bend the field outward everywhere but also tend to confine plasma to the more distant equatorial part of the flux tubes.

Beyond the point where the equatorial magnetic field pressure is comparable to the energy in the plasma rotation, the field lines start to bow out strongly at the equator in response to centrifugal stresses 2 and 3. The shell is

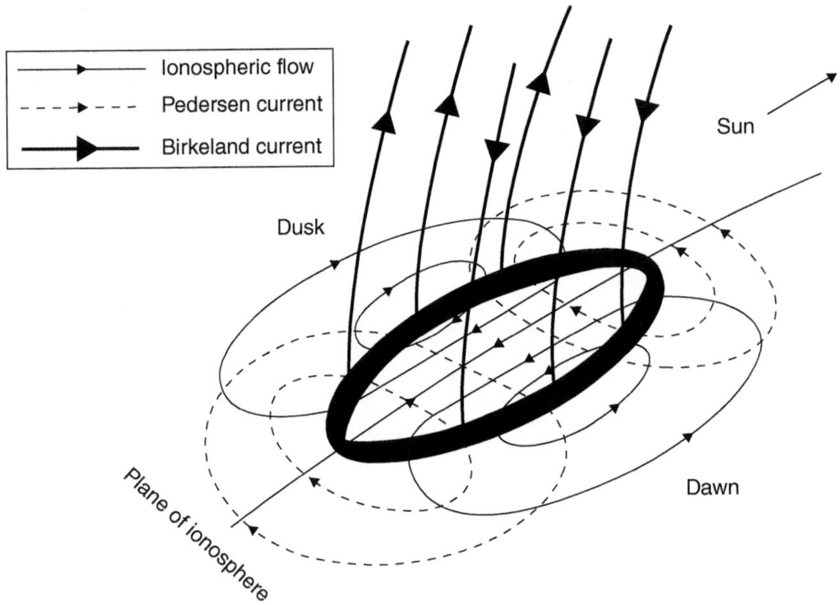

Figure 12.3 Basic double-vortex flow system of the terrestrial Dungey cycle induced by reconnection on the dayside magnetopause and in the tail. The thick dark oval corresponds to the location of where field-aligned currents enter or leave the ionosphere. From *Southwood* [1985].

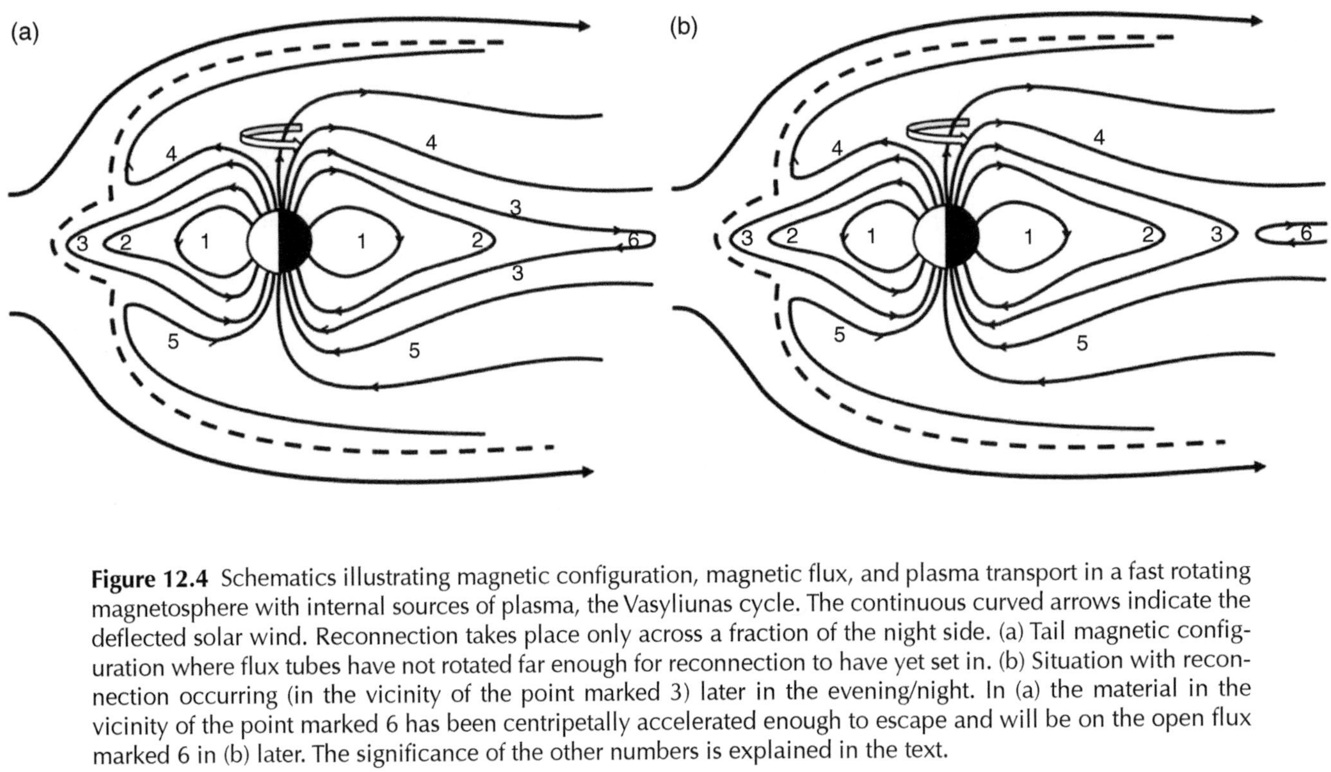

Figure 12.4 Schematics illustrating magnetic configuration, magnetic flux, and plasma transport in a fast rotating magnetosphere with internal sources of plasma, the Vasyliunas cycle. The continuous curved arrows indicate the deflected solar wind. Reconnection takes place only across a fraction of the night side. (a) Tail magnetic configuration where flux tubes have not rotated far enough for reconnection to have yet set in. (b) Situation with reconnection occurring (in the vicinity of the point marked 3) later in the evening/night. In (a) the material in the vicinity of the point marked 6 has been centripetally accelerated enough to escape and will be on the open flux marked 6 in (b) later. The significance of the other numbers is explained in the text.

more distorted by night than by day because there is no pressure at infinity at night. For this reason, a tail forms. As one moves around the planet from noon to afternoon to night, the field will become increasingly distended by the centrifugal effects. This can eventually lead to the field no longer being able to contain the plasma at large enough distance and reconnection will occur. Indeed, as long as there is an active internal source of plasma, plasma has to be lost from the system and the tail, where the pressure at infinity is very small, is the most likely escape route. As long as the field is connected to the planet, the down-tail streaming is driven by the centripetal acceleration transmitted by the field from the rotating ionosphere.

Returning to Figure 12.4, one sees that in Figure 12.4a the shell is numbered 3 by day and has two numbers (3 and 6) by night. The reason for this labeling is seen in Figure 12.4b and is a result of the escape process. Recalling that material is rotating from Figure 12.4a to 12.4b, in sketch 12.4b the continuous field lines of the shell at earlier local time are now on two separate field lines. At this point, reconnection has occurred. The field line marked 3 is closed but some of the plasma has been lost. The lower density allows it to return to rotation about the planet toward the morning side. That plasma is labeled 6 and is now on a flux tube disconnected from the planet entirely. It is important to note that, as long as the internal magnetospheric source is active, such a release must happen from time to time but it need not happen steadily or even occur on every rotation. Indeed, even the term *rotation* is ambiguous for the flux tubes of the Vasyliunas cycle as the frictional transfer of momentum from the ionosphere means there is large slippage between magnetosphere and ionosphere, as we will discuss below.

Sketching the expected magnetic configuration as a quasi-steady situation is not hard. However, sketching a steady flow is more difficult. Authors like *Kivelson and Southwood* [2005] have indeed suggested a nonsteady breaking off of material downstream into the tail is more likely in the Jupiter context. At Jupiter there is even evidence of the field-breaking process occurring and material escaping sporadically in the morning [*Southwood et al.*, 1993, *Haynes et al.*, 1994, *Leamon et al.*, 1995]. In Figure 12.5, a notional steady flow is shown based on the familiar sketch [*Vasyliunas,* 1983]. On the dusk/evening flanks there is a flow straight down the tail as plasma escapes directly. The flux tubes on which this plasma finds itself are going eventually to break, but as long as the tube is connected to the planet, there is a component of azimuthal motion. It follows that at some point across the tail a reconnection X line appears where the field breaks (change from case A to B in Figure 12.4). There is also an O line as marked. The flow arrows are shaded to indicate differential loading. The plot shows how the outer magnetosphere flow can act as a conveyor to release the ionized material transported from the inner magnetosphere.

Viewed in the frame of the rotating plasma, reconnection will occur in the tail as a result of centrifugal forces distending the equatorial field outward in a ballooning process [*Kivelson and Southwood*, 2005]. In Figure 12.5 the idea is given that interchange proceeds in a "high-*m*" mode (short east-west scale length). *Kivelson and Southwood* [2005] suggested that for Jupiter the resulting

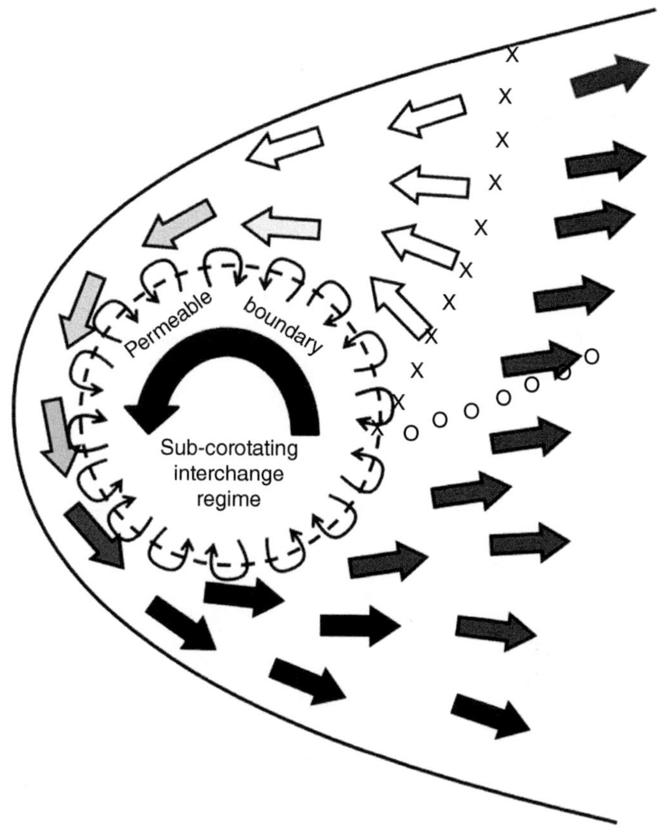

Figure 12.5 Equatorial plane view of a steady Vasyliunas cycle. The X's mark a reconnection X line, the O's mark an O line. Along the X line flux tubes break and so depleted tubes emerge and move toward the dayside. In the interior, where the dipole field is dominant, the material is sub-corotating (at around 50% of the corotation speed). In this region, there is radial mass transport due to the presence of a source deep inside (e.g., in the case of Saturn, the Enceladus torus). The arrows are designed to give a sense of the loading of flux tubes. Along the X line reconnection causes the flux tubes to break. The equatorial part moves downtail eventually releasing material from the interior of the magnetosphere into the interplanetary medium. The off-equatorial less dense parts of the flux tube close and move sunward. These newly closed depleted tubes will undergo reconnection again on the nightside and so in steady state must refill as they pass through the dayside from whatever process transports material outward in the inner region.

nightside instability could be also high *m* and likely to make the tail a mix of inward and outward moving flux.

A further complication of the Jovian system is the large azimuthal asymmetry of the internal field. This presents an additional complication to any analysis of the tail formation as the internal field asymmetry causes an inherent rocking of the system in response to the planetary rotation. In contrast, Saturn appears to present an almost impossibly simple situation. The internal field of Saturn is to this point in time regarded as symmetric about the planetary rotation axis [*Burton et al.*, 2009], Moreover, the Cassini space mission has provided an enormous wealth of data with good coverage in radial distance (beyond $4R_S$) with a variety of orbital inclinations in the nine years of operations. Accordingly, we now concentrate on describing the Saturn system.

Nonetheless, the Saturn magnetosphere is not as simple as one might have predicted. In particular, in the center of the magnetotail, the plasma sheet where material from Enceladus would be either trapped or escaping down tail is probably generally understood at a basic level. However, the relationship between the plasma on the flux tubes in the polar cap making up the low-density flux tubes of the tail and the solar wind is not yet at all clear.

12.4. SATURN'S POLAR CAP AND SOLAR WIND: DUNGEY CYCLE

Does one need to invoke the presence of a solar wind associated circulation, a Dungey cycle, at invariant latitudes poleward of the Vasyliunas cycle? In practice, it does seem likely that solar wind material would penetrate in some way on the dayside magnetopause by direct entry through the cusp, if not by reconnection. There is a strong reason to feel that there should be some solar wind interaction and stress transfer.

In the absence of some form of stress transfer such as magnetic reconnection with the solar wind on the dayside, i.e., in the pure Vasyliunas cycle outlined in the previous section, the polar cap takes a purely passive part in tail formation. Although there must be magnetic reconnection in the central tail to release the internal material down tail, the permanently open flux tubes located in the polar cap, i.e., at invariant latitudes above which the field lines are marked 4 (northern hemisphere) and 5 (southern hemisphere), transport neither plasma nor magnetic flux. The tubes would be sparsely populated and, far from the planet, the tubes would bend antisunward to form the two lobes of the tail on either side of the plasma sheet formed by material transported from the inner magnetosphere in the equatorial plane.

For our discussion here, the precise mechanism of momentum transfer is not important; we shall use the term Dungey cycle [*Cowley et al.*, 2004] for any solar wind driven momentum transfer and resultant circulation on the polar cap tubes. In our view the fundamental difference between the likely Dungey cycle in a fast rotator like Jupiter or, in particular, Saturn is that rotation will be important. In early discussions of such magnetospheres, it was even believed that rotation might be enough to generate a continuous outward flow of plasma from the planetary atmosphere accelerated centripetally by the ionosphere along the field (see, for example, the discussion of the early Pioneer spacecraft results by *Kennel and Coroniti* [1975]). Something less extreme is likely on the flux tubes immediately beyond the outermost tubes of the Vasyliunas cycle near what is the boundary of the open and closed flux. We propose that there is centripetal acceleration of plasma leading to outflow in the tail in the evening sector along these tubes. The unusual feature of the proposal is that we suggest that most of the time the material is likely not of planetary origin but to be of solar wind origin. The outflow down tail is along the field. Accordingly, if the flow starts on closed field lines, eventually flux tubes must break, and because magnetic flux needs to be conserved, such an outflow needs to be balanced by a return flow of flux, constituting a Dungey cycle.

However, if solar wind access is primary through direct entry through the cusp on an open field, return flow is not necessary. Accordingly, reconnection between the solar wind field and the magnetosphere is not an absolute requirement in what follows. The key point is that rotation driven from the ionosphere below dominates the motion on the flux tubes immediately beyond the Vasyliunas cycle. Unlike the Vasyliunas cycle where, as we will see later, there is large observed slippage between the ionosphere and magnetosphere, we suggest that the low mass density of material on these tubes means that the polar ionospheres can sustain rotation of the magnetospheric flux tubes with a period near 10.7 h. As a result, we suggest also that the 10.7 h magnetic signals seen everywhere in the Saturn system [*Andrews et al.*, 2012] originate on the tubes of the Dungey cycle. Indeed, it could even be argued that the 10.7 h periodicities are the strongest evidence for the existence of a Dungey cycle.

In two papers that perhaps have not received enough attention, *Gurnett et al.* [2010, 2011] provide evidence of where the Vasyliunas cycle ends and the Dungey cycle begins in the Saturn magnetosphere. Using the Cassini Radio and Plasma Wave Science (RPWS) instrument and other Cassini spacecraft data, they report detection of a steep density drop on flux tubes at invariant latitudes near $\Lambda = 70°–74°$. *Gurnett et al.* [2010] refer to this boundary as a high-latitude plasmapause. The term plasmapause is imported from the terrestrial context. There the plasmapause marks the boundary or streamline in steady flow inside which the flux tubes are permanently closed and just rotate about Earth and outside which flux tubes open once per cycle. In the closed field region, the latter flux tubes follow streamlines coming in from the tail on the nightside and go to the magnetopause on the dayside. Through the agent of reconnection on the magnetopause and in the tail, the flow system closes by a flow over the polar cap (cf. section 12.2.). In other words the plasmapause at Earth is the inner boundary of the flux tubes participating in the Dungey cycle. *Gurnett et al.* [2010, 2011] make it clear that they have the same notion for Saturn.

Cowley et al. [2004] place the Dungey reconnection line as an extension near dawn of the Vasyliunas line across the tail [see Figure 2, *Cowley et al.*, 2004], but they indicate there is no basic requirement for this to be so. *Jia et al.* [2012a] have published a global Saturn magnetosphere simulation including both Vasyliunas and Dungey cycles drawing a similar conclusion that the nightside Dungey cycle reconnection is located toward dawn. This is not an important issue in what follows.

The main point is that there is a circulation of a much lower mass density plasma on flux tubes at invariant latitudes beyond that of the Vasyliunas cycle. We suggest that some solar wind material is present on the flux tubes outside the Gurnett boundary and that circulation of plasma and flux occurs there. The most likely access point for solar wind to Saturn's high-latitude magnetosphere is on the dayside, in or near the cusp (in between shells 3 and 4 of Figures 12.4a and 12.4b).

The corotation speed in the outer dayside cusp is in excess of the local magnetosheath speed. Taking a spatial scale of $20R_S$ to $25R_S$ and a rotation period of ~11 h, one sees that the speed of material moving at the corotation speed is of order ~$12R_S$/h in the vicinity of the magnetopause. In contrast, the magnetosheath might be moving at ~250 km/s ≈ $5R_S$/h. Thermal proton speeds are similar or smaller. Rotation will dominate the motion of material entering at high latitude in the vicinity of the cusp once it is on Saturn field lines. The rotation is fast. Independent of the fact that material entering from the solar wind will have antisolar momentum before entry, in 3 h corotation moves particles from noon to beyond dusk. Solar wind protons entering near the cusp far off the equator are not likely to complete even a quarter bounce along the field as they move from noon to dusk and on to the nightside. Nonetheless, once in the tail, the primary velocity will be along the field. Throughout this process the particles will be subject to increasing centripetal acceleration along the field and the field itself becomes more and more radial. By the time they enter the night side, there should be an outflow down tail much as discussed earlier for the Vasyliunas cycle at lower latitudes. A flow of solar wind material then additionally accelerated will not be ultimately able to be contained by the outflowing plasma. The flux tube will break, equatorial material will escape, and a residual closed tube containing much reduced amounts of solar wind plasma will be dragged around to the morning side. The flux tubes will return to rotation and in turn reconnect and pick up new solar wind material on the dayside. In its simplest form, then, the Dungey cycle would all take place on tubes just beyond the outer limit of the Vasyliūnas cycle.

The inherent speed of rotation compared to the likely speed of plasma motion along the field means that northern and southern ionospheres act independently as flywheels on material entering off-equator near the northern and southern cusps, respectively. It has been known for some time that the Saturnian radio signals [*Gurnett et al.*, 2009] pulsed at slightly different periods in the northern and southern hemispheres. Indeed, *Gurnett et al.* [2011] report that the density gradient, the plasmapause, which they have identified oscillates with the southern radio period in the south and northern in the north. The independence of northern and southern flow systems as described here is probably a substantial part of the explanation.

12.5. CASSINI MAGNETOMETER OBSERVATIONS AT HIGH INVARIANT LATITUDE

Before returning to the question of 10.7 h modulation, we look at Cassini spacecraft magnetometer data from invariant latitudes corresponding to what we have now identified as those of the Dungey cycle.

Cassini does get to high invariant latitude, Λ, well beyond the location of the Gurnett boundary at around Λ ≈ 70°–74°. Several sequences of highly inclined orbits have allowed the spacecraft to penetrate well above Λ ~ 80° while relatively close to the planet ($12R_S$ to $20R_S$). For comparison, the cusp (the dimple evident on the dayside magnetopause in between flux tubes 3 and 4 in Figure 12.4) was directly encountered several times by Cassini in early 2007 at Λ~78° near noon. The noon meridian should normally mark the highest latitude reached by the cusp and thus by extension the outer limit of the Vasyliunas cycle.

Plots of a sequence of five identical orbits are shown in Figure 12.6a. The left-hand panel shows the orbits (each orbital overlays the others) in the (*R*, *Z*) plane, where *Z* is along the rotation axis and *R* is the axial distance from it. The model planetary dipole field lines are shown. The right-hand side shows a polar plot of the orbit based on projection of the magnetic shells based on the planetary dipole. The five short-period (~10 day) high-inclination orbits come from between 21 December 2008 and 7 February 2009. In Figure 12.6b the field observed on the orbits is compared with the model planetary dipole. The two figures show that it is not unreasonable to use the dipole parameters as a guide to the mapping of planetary flux. The departures of the actual field from the dipole are largely due to two effects. The first is the depression of the B_θ component near the nightside equatorial passages due to the planetary ring current. The equatorial passes where the spacecraft is below Λ = 78° both day and night (marked by vertical lines on the plots in Figure 12.6b). The second perturbation is due to the tail current sheet which produces a north-south symmetric distension in the radial field that is minimum near noon. It contributes to a net increase in the observed field over

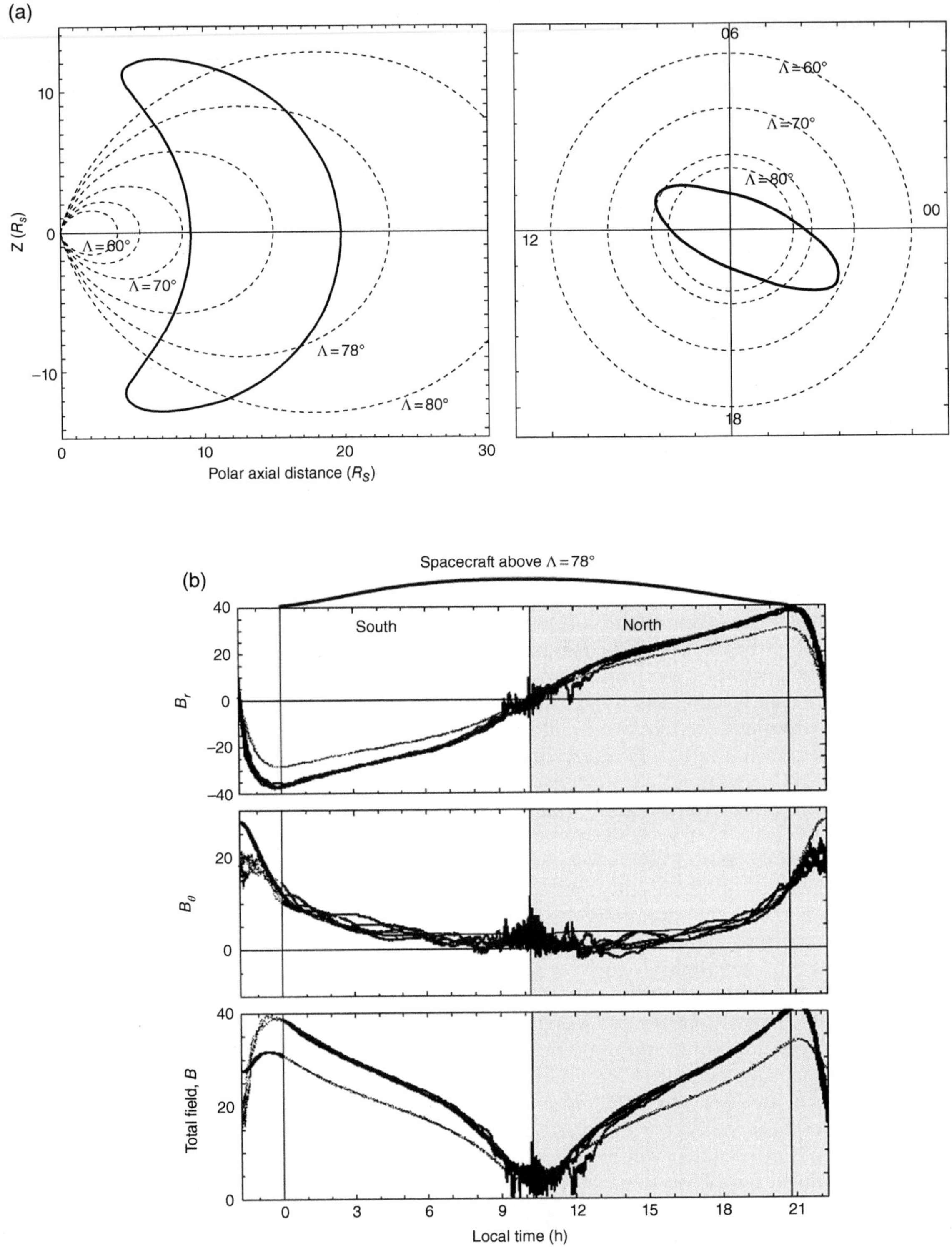

Figure 12.6 (a) Plots of five Cassini high-inclination orbits between 21 December 2008 and 7 February 2009 projected into a generalized meridian view and a polar view (looking down from the north). The relationship of the orbits to the model planetary dipole field is indicated in each panel. In the left-hand panel, field lines are shown for Λ = 60°, 65°, 70°, 75°, 78°, 80°. The segments of the orbits where the spacecraft is above invariant latitude, Λ = 78°, are shown with thicker traces in both panels. The right-hand panel is a notational polar projection of the orbits using the planetary model dipole. Circles representing magnetic shells at Λ = 60°, 70°, 78°, 80° are shown (b) Comparison of the model planetary dipole field and the measured field on the five Cassini high-inclination orbits between 21 December 2008 and 7 February 2009. Fainter traces show the model field and the observed field (in nT) plotted for meridional components B_r, $B\Theta$, $|B|$, respectively. The abscissa for each panel is spacecraft local time. The orbits are effectively identical. The spacecraft is in the polar cap field on each orbit from around midnight through to just before 2100 LT on each orbit. The fainter trace shows the model field. The discrepancy with the dipole field is less than 25% throughout.

the model value and is attributed to the centrifugal effect [*Arridge et al.*, 2008]. The external currents associated with the compression of the magnetic field at the magnetopause appear to not strongly affect the measured field on the orbits in question. Although the model invariant latitude maps well near noon, the field at the spacecraft will map to increasingly lower latitude than the indicated invariant latitude as one moves toward midnight where the discrepancy is maximum.

In Figure 12.7 we use invariant latitude as determined from the model field to order the data from the same orbits as used for Figure 12.6. The lower panel indicates spacecraft invariant latitude Λ, and in the upper panel, the magnetometer data traces of 1 min resolution of the azimuthal component from five orbits are overlaid and plotted against local time. The 10.7 h field oscillations that pervade the Saturn magnetosphere have not been removed from the data and are responsible for the shorter scale fluctuations that are visible. Ignoring the short-scale fluctuations, one sees that the azimuthal perturbation of the dipole field at invariant latitudes above ~78° varies systematically with spacecraft local time. The situation is made clearer when account is taken of the shading that shows when the spacecraft is in the southern or northern hemisphere. The local time dependence is folded with a hemispheric effect where the sign of the perturbation changes as the spacecraft moves from the northern hemisphere to south and back. The hemisphere change shows the perturbation is symmetrical about the equator. Based on this deduction, we have marked the plot "lag" and "lead" to show where the perturbation implies that the magnetic field is bent forward (i.e., in the direction of corotation) or backward out of the local dipole meridian.

The bending forward and backward with local time might be due to two effects and is probably due to a combination. First, the external pressure of the solar wind encloses the magnetospheric field and will naturally bend the planetary field out of the dipole meridians. Such an effect would be symmetrical with respect to the Sun-planet line. This is seen on the dayside where B_ϕ switches sign on the noon meridian. The second way to produce field bending out of the meridian is associated with slippage between the ionospheric part of the tube and the magnetospheric part. The nightside part of the perturbation appears to be of this nature. Lead turns to lag in midevening near 2100–2200 LT, 2–3 h before midnight

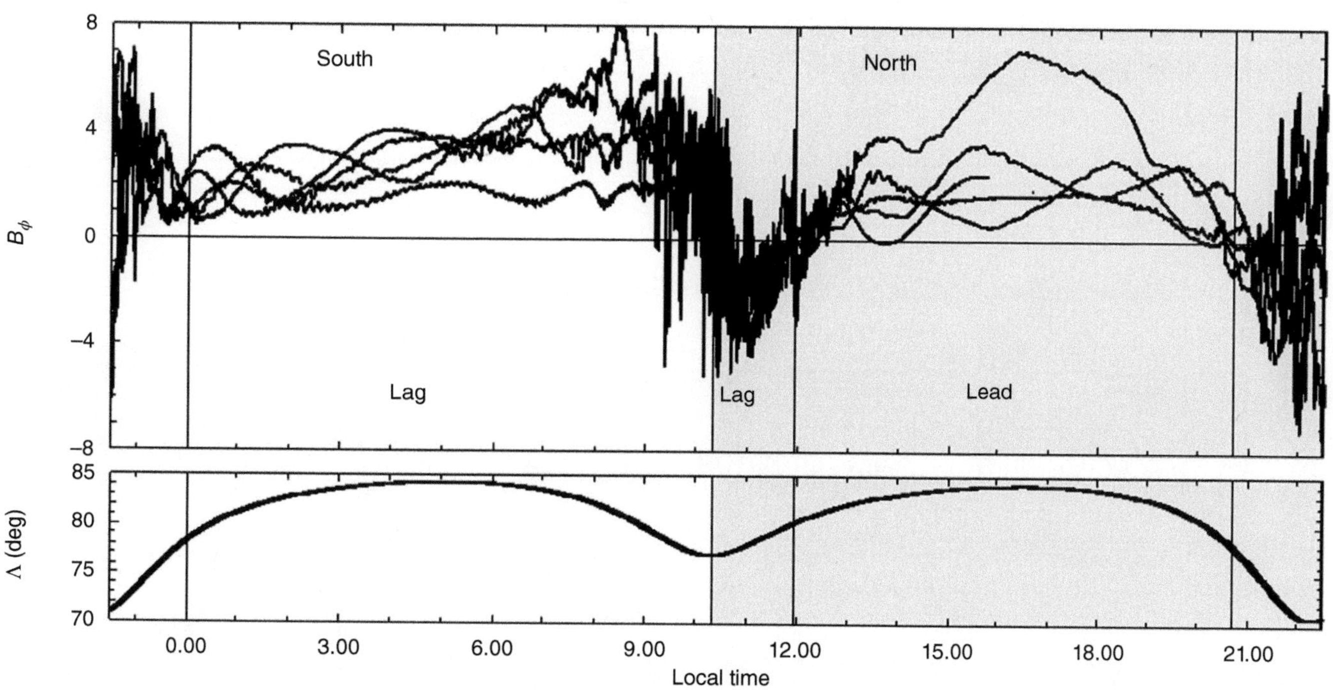

Figure 12.7 Imperial College Cassini magnetometer data. The local time variation of azimuthal field measured at high invariant latitude is shown for five orbits between 21 December 2008 and 7 February 2009. Upper panel: azimuthal magnetic field component B_ϕ (nT) versus local time (h). Lower panel: spacecraft invariant latitude (deg) versus local time (h). Vertical dashed lines mark where the magnetic configuration changes from lead (magnetosphere with respect to the ionosphere) to lag. Vertical dashed lines mark the switch between lead and lag as explained in the text. The background shading indicates when the spacecraft was in the north or the south. The labels "lag" and "lead" indicate where the azimuthal field perturbation implies that the equatorial (magnetospheric) portion of the flux tube is apparently lagging or leading the rotation of the ionospheric portion.

rather than in the midnight meridian. This shows the bending is not a purely a solar wind pressure effect. The explanation is that here one is in the potential outflow region. Once material has come past dusk, there is no pressure at large distance. If the field strength is not adequate to restrain the rotating plasma, there can be a free expansion down tail (cf. Figure 12.5). However, recall the spacecraft is beyond the invariant latitudes of the flux tubes of the Vasyliunas cycle. The plasma will have been accelerated by the ionospheric flywheel as it moves around from noon. Nonetheless, the signature is of the tubes being pulled out radially in early evening. Then further into the evening sector the reversal of B_ϕ well before midnight would occur because of the discrepancy between the tailward distension due to the outflow on the magnetospheric parts of the tubes and the continued azimuthal motion of the feet in the ionosphere.

If the tubes were originally closed, eventually the down-tail motion of the plasma will cause the field to break. The newly closed flux tubes, containing any part of the solar wind plasma that has not escaped, will continue toward around dawn.

The most extreme B_ϕ perturbations in Figure 12.7 occur around midnight to 0200 LT where the spacecraft orbit dips to its lowest invariant latitudes around Λ~70°. These perturbations can be identified as the field-aligned currents associated with the Gurnett plasmapause boundary and identified as the outer limit of the Vasyliunas cycle. On different orbits the local time of the encounters with the B_ϕ perturbation varies. The invariant latitude of the perturbations is not only where the Gurnett boundary is but also where *Southwood and Kivelson* [2007, 2009] and *Jia et al.* [2012b] located the field-aligned currents associated with 10.7 h cam magnetic perturbations in the dipole region, a point we shall return to subsequently.

12.6. MODULATION OF SATURN TAIL

The data in Figure 12.7 give a picture of what happens near the boundary between open and closed field lines (i.e., between shells 3 and 4 in Figure 12.4). Next we consider what happens in the equatorial plane in the evening local time zone. Figure 12.8 shows Cassini magnetometer data from a pass in the equatorial plane of Saturn's

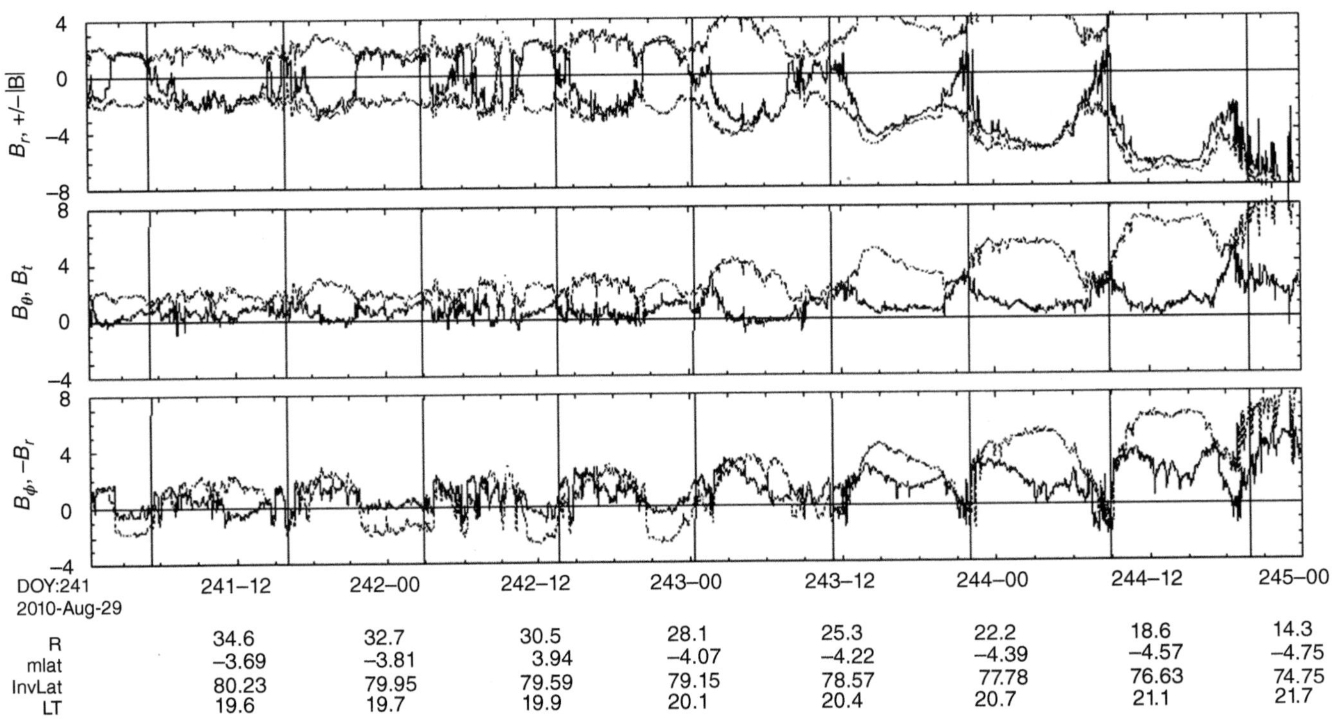

Figure 12.8 Imperial College magnetometer data from the Cassini spacecraft from a time near equinox, 29 August to 2 September 2010. The spacecraft remains close to the equatorial plane in early to midevening (19.6–21.7 LT) and covers a radial distance from 34.6 to 14.3 R_s (Saturn radii). The abscissa is day of 2010:h of day. Ancillary data giving the spacecraft position (radial distance, RS; dipole magnetic latitude, deg; invariant latitude, deg; local time, h) are shown beneath each time mark. The ordinate shows the components (nT) in Saturn polar coordinates centerd on the planetary axis. The components displayed are explained in the text. Vertical markers mark cycles of a rotating signal with 642.3 min period, the mean of northern and southern 10.7 h magnetic oscillations at this time.

magnetotail. Data shown are from 2010 around the equinox, when the Cassini spacecraft made a series of near-equatorial orbits with apo-apsis around dusk, which is normally in the solar wind. The inbound passes take the spacecraft right across the evening sector of the tail and give an opportunity to look at the magnetic field in the early evening sector where both the Vasyliunas cycle outflow should be as well as the higher latitude sector looked at in the previous section. In the data shown, the spacecraft is inbound, moving radially between $34R_S$ and $14R_S$ and local time changes from ~1900 to nearly 2200 LT. Magnetometer data from days 241–245 of 2010 are shown. In the top panel we show the radial component B_r. Also shown are the field modulus $|B|$ and $-|B|$. This presentation allows one to see that throughout the magnitude of B_r is similar to that of $|B|$ and so the field is primarily in the radial direction.

The radial field is good evidence of the distension of the field associated with the centrifugal effect in this sector, just as expected in a fast rotating magnetosphere. High-frequency fluctuations can be seen throughout and the current sheet (where B_r reverses) appears to be flapping (particularly between 0200 and 1600 on day 242).

Although on short time scales there is some evidence of turbulence which might be perhaps *Kivelson and Southwood*'s [2005] high-*m* structures, the dominant feature in Figure 12.8 is a quasi-periodic 10.7 h modulation. The vertical markers mark the periodicity. They are placed so they mark the passage over the spacecraft of a rotating signal with period of 642.3 min and an azimuthal wave number $m = 1$. It is well known that northern and southern magnetic perturbations (as well as the SKR radio signals) have slightly different periods (see, e.g., the survey by *Andrews et al.* [2012]). The 642.3 min has been chosen as it is the mean period of northern and southern 10.7 h pervasive magnetic signals at the time the data were taken. Although variations are far from sinusoidal, all three components show some ordering by the phase markers,.

The upper panel shows the radial component B_r. Also shown is a plot of the field strength $|B|$ (dotted). Also shown plotted is $-|B|$ (also dotted). This manner of plotting allows one to see immediately that the field is predominantly either radial or antiradial. The field strength varies periodically. As the spacecraft moves inward and to later local time the minima in field strength become shorter but more and more pronounced. At the same time, in the more distant regions, the field is primarily radial throughout the periodic cycle, but the radial component reverses in the lower field strength region. At the same time in the sector where the field strength is lower, the azimuthal component is close to zero. Closer to the planet, the pattern changes. Not only are the field minima shorter but the dominant field component becomes B_θ there while B_ϕ is small or even reverses.

The middle panel shows the southward component, B_θ, and $|B|$. Regular pulsing of the field strength $|B|$ is clear. Where $|B|$ is a minimum, the field is not radial but in the B_θ direction, as is shown by the coalescence of the plots in the panel. A time-dependent signature with B_θ peaking as B_r drops is characteristic of the magnetotail signature of a substorm at Earth (dipolarization). However, the signature also fits with a flapping of the current sheet. However, the B_ϕ trace shows that the pattern is not quite as symmetrical as flapping might suggest. The bottom panel shows B_ϕ. Also shown is the radial component (dotted), but it has been inverted [i.e., $-B_r$]. What is clear is that there is a general anticorrelation between the radial and azimuthal components. However, the correlation is not strict. There is a tendency for B_ϕ to not actually change sign while B_r always reverses.

The once-per-cycle dipolarization [B_r decreases, B_θ increases] was already noted. The more distant dipolarizations might appear to be due to the passage over the spacecraft of the current sheet due to a simple flapping of the sheet. This cannot be true as, although the radial components are symmetric between halves of the cycle, the field strength changes and the B_ϕ variations do not. There is a regular changing stress in the field through the cycle with there being more bending of the field out of the meridian in the part of the cycle when the magnetic field is stronger. In other words, the stress between ionosphere and magnetosphere is modulated, being greater when the field strength is maximal.

In conclusion, several features can be noted. There is an underlying periodic structure with an apparent radial variation. The periodic dipolarization was noted early on in the Cassini orbital era and at that time identified and recognized to resemble a periodic form of the terrestrial substorm [*Mitchell et al.*, 2005]. In a substorm, magnetic stress in the tail is released and the field changes from more or less radial to a much more dipolar form. At the same time, there is an injection of plasma into the magnetosphere. There seem to be similar injections of energetic charged particles in the Saturnian case [*Mitchell et al.*, 2005]. It is clear also that the structure of the magnetic field in the equatorial magnetotail at the distances probed by Cassini shows a nonsinusoidal modulation with a period of around 10.7 h. Recalling that the rotation period of the heavy material is much longer than 10.7 h, explanation of the period is important.

12.7. MAGNETIC FIELD IN POLAR CAP

In Figure 12.9, we show further Cassini magnetometer data from very high invariant latitude. The data come from an inclined orbit in late 2006 when the spacecraft is at high invariant latitude on open flux tubes in the polar cap and certainly poleward of the magnetic shell where

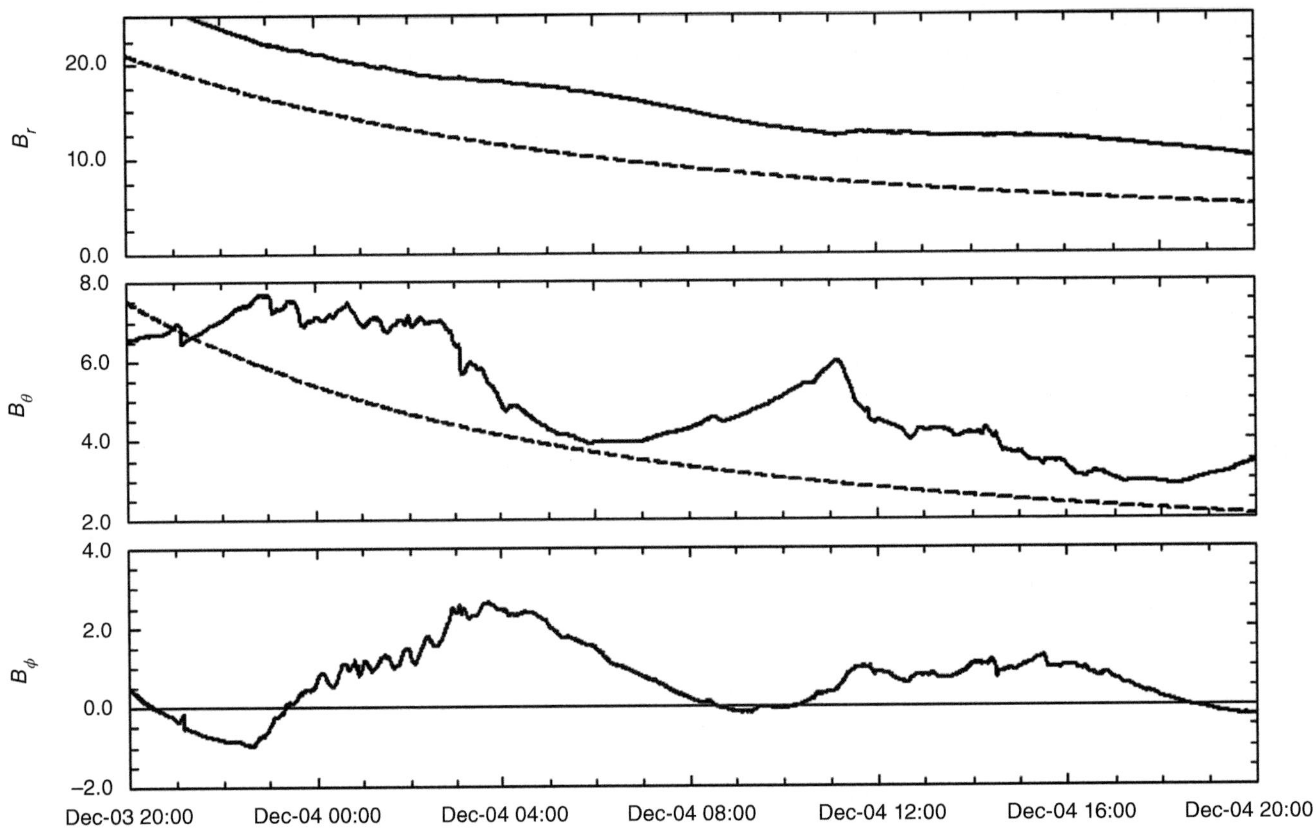

R	11.8	13.1	14.3	15.5	16.6	17.6	18.6
Lshell	34.8	39.6	42.5	44.2	45.0	45.3	45.4
InvLat	80.2	80.9	81.2	81.3	81.4	81.5	81.5
mlat	54.5	54.9	54.5	53.7	52.6	51.4	50.2
LT	20.5	21.1	21.5	21.9	22.2	22.4	22.6

Figure 12.9 Imperial College Cassini spacecraft magnetometer data from 3–4 December 2006 when the spacecraft was on polar cap field lines. One minute averaged data are shown with no further filtering. The three panels show components B_r, B_θ, B_φ (units, nT) in a coordinate system aligned with the planetary rotation/magnetic axis. The dashed traces indicate the planetary dipole field strength at the spacecraft. Below the plots are give spacecraft radial distance (R_s, local time (LT), hr) as well as magnetic L shell, magnetic latitude (deg), and invariant latitude (deg) based on the planetary dipole field.

Gurnett et al. [2010] place the plasmapause boundary. The outbound spacecraft moves from a radial distance of ~$12R_S$ to beyond $18R_S$. The spacecraft is near the transition from a dipolar field to a magnetotail configuration on the evening side of the planet. The dotted line in the figure shows the model planetary dipole field along the orbit. By the end of the segment, the background field is almost entirely radial.

The major background field component throughout is B_r and so B_θ, B_ϕ are primarily transverse. The two lower panels show these components. The 10.7h waves are clearly seen in both transverse components and amplitudes are more or less the same. A different scale has been used for the radial signal in the top panel. This obscures that there are also oscillations of similar amplitude to the transverse oscillations (~1.5 nT) in $|B|$ in antiphase with the transverse meridional θ component.

The azimuthal ϕ component is seen to be in quadrature with the θ component with B_θ leading. The transverse signals are thus close to circularly polarized and rotating in the sense of the planetary rotation. Any circularly polarized electromagnetic wave carries angular momentum. If, in this case, the wave is propagating away from the planet, when mapped back to the ionosphere, the angular momentum would be in the direction of planetary rotation. If far away, the waves are not reflected, a steady structure cannot develop, and the waves constitute the mechanism for (one-way) angular momentum transfer.

The data illustrating the signals in Figure 12.9 are not unique. They have been seen by Cassini (up to ~85° invariant latitude) throughout the cap and the phase is regular enough that it can be mapped from orbit to orbit. The steady nature of the phase and the single ($m=1$) angular wave number detected means that the field coning motion is coherent across the entire cap.

Previously *Kennel and Coroniti* [1975], *Isbell et al.* [1984], and *Milan et al.* [2005] have considered the transmission of ionospheric torque along the open field lines of the polar cap of a fast rotating magnetosphere. Each paper has suggested that there would be a symmetric winding of the field about the pole in the nonrotating inertial frame. What appears to happen in practice in the data shown in Figure 12.9 is that the field rocks in a coning motion in the inertial frame. The structure in the field is steady in a frame rotating at the rocking period. The polar cap ionosphere does indeed transmit a torque outward but through the 10.7h periodic signals. These replace the axisymmetric twist and the steady wound-up field envisaged by the previous authors. The polar cap flux is kinked (displaced transverse to itself) as well being twisted. The behavior is rather like the helical form that a jump rope takes when twisted from one end.

The compressional component in the signal is significant. If, as suggested here, the 10.7h signals in the polar cap are due to angular momentum transmission outward from the ionosphere, the compression almost certainly results from the confinement of the wave shearing motions due to the finite dimensions of the tail. In Figure 12.9, the spacecraft is ~1.8×10^6km from the planet. An order-of-magnitude estimate of the plasma spatial displacement ξ implicit in the transverse disturbances gives, from the frozen-in field theorem,

$$\xi \sim \lambda_{\parallel} \frac{b}{|B|}$$

where $\lambda_{\parallel}$ is a parallel scale length. A minimum value for $\lambda_{\parallel}$ might be approximately the spacecraft radial distance, 30 R_s, although it is likely to be larger (e.g., the parallel wavelength of an Alfvén wave could be several hundred R_s). The scale of motion is at least ~$10R_s$, comparable to the transverse dimensions of the system at the spacecraft altitude. Although more analysis is needed, it is straightforward to imagine that the transverse motion is likely to be limited by pressure exerted at the boundaries and the compressional fields result.

Whatever the source, the compressional signals are indeed seen in the polar cap. They have the distinct property of transmitting stress transverse to the field. If the 10.7h signals originate from large-scale Alfvén waves emitted each polar cap, as, for example, *Southwood and Cowley* [2014] propose, then it is the mode conversion to compressional signals that leads to the ubiquity of the signals throughout the magnetosphere.

12.8. THEORIES FOR MODULATION AT SATURN

The 10.7h modulation is seen in the central polar cap field and the tail lobe. A more complex modulation is seen at the same period in the center of the tail. The Gurnett boundary oscillates at the period. Moreover, from magnetic data from shells between the highest invariant latitudes and the central tail, the shells identified as just beyond the Vasyliunas cycle also exhibit the signals. The 10.7h modulation is a central feature of the Saturn system. There have been several theoretical ideas put forward to explain the 10.7h modulation evident in the magnetotail data shown in Figures 12.8 and 12.9. We now review them.

The first proposal was from *Goldreich and Farmer* [2007] and *Gurnett et al.* [2007]. They proposed that the 10.7h magnetic oscillations in the inner magnetosphere, which also show up in the SKR radio emission, were due to a rotating plasma convection system at Saturn. If there were a simple Vasyliunas system where the heavy plasma from the inner magnetosphere was rotating every 10.7h and diffusing slowly outward, the modulation could be explained by the circulation consisting of two separate sectors, a denser sector mainly moving outward and a less dense sector moving inward. Such a picture would fit with a two-cell rotating convection system for the whole closed field line part of the magnetosphere. The original notion of a rotating convection system was first proposed in an abstract by *Vasyliunas* [1978] referred to in a paper by *Hill et al.* [1981] which provides the first theoretical outline, albeit for the Jovian system rather than Saturn. Inward and outward flow sectors would be superimposed on a general plasma rotation at 10.7h everywhere where the heavy material of Enceladus origin is present. The bimodal tail structure forms where the material in the more dense sector is free to blow down tail.

The arguments referred to in the previous paragraph are very appealing. Nevertheless, there are fundamental theoretical and observational problems. In our view they lead to the rejection of the idea as the basis of the bimodal structure of the central tail revealed in Figure 12.8 and the periodicity in the lobe implied by Figure 12.9. First, theories by *Hill* [1979] and *Bunce et al.* [2003] show that any azimuthal flow would be sub-corotating where radial mass transport is occurring. Moreover, from an observational standpoint the evidence is that the inner and middle magnetosphere does not rotate with a period anywhere near as short as 10.7h. Typical plasma rotation speeds reported in the inner and middle magnetosphere are around half the corotation rate and there would be very substantial slippage with respect to the ionosphere at the feet of the field lines in question. The natural period associated with plasma rotation would then be more like 20–22 h [*Richardson,* 1986; *Wilson et al.*, 2009; *Thomsen et al.*, 2010; *Arridge et al.*, 2011]. Indeed, what evidence there is in the tail is that sub-corotation persists [*McAndrews et al.*, 2009] there also.

Khurana et al. [2009] provide a slightly more subtle theory than the *Gurnett et al.* [2007] and *Goldreich and Farmer* [2007] explanations of the 10.7 h modulation. They argue for a 10.7 h structure in the plasma current system that leads to a prime (denser) and nonprime (less dense) sector. A sector structure much as they discuss almost certainly exists as we indicated in our discussion above. However, their particular mechanism relies on an asymmetric force from the solar wind. Near equinox the solar wind is at right angles to the Saturn rotation axis and the asymmetry disappears. *Khurana et al.* [2009] note this and predict that the effect would disappear at equinox. It has not. Indeed, the data shown in Figure 12.8 are equinoctial and the modulation is present as strongly as ever.

Jia and Kivelson [2012] and *Jia et al.* [2012b] produced global Saturn magnetospheric simulations which well reproduce the global magnetic 10.7 h modulation and, in particular, the magnetotail structure as described here. They adapted the University of Michigan BATSRUS code global simulation of the aligned rotator Saturn magnetospheric circulation system with disturbances in circulation imposed from below and rooted in the thermosphere (neutral atmosphere), thereby avoiding the problem of the slow (20–22 h) rotation period of the material of the Vasyliunas cycle. Twin vortices with angular wave number $m = 1$ centerd on 71°S latitude and 74°N latitude rotate in the neutral polar atmosphere and carry the ionospheric ions with them, thus producing plasma vortices and field-aligned currents which transmit stress to the magnetosphere. The vortex amplitudes and location are chosen to match the magnetic amplitudes observed. However, the simulations then reproduce very well not only the magnetic field measurements of Cassini but also many other periodic features of the Saturn plasma environment.

The idea of such a set of disturbances caused by a local thermospheric heat source rotating with the planet had been first proposed by *Smith* [2011], but the Michigan work takes the idea much further forward. However, one may ask if specifying a neutral atmosphere source is necessary or even likely. Once one allows for the displacement of the planetary dipole along the symmetry to the north by $0.04R_S$, the latitudes 71°S and 74°N map to the same magnetic shell. That shell is not only where the most intense field-aligned currents are detected but also where *Gurnett et al.* [2010] observationally locate the large density gradient that they call the high-latitude plasmapause. We suggested, consistent with Gurnett's proposal, that this is the shell that marks the outermost field lines containing heavy material from the inner magnetosphere and thus the outer limit of the Vasyliunas circulation. There would be a natural azimuthal flow shear at this bounding shell as the shells poleward containing much lower density (and likely lower mass) plasma will rotate much more easily. The presence of a natural shear in the plasma flow at the required location makes it seem more likely to this author that the vortices are seated in the plasma environment. They are associated with either the shear expected across the Gurnett boundary itself or the plasma on the open field lines at higher invariant latitude. Indeed, *Gurnett et al.* [2011] report that the boundary oscillates with the period of the respective hemisphere.

In the rotating frame in which the signals are stationary, the vortices are no more than a camlike displacement of the boundary magnetic shell. In Figure 12.9, we have already shown that poleward of the Gurnett boundary, in the polar cap, there are quite sinusoidal oscillations, which we have identified as associated with outward transfer of angular momentum along the open field lines of the cap.

The highest latitude data appear to show rocking of the whole cap in Figure 12.9. As noted earlier, not only is the phase of the magnetic periodicity remarkably steady [*Andrews et al.,* 2008, *Andrews et al.,* 2010, *Provan et al.,* 2009a, 2009b, 2012; *Andrews et al.* 2012] in the polar cap, but also slightly different periods are detected in the northern and southern hemispheres [*Andrews et al.*, 2010, *Southwood*, 2011]. Importantly, *Andrews et al.* [2012] have determined that close to pristine southern or northern period signals were detected in the respective polar caps while both signals are present on the closed flux tubes. This result suggests that the signals themselves originate from each polar cap [see, e.g., *Southwood and Cowley,* 2014]. We have already provided a reason for this to be so, namely that large-scale hydromagnetic waves transmit angular momentum from the ionosphere upward along the open field lines. The observation by *Gurnett et al.* [2011] that the position of the steep density gradient appears to oscillate at the same period and with the same hemispheric dependence as the SKR radio signals and the magnetic signals confirms the idea.

What then can one say of the 10.7 h modulation of the central plane of the magnetotail so evident in Figure 12.8? It looks to be a fairly central feature of the tail structure at Saturn. The contrast between the remarkably sinusoidal signals in the polar cap field lines of the lobe (see e.g., Figure 12.9) and the irregular nonharmonic nature of the periodicity of the signals in the center plane (as illustrated in Figure 12.8) is clear. As noted above, the magnetic pressure oscillations in the cap will carry the 10.7 h signals throughout the rest of the magnetosphere.

In particular, such pressure fluctuations will cause physical motion of the central plasma and current sheet of the tail. Thus the fluctuations in the center plane are complicated not only by the presence of the plasma sheet and the likely outward flow of material there but also by motion of the plasma north and south in response to the magnetic pressure fluctuations. However, what of the analogy given by *Mitchell et al.* [2005] to "periodic substorms" and evident on Figure 12.8? The period in question is the 10.7 h period, which is the period of rotation of the solar wind material in the polar cap outside the Vasyliunas cycle. It is not the period of rotation of the heavy material from Enceladus, as we have noted. However, the field lines on days 243–245 in Figure 12.8 where the periodic signals are seen are very likely to map to the Vasyliunas cycle regime in this mid evening sector. Moreover, other particle data do show ordering by the 10.7 h periodicity [e.g., *Mitchell et al.*, 2009, or *Kennelly et al.*, 2013]. Could it be that the material that causes the faster rotation of solar wind material to have an effect on the higher invariant latitude field lines can modulate loss of material down tail and injection of energetic particles in the inner regions of the Saturn magnetosphere?

12.9. DIFFERENCES BETWEEN JUPITER AND SATURN

It is probably fair to say that the fast rotating magnetosphere of Jupiter is not as well understood as Saturn's. Both are indubitably fast rotators, but the much stronger and much more asymmetric magnetic field renders Jupiter's system different in many ways. *Kivelson and Southwood* [2005] attempt a basic picture of the Jovian magnetospheric dynamics by describing the dominant process local time sector by local sector. The basis of their description is the Vasyliunas cycle, which moves heavy iogenic ionized material from the inner magnetosphere eventually to interplanetary space, just as must occur at Saturn with material from Enceladus. There is a large dawn-dusk asymmetry in the plasma sheet with much thicker sheet in the afternoon sector, suggesting very substantial heating/acceleration there leading to perhaps continuous loss down the tail of iogenic material from the deep interior. On the morning side there is also a large region depleted of plasma at a large radial distance with no apparent analogue at Saturn, the cushion region. In this region, there is no equatorial current sheet. The region is also the site of bubbles of plasma [*Southwood et al.*, 1993; *Haynes et al.*, 1994; *Leamon et al.*, 1995], suggesting material can escape sporadically on the morning side as well as probably more continuously at night. Saturn appears to have no similar analogues. Hence, although it is fair to say that some aspects of the Saturn system do resemble the picture given by *Kivelson and Southwood* [2005], there are very significant differences. The presence of a Dungey cycle at Jupiter has been seriously brought into doubt in the Jovian magnetosphere [*McComas, and Bagenal*, 2007]. The Galileo orbiter spacecraft really did not explore the high-latitude magnetosphere at all and so as yet information only comes from observations of the aurora. However, matters should change substantially with the arrival of the NASA Juno mission in the not-too-distant future. It is important to note that Cassini's high-inclination orbits have provided much critical information in our present picture at Saturn.

12.10. CONCLUDING REMARKS

This short review has worked from the idealized Dungey model of flux and plasma circulation in the terrestrial system that gives rise to the magnetotail of Earth's magnetosphere to look at the modifications introduced in fast rotating systems such as at the giant planets. Internal sources of ionized material further complicate the fast rotators represented by Jupiter and Saturn. Jupiter still awaits important exploration in the near future and our analysis has concentrated on Saturn, using in particular the Cassini magnetometer data set. Although basic processes are almost certainly the same at Jupiter, its magnetosphere is certainly much larger and probably inherently more complex than Saturn's. Whereas the Vasyliunas cycle, long propounded to explain the basic circulation of flux and plasma at Jupiter or Saturn, is certainly part of the story, we have used Cassini magnetometer data to argue that a high-latitude Dungey cycle appears a necessary element to explain observations at Saturn. The Dungey cycle may be a little like the cycle at Earth in that, as we have pointed out, much of the solar wind material entering by day is likely to exit directly down tail on the nightside and not become trapped at all. We have argued that the 10.7 h magnetic modulation seen throughout the Saturn system is intimately linked to the Dungey cycle. Although we propose it originates from the imposition of rotation on the open flux tubes of the polar cap, and so the tubes that make up the lobes of the tail, the magnetic compression associated with the waves propagating outward modulate the magnetotail overall, including probably the escape of more massive ionized material down the center of the tail.

ACKNOWLEDGMENTS

The author acknowledges useful discussions with Xianzhe Jia, Margaret Kivelson, and Vytenis Vasyliunas. Cassini data reduction at Imperial College have been supported by the U.K. Space Agency through a contract from the European Space Agency.

REFERENCES

Andrews, D. J., E. J. Bunce, S. W. H. Cowley, M. K. Dougherty, G. Provan, and D. J. Southwood (2008), Planetary period oscillations in Saturn's magnetosphere: Phase relation of equatorial magnetic field oscillations and SKR modulation, *J. Geophys. Res.*, *113*, A09205, doi:10.1029/2007JA012937.

Andrews, D. J., A. J. Coates, S. W. H. Cowley, M. K. Dougherty, L. Lamy, G. Provan, and P. Zarka (2010), Magnetospheric period oscillations at Saturn: Comparison of equatorial and high-latitude magnetic field periods with north and south SKR periods, *J. Geophys. Res.*, *115*, A12252, doi:10.1029/2010JA015666.

Andrews, D. J., S. W. H. Cowley, M. K. Dougherty, L. Lamy, G. Provan, and D. J. Southwood (2012), Planetary period oscillations in Saturn's magnetosphere: Evolution of magnetic oscillation properties from southern summer to post-equinox, *J. Geophys. Res.*, *117*, A04224, doi:http://dx.doi.org/10.1029/2011JA017444.

Arridge, C. S., K. K. Khurana, C. T. Russell, D. J. Southwood, N. Achilleos, M. K. Dougherty, A. J. Coates, and H. K. Leinweber (2008), Warping of Saturn's magnetospheric and magnetotail current sheets, *J. Geophys. Res.*, *113*, A08217, doi:10.1029/2007JA012963.

Arridge, C. S., et al. (2011), Upstream of Saturn and Titan, *Space. Sci. Rev.*, *162*, 25–83.

Aubry, M. P., C. T. Russell, and M. G. Kivelson (1970), Inward motion of the magnetopause before a substorm, *J. Geophys. Res.*, *75*, 7018–7031, doi:10.1029/JA075i034p07018.

Axford, W. I., H. E. Petschek, and G. L. Siscoe, (1965), Tail of the magnetosphere, *J. Geophys. Res.*, *70*, 1231–1236.

Bunce, E. J., S. W. H. Cowley, and J. A. Wild (2003), Azimuthal magnetic fields in Saturn's magnetosphere: Effects associated with plasma sub-corotation and the magnetopause-tail current system, *Ann. Geophys.*, *21*, 1709.

Burton, M. E., M. K. Dougherty, and C. T. Russell (2009), Model of Saturn's internal planetary magnetic field based on Cassini observations, *Planet. Space Sci.*, *57*, 1706–1713.

Cahill, L. J., Jr., (1964) Preliminary results of magnetic field measurements in the tail of the geomagnetic cavity, *Trans. Am. Geophys. Union*, *45*, 231–235.

Cowley, S. W. H., and M. Lockwood (1992), Excitation and decay of solar wind-driven flows in the magnetosphere-ionosphere system, *Ann. Geophys.*, *10*, 103–115.

Cowley, S. W. H., E. J. Bunce, and R. Prangé, (2004), Saturn's polar ionospheric flows and their relation to the main auroral oval, *Ann. Geophys.*, *22*, 1379–1394, doi:1432-0576/ag/2004-22-1379

Dungey, J. W. (1961), Interplanetary magnetic field and the auroral zones, *Phys. Rev. Lett.*, *6*, 47–48.

Dungey, J. W. (1963), Interactions of the solar plasma with the geomagnetic field, *Planet. Space Sci.*. *10*, 233–237.

Dungey, J. W. (1965), The length of the magnetospheric tail, *J. Geophys. Res.*, *70*, 1753.

Freeman, M. P., and D. J. Southwood (1988), The effect of magnetospheric erosion on mid- and high-latitude ionospheric flows, *Planet. Space Sci.*, *36*, 509.

Goldreich, P., and A. J. Farmer (2007), Spontaneous symmetry breaking of the external field of Saturn, *J. Geophys. Res.*, *112*, A05225, doi:10.1029/2006JA012163.

Gurnett, D. A., A. M. Persoon, W. S. Kurth, J. B. Groene, T. F. Averkamp, M. K. Dougherty, and D. J. Southwood (2007), The rotation of the inner region of Saturns plasma disk, *Science*, *316*, 442–445, doi:10.1126/science.1138562.

Gurnett, D. A., A. Lecacheux, W. S. Kurth, A. M. Persoon, J. B. Groene, L. Lamy, P. Zarka, and J. F. Carbary (2009), Discovery of a north-south asymmetry in Saturn's radio rotation period, *Geophys. Res. Lett.*, *36*, L16102, doi:10.1029/2009GL039621.

Gurnett, D. A., et al. (2010), A plasmapause-like density boundary at high latitudes in Saturn's magnetosphere, *Geophys. Res.Lett.*, *37*, L16806, doi:10.1029/2010GL044466.

Gurnett, D. A., A. M. Persoon, J. B. Groene, W. S. Kurth, M. Morooka, J.-E. Wahlund, and J. D. Nichols (2011), The rotation of the plasmapause-like boundary at high latitudes in Saturn's magnetosphere and its relation to the eccentric rotation of the northern and southern auroral ovals, *Geophys. Res. Lett.*, *38*, L21203, doi:10.1029/2011GL049547.

Haynes. P. L., A. Balogh, M. K. Dougherty, D. J. Southwood, A. Fazakerley, and E. J. Smith (1994), Null fields in the outer Jovian magnetosphere: Ulysses observations, *Geophys. Res. Lett*, *21*, 405.

Hill, T. W. (1976), Interchange stability of a rapidly rotating magnetosphere, *Planet. Space Sci.*, *24*, 1151.

Hill, T. W. (1979), Inertial limit on corotation, *J. Geophys. Res.*, *84*, 6554.

Hill, T. W., A. J. Dessler, and L. J. Maher (1981), Corotating magnetospheric convection, *J. Geophys. Res.*, *86*, 9020–9028.

Isbell, J., A. J. Dessler, and J. H. Waite, Jr. (1984), Magnetospheric energization by interaction between planetary spin and the solar wind, *J. Geophys. Res.*, *89*, 10,716.

Jia, X., and M. G. Kivelson (2012), Driving Saturn's magnetospheric periodicities from the upper atmosphere/ionosphere: Magnetotail response to dual sources, *J. Geophys. Res.*, *117*, A11219, doi:10.1029/2012JA018183.

Jia, X., K. C. Hansen, T. I. Gombosi, M. G. Kivelson, G. Tóth, D. L. DeZeeuw, and A. J. Ridley (2012a), Magnetospheric configuration and dynamics of Saturn's magnetosphere: A global MHD simulation, *J. Geophys. Res.*, *117*, A05225, doi:10.1029/2012JA017575.

Jia, X., M. G. Kivelson, and T. I. Gombosi (2012b), Driving Saturn's magnetospheric periodicities from the upper atmosphere/ionosphere, *J. Geophys. Res.*, *117*, A04215, doi:10.1029/2011JA017367.

Kennel, C. F., and F. V. Coroniti (1975), Is Jupiter's magnetosphere like a pulsar's or Earth's? *Space Sci. Rev.*, *17*, 857–883.

Kennelly, T. J., J. S. Leisner, G. B. Hospodarsky, and D. A. Gurnett (2013), Ordering of injection events within Saturnian SLS longitude and local time, *J. Geophys. Res. Space Phys.*, *118*, 832–838, doi:10.1002/jgra.50152.

Khurana, K. K., D. G. Mitchell, C. S. Arridge, M. K. Dougherty, C. T. Russell, C. Paranicas, N. Krupp, and A. J. Coates, (2009), Sources of rotational signals in Saturn's magnetosphere, *J. Geophys. Res.*, *114*, A02211, doi:10.1029/2008JA013312.

Kivelson, M. G., and D. J. Southwood (2005), Dynamical consequences of two modes of centrifugal instability in Jupiter's outer magnetosphere, *J. Geophys. Res.*, *110*, A12209, doi:10.1029/2005JA011176.

Leamon, R. J., M. K. Dougherty, D. J. Southwood, and P. L. Haynes (1995), Magnetic nulls in the outer magnetosphere of Jupiter: Detections by Pioneer and Voyager spacecraft, *J. Geophys. Res.*, *100*, 1829.

McAndrews, H. J., et al. (2009), Plasma in Saturn's nightside magnetosphere and the implications for global circulation, *Planet. Space Sci.*, *57*, 1714–1722.

McComas, D. J., and F. Bagenal (2007), Jupiter: A fundamentally different magnetospheric interaction with the solar wind, *Geophys. Res. Lett.*, *34*, L20106, doi:10.1029/2007GL031078.

Milan, S. E., E. J. Bunce, S. W. H. Cowley, and C. M. Jackman (2005), Implications of rapid planetary rotation for the Dungey magnetotail of Saturn, *J. Geophys. Res.*, *110*, A03209, doi:10.1029/2004JA010716.

Mitchell, D. G., et al. (2005), Energetic ion acceleration in Saturn's magnetotail: Substorms at Saturn? *Geophys. Res. Lett.*, *32*, L20S01, doi:10.1029/2005GL022647.

Mitchell, D. G., et al. (2009), Recurrent energization of plasma in the midnight-to-dawn quadrant of Saturn's magnetosphere, and its relationship to auroral UV and radio emissions, *Planet. Space Sci.*, *57*, 1732–1742.

Ness, N. F. (1965), The Earth's magnetic tail, *J. Geophys. Res.*, *70*(13), 2989–3005, *doi:*10.1029/JZ070i013p02989.

Parker, E. N. (1958), Interaction of the solar wind with the geomagnetic field, *Phys. Fluids 1*, 171–187,

Piddington, J. H. (1960), Geomagnetic storm theory, *J. Geophys. Res.*, *65*, 93–106.

Provan, G., D. J. Andrews, C. S. Arridge, A. J. Coates, S. W. H. Cowley, S. E. Milan, M. K. Dougherty, and D. M. Wright (2009a), Polarization and phase of planetary-period magnetic field oscillations on high-latitude field lines in Saturn's magnetosphere, *J. Geophys. Res.*, *114*, A02225, doi:10.1029/2008JA013782.

Provan, G., S. W. H. Cowley, and J. D. Nichols (2009b), Phase relation of oscillations near the planetary period of Saturn's auroral oval and the equatorial magnetospheric magnetic field, *J. Geophys. Res.*, *114*, A04205, doi:10.1029/2008JA013988.

Provan, G., D. J. Andrews, C. S. Arridge, A. J. Coates, S. W. H. Cowley, G. Cox, M. K. Dougherty, and C. M. Jackman (2012), Dual periodicities in planetary-period magnetic field oscillations in Saturn's tail, *J. Geophys. Res.*, *117*, A01209, doi:10.1029/2011JA017104.

Richardson, J. D. (1986), Thermal ions at Saturn: Plasma parameters and implications, *J. Geophys. Res.*, *91*(A2), 1381–1389, doi:10.1029/JA091iA02p01381.

Russell, C. T., and R. C. Elphic, (1979), ISEE observations of flux transfer events at the dayside magnetopause, *Geophys. Res. Lett.*, *6*, 33–36, doi:10.1029/GL006i001p00033.

Smith, C. G. A. (2011), A Saturnian cam current system driven by asymmetric thermospheric heating, *Mon. Not. R. Astr. Soc.*, *410*, 2315–2328, doi: 10.1111/j.1365-2966.2010.17602.x.

Southwood, D. J. (1985), Theoretical aspects of ionosphere-magnetosphere-solar wind coupling, in Physics of Ionosphere-Magnetosphere Connection, edited by E. R. Schmerling, *Adv. Space Res.*, *5*, 4.

Southwood, D. J. (2011), Direct evidence of differences in magnetic rotation rate between Saturn's northern and southern polar regions, *J. Geophys. Res.*, *116*, A01201, doi:10.1029/2010JA016070.

Southwood D. J. and S. W. H. Cowley (2014), The origin of Saturn's magnetic periodicities: Northern and southern current systems, *J. Geophys. Res. Space Physics*, *119*, 1563–1571, doi:10.1002/2013JA019632.

Southwood, D. J., and M. G. Kivelson (1989), Magnetospheric interchange motions, *J. Geophys. Res.*, *94*, 299–308.

Southwood, D. J., and M. G. Kivelson (2007), Saturnian magnetospheric dynamics: Elucidation of a camshaft model, *J. Geophys. Res.*, *112*, A12222, doi:10.1029/2007JA012254.

Southwood, D. J., and M. G. Kivelson (2009), The source of Saturn's periodic radio emission, *J. Geophys. Res.*, *114*, A09201, doi:10.1029/2008JA013800.

Southwood, D. J., M. K. Dougherty, P. Canu, A. Balogh, and P. J. Kellogg, (1993), Correlations between magnetic field and electron density observations during the inbound Ulysses Jupiter flyby, *Planet. Space Sci.*, *41*, 919–930.

Southwood, D. J., C. J. Farrugia, and M. A. Saunders (1988), What are flux transfer events? *Planet. Space Sci.*, *36*, 503.

Thomsen, M. F., D. B. Reisenfeld, D. M. Delapp, R. L. Tokar, D. T. Young, F. J. Crary, E. C. Sittler, M. A. McGraw, and J. D. Williams (2010), Survey of ion plasma parameters in Saturn's magnetosphere, *J. Geophys. Res.*, *115*, A10220, doi:10.1029/2010JA015267.

Vasyliunas,V. M. (1978), A mechanism for plasma convection in the inner Jovian magnetosphere (abstract), in STP Symposium Abstracts, Innsbruck, Austria.

Vasyliunas, V. M. (1983), Plasma distribution and flow, in *Physics of the Jovian Magnetosphere*, edited by A. J. Dessler, pp. 395–453, Cambridge Univ. Press, New York.

Wilson, R. J., R. L. Tokar, and M. G. Henderson (2009), Thermal ion flow in Saturn's inner magnetosphere measured by the Cassini plasma spectrometer: A signature of the Enceladus torus? *Geophys. Res. Lett.*, *36*, L23104, doi:10.1029/2009GL040225.

13

Solar Wind Interaction with Giant Magnetospheres and Earth's Magnetosphere

P. A. Delamere

13.1. INTRODUCTION

The formation and detailed character of the magnetotail are fundamentally tied to the solar wind interaction with the magnetosphere, starting with dayside and flank magnetopause boundary processes (Chapter 1). Magnetospheric convection was proposed by *Axford and Hines* [1961] as a consequence of the solar wind interaction driven by a tangential drag at the magnetopause boundary. *Axford and Hines* [1961] further proposed a viscous like interaction as a possible source of tangential drag and later mentioned magnetic stress as an alternative in light of the *Dungey* [1961] model of an open magnetosphere. Large-scale magnetic reconnection in the *Dungey* [1961] model couples the solar wind via wave propagation to the planetary ionosphere on open magnetic field lines (see discussion of reconnection in Chapter 15). The viscous interaction can transport mass, momentum, and energy to/from the magnetosphere through diffusive processes and intermittent reconnection (see Chapter 14), resulting in the limiting case of a closed magnetosphere. A complete description of the solar wind interaction should, however, include large-scale Dungey reconnection (i.e., open magnetosphere) and small-scale viscous processes. This chapter will compare the solar wind interaction with Earth and with the giant magnetospheres and discuss implications for the magnetotail.

At Earth, magnetic reconnection is considered the dominant process that couples the solar wind momentum to the magnetosphere by driving ionospheric convection that circulates magnetic flux through the magnetosphere through the Dungey cycle of magnetic reconnection [*Dungey*, 1961]. While ionospheric outflows contribute plasma to the magnetospheric cavity, it is well known that the magnetospheric plasma properties correlate with properties of the solar wind. Thus, the solar wind entry mechanisms are important and the viscous interaction is sometimes considered for mass transport (see Chapter 14). In particular, shear flow-driven instabilities (i.e., Kelvin-Helmholtz) [*Lee et al.*, 1981; *Pu and Kivelson*, 1983; *Thomas and Winske*, 1991, 1993; *Huba*, 1996; *Otto and Fairfield*, 2000; *Nykyri and Otto*, 2001, 2004; *Matsumoto and Seki* , 2010; *Nakamura et al.*, 2010; *Sundberg et al.*, 2012; *Karimabadi et al.*, 2013] and other wave-particle interactions (e.g., kinetic Alfvén waves) [*Johnson and Cheng*, 1997; *Johnson et al.*, 2001; *Johnson and Cheng*, 2001] are thought to play an important role in mass transport. However, dual-lobe reconnection can also transport solar wind material to the magnetosphere [*Song and Russell* , 1992]. An advantage that magnetic reconnection has at Earth in driving magnetospheric dynamics is the relative absence of corotating magnetospheric plasma as compared with the rapidly rotating giant magnetospheres.

At the giant magnetospheres, magnetic reconnection must not only drive ionospheric convection (e.g., overcoming ionospheric flows associated with the rapid rotation) but also transfer sufficient momentum to alter magnetodisc flows if an Earth-like Dungey flux circulation is to be achieved. The standard corotation-convection models nicely illustrate the energetics of this situation [e.g., *Brice and Ioannidis,* 1970]. The combined potential, in the sun-planet rest frame, has the form

$$\phi = \frac{\Omega B_o R_P^2}{\rho} + \eta v_{sw} B_{sw} R_P y, \qquad (13.1)$$

Geophysical Institute, University of Alaska Fairbanks, Fairbanks, Alaska

Magnetotails in the Solar System, Geophysical Monograph 207. First Edition.
Edited by Andreas Keiling, Caitríona M. Jackman, and Peter A. Delamere.

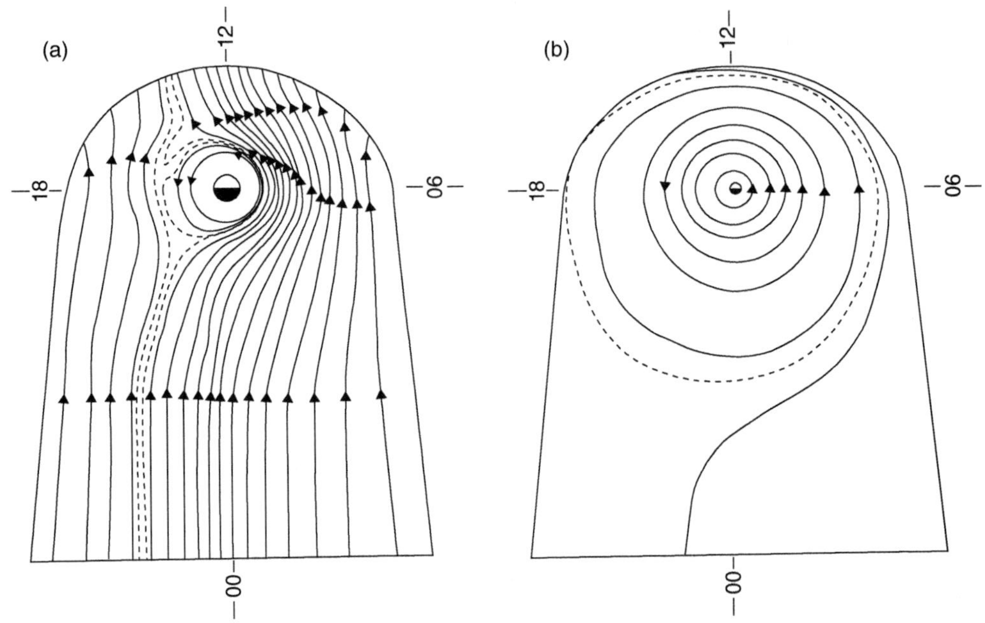

Figure 13.1 Corotation-convection models for Earth and Jupiter assuming sunward convection in the tail [*Brice and Ioannidis*, 1970]. The plasmapause boundary lies inside the magnetopause boundary for Earth and at the magnetopause boundary for Jupiter.

where the subscript sw refers to solar wind, Ω is the planet's angular velocity, B_o is the planet's surface field strength, ρ and y are the respective radial and dawn-dusk distances in units of planetary radius R_p, and η is an empirically determined factor (usually 0.1 – 0.2 for Earth, e.g., *Kennel and Coroniti* [1977]) that determines the reconnection potential associated with the Dungey cycle. Figure 13.1 compares Earth and Jupiter, showing that the corotating plasmasphere at Jupiter (i.e., in the form of a centrifugally confined magnetodisc) extends to the magnetopause boundary due to the dominance of the internal, planetary driver. Centrifugal stresses will stretch and thin the rotating magnetodisc until plasma blobs are ejected into the tail via "Vasyliunas cycle" reconnection [*Vasyliunas*, 1983]. When opening reconnection does occur, *Cowley et al.* [2005] suggested that the closing reconnection likely occurs on the dawn/tail flank and flux returns to the dayside with the corotational flows (Figure 13.2).

There has been considerable debate regarding the nature of the solar wind interaction with Jupiter's magnetosphere (see *McComas and Bagenal* [2007] and comment by *Cowley et al.* [2008] and counter response by *McComas and Bagenal* [2008]). The Dungey cycle of reconnection must be important if dayside reconnection is operating [*Cowley et al.*, 2005; *Nichols et al.*, 2006], though perhaps only under compressed conditions of the magnetosphere [*Badman and Cowley*, 2007]. However, *Mozer and Hull* [2010] argued that the energy conversion rates from reconnection are small and reconnection is less important at the giant magnetospheres. *Delamere and Bagenal* [2010] argued that the viscous interaction may in fact be the dominant mechanism for momentum and energy transport at the magnetopause boundary.

We will discuss aspects of the solar wind interaction by considering the limiting cases of momentum transfer from the solar wind to the magnetosphere (section 13.2). The momentum input to the ionosphere vs. momentum input directly to the magnetosphere via tangential stresses generated by a viscous interaction will be considered (sections 13.3 and 13.4). Aspects of the internal driver (i.e., Vasyliunas cycle at the giant magnetospheres) will be discussed in section 13.5. A brief summary of auroral signatures that may distinguish these drivers will be addressed in section 13.6. Finally, a discussion of magnetospheric scale and the structure of the magnetotail will be given in sections 13.7 and 13.8.

For the sake of clarity, a careful definition of terms is necessary in any comparative discussion of planetary magnetospheres. The term "Dungey cycle" used throughout this chapter will refer to the idealized situation where opening dayside reconnection drives large-scale (ideally two cells) ionospheric convection and closing reconnection occurs in the nightside tail region. The Dungey cycle is also a key aspect of the terrestrial substorm (see discussion of magnetotail current sheet formation in Chapter 17). Used in this context, the solar wind must be

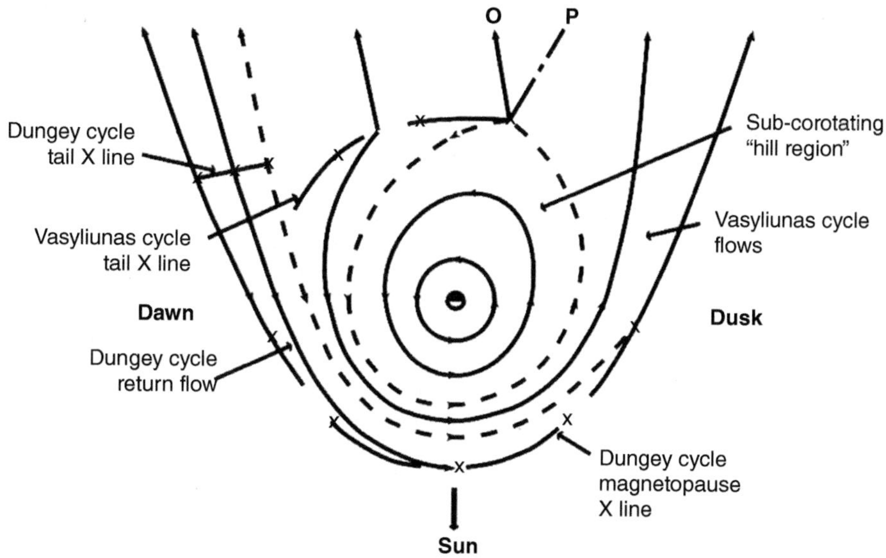

Figure 13.2 Flows in the equatorial plane for Jupiter [*Cowley et al.*, 2003] showing the relation between the Vasyliunas X line and the Dungey X line.

capable of supplying sufficient momentum to drive ionospheric convection on open and closed field lines, which requires sunward magnetospheric flows (see Figure 13.4 in Chapter 1). On the other hand, magnetic stresses due to field line threading of the magnetopause boundary can be generated by shear flow instabilities (e.g., Kelvin-Helmholtz (KH)) [*Otto*, 2007; *Ma et al.*, 2014a, 2014b]. In this context, reconnection is driven by the instability, and the validity of empirical reconnection rates for nondriven, Dungey reconnection at Earth should be questioned for the giant magnetospheres where shear flow unstable conditions exist in the subsolar region. The evolution of open field lines generated by the KH instability is difficult to determine due to twisted and entangled magnetic field lines in the KH vortices. While net magnetic flux transport is possible, closing reconnection along the flanks of the magnetosphere can also occur and on average this type of intermittent and thus small-scale reconnection contributes to tangential stresses typically considered to be "viscous."

13.2. MOMENTUM TRANSFER FROM SOLAR WIND

Quantifying the momentum transfer from the solar wind to the magnetosphere is critical to understanding magnetospheric dynamics and ultimately to the formation of the magnetotail structure. We will focus our discussion by considering the solar wind stresses generated at the magnetopause boundary. The momentum equation in the limits of nonrelativistic bulk flows, charge quasi-neutrality, and Alfvén speed, $v_A^2 << c^2$, is given by

$$\frac{\partial(\rho\mathbf{v})}{\partial t}+\nabla\cdot\left(\rho\mathbf{v}\mathbf{v}+\mathbf{P}-\frac{\mathbf{BB}}{\mu_o}+\mathbf{I}\frac{B^2}{2\mu_o}\right)=0, \tag{13.2}$$

where $\mathbf{P}$ is the pressure tensor, $\mathbf{B}$ is the magnetic field, $\mathbf{I}$ is the unit dyad, ρ is the mass density, and $\mathbf{v}$ is the bulk flow of the plasma [e.g., *Vasyliunas,* 2007]. We will limit the discussion to tangential Maxwell stresses at the magnetopause boundary, resulting from the interaction of the solar wind with the magnetosphere. Other terms (e.g., dynamic pressure and magnetic pressure) are clearly important on the dayside, but we will focus on the tailward forces generated by a draped magnetic field line (fourth term on the left-hand side), though tangential stresses also arise from the inertial and pressure terms. The momentum transfered to the magnetosphere/ionosphere system can be determined by integrating the Maxwell stresses over the surface of an appropriate magnetospheric volume. In the limiting case, the tangential stress can be estimated by the draping of the super-Alfvénic solar wind flow over the magnetospheric obstacle, or

$$\frac{dp_x}{dt}=\oint_S(\overleftrightarrow{\mathbf{T}}\cdot d\mathbf{a})_x=\frac{2B_xB_{\mathrm{IMF}}A}{\mu_o}, \tag{13.3}$$

where $\overleftrightarrow{\mathbf{T}}$ is the Maxwell stress tensor, p is momentum, B_{IMF} (the interplanetary magnetic field), for simplicity, can be in either the y or z direction (at the giant magnetospheres B_{IMF} is mostly in the y direction), and A is the surface area of the volume under consideration. We can rewrite this expression in terms of the solar wind quantities,

$$\frac{dp_x}{dt} = 2\rho_{sw} v_A v_{sw} A, \tag{13.4}$$

where v_{sw} and ρ_{sw} are the solar wind speed and mass density and v_A is the solar wind Alfvén velocity. This latter expression is the upper limit for momentum transfer and assumes a fully saturated Alfvénic interaction where the draping of the magnetic field is maximized and flux conservation requires that

$$\frac{v_{sw}}{v_A} = \frac{B_x}{B_{IMF}}. \tag{13.5}$$

A fully draped magnetic field configuration can be generated by either anchoring the magnetic footprint to the ionosphere in the case of open field lines or otherwise impeding the solar wind flow, generating a region of stagnation (e.g., a mass-loaded flow, such as a comet, generates a saturated Alfvénic interaction). In reality, there may be instances where the fully developed B_x component as defined by equation 13.5 does not exist (e.g., weak viscous stresses where the solar wind flow is not strongly impeded). But we emphasize, again, that there is an upper limit to the momentum transfer rate.

13.3. MAGNETIC RECONNECTION AND IONOSPHERIC CONVECTION

In an open magnetosphere, momentum is transferred directly from the solar wind to the ionosphere on open field lines. The solar wind exerts a tailward force on the magnetospheric system and the ionosphere must accommodate this tailward force while preserving the magnetic field through incompressible flows. The large fast mode speed in the ionosphere guarantees that the magnetic field cannot be compressed in the ionosphere. Convective circulation of the plasma flow will preserve the magnetic field strength and on a global scale this is manifested as the two-cell Dungey circulation of magnetic flux (Figure 13.3).

Earth's extensive Super Dual Auroral Radar Network (SuperDARN) has provided detailed coverage of ionospheric plasma circulation, showing that flows in the polar cap are largely antisunward, especially during southward IMF conditions [*Bristow et al.*, 2011]. However, there are instances of significant temporal variability in global flow speeds of the polar cap region. Figure 13.4 from *Bristow et al.* [2011] shows ionospheric flows (red) over the south geomagnetic pole from SuperDARN HR radar at McMurdo Station, Antarctica, on 15 March, 2010. Also plotted (black) is IMF B_z. In the interval from 16:50 UT to 21:00 UT there are significant fluctuations in B_z and corresponding ionospheric flow variations. The correspondence between the IMF variations and observed velocity fluctuations is clear. The trend is further illustrated in Figure 13.5, showing the Earth-Sun (E-S) velocity vs. IMF B_z for more than 40,000 data points with linear regression errors $\sim 1\%$. A statistically significant correlation between the Earth-Sun velocity (v_{E-S}) and B_z should not be expected for the entire data set because, for example, tail dynamics should also influence the ionospheric flows. The variability in the 15 March, 2010 case is clearly tied to solar wind conditions at the dayside magnetopause boundary (v_{E-S} correlation with $B_z \sim 0.8$) and the entire polar cap can respond rapidly (e.g., a few minutes) to variable solar wind momentum input starting from the dayside cusp region [*Bristow et al.*, 2011]. Apparently, the solar wind momentum transfer rates

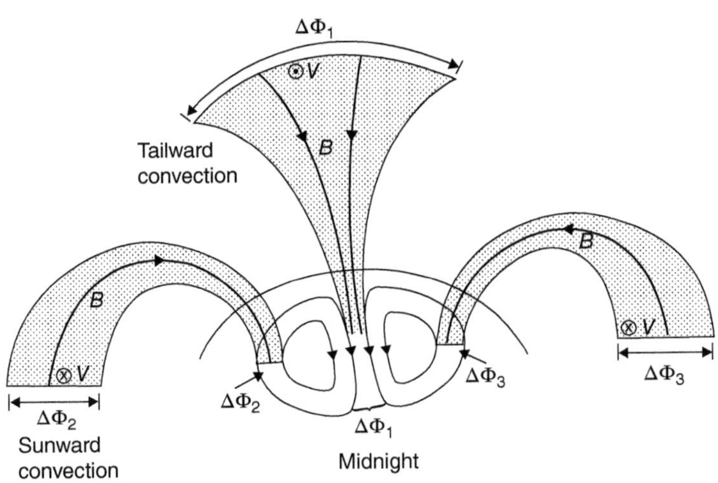

Figure 13.3 Ionospheric convection and associated cross-polar cap potentials due to the Dungey cycle of reconnection. Courtesy of A. Otto.

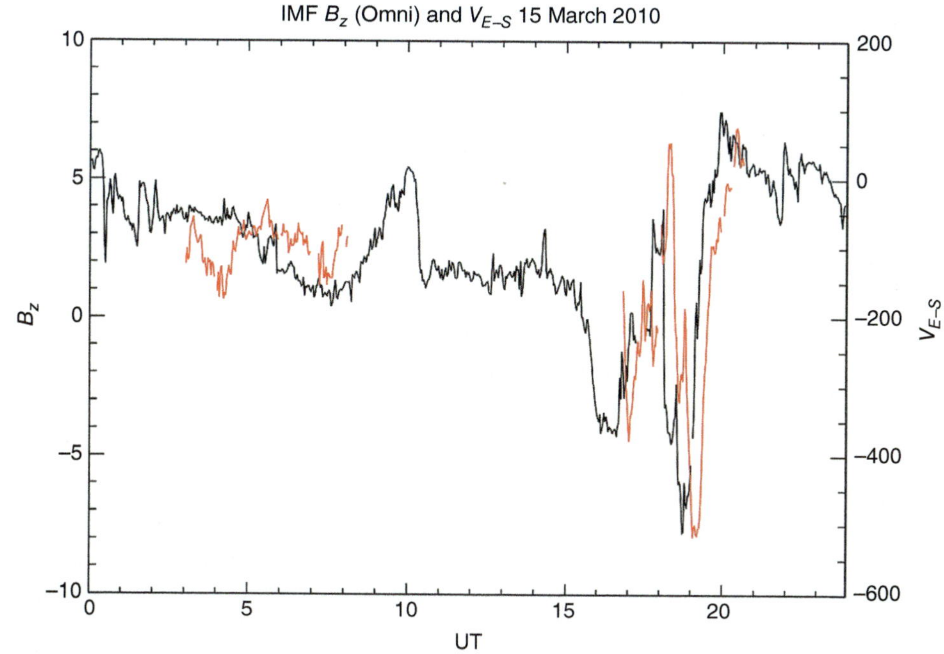

Figure 13.4 Sample observed fluctuations in the Earth-Sun velocity component (red) and IMF B_z (black) from *Bristow et al.* [2011] for recent observations of flows over the south geomagnetic pole (polar cap region) from the SuperDARN HF radar at McMurdo Station, Antarctica. Note the clear correlation between flow and IMF B_z in the interval from 16:50 UT and 21:00 UT, where the rapid changes in IMF B_z and v_{E-S} occur on time scales ~ 5 min.

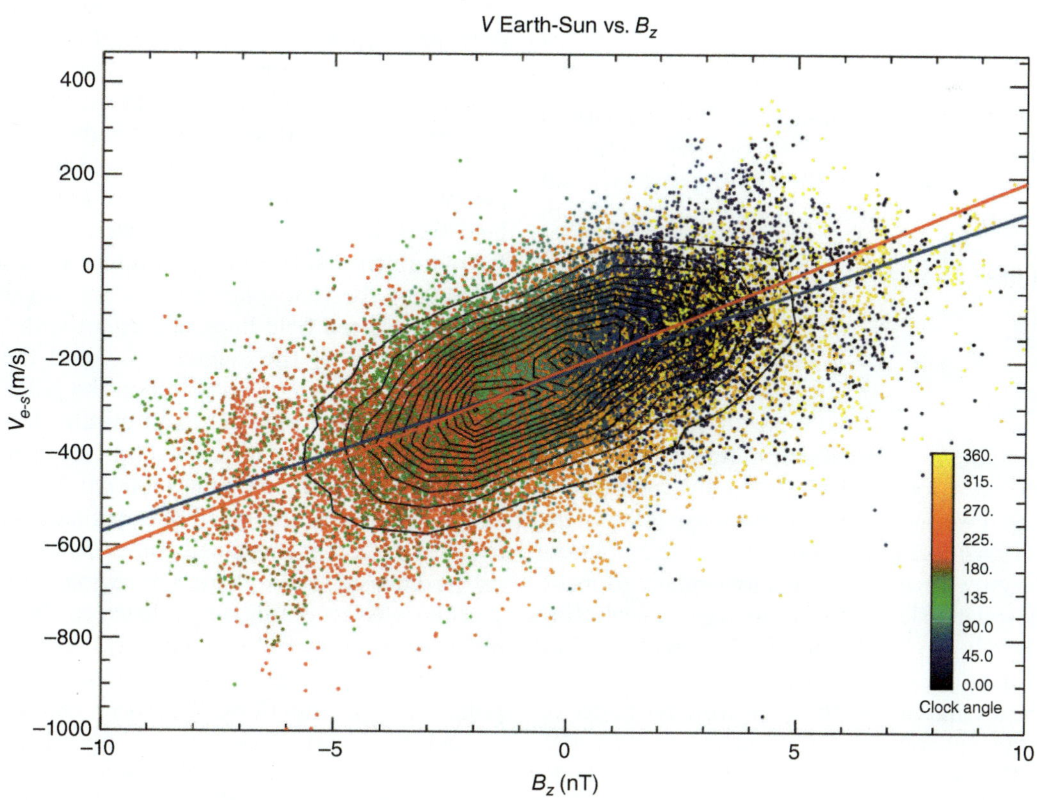

Figure 13.5 Earth-Sun velocity component vs. IMF B_z from *Bristow et al.* [2011] for recent observations of flows over the south geomagnetic pole (polar cap region) from the SuperDARN HF radar at McMurdo Station, Antarctica.

through the limited cusp region is sufficient to move the entire ionospheric slab of plasma to conform to the instantaneous solar wind conditions. This situation is contrasted with an alternate scenario where the solar wind momentum input is weak and the global flows would be best represented by the solar wind momentum transfer rates averaged over the roughly 1–4 h convection time scale of the polar cap (i.e., from dayside to nightside).

An estimate of the momentum transfer rates from the solar wind to the ionosphere on open field lines can be made by considering the polar cap as Alfvén wings as described by *Ridley* [2007] and *Kivelson and Ridley* [2008]. The upper limit for momentum transfer is given by equation 13.4. If we assume that the newly opened field lines occupy an area of $1R_E \times 0.1R_E$ in the dayside cusp region and that this area maps to $10R_E \times 1R_E$ in the solar wind, then the momentum transfer rate is

$$\frac{dp_x}{dt} = \frac{M_A B_{\text{IMF}}^2 A}{\mu_o} \approx 4\times10^4\ \text{N}, \tag{13.6}$$

where $M_A \sim 5$ is the Alfvén Mach number of the solar wind flow, $B_{\text{IMF}} \sim 5$ nT, and where we ignore the factor of 2 in equation 13.4 when considering only one hemisphere. The mass of the entire slab of plasma in the polar cap is roughly

$$m_{\text{isph}} = \rho_{\text{isph}} A_{\text{pc}}\, \Delta z, \tag{13.7}$$

where $A_{\text{pc}} \approx \pi(0.5R_E)^2$, $\rho_{\text{isph}} \approx (16\ \text{amu})(1.67\times10^{-27}\ \text{kg/amu})(2\times10^{11}\ \text{m}^{-3})$, and $\Delta z \approx 100$ km. Now consider changes in ionospheric plasma velocity for the limiting case of rapid variations in B_z. Assume that $\Delta v \sim 300$ m/s and $\Delta t \sim 5$ min (following the example fluctuations from Figure 13.4). The force required for these fluctuations is

$$F_x = m_{\text{isph}} \frac{\Delta v}{\Delta t} \approx 3\times10^4\ \text{N}, \tag{13.8}$$

which is comparable to the momentum transfer rate from the solar wind to a localized patch in the cusp region (equation 13.6). (We note that this estimate follows the formalism of *Vasyliunas* [2007], but we use the amplification of the Lorentz force in the cusp ionosphere to move the entire ionospheric slab in the polar region, neglecting momentum coupling to Earth's interior as is evident in the SuperDARN data for time scales ~5 min.) This rough calculation shows that indeed the solar wind is capable of changing the ionospheric convection on short time scales. Typically the fluctuations in B_z are more gradual (~ 10 min) and the ionospheric flows are also influenced by tail dynamics [*Cowley and Lockwood*, 1992]. As a result, the observed spread in velocities as a function of B_z, shown in Figure 13.5 from, might be expected. The coupling of the solar wind to Earth's ionosphere is of fundamental importance in the Dungey cycle.

Although ionospheric convection (as measured by the cross-polar cap potential) is largely driven by IMF B_z (e.g., see discussions by Ridley [2005] and Kivelson and Ridley [2008]), there remains a viscous contribution to the cross-polar cap potential (CPCP), Φ_V, that is poorly understood. Recent studies using the Lyon-Fedder-Mobarry (LFM) magnetohydrodynamic simulation by *Bruntz et al.* [2012] attempted to quantify the viscous potential and derived formulas for Φ_V as a function of solar wind density and velocity. *Bruntz et al.* [2012] found that the viscous potential increased with increasing solar wind density and velocities with values ranging between 10 and 80 kV. During southward B_{IMF} conditions (i.e., Dungey convection) the CPCP generally saturates at 150 kV with typical values ranging between 10s to 100 kV [Boyle *et al.*, 1997]. While the Dungey cycle appears to dominate over the viscous interaction at Earth, it should be noted that the viscous interaction is nonnegligible.

At the giant magnetospheres a comprehensive superDARN network is, obviously, not available and our knowledge of global ionospheric circulation is limited to ground-based, line-of-sight, Doppler-shifted H_3^+ emissions [*Stallard et al.*, 2001, 2003, 2004]. Nevertheless, due to the rapid rotation of these planets, the corotational flows are readily observed and deviations from corotation can be attributed to internal mass loading of the magnetodisc (on closed field lines) or to the solar wind control of the ionospheric flows. At Jupiter, the ionospheric flows are largely corotational and subcorotational with the exception of a region in the dawnside ionosphere that is held fixed in the inertial reference frame (Figure 13.6). It is assumed that this region is under solar wind control. At Saturn the ionospheric flows are largely subcorotational on closed field lines, consistent with the measured subcorotation of the magnetodisc, and the ionospheric flows in the region of open flux in the polar cap are consistent with corotation. Though the absolute mapping of open vs. closed flux is uncertain, the open polar cap does not appear to be held by the solar wind.

The incompressibility of the magnetic field at Jupiter requires at least a single-cell convection pattern localized on the dawnside ionosphere. The region is aurorally active, and there is no definitive evidence to date whether this region is on open or closed magnetic field lines. Motivated by no clear evidence of an open/closed boundary in the polar auroral emissions, *Delamere and Bagenal* [2010] argued that the dawnside convection was possibly driven by a viscous interaction on the dawn flank (see discussion of Jupiter's polar aurora in section 13.6.). On the other hand, *Cowley et al.* [2003] suggested that the ionospheric flows were instead part of a single-cell Dungey cycle. At

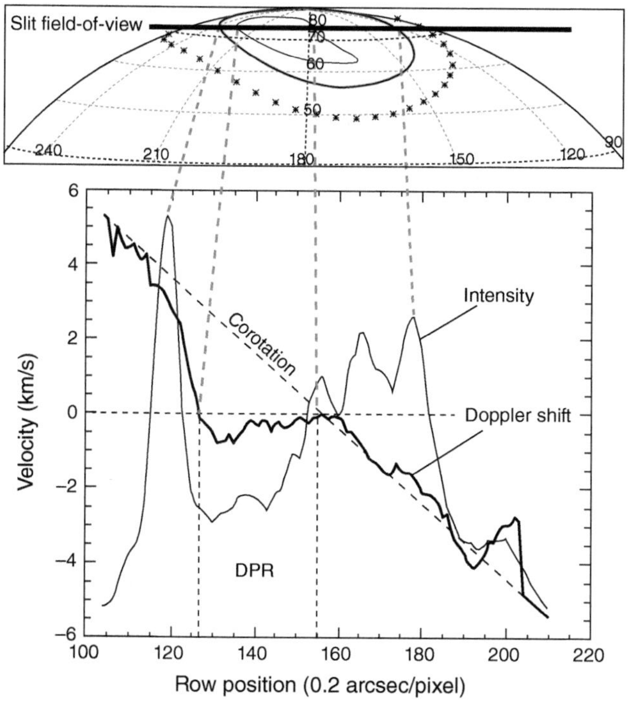

Figure 13.6 Viewing geometry (top) and Doppler-shifted H_3^+ IR emission profile (bottom) from *Stallard et al.* [2003] suggesting anti-sunward ionospheric flows (i.e., solar wind control) in the dark polar region (DPR).

Saturn, the magnetic field remains incompressible by simply corotating with the planet (trivial solution) if the influence of the solar wind and magnetosphere is negligible compared with thermospheric input.

13.4. VISCOUS INTERACTION

The effectiveness of reconnection at Jupiter and Saturn has been debated in the literature. Magnetopause plasma β gradients [*Masters et al.*, 2012; *Desroche et al.*, 2012, 2013], magnetosonic Mach number in the solar wind [*Scurry et al.*, 1994; *Grocott et al.*, 2009], and magnetopause shear flow conditions [*Desroche et al.*, 2012, 2013] have been considered as factors that may influence magnetic reconnection (see discussion for Saturn in Chapter 6). If nondriven reconnection is suppressed, then the viscous interaction may play a more important role at the giant magnetospheres. The viscous interaction transfers mass, momentum, and energy from the solar wind in the sheath to a low-latitude boundary layer on the planetward side of the magnetopause boundary (see discussion in Chapter 14), resulting in plasma circulation through the magnetosphere in a pair of viscous cells that can drive ionospheric convection [*Axford and Hines*, 1961]. The viscous interaction is complex, and for simplicity we discuss the KH instability as an example of a viscous process. This shear flow-driven instability requires super-Alfvénic shear flow, implying a plane for the wave k vector that minimizes magnetic tension across the boundary according to the instability condition [*Chandrasekhar*, 1961],

$$\left[\mathbf{k}\cdot(\mathbf{v}_1-\mathbf{v}_2)\right]^2 > \frac{\rho_1+\rho_2}{\mu_o\rho_1\rho_2}\left[(\mathbf{k}\cdot\mathbf{B}_1)^2+(\mathbf{k}\cdot\mathbf{B}_2)^2\right], \tag{13.9}$$

where $\mathbf{v}$ denotes velocity, ρ is the mass density, $\mathbf{B}$ is the magnetic field, $\mathbf{k}$ is the wave vector, and the indices refer to the two regions across the shear flow boundary.

At Earth, there is evidence of KH instabilities along the equatorial flanks of the magnetosphere [e.g., *Fairfield et al.*, 2000; *Otto and Fairfield,* 2000]. It is expected that the surface waves and vortex structures evolve symmetrically on the dawn and dusk flanks due to symmetry of the sheath flows with respect to the relatively slow magnetospheric plasma flows [*Ma et al.*, 2014a] though some asymmetry has been discussed by *Huba* [1994]. At the giant magnetospheres, on the other hand, the rapidly rotating magnetodisc generates a dawn/dusk asymmetry in the shear flow at the magnetopause boundary. In fact, the corotating flows in the subsolar region can also render the boundary KH unstable and *Desroche et al.* [2012, 2013] showed that much of Jupiter's and Saturn's magnetopause is KH unstable under all solar wind conditions.

In the subsolar region of the giant magnetospheres where the sheath flow is stagnant, a KH vortex will propagate with a velocity

$$v_{kh} = v_{ms}\frac{\rho_{ms}}{\rho_{ms}+\rho_{sh}}, \tag{13.10}$$

where ρ is the mass density and subscripts ms and sh refer to magnetosphere and magnetosheath, respectively [*Chandrasekhar*, 1961]. When the tailward sheath flow is added in the prenoon sector, then there will be a region where the vortex motion is nearly stationary if the momentum density is balanced across the boundary. On the duskside there is a small region with zero shear flow, yet based on equation 13.10, KH vortices formed in the subsolar region will be carried by the magnetospheric flow to the tail flank. *Ma et al.* [2014c] showed that indeed KH vortices will propagate through the postnoon region despite minimum shear flow and that the dawn vs. dusk development of KH vortices are highly asymmetric. Figure 13.7 shows the asymmetric evolution of KH vortices at Saturn's magnetopause boundary. In this numerical experiment the Sun is to the top, the subsolar region is in the middle, dusk is to the left, and dawn is to the right. Note the stretching of the spine connecting the vortices on the duskside.

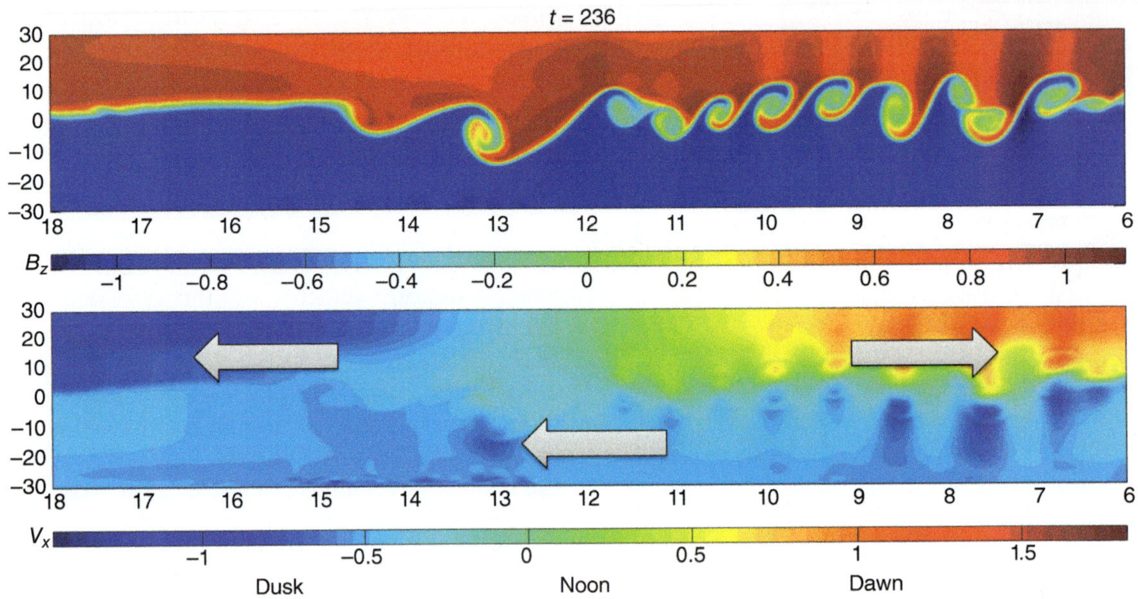

Figure 13.7 Asymmetric development of KH vortices at Saturn's magnetopause boundary. The sun is to the top, dusk is to the left, and dawn is to the right. Initially the sheath flow (top half) was initialized as an incompressible flow with increasing flow speed away from the subsolar point (middle).

A consequence of the asymmetric KH development is the KH-driven thinning of the magnetopause boundary along regions of the dusk flank. To date there is little evidence of magnetic reconnection operating at Jupiter's [*Walker and Russell*, 1985] or Saturn's [*McAndrews et al.*, 2008 *Lai et al.*, 2012] dayside magnetopause boundary. However, reconnection can be triggered by KH activity [*Ma et al.*, 2014] due to the thinning of the current sheet connecting two vortices. This is expected to occur more frequently in the postnoon sector and may be the driver of magnetic reconnection at Saturn given the difficulty of reconnection operating in the presence of super-Alfvénic flows [*La Belle-Hamer* et al., 1995] and gradients in the plasma β [*Masters et al.*, 2012].

Ultimately, the KH instability will involve many small-scale structures that form during the nonlinear vortex rollup. Mass transport has been discussed as a result of magnetic reconnection involving the in-plane components of the magnetic field [*Nykyri and Otto*, 2004]. However, *Cowee et al.* [2009] discussed superdiffusion (enhanced over classical diffusion) seen in hybrid simulations. *Delamere et al.* [2011] also confirmed the enhanced mass transport seen by *Cowee et al.* [2009] with two-dimensional hybrid simulations of Saturn's magnetopause boundary. Particle-in-cell simulations by *Karimabadi et al.* [2013] indicate that the vortex motion can lead to turbulence where the shear flow energy is driven into bulk heating of the plasma. The expected net effect of this type of viscous process is a tangential stress generated at the magnetopause boundary. If the IMF becomes entangled in this boundary mixing, then the viscous interaction can be significant and could, potentially, be the primary solar wind interaction mechanism at the giant magnetospheres [*Delamere and Bagenal*, 2010].

13.5. INTERNAL DRIVERS AND PLASMA TRANSPORT

The giant magnetospheres have the additional complexity of an internal driver. The rapid rotation introduces centrifugal stresses on the internally loaded plasma sheet, whereby plasma can be ejected down the magnetotail via a Vasyliunas cycle of magnetic reconnection [*Vasyliunas*, 1983] (Figure 13.8). However, this process is complicated by the nonadiabatic heating of the plasma sheet during radial transport. Plasma originating at the moons (Io and Enceladus) is expected to cool adiabatically during radial transport. Yet significant heating occurs instead. *Bagenal and Delamere* [2011] showed that an input power of 3–16 TW at Jupiter and 75–630 GW at Saturn is required to explain the heating. Furthermore, *Southwood and Kivelson* [2001] argued that a pressure anisotropy develops during radial transport that can result in ballooning of the magnetodisc, leading to a breakdown in containment as Alfvénic communication along the magnetic field is truncated. *Kivelson and*

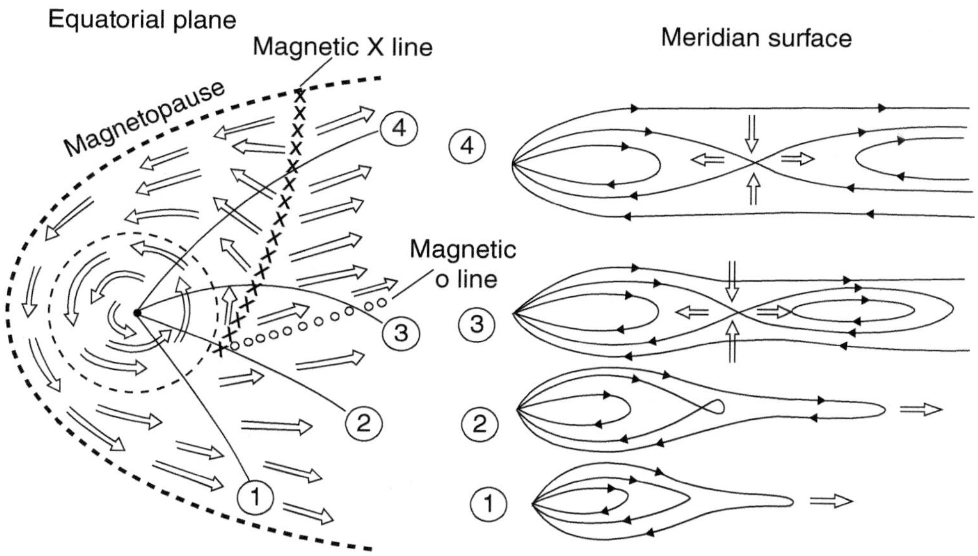

Figure 13.8 Schematic of the Vasyliunas cycle [*Vasyliunas*, 1983], showing the local time variation of the detachment of plasmoids.

Southwood [2005] proposed an alternative explanation for the Vasyliunas cycle for magnetotail flow, showing that the dynamics of streaming outflow on the dusk flank are consistent with that of an explosive ballooning instability (Figure 13.9).

Reconnection in the magnetotails of Jupiter and Saturn has been observed and attributed to the on-going Vasyliunas cycle [*Russell et al.*, 1999; *Kronberg et al.*, 2005; *Vogt et al.*, 2010; *Hill et al.*, 2008; *Jackman et al.*, 2011]. However, *Bagenal* [2007] noted that the inferred occurrence of plasmoids in Jupiter's tail from auroral signatures is insufficient to account for the roughly 1000 kg/s that must be lost from Jupiter's magnetosphere. Instead, *Bagenal* [2007] argued that most of the iogenic plasma must be lost via small-scale drizzle on the flanks. A similar conclusion was reached by *Jackman et al.* [2013] for the mass loss mechanism from Saturn's magnetosphere. The physical mechanism for drizzle could be related to the *Kivelson and Southwood* [2005] proposal for ballooning; however, the complete picture of plasma loss should also include the solar wind interaction. A few outstanding questions that remain include:

- How is solar wind-controlled reconnection assimilated into the magnetotail?
- How does the viscous interaction affect the plasma loss on the flanks of the magnetosphere?
- What is the relative importance of solar wind-driven (reconnection and viscous) vs. internally driven (Vasyliunas/ballooning) plasma loss from the giant magnetospheres?

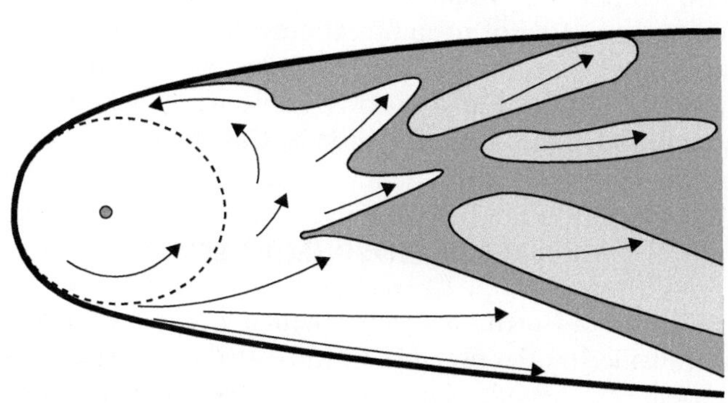

Figure 13.9 Magnetotail flow with a ballooning unstable plasma sheet [*Kivelson and Southwood*, 2005]. Plasmoids are released down the tail at nearly their release velocity and empty flux tubes begin to flow sunward on the dawn flank.

13.6. AURORAL SIGNATURES

Auroral emissions provide important clues regarding the nature of the solar wind interaction. Experience at Earth shows that open field lines are generally devoid of auroral emissions and a very rough approximation of the open flux in the polar cap can be made by estimating the magnetic flux in an area polarward of the main auroral oval [e.g., *Paschmann et al.*, 2002]. Similar assumptions have been made for the case of Saturn's main auroral emissions by *Badman et al.* [2005] and *Radioti et al.* [2011] and essentially confirmed with simultaneous Hubble Space Telescope (HST) and Cassini in situ observations [*Bunce et al.*, 2008]. Jupiter, on the other hand, is perplexing. The solar wind control of Jupiter's dawnside polar region led *Cowley et al.* [2003] to conclude that this was a region of open flux and part of a single-cell convection

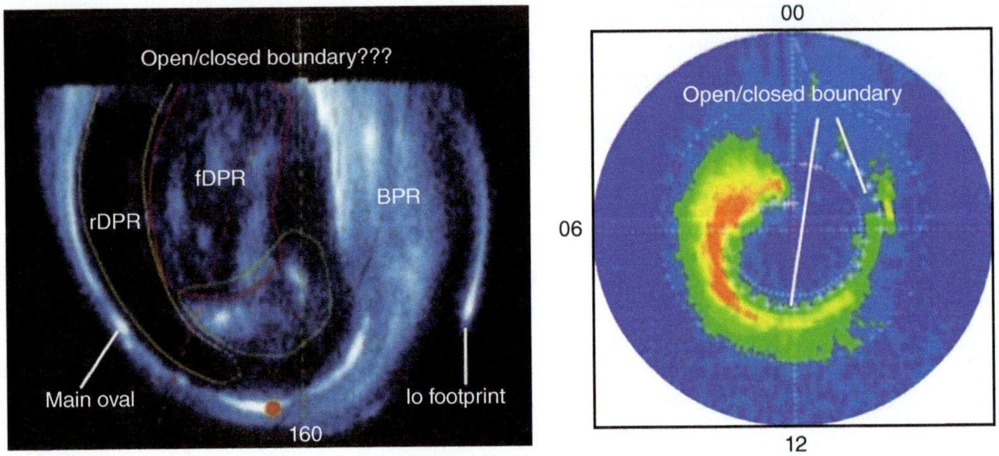

Figure 13.10 Left: Jupiter's aurora [from *Grodent et al.*, 2003], showing an aurorally active polar region. The dark polar region (DPR) and the bright polar region (BPR) follow the definitions of *Stallard et al.* [2003]. Auroral emission intensity is generally weaker in the dawn sector but rarely absent (i.e., DPR), making definition of an open/closed magnetic flux boundary with UV emissions difficult. Right: Saturn's spiral aurora from 2004-01-28 01:27:28 [from *Badman et al.*, 2005]. The open/closed boundary is indicated and verified by *Bunce et al.* [2008] with Cassini's exploration of Saturn's high-latitude magnetosphere.

that must be associated with the Dungey cycle. Yet this region is aurorally active in comparison to Earth's and Saturn's open polar cap. Figure 13.10 compares images from the HST of Jupiter's and Saturn's UV aurora. Jupiter clearly lacks a well-defined open polar cap in terms of auroral emissions, while in situ Cassini observations confirm an aurorally dark open polar cap [*Radioti et al.*, 2011; *Bunce et al.*, 2008]. The apparent solar wind control of this region from the infrared Doppler-shifted H_3^+ emissions of *Stallard et al.* [2003] combined with the auroral activity led *Delamere and Bagenal* [2010] to postulate that the magnetosphere could be closed with solar wind control enabled by a viscous interaction on the dawn flank. *Vogt et al.* [2011] developed a flux equivalence magnetic field model for Jupiter and concluded, contrary to the suggestion of *Delamere and bagenal* [2010], that Jupiter's magnetosphere contains ~700 GWb of open flux, i.e., much of the region poleward of Jupiter's main auroral oval that remains aurorally active. The much-anticipated arrival of the Juno spacecraft at Jupiter will hopefully resolve this debate.

Magnetic reconnection in Saturn's tail is inferred from observations of Saturn's auroral emissions by the HST [*Cowley et al.*, 2005]. During the January 2004 HST observing campaign, Cassini monitored the upstream solar wind conditions. Following the arrival of solar wind compression, the aurora expanded polarward on the dawnside, suggesting significant tail reconnection in response solar wind compression. *Crary et al.* [2005] concluded that solar wind dynamic pressure was the primary factor for controlling Saturn's aurora, in contrast to Earth, where the IMF B_z component plays a more important role. It is difficult to understand the response of Jupiter's auroral emissions to the solar wind given that the main oval of emissions is internally driven and the polar emissions are highly variable. However, *Nichols et al.* [2007] concluded that brightened emissions occurred during magnetospheric compression followed by an expansion.

Dayside magnetic reconnection at Saturn was inferred by *Radioti et al.* [2011] and *Badman et al.* [2013] using auroral images taken with the Cassini UVIS instrument. The images show an intensification of the main oval in the subsolar region followed by consecutive bifurcations of the main ring of emissions. The bifurcations are assumed to be transient in nature and are located poleward of the open/closed field line boundary due to tailward convection. *Radioti et al.* [2011] argued that the addition of ~10% of open flux contained within the polar cap is similar to the fraction estimated for flux transfer events (FTEs) at Earth. Such FTEs have, however, not been found in the subsolar region by Cassini in situ observations [*Lai et al.*, 2012].

13.7. MAGNETOSPHERIC SCALE

The size of Earth's magnetosphere is easily understood by considering the pressure balance at the dayside magnetopause. The subsolar distance of the magnetopause, R_{MP}, is determined by balancing the planetary magnetic field pressure with the dynamic pressure of the solar wind, where $R_{MP} \sim (\rho v^2)^{-1/6}$. The situation at the giant magnetospheres is different. The internal sources of plasma at Jupiter (Io's volcanos) and Saturn (Enceladus' water plumes) inflate the magnetospheres considerably to the point where the pressure balance at the magnetopause is dominated by the plasma pressure of the high-β magnetodisc (see discussions by *Krupp et al.* [2004], *Mitchell et al.* [2009], and *Mauk et al.* [2009]). At Jupiter, $R_{MP} \sim (\rho v^2)^{-a}$, where $\frac{1}{5} < a < \frac{1}{4}$ [*Slavin et al.*, 1985; *Huddleston et al.*, 1998; *Joy et al.*, 2002; *Alexeev and Belenkaya*, 2005]. Factor-of-10 variations in the solar wind pressure at 5 AU results in $50\,R_J < R_{MP} < 100\,R_J$ at Jupiter [*McComas et al.*, 2014].

Delamere and Bagenal [2013] showed that the loss of plasma from Jupiter (500–1000 kg/s) and Saturn (20–50 kg/s) is consistent with magnetospheric scale. *Delamere and Bagenal* [2013] argued that the magnetospheric scale must fundamentally be determined by the ability of the solar wind to couple sufficient momentum from the solar wind through magnetic tangential stresses (equation 13.4) to pick up the magnetospheric plasma. If the momentum transfer rates are insufficient, then the magnetodisc must expand, inflating the magnetosphere. (In reality, the magnetospheric plasma sources are relatively constant, and the magnetospheric size changes according to solar wind conditions.) In addition, *Delamere and Bagenal* [2010] showed that the location of the magnetic X line in the tail is consistent with magnetic stress balance between the solar wind and the planetary magnetic field, suggesting that globally the solar wind interaction must be generating tangential stresses (perhaps viscous) at the magnetopause boundary. It should be emphasized that a description of the solar wind interaction that does not account for the Dungey cycle, the Vasyliunas cycle, and the viscous interaction is incomplete.

13.8. SOLAR WIND INTERACTION AND MAGNETOTAIL STRUCTURE

The magnetotail is somewhat of an enigma (see Chapter 1). In an open magnetosphere, the tail can be viewed as the draping of the open magnetic field lines in the polar cap due to the solar wind flow. The length of the tail can be estimated by considering the convection time of the ionospheric magnetic footprint across the polar cap multiplied by the solar wind speed [e.g. *Kennel and Coroniti,* 1977]. Such estimates yield a tail length of roughly 1 AU (for the case of Jupiter). Yet, Jupiter's magnetotail was observed 4 AU downstream while the Voyager spacecraft was approaching Saturn [*Scarf et al.*, 1981; *Kurth et al.*, 1981]. The identification of the magnetotail was based on observations of continuum radiation generated within Jupiter's magnetosphere, requiring a low-density conduit from Jupiter to the spacecraft. A cylindrical tail structure devoid of plasma would facilitate such an observation. On the other hand, *Kurth et al.* [1981] suggested that the continuum radiation observations may also be consistent with a cometlike tail disruption. Such disruptions have been observed and even suggested for Earth's magnetotail due to velocity shears in the solar wind [*Borovsky*, 2012]. It is therefore worthwhile comparing magnetotail structures with that of comets (see discussion of cometary interactions in Chapter 10).

For the giant magnetospheres, the possible small polar caps (particularly at Jupiter where a definitive identification of the open polar cap via auroral emissions is difficult) may have limited application to the overall magnetotail structure. Wings of closed flux along the flanks are generated by global MHD simulations of Earth's and Saturn's magnetospheres [*Song et al.*, 2001; *Jia et al.*, 2012]. Such wing structures contribute to the tail structure, and when combined with the viscous entanglement of the IMF along the magnetopause boundary, the cometary model of the magnetotail is perhaps entirely relevant. In the event of a strong viscous interaction, the draping of the IMF would approach the momentum transfer limit given by equation 13.4. Indeed, at high Alfvén Mach number the interaction would resemble that of a cometary tail. At Earth, the relatively large open polar cap can provide a basis for magnetotail structure with modest contributions from the viscous interaction.

The New Horizons spacecraft, bound for Pluto, made plasma measurements in Jupiter's magnetotail [*McNutt et al.*, 2007; *McComas et al.*, 2007]. These unique data sets (previous spacecraft at Earth, Jupiter, and Saturn have covered the tail region to within an Earth-normalized value of $30\,R_E$ [*Jackman and Arridge*, 2011]) from the Solar Wind at Pluto (SWAP) and Pluto Energetic Particle Spectrometer Science Investigation (PEPPSI) revealed an unexpected tail structure that led *McComas and Bagenal* [2007] to conclude that Jupiter's magnetosphere is largely closed and is fundamentally different from Earth's. The spacecraft moved down the duskside of the tail and exited the tail $> 2000\,R_J$ on the dusk flank. The SWAP instrument (35 eV/q to 7.5 keV/q) initially showed the hot plasma magnetodisc, but in the tail the flows became increasingly more variable in flux and mean energy. Large-scale, quasi-periodic intensity enhancements between $400\,R_J$ and $1600\,R_J$ were interpreted as plasmoids of magnetospheric origin moving down the tail. The PEPPSI instrument observed a series of dispersion events in the deep tail (i.e., $>500\,R_J$) and inferred their origin to be roughly $100\,R_J$ to $1000\,R_J$ from the planet [*Hill et al.*, 2008]. One possible interpretation is that the bursts were associated with reconnection. Yet these bursts span a range that is much larger than the reconnection X line distance of $100\,R_J$ to $150\,R_J$ observed by Galileo magnetometer data and energetic particle data [*Woch et al.*, 2002; *Kronberg et al.*, 2005; *Vogt et al.*, 2010; *Kasahara et al.*,

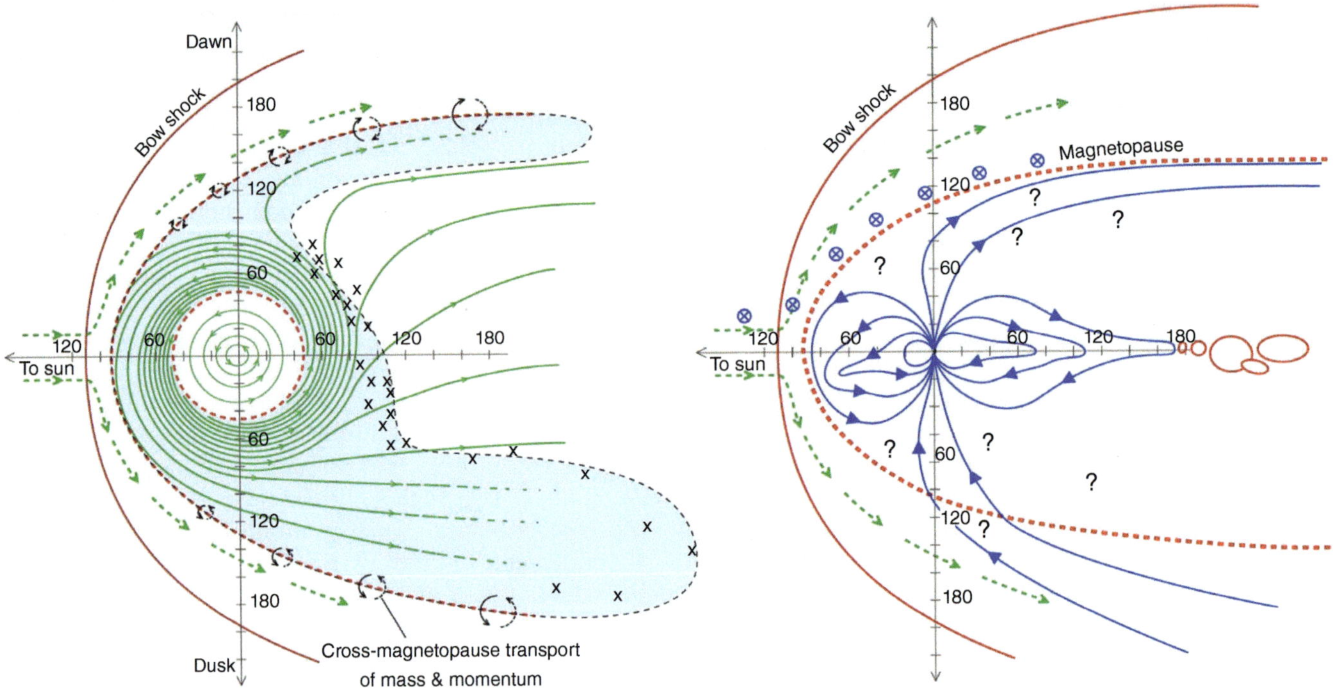

Figure 13.11 Schematic of the equatorial (left) noon-midnight (right) meridians for Jupiter's magnetosphere. The left panel is adapted from *Delamere and Bagenal* [2010] showing the average plasma flow (green), the location of magnetic X line(s), and the regions of closed magnetic flux (light blue). The wings of closed flux on the dawn and dusk flanks are taken from global MHD simulations [*Song et al.*, 2001; *Jia et al.*, 2012] and based on New Horizons dispersive events [*McNutt et al.*, 2007; *Hill et al.*, 2009]. The flux content of the wings could be comparable with the open flux content and therefore must be considered in the overall magnetotail structure. The magnetic field (IMF) is shown into the page on the right and emphasizes that the magnetic field flows preferentially over the poles for a polar-flattened magnetosphere [*Desroche et al.*, 2012] so that flux that may open on the dayside may not drive a Dungey cycle but instead will get carried back to a distance where it no longer communicates with the planet and becomes entangled in boundary layer mixing.

2013]. The analyses of these observations are ongoing but certainly provoke a number of such questions as:

• How is plasma lost from the magnetosphere on the dusk flank? Does plasma "drizzle" from the magnetosphere?

• Do pressure anisotropies contribute to the evolution (e.g., ballooning) of magnetic flux tubes on the dusk flank?

• How large is the closed flux region on the dusk flank and what is the magnetic field topology?

• How is current sheet thinning achieved for reconnection to occur at the giant magnetospheres (see Chapter 17)?

The New Horizons spacecraft did not carry a magnetometer, so some of these questions cannot be answered. But much can be done with the plasma measurements and modeling studies that address kinetic aspects involving high-β plasmas. In the meanwhile, we can address limiting aspects of the solar wind interaction (i.e., closed, viscously driven magnetospheres, to open, viscous-free interactions) to explore possible scenarios that can address the New Horizons observations. *Delamere and Bagenal* [2013] suggested that the flanks of the magnetotail contain closed wings of magnetic flux (see Figure 13.11) that extend down the flanks for considerable distances. This interpretation follows the closed-magnetosphere limit. Ultimately, models of the rapidly rotating, giant magnetosphere should consider all aspects of the solar wind interaction for understanding magnetotail structure.

13.9. CONCLUSIONS

Magnetotail structures are common throughout the solar system. While the overall topology of a draped magnetic field in a super-Alfvénic solar wind flow is similar between comets and magnetospheres, the basic mechanisms that cause the interplanetary magnetic field to drape are different. Magnetic reconnection couples the solar wind flow to a conductive ionosphere/atmosphere. Escaping cometary atmospheres mass load the solar wind flow via photoionization and charge exchange processes and the resulting pickup of cometary material into the solar wind flow leads to tail formation. Viscous interactions which include small-scale, intermittent reconnection at the magnetopause boundary lead to mass, momentum,

Table 13.1 Characteristics of solar wind interaction with Earth, Jupiter, and Saturn

	Earth	Jupiter	Saturn
In situ evidence of dayside reconnection (FTEs)	Yes [*Russell*, 1984]	~ Few [*Walker and Russell*, 1985]	Uncertain [*Lai et al.*, 2012]
Auroral evidence of dayside reconnection	Poleward moving auroral forms [*Fasel et al.*, 1995]	Quasi-periodic polar flares [*Bonfond et al.*, 2011	Noon-sector auroral bifurcations [*Radioti et al.*, 2011]
Auroral evidence of open flux	Yes	Intermittent [*Pallier and Prangè*, 2001]	Yes [*Badman et al.*, 2005; *Bunce et al.*, 2008]
Open flux content (GWb)	0.2–0.9 [e.g. *Milan et al.*, 2004]	500–1000 [*Nichols et al.*, 2006; *Vogt et al.*, 2011]	15–50 [*Badman et al.*, 2005]
Distance to subsolar magnetopause boundary, R_{MP} (Planetary radii)	10	75	22
Expected R_{MP} for planetary dipole only	10	30–40	15–18
Estimate of open flux content[a] (GWb)	0.76	156	22
In situ evidence of Kelvin-Helmholtz at magnetopause boundary	Yes [e.g., *Fairfield et al.*, 2000]	Uncertain	Yes [*Masters et al.*, 2010; *Delamere et al.*, 2013]
In situ evidence of tail reconnection	Yes [*Nishida*, 1984]	Many cases [*Russell et al.*, 1999; *Kronberg et al.*, 2005; *Vogt et al.*, 2011]	~Few [*Jackman et al.*, 2007; *Hill et al.*, 2008]

[a]Open flux content estimated by integrating the flux of the dipole field through the equatorial plane from infinity to the subsolar magnetopause standoff distance following *Vasyliuas* [2009].

and energy transport and this, too, can deflect the IMF into a tail configuration.

Little data are available in the distant magnetotails of Earth and Saturn. But Jupiter's distant magnetotail has been explored by two separate spacecraft (Voyager and New Horizons). The long tail (~>4 AU) is surprising and a complete description of its structure and topology is perplexing. Understanding the basic processes that govern the solar wind interaction is crucial for establishing the conditions for tail formation and stability. We summarize by comparing and contrasting the solar wind interaction at Earth, Jupiter, and Saturn. Table 13.1 provides a summary of observations of key aspects of the solar wind interaction. While many similar characteristics are present, the relative importance of different physical processes cannot be clearly distinguished. For example, in the case of tail reconnection, the release of plasmoids down the tails of the giant magnetospheres cannot distinguish between a Dungey cycle of reconnection and the Vasyliunas cycle. Estimates of the open flux content of Jupiter's polar cap seem excessive compared with the expected maximum amount of open flux. If Jupiter's magnetosphere contains so much open flux, then how are the polar auroral emissions generated? If Saturn's polar aurora show significant dayside reconnection activity, then why do definitive in situ FTEs elude detection? These are some of the issues that will hopefully be addressed during the end-of-mission polar orbit of the Cassini spacecraft and the arrival of the polar-orbiting Juno spacecraft at Jupiter.

REFERENCES

Alexeev, I. I., and E. S. Belenkaya (2005), Modeling of the Jovian Magnetosphere, *Ann. Geophys.*, *23*, 809–826, doi:10.5194/angeo-23-809-2005.

Axford, W. I., and C. O. Hines (1961), A unifying theory of high latitude geophysical phenomena and geomagnetic storms, *Can. J. Phys.*, *39*, 1433–1464.

Badman, S. V., and S. W. H. Cowley (2007), Significance of Dungey-cycle flows in Jupiter's and Saturn's magnetospheres, and their identification on closed equatorial field lines, *Ann. Geophys.*, *25*, 941–951.

Badman, S. V., E. J. Bunce, J. T. Clarke, S. W. H. Cowley, J.-C. Gérard, D. Grodent, and S. E. Milan (2005), Open flux estimates in Saturn's magnetosphere during the January 2004 Cassini-HST campaign, and implications for reconnection rates, *J. Geophys. Res. (Space Phys.)*, *110*, A11,216, doi:10.1029/2005JA011240.

Badman, S. V., A. Masters, H. Hasegawa, M. Fujimoto, A. Radioti, D. Grodent, N. Sergis, M. K. Dougherty, and A. J. Coates (2013), Bursty magnetic reconnection at Saturn's magnetopause, *Geophys. Res. Lett.*, *40*, 1027–1031, doi:10.1002/grl.50199.

Bagenal, F. (2007), The magnetosphere of Jupiter: Coupling the equator to the poles, *J. Atmos. Solar-Terrest. Phys.*, *69*, 387–402, doi:10.1016/j.jastp.2006.08.012.

Bagenal, F., and P. A. Delamere (2011), Flow of mass and energy in the magnetospheres of Jupiter and Saturn, *J. Geophys. Res. (Space Phys.)*, *116*, A05209, doi:10.1029/2010JA016294.

Bonfond, B., M. F. Vogt, J.-C. Gérard, D. Grodent, A. Radioti, and V. Coumans (2011), Quasi-periodic polar flares at Jupiter: A signature of pulsed dayside reconnections?, *Geophys. Res. lett.*, *38*, L02104, doi:10.1029/2010GL045981.

Borovsky, J. E. (2012), The effect of sudden wind shear on the Earth's magnetosphere: Statistics of wind shear events and CCMC simulations of magnetotail disconnections, *J. Geophys. Res. (Space Phys.)*, *117*, A06224, doi:10.1029/2012JA017623.

Boyle, C. B., P. H. Reiff, and M. R. Hairston (1997), Empirical polar cap potentials, *102*, 111–126, doi:10.1029/96JA01742.

Brice, N. M., and G. A. Ioannidis (1970), The magnetospheres of Jupiter and Earth, *Icarus*, *13*, 173–, doi:10.1016/0019-1035(70)90048-5.

Bristow, W. A., J. Spaleta, and R. T. Parris (2011), First observations of ionospheric irregularities and flows over the south geomagnetic pole from the Super Dual Auroral Radar Network (SuperDARN) HF radar at McMurdo Station, Antarctica, *J. Geophys. Res. (Space Phys.)*, *116*, A12325, doi:10.1029/2011JA016834.

Bruntz, R., R. E. Lopez, S. K. Bhattarai, K. H. Pham, Y. Deng, Y. Huang, M. Wiltberger, and J. G. Lyon (2012), Investigating the viscous interaction and its role in generating the ionospheric potential during the Whole Heliosphere Interval, *J. Atmos. Solar-Terrest. Phys.*, *83*, 70–78, doi:10.1016/j.jastp.2012.03.016.

Bunce, E. J., et al. (2008), Origin of Saturn's aurora: Simultaneous observations by Cassini and the Hubble Space Telescope, *J. Geophys. Res. (Space Phys.)*, *113*, A09209, doi:10.1029/2008JA013257.

Chandrasekhar, S. (1961), *Hydrodynamic and Hydromagnetic Stability*, Dover Publications Inc., New York.

Cowee, M. M., D. Winske, and S. P. Gary (2009), Two-dimensional hybrid simulations of superdiffusion at the magnetopause driven by Kelvin-Helmholtz instability, *J. Geophys. Res. (Space Phys.)*, *114*, 10,209, doi:10.1029/2009JA014222.

Cowley, S. W. H., and M. Lockwood (1992), Excitation and decay of solar wind-driven flows in the magnetosphere-ionosphere system, *Ann. Geophys.*, *10*, 103–115.

Cowley, S. W. H., E. J. Bunce, T. S. Stallard, and S. Miller (2003), Jupiter's polar ionospheric flows: Theoretical interpretation, *Geophys. Res. Lett.*, *30*, 5, 1220, doi:10.1029/2002GL016030.

Cowley, S. W. H., S. V. Badman, E. J. Bunce, J. T. Clarke, J. Gérard, D. Grodent, C. M. Jackman, S. E. Milan, and T. K. Yeoman (2005), Reconnection in a rotation-dominated magnetosphere and its relation to Saturn's auroral dynamics, *J. Geophys. Res. (Space Phys.)*, *110*, 2201, doi:10.1029/ 2004JA010796.

Cowley, S. W. H., S. V. Badman, S. M. Imber, and S. E. Milan (2008), Comment on "Jupiter: A fundamentally different magnetospheric interaction with the solar wind" by D. J. McComas and F. Bagenal, *Geophys. Res. Lett.*, *35*, 10,101, doi:10.1029/2007GL032645.

Crary, F. J., et al. (2005), Solar wind dynamic pressure and electric field as the main factors controlling Saturn's aurorae, *Nature*, *433*, 720–722, doi:10.1038/nature03333.

Delamere, P. A., and F. Bagenal (2010), Solar wind interaction with Jupiter's magnetosphere, *Journal of Geophysical Research (Space Physics)*, *115*, 10,201–+, doi:10.1029/2010JA015347.

Delamere, P. A., and F. Bagenal (2013), Magnetotail structure of the giant magnetospheres: Implications of the viscous interaction with the solar wind, *J. Geophys. Res. (Space Phys.)*, *118*, 7045–7053, doi:10.1002/2013JA019179.

Delamere, P. A., R. J. Wilson, and A. Masters (2011), Kelvin-Helmholtz instability at Saturn's magnetopause: Hybrid simulations, *J. Geophys. Res. (Space Phys.)*, *116*, A10222, doi:10.1029/2011JA016724.

Delamere, P. A., R. J. Wilson, S. Eriksson, and F. Bagenal (2013), Magnetic signatures of Kelvin-Helmholtz vortices on Saturn's magnetopause: Global survey, *J. Geophys. Res. (Space Phys.)*, *118*, 393–404, doi:10.1029/2012JA018197.

Desroche, M., F. Bagenal, P. A. Delamere, and N. Erkaev (2012), Conditions at the expanded Jovian magnetopause and implications for the solar wind interaction, *J. Geophys. Res. (Space Phys.)*, *117*, A07202, doi:10.1029/2012JA017621.

Desroche, M., F. Bagenal, P. A. Delamere, and N. Erkaev (2013), Conditions at the magnetopause of Saturn and implications for the solar wind interaction, *J. Geophys. Res. (Space Phys.)*, *118*, 3087–3095, doi:10.1002/jgra.50294.

Dungey, J. W. (1961), Interplanetary magnetic field and the auroral zones, *Phys. Rev. Lett.*, *6*(2), 47–48, doi:10.1103/PhysRevLett.6.47.

Fairfield, D. H., A. Otto, T. Mukai, S. Kokubun, R. P. Lepping, J. T. Steinberg, A. J. Lazarus, and T. Yamamoto (2000), Geotail observations of the Kelvin-Helmholtz instability at the equatorial magnetotail boundary for parallel northward fields, *J. Geophys. Res.*, *105*, 21,159–21,173.

Fasel, G. J., L. C. Lee, and R. W. Smith (1995), Dayside poleward-moving auroral forms: A brief review, *AGu Geophys. Monogr. Ser.*, *90*, 439.

Grocott, A., S. V. Badman, S. W. H. Cowley, S. E. Milan, J. D. Nichols, and T. K. Yeoman (2009), Magnetosonic Mach number dependence of the efficiency of reconnection between planetary and interplanetary magnetic fields, *J. Geophys. Res. (Space Phys.)*, *114*, A07219, doi:10.1029/2009JA014330.

Grodent, D., J. T. Clarke, J. H. Waite, S. W. H. Cowley, J. Gérard, and J. Kim (2003), Jupiter's polar auroral emissions, *J. Geophys. Res. (Space Phys.)*, *108*, 1366, doi:10.1029/2003JA010017.

Hill, M. E., D. K. Haggerty, R. L. McNutt, and C. P. Paranicas (2009), Energetic particle evidence for magnetic filaments in Jupiter's magnetotail, *J. Geophys. Res. (Space Phys.)*, *114*, 11,201, doi:10.1029/2009JA014374.

Hill, T. W., et al. (2008), Plasmoids in Saturn's magnetotail, *J. Geophys. Res. (Space Phys.)*, *113*, A01214, doi:10.1029/2007JA012626.

Huba, J. D. (1994), Hall dynamics of the Kelvin-Helmholtz instability, *Phys. Rev. Lett.*, *72*, 2033–2036, doi:10.1103/PhysRevLett.72.2033.

Huba, J. D. (1996), The Kelvin-Helmholtz instability: Finite Larmor radius magnetohydrodynamics, *Geophys. Res. Lett.*, *23*, 2907–2910, doi:10.1029/96GL02767.

Huddleston, D. E., C. T. Russell, M. G. Kivelson, K. K. Khurana, and L. Bennett (1998), Location and shape of the

Jovian magnetopause and bow shock, *J. Geophys. Res.*, *103*, 20,075–20,082, doi:10.1029/98JE00394.

Jackman, C. M., and C. S. Arridge (2011), Statistical properties of the magnetic field in the Kronian magnetotail lobes and current sheet, *J. Geophys. Res. (Space Phys.)*, *116*, A05224, doi:10.1029/2010JA015973.

Jackman, C. M., C. T. Russell, D. J. Southwood, C. S. Arridge, N. Achilleos, and M. K. Dougherty (2007), Strong rapid dipolarizations in Saturn's magnetotail: In situ evidence of reconnection, *Geophys. Res. Lett.*, *34*, L11203, doi:10.1029/2007GL029764.

Jackman, C. M., J. A. Slavin, and S. W. H. Cowley (2011), Cassini observations of plasmoid structure and dynamics: Implications for the role of magnetic reconnection in magnetospheric circulation at Saturn, *J. Geophys. Res. (Space Phys.)*, *116*, A10212, doi:10.1029/2011JA016682.

Jackman, C. M., , J. A. Slavin, M. Kivelson, D. J. Southwood, N. Achilleos, M. Thomsen, G. A. DiBraccio, J. Eastwood, M. Freeman, and M. Dougherty (2013), Properties of reconnection signatures in Saturn's tail: Local and global effects, *Geophys. Res. Lett.*., submitted.

Jia, X., K. C. Hansen, T. I. Gombosi, M. G. Kivelson, G. Tóth, D. L. DeZeeuw, and A. J. Ridley (2012), Magnetospheric configuration and dynamics of Saturn's magnetosphere: A global MHD simulation, *J. Geophys. Res. (Space Phys.)*, *117*, A05225, doi:10.1029/2012JA017575.

Johnson, J. R., and C. Z. Cheng (1997), Kinetic Alfvén waves and plasma transport at the magnetopause, *Geophys. Res. Lett.*, *24*, 1423–1426, doi:10.1029/97GL01333.

Johnson, J. R., and C. Z. Cheng (2001), Stochastic ion heating at the magnetopause due to kinetic Alfvén waves, *Geophys. Res. Let*, *28*(23), 4421–4424, doi:10.1029/2001GL013509.

Johnson, J. R., C. Z. Cheng, and P. Song (2001), Signatures of mode conversion and kinetic Alfvén waves at the magnetopause, *Geophys. Res. Lett.*, *28*(2), 227–230, doi:10.1029/2000GL012048.

Joy, S. P., M. G. Kivelson, R. J. Walker, K. K. Khurana, C. T. Russell, and T. Ogino (2002), Probabilistic models of the Jovian magnetopause and bow shock locations, *J. Geophys. Res. (Space Phys.)*, *107*, 1309, doi:10.1029/2001JA009146.

Karimabadi, H., et al. (2013), Coherent structures, intermittent turbulence, and dissipation in high-temperature plasmas, *Phys. Plasmas*, *20*,01,2303.

Kasahara, S., E. A. Kronberg, T. Kimura, C. Tao, S. V. Badman, A. Masters, A. Retinò, N. Krupp, and M. Fujimoto (2013), Asymmetric distribution of reconnection jet fronts in the Jovian nightside magnetosphere, *J. Geophys. Res. (Space Phys.)*, *118*, 375–384, doi:10.1029/2012JA018130.

Kennel, C. F., and F. V. Coroniti (1977), Possible origins of time variability in Jupiter's outer magnetosphere. 2. Variations in solar wind magnetic field., *Geophys. Res. Lett.*, *4*, 215–218, doi:10.1029/GL004i006p00215.

Kivelson, M. G., and A. J. Ridley (2008), Saturation of the polar cap potential: Inference from Alfvén wing arguments, *J. Geophys. Res. (Space Phys.)*, *113*, 5214, doi:10.1029/2007JA012302.

Kivelson, M. G., and D. J. Southwood (2005), Dynamical consequences of two modes of centrifugal instability in Jupiter's outer magnetosphere, *J. Geophys. Res. (Space Phys.)*, *110*, 12,209, doi:10.1029/2005JA011176.

Kronberg, E. A., J. Woch, N. Krupp, A. Lagg, K. K. Khurana, and K. Glassmeier (2005), Mass release at Jupiter: Substorm-like processes in the Jovian magnetotail, *J. Geophys. Res. (Space Phys.)*, *110*, 3211, doi:10.1029/2004JA010777.

Krupp, N., et al. (2004), The Dynamics of the Jovian magnetosphere. In *Jupiter: The Planet, Satellites and Magnetosphere,* edited by Bagenal, F., T. Dowling, and W. McKinnon, pp. 617–638, Cambridge University Press.

Kurth, W. S., D. A. Gurnett, F. L. Scarf, R. L. Poynter, and J. D. Sullivan (1981), Voyager observations of Jupiter's distant Magnetotail, *J. Geophys. Res. 86*, 8402–8412, doi:10.1029/JA086iA10p08402.

La Belle-Hamer, A. L., A. Otto, and L. C. Lee (1995), Magnetic reconnection in the presence of sheared flow and density asymmetry: Applications to the Earth's magnetopause, *J. Geophys. Res.*, *100*, 11,875.

Lai, H. R., H. Y. W. C. T. Russell, C. S. Arridge, and M. K. M. K. Dougherty (2012), Reconnection at the magnetopause of saturn: Perspective from tte occurrence and magnetosphere size, *J. Geophys. Res.*, *117*(A5), A05,222.

Lee, L. C., R. K. Albano, and J. R. Kan (1981), Kelvin-Helmholtz instability in the magnetopause-boundary layer region, *J. Geophys. Res.*, *86*, 54–58, doi:10.1029/JA086iA01p00054.

Ma, X., A. Otto, and P. A. Delamere (2014a), Interaction of magnetic reconnection and Kelvin-Helmholtz modes for large magnetic shear: 1. Kelvin-Helmholtz trigger, *J. Geophys. Res. (Space Phys.)*, *119*, 781–797, doi:10.1002/2013JA019224.

Ma, X., A. Otto, and P. A. Delamere (2014b), Interaction of magnetic reconnection and Kelvin-Helmholtz modes for large magnetic shear: 2. Reconnection Trigger, *J. Geophys. Res. (Space Phys.)*, *119*, 808–820, doi:10.1002/2013JA019225.

Ma, X., P. Delamere, B. Stauffer, and A. Otto (2014c), Asymmetric Kelvin-Helmholtz evolution at Saturn's magnetopause boundary, *J. Geophys. Res.*, submitted .

Masters, A., et al. (2010), Cassini observations of a Kelvin-Helmholtz vortex in Saturn's outer magnetosphere, *J. Geophys. Res. (Space Phys.)*, *115*, A07225, doi:10.1029/2010JA015351.

Masters, A., J. P. Eastwood, M. Swisdak, M. F. Thomsen, C. T. Russell, N. Sergis, F. J. Crary, M. K. Dougherty, A. J. Coates, and S. M. Krimigis (2012), The importance of plasma β conditions for magnetic reconnection at Saturn's magnetopause, *Geophys. Res. Lett.*, *39*, L08103, doi:10.1029/2012GL051372.

Matsumoto, Y., and K. Seki (2010), Formation of a broad plasma turbulent layer by forward and inverse energy cascades of the Kelvin-Helmholtz instability, *J. Geophys. Res. (Space Phys.)*, *115*, 10,231, doi:10.1029/2009JA014637.

Mauk, B. H., et al. (2009), *Fund. Plasma Process. Saturn's Magnetosphere*, p. 281, doi:10.1007/978-1-4020-9217-6-11.

McAndrews, H. J., C. J. Owen, M. F. Thomsen, B. Lavraud, A. J. Coates, M. K. Dougherty, and D. T. Young (2008), Evidence for reconnection at Saturn's magnetopause, *J. Geophys. Res. (Space Phys.)*, *113*, A04210, doi:10.1029/2007JA012581.

McComas, D. J., and F. Bagenal (2007), Jupiter: A fundamentally different magnetospheric interaction with the solar wind, *J. Geophys. Res.*, *34*, 20,106, doi:10.1029/2007GL031078.

McComas, D. J., and F. Bagenal (2008), Reply to comment by S. W. H. Cowley et al. on "Jupiter: A fundamentally different magnetospheric interaction with the solar wind," *Geophys. Res. Lett.*, *35*, 10,103, doi:10.1029/2008GL034351.

McComas, D. J., F. Allegrini, F. Bagenal, F. Crary, R. W. Ebert, H. Elliott, A. Stern, and P. Valek (2007), Diverse plasma populations and structures in Jupiter's magnetotail, *Science*, *318*, 217, doi:10.1126/science.1147393.

McComas, D. J., F. Bagenal, and R. Ebert (2014), Bimodal size of Jupiter's magnetosphere, *J. Geophys. Res. (Space Phys.)*, doi:10.1002/2013JA019660.

McNutt, R. L., et al. (2007), Energetic particles in the Jovian magnetotail, *Science*, *318*, 220, doi:10.1126/science.1148025.

Milan, S. E., S. W. H. Cowley, M. Lester, D. M. Wright, J. A. Slavin, M. Fillingim, C. W. Carlson, and H. J. Singer (2004), Response of the magnetotail to changes in the open flux content of the magnetosphere, *J. Geophys. Res. (Space Phy.)*, *109*, A04220, doi:10.1029/2003JA010350.

Mitchell, D. G., J. F. Carbary, S. W. H. Cowley, T. W. Hill, and P. Zarka (2009), *The Dynamics of Saturn's Magnetosphere*, p. 257, Springer, Dortrecht Heidelberg London New York doi:10.1007/978-1-4020-9217-6-10.

Mozer, F. S., and A. Hull (2010), Scaling the energy conversion rate from magnetic field reconnection to different bodies, *Phys. Plasmas*, *17*, 102,906.

Nakamura, T. K. M., H. Hasegawa, and I. Shinohara (2010), Kinetic effects on the Kelvin-Helmholtz instability in ion-to-magnetohydrodynamic scale transverse velocity shear layers: Particle simulations, *Phys. Plasmas*, *17*(4), 042,119, doi:10.1063/1.3385445.

Nichols, J. D., S. W. H. Cowley, and D. J. McComas (2006), Magnetopause reconnection rate estimates for Jupiter's magnetosphere based on interplanetary measurements at˜5AU, *Ann. Geophys.*, *24*, 393–406, doi:10.5194/angeo-24-393-2006.

Nichols, J. D., E. J. Bunce, J. T. Clarke, S. W. H. Cowley, J.-C. Gérard, D. Grodent, and W. R. Pryor (2007), Response of Jupiter's UV auroras to interplanetary conditions as observed by the Hubble Space Telescope during the Cassini flyby campaign, *J. Geophys. Res. (Space Phys.)*, *112*, 2203, doi:10.1029/2006JA012005.

Nishida, A. (1984), Reconnection in Earth's Magnetotail: An overview, in *Magnetic Reconnection in Space and Laboratory Plasmas*, Hones, E. W., Jr. Monograph, AGU, Washington, D.C.

Nykyri, K., and A. Otto (2001), Plasma transport at the magnetospheric boundary due to reconnection in Kelvin-Helmholtz vortices, *Geophys. Res. Lett*, *28*, 3565–3568, doi:10.1029/2001GL013239.

Nykyri, K., and A. Otto (2004), Influence of the Hall term on KH instability and reconnection inside KH vortices, *Ann. Geophys.*, *22*, 935–949.

Otto, A. (2007), Plasma entry and Kelvin-Helmholtz modes at the flanks of the magnetosphere, *Fall Meet. Abstracts*, p. B7, AGu, Washington, D.C.

Otto, A., and D. H. Fairfield (2000), Kelvin-Helmholtz instability at the magnetotail boundary: MHD simulation and comparison with geotail observations, *J. Geophys. Res.*, *105*, 175.

Pallier, L., and R. Prangé (2001), More about the structure of the high latitude Jovian aurorae, *Planet. Space Sci.*, *49*, 1159–1173.

Paschmann, G., S. Haaland, and R. Treumann (2002), Auroral plasma physics, *Space Sci. Rev.*, *103*, doi:10.1023/A:1023030716698.

Pu, Z.-Y., and M. G. Kivelson (1983), Kelvin-Helmholtz instability at the magnetopause. I. Solution for compressible plasmas. II. Energy flux into the magnetosphere, *J. Geophys. Res.*, *88*, 841–861, doi:10.1029/JA088iA02p00841.

Radioti, A., D. Grodent, J.-C. Gérard, S. E. Milan, B. Bonfond, J. Gustin, and W. Pryor (2011), Bifurcations of the main auroral ring at Saturn: Ionospheric signatures of consecutive reconnection events at the magnetopause, *J. Geophys. Res. (Space Phys.)*, *116*, A11209, doi:10.1029/2011JA016661.

Ridley, A. J. (2005), A new formulation for the ionospheric cross polar cap potential including saturation effects, *Ann. Geophys.*, *23*, 3533–3547.

Ridley, A. J. (2007), Alfven wings at earth's magnetosphere under strong interplanetary magnetic fields, *Ann. Geophys.*, *25*, 533–542.

Russell, C. T. (1984), Reconnection at the Earth's magnetopause: Magnetic field observations and flux transfer events, in *Magnetic Reconnection in Space and Laboratory Plasmas*, Monograph, AGU, Washington, D.C.

Russell, C. T., M. G. Kivelson, K. K. Khurana, and D. E. Huddleston (1999), Magnetic fluctuations close to Io: Ion cyclotron and mirror mode wave properties, *Planet. Space Sci.*, *47*, 143.

Scarf, F. L., W. S. Kurth, D. A. Gurnett, H. S. Bridge, and J. D. Sullivan (1981), Jupiter tail phenomena upstream from Saturn, *Nature*, *292*, 585, doi:10.1038/292585a0.

Scurry, L., C. T. Russell, and J. T. Gosling (1994), Geomagnetic activity and the beta dependence of the dayside reconnection rate, *J. Geophys. Res*, *99*, 14,811, doi:10.1029/94JA00794.

Slavin, J. A., E. J. Smith, J. R. Spreiter, and S. S. Stahara (1985), Solar wind flow about the outer planets: Gas dynamic modeling of the Jupiter and Saturn bow shocks, *J. Geophys. Res.*, *90*, 6275–6286, doi:10.1029/JA090iA07p06275.

Song, P., and C. T. Russell (1992), Model of the formation of the low-latitude boundary layer for strongly northward interplanetary magnetic field, *J. Geophys. Res.*, *97*, 1411–1420, doi:10.1029/91JA02377.

Song, P., D. L. Dezeeuw, T. I. Gombosi, J. U. Kozyra, and K. G. Powell (2001), Global MHD simulations for southward IMF: A pair of wings in the flanks, *Adv. Space Res.*, *28*, 1763–1771, doi:10.1016/S0273-1177(01)00544-0.

Southwood, D. J., and M. G. Kivelson (2001), A new perspective concerning the influence of the solar wind on the Jovian magnetosphere, *J. Geophys. Res*, *106*, 6123–6130, doi:10.1029/2000JA000236.

Stallard, T., S. Miller, G. Millward, and R. D. Joseph (2001), On the dynamics of the Jovian ionosphere and thermosphere. I. The measurement of ion winds, *Icarus*, *154*, 475–491, doi:10.1006/icar.2001.6681.

Stallard, T. S., S. Miller, S. W. H. Cowley, and E. J. Bunce (2003), Jupiter's polar ionospheric flows: Measured intensity and velocity variations poleward of the main auroral oval, *Geophys. Res. Lett.*, *30*(5), 1221, doi:10.1029/2002GL016031.

Stallard, T. S., S. Miller, L. M. Trafton, T. R. Geballe, and R. D. Joseph (2004), Ion winds in Saturn's southern auroral/polar region, *Icarus*, *167*, 204–211, doi:10.1016/j.icarus.2003.09.006.

Sundberg, T., S. A. Boardsen, J. A. Slavin, B. J. Anderson, H. Korth, T. H. Zurbuchen, J. M. Raines, and S. C. Solomon (2012), MESSENGER orbital observations of large-amplitude Kelvin-Helmholtz waves at Mercury's magnetopause, *J. Geophys. Res (Space Phys.)*, *117*, A04216, doi:10.1029/2011JA017268.

Thomas, V. A., and D. Winske (1991), Kinetic simulation of the Kelvin-Helmholtz instability at the Venus ionopause, *Geophys. Res. Lett.*, *18*, 1943–1946, doi:10.1029/91GL02552.

Thomas, V. A., and D. Winske (1993), Kinetic simulations of the Kelvin-Helmholtz instability at the magnetopause, *J. Geophys. Res. Lett.*, *98*, 11,425, doi:10.1029/93JA00604.

Vasyliunas, V. M. (1983), Plasma distribution and flow, in *Physics of the Jovian Magnetosphere*, edited by Dessler, A. J., pp. 395–453, Cambridge Univ. Press.

Vasyliunas, V. M. (2007), The mechanical advantage of the magnetosphere: solar-wind-related forces in the magnetosphere-ionosphere-earth system, *Ann. Geophys.*, *25*, 255–269.

Vasyliunas, V. M. (2009), Fundamentals of planetary magnetospheres, in *Heliophysics: Plasma Physics of the Local Cosmos*, edited by Schrijver, C. J., and. G. L. Siscoe, Cambridge Univ. Press.

Vogt, M. F., M. G. Kivelson, K. K. Khurana, S. P. Joy, and R. J. Walker (2010), Reconnection and flows in the Jovian magnetotail as inferred from magnetometer observations, *J. Geophys. Res. (Space Phys.)*, *115*, A06,219, doi:10.1029/2009JA015098.

Vogt, M. F., M. G. Kivelson, K. K. Khurana, R. J. Walker, B. Bonfond, D. Grodent, and A. Radioti (2011), Improved mapping of Jupiter's auroral features to magnetospheric sources, *J. Geophys. Res. (Space Phys.)*, *116*, A03,220, doi:10.1029/2010JA016148.

Walker, R. J., and C. T. Russell (1985), Flux transfer events at the Jovian magnetopause, *J. Geophys. Res.*, *90*, 7397–7404, doi:10.1029/JA090iA08p07397.

Woch, J., N. Krupp, and A. Lagg (2002), Particle bursts in the Jovian magnetosphere: Evidence for a near-Jupiter neutral line, *Geophys. Res. Lett.*, *29*(7), 42-1–42-4, doi:10.1029/2001GL014080.

14

Solar Wind Entry Into and Transport Within Planetary Magnetotails

Simon Wing[1] and Jay R. Johnson[2]

14.1. INTRODUCTION

Earth's magnetosphere on the nightside is stretched out in an elongated fashion that resembles a taillike configuration, which is often referred to as a magnetotail. The magnetotail is fundamentally important for understanding dynamical processes in the magnetosphere such as geomagnetic storms and substorms. A region in the magnetotail that is usually characterized by high plasma density ($n > \sim 0.5\,\mathrm{cm}^{-3}$) and plasma $\beta \sim 1$ is called the plasma sheet, which is an important source population for ring current, energetic outer radiation belt particles, and auroral precipitation.

It is believed that there are two plasma sources for the plasma sheet, namely, the solar wind and the ionosphere. Their relative importance depends on solar wind and geomagnetic conditions. It is well established that plasma sheet properties such as density, pressure, and temperature are correlated with solar wind conditions [e.g., *Borovsky et al.*, 1998a; *Terasawa et al.*, 1997; *Wing and Newell*, 2002; *Nagata et al.*, 2007]. However, the path the solar wind takes to reach the plasma sheet and the subsequent plasma transport within the plasma sheet are not always clear.

During southward interplanetary magnetic field (IMF), reconnection can occur at the low-latitude magnetopause on the dayside where the IMF and Earth's magnetic field line are nearly antiparallel [*Crooker*, 1979]. After reconnection, the open field lines would convect to the nightside toward the lobe region. Because of the solar wind duskward electric field, these open field lines would $\mathbf{E} \times \mathbf{B}$ drift to the equatorial region where they would once again reconnect [e.g., *Pedersen et al.*, 1985; *Keiling et al.*, 2004]. This process is depicted in Figure 14.1. Once reconnected, the field line, which is filled with magnetosheath plasma, becomes part of the plasma sheet. In essence, the magnetosheath plasma is now captured in the plasma sheet. Subsequently, this newly closed field line would $\mathbf{E} \times \mathbf{B}$ convect sunward to the dayside, returning the magnetic flux back to the dayside. This process is known as the Dungey cycle [*Dungey*, 1961], although the solar wind entry into the plasma sheet during southward IMF can be more complicated than the Dungey cycle and can involve other processes.

In the antiparallel merging model [*Crooker*, 1979], the reconnection would occur at high latitude poleward of the cusp in the lobe during northward IMF. After the reconnection, the footpoint of the newly reconnected field line would initially move sunward before being swept away to the nightside, but the solar wind electric field is dawnward instead of duskward for the southward IMF case. So, the entry mechanism for northward IMF may not be as simple and straightforward as the scenario portrayed in the Dungey cycle. The intrigue and the difficulty only deepened when it was discovered that compared to southward IMF conditions, more solar wind particles enter the magnetotail during northward IMF conditions, which can lead to the formation of cold, dense plasma sheet (temperature < 1 keV and density $> 1\,\mathrm{cm}^{-3}$) [e.g., *Terasawa et al.*, 1997; *Fujimoto et al.*, 1998; *Wing and Newell*, 2002; *Stenuit et al.*, 2002; *Øieroset et al.*, 2008].

It turns out that reconnection can still play important roles in transporting solar wind plasma into the magnetotail during periods of northward IMF. A special case of high-latitude reconnection, where Earth's field lines are reconnected at both hemispheres, forms a newly

[1] *Applied Physics Laboratory, The Johns Hopkins University, Laurel, Maryland*

[2] *Princeton Plasma Physics Laboratory, Princeton University, Princeton, New Jersey*

Magnetotails in the Solar System, Geophysical Monograph 207. First Edition.
Edited by Andreas Keiling, Caitríona M. Jackman, and Peter A. Delamere.

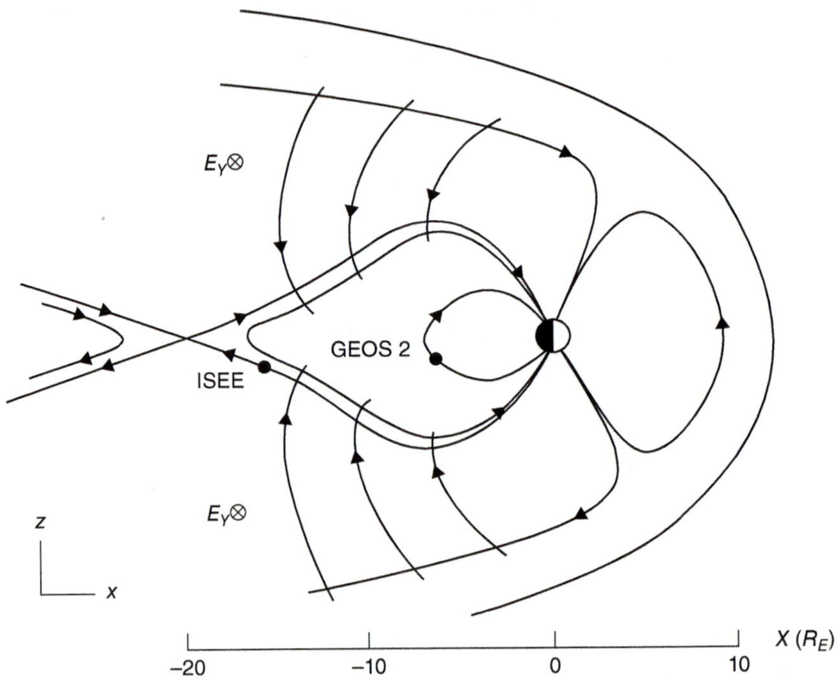

Figure 14.1 Electric and magnetic field lines on the nightside for southward IMF. After reconnection on the dayside, the open field line convects over the pole to the nightside where, due to the duskward electric field, the field line would $\mathbf{E} \times \mathbf{B}$ toward the equatorial plane where it would reconnect once more to form closed field line that is part of the plasma sheet. from Figure 11 of *Pedersen et al.* [1985].

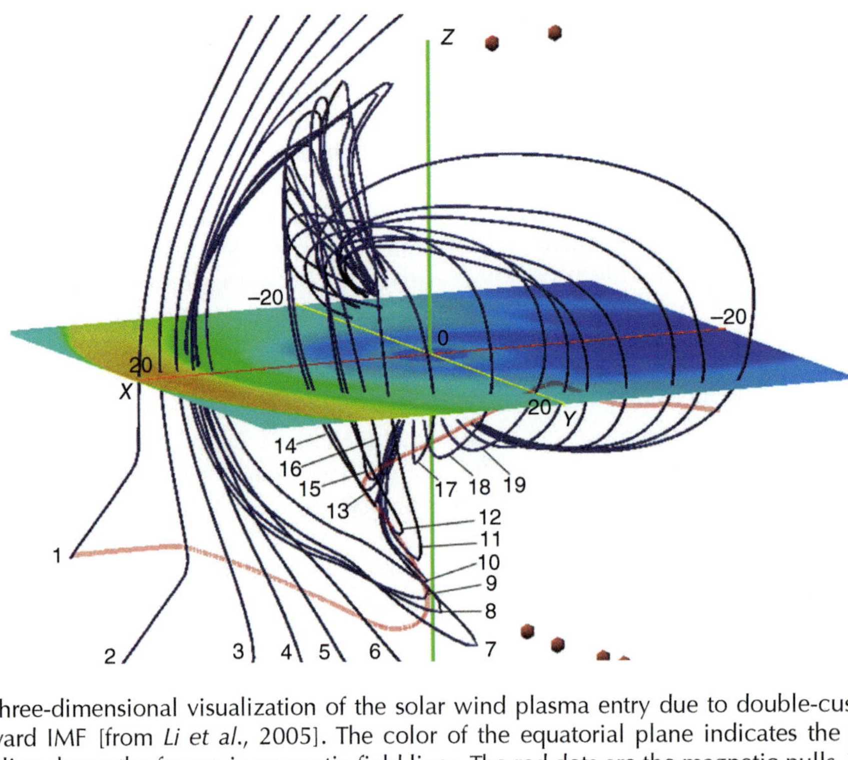

Figure 14.2 Three-dimensional visualization of the solar wind plasma entry due to double-cusp reconnection during northward IMF [from *Li et al.*, 2005]. The color of the equatorial plane indicates the plasma density. The dark blue line shows the frozen-in magnetic field lines. The red dots are the magnetic nulls. The red, yellow, and green axes are Geocentric Solar Ecliptic (GSE) *X*, *Y*, and *Z* axes, respectively. The numbers 1–19 indicate the progression of the field line in time.

closed field line filled with magnetosheath plasma ("double-cusp reconnection") [*Song and Russell*, 1992; *Le et al.*, 1996; *Sandholt et al.*, 1999; *Fuselier et al.*, 2002]. Magnetohydrodynamic (MHD) simulations have been performed to show that this process can effectively bring magnetosheath plasma into the plasma sheet [*Raeder et al.*, 1995; 1997]. This process is depicted in Figure 14.2, which can be contrasted with Figure 14.1. Figure 14.2, from an MHD simulation [*Li et al.*, 2005], shows the time sequence of a field line as it convects with the solar wind, crosses the bow shock, and then reconnects sequentially in short order with lobe field lines in each hemisphere. After that, the field line, along with the captured magnetosheath plasma, slowly (on a time scale of 1–3 h) convects into the tail. Initially, the newly closed dayside field line would move sunward and then it would move tailward to the flanks. *Øieroset et al.* [2005] reported a fairly good agreement in the plasma sheet temperature and density between an MHD simulation in which double-cusp reconnection fills the plasma sheet and Cluster satellite observations in a case study. MHD simulations for northward IMF conditions replicate the development of a low-latitude boundary layer detected by the five THEMIS spacecraft as they encountered the magnetopause boundary [*Li et al.*, 2009].

Another possible solar wind entry mechanism is Kelvin-Helmholtz instability (KHI). Kelvin-Helmholtz (KH) waves grow along an inhomogeneous velocity shear layer (as found on the flank magnetopause) and eventually develop a rolled-up vortex pattern in density and magnetic field, as illustrated in Figure 14.3. KHI has been successfully simulated with MHD code [e.g., *Otto and Fairfield*, 2000; *Nakamura and Fujimoto*, 2005; *Nykyri and Otto*, 2001; *Nykyri et al.*, 2006; *Claudepierre et al.*, 2008; *Guo et al.*, 2010; *Merkin et al.*, 2013]. Signatures of these vortices have been found on the dusk and dawn flanks of the magnetosphere [e.g., *Fairfield et al.*, 2000; *Fujimoto et al.*, 2003; *Hasegawa et al.*, 2006]. Moreover, reconnection in the nonlinear stage of the KHI could lead to the detachment of plasma from the vortex structures, leading to significant magnetopause transport [e.g., *Otto and Fairfield*, 2000; *Nykyri et al.*, 2006]. Hybrid simulations have indicated that ion blobs could become detached from the vortex structure of the KHI and could provide filaments, producing a mixing of plasma in the shear layer [*Thomas and Winske*, 1991, 1993; *Fujimoto and Terasawa*, 1994, 1995]. *Hasegawa et al.* [2006] noted that there is an inverse relationship between the magnetospheric plasma density and tailward flow speed in KH vortices, leading to the development of an algorithm for detecting rolled-up KH vortices with a single spacecraft.

Large-amplitude Alfvén waves have also been observed on the magnetospheric boundary [e.g., *Tsurutani et al.*, 1982; *LaBelle and Treumann*, 1988; *Anderson and Fuselier*, 1994]. There is evidence that the waves could be the result of mode conversion of magnetosheath compressions in the sharp magnetopause gradients at the magnetopause [*Lee et al.*, 1994; *Johnson and Cheng*, 1997; *Johnson et al.*, 2001]. Because the wavelength of the mode-converted waves are on the order of the ion gyroradius, they can

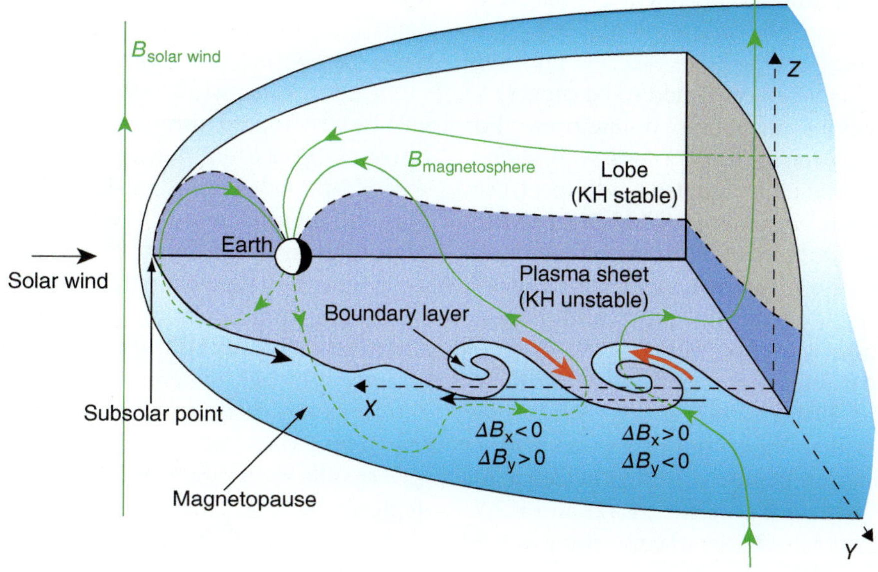

Figure 14.3 Three-dimensional view of the magnetosphere showing the KH vortices at the duskside magnetopause. From *Hasegawa et al.* [2004b].

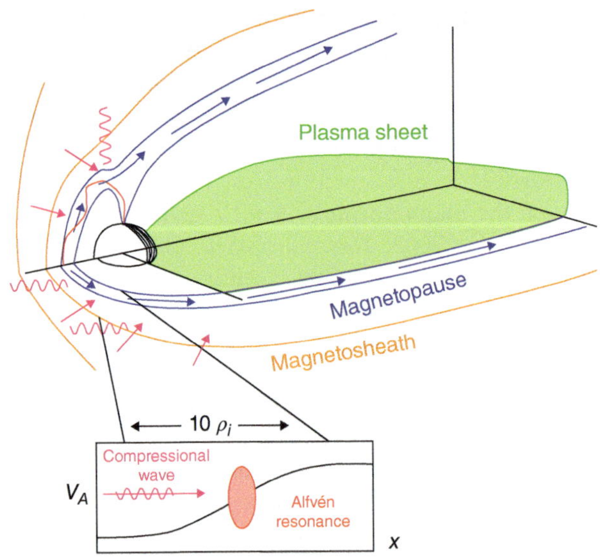

Figure 14.4 Three-dimensional view of the magnetosphere showing kinetic Alfvén waves in the magnetosheath leading to solar wind/magnetosheath plasma entry into the magnetosphere.

lead to efficient convective and diffusive transport of ions across the magnetopause, as illustrated in Figure 14.4 [*Johnson and Cheng*, 1997; *Chen*, 1999; *Chaston et al.*, 2008]. The mode conversion process has been simulated with 2D hybrid simulations [*Lin et al.*, 2010] showing linear mode conversion and 3D simulations [*Lin et al.*, 2012] showing nonlinear decay of the mode-converted waves such that transport is greatly enhanced because of the development of modes with large azimuthal wave number. Kinetic Alfvén waves (KAWs) also lead to nonlinear plasma heating because of the large amplitudes observed at the magnetopause [*Johnson and Cheng*, 2001; *Chaston et al.*, 2008].

These three processes are not necessarily mutually exclusive and have all been demonstrated to be capable to transport plasma across magnetic boundaries. For example, KH waves can excite KAWs [*Sibeck et al.*, 1999]. KAWs have been observed in conjunction with reconnection and KH structures [*Chaston et al.*, 2005, 2007, 2009]. *Nishino et al.* [2007] also observed bidirectional electrons and cold protons inside a KH vortical structure, which they interpreted as a signature reconnection. Additionally, they observed electron and ion heating, which they attributed to wave-particle interactions such as KAW heating [*Johnson and Cheng*, 1997, 2001]. *Taylor and Lavraud* [2008] found that there could be three distinct ion populations, which can be interpreted as evidence for KAW, KHI, and reconnection processes. KHI and KAW can also play significant roles in the plasma entry during southward IMF.

It has been a challenge to distinguish which of these three mechanisms can play a more significant role and under what conditions. Recent observations have attempted to identify several potential discriminators based on both in situ and remote sensing methods. One of the potential discriminators is the filling rate. Each entry mechanism may lead to a different entry or filling rate. Another potential discriminator is the dawn-dusk asymmetries. Dawn-dusk asymmetries are particularly interesting because they may be related to how upstream boundary conditions (e.g., Parker spiral magnetic field orientation) affect the various entry mechanisms, which can be tested by observation and theory. Another potential discriminator is the specific entropy or entropy per unit volume ($s=p/\rho^{\gamma}$, where p=plasma perssure, ρ=mass density, and $\gamma=5/3$ is the polytropic index). Changes in the entropy profiles may be indicators that nonadiabatic processes are operating in conjunction with plasma transport, and how the entropy changes is likely related to the entry mechanism.

While cold plasma is preferentially observed in the plasma sheet during periods of northward IMF, hot plasma is preferentially observed during periods of southward IMF. However, both hot and cold plasma have been observed to coexist in the same region/fieldline under all IMF conditions. The mechanism for heating cold plasma, either solar wind or ionospheric origin, to a hot-plasma population has not been firmly established, but it does appear that the ion-to-electron temperature ratio (T_i/T_e) is roughly conserved by the entry, energization, heating, and transport processes. This intriguing and seemingly "conserved" property deserves close and careful examination. Unlike entropy, there has not been a physical argument for the conservation of the T_i/T_e ratio. It is not clear what causes the T_i/T_e ratio conservation, if it is conserved, and if the T_i/T_e ratio can really serve as a constraint on the plasma transport across the magnetopause and within the plasma sheet.

Once the solar wind plasma enters the magnetotail, it will be distributed throughout the plasma sheet by the transport processes within the magnetotail or plasma sheet. The plasma can get heated and energized during these processes. Some of the transport processes, such as curvature and gradient drifts, introduce dawn-dusk asymmetries in the particles, especially for the hot component. Transport within the plasma sheet has been described with MHD models, which conserve total entropy ($S=pV^{\gamma}$, where V=flux tube volume). Total-entropy conservation provides important constraints on the accessibility of plasma to the plasma sheet and governs the redistribution of plasma when flux tubes are depleted of total entropy due to nonadiabatic processes such as reconnection, plasma diffusion, and wave-particle interactions.

During active times, ion outflow rates increase [e.g., *Wilson et al.*, 2001; *Tung et al.*, 2001] and so the plasma sheet heavy ion population can be significant [*Chappell et al.*, 1987; *Schriver and Ashour-Abdalla*, 1990; *Kistler et al.*, 2006, 2010]. *Bouhram et al.* [2005] analyzed Cluster data and found that O^+ density is significantly higher on the duskside than the dawnside of the magnetosphere near the magnetopause. Ion outflows or more precisely the presence of the heavy ions may affect the transport processes within the plasma sheet such as magnetotail convection and reconnection [e.g., *Winglee et al.*, 2002]. It may also affect KHI and solar wind entry at the magnetotail flanks. The effects of heavy ions on solar wind entry are actually very significant to other planetary magnetospheres as discussed in *Johnson et al.* [2014]. Ion outflows and their effects on magnetosphere-ionosphere coupling are discussed in Chapter 22 (*Wiltberger*, in this volume).

In the following sections, we discuss and review (a) plasma sheet filling rate; (b) dawn-dusk asymmetry; (c) entropies (s and S); and (d) T_i/T_e ratio and how they can discriminate the three entry mechanisms. Then, we discuss the transport within the plasma sheet. Finally, we provide a summary and conclusion.

14.2. PLASMA TRANSPORT ACROSS MAGNETOPAUSE (SOLAR WIND ENTRY)

14.2.1. Plasma Sheet Filling Rate

In the discussion of which entry mechanism(s) are significant in filling of the plasma sheet with solar wind plasma, it is useful to establish first whether each entry mechanism is able to explain the observed plasma sheet filling rate. In other words, how much of the observed filling rate can be attributed to each entry mechanism.

In an attempt to do this, *Wing et al.* [2006] examined a northward IMF interval in which the plasma sheet ion filling rate is about $0.5\,cm^{-3}h^{-1}$ as observed locally by Geotail and remotely by DMSP satellites. They performed an order-of-magnitude calculation of the magnetosheath entry rate for (1) double-cusp reconnection [e.g., *Song and Russell*, 1992], (2) KHI [e.g., *Otto and Fairfield*, 2000], and (3) KAW [e.g., *Johnson and Cheng*, 1997] and found that each of the three mechanisms is capable to fill the plasma sheet at the observed filling rate. This result is consistent with double-cusp reconnection studies [e.g., *Øieroset et al.*, 2005; *Li et al.*, 2008] and KHI studies [e.g., *Nykyri and Otto*, 2004; *Nykyri et al.*, 2006], which showed that double-cusp reconnection or KHI is sufficient to fill the plasma sheet with cold, dense ions, respectively. The diffusive entry rate (either KHI or KAW) scales with solar wind density (n_{sw}) [*Wing et al.*, 2006] while the double-cusp reconnection entry rate depends weakly and nonmonotonically with n_{sw} [*Li et al.*, 2008]. Some of the magnetospheric plasma is lost to the solar wind, inner magnetosphere, and ionosphere. Hence, the net plasma sheet filling rate would need to take into account not just the entry rate but also the loss rate.

14.2.2. Dawn-Dusk Asymmetries

Global asymmetries in plasma sheet profiles can provide a significant constraint on plasma entry. *Fujimoto et al.* [1998] showed that in Geotail observations, during northward IMF, ions frequently have two distinct components (a hot and a cold component) on the dusk flanks but only one component with a broad peak on the dawn flank, as shown in Figure 14.5. Moreover, *Hasegawa et al.* [2003] reported that Geotail observations show that the ion hot component is frequently nearly isotropic on both the dusk and dawn flanks. However, the cold component, which is similar to the magnetosheath ion population, is heated in the direction perpendicular to the magnetic field on the dawnside while it remains isotropic on the duskside, as shown in Figures 14.6a and 14.6b [*Hasegawa et al.*, 2003]. In their follow-up study, *Hasegawa et al.* [2004a] performed a statistical study on the ion properties on both flanks. The results are shown in Figure 14.7. When the ions have two clear components, one component with $T_h > 3\,keV$ and one with $T_c < 3\,keV$, they are plotted as red and green dots, respectively. However, when the ions have just one peak, then they are plotted as black dots. Figure 14.7 shows statistically that ions frequently have two components on the duskside but only one component on the dawnside. The cold component of the electron (temperature less than a few hundred electron-volts) has been heated in the direction parallel to the magnetic field line [*Hasegawa et al.*, 2003; *Fujimoto et al.*, 1998].

Wing et al. [2005] used DMSP satellites to infer plasma sheet ion temperatures and densities during periods of northward IMF. One advantage of using DMSP satellites is that these satellites have low-altitude orbits with orbital periods of about 100 min. So, in about 2 h, each DMSP satellite can cut through the footprint of the entire plasma sheet twice. *Wing et al.* [2005] fitted the ion spectra to one-component Maxwellian, two-component Maxwellian, and kappa (κ) distributions and selected the best fits.

Their cold-component density and temperature profiles are displayed in Figures 14.6c and 14.6d, respectively. The cold-component density profile has peaks at dawn and dusk flanks, as shown in Figure 14.6c. This suggests that the source of the cold component is magnetosheath or low latitude boundary layer (LLBL). In contrast, the hot-component ion density profile does not have peaks at the flanks (see Figure 8 in *Wing et al.* [2005]), which would be expected if the source of the hot component is not the solar wind. Interestingly, the density of the cold component is higher on the dawnside than that on the duskside, suggesting perhaps there is

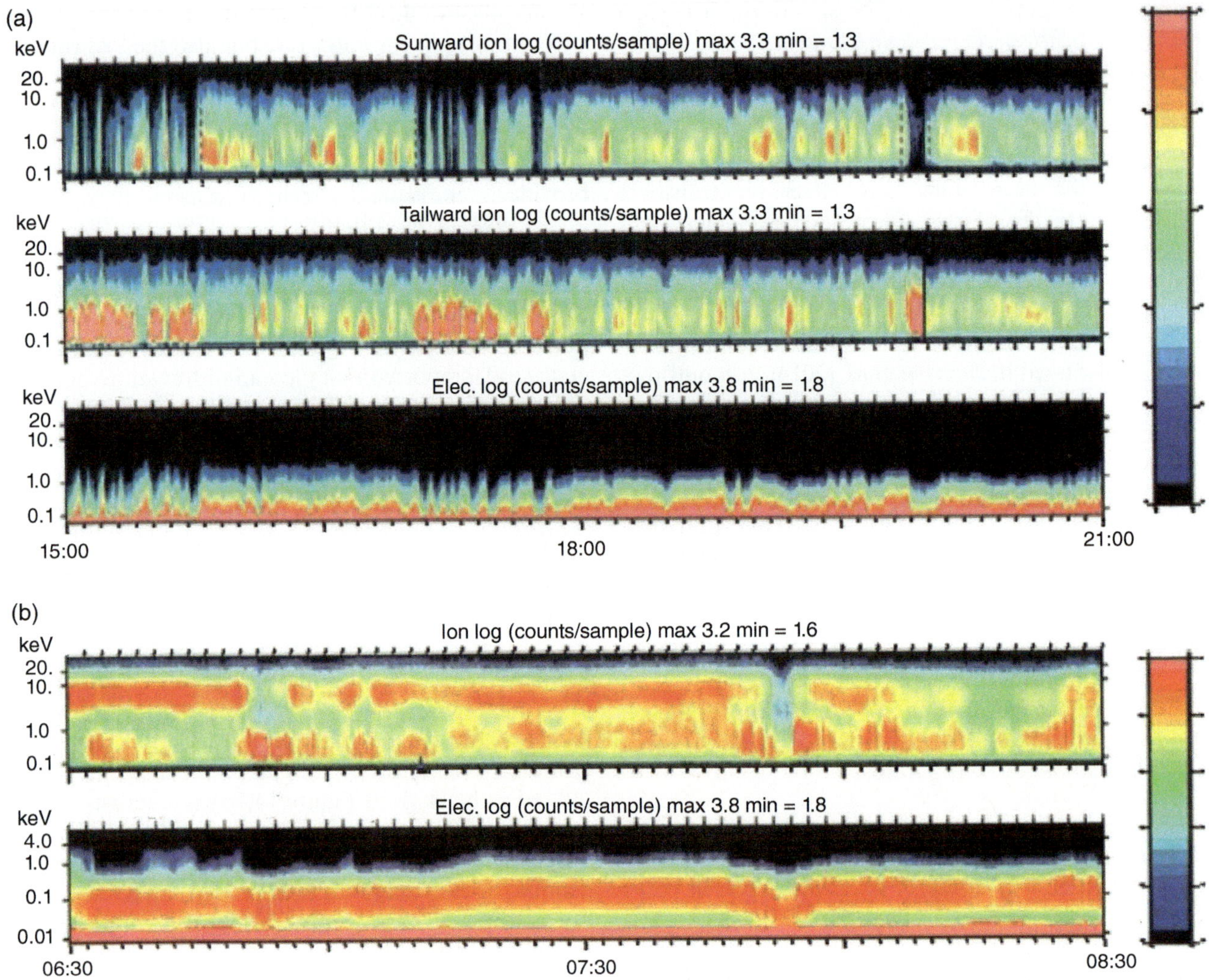

Figure 14.5 Spectrograms showing Geotail observations during northward IMF in the magnetotail just inside the magnetopause on (a) the dawnside and (b) the duskside. On the dawnside (a) the ions appear as having one component with a broad peak, but on the duskside the ions have two components: (1) hot and (2) cold. (a) Three panels show sunward ions, tailward ions, and omnidirectional electrons. Color codes are assigned according to logarithms of counts per sample with maximum=red and minimum=dark blue. (b) Top panel shows omnidirectional ions depicting mixing of hot and cold components and bottom panel shows omnidirectional electrons. From *Fujimoto et al.* [1998].

more solar wind ion entry on the dawnside than on the duskside of the magnetopause (see also *Guild et al.* [2008] and *Nagata et al.* [2007]). Figure 14.6d shows that the cold-component temperatures are higher on the dawnside than the duskside, consistent with *Hasegawa et al.* [2003]. This observation suggests that the magnetosheath ions have been heated in the entry process on the dawnside. The dawnside cold-ion temperature is about 30–40 % higher than that on the duskside (see Figure 7 in *Wing et al.* [2005]).

In agreement with *Fujimoto et al.* [1998] and *Hasegawa et al.* [2003, 2004a], *Wing et al.* [2005] found that two components are more frequently found on the dusk flank than on the dawn flank. The hot component has higher temperatures on the dusk flank than on the dawn flank, which can be attributed to the curvature and gradient drifts. The cold component, on the other hand, has higher temperatures on the dawn flank than on the dusk flank. Hence, on the dusk flank, the cold- and hot-component temperatures frequently have distinct separations and the ions can be identified as having two components. In contrast, on the dawn flank, the cold- and hot-component temperatures frequently are closer to each other, forming one component having a broad peak.

It is not clear what causes the temperature asymmetry in the cold-ion profile. For the solar wind Parker spiral,

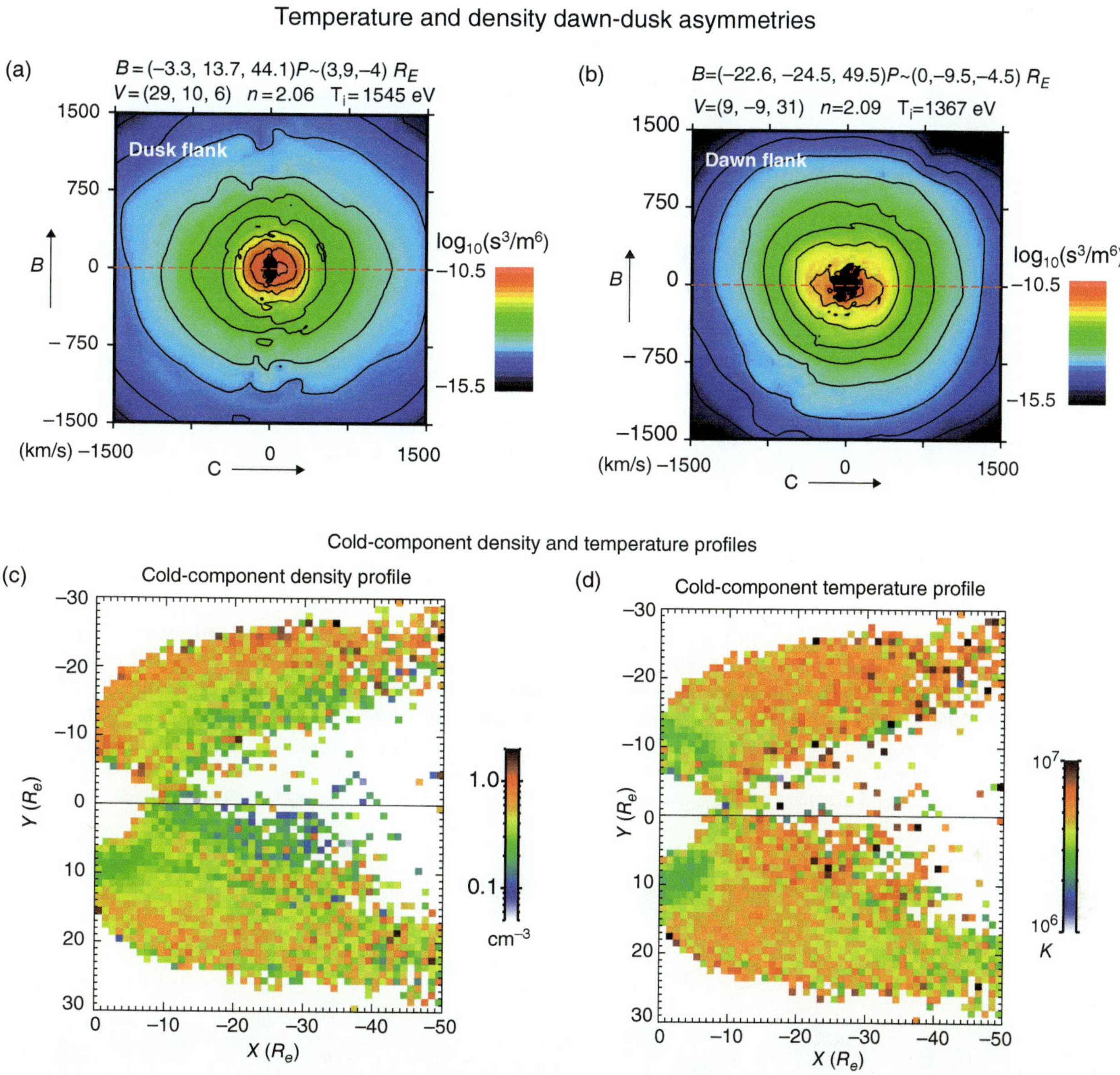

Figure 14.6 Ion distribution functions in the mixed region where cold magnetosheath-like and hot magnetospheric ions are simultaneously observed in (a) the dusk flank and (b) the dawn flank during northward IMF. Shown are the slices of the flow in the *BC* planes, where the *Y* axis = *B* = aligned with the magnetic field and the *X* axis = *C* = perpendicular to the magnetic field. The cold and hot components in the dusk flank (a) are fairly isotropic. However, in the dawn flank (b), while the hot component is fairly isotropic, the cold component shows evidence of heating in the direction perpendicular to the magnetic field (B = magnetic field, nT; P = satellite coordinate, R_E; V = ion velocity, km/s; n = ion density, cm^{-3}; T_i = ion temperature, eV) [from *Hasegawa et al.*, 2003]. (c) Density and (d) temperature profiles of the cold component of the two-component Maxwellian distribution of the plasma sheet ions during northward IMF. Note the dawn-dusk asymmetry in the temperature profile, with the dawn flank ions having higher temperatures than the dusk flank ions [from *Wing et al.*, 2005].

which is the dominant solar wind orientation, the magnetosheath magnetic field has been shown to be more turbulent downstream of the quasi-parallel shock (the dawnside magnetosheath) than downstream of the quasi-perpendicular shock (the duskside magnetosheath) [e.g., *Petrinec*, 2013]. An example of this for solar wind Parker spiral orientation can be seen in Figure 14.8, which shows Geotail magnetosheath magnetic field

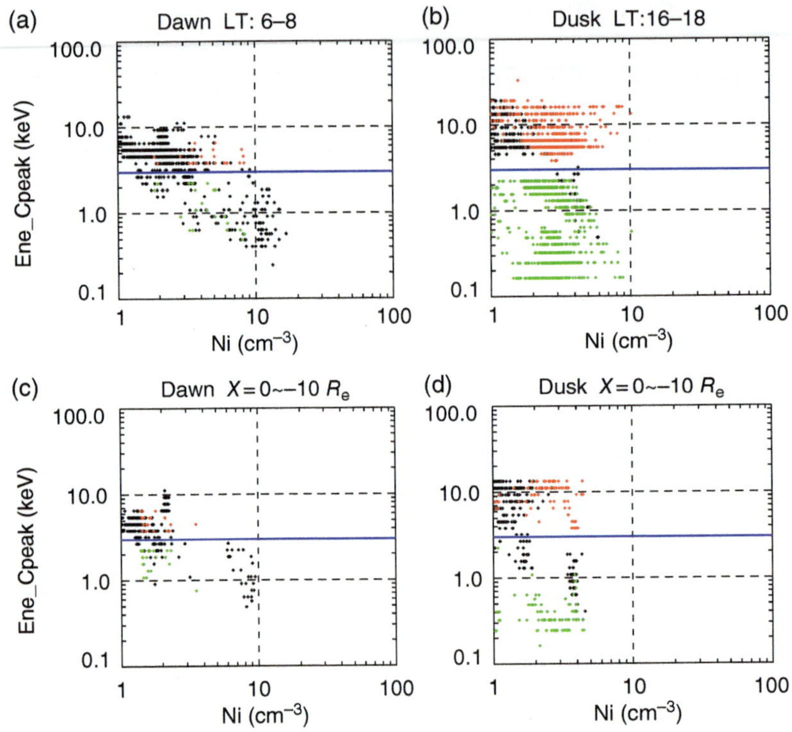

Figure 14.7 Ion energy spectrum behavior in the mixed region during extended northward IMF periods in [(a), (b)] dawn and [(c),(d)] dusk flanks. For each data sample, if the ion energy spectrum has a count rate (energy flux) peak both below and above 3 keV, the higher and lower peak energies are plotted as red and green dots, respectively. Otherwise, the energy at which the count rate reaches maximum is plotted as a black dot. There is a clear dawn-dusk asymmetry. There are more two-component ions on the duskside than on the dawnside. Conversely, there are more one-component ions on the dawnside than on the duskside. On the dawnside, the hot and cold components (red and green dots) are closer together in energy/temperature space. From *Hasegawa et al.* [2004a].

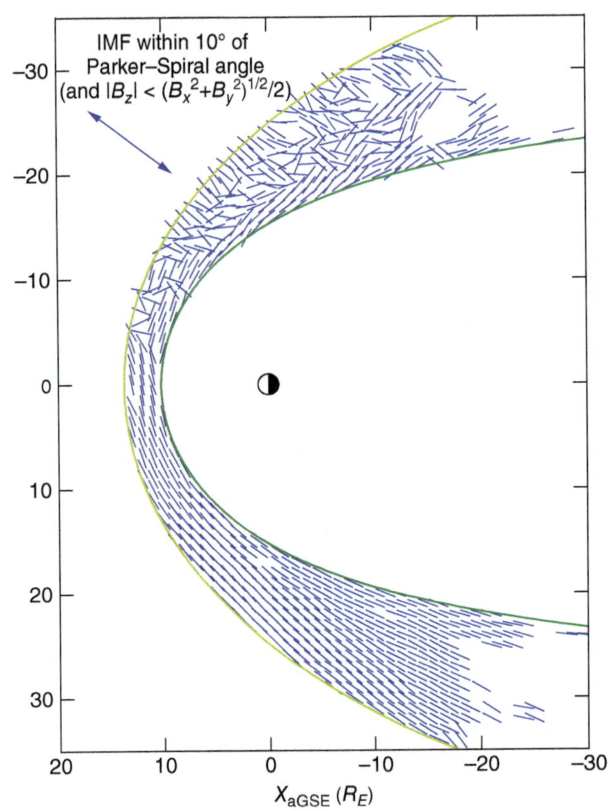

Figure 14.8 Median magnetic field vector orientation in $1 \times 1 R_E$ bins in the equatorial plane when IMF is within 10° of Parker spiral angle. Adapted from *Petrinec* [2013, Figure 6].

observations. The fluctuations in the magnetic field could indicate the presence of the compressional waves. KAWs can be generated when the compressional waves interact with the magnetospheric boundary, which can lead to diffusive transport of the solar wind/magnetosheath ions across the magnetopause boundary [*Johnson and Cheng*, 1997]. The waves have a parallel electric field, which can heat electrons in the direction of the magnetic field [*Hasegawa and Chen*, 1975; *Hasegawa and Mima*, 1978]. When the waves have large amplitudes, they can also heat ions in the direction perpendicular to the magnetic field [*Johnson and Cheng*, 2001]. Because the source of compressions could be compressional instabilities generated in the magnetosheath or compressions driven directly by the solar wind, in a statistical sense, there can be a dawn-dusk asymmetry in the KAWs and ion heating during entry. A recent survey of wave power in the sheath and magnetopause suggests that the wave power associated with tranvsverse KAWs is enhanced along the dawn flank [*Yao et al.*, 2011], which would provide enhanced transport. The study of *Chaston et al.* [2008] also provides observational evidence of stochastic heating of ions by KAWs as predicted by *Johnson et al.* [2001].

It is not clear what causes the dawn-dusk asymmetry in the cold-ion density profile either. The dawn-dusk density asymmetry can be caused by the dawn-dusk asymmetry in the source, i.e., magnetosheath population itself, or by a dawn-dusk asymmetry in the entry rate. Several magnetosheath studies showed that the magnetosheath n is higher on the dawnside than the duskside [e.g., *Paularena et al.*, 2001, *Nemecek et al.*, 2002]. More interestingly, *Walsh et al.* [2012] found that in THEMIS observations, near noon, the magnetosheath density is essentially dawn-dusk symmetrical, $n_{dawn}/n_{dusk} \sim 1$, but n_{dawn}/n_{dusk} grows larger away from noon, reaching ~1.2 at 5h away from noon. They also found similar patterns in BATS-R-US MHD, except that the asymmetry is even more pronounced, $\sim n_{dawn}/n_{dusk} \sim 1.5$, near the dawn and dusk flanks. This new result may put an additional constraint on the entry location of the magnetosheath ions. That is, if the source of the magnetospheric cold-ion dawn-dusk density asymmetry is magnetosheath, then the magnetosheath source would have to be located several hours away from noon rather than close to noon.

The dawn-dusk asymmetry in the cold-ion density may be attributed to the dawn-dusk asymmetry in the entry rate or entry process, i.e., the entry rate at the dawn flank magnetopause is higher than that at the dusk flank. *Hasegawa et al.* [2006] examined the occurrence rate of the KHI on the dawn and dusk flanks and found that the occurrence rate is roughly equal. Interestingly, *Wang et al.* [2007] reported that the magnetosheath ion entry from the dawn flank is more efficient with decreasing solar wind velocity (V_{sw}), but no significant V_{sw} effect is seen on the dusk flank.

The asymmetric heating of the cold-component/magnetosheath ions may suggest an asymmetry from the KAW-led entry, i.e., more KAW-led entry on the dawnside than the duskside magnetopause. The asymmetry may also result in KH waves because of differences in the diamagnetic drift on the two flanks [*Wilber and Winglee*, 1995], differences of the velocity shears at the two flanks, the effects of heavy ions [*Johnson et al.*, 2014], or by differences in field line draping at the flanks.

Olson and Pfitzer [1985] proposed that gradient drift entry (GDE) would allow magnetosheath ions to enter along the dawn flank and the electrons along the dusk flank. They argued that this entry mechanism can lead to the observed field-aligned current (FAC) polarity and the dawn-dusk current in the magnetotail, at least qualitatively. However, it is not clear if GDE alone can account for the observed filling rate or the cold dense state of the plasma sheet. *Treumann and Baumjohann* [1988] calculated that only 5% of the magnetosheath particles that come in contact with the magnetopause can be trapped in the magnetosphere in their model. The difficulty here is that the number of particles that come in contact with the magnetopause is unknown in any realistic magnetopause model. Hence, the efficiency of GDE is unkown. *Richard et al.* [1994] traced thousands of ions in fixed MHD electric and magnetic fields and found that most solar wind ions enter the magnetosphere through double-cusp-type reconnection [*Song and Russell*, 1992] and only a small fraction enters with the GDE mechanism. More recently, *Zhou et al.* [2007] used a simple model to calculate GDE efficiency and found that only electrons and ions having energies larger than several kilo-electron-volts can gradient drift into the magnetosphere during southward IMF. However, during northward IMF, lower-energy particles can enter, but their efficiency was not calculated. In any case, GDE cannot explain the ion and electron heating in the entry process.

The dawn-dusk asymmetry in the solar wind entry rate may also be attributed to the asymmetry in the magnetotail flank reconnection. *Peroomian and El-Alaoui* [2008] performed large-scale kinetic (LSK) simulations that show that most solar wind ions enter along the dawn flank primarily through reconnection. They also found that the entry locations can be modulated by IMF B_y. When IMF B_y is positive, the solar wind ion entrees along the dusk flank can increase. Similarly, using the same technique, *Peroomian et al.* [2011] showed that a large number of solar wind ions enter along the dawn flank through reconnection and get energized nonadiabatically before reaching the dawnside ring current.

Li et al. [2008] studied the double-cusp reconnection entry mechanism with MHD and found that ionospheric conductance affects the convection of the newly closed

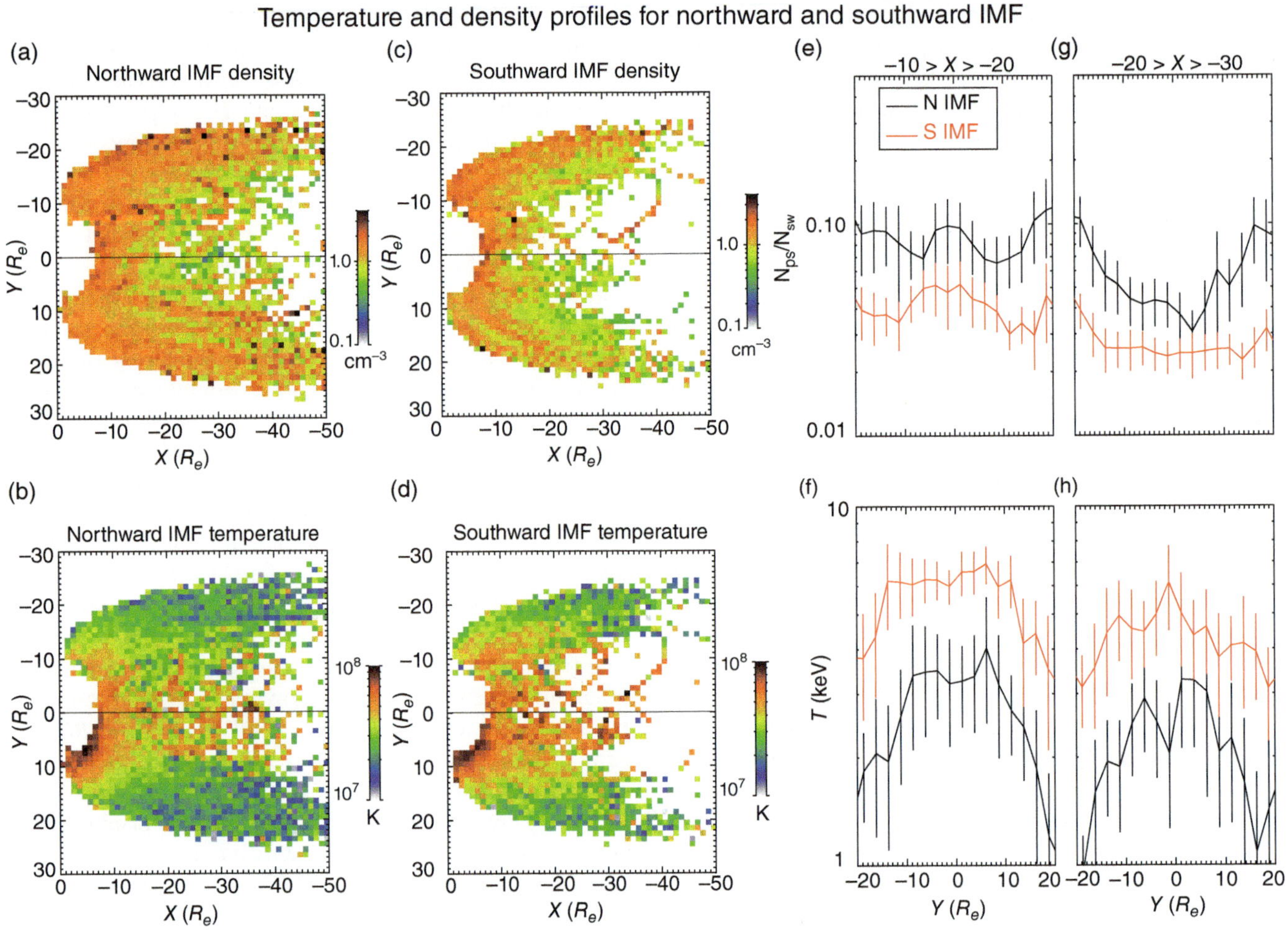

Figure 14.9 The plasma sheet ion density 2D profiles inferred from DMSP observations for the northward and southward IMF are shown in (a) and (c), respectively, and their corresponding temperature profiles are shown in (b) and (d), respectively [from *Wing and Newell*, 2002]. The density and temperature 1D profiles obtained from Geotail observations at $-10 > X > -20R_E$ are plotted in (e) and (f), respectively, and at $-20 > X > -30R_E$ in (g) and (h), respectively. From Wang et al., PET09 Workshop, Fairbanks, Alaska, March 2009; the method is described in *Wang et al.* [2010].

field lines and can introduce asymmetry in the entry rate. However, this mechanism may not introduce dawn-dusk asymmetry in the cold-component temperature, although asymmetries in MHD simulations may also result from local time differences in ionospheric conductances [*Zhang et al.*, 2012].

During southward IMF, there is evidence that solar wind particles can still enter from the flank magnetopause albeit in smaller quantities. Figure 14.9 panels a – d show plasma sheet ion n and T profiles for northward and southward IMF obtained from DMSP observations (note here the hot and cold components are combined) [*Wing and Newell*, 2002]. Figures 14.9a and 14.9b show that during northward IMF there are density maxima and temperature minima along both flanks, which can be interpreted as a signature for solar wind ion entry, as discussed above. However, during southward IMF, there is also evidence for solar wind ion entry along the flank, as suggested by the density maxima and temperature minima along both flanks in Figures 14.9c and 14.9d, albeit the density maxima and the temperature minima are smaller compared to those for northward IMF. These DMSP observations are corroborated by Geotail observations in Figure 14.9 panels e – h, which show that during southward IMF there are density peaks and temperature minima on both flanks, but the flank source is weaker [*Wang et al.*, 2006]. Presently, it is not firmly established what is/are the entry mechanism(s) for solar wind ions along the flanks during southward IMF, but KHI and KWA, in principle, can still operate under southward IMF, although KHI may be less favorable for southward IMF [*Miura*, 1995]. Recently, *Hwang et al.*

[2011] observed KH waves under southward IMF on the dawnside, but the KH vortices can be irregular and intermittent.

14.2.3. Entropy

For studies of plasma transport into and within the plasma sheet, it is often useful to consider what parameters are conserved and under what conditions. One fundamental quantity is the entropy, which is conserved both locally and globally when the plasma response is adiabatic [*Birn et al.*, 2006b]. Locally, the entropy per unit mass in an ideal, isotropic plasma is $\sigma = C\ \log(p^{1/\gamma}/\rho)$ [*Baumjohann*, 1993]. For an entropy-conserving process, σ or, equivalently, $s = p/\rho^{\gamma}$ (sometimes termed specific entropy or local entropy) is conserved. In the ideal MHD description, the pressure response is typically prescribed by an adiabatic equation of state such that $ds/dt = 0$. In a multifluid description, specific entropies for each species are also conserved as long as pressures are basically isotropic and there is no heat exchange between species, heat fluxes, or other dissipative processes.

The total entropy of a flux tube of unit magnetic flux is obtained by integrating the entropy per unit mass over the flux tube volume $\Sigma = \int \rho s\ dl/B$, where dl = infinitesimal length along B and B is the magnetic field. Equivalently, the pressure equation implies an integral constraint on the total mass content of a flux tube, $M = \int \rho\ dl/B$, is proportional to $S = \int p^{1/\gamma} dl/B$. If the plasma in a flux tube is in quasi-equilibrium, the pressure is constant and $S = p^{1/\gamma} V$ is conserved (i.e., pV^{γ} = const). While S is not exactly the entropy of the flux tube, it is entropylike in nature [*Birn et al.*, 2006b] and is sometimes referred to as the flux tube entropy parameter, flux tube entropy, total entropy, or global entropy. The specific entropy is an intensive variable whereas the total entropy is extensive.

More detailed discussion and application of entropy can be found in the 2009 Journal of Geophysical Research special section on "Entropy Properties and Constraints Related to Space Plasma Transport" (see *Wing and Johnson* [2010] and references therein).

Specific entropy s places significant constraints on plasma entry mechanisms. Figure 14.10a shows that statistically the magnetosheath ion s is 2–3 orders of magnitude lower than the plasma sheet ion s. Hence, it would be expected that a signature of the solar wind entry is to reduce s in the plasma sheet. This is indeed observed, as shown in Figure 14.10b, which shows that s is lower in both flanks, where there is a higher concentration of cold (solar wind) ions [*Wing et al.*, 2005], than that near the midnight meridian. Figure 14.10c suggests that there is a duskward heat flux in the hot-ion entropy profile, as would be expected from the curvature and gradient drifts. Figure 14.10 illustrates two interesting and relevant points: (1) during entry solar wind ion specific entropy increases by 2–3 orders of magnitude (Figure 14.10a) and (2) upon entry the transport process that moves cold ions of solar wind origin from the flanks to near midnight increases the specific entropy further by a factor of ~5 (Figure 14.10d). Point 2 can be crucial in resolving another key issue: what process transports this cold plasma from the flanks to the midnight meridian. Figure 14.10d suggests that whatever the process may be it would need to increase the specific entropy by a factor of ~5.

Changes in specific entropy may be associated with wave-particle interactions, heat fluxes of captured populations, and other dissipative mechanisms. Three-dimensional simulations of shear flow instabilities show that magnetic reconnection can mix magnetosheath and magnetosphere populations leading to reduced entropy on closed field line regions. Eddy diffusion and kinetic drifts can also affect the entropy profiles in the plasma sheet.

The total entropy, $S = p^{1/\gamma}\ V$, provides an important constraint on the source of plasma sheet material because, once material is captured on closed flux tubes, it should satisfy the entropy constraint if it is convected adiabatically into the plasma sheet. The value of this constraint, S, can be estimated from solar wind conditions and assumptions about the plasma capture process and compared with typical values in the plasma sheet. Total entropy does not increase significantly for double-cusp reconnection [*Johnson and Wing*, 2009], but it may increase significantly for reconnection in KH vortices. Moreover, the observed increase of plasma temperature tailward along the flanks is also inconsistent with specific entropy conservation of plasma captured by double-cusp reconnection, which implies that nonadiabatic processes leading to plasma loss and/or heating are required in order for plasma captured by double-cusp reconnection to reach the plasma sheet [*Johnson and Wing*, 2009].

14.2.4. Ion-to-Electron Temperature Ratio

Another important observational "constraint" on solar wind entry and plasma transport processes in the plasma sheet is the rough, not exact, preservation of the temperature ratio of ions and electrons, T_i/T_e. For example, *Baumjohann et al.* [1989] reported $5.5 < T_i/T_e < 11$ in the plasma sheet. The ratio T_i/T_e has been shown to have a dependence on solar wind Alfvén Mach number (M_A). For example, *Lavraud et al.* [2009] examined an event during which the solar wind M_A is low, ~3, and found that T_i/T_e is also unusually low, ~3 in both the magnetosheath and magnetopause boundary layers (i.e., inside the magnetopause). *Wang et al.* [2012] found a similar dependence on solar wind velocity.

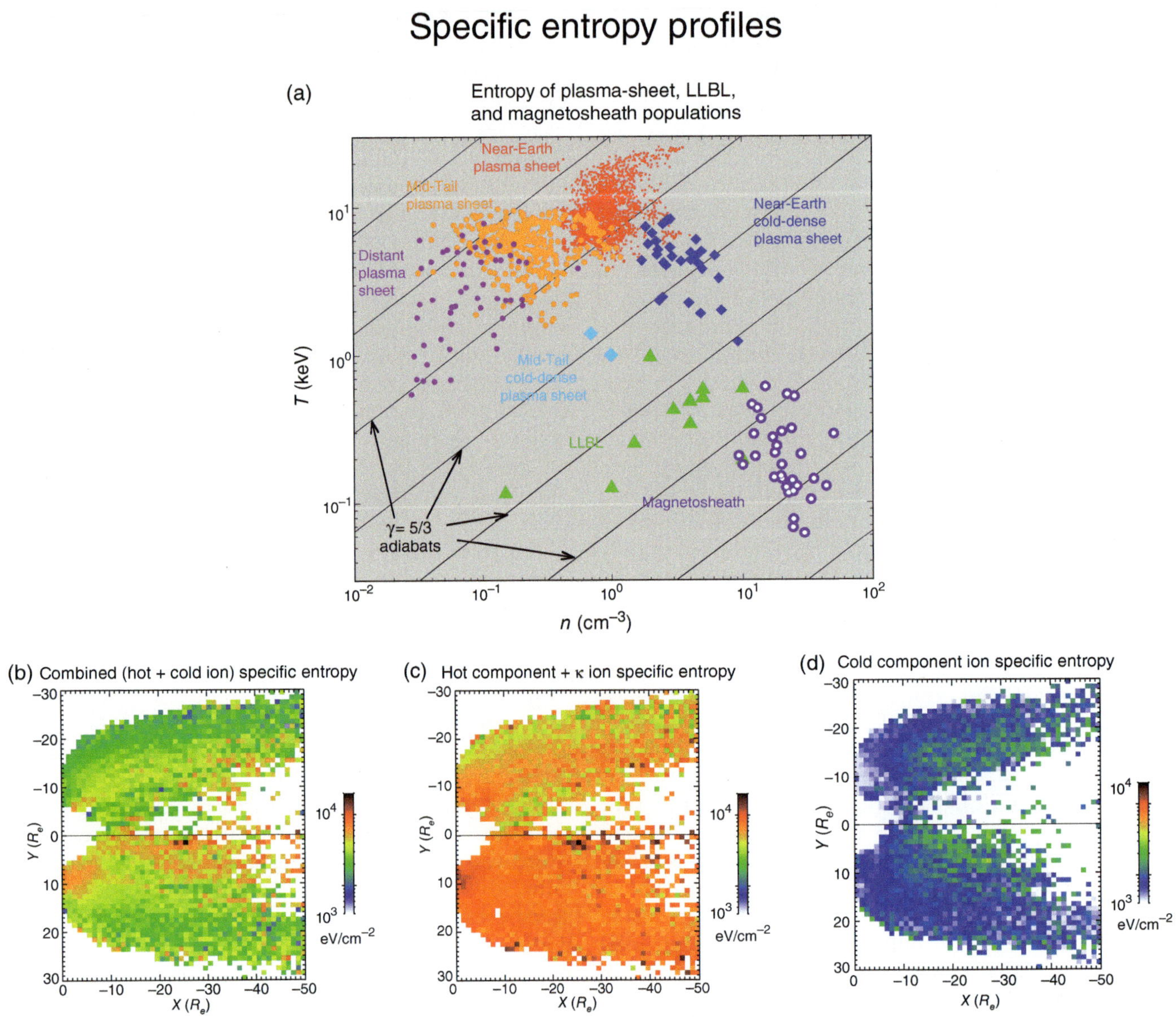

Figure 14.10 (a) Entropy density of plasma sheet, LLBL, and magnetosheath ions in the *n–T* log-log space. The ions in the distant, midtail, and near-Earth plasma sheet tend to have the same adiabats, indicated by the diagonal lines with slope $\gamma = 5/3$, although the scatter is large. The magnetosheath entropy is lower than that of the LLBL, which is lower than that of the cold, dense plasma sheet. The data source and treatment for this figure are described by *Borovsky et al.* [1998b]. (Figure courtesy of Joe Borovsky.) Plasma sheet specific entropy profiles for (b) combined hot and cold components, (c) hot component, and (d) cold component. The hot-component profiles shows a dawn-dusk gradient, while the cold-component profile shows *s* approximately conserved in the *x* direction and a gradient toward the midnight meridian [from *Johnson and Wing*, 2009].

The ratio T_i/T_e changes significantly at the bow shock and is determined by solar wind parameters. However, it is particularly interesting that the solar wind entry process appears to roughly conserve the T_i/T_e of the magnetosheath particles during the transport across the magnetopause [*Lavraud et al.*, 2009]. This property can be seen in Figure 14.11a, which shows $5 < T_i/T_e < 10$ in the magnetosheath and plasma sheet [*Wang et al.*, 2012]. Within the magnetosheath, T_i/T_e remains the same as the particles flow downstream and cool adiabatically, as would be expected. Figures 14.11b and 14.11c show two events in which THEMIS crossed from the magnetosheath to the plasma sheet. Both figures show that T_i/T_e of the magnetosheath (green line in the third panel

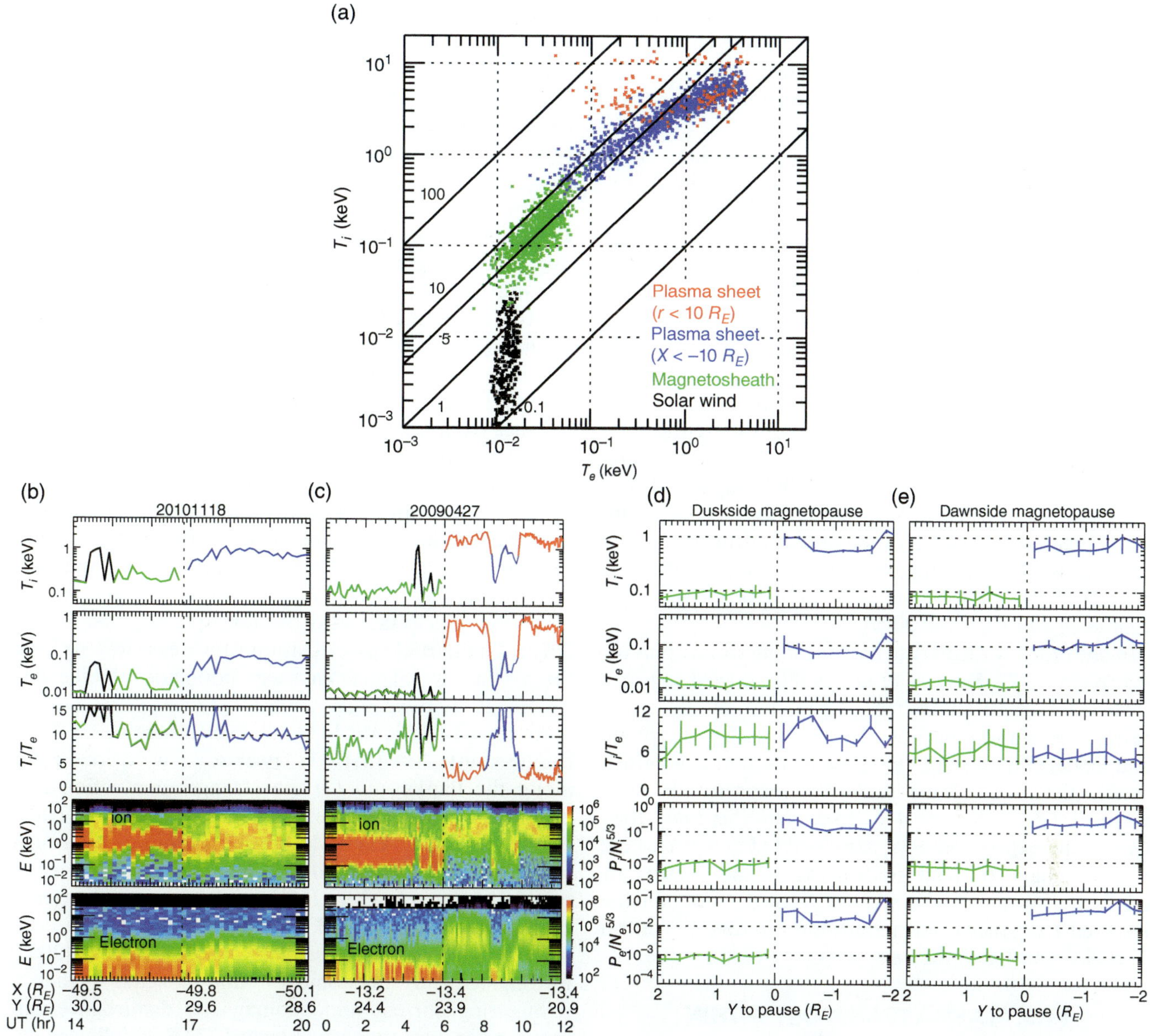

Figure 14.11 (a) Scatter diagram of ion temperature (T_i) versus electron temperature (T_e) in the solar wind, magnetosheath, and plasma sheet; $T_i/T_e = 0.1$, 1, 5, 10, 100 indicated by the black lines; T_i, T_e, T_i/T_e and energy spectrums of ion and electron energy fluxes (eV/(s-sr-cm^2-eV)) observed by THEMIS-b as it crossed the magnetopause (indicated by the dotted vertical lines) on (b) 20090423 and (c) 20090427. The green curves indicate magnetosheath plasma while the black curves indicate the hot plasma leaking out from the magnetosphere. The cool (warm) plasma sheet plasma are indicated by the blue (red) curves. The T_i, T_e, T_i/T_e, ion specific entropy $P_i/N_i^{5/3}$(nPa-cm^5), and electron specific entropy $P_e/N_e^{5/3}$(nPa-cm^5) versus the Y distance to the magnetopause (positive Y for the magnetosheath and negative Y for the plasma sheet) at (d) the duskside magnetopause and (e) the dawnside magnetopause. From *Wang et al.* [2012].

from the top) appears to be similar to the T_i/T_e of the cool plasma (blue line in the third panel from the top). Figures 14.11d and 14.11e show statistically T_i/T_e of the magnetosheath and the cool plasma (not necessarily cold component, although it should be dominated by the cold component) on both sides of the dusk and dawn magnetopause, respectively, at $X < -40 R_E$. Figures 14.11d and 14.11e suggest that when ions and electrons cross the magnetopause boundary from the magnetosheath to the magnetosphere, T_i and T_e increase by a factor of

~6–10 and specific entropies s_i and s_e increase by a factor of ~20. *Wang et al.* [2012] interpreted these changes as the signature of a nonadiabatic heating process at the boundary that heats the ions and electrons by the same proportion.

Within the plasma sheet, there is a strong dawn-dusk asymmetry in T_i/T_e which can be attributed to the curvature and gradient drifts such that T_i (T_e) is higher at the dusk (dawn) [e.g., *Wang et al.*, 2012]. Hence, T_i/T_e would have dependence on locations. *Kaufmann et al.* [2005] found that T_i/T_e decreases earthward near the midnight meridian. They offered several possible explanations: (1) electrons proportionally gain more energy than ions, (2) the ions lose energy through parallel heat flux to the ionosphere by radiating Alfvén waves, and (3) magnetic reconnection can heat ions and electrons differently. Fast flows can reduce ion density but increase ion temperature. Because of these, *Kaufmann et al.* [2005] suggested that the factors influencing T_i/T_e variability can be more complicated than just curvature and gradient drifts. The nearly constant T_i/T_e suggests that ions absorb 5–10 times as much heat as the electrons in the plasma sheet [e.g., *Kaufmann and Paterson*, 2009].

In one of a few theoretical and modeling studies that try to explain T_i/T_e, *Schriver et al.* [1998] modeled ion and electron acceleration in the magnetotail by following trajectories of thousands of particles in the T89 magnetic field model [*Tsyganenko*, 1989] and a constant dawn-dusk electric field of 0.25 mV/m in the magnetotail. They obtained T_i/T_e values of 4–6. To explain this ratio, they derived the expression $T_i/T_e = (m_i/m_e)^{1/3} (B'_e/B'_i)^{2/3}$, where B' is the magnetic field gradient where the electron or ion first crosses the neutral sheet. Hence, to obtain the observed T_i/T_e, they would require ions to cross the neutral sheet closer to Earth than electrons.

14.3. PLASMA TRANSPORT WITHIN PLASMA SHEET

14.3.1. E × B and Curvature and Gradient Drifts

Three transport mechanisms that have been shown to play significant roles in the plasma sheet and that have successfully explained large-scale features in the plasma sheet are (1) **E** × **B** and (2) curvature and gradient drifts. For example, Figures 14.9a–14. 9d show that the quiet time or northward IMF ion T and n profiles have the following large-scale features: (1) density peaks along both flanks and the inner edge of the plasma sheet, (2) temperature peak at the dusk-midnight sectors at the inner edge of the plasma sheet ($r = 8–10 R_E$), and (3) temperature minima along both flanks [*Wing and Newell*, 1998, 2002; *Wing et al.*, 2005; *Johnson and Wing*, 2009]. These large-scale features have also been found with in situ observations [*Wang et al.*, 2001, 2006]. These features can be attributed to two plasma sources, cold ions from the LLBL and hot ions from deep tail, and three transport mechanisms, **E** × **B** and curvature and gradient drifts [e.g., *Spence and Kivelson*, 1993; *Wang et al.*, 2001, 2006; *Wing and Newell*, 1998, 2002; *Wing et al.*, 2005]. Cold ions that enter the plasma sheet from the flanks mostly **E** × **B** drift sunward, resulting in temperature minima and density maxima along both flanks. The hot ions **E** × **B** drift sunward while curvature and gradient drift westward, especially in the near-Earth region where the magnetic field is stronger ($r < 10 R_E$), leading to the temperature peak at the dusk-midnight sector at the inner edge of the plasma sheet. The importance of the magnetic drift increases whereas the importance of **E** × **B** drift decreases with decreasing distance from Earth. By the time the particles reach the inner magnetosphere, **E** × **B** drift is negligible as curvature and gradient drifts dominate [*Wang et al.*, 2004]. As a result, the ions and electrons mainly move azimuthally, with the ions moving westward while the electrons move eastward. As they circle Earth, some of the ions and electrons leave the magnetosphere through magnetopause and some precipitate into the ionosphere [*Wing et al.*, 2013]. The particle precipitation provides a way to infer (or a window to) plasma sheet population [*Wing and Newell*, 1998; *Wing et al.*, 2005].

As shown in Figure 14.9, relative to northward IMF, during southward IMF (1) fewer solar wind ions enter along the flanks and (2) the stronger convection and dawn-dusk electric field leads to larger **E** × **B** drift. These differences in drifts and plasma sheet properties between northward and southward IMF give rise to the question of what happens when IMF turns southward after a prolonged northward IMF and the plasma sheet has been loaded with cold and dense plasma. *Thomsen et al.* [2003] investigated this phenomenon and concluded that a strong ring current and geomagnetic disturbance can ensue when prolonged northward IMF is followed by strong convection due to southward IMF or a sudden increase in solar wind dynamic pressure. Basically, when IMF turns southward, the massive preloaded material in the plasma sheet is delivered to the inner magnetosphere, resulting in a strong ring current and a large geomagnetic disturbance [*Denton et al.*, 2005; *Lavraud et al.*, 2006a]. This has been referred to as "the calm before the storm" for the special case of corotating interaction regions [*Borovsky and Steinberg*, 2006], but the term seems appropriate for other cases mentioned here. *Lavraud et al.* [2006b] also studied the effect of prolonged northward IMF followed by southward IMF on inner magnetosphere plasma properties. They found two ion populations at the geosynchronous orbit: (1) midnight and (2) dawnside. The dawnside cold dense ions at the near-Earth plasma sheet during active times or substorm growth phase have been previously observed [*Wing and Johnson*, 2009; *Wing and Newell*, 1998; 2002; *Korth et al.*, 1999; *Denton et al.*, 2005] and modeled [*Wang et al.*, 2004]. It has been attributed to the stagnation point where $\mathbf{E} \times \mathbf{B}$ drift and corotation is nearly balanced by the curvature and gradient drifts [*Friedel et al.*, 2001] and solar wind entry from the dawn flank.

Interestingly, *Ashour-Abdalla et al.* [2010] found that during northward IMF the plasma sheet can temporarily split into dawn and dusk halves in LSK simulation. They attributed this division to the presence of flux ropes in the center of the tail possibly due to the localized near-Earth reconnection.

Ion outflows during active times can significantly increase plasma sheet O^+ population, which can at times carry much of the energy (e.g., storm-time ring current). Escaping O^+ has higher fluxes in the 00–12 MLT (midnight-dawn-noon) sector than 12–24 MLT (noon-dusk-midnight) sector [*Redmon et al.*, 2012; *Collin et al.*, 1988]. This may be related to the presence of cold, dense ions in the plasma sheet postmidnight sector after substorm onset [*Wing et al.*, 2007; *Wing and Johnson*, 2009]. *Peterson et al.* [2009] argued the importance of in-transit populations for ionospheric outflows for the quicker-than-expected arrival time of the outflows in the plasma sheet.

These outflows (combined with associated energization processes) could significantly modify the background plasma and magnetospheric $\mathbf{E} \times \mathbf{B}$ convection [e.g., *Winglee et al.*, 2002]. Several important physical effects might be expected. The study of *Karimabadi et al.* [2011] suggests that heavy ions only affect the reconnection rate when they are the primary current carrier but in the nonlinear stage a proton-dominated current sheet can be replaced by heavy ions in the surrounding plasma, leading to reduced reconnection efficiency, reduction in the production of secondary islands, and broadening of the reconnection structures. MHD simulations show that dayside O^+ outflows would convect to the nightside and may move the magnetotail X line closer to Earth in order to maintain the same reconnection rate on the dayside [e.g., *Brambles et al.*, 2010; *Wiltberger et al.*, 2010]. On the other hand, nightside O^+ outflows may have the opposite effect, moving the magnetotail X line further from Earth due to the increased plasma pressure [e.g., *Garcia et al.*, 2010].

14.3.2. Turbulence

Once the solar wind plasma enters the plasma sheet, it can be distributed throughout the plasma sheet. Observations show that the cold, dense ions can be found throughout the entire plasma sheet during northward IMF [e.g., *Terasawa et al.*, 1997; *Øieroset et al.*, 2005; *Wing and Newell*, 2002; *Wing et al.*, 2005, 2006; *Johnson and Wing*, 2009]. However, the mechanism for transporting cold, dense plasma from the flanks to the midnight meridian (center of the plasma sheet) has not been clearly established or uniformly accepted.

The plasma sheet has been observed to be turbulent, which can play some roles in the plasma transport [*Borovsky et al.*, 1997; *Borovsky and Funsten*, 2003; *Weygand et al.*, 2005]. One source of the turbulence may be velocity shear instabilities [*Borovsky et al.*, 1997]. *Wang et al.* [2010] performed a simulation that shows that diffusive transport due to turbulence can move cold particles from the flank to the midnight meridian. As such, turbulence is an important process that is intrinsic to the quasi-stable state of the tail and boundary layer [*Antonova*, 2005] as well as its stability [*Stepanova et al.*, 2011].

Ionospheric velocity measurements inferred from SuperDARN also suggest that even under steadily driven conditions there are significant ionospheric velocity fluctuations [*Bristow*, 2008]. They occur for all local times, although they are largest in the afternoon sector at auroral latitudes, and the prenoon polar cap. The level of fluctuation seems to be independent of substorm/nonsubstorm conditions. The average time for a significant fluctuation is on the order of 10 min for all latitudes and local times, and they obey a power law distribution for at least part of its range. The ionospheric fluctuations could be a remote indicator of plasma sheet turbulence and be useful for studies of plasma sheet transport.

14.3.3. Entropy

Section 14.2.3. discusses the concept of total entropy S and specific entropy s and their importance to the solar wind plasma entry problem. It turns out these two entropy parameters are also relevant for the study of the plasma transport within the plasma sheet. Total entropy S is an important parameter to identify when and where the magnetospheric plasma is interchange unstable. In particular, flux tubes with reduced S relative to the surrounding plasma may be interchange unstable, which can lead to earthward plasma transport at high speed, commonly known as fast flows [e.g., *Pontius and Wolf*, 1990] that redistribute plasma into a more stable magnetospheric configuration [*Gold*, 1959; *Sonnerup and Laird*, 1963; *Hill*, 1976; *Chen and Wolf*, 1999].

Statistical studies have shown fairly definitively that in the magnetotail S is not conserved along the earthward convective path; S decreases with decreasing distance from Earth [*Erickson and Wolf*, 1980; *Kaufmann and Paterson*, 2009; *Wing and Johnson*, 2009; *Wang et al.*, 2009; *Garner et al.*, 2003]. Nonconservation of S can occur in nonadiabatic processes such as wave-particle interactions, heat flux, gradient/curvature drifts [*Tsyganenko*, 1982; *Kivelson and Spence*, 1988; *Wang et al.*, 2001, 2004; *Wolf et al.*, 2009; *Wing and Johnson*, 2009; *Kaufmann and Paterson*, 2008], and non-earthward $\mathbf{E} \times \mathbf{B}$ drift [e.g., *Kaufmann et al.*, 2004]. Nonconservation of S can also occur when there is a loss of plasma content in the flux tube [e.g., *Pontius and Wolf*, 1990; *Chen and Wolf*, 1993; *Wing and Johnson*, 2009; *Borovsky et al.*, 1998b]. In addition, S and s can change when plasma is added to the flux tube, e.g., through solar wind plasma entry or ionospheric outflow [e.g., *Johnson and Wing*, 2009]. If the localized process leads to field line reconfiguration (reconnection), then the change of S is more related to the change of flux tube volume than to changes in s within that volume.

Birn et al. [2006a, 2009] and *Wing and Johnson* [2009] discussed S changes in flux tube reconfiguration in magnetic field reconnection process in which the closed field line splits into two parts: (1) a newly closed field line albeit with reduced length and (2) a tailward convecting plasmoid. The newly closed field line has much lower S and V compared to those in the original flux tube, while the remainder resides in the newly formed plasmoid.

An earthward fast flow can attain velocity greater than 400 km/s and has been associated with a bubble, which is a magnetotail flux tube having lower density, pressure, and S than surrounding flux tubes [e.g., *Sergeev et al.*, 1996; *Nakamura et al.*, 2005; *Angelopoulos et al.*, 1994]. Herein, the term *fast flow* includes bubble, bursty bulk flow (BBF), and flow burst (shorter duration than BBF). Fast earthward flows can be more frequently observed during magnetically active times and have been associated with substorm onsets, and their occurrence frequency reaches a peak at the start of the recovery phase [e.g., *Nakamura et al.*, 2002; *Angelopoulos et al.*, 1994; *Baumjohann et al.*, 1990; *McPherron et al.*, 2011; *Sergeev et al.*, 2008; *Slavin et al.*, 2002]. Fast-flow observations and simulations often show two parts: (1) the head where B_z is large (the dipolarization front) and (2) the tail. The pressure inside the bubble is smaller than that outside the bubble. This pressure gradient leads to the field-aligned current with region 1 polarity whose particle precipitation signature in optical images often appears as an auroral streamer. Thus, many studies pair auroral streamers with fast flows [e.g., *Sergeev et al.*, 2004; *Nakamura et al.*, 2001; *Lyons et al.*, 1999; *Sanchez et al.*, 2012].

There has been no clear picture of the earthward fast flows $r < 10R_E - 12R_E$. It appears that many fast flows slow down at $r < 10R_E - 12R_E$ [e.g., *McPherron et al.*, 2011; *Panov et al.*, 2010a] and many do not reach the inner magnetosphere or geosynchronous orbit [e.g., *Ohtani et al.*, 2006]. Observations and theory suggest that in the final stage of the bubble earthward motion the filamentary bubble executes a damped oscillation about an equilibrium (interchange oscillation) as the bubble slows down and eventually comes to a stop (interchange oscillation) [*Wolf et al.*, 2012; *Panov et al.*, 2010b]. In Rice Convection Model-Equilibrium (RCM-E) simulations, flux tubes do not generally reach in the inner magnetosphere unless their total entropies (S) have been reduced [*Lemon et al.*, 2004; *Zhang et al.*, 2008].

In the *Birn et al.* [2009] simulations, after the reconnection, the newly closed field line, which has smaller entropy and flux tube compared to the surrounding flux tubes, can initiate the unstable growth of ballooning and interchange instability, resulting in the earthward propagation of the flux tube. These bubbles would move earthward until their S equals that of their neighbors [*Wolf et al.*, 2009]. The more depleted the flux tube, the closer to Earth it can penetrate. This view has been supported in some recent observational studies [e.g., *Dubyagin et al.*, 2011].

Recently, *Birn et al.* [2011] showed that as bubbles move earthward vortices of return flow develop and the bubbles also fragment, developing cross-tail structuring in the later stages. This process is displayed in Figure 14.12. Initially, reconnection at $X = -8$ (simulation scale; $X = -10$ corresponds to $X_{GSM} = -23R_e$) leads to fast earthward flows through interchange instability generated by buoyancy forces. In Figures 14.12a and 14.12d, the blue patch that is aligned with the X axis near midnight corresponds to fast earthward flow of flux tube with depleted S (BBF), which brakes in the inner magnetosphere at about $X = -2$ (simulation scale). Sometime later, the earthward flow near midnight is reversed, as shown in Figures 14.12b and 14.12e. Additionally, two secondary earthward fast-flow channels develop eastward and westward of the initial fast flow. Small-scale structuring develops. Finally, the flow subsides and the entire region in the inner magnetosphere has overwhelmingly low entropy plasma (blue), as shown in Figure 14.12c. This final configuration suggests that the BBFs have successfully carried the low-entropy plasma from the magnetotail into a large region in the inner magnetosphere.

Fast flows and S may be relevant to plasma transport in other planetary magnetospheres as well. For example, many studies found that the plasma injection at Saturn's magnetopshere is often characterized as having "deep density cavity" where the ions (both protons and water group) and electrons in the injected flux tube have lower densities than those of the surrounding plasma [e.g., *Sittler et al.*, 2006; *Rymer et al.*, 2009; *Burch et al.*, 2005; *Hill et al.*, 2005; *Sittler et al.*, 2005]. This is exactly what is observed with fast-flow and fast-flow-led plasma injection at Earth's magnetosphere.

At Saturn, distant reconnection, either Vasyliunas- or Dungey-type reconnection, can lead to the reduction of the flux tube volume and hence S. As at Earth, the entropy-depleted flux tube can lead to interchange instability [e.g., *Young et al.*, 2005; *Hill et al.*, 2005] and move inward where the background S is lower (even though the density can be higher) [e.g., *Thomsen et al.* 2010, *McAndrews et al.*, 2009]. Some Saturn studies have inter-

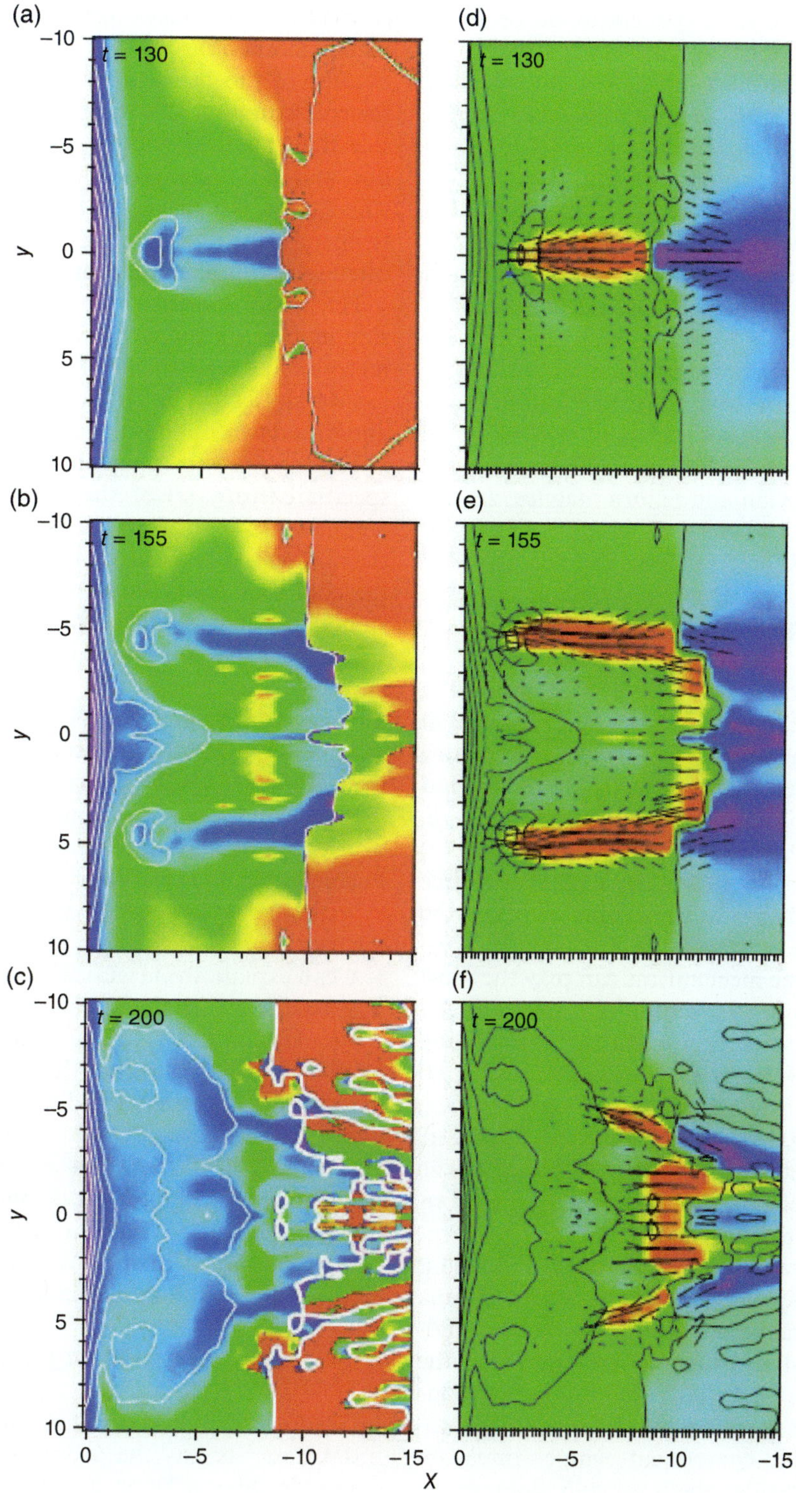

Figure 14.12 Total entropy S (a, b, c) and flow in the X-Y plane (d, e, f) at three different times by *Birn et al.* [2011] MHD simulation.

preted the density-depleted flux tube as a bubble [e.g., *André et al.*, 2007]. As at Earth, the bubble would seek a path to the region where the background S is the same as the bubble's, which means that it would move inward. An interesting variation of the process for Saturn may be that as a bubble moves inward in the inner magnetosphere it encounters colder neutrals and ions may undergo charge exchange. This may reduce the entropy of the bubble even more and perhaps allow it to penetrate deeper into Saturn's inner magnetosphere. The charge exchange makes it possible to study the bubbles with an energetic neutral atom (ENA) imager. Such studies have indeed been carried out. For example, *Mitchell et al.* [2005, 2009] reported that Cassini ENA observations frequently show brightening near midnight, which can be associated with reconnections tailward of 15 Saturn radii (R_s), which lead to flux tube interchange and flux tube planetward motion to $8R_s$. The particles in the flux tube are heated adiabatically as the flux tube moves to the inner magnetosphere where they encounter higher neutral density leading to higher ENA emission. A direct association between plasmoids in Saturn's magnetotail and ENA image brightening has been made by *Hill et al.* [2008].

More discussion on Jovian and Saturn magnetospheres can be found in Chapter 5 (*Krupp et al.*, in this volume) and Chapter 6 (*Jackman et al.*, in this volume), respectively.

14.4. SUMMARY AND CONCLUSION

This chapter reviews our current understanding of the plasma transport from the solar wind to the magnetotail and transport within the magnetotail. Three solar entry mechanisms are examined for northward IMF solar wind entrees: (1) double-cusp reconnection [*Song and Russell*, 1992], (2) KHI [e.g., *Otto and Fairfield*, 2000], and (3) KAW [*Johnson and Cheng*, 1997]. Each of these three mechanisms has been shown to be capable of filling the plasma sheet with cold, dense plasma. In order to determine which of these three mechanisms can play the most significant roles and under what conditions, we examined four observational constraints: (1) plasma sheet filling rate, (2) dawn-dusk asymmetries, (3) specific and total entropies, and (4) T_i/T_e.

It turns out that all three mechanisms are capable of filling the plasma sheet with cold, dense plasma. The entry rate from each mechanism is quite similar and consistent with the observed filling rate.

In situ and remote observations of the cold-component (magnetosheath/solar wind origin) ions at the plasma sheet flanks near the magnetopause reveal that the cold-component density and temperature are higher on the dawn flank than on the dusk flank. In situ observations show that while, on the duskside, the ion hot- (magnetospheric origin) and cold-component temperatures are isotropic, on the dawnside, the ion cold component appears to have been heated in the direction perpendicular to the magnetic field and the electrons have been heated in the direction parallel to the magnetic field. These observations are consistent with the expected signature of KAW interaction with particles. On the duskside, the ions are frequently observed to have two components whereas on the dawnside the ions are frequently observed to have one component. The hot component has higher temperature on the duskside than the dawnside because of the curvature and gradient drifts. In contrast, the cold component has higher temperature on the dawnside than the duskside. As a result, on the duskside, the hot and cold components are often easily identified whereas on the dawnside the hot- and cold-component temperatures are often close to each other and the ions are often seen as having one component with a broad peak.

The higher density of the cold component on the dawnside magnetotail may result from the dawn-dusk asymmetry of the magnetosheath ion density (magnetosheath ion density is higher on the dawnside than the duskside) and/or a dawn-dusk asymmetry of the entry rate.

The solar wind/magnetosheath entrees reduce the specific entropy (s) of the plasma sheet ions. The cold component s increases by a factor of 5 from the flanks to the midnight meridian. So, the mechanism that transports the cold ions from the flanks to the midnight meridian should also have the effect of increasing s.

The ratio T_i/T_e appears to be roughly conserved when the magnetosheath particles are transported across the magnetopause into plasma sheet even though T_i, T_e, s_i, and s_e all increase significantly. If the adiabatic equation is $TV^{(\gamma-1)}$, then T_i and T_e would increase according to the flux tube volume V and T_i/T_e can be conserved. However, the increases of s_i and s_e suggest nonadiabatic heating during entry. The fluctuation in T_i/T_e also indicates nonadiabatic processes.

Within the magnetotail, $\mathbf{E}\times\mathbf{B}$ and curvature and gradient drifts play important roles in the plasma transport and can explain the large-scale observed features of the n, T, p profiles. Turbulence can also play a significant role, particularly in transporting cold plasma from the flanks to the midnight meridian. Ion outflows can change the magnetotail convection and reconnection location. Total entropy S and fast flows can play significant roles in the earthward transport. Many fast flows do not reach the inner magnetosphere, but simulations show that fast flows would only travel inward until its S matches the background S. Total entropy S can also play a significant role in transport in other planetary magnetospheres.

ACKNOWLEDGMENTS

Gordon Wilson has been helpful in acquisition of DMSP SSJ/4 data, as has the World Data Center in Boulder, Colorado. Simon Wing gratefully acknowledges support from NSF grants ATM-0802715 and AGS-1058456 and NASA grant NNX13AE12G. Jay Johnson was funded by NASA grants (NNH09AM53I, NNH09AK63I, and NNH11AR07I), NSF grants ATM0902730 and AGS-1203299, and DOE contract DE-AC02-09CH11466. We acknowledge the support of the International Space Science Institute (ISSI) Bern, Switzerland. Last but not least, we also thank NSF GEM for supporting Plasma entry and transport into and within the magnetotail (PET) Focus Group, which provided a forum for fruitful discussions of the topics covered here.

REFERENCES

Anderson, B. J., and S. A. Fuselier (1994), Response of thermal ions to electromagnetic ion cyclotron waves, *J. Geophys. Res.*, *99*, 19,413–19,425.

André, N., et al. (2007), Magnetic signatures of plasma-depleted flux tubes in the Saturnian inner magnetosphere, *Geophys. Res. Lett.*, *34*, L14108, doi:10.1029/2007GL030374.

Angelopoulos, V., et al. (1994), Statistical characteristics of bursty bulk flow events, *J. Geophys. Res.*, *99*, 21,257–21,280.

Antonova, E. E. (2005), The structure of the magnetospheric boundary layers and the magnetospheric turbulence, *Planet Space Sci.*, *53*, 161–168, doi:10.1016/j.pss.2004.09.041.

Ashour-Abdalla, M., J. M. Bosqued, M. El-Alaoui, V. Peroomian, and R. Walker (2010), Observations and simulations of a highly structured plasma sheet during northward IMF, *J. Geophys. Res.*, *115*, A10227, doi:10.1029/2009JA015135.

Baumjohann, W. (1993), The near-Earth plasma sheet: An AMPTE/IRM perspective, *Space Sci. Rev.*, *64*, 141–163.

Baumjohann, W., G. Paschmann, and C. A. Cattell (1989), Average plasma properties in the central plasma sheet, *J. Geophys. Res.*, *94*, 6597.

Baumjohann, W., G. Paschmann, and H. Luhr (1990), Pressure balance between lobe and plasma sheet, *Geophys. Res. Lett.*, *17*, 45–48.

Birn J., M. Hesse, and K. Schindler (2006a), On the role of entropy conservation and entropy loss governing substorm phases, in *Proceedings of the Eighth International Conference of Substorms*, edited by M. Syrjäsuo, and E. Donovan, pp. 19–24, Univ. of Calgary, Calagary, Canada.

Birn J., M. Hesse, and K. Schindler (2006b), Entropy conservation in simulations of magnetic reconnection, *Phys. Plasmas*, *13*, 092117, doi:10.1063/1.2349440.

Birn, J., M. Hesse, K. Schindler, and S. Zaharia (2009), Role of entropy in magnetotail dynamics, *J. Geophys. Res.*, *114*, A00D03, doi:10.1029/2008JA014015.

Birn, J., R. Nakamura, E. V. Panov, and M. Hesse (2011), Bursty bulk flows and dipolarization in MHD simulations of magnetotail reconnection, *J. Geophys. Res.*, *116*, A01210, doi:10.1029/2010JA016083.

Bouhram, M., B. Klecker, G. Paschmann, S. Haaland, H. Hasegawa, H. Reme, J.-A. Sauvaud, L. M. Kistler, and A. Balogh (2005), Survey of energetic O^+ near the dayside mid-latitude magnetopause with Cluster, *Ann. Geophys.*, *23*, 1281–1294.

Borovsky, J. E., and H. O. Funsten (2003), MHD turbulence in the Earth's plasma sheet: Dynamics, dissipation, and driving, *J. Geophys. Res.*, *108*, 1284, doi:10.1029/2002JA009625.

Borovsky, J. E., and J. T. Steinberg (2006), The "calm before the storm" in CIR/magnetosphere interactions: Occurrence statistics, solar wind statistics, and magnetospheric preconditioning, *J. Geophys. Res.*, *111*, A07S10, doi:10.1029/2005JA011397.

Borovsky, J. E., R. C. Elphic, H. O. Funsten, and M. F. Thomsen (1997), The Earth's plasma sheet as a laboratory for turbulence in high-beta MHD, *J. Plasma Phys.*, *57*, 1–34.

Borovsky, J. E., M. F. Thomsen, and R. C. Elphic (1998a), The driving of the plasma sheet by the solar wind, *J. Geophys. Res.*, *103*, 17,617–17,639.

Borovsky, J. E., M. F. Thomsen, R. C. Elphic, T. E. Cayton, and D. J. McComas (1998b), The transport of plasma sheet material from the distant tail to geosynchronous orbit, *J. Geophys. Res.*, *103*, A9, 20,297–20,331.

Brambles, O. J., W. Lotko, P. A. Damiano, B. Zhang, M. Wiltberger, and J. Lyon (2010), Effects of causally driven cusp O^+ outflow on the storm time magnetosphere-ionosphere system using a multifluid global simulation, *J. Geophys. Res.*, *115*, A00J04, doi:10.1029/2010JA015469.

Bristow, W. (2008), Statistics of velocity fluctuations observed by SuperDARN under steady interplanetary magnetic field conditions, *J. Geophys. Res.*, *113*, A11202, doi:10.1029/2008JA013203.

Burch, J. L., J. Goldstein, T. W. Hill, D. T. Young, F. J. Crary, A. J. Coates, N. Andre, W. S. Kurth, and E. C. Sittler, Jr. (2005), Properties of local plasma injections in Saturn's magnetosphere, *Geophys. Res. Lett.*, *32*(14), L14S02.

Chappell, C. R., T. E. Moore, and J. H. Waite Jr. (1987), The ionosphere as a fully adequate source of plasma for the Earth's magnetosphere, *J. Geophys. Res.*, *92*(A6), 5896–5910, doi:10.1029/JA092iA06p05896.

Chaston, C. C., T. D. Phan, J. W. Bonnell, F. S. Mozer, M. Acun, M. L. Goldstein, A. Balogh, M. Andre, H. Reme, and A. Fazakerley (2005), Drift-kinetic Alfven waves observed near a reconnection X line in the Earth's magnetopause, *Phys. Rev. Lett.*, *95*(6), 065,002.

Chaston, C. C., M. Wilber, F. S. Mozer, M. Fujimoto, M. L. Goldstein, M. Acuna, M. H. Reme, and A. Fazakerley (2007), Mode conversion and anomalous transport in Kelvin-Helmholtz vortices and kinetic Alfven waves at the Earth's magnetopause, *Phys. Rev. Lett.*, *99*, 175,004.

Chaston, C., et al. (2008), Turbulent heating and cross-field transport near the magnetopause from THEMIS, *Geophys. Res. Lett.*, *35*, L17S08, doi:10.1029/2008GL033601.

Chaston, C. C., J. R. Johnson, M. Wilber, M. Acuna, M. L. Goldstein, and H. Reme (2009), Kinetic Alfven wave turbulence and transport through a reconnection diffusion region, *Phys. Rev. Lett.*, *102*, 015,001.

Chen, C. X., and R. A. Wolf (1993), Interpretation of high-speed flows in the plasma sheet, *J. Geophys. Res.*, *98*, 21,409–21,419.

Chen, C. X., and R. A. Wolf (1999), Theory of thin-filament motion in Earth's magnetotail and its application to bursty bulk flows, *J. Geophys. Res.*, *104*(A7), 14,613–14,626, doi:10.1029/1999A900005.

Chen, L. (1999), Theory of plasma transport induced by low-frequency hydromagnetic waves, *J. Geophys. Res.*, *104*(A2), 2421–2427, doi:10.1029/1998JA900051.

Claudepierre, S. G., S. R. Elkington, and M. Wiltberger (2008), Solar wind driving of magnetospheric ULF waves: Pulsations driven by velocity shear at the magnetopause, *J. Geophys. Res.*, *113*, *A05218*, doi:10.1029/2007JA012890.

Collin, H. L., W. K. Peterson, J. F. Drake, and A. W. Yau (1988), The helium components of energetic terrestrial ion upflows: Their occurrence, morphology, and intensity, *J. Geophys. Res.*, *93*, 7558–7564, doi:10.1029/JA093iA07p07558.

Crooker, N. U. (1979), Dayside merging and cusp geometry, *J. Geophys. Res.*, *84*(A3), 951–959, doi:10.1029/JA084iA03p00951.

Denton, M. H., M. F. Thomsen, H. Korth, S. Lynch, J. C. Zhang, and M. W. Liemohn (2005), Bulk plasma properties at geosynchronous orbit, *J. Geophys. Res.*, *110*, A07223, doi:10.1029/2004JA010861.

Dubyagin, S., V. Sergeev, S. Apatenkov, V. Angelopoulos, A. Runov, R. Nakamura, W. Baumjohann, J. McFadden, and D. Larson (2011), Can flow bursts penetrate into the inner magnetosphere? *Geophys. Res. Lett.*, *38*, L08102, doi:10.1029/2011GL047016.

Dungey, J. W. (1961), Interplanetary magnetic field and the auroral zone, *Phys. Res. Lett.*, *6*, 47.

Erickson, G. M., and R. A. Wolf (1980), Is steady convection possible in the Earth's magnetotail? *Geophys. Res. Lett.*, *7*, 897–900, doi:10.1029/GL007i011p00897.

Fairfield, D. H., A. Otto, T. Mukai, S. Kokubun, R. P. Lepping, J. T. Steinberg, A. J. Lazarus, and T. Yamamoto (2000), Geotail observations of the Kelvin-Helmholtz instability at the equatorial magnetotail boundary for parallel northward fields, *J. Geophys. Res.*, *105*, 21,159–21,173.

Friedel, R. H. W., H. Korth, M. G. Henderson, M. F. Thomsen, and J. D. Scudder (2001), Plasma sheet access to the inner magnetosphere, *J. Geophys. Res.*, *106*(A4), 5845–5858, doi:10.1029/2000JA003011.

Fujimoto, M., and T. Terasawa (1994), Anomalous ion mixing within an MHD scale Kelvin-Helmholtz vortex, *J. Geophys. Res.*, *99*, 8601–8614.

Fujimoto, M., and T. Terasawa (1995), Anomalous ion mixing within an MHD scale Kelvin-Helmholtz vortex, 2, Effects of inhomogeneity, *J. Geophys. Res.*, *100*, 12,025–12,034.

Fujimoto, M., T. Terasawa, T. Mukai, Y. Saito, T. Yamamoto, and S. Kokubun (1998), Plasma entry from the flanks of the near-Earth magnetotail: Geotail observations, *J. Geophys. Res.*, *103*(A3), 4391–4408, doi:10.1029/97JA03340.

Fujimoto, M., T. Tonooka, and T. Mukai (2003), Vortex-like fluctuations in the magnetotail flanks and their possible roles in plasma transport, in *Earth's Low-Latitude Boundary Layer*, edited by P. Newell, and T. Onsager, *Geophys. Monogr.*, pp. 241–251, AGU, Washington, D.C.

Fuselier, S. A., J. Berchem, K. J. Trattner, and R. Friedel (2002), Tracing ions in the cusp and low-latitude boundary layer using multispacecraft observations and a global MHD simulation, *J. Geophys. Res.*, *107*(A9), 1226, doi:10.1029/2001JA000130.

Garcia, K. S., V. G. Merkin, and W. J. Hughes (2010), Effects of nightside O^+ outflow on magnetospheric dynamics: Results of multifluid MHD modeling, *J. Geophys. Res.*, *115*, A00J09, doi:10.1029/2010JA015730.

Garner, T. W., R. A. Wolf, R. W. Spiro, M. F. Thomsen, and H. Korth (2003), Pressure balance inconsistency exhibited in a statistical model of magnetospheric plasma, *J. Geophys. Res.*, *108*, 1331, doi:10.1029/2003JA009877.

Gold, T. (1959), Motions in the magnetosphere of the earth, *J. Geophys. Res.*, *64*, 1219.

Guild, T. B., H. E. Spence, E. L. Kepko, V. Merkin, J. G. Lyon, M. Wiltberger, and C. C. Goodrich (2008), Geotail and LFM comparisons of plasma sheet climatology: 1. Average values, *J. Geophys. Res.*, *113*, A04216, doi:10.1029/2007JA012611.

Guo, X. C., C. Wang, and Y. Q. Hu (2010), Global MHD simulation of the Kelvin-Helmholtz instability at the magnetopause for northward interplanetary magnetic field, *J. Geophys. Res.*, *115*, A10218, doi:10.1029/2009JA015193.

Hasegawa, A., and L. Chen (1975), Kinetic process of the plasma heating by Alfven wave, *Phys. Rev. Lett.*, *35*, 370.

Hasegawa, A., and K. Mima (1978), Anomalous transport produced by kinetic Alfvén wave turbulence, *J. Geophys. Res.*, *83*(A3), 1117–1123, doi:10.1029/JA083iA03p01117.

Hasegawa, H., M. Fujimoto, K. Maezawa, Y. Saito, and T. Mukai (2003), Geotail observations of the dayside outer boundary region: Interplanetary magnetic field control and dawn-dusk asymmetry, *J. Geophys. Res.*, *108*(A4), 1163, doi:1029/2002JA009667.

Hasegawa, H., M. Fujimoto, Y. Saito, and T. Mukai (2004a), Dense and stagnant ions in the low-latitude boundary region under northward interplanetary magnetic field, *Geophys. Res. Lett.*, *31*, L06802, doi:10.1029/2003GL19120.

Hasegawa, H., M. Fujimoto, T.-D. Phan, H. Rème, A. Balogh, M. W. Dunlop, C. Hashimoto, and R. TanDokoro (2004b), Transport of solar wind into Earth's magnetosphere through rolled-up Kelvin-Helmholtz vortices, *Nature*, *430*, 755–758.

Hasegawa, H., M. Fujimoto, K. Takagi, Y. Saito, T. Mukai, and H. Rème (2006), Single-spacecraft detection of rolled-up Kelvin-Helmholtz vortices at the flank magnetopause, *J. Geophys. Res.*, *111*, A09203, doi:10.1029/2006JA011728.

Hill, T. W. (1976), Interchange stability of a rapidly rotating magnetosphere, *Planet. Space Sci.*, *24*, 1151.

Hill, T. W., et al. (2005), Evidence for rotationally driven plasma transport in Saturn's magnetosphere, *Geophys. Res. Lett.*, *32*, L14S10.

Hill, T. W., et al. (2008), Plasmoids in Saturn's magnetotail, *J. Geophys. Res.*, *113*, A01214, doi:10.1029/2007JA012626 .

Hwang, K.-J., M. M. Kuznetsova, F. Sahraoui, M. L. Goldstein, E. Lee, and G. K. Parks (2011), Kelvin-Helmholtz waves under southward interplanetary magnetic field, *J. Geophys. Res.*, *116*, A08210, doi:10.1029/2011JA016596.

Johnson, J. R., and C. Z. Cheng (1997), Kinetic Alfven waves and plasma transport at the magnetopause, *Geophys. Res. Lett.*, *24*, 1423–1426.

Johnson, J. R., and C. Z. Cheng (2001), Stochastic ion heating at the magnetopause due to kinetic Alfven waves, *Geophys. Res. Lett.*, *28*, 4421–4424.

Johnson, J. R., and S. Wing (2009), Northward interplanetary magnetic field plasma sheet entropies, *J. Geophys. Res.*, *114*, A00D08, doi:10.1029/2008JA014017.

Johnson, J. R., C. Z. Cheng, and P. Song (2001), Signatures of mode conversion and kinetic Alfven waves at the magnetopause, *Geophys. Res. Lett.*, *28*(2), 227–230.

Johnson, J. R., S. Wing, and P. Delamere (2014), Kelvin Helmholtz Instability in Planetary Magnetospheres, *Space Sci. Rev.*, 1–31, doi:10.1007/s11214-014-0085-z.

Karimabadi, H., V. Roytershteyn, C.G. Mouikis, L.M. Kistler, W. Daughton (2011), Flushing effect in reconnection: Effects of minority species of oxygen ions, *Planet. Space Sci.*, *59*, 526–536, doi:10.1016/j.pss.2010.07.014.

Kaufmann, R. L., and W. R. Paterson (2008), Ion heat flux and energy transport near the magnetotail neutral sheet, *J. Geophys. Res.*, *113*, A05207, doi:10.1029/2007JA012929.

Kaufmann, R. L., and W. R. Paterson (2009), Boltzmann *H* function and entropy in the plasma sheet, *J. Geophys. Res.*, *114*, A00D04, doi:10.1029/2008JA014030.

Kaufmann, R. L., W. R. Paterson, and L. A. Frank (2004), Pressure, volume, density relationships in the plasma sheet, *J. Geophys. Res.*, *109*, A08204, doi:10.1029/2003JA010317.

Kaufmann, R. L., W. R. Paterson, and L. A. Frank (2005), Relationships between the ion flow speed, magnetic flux transport rate, and other plasma sheet parameters, *J. Geophys. Res.*, *110*, A09216, doi:10.1029/2005JA011068.

Keiling, A., et al. (2004), New properties of energy-dispersed ions in the plasma sheet boundary layer observed by Cluster, *J. Geophys. Res.*, *109*, A05215, doi:10.1029/2003JA010277.

Kistler, L. M., et al. (2006), Ion composition and pressure changes in storm time and nonstorm substorms in the vicinity of the near-Earth neutral line, *J. Geophys. Res.*, *111*, A11222, doi:10.1029/2006JA011939.

Kistler, L. M., C. G. Mouikis, B. Klecker, and I. Dandouras (2010), Cusp as a source for oxygen in the plasma sheet during geomagnetic storms, *J. Geophys. Res.*, *115*, A03209, doi:10.1029/2009JA014838.

Kivelson, M. G., and H. E. Spence (1988), On the possibility of quasistatic convection in the quiet magnetotail, *Geophys. Res., Lett.*, *15*, 1541–1544.

Korth, H., M. F. Thomsen, J. E. Borovsky, and D. J. McComas (1999), Plasma sheet access to geosynchronous orbit, *J. Geophys. Res.*, *104*(A11), 25,047–25,061, doi:10.1029/1999 JA900292.

LaBelle, J. , and R. A. Treumann (1988), Plasma waves at the dayside magnetopause, *Space Sci. Rev.*, *47*, 175.

Lavraud, B., M. F. Thomsen, J. E. Borovsky, M. H. Denton, and T. I. Pulkkinen (2006a), Magnetosphere preconditioning under northward IMF: Evidence from the study of coronal mass ejection and corotating interaction region geoeffectiveness, *J. Geophys. Res.*, *111*, A09208, doi:10.1029/2005JA011566.

Lavraud, B., M. F. Thomsen, S. Wing, M. Fujimoto, M. H. Denton, J. E. Borovsky, A. Aasnes, K. Seki, and J. Weygand (2006b), Observations of two distinct cold, dense ion populations at geosynchronous orbit: Local time asymmetry, solar wind dependence and origin, *Ann. Geophys.*, *24*(12), 3451–3465.

Lavraud, B., et al. (2009), Tracing solar wind plasma entry into the magnetosphere using ion-to-electron temperature ratio, *Geophys. Res. Lett.*, *36*, L18109, doi:10.1029/2009GL039442.

Le, G., C. T. Russell, J. T. Gosling, and M. F. Thomsen (1996), ISEE observations of low-latitude boundary layer for northward interplanetary magnetic field: Implications for cusp reconnection, *J. Geophys. Res.*, *101*, 27,239.

Lee, L. C., J. R. Johnson, and Z. W. Ma (1994), Kinetic Alfvén waves as a source of plasma transport at the dayside magnetopause, *J. Geophys. Res.*, *99*, 17,405–17,411.

Lemon, C., R. A. Wolf, T. W. Hill, S. Sazykin, R. W. Spiro, F. R. Toffoletto, J. Birn, and M. Hesse (2004), Magnetic storm ring current injection modeled with the Rice Convection Model and a self-consistent magnetic field, *Geophys. Res. Lett.*, *31*, L21801, doi:10.1029/2004GL020914.

Li, W., J. Raeder, J. Dorelli, M. Øieroset, and T. D. Phan (2005), Plasma sheet formation during long period of northward IMF, *Geophys. Res. Lett.*, *32*, L12S08, doi:10.1029/2004GL021524.

Li, W., J. Raeder, M. F. Thomsen, and B. Lavraud (2008), Solar wind plasma entry into the magnetosphere under northward IMF conditions, *J. Geophys. Res.*, *113*, A04204, doi:10.1029/2007JA012604.

Li, W., J. Raeder, M. Øieroset, and T. D. Phan (2009), Cold dense magnetopause boundary layer under northward IMF: Results from THEMIS and MHD simulations, *J. Geophys. Res.*, *114*, A00C15, doi:10.1029/2008JA013497.

Lin, Y., J. R. Johnson, and X. Y. Wang (2010), Hybrid simulation of mode conversion at the magnetopause, *J. Geophys. Res.*, *115*, A04208, doi:10.1029/2009JA014524.

Lin, Y., J. R. Johnson, and X. Wang (2012), Three-dimensional mode conversion associated with kinetic Alfven waves, *Phys. Rev. Lett.*, *109*, 125,003, doi:10.1103/PhysRevLett.109.125003.

Lyons, L. R., et al. (1999), Association between geotail plasma flows and auroral poleward boundary intensifications observed by canopus photometer, *J. Geophys. Res.*, *104*(A3), 4484–4500, doi:10.1029/1998JA900140.

McAndrews, H. J., et al. (2009), Plasma in Saturn's nightside magnetosphere and the implications for global circulation, *Planet. Space Sci.*, *57*, 1714–1722, doi:10.1016/j.pss.2009.03.003.

McPherron, R. L., T.-S. Hsu, J. Kissinger, X. Chu, and V. Angelopoulos (2011), Characteristics of plasma flows at the inner edge of the plasma sheet, *J. Geophys. Res.*, *116*, A00133, doi:10.1029/2010JA015923.

Merkin, V. G., J. G. Lyon, and S. G. Claudepierre (2013), Kelvin-Helmholtz instability of the magnetospheric boundary in a three-dimensional global MHD simulation during northward IMF conditions, *J. Geophys. Res. Space Phys.*, *118*, 5478–5496, doi:10.1002/jgra.50520.

Mitchell, D. G., et al. (2005), Energetic ion acceleration in Saturn's magnetotail: Substorms at Saturn? *Geophys. Res. Lett.*, *32*, L20S01, doi:10.1029/2005GL022647.

Mitchell, D. G., et al. (2009), Recurrent energization of plasma in the midnight-to-dawn quadrant of Saturn's magnetosphere, and its relationship to auroral UV and radio emissions, *Planet. Space Sci.*, *57*, 1732–1742, doi:10.1016/j.pss.2009.04.002.

Miura, A. (1995), Dependence of the magnetopause Kelvin-Helmholtz instability on the orientation of the magnetosheath magnetic field, *Geophys. Res. Lett.*, *22*, 2993–2996.

Nagata, D., S. Machida, S. Ohtani, Y. Saito, and T. Mukai (2007), Solar wind control of plasma number density in the near-Earth plasma sheet, *J. Geophys. Res.*, *112*, A09204, doi:10.1029/2007JA012284.

Nakamura, T. K. M., and M. Fujimoto (2005), Magnetic reconnection within rolled-up MHD-scale Kelvin-Helmholtz vortices: Two-fluid simulations including finite electron inertial effects, *Geophys. Res. Lett.*, *32*, L21102, doi:10.1029/2005GL023362.

Nakamura, R., W. Baumjohann, M. Brittmacher, V. A. Sergeev, M. Kubyshkina, T. Mukai, and K. Liou (2001), Flow bursts and auroral activations, *J. Geophys. Res.*, *106*(A6), 10,777–10,789.

Nakamura, T. K., et al. (2002), Fast flow during current sheet thining, *Geophys. Res. Lett.*, *29*(23), 2140, doi:10.1029/2002GL016200.

Nakamura, R., et al. (2005), Localized fast flow disturbance observed in the plasma sheet and in the ionosphere, *Ann. Geophys.*, *23*(2), 553–566.

Nemecek, Z., J. Safrankova, G. N. Zastenkar, P. Pisoft, and K. I. Paularena (2002), Spatial distribution of the magnetosheath ion flux, *Adv. Space. Res.*, *30*, 2751–2756, doi:10.1016/S0273-1177(02)80402-1.

Nishino, M. N., M. Fujimoto, T. Terasawa, G. Ueno, K. Maezawa, T. Mukai, and Y. Saito (2007), Geotail observations of temperature anisotropy of the two-component protons in the dusk plasma sheet, *Ann. Geophys.*, *25*, 769–777.

Nykyri, K., and A. Otto (2001), Plasma transport at the magnetospheric boundary due to reconnection in Kelvin-Helmholtz vortices, *Geophys. Res. Lett.*, *28*, 3565–3568.

Nykyri, K., and A. Otto (2004), Influence of the Hall term on KH instability and reconnection inside KH vortices, *Ann. Geophys.*, *22*, 935–949.

Nykyri, K., A. Otto, B. Lavraud, C. Mouikis, L. M. Kistler, A. Balogh, and H. Reme (2006), Cluster observations of reconnection due to the Kelvin-Hemholtz instability at the dawnsdie magnetospheric flank, *Ann. Geophys.*, *24*, 2619–2643.

Ohtani, S., H. J. Singer, and T. Mukai (2006), Effects of the fast plasma sheet flow on the geosynchronous magnetic configuration: Geotail and GOES coordinated study, *J. Geophys. Res.*, *111*, A01204, doi:10.1029/2005JA011383.

Øieroset, M., J. Raeder, T. D. Phan, S. Wing, J. P. McFadden, W. Li, M. Fujimoto, H. Rème, and A. Balogh (2005), Global cooling and densification of the plasma sheet during an extended period of purely northward IMF on October 22–24, 2003, *Geophys. Res. Lett.*, *32*, L12S07, doi:10.1029/2004GL021523.

Øieroset, M., T. D. Phan, V. Angelopoulos, J. P. Eastwood, J. McFadden, D. Larson, C. W. Carlson, K.-H. Glassmeier, M. Fujimoto, and J. Raeder (2008), THEMIS multi-spacecraft observations of magnetosheath plasma penetration deep into the dayside low-latitude magnetosphere for northward and strong By IMF, *Geophys. Res. Lett.*, *35*, L17S11, doi:10.1029/2008GL033661.

Olson, W. P., and K. A. Pfitzer (1985), Magnetospheric responses to the gradient drift entry of solar wind plasma, *J. Geophys. Res.*, *90*(A11), 10,823–10,833, doi:10.1029/JA090iA11p10823.

Otto, A., and D. H. Fairfield (2000), Kelvin-Helmholtz instability at the magnetotail boundary: MHD simulation and comparison with Geotail observations, *J. Geophys. Res.*, *105*, 21,175–21,190.

Panov, E. V., et al. (2010a), Plasma sheet thickness during a bursty bulk flow reversal, *J. Geophys. Res.*, *115*, A05213, doi:10.1029/2009JA014743.

Panov, E. V., et al. (2010b), Multiple overshoot and rebound of a bursty bulk flow, *Geophys. Res. Lett.*, *37*, L08103, doi:10.1029/2009GL041971.

Paularena, K. I., J. D. Richardson, M. A. Kolpak, C. R. Jackson, and G. L. Siscoe (2001), A dawn-dusk density asymmetry in Earth's magnetosheath, *J. Geophys. Res.*, *106*(A11), 25,377–25,394, doi:10.1029/2000JA000177.

Pedersen, A., C. A. Cattell, C.-G. Fälthammar, K. Knott, P.-A. Lindqvist, R. H. Manka, and F. S. Mozer (1985), Electric fields in the plasma sheet and plasma sheet boundary layer, *J. Geophys. Res.*, *90*(A2), 1231–1242, doi:10.1029/JA090iA02p01231.

Peroomian, V., and M. El-Alaoui (2008), The storm-time access of solar wind ions to the nightside ring current and plasma sheet, *J. Geophys. Res.*, *113*, A06215, doi:10.1029/2007JA012872.

Peroomian, V., M. El-Alaoui, and P. C. Brandt (2011), The ion population of the magnetotail during the 17 April 2002 magnetic storm: Large-scale kinetic simulations and IMAGE/HENA observations, *J. Geophys. Res.*, *116*, A05214, doi:10.1029/2010JA016253.

Peterson, W. K., L. Andersson, B. Callahan, S. R. Elkington, R. W. Winglee, J. D. Scudder, and H. L. Collin (2009), Geomagnetic activity dependence of O^+ in transit from the ionosphere, *J. Atmos. Solar-Terres. Phys.*, *71*, 1623–1629, doi:10.1016/j.jastp.2008.11.003.

Petrinec, S. M. (2013), On the magnetic field configuration of the magnetosheath, *Terr. Atmos. Ocean. Sci.*, *24*, 265–272, doi:10.3319/TAO.2012.10.17.02(SEC).

Pontius, D. H.., and R. A. Wolf (1990), Transient flux tubes in the terrestrial magnetoshere, *Geophys. Res. Lett.*, *17*(1), 49–52.

Raeder, J., R. J. Walker, and M. Ashour-Abdalla (1995), The structure of the distant geomagnetic tail during long periods of northward IMF, *Geophys. Res. Lett.*, *22*, 349.

Raeder, J., et al. (1997), Boundary layer formation in the magnetotail: Geotail observations and comparisons with a global MHD simulation, *Geophys. Res. Lett.*, *24*, 951–954, doi:10.1029/97GL00218.

Redmon, R. J., W. K. Peterson, L. Andersson, and P. G. Richards (2012), Dawnward shift of the dayside O^+ outflow distribution: The importance of field line history in O^+ escape from the ionosphere, *J. Geophys. Res.*, *117*, A12222, doi:10.1029/2012JA018145.

Richard, R. L., R. J. Walker, and M. Ashour-Abdalla (1994), The population of the magnetosphere by solar winds ions when the interplanetary magnetic field is northward, *Geophys. Res. Lett.*, *21*, 2455–2458, doi:10.1029/94GL01427.

Rymer, et al. (2009), Cassini evidence for rapid interchange transport at Saturn, *Planet. Space Sci.*, *57*, 1779–1784.

Sanchez, E. R., S. Wing, E. Spanswick, and E. Donovan (2012), Entropy conservation and rate of propagation of bubbles in the Earth's magnetotail: A case study, *J. Geophys. Res.*, *117*, A05226, doi:10.1029/2011JA017287.

Sandholt, P. E., C. J. Farrugia, S. W. H. Cowley, W. F. Denig, M. Lester, J. Moen, and B. Lybekk (1999), Capture of magnetosheath plasma by the magnetosphere during northward IMF, *Geophys. Res. Lett.*, *26*, 2833–2836.

Schriver, D., and M. Ashour-Abdalla (1990), Cold plasma heating in the plasma sheet boundary layer: Theory and simulations, *J. Geophys. Res.*, *95*(A4), 3987–4005, doi:10.1029/JA095iA04p03987.

Schriver, D., M. Ashour-Abdalla, and R. L. Richard (1998), On the origin of the ion-electron temperature difference in the plasma sheet, *J. Geophys. Res.*, *103*(A7), 14,879–14,895, doi:10.1029/98JA00017.

Sergeev, V. A., V. Angelopoulos, J. T. Gosling, C. A. Cattell, and C. T. Russell (1996), Detection of localized plasma-depleted flux tubes or bubbles in the midtail plasma sheet, *J. Geophys. Res.*, *101*, 10,817–10,826, doi:10.1029/96JA00460.

Sergeev, V. A., K. Liou, P. T. Newell, S.-I. Ohtani, M. R. Hairston, and F. Rich (2004), Auroral streamers: Characteristics of associated precipitation, convection and field-aligned currents, *Ann. Geophys.*, *22*, 537–548.

Sergeev, V. A., S. V. Apatenkov, V. Angelopoulos, J. P. McFadden, D. Larson, J. W. Bonnell, M. Kuznetsova, N. Partamies, and F. Honary (2008), Simultaneous THEMIS observations in the near-tail portion of the inner and outer plasma sheet flux tubes at substorm onset, *J. Geophys. Res.*, *113*, A00C02, doi:10.1029/2008JA013527.

Sibeck, D. G., et al. (1999), Comprehensive study of the magnetospheric response to a hot flow anomaly, *J. Geophys. Res.*, *104*(A3), 4577–4593, doi:10.1029/1998JA900021.

Sittler, E. C., Jr., et al. (2005), Preliminary results on Saturn's inner plasmasphere as observed by Cassini: Comparison with Voyager, *Geophys. Res. Lett.*, *32*, L14S07, doi:10.1029/2005GL022653.

Sittler, E. C., Jr., et al. (2006), Cassini observations of Saturn's inner plasmasphere: Saturn orbit insertion results, *Planet. Space Sci.*, *54*, 1197–1210.

Slavin, J. A., et al. (2002), Simultaneous observations of earthward flow bursts and plasmoid ejection during magnetospheric substorms, *J. Geophys. Res.*, *107*(A7), doi:10.1029/2000JA003501.

Song, P., and C. T. Russell (1992), Model of the formation of the low-latitude boundary layer for strongly northward interplanetary magnetic field, *J. Geophys. Res.*, *97*, 1411.

Sonnerup, B. U. Ö., and M. J. Laird (1963), On magnetospheric interchange instability, *J. Geophys. Res.*, *68*(1), 131–139, doi:10.1029/JZ068i001p00131.

Spence, H. E., and M. G. Kivelson (1993), Contributions of the low-latitude boundary layer to the finite width magnetotail convection model, *J. Geophys. Res.*, *98*(A9), 15,487–15,496, doi:10.1029/93JA01531.

Stenuit, H., et al. (2002), Multi-spacecraft study oin the dynamics of the dusk-flank magnetosphere under northward IMF: January 10–11, 1997, *J. Geophys. Res.*, *107*(A10), 1333, doi:10.1029/2002JA9009246.

Stepanova, M., V. Pinto, J. A. Valdivia, and E. E. Antonova (2011), Spatial distribution of the eddy diffusion coefficients in the plasma sheet during quiet time and substorms from THEMIS satellite data, *J. Geophys. Res.*, *116*, A00I24, doi:10.1029/2010JA015887.

Taylor, M. G. G. T., and B. Lavraud (2008), Observation of three distinct ion populations at the Kelvin-Helmholtz-unstable magnetopause, *Ann. Geophys.*, *26*, 1559–1566.

Terasawa, T., et al. (1997), Solar wind control of density and temperature in the near-Earth plasma sheet: WIND/GEOTAIL collaboration, *Geophys. Res. Lett.*, *24*, 935–938.

Thomas, V. A., and D. Winske (1991), Kinetic simulation of the Kelvin-Helmholtz instability at the Venus ionopause, *Geophys. Res. Lett.*, *18*, 1943–1946.

Thomas, V. A., and D. Winske (1993), Kinetic simulations of the Kelvin-Helmholtz instability at the magnetopause, *J. Geophys. Res.*, *98*, 11,425–11,438.

Thomsen, M. F., J. E. Borovsky, R. M. Skoug, and C. W. Smith (2003), Delivery of cold, dense plasma sheet material into the near-Earth region, *J. Geophys. Res.*, *108*, 1151, doi:10.1029/2002JA009544.

Thomsen, M. F., D. B. Reisenfeld, D. M. Delapp, R. L. Tokar, D. T. Young, F. J. Crary, E. C. Sittler, M. A. McGraw, and J. D. Williams (2010), Survey of ion plasma parameters in Saturn's magnetosphere, *J. Geophys. Res.*, *115*, A10220, doi:10.1029/2010JA015267.

Treumann, R. A., and W. Baumjohann (1988), Particle trapping at a tangential discontinuity: Multiple incidence, *Planet. Space Sci.*, *36*, 1477–1484.

Tsurutani, B. T., E. J. Smith, R. R. Anderson, K. W. Ogilvie, J. D. Scudder, D. N. Baker, and S. J. Bame (1982), Lion roars and nonoscillatory drift mirror waves in the magnetosheath, *J. Geophys. Res.*, *87*, 6060–6072.

Tsyganenko, N. A. (1982), On the convective mechanism for formation of the plasma sheet in the magnetospheric tail, *Planet. Space Sci.*, *30*, 1007–1012.

Tsyganenko, N. A. (1989), A magnetospheric magnetic field model with a warped tail current sheet, *Planet. Space Sci.*, *37*, 5–20.

Tung, Y.-K., C. W. Carlson, J. P. McFadden, D. M. Klumpar, G. K. Parks, W. J. Peria, and K. Liou (2001), Auroral polar cap boundary ion conic outflow observed on FAST, *J. Geophys. Res.*, *106*(A3), 3603–3614, doi:10.1029/2000JA900115.

Walsh, B. M., D. G. Sibeck, Y. Wang, and D. H. Fairfield (2012), Dawn-dusk asymmetries in the Earth's magnetosheath, *J. Geophys. Res.*, *117*, A12211, doi:10.1029/2012JA018240.

Wang, C.-P., L. R. Lyons, M. W. Chen, and R. A. Wolf (2001), Modeling the quiet time inner plasma sheet protons, *J. Geophys. Res.*, *106*(A4), 6161–6178, doi:10.1029/2000JA000377.

Wang, C.-P., L. R. Lyons, M. W. Chen, and F. R. Toffoletto (2004), Modeling the transition of the inner plasma sheet from weak to enhanced convection, *J. Geophys. Res.*, *109*, A12202, doi:10.1029/2004JA010591.

Wang, C.-P., L. R. Lyons, J. M. Weygand, T. Nagai, and R. W. McEntire (2006), Equatorial distributions of the plasma sheet ions, their electric and magnetic drifts, and magnetic fields under different interplanetary magnetic field Bz conditions, *J. Geophys. Res.*, *111*, A04215, doi:10.1029/2005JA011545.

Wang, C.-P., L. R. Lyons, T. Nagai, J. M. Weygand, and R. W. McEntire (2007), Sources, transport, and distributions of plasma sheet ions and electrons and dependences on interplanetary parameters under northward interplanetary magnetic field, *J. Geophys. Res.*, *112*, A10224, doi:10.1029/2007JA012522.

Wang, C.-P., L. R. Lyons, R. A. Wolf, T. Nagai, J. M. Weygand, and A. T. Y. Lui (2009), Plasma sheet $PV^{5/3}$ and nV and associated plasma and energy transport for different convection strengths and AE levels, *J. Geophys. Res.*, *114*, A00D02, doi:10.1029/2008JA013849.

Wang, C.-P., L. R. Lyons, T. Nagai, J. M. Weygand, and A. T. Y. Lui (2010), Evolution of plasma sheet particle content under different interplanetary magnetic field conditions, *J. Geophys. Res.*, *115*, A06210, doi:10.1029/2009JA015028.

Wang, C.-P., M. Gkioulidou, L. R. Lyons, and V. Angelopoulos (2012), Spatial distributions of the ion to electron temperature ratio in the magnetosheath and plasma sheet, *J. Geophys. Res.*, *117*, A08215, doi:10.1029/2012JA017658.

Weygand, J. M., et al. (2005), Plasma sheet turbulence observed by Cluster II, *J. Geophys. Res.*, *110*, A01205, doi:10.1029/2004JA010581.

Wilber, M., and R. M. Winglee (1995), Dawn-dusk asymmetries in the low-latitude boundary layer arising from the Kelvin-Helmholtz instability: A particle simulation, *J. Geopshys. Res.*, *100*, 1883–1898.

Wilson, G. R., D. M. Ober, G. A. Germany, and E. J. Lund (2001), The relationship between suprathermal heavy ion outflow and auroral electron energy deposition: Polar/Ultraviolet Imager and Fast Auroral Snapshot/Time-of-Flight Energy Angle Mass Spectrometer observations, *J. Geophys. Res.*, *106*(A9), 18,981–18,993, doi:10.1029/2000JA000434.

Wiltberger, M., W. Lotko, J. G. Lyon, P. Damiano, and V. Merkin (2010), Influence of cusp O^+ outflow on magnetotail dynamics in a multifluid MHD model of the magnetosphere, *J. Geophys. Res.*, *115*, A00J05, doi:10.1029/2010JA015579.

Wing, S., and J. R. Johnson (2009), Substorm entropies, *J. Geophys. Res.*, *114*, A00D07, doi:10.1029/2008JA013989.

Wing, S., and J. R. Johnson (2010), Introduction to special section on entropy properties and constraints related to space plasma transport, *J. Geophys. Res.*, *115*, A00D00, doi:10.1029/2009JA014911.

Wing, S., and P. T. Newell (1998), Central plasma sheet ion properties as inferred from ionospheric observations, *J. Geophys. Res.*, *103*, 6785–6800.

Wing, S., and P. T. Newell (2002), 2D plasma sheet ion density and temperature profiles for northward and southward IMF, *Geophys. Res. Lett.*, *29*, doi:10.1029/2001GL013950.

Wing, S., J. R. Johnson, P. T. Newell, and C.-I. Meng (2005), Dawn-dusk asymmetries, ion spectra, and sources in the northward interplanetary magnetic field plasma sheet, *J. Geophys. Res.*, *110*, A08205, doi:10.1029/2005JA011086.

Wing, S., J. R. Johnson, and M. Fujimoto (2006), Timescale for the formation of the cold-dense plasma sheet: A case study, *Geophys. Res. Lett.*, *33*, L23106, doi:10.1029/2006GL027110.

Wing, S., J. W. Gjerloev, J. R. Johnson, and R. A. Hoffman (2007), Substorm plasma sheet ion pressure profiles, *Geophys. Res. Lett.*, *34*, L16110, doi:10.1029/2007GL030453.

Wing, S., M. Gkioulidou, J. R. Johnson, P. T. Newell, and C.-P. Wang (2013), Auroral particle precipitation characterized by the substorm cycle, *J. Geophys. Res. Space Phys.*, *118*, 1022–1039, doi:10.1002/jgra.50160.

Winglee, R. M., D. Chua, M. Brittnacher, G. K. Parks, and G. Lu (2002), Global impact of ionospheric outflows on the dynamics of the magnetosphere and cross-polar cap potential, *J. Geophys. Res.*, *107*(A9), 1237, doi:10.1029/2001JA000214.

Winske, D., and N. Omidi (1995), Diffusion at the magnetopause: Hybrid simulations, *J. Geophys. Res.*, *100*(A7), 11,923–11,933, doi:10.1029/94JA02730.

Wolf, R. A., Wan, X. Xing, J. Zhang, and S. Sazykin (2009), Entropy and plasma sheet transport, *J. Geophys. Res.*, *114*, A00D05, doi:10.1029/2009JA014044.

Wolf, R. A., C. X. Chen, and F. R. Toffoletto (2012), Thin filament simulations for Earth's plasma sheet: Interchange oscillations, *J. Geophys. Res.*, *117*, A02215, doi:10.1029/2011JA016971.

Yao, Y., C. C. Chaston, K.-H. Glassmeier, and V. Angelopoulos (2011), Electromagnetic waves on ion gyro-radii scales across the magne- topause, *Geophys. Res. Lett.*, *38*, L09102, doi:10.1029/ 2011GL047328.

Young, et al. (2005), Composition and dynamics of plasma in saturn's magnetosphere, *Science*, *307*, 1262.

Zhang, B., W. Lotko, O. Brambles, P. Damiano, M. Wiltberger, and J. Lyon (2012), Magnetotail origins of auroral Alfvénic power, *J. Geophys. Res.*, *117*, A09205, doi:10.1029/2012JA017680.

Zhang, J.-C., R. A. Wolf, S. Sazykin, and F. R. Toffoletto (2008), Injection of a bubble into the inner magnetosphere, *Geophys. Res. Lett.*, *35*, L02110, doi:10.1029/2007GL032048.

Zhou, X.-Z., Z. Y. Pu, Q.-G. Zong, and L. Xie (2007), Energy filter effect for solar wind particle entry to the plasma sheet via flank regions during southward interplanetary magnetic field, *J. Geophys. Res.*, *112*, A06233, doi:10.1029/2006JA012180.

15

Magnetic Reconnection in Different Environments: Similarities and Differences

Michael Hesse,[1] Nicolas Aunai,[1] Masha Kuznetsova,[1] Seiji Zenitani,[2] and Joachim Birn[3,4]

15.1. INTRODUCTION

Magnetic reconnection is broadly recognized as a key plasma transport and energy conversion process in space plasmas. Ranging from solar activity [e.g., *Cargill and Klimchuk*, 1997; *Antiochos et al.*, 1999] through interplanetary space [e.g., *Gosling et al.*, 1995, 2005] to the boundaries [e.g., *Paschmann et al.*, 1979; *Sonnerup et al.*, 1981] and interiors of planetary magnetospheres [e.g., Hones, 1977], magnetic reconnection facilitates highly time dependent, eruptive behavior and it transports mass, momentum, and energy into regions featuring otherwise nearly impenetrable boundary layers.

A key feature of magnetic reconnection is that it primarily converts magnetic energy in the two inflow regions into internal and kinetic energy. The efficacy of this energy conversion process depends strongly on plasma parameters in the two inflow regions [e.g., *Aunai et al.* 2011]. For example, high plasma $\beta = 2\mu_0 p/B^2$ [*Swisdak et al.*, 2003] or low Alfvén speed [*Hesse et al.*, 1996] in the inflow slows down the reconnection process.

In Earth's magnetosphere, magnetic reconnection operates both at the magnetopause, the boundary between solar-wind-dominated plasma and the region dominated by Earth's magnetic field, and on the nightside, within Earth's magnetotail. In the former situation, reconnection merges magnetic fields and plasmas of solar origin on one side with magnetospheric magnetic field and plasma on the other in a usually very asymmetric geometry [e.g., *Mozer et al* 2008]. Furthermore, reconnecting magnetic fields are typically not coplanar, implying, in the reconnection plane, the existence of an out-of-plane magnetic field. On the nightside, however, reconnection is believed to involve considerably more symmetric environments, with usually relatively small guide magnetic fields. It is expected that similar situations prevail at the other magnetized planets. Given the varied nature of the environments magnetic reconnection operates in, it is therefore of great interest to understand how reconnection adjusts to plasma parameters in both inflow regions.

Owing to its importance, reconnection has therefore been studied by means of spacecraft observations [e.g., *Eastwood et al.*, 2007, 2010; *Paschmann*, 2008], by analytic theory [*Cassak and Shay*, 2007; *Birn et al.*, 2011; *Hesse et al.*, 2011a], and by numerical modeling. Recent work has indicated that reconnection changes substantially if the inflow sides are unequal [*Swisdak et al.*, 2003; *Mozer and Pritchett*, 2011; *Hesse et al.*, 2013] or if a guide magnetic field is being introduced [e.g., *Hesse et al.*, 2011b]. Kinetic, hybrid, and Hall-MHD simulations consistently exhibit a reconnection rate much larger than typical Sweet-Parker rates [*Birn et al.*, 2001]; however, recent evidence suggests that island formation may impact reconnection rates in kinetic models [*Daughton et al.*, 2006, 2009, *Karimabadi et al.*, 2007], and it may play a critical role in enabling fast reconnection even in MHD systems [*Huang and Bhattacharjee*, 2013].

This chapter therefore addresses two important aspects of magnetic reconnection: time-dependent rates affected by island formation and the preferred direction of the

[1] *Heliophysics Science Division, NASA Goddard Space Flight Center, Greenbelt, Maryland*
[2] *National Astronomical Observatory of Japan, Tokyo, Japan*
[3] *Los Alamos National Laboratory, Los Alamos, New Mexico*
[4] *Space Science Institute, Boulder, Colorado*

Magnetotails in the Solar System, Geophysical Monograph 207. First Edition.
Edited by Andreas Keiling, Caitríona M. Jackman, and Peter A. Delamere.

reconnection line if merging magnetic field components are not antiparallel. The former presents new results, whereas the latter will summarize recent findings [*Hesse et al.*, 2013]. The two aspects of the analysis will be prefaced by a basic discussion of the impact magnetic reconnection has on energy transport across topological boundaries.

15.2. ENERGY TRANSPORT

While it is widely known that magnetic reconnection facilitates mass, momentum, and energy transport in plasmas, it is illustrative to consider an analytical approach to describing the efficacy of reconnection as a transport mechanism. For simplicity, we consider a simple, two-dimensional geometry, where an X-shaped reconnection region forms the intersection of two separatrices. The geometry is depicted in Figure 15.1.

Here, V denotes the volume enclosed by the two separatrices, ∂V its boundary (the separatrices), $\vec{n}$ the local boundary normal, $\vec{v}$ the plasma velocity, and $\vec{v}_s$ the velocity of the separatrix surface itself. Note that the normal here is represented by the blue arrow in Figure 15.1 and must not be confused with the normal direction to the reconnecting current sheet.

Integrating the MHD energy equation over the volume V then leads to

$$\begin{aligned}\frac{d}{dt}\int_V dV\left[\rho\frac{v^2}{2}+\frac{B^2}{2\mu_o}+\frac{p}{\gamma-1}\right]\\ =\oint_{\partial V} dS\,\vec{n}\cdot\left[-\frac{1}{\mu_o}\vec{R}\times\vec{B}+\frac{B^2}{\mu_o}\left(\frac{\vec{v}_s}{2}-\vec{v}\right)+\rho\frac{v^2}{2}(\vec{v}_s-\vec{v})\right.\\ \left.+\frac{p}{\gamma-1}(\vec{v}_s-\gamma\vec{v})\right].\end{aligned} \tag{15.1}$$

Here we have assumed the generic relation

$$\vec{E}+\vec{v}\times\vec{B}=\vec{R}, \tag{15.2}$$

where $\vec{R}\neq 0$ only inside the red-shaded region in Figure 15.1, implying that the first term on the right-hand side (RHS) of (1) can be ignored. Assuming a strictly

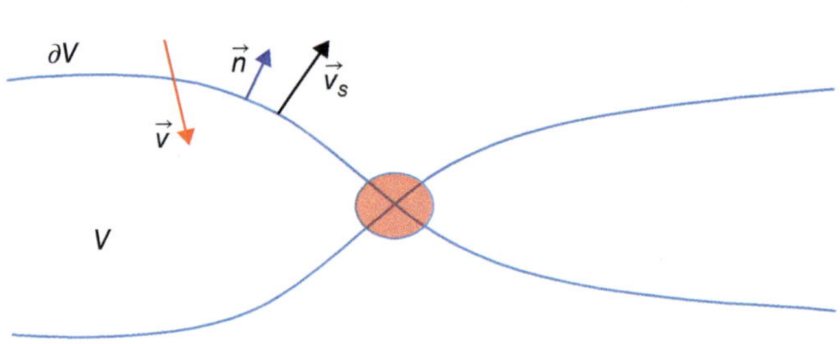

Figure 15.1 Schematic of the reconnection geometry.

two-dimensional system, we can represent the magnetic field through a flux function A such that

$$\vec{B}=\nabla A\times\vec{e}_y, \tag{15.3}$$

where $\vec{e}_y$ is the unit vector in the direction of the ignorable coordinate y. In the Coulomb gauge, the time evolution of the magnetic field can then be described by

$$\frac{\partial A}{\partial t}+\vec{v}\cdot\nabla A=-R_y. \tag{15.4}$$

At the X point proper, the gradient of A vanishes. At the same time, the numerical value A_x of the flux function at the X point is constant along the evolving separatrix surfaces, implying

$$\left.\frac{\partial A}{\partial t}\right|_{\text{separatrix}}=-R_y\big|_{X\text{ point}}. \tag{15.5}$$

Away from the diffusion region, the magnetic field is frozen into the plasma such that

$$\frac{\partial A}{\partial t}+\vec{v}\cdot\nabla A=0. \tag{15.6}$$

This is the change of A seen by a moving plasma element. If, at the separatrix, this change is the same as the one defined by (15.5), plasma does not cross the separatrix, and magnetic reconnection does not occur.

For reconnection, the surface and plasma velocities hence satisfy, on the separatrix,

$$(\vec{v}_s-\vec{v})\cdot\nabla A\big|_{\text{separatrix}}=(\vec{v}_s-\vec{v})\cdot\vec{n}B=-R_y\big|_{X\text{ point}}=-E_r. \tag{15.7}$$

We can insert this result into the energy balance to yield

$$\begin{aligned}\frac{d}{dt}\int_V dV\left[\rho\frac{v^2}{2}+\frac{B^2}{2\mu_o}+\frac{p}{\gamma-1}\right]\\ =\oint_{\partial V} dS\left[\left(\rho\frac{v^2}{2}+\frac{B^2}{2\mu_o}+\frac{p}{\gamma-1}\right)\frac{E_r}{B_{\text{separatrix}}}-\vec{n}\cdot\vec{v}_s\left(\frac{B^2}{2\mu_o}+p\right)\right].\end{aligned} \tag{15.8}$$

Equation (15.8) describes two contributions to the change of overall energy content of the volume enclosed by the separatrix: a compression or expansion contribution (second term on RHS) and a net energy transport facilitated by the reconnection electric field.

Similar arguments are readily made for mass and momentum transport. The analysis presented here shows that reconnection is a key transport mechanism, which, by opening otherwise closed magnetic surfaces, allows for significant changes to energy content. This feature is a critically important provider of energy sources for the multitude of physical processes occurring inside the reconnected plasma.

In the following sections, we focus on the kinetic essentials, which render reconnection possible in collisionless plasmas. Arguably, the most important aspect here is the physics of the diffusion region, which supports the electric field that regulates the overall transport. We will present aspects of the collisionless physics of the diffusion region for two different environments: symmetric and asymmetric geometries.

15.3. TIME-DEPENDENT RECONNECTION

Time dependence and plasmoid formation have attracted considerable attention [e.g., *Daughton et al.*, 2006, 2009; *Drake et al.*, 2006; *Karimabadi et al.*, 2007]. We here perform a detailed analysis of the correlation between diffusion region dimensions and reconnection rate. For this purpose, we employ open boundary condition calculations of a continuously driven reconnecting system. The employed code is a 2.5D variant of our proven 3D particle-in-cell code [*Hesse et al.*, 1999], where periodic boundary conditions have been replaced by open boundaries, where normal derivatives of density, velocity, and isotropic pressure are assumed to vanish. Here and throughout the remainder of this chapter we normalize densities by a typical density n_0 in the current sheet, the magnetic field by the asymptotic value B_0 of the in-plane magnetic field. Ions are assumed to be protons (mass m_p) throughout, and length scales by the ion inertial length c/ω_i, where the ion plasma frequency $\omega_i = \sqrt{e^2 n_0/\varepsilon_0 m_p}$ is evaluated for the reference density. Velocities are measured in units of the ion Alfvén velocity $v_A = B_0/\sqrt{\mu_0 m_p n_0}$ based on the reference magnitudes of magnetic field and density. The electric field is measured in units of $E_0 = v_A B_0$ and the pressure in units of $p_0 = B_0^2/\mu_0$. The above length scale and Ampere's law imply that the current density is normalized to $j_0 = \omega_i B_0/c\mu_0$.

The initial condition consists of a poloidal magnetic field, a modified Harris sheet [*Harris*, 1962] with a current sheet half width of half the ion inertial length. The initial magnetic field is coplanar, i.e., does not contain a guide field component. In a coordinate system where the x axis is in the direction of the initial magnetic field, the y axis is in the initial current direction, and the z direction completes a right-handed coordinate system, it is of the following form:

$$B_x = \tanh(2z) + a_0\pi/L_z \cos(2\pi x/L_x)\sin(\pi z/L_z), \quad (15.9a)$$

$$B_z = -a_0 2\pi/L_z \sin(2\pi x/L_x)\cos(\pi z/L_z). \quad (15.9b)$$

The perturbation amplitude $a_0 = 0.1$ leads to an initial value of the normal magnetic field of about 3% of B_0. The system size is $L_x = 100$ and $L_z = 25$. The ion-electron mass ratio is chosen to be $m_i/m_e = 25$. A total of 2×10^8 macroparticles are moved on a 1600×800 grid, with an electron/ion temperature ratio of $T_e/T_i = 0.2$. The system is, at its upper boundary, subjected to a constant driving electric field of $E_y = 0.185$. Particles are fed into the calculation to maintain, at the upper and lower boundaries, a density of $n_b = 0.2$ throughout the calculation and removed as they leave the boundaries at $x = 0$ and $x = 100$. The calculation extends for 1000 Alfvén times. Figure 15.2 displays two typical times of the simulation.

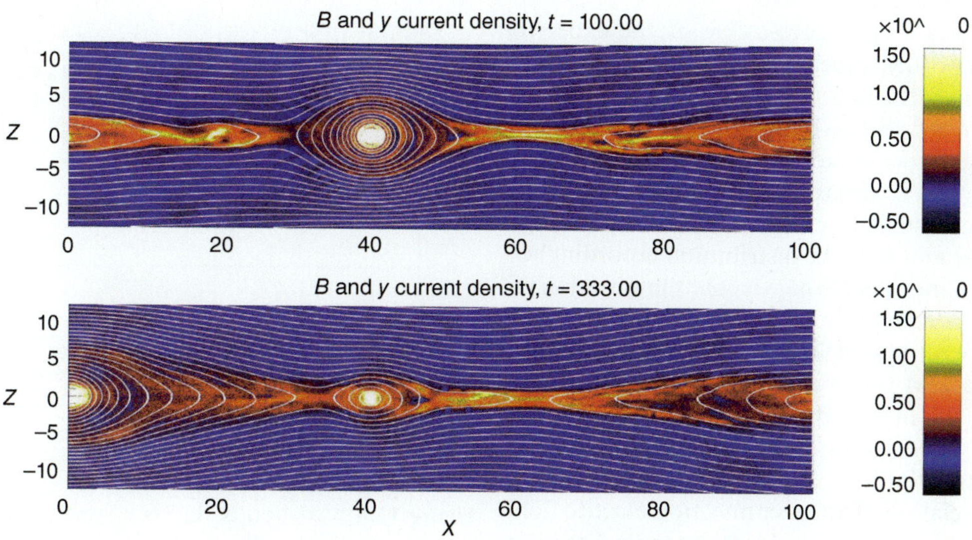

Figure 15.2 Two times of the simulation showing the common presence of magnetic islands at different times of the simulation. Note that the island shown in the bottom panel formed newly and did not evolve from the island shown in the top panel.

In order to assess the effect of island formation on magnetic reconnection, we need to define the dimensions of the diffusion region. The electric field is given by

$$\vec{E} + \vec{v}_e \times \vec{B} = -\frac{1}{en_e}\nabla\cdot\vec{P}_e - \frac{m_e}{e}\frac{d\vec{v}_e}{dt} = \vec{R}_e. \quad (15.10)$$

Here the index e denotes electron quantities. At each X point, the electric field is given entirely by the nonideal terms on the RHS of (10). When deviating either horizontally or vertically from the diffusion region location, the contribution to the electric field by the convective term (second term on the left-hand side (LHS)) begins to increase, whereas the nonideal contribution decreases. The horizontal and vertical edges of the diffusion region are now defined to be the location of crossover between ideal and nonideal terms, i.e.,

$$-\left(\vec{v}_e \times \vec{B}\right)\cdot\vec{e}_y = R_{ey}. \quad (15.11)$$

Equation 15.11 defines one location each to the right and left and top and bottom of the X point. The distance between the horizontal crossover points defines the length λ_x and the corresponding vertical distance λ_z is defined similarly. We perform this analysis for the dominant X point for each inverse ion cyclotron period during the simulation. The dominant X point is defined to be the one whose associated separatrix encloses the larges amount of magnetic flux.

We expect a dependence of the reconnection electric field on the ratio of the two quantities. This expectation is demonstrated in Figure 15.3, which plots the instantaneous electric field at the dominant X point against the diffusion region aspect ratio.

The figure shows a clear correlation of the expected kind. It is noteworthy, albeit not surprising, that the electric field values fluctuate about the applied electric field. The mean reconnection electric field strength equals the driving value; however, reconnection remains fast at all times. Applying a lower or larger driving electric field will likely lead to a smaller or larger, respectively, average reconnection rate and shift the distribution accordingly.

The point we want to make here is that the variation of the electric field is essentially due only to the variation of the diffusion region length in the x direction. For this purpose, Figure 15.4 illustrates the dependence of the electric field on the inverse diffusion region length in x, and Figure 15.5 depicts a histogram of encountered scale length in the z direction.

The clear correlation of the reconnection electric field with the inverse diffusion region length in the x direction combined with the very limited variation of the diffusion region thickness, implies the variation of the

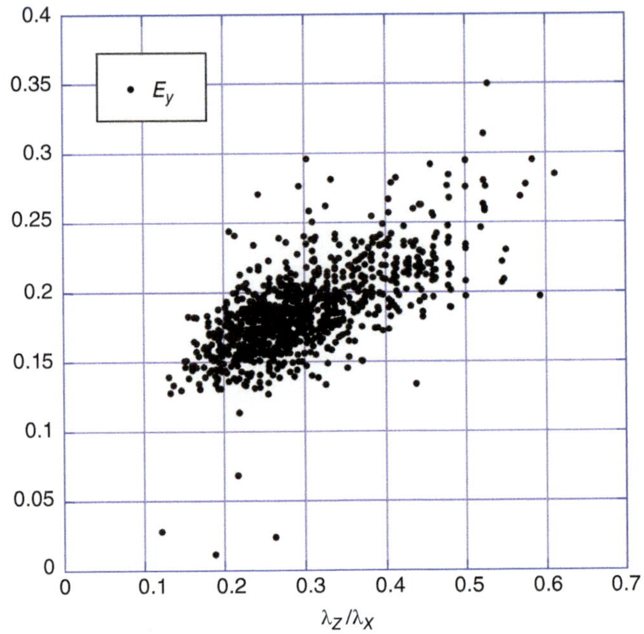

Figure 15.3 Scatter plot of the instantaneous electric field at the major X point against the aspect ratio of the diffusion region. The outliers in the reconnection rates are caused by unusually strongly fluctuating electric fields and plasma properties, which render an analysis very difficult.

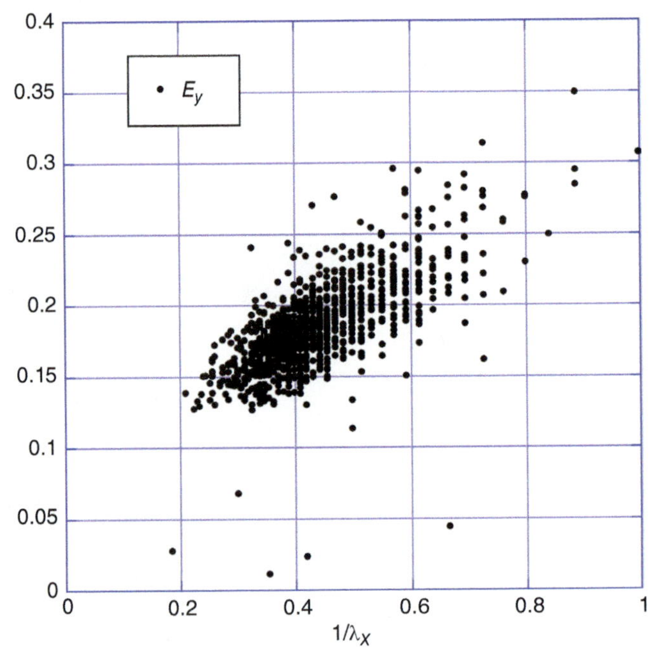

Figure 15.4 Variation of the reconnection electric field at the dominant X point with inverse diffusion region scale length in the x direction. The distribution of inverse diffusion region lengths has a mean of 0.54 and a standard deviation of 0.27.

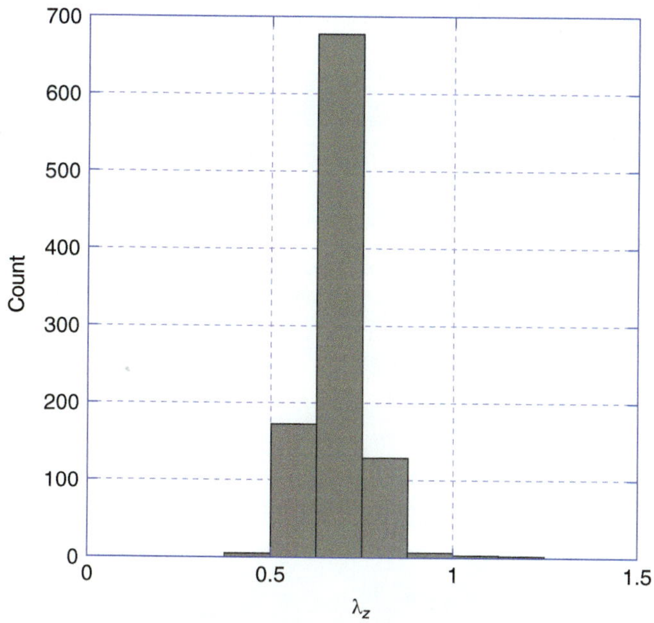

Figure 15.5 Histogram of diffusion region thicknesses throughout the calculation. The figure demonstrates only very small variations of this parameter. The distribution has a mean of 0.67 with a standard deviation of 0.09.

reconnection electric field is primarily due to variations of the diffusion region dimensions in the x direction. This latter variation is caused by changes to the electron magnetization in the horizontal direction, which extends or contracts the extent of the diffusion region. Electron magnetization is primarily affected by the strength of the magnetic field.

It appears evident that the growth of tearing mode islands, which form in the reconnecting current layer, are the prime cause of the variation of this magnetization. As a tearing mode island becomes nonlinear, it will begin to magnetize electrons and thus shorten the electron bounce width in the x direction. This leads to an effective shortening of λ_x, which, as we have seen above, will typically lead to a faster reconnection rate. Therefore, we conclude that island formation in the current layer associated with an active X point can increase the rate of magnetic reconnection. However, we note that reconnection can be fast for longer diffusion regions, presumably related to the absence of islands in the proximity of the X point under consideration. Figure 15.4 clearly shows that the rate virtually never drops below $E=0.12$. We point out that the applied electric field exceeds the empirical minimum values. Hence it is conceivable that plasmoid formation may be necessary to raise the average reconnection rate to that required for the flux transport prescribed by the external driving.

15.4. RECONNECTION LINE ORIENTATION FOR ASYMMETRIC RECONNECTION

In this section, we review recent results [*Hesse et al.*, 2013] pertaining to another aspect of magnetic reconnection: merging of plasmas from asymmetric inflow regions. Such asymmetry is commonly found at the magnetopause of Earth as well as on other planets. Typically, the shear angle of the magnetic field is different from the 180° modeled above, which, in comparison, is represented by the presence of a guide magnetic field. Here a key question is which direction a reconnection line will follow: Will it be normal to the plane wherein the guide field is constant [*Sonnerup*, 1974] or will it choose a different direction [*Gonzales and Mozer*, 1974; *Cowley*, 1976]? In order to address this question, we model a sample system with parameters set up similar to *Pritchett* [2008]. The poloidal magnetic field is of the form

$$B_x = 0.5 + tanh\left[z/l\right] \tag{15.12}$$

with a guide magnetic field

$$B_y = B_{y0} = 1. \tag{15.13}$$

The initial ion and electron densities are chosen to exhibit asymptotic values of $n=1$ for $z>0$ and $n=0.333$ for $z<0$:

$$n_i = n_e = 1 - tanh\left[z/l\right]/3. - tanh^2\left[z/l\right]/3. \tag{15.14}$$

The initial current layer half-width is $l=0.5$. A temperature of $T_e/T_i=0.2$ and a temperature of $T=T_i+T_e=1.5$ are employed. A small, X-type, initial perturbation is introduced into current density and magnetic fields with an amplitude of $\delta B=0.1$, leading, e.g., to a form of B_z similar to (9b).

Particles, initialized as drifting Maxwellians, are integrated on a grid of 1000×800 cells representing a system of size $L_x=64$ and $L_z=25.6$. Periodic boundary conditions are employed in the x direction, whereas the particles are specularly reflected at the upper and lower boundaries. The ion-electron mass ratio is chosen to be $m_i/m_e=25$. A total of 1.6×10^9 macro particles are employed for the simulations.

In addition to investigating the reference case above, we further study the guide field model in coordinate systems of different orientation, derived from the one above by rotation around the z axis. The rotation angle α is defined in Figure 15.6. For each simulation in a rotated coordinate system, the entire set of vector quantities defined above is transformed into the rotated coordinate system, and the numerical model is applied to the new initial condition.

The overall evolution of the reference system, with uniform guide field of unity (i.e., $\alpha=0$), is illustrated for

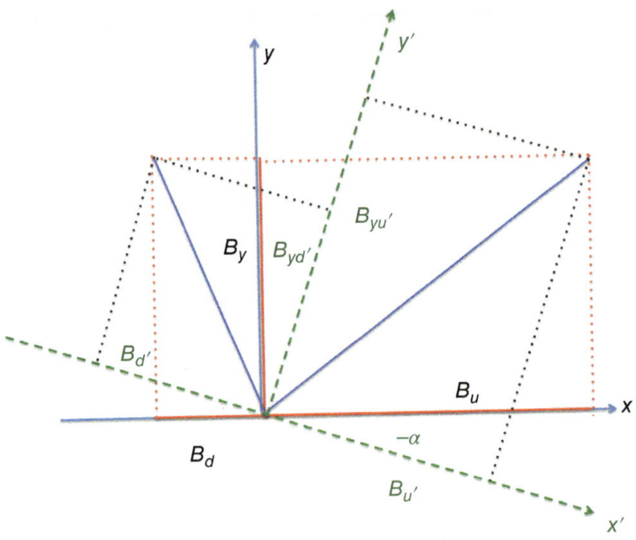

Figure 15.6 Graphics of the asymptotic magnetic field values in a coordinate system rotated by an angle α about the z axis. The indices u (up) and d (down) refer to asymptotic values above and below the current layer, respectively. The globally constant guide field $B_y = 1$ for $\alpha = 0$ becomes spatially dependent after rotation. The asymptotic values are shown on the y' axis.

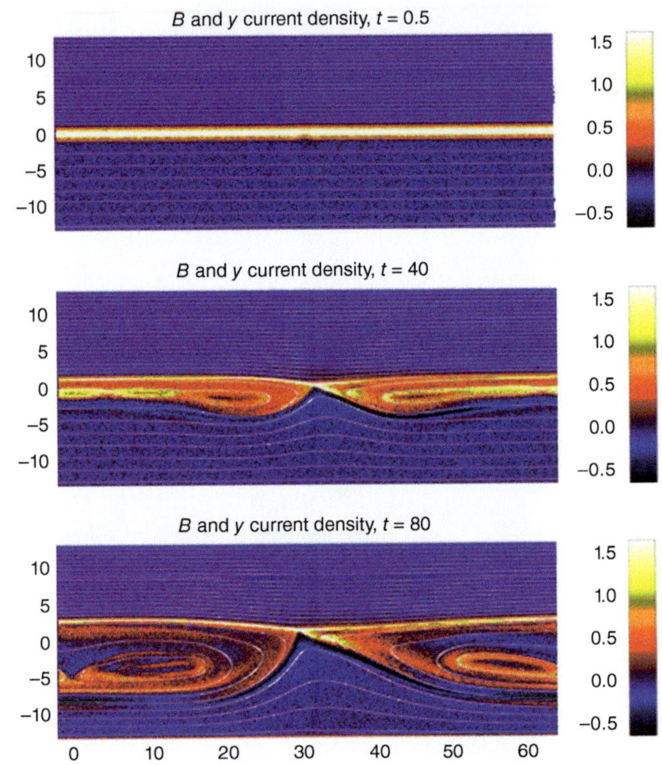

Figure 15.7 Magnetic field (white lines) and current density (color) evolution for the reference run, for which the initial guide field is uniform and of unit value.

three different times in Figure 15.7. The figure shows a transition from initially symmetric behavior to the development of a significant left-right asymmetry toward the end of the simulation period ($t = 80$).

The temporal evolution of the reference reconnection rate is shown in Figure 15.8. The figure demonstrates that *Sonnerup*'s [1974] prediction of fast reconnection in the presence of a guide field is correct: The peak reconnection rate is still about 0.05. The simulation never attains a steady state, a consequence of both periodic boundary conditions in the x direction and the closed top and bottom boundaries, which imply limited amount of magnetic flux and energy to tap into. Open boundary conditions for asymmetric systems are difficult to implement, which led us to choose the present implementation.

A reasonable assumption has the X line naturally orienting itself in such a way to obtain the fastest reconnection rate. The most important question is therefore whether faster reconnection can be obtained if $\alpha \neq 0$. Conducting a set of simulations, we find substantial variations of the reconnection rates depending on the orientation of the coordinate system. Figure 15.9 displays the time evolution of the reconnection rates for all runs obtained from the guide field reference model by coordinate system rotation and illustrates this strong dependence.

Figure 15.9 exhibits a maximum reconnection rate for $\alpha = -14.87°$. Adjacent values of the rotation angle yield almost identical reconnection rates, indicating that the maximum is not strongly localized. Extending rotation

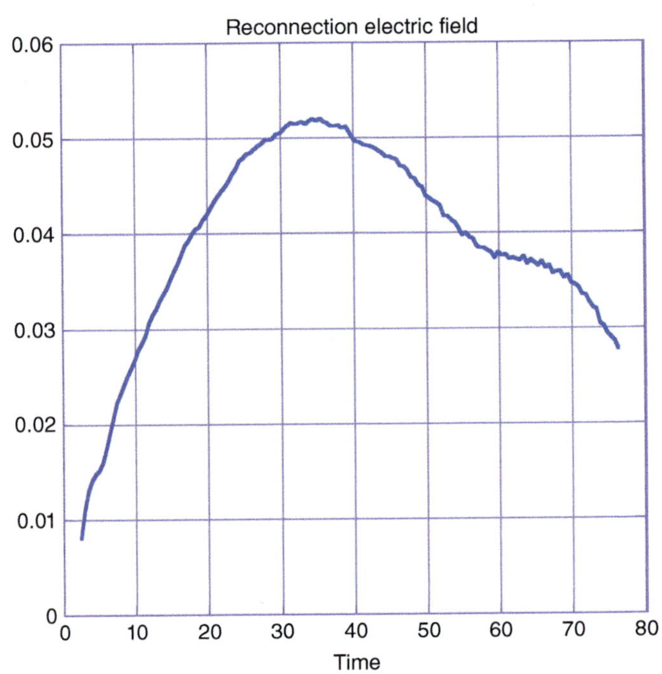

Figure 15.8 Time evolution of the reconnection electric field for the reference run for which the initial guide field is uniform and of unit value.

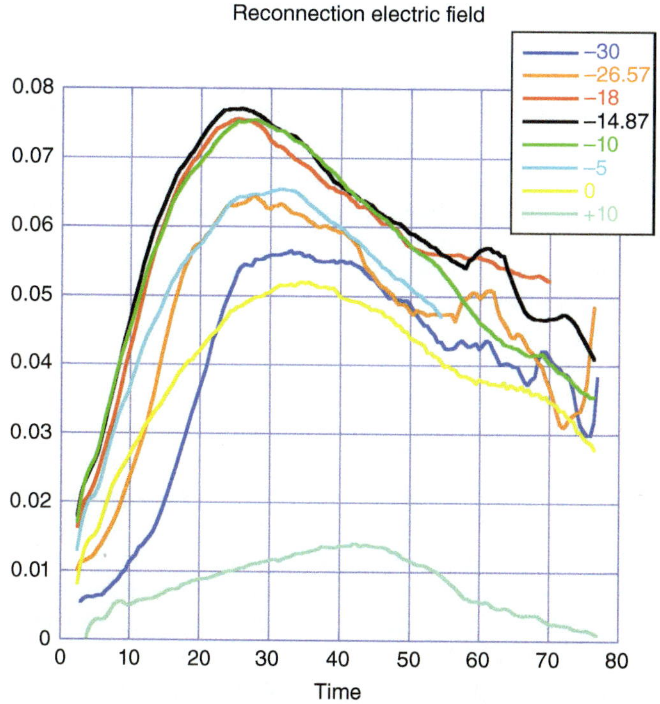

Figure 15.9 Time evolution of the reconnection electric field for the entire set of runs derived from rotating the frame of the guide field calculation by an angle α. The different colors denote different runs, and the angles are denoted in the figure.

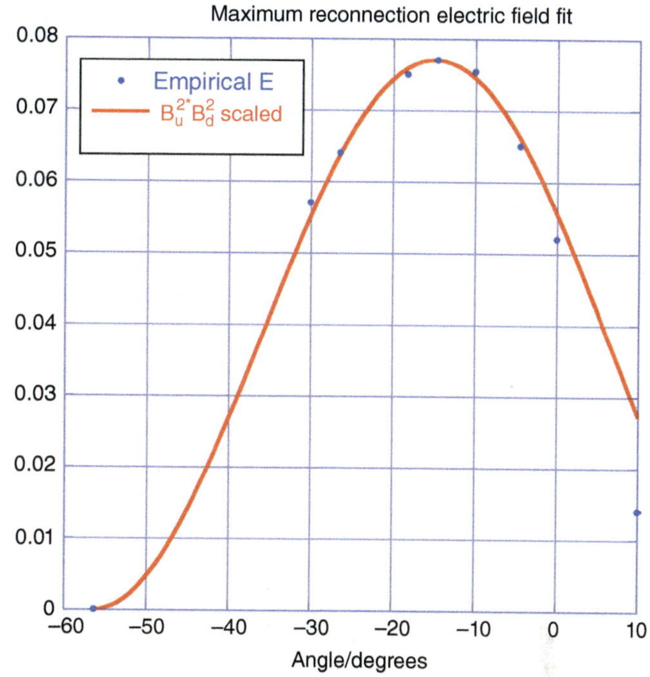

Figure 15.10 Plot of peak electric field values and predictions based on the magnetic energy available for magnetic reconnection. The magnetic-energy-based prediction exhibits an excellent match.

angles more significantly to either side, however, yields rapidly decreasing values of the reconnection electric fields. It is noteworthy that the reconnection evolution for $\alpha = -26.57°$, for which the reconnection magnetic fields are equal on both sides, still leads to considerably slower reconnection rates than the runs for smaller rotation angles. Therefore, symmetry of the in-plane magnetic field does not lead to the fastest reconnection rates.

Furthermore, the reconnection rate has to vanish for rotations, for which one of the two rotated magnetic field components

$$B_{u'} = B_u \cos(\alpha) + B_y \sin(\alpha), \quad (15.15)$$

$$B_{d'} = B_d \cos(\alpha) + B_y \sin(\alpha) \quad (15.16)$$

vanishes. For the system investigated here, $B_u = 1.5$ and $B_d = -0.5$. We find $B_u{}' = 0$ for the rotation value $\alpha \approx -57°$.

In order to develop a scaling relation for the reconnection electric field, we consider the maximum value of the entire time evolution for a fixed rotation angle. The key question is how the maximum values shown here relate to physical parameters. We will approach this issue from two different angles. First, we will make a reasonable assumption: reconnection rates should scale with the available magnetic energy. Because of the constancy of the total volume, this assumption implies that reconnection rates should be proportional to the magnetic energy density. Since the energy densities available for reconnection can differ across the current layer, our assumed proportionality implies

$$E_r = \psi_1 B_{u'}^2 B_{d'}^2, \quad (15.17)$$

where the factor ψ_I should depend on parameters such as density and total magnetic field, which are held constant in the present investigation. Figure 15.10 demonstrates that the model (17) does an excellent job representing the maxima of the modeled reconnection rates.

We can use our model (17) to derive the rotation angle, for which this energy product assumes its maximum. Using the rotated magnetic field expressions (15) and (16), it is a straightforward exercise to derive the equation

$$\cot^2 \alpha_{\max} - 2\frac{B_d B_u - B_y^2}{B_y (B_u + B_d)} \cot \alpha_{\max} - 1 = 0 \quad (15.18)$$

for the angle $\alpha_{\max}$, for which (17) attains its maximum. The relevant solution is

$$\alpha_{\max} = -14.87°,$$

matching our empirical determination.

It is further interesting to note that equation 15.18 also describes exactly the angle which bisects the angle between the asymptotic magnetic field on both sides of the current layer, a result readily derived from geometric considerations. This result applies to more than the system modeled here and is hence quite generic: The half-angle direction maximizes the magnetic energy available for reconnection.

In summary, our results indicate that the magnetic reconnection line in asymmetric systems is preferentially oriented in such a way that it bisects the direction of the asymptotic magnetic field direction on both inflow sides. This orientation is identical to the one for which the product of available magnetic energy is maximized.

15.5. SUMMARY

In this chapter, we presented results pertaining to three different aspects of magnetic reconnection. The first aspect, transport of magnetic flux, mass, momentum, and energy, is generic to all geometries and environments in which reconnection is operating. Focusing on energy transport, we could show that magnetic reconnection enables effective energy transport beyond what is possible by compression or expansion of a volume enclosed by a separatrix. We showed that the critical quantity providing the amplitude of the transport rate is exactly the reconnection electric field, measured at the X point. While not developed here, it is straightforward to develop similar analyses for mass and momentum transport.

Having demonstrated that effective energy transport is impossible without magnetic reconnection, we embarked on a discussion of the second topic: time dependence. Using our open boundary condition models, we executed an extended calculation of a driven, continuously reconnecting, symmetric current sheet.

For the purpose of the analysis of the model run, we defined diffusion region dimension through the region, where nonideal contributions to the electric field exceed the electric field contribution by electron convection. The magnitude crossover points were identified with the horizontal and vertical diffusion region edges. We developed an automated algorithm to calculate, for each output time step, these dimensions and take note of the instantaneous value of the reconnection electric field.

We then used the results for a statistical correlation. Not surprisingly, we found that rates tend to correlate well with the aspect ratio of the diffusion region. More surprisingly, we discovered that the aspect ratio variation was almost exclusively due to variation of diffusion region lengths, whereas diffusion region width showed relatively limited variations. We explained this phenomenon by the repetitive formation and growth of plasmoids, which, when grown past the linear stage, begin to magnetize electrons. This magnetization changes the crossover location between nonideal and convection electric fields and renders the diffusion region shorter. Thus, the aspect ratio becomes larger, and, on average, the reconnection electric field grows. This modulation is of the order of 50%, implying that reconnection is fast even in the temporary absence of magnetic islands. Therefore, our conclusion is that island formation is not necessary for fast reconnection, but it typically serves to enhance the reconnection rate substantially and in a very time-dependent manner.

As the third part of this chapter, we briefly reviewed recent results pertaining to the direction of the reconnection line in asymmetric magnetic reconnection. Starting with an asymmetric configuration with a guide magnetic field, which, in many ways, is typical for reconnection geometries expected at a planetary magnetopause, we performed a set of kinetic simulations of this system in rotated coordinate systems. We found a surprising variation in reconnection rates, even more surprising since neither constant guide field nor symmetric in-plane magnetic fields led to the fastest reconnection rate. Instead, we found empirically that the reconnection rates scaled well with the product of energy densities of both inflow regions. The direction which maximizes this product turned out to be also halving the angle between the asymptotic magnetic field directions on both sides of the current layer. Clearly, this result requires further testing, including for different density asymmetries but it is highly suggestive that the X line will orient itself such that it bisects the directions on the magnetic fields in the two inflow regions.

Finally, one might ask the question of how these results pertain to the magnetospheres of planets other than Earth. The general results presented in section 15.2. are universal, i.e., they apply, at least in principle, also to planetary magnetospheres. As the analysis in subsequent sections has been conducted in dimensionless units, a large variety of applications can be developed by employing appropriate dimensions in the interpretation of the results. This freedom is an application of the "Vlasov scaling" [*Schindler*, 1969], whereby Vlasov scaling laws permit rescaling of solution to different applications. Such rescaling will greatly extend the applicability of the results presented here; however, just as in the terrestrial magnetosphere, rescaling will fall short of covering all relevant scenarios. Therefore, much more research needs to be conducted, and many more exciting and surprising discoveries lie in wait.

ACKNOWLEDGMENTS

This work was supported by NASA's Magnetospheric MultiScale Mission. One of us (N.A.) gratefully acknowledges support from the NASA Postdoctoral fellowship program.

REFERENCES

Antiochos, S. K., C. R. DeVore, and J. A. Klimchuk (1999), A model for solar coronal mass ejections, *Astrophys. J.*, *510*, 485–493.

Aunai, N., G. Belmont, and R. Smets (2011), Energy budgets in collisionless magnetic reconnection: Ion heating and bulk acceleration, *Phys. Plasmas*, *18*, 2901.

Birn, J., J. F. Drake, M. A. Shay, B. N. Rogers, R. E. Denton, M. Hesse, M. Kuznetsova, Z. W. Ma, A. Bhattacharjee, A. and Otto, and P. L. Pritchett (2001), GEM magnetic reconnection challenge, *J. Geophys. Res.*, *106*, 3715.

Birn, J., M. Hesse, and S. Zenitani, (2011), Reconnection in compressible plasmas: Extended conversion region, *Phys. Plasmas*, *18*, 111,202.

Cargill, P. A., and J. A. Klimchuk (1997), A nanoflare explanation for the heating of coronal loops observed by Yohkoh, *Astrophys. J.*, *478*, 799.

Cassak, P. A., and M. A. Shay (2007), Scaling of asymmetric magnetic reconnection: General theory and collisional simulations, *Phys. Plasmas*, *14*, 102,114.

Cowley, S. W. (1976), Comments on the merging of nonantiparallel magnetic fields, *J. Geophys. Res.*, *81*, 3455.

Drake, J. F., M. Swisdak, K. M. Schoeffler, B. N. Rogers, and S. Kobayashi (2006), Formation of secondary islands during magnetic reconnection, *Geophys. Res. Lett.*, *33*(13), L13105, doi:10.1029/2006GL025957.

Daughton, W., J. Scudder, and H. Karimabadi, (2006), Fully kinetic simulations of undriven magnetic reconnection with open boundary conditions, *Phys. Plasmas*, *13*, 2101.

Daughton, W, V. Roytershteyn, B. Albright, H. Karimabadi, L. Yin, and K. J. Bowers (2009), Transition from collisional to kinetic regimes in large-scale reconnection layers, *Phys. Rev. Lett.*, *103*, 65,004.

Eastwood, J. P., T.-D. Phan, F. S. Mozer, M. A. Shay, M. Fujimoto, A. Retinò, M. Hesse, A. Balogh, E. A. Lucek, and I. Dandouras (2007), Multi-point observations of the Hall electromagnetic field and secondary island formation during magnetic reconnection, *J. Geophys. Res.*, *112*, A06235, doi:10.1029/2006JA012158.

Eastwood, J., T. Phan, M. Øieroset, and M. Shay (2010), Average properties of the magnetic reconnection ion diffusion region in the Earth's magnetotail: The 2001–2005 Cluster observations and comparison with simulations, *J. Geophys. Res.*, *115*, 08215.

Gonzales, W. D., and F. S. Mozer (1974), A quantitative model for the potential resulting from reconnection with an arbitrary interplanetary magnetic field, *J. Geophys. Res.*, *79*, 4186.

Gosling, J. T., J. Birn, and M. Hesse (1995), Three-dimensional magnetic reconnection and the magnetic topology of coronal mass ejection events, *Geophys. Res. Lett.*, *22*, 869.

Gosling, J. T., R. M. Skoug, D. J. McComas, and C. W Smith (2005), Direct evidence for magnetic reconnection in the solar wind near 1 AU, *J. Geophys. Res.*, *110*, 2156.

Harris, E.G. (1962), On a plasma sheath separating regions of oppositely directed magnetic field. *Nuovo Cimento*, *23*, 115.

Hesse, M., J. Birn, D. N. Baker, and J. A. Slavin (1996), MHD simulations of the transition of magnetic reconnection from closed to open field lines, *J. Geophys. Res.*, *101*, 10,805.

Hesse, M., K. Schindler, J. Birn, and M. Kuznetsova (1999), The diffusion region in collisionless magnetic reconnection, *Phys. Plasmas*, *6*, 1781.

Hesse, M., J. Birn, and S. Zenitani (2011a), Magnetic reconnection in a compressible MHD plasma, *Phys. Plasmas*, *18*, 042,104.

Hesse, M., T. Neukirch; K. Schindler, M. Kuznetsova, and S. Zenitani (2011b), The diffusion region in collisionless magnetic reconnection, *Space Sci. Rev.*, *160* (1–4) 3.

Hesse, M., N. Aunai, S. Zenitani, M. Kuznetsova, and J. Birn (2013), Aspects of collisionless magnetic reconnection in asymmetric systems, *Phys. Plasmas*, *20*, 1210.

Hones, E. W., Jr. (1977), Substorm processes in the magnetotail: Comments on "On hot tenuous plasma fireballs and boundary layers in the earth's magnetotail" by L. A. Frank, K. L. Ackerson, and R. P. Lepping, *J. Geophys. Res.*, *82*, 5633.

Huang, Y.-M., and A. Bhattacharjee (2013), Plasmoid instability in high-Lundquist-number magnetic reconnection, *Phys. Plasmas*, *20*, 055,702.

H. Karimabadi, W. Daughton, and J. D. Scudder (2007), Multiscale structure of the electron diffusion region, *Geophys. Res. Lett.*, L13104, doi:10.1029/2007GL030306.

Mozer, F., V. Angelopoulos, J. Bonnell, K.-H. Glassmeier, and J. McFadden (2008), THEMIS observations of modified Hallfields in asymmetric magnetic field reconnection, *Geophys. Res. Lett.*, *35*, L17S04, doi:10.1029/2007GL033033.

Mozer, F. S., and P. L. Pritchett (2011), Electron physics of asymmetric magnetic field reconnection, *Space Sci. Rev.*, *158*, 119.

Paschmann, G. (2008), Recent in-situ observations of magnetic reconnection in near-Earth space, *Geophys. Res. Lett.*, *35*, 19,109.

Paschmann, G., et al. (1979), Plasma acceleration at the earth's magnetopause: Evidence for reconnection, *Nature*, *282*, 243.

Pritchett, P. L. (2008), Collisionless magnetic reconnection in an asymmetric current sheet, *J. Geophys. Res.*, *113*, A06210, doi:10.1029/2007JA012930.

Schindler, K. (1969), Laboratory experiments related to the solar wind and the magnetosphere, *Rev. Geophys.*, *7*, 51.

Sonnerup, B. U. O. (1974), Magnetopause reconnection rate, *J. Geophys. Res.*, *79*, 1546.

Swisdak, M., B. N. Rogers, J. F. Drake, and M. A. Shay (2003), Diamagnetic suppression of component magnetic reconnection at the magnetopause, *J. Geophys. Res.*, *108*, doi:10.1029/2002JA009726.

16

Origin and Evolution of Plasmoids and Flux Ropes in the Magnetotails of Earth and Mars

J. P. Eastwood[1] and S. A. Kiehas[2,3]

16.1. INTRODUCTION

Magnetic reconnection is one of the most important processes determining the behavior of planetary magnetotails. It controls the rates at which plasma and magnetic flux circulate, and, as described elsewhere in this volume, has been observed at all the planets where extended in situ magnetotail observations have been made: Mercury (see Chapter 2), Venus and Mars (Chapter 3), Earth (Chapter 4), Jupiter (Chapter 5), and Saturn (Chapter 6). In general, the onset of magnetic reconnection across the magnetotail current sheet leads to the formation of oppositely directed jets of heated plasma, confined to the current sheet, which are energized at the expense of energy previously stored in the magnetotail lobes. Magnetotails often exhibit time-dependent behavior, and this is intimately related to the time-dependent processes of energy storage, transfer, and release that are enabled by magnetic reconnection (Chapter 15). As such, reconnection lies at the heart of many dynamic magnetotail phenomena, such as substorms at Earth (Chapter 18).

In addition to releasing stored magnetic energy and creating jets of heated plasma, reconnection also produces a variety of plasma structures such as islands, plasmoids, flux ropes, bubbles, and dipolarizations. These structures are very important for understanding on a more quantitative level how magnetotails work and the time scales and pathways by which magnetic reconnection changes magnetotail topologies and redistributes energy. For example, at Earth substorm activity is closely related to plasmoid formation [e.g., *Moldwin and Hughes*, 1992; *Slavin et al.*, 1993]. It is therefore very important to understand the underlying physics controlling the formation of each type of structure and to ascertain for each their typical properties, such as location, occurrence rate, size, and motion, in order to establish a deeper overall understanding of magnetospheric dynamics.

At Earth, thanks to comprehensive multipoint observations, we now have a relatively good understanding of magnetotail behavior which allows both qualitative and quantitative studies of reconnection and the structure it generates. In this review we first discuss the origin and evolution of plasmoids and flux ropes in Earth's magnetotail, providing an overview of our current understanding based on recent multipoint and multimission data analysis. Plasmoids, flux ropes, and other similar structures have been extensively studied, and so we encourage the interested reader to also consult earlier reviews and the references contained therein [e.g., *Hesse and Kivelson*, 1998; *Slavin*, 1998; *Sharma et al.*, 2008].

Recent activity in planetary magnetotail research has seen an increased interest in comparative studies, in part due to the increasing availability of data from different planets. Such comparative studies are important for two reasons. First, planetary missions, owing to their more general scientific objectives (particularly surface imaging at a variety of wavelengths), often cannot accommodate a full suite of plasma instrumentation. In particular, while planetary missions often make very good magnetic field measurements, complete measurements of the plasma distribution functions are difficult. Since flux ropes, plasmoids, etc., can be observed in magnetic field data alone, transfer of models and knowledge from measurements

[1] *The Blackett Laboratory, Imperial College London, United Kingdom*

[2] *Space Research Institute, Austrian Academy of Sciences, Graz, Austria*

[3] *Department of Earth, Planetary, and Space Sciences, University of California, Los Angeles, California*

Magnetotails in the Solar System, Geophysical Monograph 207. First Edition.
Edited by Andreas Keiling, Caitríona M. Jackman, and Peter A. Delamere.

made at Earth by dedicated space plasma satellites can prove crucial to properly interpreting and understanding planetary magnetotail observations. Conversely, different planetary magnetotails provide access to different regions of parameter space which cannot be routinely achieved at Earth. Such observations enable the development of a more general understanding of collisionless plasma processes.

To illustrate this, we also present recent results concerning observations of flux ropes in the vicinity of Mars: much of our knowledge about flux ropes at Mars is informed and influenced by what has previously been learned at Earth. The martian magnetosphere is particularly interesting because although it has no global dipole, there are regions of strong crustal fields which participate in its interaction with the solar wind. Understanding the Mars solar wind interaction is very important for determining its atmospheric history, and recent discoveries show that magnetic reconnection-generated structures may play a significant role, particularly in the vicinity of the crustal field regions.

Before proceeding further, we note that even a cursory glance at the literature reveals a profusion of terminology which can be somewhat confusing, especially for those new to the field. We will therefore now briefly discuss some of the different terms used to describe reconnection-generated structures.

16.1.1. Quasi-Two-Dimensional Structure

We first consider the simplest configuration, shown in Figure 16.1a, where a current sheet separates two regions of frozen-in plasma with antiparallel magnetic fields. Since the fields are antiparallel, the current sheet is also a neutral sheet. Such a current sheet can be unstable to the tearing mode instability, which leads to the formation of a series of reconnection *X points* [see, e.g., *Priest and Forbes*, 2000]. A diffusion region is centered on each X point, where the magnetic field decouples from the plasma and reconnection occurs. The plasmas from above and below the current sheet mix and are ejected from each X point to the left and right in heated jets confined to the current sheet. In this perfectly antiparallel configuration the magnetic field is zero at the X point, and the X points are separated by *O points*, where the magnetic field is also zero. Closed loops of magnetic flux are centered on the O points, containing plasma ejected by the X points on either side of the O point. These closed loops of magnetic flux containing reconnected plasma are termed *islands*. In three dimensions, the X and O points extend out of the page to form *X lines* (or *neutral lines*) and *O lines*. The application of the tearing mode to the magnetotail has received considerable attention [e.g., *Coppi et al.*, 1966; *Schindler*, 1974; *Sitnov and Schindler*, 2010]. One consequence of such multiple X line formation is the production of an island chain: if one X line ultimately dominates, then the plasma flow from this X line causes the ejection of a chain of islands, which is observed, for example, in Earth's magnetotail [*Slavin et al.*, 1993, 2005; *Eastwood et al.*, 2005; *Imber et al.*, 2011].

Plasmoids can be considered as a particular type of island specific to magnetotails. At Earth they are generated between the distant X line and the near-Earth X line (see Figure 16.1b) [*Hones*, 1977; *Baker et al.*, 1996]. In a simple antiparallel magnetic field geometry, as might be expected in the magnetotail in the absence of any other complicating factors, this leads to the formation of a looplike closed field region, a *plasmoid*, which pinches off from the preexisting plasma sheet. As such, the name plasmoid was introduced to describe a geophysical phenomenon. The formation of a plasmoid also leads to the production of a *traveling compression region (TCR)*, which corresponds to the compressed lobe magnetic field created by the passage of the plasmoid [*Slavin et al.*, 1984].

A feature also created by reconnection, but on a much smaller scale, is the *secondary island*. Kinetic simulations show that the reconnection current layer can elongate into a thin current sheet which becomes unstable to the formation of islands [*Drake et al.*, 2006a; *Karimabadi et al.*, 2007]. Once formed by this secondary instability, the islands are periodically ejected in the reconnection outflow (see Figure 16.1c). Such periodic ejection could also create a chain of islands; this has been suggested to explain observations of flux rope chains in the magnetotails of Mars and Earth where the system size was not found to be commensurate with the chain size and duration [*Eastwood et al.*, 2012a; *Hietala et al.*, 2013]. Again Figure 16.1 shows antiparallel reconnection, and so the secondary island consists of closed loops. However, observations (discussed in more detail in section 16.2.5.) show that even under antiparallel reconnection the field lines are not closed and the island is more correctly described as a flux rope [*Eastwood et al.*, 2007].

16.1.2. Flux Ropes

Thus far, we have described the production of islands, plasmoids, and secondary islands by antiparallel reconnection. However, in general, the reconnecting fields may not be antiparallel, and this leads to a fundamental change in the magnetic topology, as shown in Figure 16.2. If the shear angle between the reconnecting magnetic fields is less than 180°, then the magnetic field can be decomposed into a reconnecting field and a guide field.

Any B_y present in the magnetotail will lead to a shear of magnetic field lines in the northern and southern tail hemispheres. This means that the northern part of a closed plasma sheet field line will not be perfectly aligned with the southern part of the same field line. Instead, the northern and southern parts of the same field line exhibit

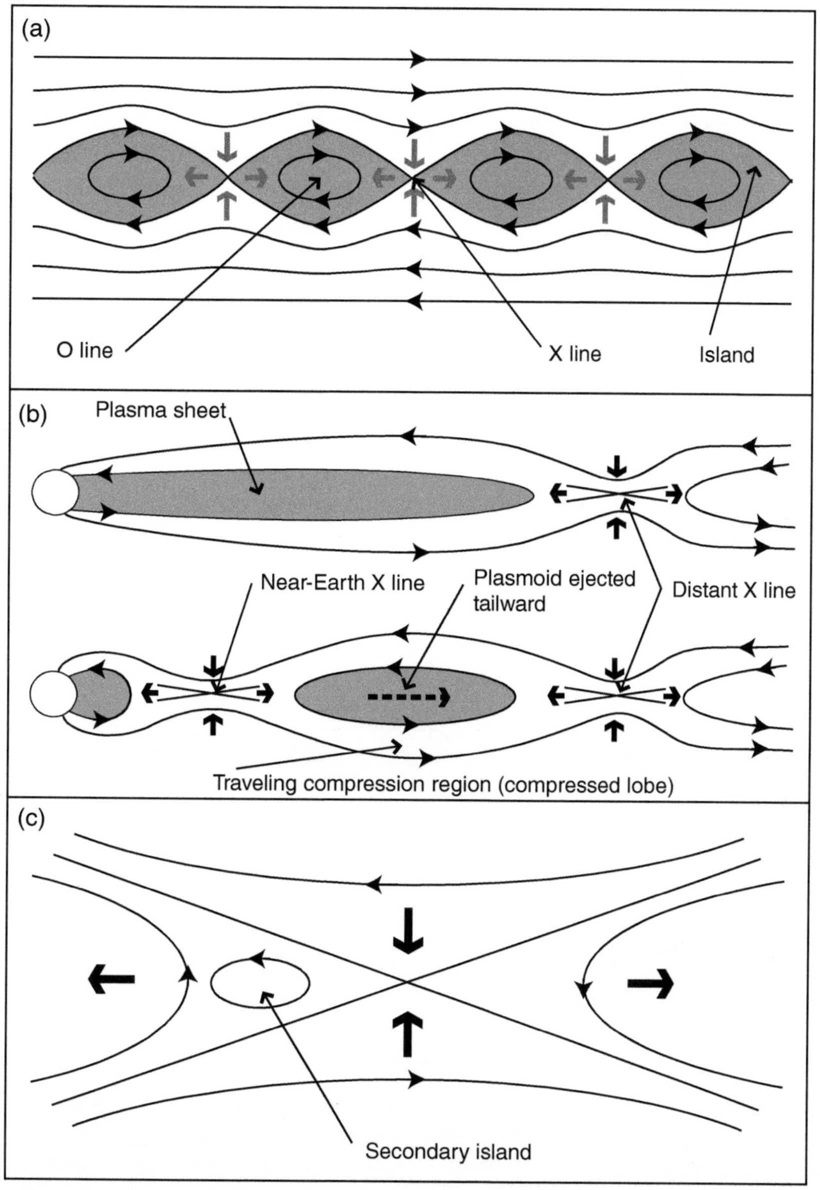

Figure 16.1 (a) A current sheet reconnects via, e.g., a tearing mode instability to form a series of X lines separated by O lines. The closed field regions between the X lines represent islands. Note that this is a fundamentally 2D picture, in that the magnetic fields above and below the current sheet are perfectly antiparallel. (b) A plasmoid is created between the distant and near-Earth X lines, or neutral lines. Originally conceived as a closed loop in a simple antiparallel magnetic field geometry. (c) A secondary island is generated by a secondary instability once reconnection has commenced. They are typically small compared to islands and plasmoids.

a shear and the northern part of one field line is actually adjacent to the southern part of a different field line. Consequently, the northern part of one field line will reconnect not with its southern part but with the southern part of a different field line. As a result, a three-dimensional (3D) *flux rope* with helical field lines is formed rather than a quasi–two–dimensional (2D) plasmoid with looplike field lines. The typical plasmoid picture is obtained by projecting the flux rope in the meridional plane, and the flux rope field lines appear to close on themselves. In addition to the looplike magnetic field, there is a "core field", here pointing along the $-Y$ direction since $B_y < 0$. There is a net magnetic flux along the axis of the flux rope.

Various flux rope models have been developed; the simplest is the force-free model [*Lundquist*, 1950; *Burlaga*, 1988; *Lepping et al.*, 1990]. If the plasma is force free and low β, then $J \times B = 0$, a solution of which is a helical magnetic field whose flux content can be calculated

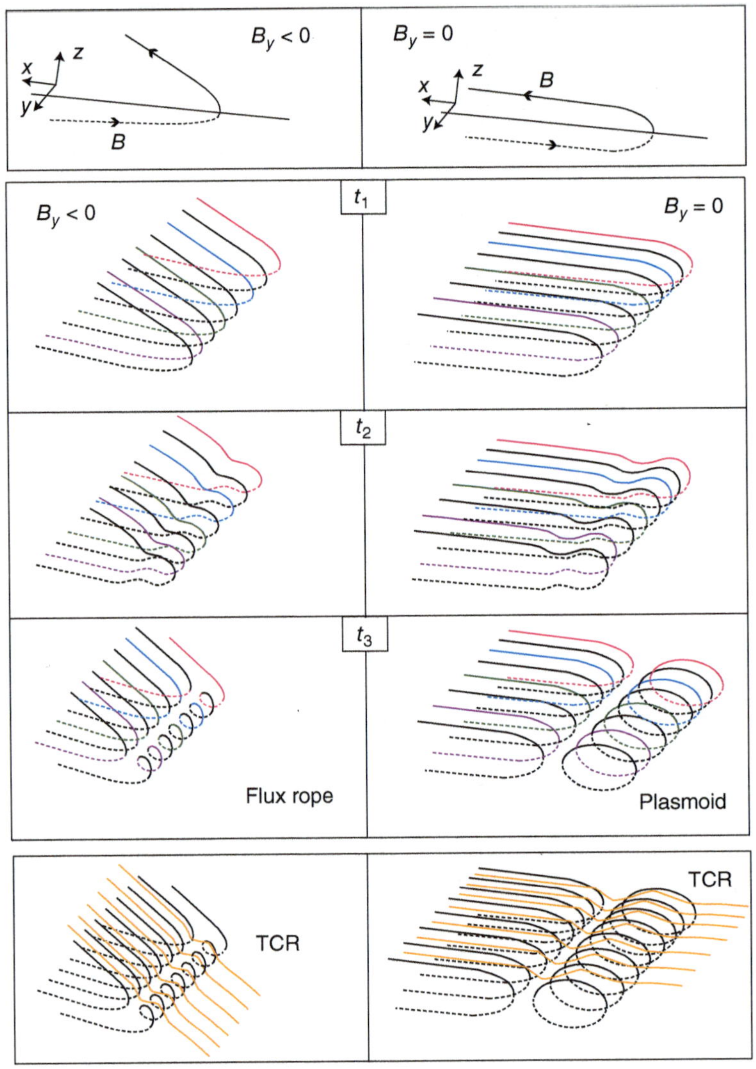

Figure 16.2 Formation of a flux rope (left panels) and plasmoid (right panels). Earth is located at the left and all panels depict the same perspective view. Top panels: The solid straight line symbolizes the current sheet. The dashed (solid) part of the field line is located in the southern (northern) hemisphere. For the formation of a flux rope a nonzero background B_y is assumed (left top panel, we choose B_y<0), while B_y=0 for the formation of a plasmoid (right top panel). For the formation of a plasmoid, the northern and southern parts of the sample field line are within the noon–midnight–meridional plane. For the formation of a flux rope, the northern and southern parts of the sample field line are not aligned within the noon–midnight–meridional plane. Instead, the northern part of the field line is angled into the postmidnight sector (-*y*), while the southern part is angled into the premidnight sector (+*y*). Middle panels: To distinguish between field lines, they are colored differently. At t_0 the initial configuration of a set of field lines are crossing the current sheet at the same *y* location. In the B_y<0 case, a shear of the field lines in the northern and southern hemispheres occurs. In the B_y=0 case the field lines are symmetric with respect to the current sheet. At t_2 we assume a reconnection line elongated in the *y* direction. Inflowing plasma convects field lines from both hemispheres toward the current sheet, symbolized by a dent in the drawn field lines. At t_3 field lines from both hemispheres reconnect at the reconnection line. For B_y<0 the northern part of one field line reconnects with the southern part of a different field line due to the shear of the northern and southern field lines. Hence, a flux rope is formed. For B_y=0 field lines from the northern and southern hemispheres reconnect with themselves, forming a plasmoid. Bottom panels: In both cases non–reconnected field lines (orange) drape around the flux rope/plasmoid, leading to a TCR. [From *Kiehas et al.*, 2013.]

analytically (am equation for this is given in, e.g., *Eastwood et al.* [2012b]). Although relatively simple and idealized, the force-free model has proved to be a very useful tool in studying flux ropes and is used in many of the studies that are discussed in more detail below.

While plasmoids/islands and flux ropes can easily be distinguished in terms of their magnetic field line configuration (flux ropes as 3D, helical magnetic field line structures with an axial core field and plasmoids as quasi–2D features with closed-loop magnetic field lines), the terms "plasmoid" and "flux rope" are sometimes used interchangably in the literature. In some publications the term "plasmoid" is not only used for a closed magnetic field line structure (loops) but also for a helical field line structure (flux rope). For instance, *Ieda et al.* [1998] define plasmoids as structures with both helical *and* loop magnetic field line configurations. Also, *Moldwin and Hughes* [1992] discuss the formation and evolution of plasmoids but find them to have a magnetic core field and being "open" magnetic structures. We would argue that islands, plasmoids, and secondary islands can be seen as idealized subsets of flux ropes that are generated if no shear angle between the reconnecting magnetic field lines is present.

16.1.3. Other Structure

The interaction of earthward directed reconnection jets with the preexisting magnetotail plasma can also lead to the formation of structure. In particular, the onset of reconnection can lead to the pileup of magnetic flux and an increase in the north-south component of the magnetic field [e.g., *Sitnov et al.*, 2013]. This dipolarization can be localized in a thin *dipolarization front* which can propagate earthward while maintaining its structure [e.g., *Runov et al.*, 2009]. Dipolarization fronts are associated with depleted entropy regions, or *bubbles* [*Birn et al.*, 2011]. It is thought that the low-entropy flux tubes associated with the dipolarization front can efficiently contract as they move earthward, thus "rising" to the near-Earth tail region [e.g., *Wolf et al.*, 2009, and references therein]. Recent 3D MHD simulations show that earthward moving bubbles are associated with tailward moving entropy-enhanced *blobs* and field-aligned currents [*Birn et al.*, 2013]. For a comprehensive overview of recent work in this area, see the review by *Sergeev et al.* [2012].

16.2. OBSERVATIONS AT EARTH

16.2.1. Plasmoids

As mentioned in the previous section, the transient phenomenon of looplike magnetic fields associated with fast plasma flows was sketched by *Hones* [1977] to link IMP–8 magnetic field and plasma observations at $X_{GSM} = -34\,R_E$ with the global reconfiguration of the magnetotail during substorms. In this picture, a near-Earth neutral line (NENL) starts to reconnect plasma sheet field lines, forming closed magnetic loops tailward of the reconnection site. As the NENL progresses to reconnect higher-latitude plasma sheet field lines, eventually the last closed field line, which separates the plasma sheet from the lobe and hence crosses the distant X line, will reconnect. At this point, the last closed field line no longer connects the distant X line to the ionosphere but instead connects with the NENL. Hence, the entire plasma sheet tailward of the NENL is disconnected from the ionosphere. Instead, it is captured between the near-Earth and distant neutral line and this structure of magnetic loops and plasma, which has been termed *plasmoid* [*Hones*, 1976], is released and moves tailward [*Baker et al.*, 1996].

This simple concept of plasmoid release is based on the dominance of tailward or earthward acting forces: as long as plasma sheet field lines envelop the plasmoid, it cannot be released since the earthward directed magnetic tension force of nonreconnected plasma sheet field lines and dynamic pressure force of earthward accelerated plasma from the distant X line act against the tailward directed dynamic pressure force of the accelerated plasma from the NENL [*Nishida et al.*, 1986]. As the NENL reconnects more and more plasma sheet field lines, the flux of enveloping plasma sheet field lines decreases and the earthward directed magnetic tension force weakens. As the last closed field line reconnects at the NENL, the tailward directed dynamic pressure overcomes the earthward directed tension forces and the plasmoid propagates tailward.

After the reconnection of plasma sheet field lines, the NENL may start to reconnect lobe field lines. The outflow speed of plasma during reconnection is linked to the Alfvén speed in the inflow region [e.g., *Priest and Forbes*, 2000]. Since lobe plasma has a lower density than plasma sheet plasma, it is this reconnection of lobe (open) field lines that generates faster outflow than the previous reconnection of plasma sheet (closed) field lines and accelerates the structure away from the reconnection site.

The development of global magnetohydrodynamic (MHD) codes enables the investigation of plasmoids in a global magnetosphere. With the aid of a GUMICS-4 simulation, *Honkonen et al.* [2011] showed that plasmoid formation is associated with sufficiently large and fast rotations of the interplanetary magnetic field (IMF) clock angle and initiated during lobe reconnection after a fast northward IMF rotation. In an earlier global MHD simulation *Slinker et al.* [1995] observed plasmoid formation after the IMF turned southward from a prolonged steady northward orientation. It should be mentioned that Hsu and McPherron [2002] showed that half of all substorms are triggered by a northward turning of the IMF but did not take simultaneous changes in B_y into account. We note that a current area of active research appears to be the incorporation of kinetic-scale physics into MHD codes so as to more accurately understand onset behavior.

16.2.2. Flux Ropes

The 2D picture of plasmoids needed to be extended after strong core fields with a flux-rope-like structure were observed [e.g., *Slavin et al.*, 1989a; *Moldwin and Hughes*,

1991; *Machida et al.*, 1994; *Frank et al.*, 1995; *Slavin et al.*, 1995]. *Moldwin and Hughes* [1991] showed that previously observed structures that were interpreted as plasmoids were better described as 3D flux ropes. *Hughes and Sibeck* [1987] described how a shear in the northern and southern tail hemispheres leads to the formation of a flux rope rather than a plasmoid. This was also shown by early simulations [*Birn et al.*, 1989]. MHD simulations also show that under the presence of a finite B_y the resulting flux ropes resemble complex structures of both open and closed field lines [e.g., *Farr et al.*, 2008]. The connectivity of flux rope field lines with the ionosphere was investigated by *Kivelson et al.* [1996], who found in their simulation closed field lines connecting the northern morning ionosphere with the southern evening ionosphere for a negative IMF B_y. This situation would be vice versa for a positive IMF B_y.

Finally, we note that the possible connection of flux ropes to the ionosphere and the mechanism for release are still open questions. From our current understanding, a flux rope can move tailward as it forms, but it cannot be ejected tailward since the ends of the flux rope are still connected to the ionosphere. Hence, lobe reconnection on the flanks of the flux rope is required to release the flux rope. Observational evidence for the release mechanism and the role of lobe reconnection is still missing.

16.2.2.1. Core Field. The magnetic field structure inside an idealized flux rope is characterized by a purely azimuthal field at its edges which rotates gradually toward a purely axial field at its center. The axial field (core field) is strong, with a magnetic field strength that can exceed the lobe field strength by as much as 10–20% [e.g., *Slavin et al.*, 1989a, 1995]. While *Slavin et al.* [1989a] and *Mukai et al.* [1998] suggested that the core field originates from compression due to draping of field lines, model results by *Hesse et al.* [1996] yield an enhancement of the axial field due to a decrease of the plasma pressure. In this model, the flux ropes grow in the cross–tail direction until they reach the low–latitude boundary layer or the magnetosheath. In *Hesse et al.*'s [1996] simulation, the hot flux rope plasma is magnetically connected to the cold magnetosheath plasma, and the resulting heat exchange leads to a decrease in the plasma pressure inside the flux rope. With the plasma pressure weakened, the magnetic tension forces will lead to a compression of the flux rope. This will take place primarily in the center of the flux rope, because the magnetic field lines are shorter near the axis and hence the heat exchange will take place faster there. Due to the compression, the core field strength increases until the magnetic pressure force balances the tension force and a force-free state is reached. This scenario requires that flux ropes are magnetically connected to the flanks of the magnetotail. However, recent dual–probe ARTEMIS observations show that flux ropes may be spatially confined and do not necessarily extend over the entire cross section of the tail (see section 16.2.3.). Certainly, any other magnetic connectivity leading to a loss of plasma or heat exchange resulting in a decrease of the plasma pressure in the center of the flux rope will result in an intensification of the core field according to *Hesse et al.*'s [1996] scenario.

It should be noted that plasmoids with apparently closed magnetic loops can exhibit enhanced internal magnetic fields. *Slavin et al.* [1989a] analyzed two plasmoid events and concluded that they exhibit closed-loop field lines based on the evaluation of a minimum variance analysis and the corresponding hodograms. The two events were characterized by an increase in the total magnetic field strength in the center of the plasmoid. To be precise, the total magnetic field exhibited a dip-increase-dip variation as the ISEE–3 spacecraft crossed the plasmoid. This magnetic field enhancement beyond the lobe field strength in the center of the plasmoid is suggested to be due to a plasmoid–associated flaring of the magnetopause with the amount of compression varying with the impact parameter of the spacecraft trajectory relative to the central axis of the plasmoid. The field strength enhancement is due to a compression of the plasmoid normal to the current sheet and is expressed primarily in B_x. The spacecraft would observe this enhancement only at the edge of the plasmoid, not during a passage at the center, as sketched in Figure 12 of *Slavin et al.* [1989a]. The same effect can also appear for flux ropes, as shown by *Moldwin and Hughes* [1991]. In the statistical survey by *Moldwin and Hughes* [1992], 23% of the bipolar B_x events had their core field predominantly in the Y direction, while 24% had their enhanced field approximately equally divided between the X and Y directions. Recent ARTEMIS observations show the same effect. Figure 16.3 shows observations of two different flux ropes made by the two ARTEMIS probes [*Kiehas et al.*, 2012]. P1 is located closer to the current sheet and passes through the center of the flux rope, observing the core field B_y. P2 is located further away from the current sheet and crosses the flux rope near its edge, observing the dip–increase–dip variation in B_x and B_y with the centered increase being dominant in B_x.

It is possible that the transverse pressure resulting from the compression of the flux rope "squeezes" plasma out of the flux rope [*Zwan and Wolf*, 1976; *Slavin et al.*, 1989a]. As a result, the decrease in the plasma pressure will lead to an intensification of the core field, as modeled by *Hesse et al.* [1996] and described above.

In previous studies, a correlation between the direction of the axial field inside flux ropes and the IMF B_y has been found [*Hughes and Sibeck*, 1987; *Lepping et al.*, 1995; *Slavin et al.*, 1995]. *Moldwin and Hughes* [1992] (*Slavin et al.* [2003a]) found that the B_y polarity of the

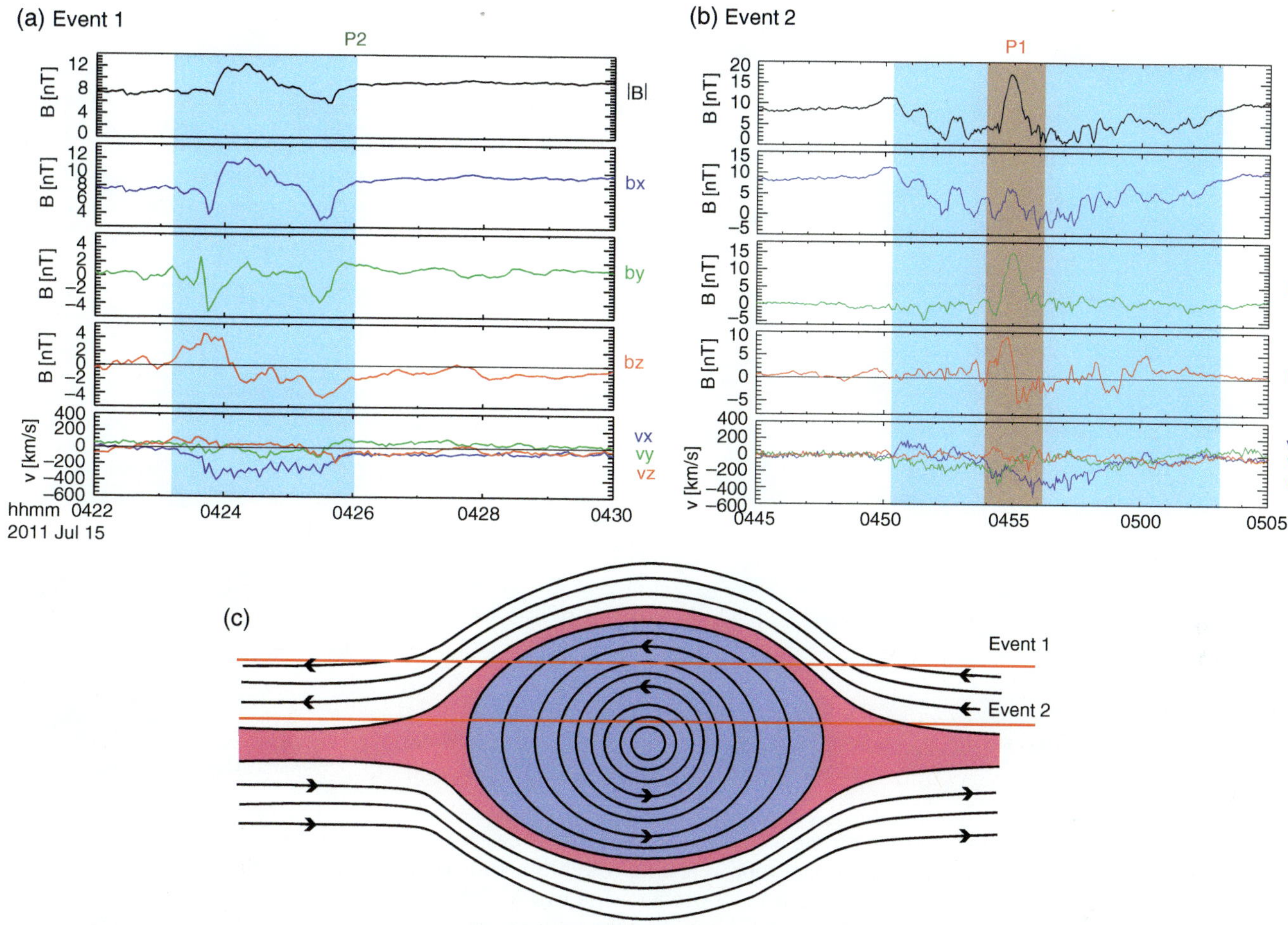

Figure 16.3 (a,b) Total magnetic field $|B|$, B_x, B_y, B_z and ion velocity for two independent flux rope encounters by ARTEMIS P1 and P2. The observations are very different since the probes were located on different latitudes, hence crossing the flux rope on different levels. (c) Possible trajectories of the spacecraft through a flux rope for events 1 and 2. [From *Kiehas et al.*, 2013].

core field was identical with the polarity of the IMF B_y in 87% (79%) of 39 (28) events. This relation sounds reasonable, since the IMF B_y penetrates into the magnetosphere according to the model of *Cowley* [1981] and the correlation between a plasma sheet B_y with the IMF B_y has been observed in the past [*Fairfield*, 1979; *Lui*, 1984; *Tsurutani et al.*, 1984]. However, *Borg et al.* [2012] could not find a clear polarity correlation between the flux rope core field and the IMF B_y for the 27 flux ropes observed with Cluster. For 13 of the 27 flux ropes the polarity of the core field was opposite of the polarity of the IMF B_y, and so the formation of the flux rope core field is not completely resolved. Finally, if the spacecraft does not cross the flux rope at its center, it may not observe its core field, and the flux rope may be misinterpreted as a plasmoid; a thorough analysis is important [*Shen et al.*, 2007; *Borg et al.*, 2012].

16.2.2.2. Tilting of Flux Ropes. A tilting of the flux rope axis in the $X-Y$ plane has been observed during several events in Earth's magnetotail [*Slavin et al.*, 2003a; *Shen et al.*, 2003; *Walsh et al.*, 2009; *Kiehas et al.*, 2012; *Hietala et al.*, 2014]. This could be explained by differential reconnection due to a gradual extension of the reconnection line over the Y direction ([*Kiehas et al.*, 2012]; see Figure 16.4, left column). Localized reconnection leads to the initial formation of a flux rope of limited azimuthal extent in the center of the magnetotail. This first and central part of the flux rope moves tailward while the reconnection line spreads farther into the dusk and dawn sectors and adds more twists to the already established flux rope. Since the previously formed flux rope has already been transported a certain distance downtail, these newly formed ends of the flux rope trail the previously formed central part. This leads to a boomerang-shaped flux rope, exhibiting a nonuniform shape with inclined "wings."

Alternatively, since reconnection at the dawn- and duskside flanks need not necessarily happen at the same

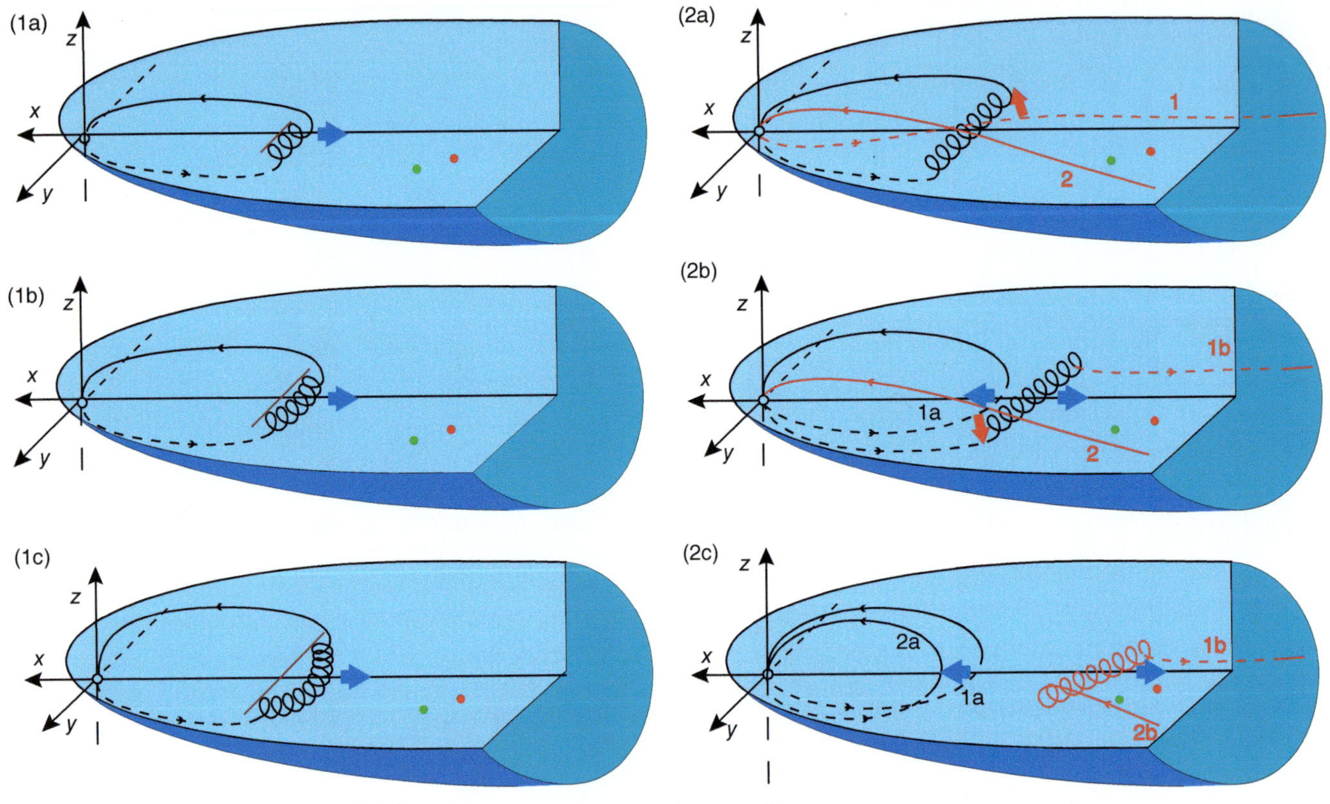

Figure 16.4 Two possible scenarios for the formation of tilted flux ropes in Earth's magnetosphere (blue). In the initial state of this idealized sketch, the flux rope is resting within the equatorial plane, exhibiting a dawn-dusk extent and is connected at its ends to the Earth's ionosphere. The two ARTEMIS spacecraft P1 and P2 are symbolized by red and green dots, respectively. Left column: Reconnection starts near midnight and forms the central segment of the flux rope (panel 1a). The reconnection line is indicated by a brown line. As an assumption, the reconnection line spreads along the east–west direction and adds additional twists to the flux rope. Since the central part (formed during step 1a) has moved farther downtail, the newly added twists are trailing the central part (panel 1b). As the reconnection line spreads farther along the east-west direction, the flux rope forms a non-uniform, boomerang-like shape (panel 1c). This also implies that the flux rope is thicker at its center than at its flanks. For simplicity, only one flux rope field line is drawn. Right column: Open field line 1 (located in the southern lobe in the dawn sector) reconnects on the dawnside end of the flux rope (panel 2a). Now, this end of the flux rope is not connected to the ionosphere any more and moves into the tail. This leads to a tilting of the flux rope, since the duskside end is still connected to the ionosphere until open field line 2 (located in the northern lobe in the dusk sector) reconnects at this end (panel 2b). With no connection to the ionosphere, the flux rope now propagates into the tail (panel 2c). [From *Kiehas et al.*, 2012].

time, this might also lead to a tilting of the flux rope axis (see Figure 16.4, right column). While one end is already-reconnected to an open field line, the other end is still connected to the ionosphere. Hence, the dawnside end will be pulled downtail by the open field line, while the duskside end is still connected to Earth. This differential motion leads to a tilt in the flux rope. When the other end of the flux rope is finally reconnected to an open field line as well, the flux rope is no longer connected to the ionosphere and is free to move downtail, exhibiting a tilt [*Kiehas et al.*, 2012]. Certainly, any tilt should be seen with respect to the aberrated magnetosphere, rather than with respect to the GSM coordinate system.

16.2.3. Dimensions of Plasmoids and Flux Ropes

Investigations into the scale size of plasmoids and flux ropes during the pre-Cluster/THEMIS/ARTEMIS era relied on single-spacecraft measurements and hence need to be taken with care. One–point measurements of a 3D structure can only yield very rough estimates of its dimensions, if at all. The extent of flux ropes in X can easily be estimated with single-spacecraft measurements if the duration of the flux rope passage and the propagation speed are known. However, due to the circular cross section of flux ropes, the result will also depend on the impact parameter, i.e., whether the spacecraft crosses the

Table 16.1 Observations of plasmoids

Reference	No.	s/c	Observational Domain	Plasmoid Length	Comment
Hones et al. [1984]	1	ISEE–3	$-220R_E$	$75R_E$ to $110R_E$	
Scholer et al. [1984]	20	ISEE–3	$-80R_E$ to $-218R_E$	$50R_E$ to $100R_E$	
Richardson et al. [1987]	37	ISEE–3		$20R_E$ to $60R_E$	
Moldwin and Hughes [1992]	366	ISEE–3	$-50R_E$ to $-250R_E$	$16.7 \pm 13R_E$	Bipolar B_z or B_y
Slavin et al. [1993]	116	ISEE–3	$-50R_E$ to -250	$35R_E$	TCRs
Ieda et al. [1998]	824	Geotail	$-16R_E$ to $-210R_E$	$4R_E$ to $10R_E$	
Slavin et al. [2003a]	73	Geotail	$-14R_E$ to $-30R_E$	$2R_E$ to $5R_E$	
Slavin et al. [2005]	148	Cluster	$-11R_E$ to $-19R_E$	$4R_E$	TCRs

s/c, spacecraft.

flux rope at its center or the edge. Furthermore, constraints in the plasma instrumentation may lead to uncertainties in the velocity determination and hence to errors in spatial calculations, especially for fast, transient phenomena. Also, the flux rope boundaries cannot always be determined with great precision from the available instruments and definitions of these boundaries may differ.

Scholer et al. [1984] evaluated 20 plasmoid events observed by ISEE–3 between $80R_E$ and $218R_E$ downtail and found their length in X to be between $50R_E$ and $100R_E$ with a mean value of $80R_E$. However, *Scholer et al.* [1984] used energetic (≥35 keV) ion bursts as a selection criterion for the plasmoid events and therefore tended to select longer duration events due to the 64 and 128 s resolution of the ultralow–energy charge analyzer instrument [*Moldwin and Hughes*, 1992]. *Moldwin and Hughes* [1992] identified 366 plasmoids during ISEE–3's magnetotail crossings between ~ $50R_E$ and ~ $250R_E$ downtail. This large sample is due to their selection criterion, which included all events with a bipolar variations in B_z *or* B_y. They found the average length of the plasmoids to be 16.7 ± $13R_E$, with variations depending on the downtail location and therefore evolution (see section 16.2.4.).

From the same data set, *Slavin et al.* [1993] identified 116 TCRs and used indirect speed determinations and assumptions of the relation between the size of the TCRs and the underlying flux ropes to estimate the flux rope size in X to be about $35R_E$. The size of flux ropes in Y and Z was estimated by *Slavin et al.* [1993] using considerations of flux conservation and the spatial distribution of TCR observations, which yields flux rope dimensions in Y and Z of about $15R_E$. Hence, the estimated average flux rope size of $35R_E$ (length) × $15R_E$ (width) × $15R_E$ (height) is a very rough estimate, based on indirect observations.

Similarly, *Ieda et al.* [1998] also estimated the azimuthal size (in Y) of the observed plasmoids, based on the Y_{GSM} range over which plasmoids were observed and not on direct or indirect width measurements, to be $20R_E$ at $X_{GSM} = -30R_E$ and increased to ~40 R_E at $X_{GSM} \sim 80R_E$. The thickness of plasmoids was estimated via the same flux conservation considerations as in *Slavin et al.* [1993] and found to be about $10R_E$. The azimuthal size of flux ropes is of special interest, since it indicates if flux ropes might extend over the entire tail, allowing heat exchange with the magnetosheath. In summary, an overview of the spatial dimensions from the literature is shown in Table 16.1.

The estimate by *Slavin et al.* [1993] of $15R_E$ azimuthal extent would mean that flux ropes cover a significant portion of the dawn–dusk magnetotail width. With an azimuthal extent of $40R_E$, as argued by *Ieda et al.* [1998], flux ropes would cover the better part of the cross–tail width. However, as mentioned before, estimating the azimuthal size of flux ropes with single-spacecraft observations may not be very reliable. Multi spacecraft missions like Cluster, THEMIS, and ARTEMIS are therefore of considerable use in identifying the cross–tail extent of flux ropes.

Recently, *Kiehas et al.* [2013] investigated two events where flux rope signatures were only seen by one ARTEMIS probe, while the other probe, separated primarily in Y by ~$11R_E$, did not observe any flux rope or TCR signatures, even though it was located near the current sheet for the first event and near the plasma sheet for the second one. *Li et al.* [2014] conducted a statistical study of 74 ARTEMIS flux rope observations and found the typical azimuthal flux rope size based on two–spacecraft correlation to be about $6R_E$. Flux ropes with more than $9R_E$ azimuthal size are observed during times of enhanced geomagnetic activity (AE > 400 nT) [see also *Hietala et al.*, 2014]. This indicates that contrary to previous investigations, flux ropes may be spatially confined in the dusk–dawn direction. There are several possible reasons as to why ARTEMIS observations yield smaller azimuthal flux rope scales near lunar orbit than previous missions which ventured deeper into the tail: (a) uncertainties in the determination of the spatial extent of a 3D structure with single-spacecraft observations; (b) growth of flux ropes across the tail as they propagate further downtail; and (c) coalescence of several flux ropes, as suggested, e.g., by *Hietala et al.* [2014]. This requires further investigation.

Since magnetic reconnection is responsible for the formation of flux ropes, investigations of X line length provide additional information about the initial azimuthal extent of flux ropes just after their formation. *Nagai et al.* [2013] concluded the length of the X line to be probably less than $8R_E$ from the high occurrence of reconnection in the region $Y=0$ to $Y=8R_E$. The length of the X line can also be estimated from the spatial scale of reconnection flows, which were determined via multispacecraft studies by various authors to be $1R_E$ to $3R_E$ [*Sergeev et al.*, 1996], less than $3R_E$ [*Angelopoulos et al.*, 1997; *Kauristie et al.*, 2000], and $3R_E$ to $5R_E$ [*Nakamura et al.*, 2001].

16.2.4. Evolution of Plasmoids and Flux Ropes

Even though recent dual–spacecraft observations reveal smaller scales than previous single-spacecraft observations, distant and midtail flux ropes are still larger in size than near–Earth flux ropes. This might indicate that flux ropes grow during their propagation downtail or, alternatively, coalesce with other flux ropes. While *Richardson et al.* [1987] expected an adiabatic expansion of flux ropes during their downtail propagation, *Slavin et al.* [1989a] suggested a contraction in X and expansion in Z should occur due to the higher tension of the field lines near the equatorial plane. On the other hand, ISEE–3 observations of TCRs over the range from $50R_E$ to $250R_E$ downtail [*Slavin et al.*, 1993] suggest that flux ropes maintain approximately constant shape and volume as they propagate downtail since the TCR amplitude and duration were found to be independent from the location. *Moldwin and Hughes* [1992] performed a statistical survey of 366 plasmoid events detected by ISEE–3 and found that plasmoids beyond $100R_E$ downtail are relatively stable structures since their size, velocity, magnetic core strength, and B_z amplitude do not depend on distance beyond 100 R_E. However, the size of the observed plasmoids increases from $\sim 10 \pm 6R_E$ in the neartail to $\sim 16.5 \pm 15R_E$ in the midtail to $\sim 19 \pm 13 R_E$ in the far tail. These observations support the growth of plasmoids during their downtail propagation. *Ieda et al.* [1998] found the dimension of plasmoids to be about $4R_E$ near $X_{GSM} = -30R_E$ and about $10 R_E$ beyond $X_{GSM} = -90 R_E$, indicating a growth of plasmoids as well (Figure 16.5). The acceleration of plasmoids is interpreted as being due to the field line draping and dynamic pressure of faster plasma behind the plasmoid (i.e., in the postplasmoid plasma sheet, which contains fast plasma originating from lobe reconnection). Beyond 100 R_E downtail, a slight deceleration of the plasmoid can be seen.This might be due to the decrease of the field tension force with increasing distance to the reconnection site and the interaction of the plasmoid with ambient plasma, which may exceed the acceleration force [*Mukai et al.*, 1998]. The deceleration is also more pronounced in the flank region, which might be due to the interaction with cold, dense plasma. Recent comparative observations directly reveal plasmoid growth [*Li et al.*, 2013]: in this

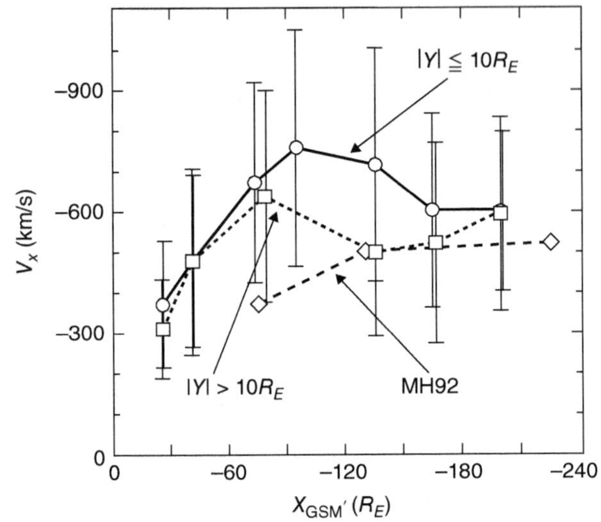

Figure 16.5 Dependence of tailward plasmoid velocities on downtail distance. The solid line represents the average values in the central region($|Y_{GSM}| \leq 10R_E$). The dotted line corresponds to the average values in the flank region ($|Y_{GSM}| > 10R_E$). Bars show the standard deviation. The results of *Moldwin and Hughes* [1992] are shown by a dashed line. [From *Ieda et al.* 1998.]

study the two ARTEMIS probes were separated mainly in X by $\sim 4 R_E$ near $60R_E$ downtail. Due to this alignment, the spacecraft could observe the same plasmoid from two different positions as it was moving downtail and consecutively swept over the two probes. The size of the plasmoid in X was determined by using the propagation velocity from timing analysis and was found to be $9R_E$, which is close to the statistical result obtained by *Ieda et al.* [1998]. As the plasmoid propagated downtail, it expanded in X from the P1 location to P2 at an expansion rate of $\sim 7R_E$/min.

16.2.5. Near–Earth Flux Ropes and Islands

So far, we have only discussed plasmoids and flux ropes moving tailward. However, earthward moving flux ropes have also been observed [*Elphic et al.*, 1986]. *Hughes and Sibeck* [1987] and *Moldwin and Hughes* [1994] interpreted these in the frame of *Nishida et al.*'s [1986] stagnant plasmoid concept. If reconnection at the NENL stops before the lobe flux gets reconnected, then the flux rope, which is still attached to the ionosphere, would convect earthward as a "protoplasmoid." Since these flux ropes were formed only during a short time, they are only a few R_E in size, or even less than $1R_E$. However, *Eastwood et al.* [2005] showed that reconnection can occur simultaneously at different points in the near–Earth current sheet. An earthward moving flux rope was bounded by two active X lines, validating the multiple X line theory on the mesoscale. Hence, the observed flux rope was the product of multiple reconnection sites at the NENL and not a convecting protoplasmoid. *Slavin et al.* [2003a] investigated near–Earth ($X > -30R_E$) flux ropes and found 48% of the observed 73 flux ropes moving earthward. By applying a constant-α flux rope model, 60% of the total flux ropes could be described as force free. The model-derived flux rope diameters were $1.4R_E$ and $4.4R_E$ for earthward and tailward moving flux ropes, respectively. The small scale of near–Earth flux ropes was also confirmed by (a) observations of TCRs with a mean scale size of 4.3 R_E at $X \sim -11R_E$ to $-19R_E$ [*Slavin et al.*, 2005], (b) applying a magnetic field rotation analysis (~1 R_E [*Shen et al.*, 2007]), and (c) the Grad-Shafranov reconstruction technique (~$1R_E$ [Möstl et al., 2009]). *Walsh et al.* [2009] reported simultaneous observations of two flux ropes more than $6R_E$ apart, indicating the activity of multiple X lines also during this event.

Slavin et al. [2003b] determined the current density inside an earthward moving flux rope observed with Cluster by both force–free modeling and the curlometer technique. While it was found that the flux rope is nearly force free, the perpendicular current dominated the parallel current in the leading part of the flux rope, indicating a non-force-free leading part. The authors argued that this deviation may be due to the interaction of the flux rope with the surrounding medium. In particular, pressure gradients resulting from the collision of the flux rope with the inner magnetosphere might compress the flux rope's leading part and result in a deviation from the ideal force–free state.

Finally, we note that secondary islands have now also been observed entrained in reconnection jets in the ion diffusion region [*Eastwood et al.*, 2007]. This event was encountered by the four Cluster satellites and was small enough such that only one satellite encountered the center of the island (the other satellites observed small-scale TCR signatures). This was confirmed by Grad-Shafranov modeling of the four-satellite data, which was used to derive a 2D magnetic field map [*Teh et al.*, 2011]. In this case the secondary island had a flux rope morphology and a strong core field, even though the overall reconnection geometry appeared to be essentially antiparallel. Secondary islands have also been observed during guide field reconnection [*Wang et al.*, 2012] and have been theorized to play an important role in the production of energetic particles [*Drake et al.*, 2006b]. Observations have shown enhancement of energetic electrons in the vicinity of secondary islands [*Chen et al.*, 2008; *Wang et al.*, 2010], possibly because of the strong electric fields which have been observed in their interior [*Eastwood et al.*, 2007] and modeled in simulations [*Zhou et al.*, 2012; *Chen et al.*, 2012].

16.3. OBSERVATIONS AT MARS

16.3.1. Introduction

Mars is unique among the planets in the solar system because while it has no global dipole field, there are strong regions of crustal magnetic anomalies which affect its interaction with the solar wind [*Acuña et al.*, 1998; *Connerney et al.*, 2001]. The existence of crustal fields indicates that Mars once had a global field and therefore a magnetosphere; analysis of magnetization in large impact basins constrains the age at which the dipole ceased to operate to ~4 billion years ago [*Acuña et al.*, 1999]. The history of the Mars magnetic field and magnetosphere is intimately related to conditions on the surface, and in particular evidence suggesting that earlier in its history Mars had a considerably thicker atmosphere and a wetter surface [e.g., *Fassett and Head*, 2011]. It is thought that the cessation of the dynamo and consequent disappearance of the global magnetosphere would have allowed solar wind sputtering to significantly increase the atmospheric loss rate [*Jakosky and Phillips*, 2001]. As such, understanding the interaction of the solar wind with Mars and its crustal fields is crucial to deciphering the history of atmospheric mass and volatile loss.

Draping of the IMF over the ionosphere of Mars leads to the formation of a magnetotail [*Yeroshenko et al.*, 1990; *Nagy et al.*, 2004], as shown in Figure 16.6. Shocked solar wind protons do not penetrate below the magnetic pileup boundary (MPB), which is somewhat analogous to the magnetopause at Earth. However, the IMF and solar wind electrons are found in the magnetic pileup region below the MPB. The photoelectron boundary (PEB) bounds this region below, and the ionosphere is unmagnetized below this boundary. Ionospheric flux ropes are created when magnetized plasma crosses the ionopause and penetrates into the unmagnetized ionosphere [*Cloutier et al.*, 1999; *Vignes et al.*, 2004; *Briggs et al.*, 2011]. However, since we are primarily concerned with magnetotail phenomena, we shall not consider them further here.

A significant amount of atmospheric material is lost through the magnetotail in a comet-like interaction, where low-energy ions in the dayside ionosphere are transported to the nightside and then away from the planet [*Lundin et al.*, 2008]. This suggests a relatively steady-state interaction (modulated only by solar forcing). However, the crustal fields exert pressure that locally affects the pressure balance, even exceeding the ionospheric thermal pressure, and they can directly interact with the solar wind plasma [*Brain*, 2006]. Evidence for other loss processes came with the discovery of aurora at Mars [*Bertaux et al.*, 2005]. The auroral emission was found to be associated with the crustal magnetic fields

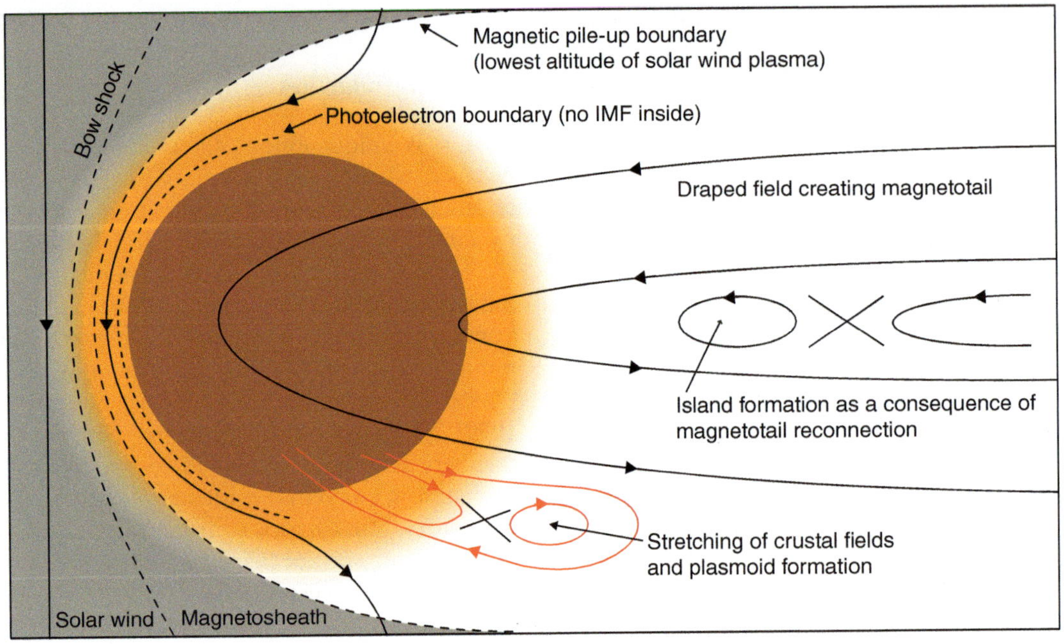

Figure 16.6 Morphology of Mars interaction with the solar wind. Draped IMF lines form a two-lobed magnetotail, and crustal magnetic fields are strong enough to directly interact with the shocked solar wind plasma.

and the cusp regions that form between them. In situ measurements showed that "auroral" electron distributions were observed at the edges of crustal field regions in the southern hemisphere [*Brain et al.*, 2006]. Prior to these observations *Krymskii et al.* [2002] had proposed that reconnection could occur between the crustal fields and the IMF, thus enabling solar wind plasma to precipitate along open field lines directly into the ionosphere [*Krymskii et al.*, 2002]. Indirect evidence for the presence of reconnection came from maps of open and closed field lines under different IMF orientation which showed that the location and distribution of open fields did depend on the IMF orientation, although this could be due to reconnection or large-scale motion and distortion of the closed field regions [*Brain*, 2006]. Recent observations of flux ropes at Mars now appear to show that reconnection occurs both in the vicinity of crustal field regions and in the magnetotail current sheet; evidence for this is presented in the next two sections.

16.3.2. Flux Ropes Associated with Crustal Fields

Flux ropes directly associated with crustal fields were first detected using data from the Mars Global Surveyor (MGS) magnetometer/electron reflectometer (MAG/ER) experiment [*Brain et al.*, 2010], as shown in Figures 16.7a and 16.7b. Between 04:51 and 04:59 UT on 9 March 2002, a substantial increase in magnetic field strength was observed (~180 nT). The red line shows the predicted crustal magnetic field strength [*Cain et al.*, 2013]; crustal fields would not be expected to be observed at this time (in fact, crustal fields are observed shortly after). There is a bipolar variation in B_z and a unipolar variation in B_y centered on the increase in field strength which is well modeled as a force-free flux rope. Figure 16.7c shows the orbit of MGS and Figures 16.7d and 16.7e illustrate possible fly-through geometries. The observations are consistent with the sampling of stretched crustal magnetic field lines, but it is not clear as to whether the flux rope has detached. However, stretching of field lines will lead to a thin current sheet where reconnection would be likely to occur, and in other similar events evidence for current sheets separating flux ropes from the crustal fields has been identified.

Of key importance for understanding ionospheric loss is the plasma content of the flux rope. However, MGS can only measure electrons via the MAG/ER instrument. The electron measurements are consistent with the plasma being of ionospheric origin before, during, and after the encounter, but additional plasma is present in the center of the flux rope. This plasma is more isotropic and thought to be magnetically disconnected from the surrounding region. Although estimates of the plasma density are difficult to make because MAG/ER only measures electron energies above 10 eV, the density could be of the order of 5–10 cm^{-3}. Similarly, the volume is uncertain because MAG/ER cannot measure electron velocity, so the relative motion of satellite and flux rope is unknown. If one assumes the plasma flow is the expected 5–15 km/s^{-1} [*Fränz et al.*, 2010] and that the flux rope is entrained in this flow, then the plasma content of the flux

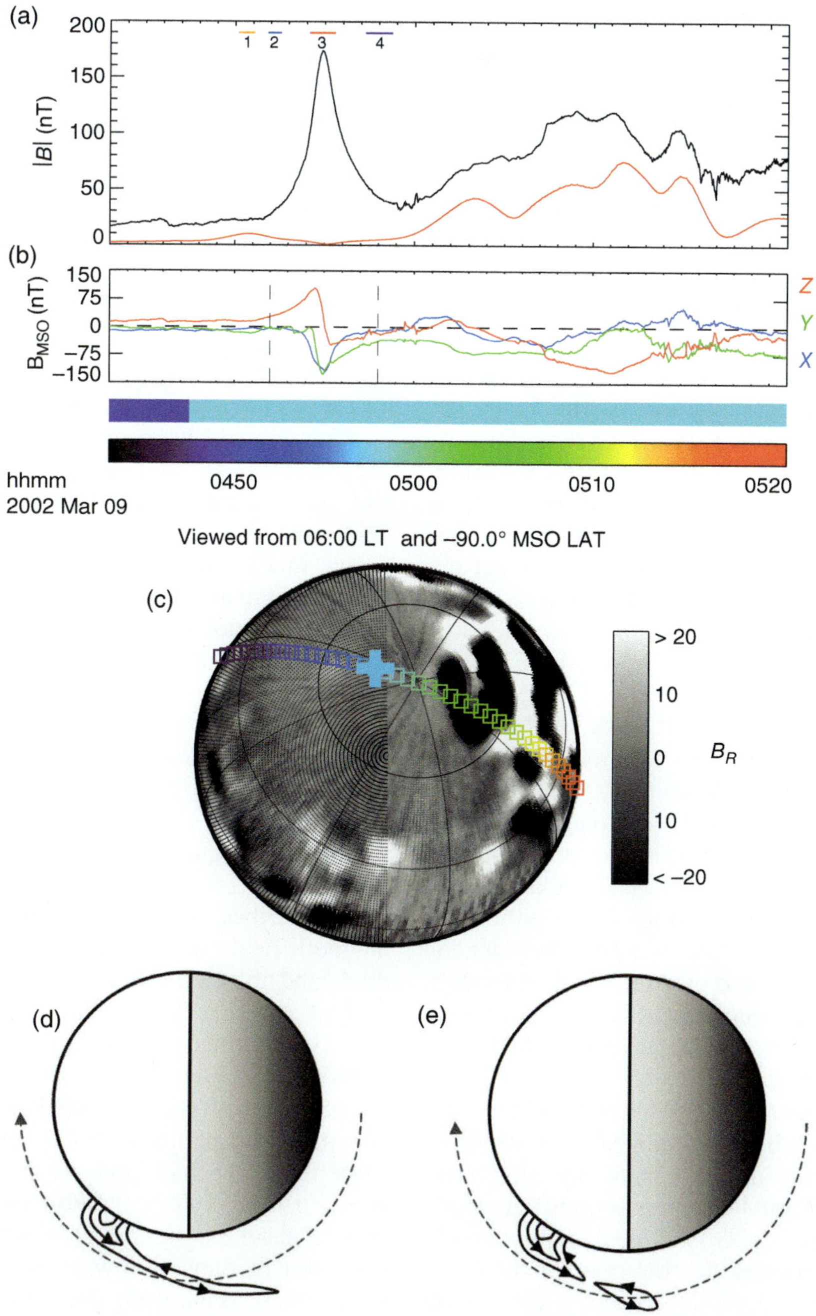

Figure 16.7 Observations of a magnetic flux rope generated by the interaction of the solar wind with Mars crustal magnetic fields. (a, b) Magnetic field strength (red: predicted crustal field strength) and components. (c) Orbit of MGS from a viewpoint looking toward the south pole with the Sun to the right. MGS was moving from the nightside over the terminator, and so the crustal fields (represented by the black and white patterned region) are sunward of the observed enhancement in magnetic field strength. (d, e) Possible MGS trajectories through the flux rope. Adapted from *Brain et al.* [2010].

rope is 7.3×10^{26} ions (the volume being derived from the duration and geometry of the encounter), giving a crude estimate that this could contribute 5–10% of the total ion loss [*Brain et al.*, 2010]. In their study, *Brain et al.*, [2010] found 190 other examples in four weeks (340 orbits) of data. *Briggs et al.* [2011] further surveyed the MGS data for flux ropes, examining four separate weeks (~84 orbits) of data evenly spaced over one year, in order to characterize secular variation in occurrence rates and properties. They found flux ropes of all types (ionospheric, magnetotail, and plasmoid) were more likely to form during intervals of low solar wind pressure.

The mapping orbit of MGS means that it returns to approximately the same location over the surface every 88 orbits (~7 Mars days). *Beharrell and Wild* [2012] examined the context of the 9 March 2002 flux rope by analyzing the magnetic field observed by MGS in the weeks before and after when similar viewing geometries were encountered. They found that flux rope signatures were also observed at these times, although the 9 March signature was by far the strongest. Applying essentially the same force-free flux rope fitting procedure to observations made two 88 orbit cycles earlier, they showed that the different signature was due to a different grazing fly-through. Given the repeated nature of the signature, they proposed that the flux ropes, including the one observed by *Brain et al.* [2010], are a stationary feature, with the crustal field being stretched and only reconnecting in the middle of the crustal field arcade but remaining connected to the planet at each end, resulting in the formation of a static flux rope signature sitting downstream of the crustal field region.

A follow-up study using Mars Express (MEX) and MGS was performed by *Morgan et al.* [2011]. In the absence of a magnetometer onboard MEX, the MARSIS (Mars Advanced Radar for Subsurface and Ionosphere Sounding) experiment on MEX [*Jordan et al.*, 1999] can be used to derive the magnetic field strength near the spacecraft. MARSIS is an ionospheric radar sounder and the sounding pulse triggers electron plasma and cyclotron waves which can be used to calculate the local electron density and magnetic field strength. The magnetic field strength can be compared with crustal field models, and localized, transient enhancements of the field were found outside of time intervals where crustal fields would be expected [*Morgan et al.*, 2011].

In particular, *Morgan et al.* [2011] found two flux ropes that were observed both by MEX and MGS. Both satellites observed an increase in the magnetic field suggestive of a localized structure, and both observations were made downstream of the strong crustal field region in the southern hemisphere, as was also the case for the *Brain et al.* [2010] study. Using minimum variance analysis on the MGS magnetic field data, it was found that the axis of the flux rope connected the locations where the peak fields were observed by MGS and MEX. ASPERA-3 plasma data from MEX showed that the electron plasma was wholly ionospheric. Interestingly, the observations at MEX and MGS were separated in time by ~30 min. It was concluded that the flux rope had therefore existed in quasi-steady-state for at least this time, suggesting that it had not yet detached from the crustal fields.

The considerable rotational asymmetry of the planetary crustal fields means that it is difficult to build up a complete global picture of the Mars overall solar wind–magnetosphere interaction. As such, global simulations are very important. *Ma et al.* [2002] combined a multi fluid MHD simulation with a crustal field model, finding that the crustal fields distorted the ionopause when on the dayside. They also observed a flux rope forming on the nightside of the planet. More recent simulations by *Harnett* [2009] concentrated on the dynamics of flux rope formation using a 3D multifluid code again with a crustal field model. The grid size was sufficiently small so that thin current sheets could form, and in the course of the simulation, several flux ropes were observed to sequentially develop near the planet, subsequently being transported into the magnetotail. Figures 16.8e and 16.8f shows the magnetic field in the magnetotail as viewed from the dusk terminator (field lines are colored only to aid visibility). Figures 16.8a – 16. 8d show the simulated oxygen, hydrogen, and electron energy spectra together with the magnetic field strength and current density along the purple line shown in Figures 16.8e and 16.8f. The flux ropes traveled downtail at speeds of 20–40 km/s, with core fields of 50–100 nT, and were a few thousand kilometers in size. Although the data from MGS and MEX have provided important insight, more comprehensive measurements are now required to test these predictions and fully understand the properties and importance of such flux ropes in the Mars magnetotail.

16.3.3. Reconnection in the Magnetotail Current Sheet

As described above, the draping of the interplanetary magnetic field over the planet leads to the formation of a dual-lobe magnetotail with an embedded magnetotail current sheet. It had previously been proposed that at comets bulk removal could be enhanced by reconnection across the tail current sheet [*Russell et al.*, 1986]; however, the geometry of magnetic field draping implies that both cometary and planetary tail current sheets are expected to be relatively thick and stable to reconnection [*Slavin et al.*, 1986a, 1986b, 1989b]. In fact, the magnetotail current sheet can be thin [*Halekas et al.*, 2006], and evidence for reconnection was first presented by *Eastwood et al.* [2008], again using data from MGS. A key element of the data was a magnetic island observed just prior to the current sheet crossing. When compared to a particle-in-cell simuation of magnetic reconnection, the data were consistent with an encounter of the ion diffusion region, and the island was identified as a secondary island. The observations were again made just downstream of crustal field regions, and so it was difficult to establish the exact relationship between the observed magnetotail current sheet reconnection and the crustal magnetic fields. *Halekas et al.* [2009] found that 0.25% of MGS thin magnetotail current sheet crossings contained magnetic field signatures consistent with an ion diffusion region encounter, and several examples of associated flux ropes

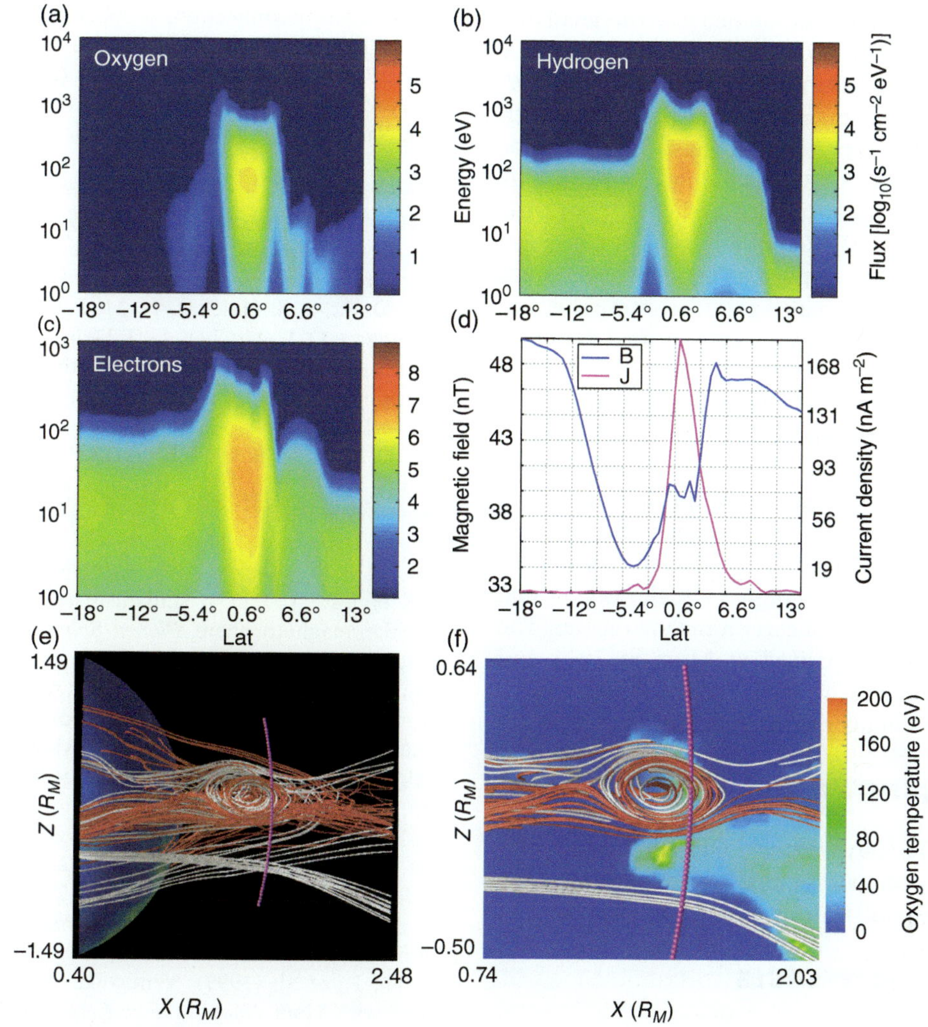

Figure 16.8 Results from a global simulation of Mars interaction with the solar wind leading to the formation of flux ropes. (a–d) Simulated oxygen, hydrogen, and electron energy spectra together with the magnetic field strength and current density along the purple line shown in panels (e) and (f). (e, f) Magnetotail magnetic field lines (color used to highlight connectivity only). [From *Harnett*, 2009].

were identified. In their study, [*Briggs et al.* 2011] concluded that flux ropes were a relatively common feature of the Mars magnetotail current sheet as observed by MGS. Finally, we also note the recent detection of a plasmoid in the Venus magnetotail by Venus Express [*Zhang et al.*, 2012]; why such features were not observed earlier by Pioneer Venus remains unclear.

Chains of flux ropes have also been observed in the magnetotail current sheet by MGS [*Eastwood et al.*, 2012a]; for example, between 14:45 and 14:55 UT on 14 January 2000, MGS crossed the magnetotail current sheet and observed a series of three flux rope signatures. Each flux rope was moving tailward, and a very good fit to a force-free flux rope model was obtained for the first and second structures. If the flux ropes were convecting with speeds given by the simulation of *Harnett* [2009], then their size was of the order of the altitude of MGS. Therefore, it would appear that they were convected around the flank of the planet from the vicinity of the crustal field region. Furthermore, it would not be possible to fit the three flux ropes into the region between the satellite and the crustal fields, and so this suggested that they were produced sequentially. As mentioned above, more comprehensive measurements are also required to fully determine the properties and importance of these flux ropes and their association with the crustal fields.

16.4. CONCLUSIONS

Magnetic reconnection generates a variety of structures, which is very important for understanding the circulation of plasma and magnetic flux through planetary magnetotails.

At Earth in particular this can be studied in great detail due to the availability of several multipoint satellite missions. A key result at Earth is that the presence of even a small B_y naturally leads to the formation of 3D flux ropes with helical magnetic field lines rather than quasi–2D plasmoids with closed magnetic loops, as depicted in Figure 16.2. A Plasmoid can thus be seen as an idealized structure that is formed in the absence of B_y. Recent advances in computer simulations, observational analysis techniques, and multi spacecraft missions have revealed that the internal structure of flux ropes is somewhat more complicated than previously thought, and this structure can change as the flux rope propagates through the magnetotail. As such, a magnetotail constellation mission would provide considerable further insight into how magnetotails work. At Mars, while significant progress has been made, this progress has been constrained by the lack of simultaneous plasma and magnetic field measurements. MGS, while measuring the magnetic field structure, could not measure the ion plasma, and conversely MEX does not carry a magnetometer. The MAVEN mission, which arrived at Mars in 2014 and carrying a comprehensive space plasma instrument package, is likely to stimulate further progress.

ACKNOWLEDGMENTS

JPE holds an STFC Advanced Fellowship, grant ST/G00725X/1. This work was supported by the Austrian Science Fund (FWF): J3041-N16.

REFERENCES

Acuña, M. H., et al. (1998), Magnetic field and plasma observations at Mars: Initial results of the Mars Global Surveyor mission, *Science*, *279*, 1676–1680.

Acuña, M. H., et al. (1999), Global distribution of crustal magnetization discovered by the Mars Global Surveyor MAG/ER experiment, *Science*, *284*, 790–793.

Angelopoulos, V., et al. (1997), Magnetotail flow bursts: Association to global magnetospheric circulation, relationship to ionospheric activity and direct evidence for localization, *Geophys. Res. Lett.*, *24*, 2271–2274.

Baker, D. N., T. I. Pulkkinen, V. Angelopoulos, W. Baumjohann, and R. L. McPherron (1996), Neutral line model of substorms: Past results and present view, *J. Geophys. Res.*, *101*, 12,975–13,010.

Beharrell, M. J., and J. A. Wild (2012), Stationary flux ropes at the southern terminator of Mars, *J. Geophys. Res.*, *117*, A12212, doi:10.1029/2012JA017738.

Bertaux, J.-L., F. Leblanc, O. Witasse, E. Quemerais, J. Lilensten, S. A. Stern, B. Sandel, and O. Korablev (2005), Discovery of an aurora on Mars, *Nature*, *435*, 790–794.

Birn, J., M. Hesse, and K. Schindler (1989), Filamentary structure of a three-dimensional plasmoid, *J. Geophys. Res.*, *94*, 241–251.

Birn, J., R. Nakamura, E. V. Panov, and M. Hesse (2011), Bursty bulk flows and dipolarization in MHD simulations of magnetotail reconnection, *J. Geophys. Res.*, *116*, A01210, doi:10.1029/2010JA016083.

Birn, J., R. Nakamura, and M. Hesse (2013), On the propagation of blobs in the magnetotail: Mhd simulation, *J. Geophys. Res. Space Phys.*, *118*, 5497–5505, doi:10.1002/jgra.50521.

Borg, A. L., M. G. G. T. Taylor, and J. P. Eastwood (2012), Observations of flux ropes during magnetic reconnection in the Earth's magnetotail, *Ann. Geophys.*, *30*, 761–773.

Brain, D. A. (2006), Mars global surveyor measurements of the martian solar wind interaction, *Space Sci. Rev.*, *126*, 77–112.

Brain, D. A., J. S. Halekas, L. M. Peticolas, R. P. Lin, J. G. Luhmann, D. L. Mitchell, G. T. Delory, S. W. Bougher, M. H. Acuña, and H. Reme (2006), On the origin of aurorae on Mars, *Geophys. Res. Lett.*, *33*, L01201, doi:10.1029/2005GL024782.

Brain, D. A., A. H. Baker, J. Briggs, J. P. Eastwood, J. S. Halekas, and T.-D. Phan (2010), Episodic detachment of martian crustal magnetic fields leading to atmospheric plasma escape, *Geophys. Res. Lett.*, *37*, L14108, doi:10.1029/2010GL043916.

Briggs, J. A., D. A. Brain, M. L. Cartwright, J. P. Eastwood, and J. S. Halekas (2011), A statistical study of flux ropes in the martian magnetosphere, *Planet. Space Sci.*, *59*, 1498–1505.

Burlaga, L. F. (1988), Magnetic clouds and force-free fields with constant alpha, *J. Geophys. Res.*, *93*(A7), 7217–7224.

Cain, J. C., B. B. Ferguson, and D. Mozzoni (2003), An $n = 90$ internal potential function of the Martian crustal magnetic field, *J. Geophys. Res.*, *108*(E2), 5008, doi:10.1029/2000JE001487.

Chen, L.-J., et al. (2008), Observation of energetic electrons within magnetic islands, *Nature Phys.*, *4*, 19–23.

Chen, L. J., W. Daughton, A. Bhattacharjee, R. B. Torbert, V. Roytershteyn, and N. Bessho (2012), In-plane electric fields in magnetic islands during collisionless magnetic reconnection, *Phys. Plasmas*, *19*, 112902.

Cloutier, P., et al. (1999), Venus-like interaction of the solar wind with Mars, *Geophys. Res. Lett.*, *26*(17), 2685–2688.

Connerney, J. E. P., M. H. Acuña, P. J. Wasilewski, G. Kletetschka, N. F. Ness, H. Reme, R. P. Lin, and D. L. Mitchell (2001), The global magnetic field of Mars and implications for crustal evolution, *Geophys. Res. Lett.*, *28*(21), 4015–4018.

Coppi, B., G. Laval, and R. Pellat (1966), Dynamics of the geomagnetic tail, *Phys. Rev. Lett.*, *16*(26), 1207–1210.

Cowley, S. W. H. (1981), Magnetospheric asymmetries associated with the *y*-component of the IMF, *Planet. Space Sci.*, *29*, 79–96, doi:10.1016/0032-0633(81)90141-0.

Drake, J. F., M. Swisdak, K. M. Schoeffler, B. N. Rogers, and S. Kobayashi (2006a), Formation of secondary islands during magnetic reconnection, *Geophys. Res. Lett.*, *33*, L13105, doi:10.1029/2006GL025957.

Drake, J. F., M. Swisdak, H. Che, and M. A. Shay (2006b), Electron acceleration from contracting magnetic islands during reconnection, *Nature*, *443*, 553–556.

Eastwood, J. P., D. G. Sibeck, J. A. Slavin, M. L. Goldstein, B. Lavraud, M. Sitnov, S. Imber, A. Balogh, E. A. Lucek, and I. Dandouras (2005), Observations of multiple X-line structure in the Earth's magnetotail current sheet: A Cluster case study, *Geophys. Res. Lett.*, *32*, L11105, doi:10.1029/2005GL022509.

Eastwood, J. P., T.-D. Phan, F. S. Mozer, M. A. Shay, M. Fujimoto, A. Retinò, M. Hesse, A. Balogh, E. A. Lucek, and I. Dandouras (2007), Multi-point observations of the Hall electromagnetic field and secondary island formation during magnetic reconnection, *J. Geophys. Res.*, *112*, A06235, doi:10.1029/2006JA012,158.

Eastwood, J. P., D. A. Brain, J. S. Halekas, J. F. Drake, T. Phan, M. Øieroset, D. G. Mitchell, R. P. Lin, and M. H. Acuña (2008), Evidence for collisionless reconnection at Mars, *Geophys. Res. Lett.*, *35*, L02106, doi:10.1029/2007GL032289.

Eastwood, J. P., J. J. H. Videira, D. A. Brain, and J. S. Halekas (2012a), A chain of magnetic flux ropes in the magnetotail of Mars, *Geophys. Res. Lett.*, *39*, L03104, doi:10.1029/2011GL050444.

Eastwood, J. P., T. D. Phan, R. C. Fear, D. G. Sibeck, V. Angelopoulos, M. Øieroset, and M. A. Shay (2012b), Survival of flux transfer event (fte) flux ropes far along the tail magnetopause, *J. Geophys. Res.*, *117*, A08222, doi:10.1029/2012JA017722.

Elphic, R. C., C. A. Cattell, K. Takahashi, S. J. Bame, and C. T. Russell (1986), ISEE-1 and 2 observations of magnetic flux ropes in the magnetotail: FTE's in the plasma sheet? *Geophys. Res. Lett.*, *13*(7), 648–651.

Fairfield, D. H. (1979), On the average configuraton of the geomagnetic tail, *J. Geophys. Res.*, *84*, 1950–1958, doi:10.1029/JA084iA05p01950.

Farr, N. L., D. N. Baker, and M. Wiltberger (2008), Complexities of a 3–D plasmoid flux rope as shown by an MHD simulation, *J. Geophys. Res.*, *113*, A12202, doi:10.1029/2008JA013,328.

Fassett, C. I., and J. W. Head (2011), Sequence and timing of conditions on early Mars, *Icarus*, *211*, 1204–1214.

Frank, L. A., W. R. Paterson, S. Kokubun, T. Yamamoto, and R. P. Lepping (1995), Direct detection of the current in a magnetotail flux rope, *Geophys. Res. Lett.*, *22*, 2697–2700.

Fränz, M., E. Dubinin, E. Nielsen, J. Woch, S. Barabash, R. Lundin, and A. Fedorov (2010), Transterminator ion flow in the martian ionosphere, *Planet. Space Sci.*, *58*, 1442–1454, doi:10.1016/j.pss.2010.06.009.

Halekas, J. S., D. A. Brain, R. J. Lillis, M. O. Fillingim, D. L. Mitchell, and R. P. Lin (2006), Current sheets at low altitudes in the martian magnetotail, *Geophys. Res. Lett.*, *33*, L13101, doi:10.1029/2006GL026229.

Halekas, J. S., J. P. Eastwood, D. A. Brain, T.-D. Phan, M. Øieroset, and R. P. Lin (2009), In situ observations of reconnection hall magnetic fields at Mars: Evidence for ion diffusion region encounters, *J. Geophys. Res.*, *114*, A11204, doi:10.1029/2009JA014544.

Harnett, E. M. (2009), High-resolution multifluid simulations of flux ropes in the martian magnetosphere, *J. Geophys. Res.*, *114*, A01208, doi:10.1029/2008JA013648.

Hesse, M., and M. G. Kivelson (1998), The formation and structure of flux ropes in the magnetotail, in *New Perspectives on the Earth's Magnetotail,* Geophys. Monogr. Ser., vol. 105, edited by D. N. B. A. Nishida and S. W. H. Cowley, AGU, Washington, D. C.

Hesse, M., J. Birn, M. M. Kuznetsova, and J. Dreher (1996), A simple model of core field generation during plasmoid evolution, *J. Geophys. Res.*, *101*, 10,797–10,804.

Hietala, H., J. P. Eastwood, and A. Isavnin (2014), Sequentially released tilted flux ropes in the earth's magnetotail, *Plasma Phys. Control. Fus.*, *56*, 064011.

Hones, E. W. (1976), The magnetotail: Its generation, and dissipation, in *Physics of Solar Planetary Environments,* vol. II, edited by D. J. Williams, p. 558,571, AGU, Washington, D. C.

Hones, E. W. (1977), Substorm processes in the magnetotail: comments on "On hot tenuous plasmas, fireballs, and boundary layers in the Earth's magnetotail" by L. A. Frank, K. L. Ackerson and R. P. Lepping, *J. Geophys. Res.*, *82*(35), 5633–5640.

Hones, E. W., J. Birn, D. N. Baker, S. J. Bame, W. C. Feldman, D. J. McComas, R. D. Zwickl, J. A. Slavin, E. J. Smith, and B. T. Tsurutani (1984), Detailed examination of a plasmoid in the distant magnetotail with ISEE 3, *Geophys. Res. Lett.*, *11*, 1046—1049, doi:10.1029/GL011i010p01046.

Honkonen, I., M. Palmroth, T. Pulkkinen, P. Janhunen, and A. Aikio (2011), On large plasmoid formation in a global magnetohydrodynamic simulation, *Ann. Geophys.*, *29*, 167–179, doi:10.1029/2008JA013328.

Hsu, T. S., and R. L. McPherron (2002), An evaluation of the statistical significance of the association between northward turnings of the interplanetary magnetic field and substorm expansion onsets, *J. Geophys. Res.*, *107*, 1398, doi:10.1029/2000JA000125.

Hughes, W. J., and D. G. Sibeck (1987), On the 3-dimensional structure of plasmoids, *Geophys. Res. Lett.*, *14*, 636–639.

Ieda, A., S. Machida, T. Mukai, Y. Saito, T. Yamamoto, A. Nishida, T. Terasawa, and S. Kokubun (1998), Statistical analysis of the plasmoid evolution with Geotail observations, *J. Geophys. Res.*, *103*, 4453–4466.

Imber, S. M., J. A. Slavin, H. U. Auster, and V. Angelopoulos (2011), A THEMIS survey of flux ropes and traveling compression regions: Location of the near-Earth reconnection site during solar minimum, *J. Geophys. Res.*, *116*, A02201, doi:10.1029/2010JA016026.

Jakosky, B. M., and R. J. Phillips (2001), Mars' volatile and climate history, *Nature*, *412*, 237–244.

Jordan, R., et al. (1999), The Mars Express MARSIS sounder instrument, *Planet. Space Sci.*, *57*, 1975–1986.

Karimabadi, H., W. Daughton, and J. Scudder (2007), Multiscale structure of the electron diffusion region, *Geophys. Res. Lett.*, *34*, L13104, doi:10.1029/2007GL030306.

Kauristie, K., V. A. Sergeev, M. Kubyshkina, T. I. Pulkkinen, V. Angelopoulos, T. Phan, R. P. Lin, and J. A. Slavin (2000), Ionospheric current signatures of transient plasma sheet flows, *J. Geophys. Res.*, *105*, 10,677–25,904, doi:10.1029/1999JA900487.

Kiehas, S. A., V. Angelopoulos, A. Runov, M. B. Moldwin, and C. Möstl (2012), On the formation of tilted flux ropes in the Earth's magnetotail observed with ARTEMIS, *J. Geophys. Res.*, *117*, A05231, doi:10.1029/2011JA017377.

Kiehas, S. A., V. Angelopoulos, A. Runov, and S.-S. Li (2013), On the azimuthal size of flux ropes near lunar orbit, *J. Geophys. Res.*, *118*, doi:10.1002/jgra.50425.

Kivelson, M. G., K. K. Khurana, R. J. Walker, L. Kepko, and D. Xu (1996), Flux ropes, interhemispheric conjugacy, and magnetospheric current closure, *J. Geophys. Res.*, *101*(A12), 27,341–27,350, doi:10.1029/96JA02220.

Krymskii, A. M., T. K. Breus, N. F. Ness, M. H. Acuña, J. E. P. Connerney, D. H. Crider, D. L. Mitchell, and S. J. Bauer (2002), Structure of the magnetic field fluxes connected with crustal magnetization and topside ionosphere at Mars, *J. Geophys. Res.*, *107*(A9), 1245, doi:10.1029/2001JA000239.

Lepping, R. P., J. A. Jones, and L. F. Burlaga (1990), Magnetic field structure of interplanetary magnetic clouds at 1 au, *J. Geophys. Res.*, *95*(8), 11,957–11,965.

Lepping, R. P., D. H. Fairfield, J. Jones, L. A. Frank, W. R. Paterson, S. Kokubun, and T. Yamamoto (1995), Cross-tail magnetic flux ropes as observed by the GEOTAIL spacecraft, *Geophys. Res. Lett.*, *22*, 1193–1196.

Li, S.-S., V. Angelopoulos, A. Runov, S. A. Kiehas, and X.-Z. Zhou (2013), Plasmoid growth and expulsion revealed by two-point artemis observations, *J. Geophys. Res.*, *118*, 2133–2144, doi:10.1002/jgra.50105.

Li, S.-S., V. Angelopoulos, A. Runov, and S. A. Kiehas (2014), On the azimuthal scale size of mid-tail plasmoids from two-point ARTEMIS observations at the earth-moon lagrange points, *J. Geophys. Res. Space Physics*, *119*, 1781–1796, doi:10.1002/2013JA019292.

Lui, A. T. Y. (1984), Characteristics of the cross-tail current in the Earth's magnetotail, in *Magnetospheric Currents,* AGU Geophys. Monogr. 28, edited by T. Potemra, pp. 158–170, AGU, Washington, D. C.

Lundin, R., S. Barabash, M. Holmström, H. Nilsson, M. Yamauchi, M. Fraenz, and E. Dubinin (2008), A comet-like escape of ionospheric plasma from Mars, *Geophys. Res. Lett.*, *35*, L18203, doi:10.1029/2008GL034811.

Lundquist, S. (1950), Magnetohydrostatic fields, *Ark. Fys.*, *2*, 361.

Ma, Y., A. F. Nagy, K. C. Hansen, D. L. DeZeeuw, and T. I. Gombosi (2002), Three-dimensional multispecies mhd studies of the solar wind interaction with Mars in the presence of crustal fields, *J. Geophys. Res.*, *107*(A10), 1282, doi:10.1029/2002JA009293.

Machida, S., T. Mukai, Y. Saito, T. Obara, T. Yamamoto, A. Nishida, M. Hirahara, T. Teresawa, and S. Kokubun (1994), GEOTAIL low energy particle and magnetic field observations of a plasmoid at $x_{GSM} = -142r_e$, *Geophys. Res. Lett.*, *21*, 2995–2998, doi:10.1029/94GL02241.

Moldwin, M. B., and W. J. Hughes (1991), Plasmoids as magnetic flux ropes, *J. Geophys. Res.*, *96*, 14,051–14,064.

Moldwin, M. B., and W. J. Hughes (1992), On the formation and evolution of plasmoids: A survey of ISEE 3 Geotail data, *J. Geophys. Res.*, *97*, 19,259–19,282, doi:10.1029/92JA01598.

Moldwin, M. B., and W. J. Hughes (1994), Observations of Earthward and tailward propagating flux rope plasmoids: Expanding the plasmoid model of geomagnetic substorms, *J. Geophys. Res.*, *99*, 183–198.

Morgan, D. D., D. A. Gurnett, F. Akalin, D. A. Brain, J. S. Leisner, F. Duru, R. A. Frahm, and J. D. Winningham (2011), Dual-spacecraft observation of large-scale magnetic flux ropes in the martian ionosphere, *J. Geophys. Res.*, *116*, A02319, doi:10.1029/2010JA016134.

Möstl, C., C. Farrugia, H. K. Biernat, S. A. Kiehas, R. Nakamura, V. V. Ivanova, and Y. Khotyaintsev (2009), The structure of an earthward propagating magnetic flux rope early in its evolution: Comparison of methods, *Ann. Geophys.*, *27*, 2215–2224, doi:10.5194/angeo-27-2215-2009.

Mukai, T., T. Yamamoto, and S. Machida (1998), Dynamics and kinetic properties of plasmoids and flux ropes: GEOTAIL observations, in New *Perspectives on the Earth's Magnetotail*, edited by A. Nishida, D. N. Baker, and S. W. H. Cowley, AGU, Washington, D. C., doi:10.1029/GM105p0117.

Nagai, T., I. Shinohara, S. Zenitani, R. Nakamura, T. K. M. Nakamura, M. Fujimoto, Y. Saito, and T. Mukai (2013), Three–dimensional structure of magnetic reconnection in the magnetotail from geotail observations, *J. Geophys. Res.*, *118*, 1667–1678, doi:10.1002/jgra.50247.

Nagy, A. F., et al. (2004), The plasma environment of Mars, *Space Sci. Rev.*, *111*, 33–114.

Nakamura, R., W. Baumjohann, R. Schödel, M. Brittnacher, V. A. Sergeev, M. Kubyshkina, T. Mukai, and K. Liou (2001), Earthward flow bursts, auroral streamers, and small expansions, *J. Geophys. Res.*, *106*, 10,791–10,802.

Nishida, A., M. Scholer, T. Terasawa, S. J. Bame, G. Gloeckler, E. J. Smith, and R. D. Zwickl (1986), Quasi–stagnant plasmoid in the middle tail: A new preexpansion phase phenomenon, *J. Geophys. Res.*, *91*, 4245–4255, doi:10.1029/JA091iA04p04245.

Priest, E., and T. Forbes (2000), *Magnetic Reconnection*, Cambridge Univ. Press, Cambridge.

Richardson, I. G., S. W. H. Cowley, E. W. Hones, Jr., and S. J. Bame (1987), Plasmoid-associated energetic ion bursts in the deep geomagnetic tail: Properties of plasmoids and the post-plasmoid plasma sheet, *J. Geophys. Res.*, *92*, 9997–10,013.

Runov, A., V. Angelopoulos, M. I. Sitnov, V. A. Sergeev, J. Bonnell, J. P. McFadden, D. Larson, K.-H. Glassmeier, and U. Auster (2009), THEMIS observations of an earthward-propagating dipolarization front, *Geophys. Res. Lett.*, *36*, L14106, doi:10.1029/2019GL038980.

Russell, C. T., M. A. Saunders, J. L. Phillips, and J. A. Fedder (1986), Near-tail reconnection as the cause of cometary tail disconnections, *J. Geophys. Res.*, *91*(A2), 1417–1423.

Schindler, K. (1974), A theory of the substorm mechanism, *J. Geophys. Res.*, *79*(19), 2803–2810.

Scholer, M., G. Gloeckler, D. Hovestadt, B. Klecker, and F. M. Ipavich (1984), Characteristics of plasmoidlike structures in the distant magnetotail, *Geophys. Res. Lett.*, *89*, 8872–8876, doi:10.1029/JA089iA10p08872.

Sergeev, V. A., V. Angelopulous, J. T. Gosling, C. A. Cattell, and C. T. Russell (1996), Detection of localized, plasma-depleted flux tubes or bubbles in the midtail plasma sheet, *J. Geophys. Res*, *101*, 10,817–10,826.

Sergeev, V. A., V. Angelopoulos, and R. Nakamura (2012), Recent advances in understanding substorm dynamics, *Geophys. Res. Lett.*, *39*, L05101, doi:10.1029/2012GL050859.

Sharma, A. S., et al. (2008), Transient and localised processes in the magnetotail: A review, *Ann. Geophys.*, *26*, 955–1006.

Shen, C., X. Li, M. Dunlop, Z. X. Liu, A. Balogh, D. N. Baker, M. Hapgood, and X. Wang (2003), Analysis on the geometrical structure of magnetic field in the current sheet based on Cluster measurements, *J. Geophys. Res.*, *108*, 1168, doi:10.1029/2002JA009612.

Shen, C., X. Li, M. Dunlop, Q. Q. Shi, Z. X. Liu, E. Lucek, and Z. Q. Chen (2007), Magnetic field rotation analysis and the applications, *J. Geophys. Res.*, *112*, A06211, doi:10.1029/2005JA011584.

Sitnov, M. I., and K. Schindler (2010), Tearing stability of a multiscale magnetotail current sheet, *Geophys. Res. Lett.*, *37*, L08102, doi:10.1029/2010GL042961.

Sitnov, M. I., N. Buzulukova, M. Swisdak, V. G. Merkin, and T. E. Moore (2013), Spontaneous formation of dipolarization fronts and reconnection onset in the magnetotail, *Geophys. Res. Lett.*, *40*, 1–6, doi:10.1029/2012GL054701.

Slavin, J. A. (1998), Traveling compression regions, in *New Perspectives on the Earth's Magnetotail,* Geophys. Monogr. Ser., vol. 105, edited by D. N. B. A. Nishida and S. W. H. Cowley, AGU, Washington, D. C.

Slavin, J. A., E. J. Smith, B. T. Tsurutani, D. G. Sibeck, H. J. Singer, D. N. Baker, J. T. Gosling, E. W. Hones, and F. L. Scarf (1984), Substorm associated traveling compression regions in the distant tail: ISEE-3 geotail observations, *Geophys. Res. Lett.*, *11*(7), 657–660.

Slavin, J. A., E. J. Smith, B. T. Tsurutani, G. L. Siscoe, D. E. Jones, and D. A. Mendis (1986a), Giacobini-Zinner magnetotail: ICE magnetic field observations, *Geophys. Res. Lett.*, *13*, 283–286.

Slavin, J. A., B. A. Goldberg, E. J. Smith, D. J. McComas, S. J. Bame, M. A. Strauss, and H. Spinrad (1986b), The structure of a cometary type i tail: Ground based and ICE observations of P/Giacobini-Zinner, *Geophys. Res. Lett.*, *13*, 1085–1088.

Slavin, J. A., D. N. Baker, D. H. Fairfield, J. D. Craven, L. A. Frank, R. C. Elphic, A. B. Galvin, W. J. Hughes, R. H. Manka, and E. J. Smith (1989a), CDAW-8 observations of plasmoid signatures in the geomagnetic tail: An assessment, *J. Geophys. Res.*, *94*, 15,153–15,175.

Slavin, J. A., D. S. Intriligator, and E. J. Smith (1989b), Pioneer Venus Orbiter magnetic field and plasma observations in the Venus magnetotail, *J. Geophys. Res.*, *94*, 2383–2398.

Slavin, J. A., M. F. Smith, E. L. Mazur, D. N. Baker, E. W. Hones, Jr., T. Iyemori, and E. W. Greenstadt (1993), ISEE 3 observations of traveling compression regions in the Earth's magnetotail, *J. Geophys. Res.*, *98*, 15,425–15,446.

Slavin, J. A., C. J. Owen, M. M. Kuznetsova, and M. Hesse (1995), ISEE 3 observations of plasmoids with flux rope magnetic topologies, *Geophys. Res. Lett.*, *22*, 2061–2064.

Slavin, J. A., R. P. Lepping, J. Gjerloev, D. H. Fairfield, M. Hesse, C. J. Owen, M. B. Moldwin, T. Nagai, A. Ieda, and T. Mukai (2003a), Geotail observations of magnetic flux ropes in the plasma sheet, *J. Geophy. Res.*, *108*, 1015, doi:10.1029/2002JA009557.

Slavin, J. A., et al. (2003b), Cluster measurements of electric current density within a flux rope in the plasma sheet, *Geophys. Res. Lett.*, *30*, 1362, doi:10.1029/2003GL016411.

Slavin, J. A., E. I. Tanskanen, M. Hesse, C. J. Owen, M. W. Dunlop, S. Imber, E. A. Lusek, A. Balogh, and K.-H. Glassmeier (2005), Cluster observations of travelling compression regions in the near-tail, *J. Geophys. Res.*, *30*, A06207, doi:10.1029/2004JA010878.

Slinker, S. P., J. A. Fedder, and J. G. Lyon (1995), Plasmoid formation and evolution in a numerical simulation of a substorm, *Geophys. Res. Lett.*, *22*, 859–862, doi:10.1029/95GL00300.

Teh, W. L., R. Nakamura, B. U. O. Sonnerup, J. P. Eastwood, M. Volwerk, A. N. Fazakerley, and W. Baumjohann (2011), Evidence of the origin of the Hall magnetic field for reconnection: Hall MHD reconstruction results from Cluster observations, *J. Geophys. Res.*, *116*, A11218, doi:10.1029/2011JA016991.

Tsurutani, B. T., D. E. Jones, R. P. Lepping, E. J. Smith, and D. G. Sibeck (1984), The relationship between the imf by and the distant tail (150–238 re) lobe and plasmasheet by fields, *J. Geophys. Res.*, *11*, 1082–1085, doi:10.1029/GL011i010p01082.

Vignes, D., M. H. Acuña, J. E. P. Connerney, D. H. Crider, H. Reme, and C. Mazelle (2004), Magnetic flux ropes in the martian atmosphere: Global characteristics, *Space Sci. Rev.*, *111*(1), 223–231.

Walsh, A. P., A. N. Fazakerley, R. J. Wilson, I. V. Alexeev, P. D. Henderson, C. J. Owen, E. Lucek, C. Carr, and I. Dandouras (2009), Near-simultaneous magnetotail flux rope observations with Cluster and Double Star, *Ann. Geophys.*, *27*, 725–743.

Wang, R. S., Q. M. Lu, A. M. Du, and S. Wang (2010), In situ observations of a secondary magnetic island in an ion diffusion region and associated energetic electrons, *Phys. Rev. Lett.*, *104*, 175,003.

Wang, R. S., et al. (2012), Asymmetry in the current sheet and secondary magnetic flux ropes during guide field magnetic reconnection, *J. Geophys. Res.*, *117*, A07223, doi:10.1029/2011JA017384.

Wolf, R. A., Y. Wan, X. Xing, J.-C. Zhang, and S. Sazykin (2009), Entropy and plasma sheet transport, *J. Geophys. Res.*, *114*, A00D05, doi:10.1029/2009JA014044.

Yeroshenko, Y., W. Riedler, K. Schwingenschuh, J. G. Luhmann, M. Ong, and C. T. Russell (1990), The magnetotail of Mars: Phobos observations, *Geophys. Res. Lett.*, *17*(6), 885–888.

Zhang, T. L., et al. (2012), Magnetic reconnection in the near venusian magnetotail, *Science*, *336*, 567–570.

Zhou, M., X. H. Deng, and S. Y. Huang (2012), Electric field structure inside the secondary island in the reconnection diffusion region, *Phys. Plasmas*, *19*, 042,902.

Zwan, B. J., and R. A. Wolf (1976), Depletion of solar wind plasma near a planetary boundary, *J. Geophys. Res.*, *81*, 1636—1648, doi:10.1029/JA081i010p01636.

17

Current Sheets Formation in Planetary Magnetotail

Antonius Otto, Min-Shiu Hsieh, and Fred Hall IV

17.1. INTRODUCTION

Current sheet (CS) formation and thinning are common and fundamentally important processes in space and laboratory plasmas. CSs are always present in the interaction of plasmas of different origins and their presence is a necessary requirement for the onset of magnetic reconnection. Here the terminology thin CS is used to address CSs much thinner than typical with the expectation that they become sufficiently thin and intense to cause dissipation (based on kinetic processes) and the onset of reconnection. The formation of thin CSs is particularly expected for fast dynamic and nonlinear plasma processes such as Kelvin-Helmholtz or Rayleigh-Taylor instabilities. However, the process of CS thinning through slow configurational changes typical for magnetotail physics is less well understood.

In order to illustrate CS formation by fast instability processes, Figure 17.1 shows a snapshot of the nonlinear evolution of a gravitationally driven Rayleigh-Taylor instability with an initially small and constant horizontal magnetic field. The initial pressure scale height is 10 in simulation units and the initial plasma β varies from about 100 at the base of the simulation to about 20 at the top. The driving force is the free energy associated with the heavier and lower entropy plasma on top of a rarefied plasma. The result in Figure 17.1 shows plasma velocity, magnetic field, and current density, and the plot on the right presents p/n^γ (where p is pressure, n is density, and γ is the ratio of specific heat), which is a measure of the plasma entropy. The result is obtained by a simple two-dimensional, high-resolution magnetohydro dynamic (MHD) simulation based on a second-order finite difference method [*Otto*, 1990; *Ma and Otto*, 2013].

The figure demonstrates the formation of multiple thin CS layers and combines three fundamental plasma processes associated with or responsible for CS generation. The primary instability is the Rayleigh-Taylor instability giving rise to the downward jetting plasma. The large velocity difference between the downward jet and the ambient plasma causes nonlinear Kelvin-Helmholtz waves as a secondary instability (illustrated by the many velocity vortices). The fast nonlinear evolution of these instabilities concentrate and amplify the magnetic field by almost two orders of magnitude and thereby generate strong and thin CSs. The significance of these CSs is their susceptibility to magnetic reconnection. Evidence for reconnection is the presence of multiple small magnetic islands. Asterisks represent plasma elements which were uniformly distributed along a single magnetic field line initially. Clearly this is no longer the case, confirming the presence of magnetic reconnection.

In contrast, CS formation in magnetotails is typically a very different physical process because it occurs through a slow force-balanced evolution. This is sometimes addressed as a quasi-static evolution because the evolution goes through a sequence of equilibrium configurations [*Schindler and Birn*, 1982]. A classic example is the CS thinning and evolution during the growth phase of geomagnetic substorms. Geomagnetic substorms [*Akasofu*, 1964; *McPherron*, 1979] represent a major mode of energy storage and release of Earth's and similar magnetospheres [e.g., *Kan et al.*, 1991; *Sharma et al.*, 2003]. During substorms a large amount of magnetic flux is opened and reclosed, implying a significant impact on the magnetic flux budget of the magnetosphere [e.g., *Boakes et al.*, 2009; *Milan et al.*, 2009]. In recent years, particularly the THEMIS mission with combined multi-

Geophysical Institute, University of Alaska Fairbanks, Fairbanks, Alaska

Magnetotails in the Solar System, Geophysical Monograph 207. First Edition.
Edited by Andreas Keiling, Caitríona M. Jackman, and Peter A. Delamere.

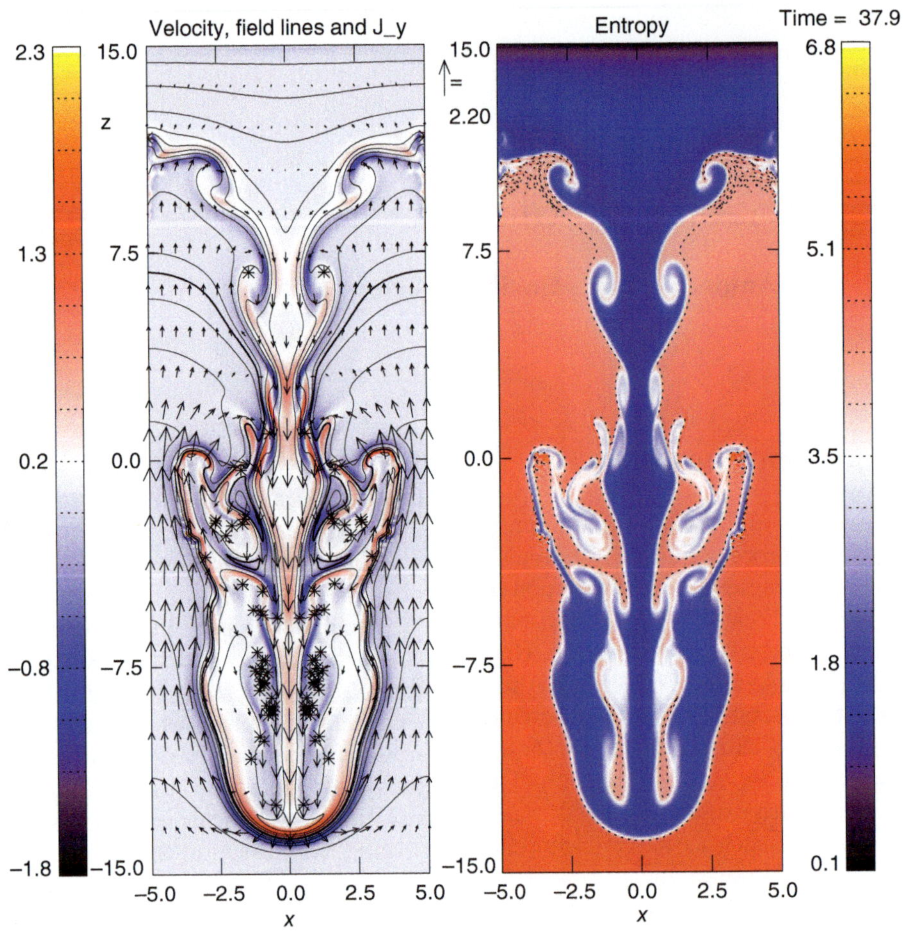

Figure 17.1 Left: Velocity (arrows), magnetic field lines (lines), and current density (color) from a Rayleigh-Taylor instability. Right: Entropy distribution for the same state.

spacecraft [*Angelopoulos*, 2008] and ground observations [*Mende et al.*, 2008] has provided a wealth of new insight.

In comparison to substorm expansion, the growth phase received less attention and central aspects such as the physics of the tail CS thinning represent a fundamentally important unresolved problems. This is critical because the state of the magnetotail at the end of the growth phase determines the conditions for the expansion-phase onset.

Typical for the substorm growth phase is an initial southward turning of the interplanetary magnetic field (IMF). The southward IMF increases dayside reconnection, and magnetic flux is removed from the dayside and accumulated in the lobes, thereby increasing magnetic energy in the magnetotail. The typical duration for the growth phase is 30–60 min, and a highly characteristic feature is the "stretching" of the near-Earth magnetic field between 6 R_E and 15 R_E from a mostly dipolar to a more taillike geometry and the evolution of a thin near-Earth CS [e.g., *McPherron et al.*, 1987; *Sergeev et al.*, 1993, 2012; *Sanny et al.*, 1994; *Ohtani et al.*, 2000; *Asano et al.*, 2004; *Petrukovich et al.*, 2007; *Zhou et al.*, 2009]. Other typical observations include (a) intensification of ionospheric convection with typical polar cap potential of 60 kV and higher; (b) gradual increase of the size of the polar cap and equatorward expansion of the so-called isotropy boundaries [e.g., *Sergeev et al.*, 1983, 2012; *Shevchenko et al.*, 2010]; (c) slow intensification of the field-aligned current (FAC) systems; and (d) convergent motion of the most equatorward discrete arc and the diffuse aurora equatorward of the energetic particle isotropy boundaries.

Most of the growth-phase properties are qualitatively consistent with enhanced dayside reconnection after a southward turning of the IMF, an enhanced transport of open magnetic flux into the lobes of the magnetotail, and an increase of the amount of open magnetic flux in the polar cap.

Only a few mechanisms have been proposed for the critical aspect of CS thinning in Earth's magnetotail. All of these have some difficulties in explaining important aspects of the CS thinning, such as the duration of the growth phase and the location and spatial extent of the thin CS formation.

Substormlike signatures have been observed in Mercury's magnetotail. MESSENGER and Mariner observations show clear signatures of magnetic reconnection [*Slavin et al.*, 2010, 2012] and dipolarization events [*Sundberg et al.*, 2012] during which the B_z component rapidly increases after a period of a slow decrease often to values close to zero. Particularly this slow decrease is indicative of the stretching and thinning of the near-planetary tail current layer.

There is some discussion whether the so-called substorm or Dungey cycle operates in the giant magnetospheres [*Cowley et al.*, 2005; *Jackman and Cowley*, 2006; *Badman and Cowley*, 2007; *McComas and Bagenal*, 2007] of Saturn and Jupiter. Both the Jovian and the Kronian inner and middle magnetospheres are strongly influenced by internal mass loading [*Vasyliunas*, 1983; *Delamere and Bagenal*, 2010; *Bagenal and Delamere*, 2011]. There is considerable evidence that magnetic reconnection operates in the magnetodisc and the magnetotail of these magnetospheres [*Woch et al.*, 2002; *Hill et al.*, 2005; *Kronberg et al.*, 2008; *Ge et al.*, 2010; *Kasahara et al.*, 2013; *Jackman et al.*, 2011]. The large mass loading requires a continuous radial transport of large amounts of plasma, which is most efficiently achieved by magnetic reconnection.

Tail CSs of Uranus and Neptune are poorly understood because of the rather limited in-situ observations and because of the very different magnetospheres compared to other planets. Both Neptune and Uranus have large magnetic dipole tilt, and the Uranian rotation axis is tilted by 97°. Therefore, the magnetotails of these planets undergo major reconfigurations on the rotation period, which likely prevents quiescent periods of time typical for other magnetospheres [*Vasyliunas*, 2011].

The following section reviews different mechanisms for CS thinning through slow adiabatic evolution. Particular emphasis is given to a new mechanism based on magnetic flux depletion (MFD) in section 17.3, which examines the physical conditions that can generate MFD and the properties of the associated CS evolution. This process is applicable to the growth phase of the terrestrial magnetosphere and possibly to the giant magnetospheres provided that a Dungey cycle exists. Section 17.4 addresses the effect of flux tube entropy changes on CS formation, and the last section presents a summary and discussion.

17.2. CURRENT SHEET THINNING PHYSICS

The thinning of the near-Earth CS is highly important because it represents a major reconfiguration of the magnetotail from a mostly dipolar to a highly stretched magnetic field, which determines the conditions for substorm onset. Thin CSs have been observed at distances from $\sim 9\,R_E$ [*Sergeev et al.*, 1990] to $\sim 18.7\,R_E$ [*Thompson et al.*, 2005]. These CSs have attained thicknesses as small as $\sim 0.1\,R_E$ [*Sergeev et al.*, 1990] compared to typical values of $\sim 5\,R_E$ an hour before substorm onset [*McPherron et al.*, 1987; *Sanny et al.*, 1994]. The corresponding current density increases by a factor of 5–10 during the growth phase [*Sergeev et al.*, 1993]. In the late growth phase and shortly before substorm onset, CS thinning appears to occur on a faster time scale. These results are confirmed and detailed by more recent Cluster and THEMIS observations [e.g., *Petrukovich et al.*, 2007; *Kubyshkina et al.*, 2011; *Sergeev et al.*, 2012].

The most basic process for CS thinning is slow adiabatic compression of the entire CS. However, entropy conservation requires $p/n^{\gamma} = \text{const}$ with $\gamma = \frac{5}{3}$ for adiabatic motion. Depending on the actual plasma properties and including the compressibility of the magnetic field, the effective value of γ is between 1 and $\frac{5}{3}$. Therefore, an increase of the cross-tail current density by a factor of 9 requires the total pressure increase of at least a factor of 9 corresponding to a lobe field increase of the factor of 3. This is much higher than observed for typical substorms.

Several models have been demonstrated to generate thin CSs through slow plasma evolution. Considerable work has focused on the evolution of CSs as a result of changing boundary conditions that do not permit an analytic equilibrium, i.e., solutions develop a discontinuous change in the magnetic field [e.g., *Schindler and Birn*, 1993; *Birn et al.*, 1998, 2004a]. The location of the evolving CS depends in part on the boundary perturbations, but a natural location appears to be the location with the largest gradient in magnetic flux tube volume [*Wiegelmann and Schindler*, 1995; *Becker et al.*, 2001]. Models which employ an electric field at the lobe boundary indicate that the electric field profile can focus the location of the CS thinning.

Mechanisms to explain the CS thinning also include the work of *Lee et al.* [1995, 1998], who proposed a local diffusion of thermal pressure at the CS boundary ("entropy antidiffusion"). To understand this mechanism, it is helpful to note that a two-dimensional equilibrium with $\boldsymbol{B} = \nabla A_y \times \boldsymbol{e}_y$ implies the equilibrium condition $p = p(A_y)$ and $j_y \sim dp/dA_y$ such the pressure p is constant on magnetic field lines. Here A_y is the flux function or y component of the vector potential and j_y is the y or cross-tail component of the current density. For a Harris sheet $p = c_0 \exp(c_1 A_y)$, where c_0 and c_1 are constants. The pressure diffusion modifies the pressure by introducing a local perturbation $\delta p(A_y)$, thereby introducing a modification of the current density $\delta j_y \sim d\delta\, p/dA_y$ such that the associated two-dimensional equilibrium displays a modified current density $j_y + \delta j_y$. Pressure diffusion also redistributes entropy as a function of the magnetic flux A_y with localized steepened and flattened profiles. The cause for the pressure diffusion is attributed to wave-particle interaction of the lower hybrid drift instability which is stabilized by the high plasma beta in the central plasma sheet but can in principle operate at the plasma sheet boundary. In a global equilibrium solution *Zaharia and Cheng* [2003] constructed a thin CS configuration based on the outer boundary conditions and specific choices of the pressure profile.

It is important to consider if and how these models can explain typical properties of CS thinning during the growth phase. Although it is possible to adjust model parameters to account for some properties, none of the models can naturally account for all of these observations, specifically the near-Earth location, the duration of the growth phase, the correlation to southward IMF conditions, and the strong decrease of the near-Earth magnetic field in the equatorial plane. The following section introduces a new CS thinning mechanism based on MFD in the near-Earth magnetotail which is examined with respect to typical growth-phase properties.

17.3. MIDNIGHT MFD AND CS THINNING

17.3.1. Basic Physics of MFD and CS Thinning

After a southward turning of the IMF, reconnection increases at the dayside magnetosphere, and a large amount of closed magnetic flux is opened and added to the magnetotail lobes. Initially the magnetopause moves inward by about 0.5 R_E to 1 R_E due to the reduction of closed flux in the dayside magnetosphere for about 15–20 min. However, after this period the earthward motion of the magnetopause stops, indicating the transport of closed magnetic flux into the dayside reconnection region to balance the dayside reconnection rate. This magnetic flux transport cannot occur in a steady state because the process strongly reduces the amount of closed magnetic flux in the magnetosphere. The most obvious magnetic flux reservoir for this transport is the near-Earth magnetotail. The configurational changes of the magnetosphere are slow, indicating an adiabatic or quasi-static evolution where static force balance applies throughout this evolution. This is consistent with the absence of fast flows at most times during the growth phases. Therefore, this transport should occur mostly along contours of constant flux tube entropy defined by the line integral

$$H = \int_{\text{ft}} \frac{p^{1/\gamma} ds}{B} \tag{17.1}$$

along a magnetic flux tube (ft), where B is the magnetic field magnitude and ds is the differential line element along the vector magnetic field **B**.

Figure 17.2 shows contour lines of flux tube entropy in the equatorial plane of the magnetosphere [*Hall*, 2006]. The result uses a typical Tsyganenko magnetic field (T96, *Tsyganenko and Stern*, 1996) for average solar wind conditions with no dipole tilt and a small 1 nT southward IMF. The plasma pressure in the equatorial plane is determined by integrating the force balance equation in the equatorial plane using the Tsyganenko magnetic field. Assuming a constant pressure along magnetic field lines, equation 17.1 yields the flux tube entropy. Although this is not a global equilibrium, it is a good approximation for the flux tube entropy distribution because deviations from force balance are small and the radial gradient of the flux tube entropy is large.

The result in Figure 17.2 demonstrates that the dayside magnetopause is connected to the near-Earth tail region at about 10 R_E by contours of constant flux tube entropy. The Figure also illustrates the large gradient in flux tube entropy along the magnetotail. This gradient combined with conservation of magnetic flux tube entropy prohibits significant magnetic flux transport from the mid or far tail into the near-Earth region [*Erickson and Wolf*, 1980; *Schindler and Birn*, 1982; *Baumjohann*, 2002]. Therefore the only reservoir of closed magnetic flux for the transport to the dayside is located in the near-Earth tail around $x=-8$ R_E to $\mu=-15$ R_E. The equatorial convection, implied by the result in Figure 17.2, converges on the dayside (where reconnection erodes magnetic flux) and diverges in the midnight tail region. Thus the near-Earth tail region is expected to be strongly depleted of closed magnetic flux over the course of the growth phase.

Observations [*Borovsky et al.*, 1998] confirm the large tailward gradient of the flux tube entropy and of the plasma content such that significant transport by earthward convection from the midtail is unlikely because it would require an excessive loss of particles and thermal energy from a flux tube convecting into the near-Earth tail. Early ideas that azimuthal flux transport may be important for the tail evolution go back to *Coroniti and Kennel* [1973] and *Kan* [1990] but have never been examined in detail. *Coroniti and Kennel* [1973] suggested the propagation of fast expansion wave in the magnetosphere converging in the midnight region. These waves are indeed important to transport the information of dayside reconnection and switch on the convection pattern; however, the physics of this convection and the quasi-static evolution have not been part of their proposal.

MFD implies a change from a more dipolar to a stretched or Harris sheetlike configuration because of the strong reduction of B_z in the near-Earth tail around $x=-8$ R_E to $x=-15$ R_E. This is exactly the region where CS formation is observed during the growth phase. Therefore MFD can predict the correct location of the CS thinning without any additional assumptions on model properties provided that the reduction of B_z indeed leads to CS formation. However, does this model also predict the correct

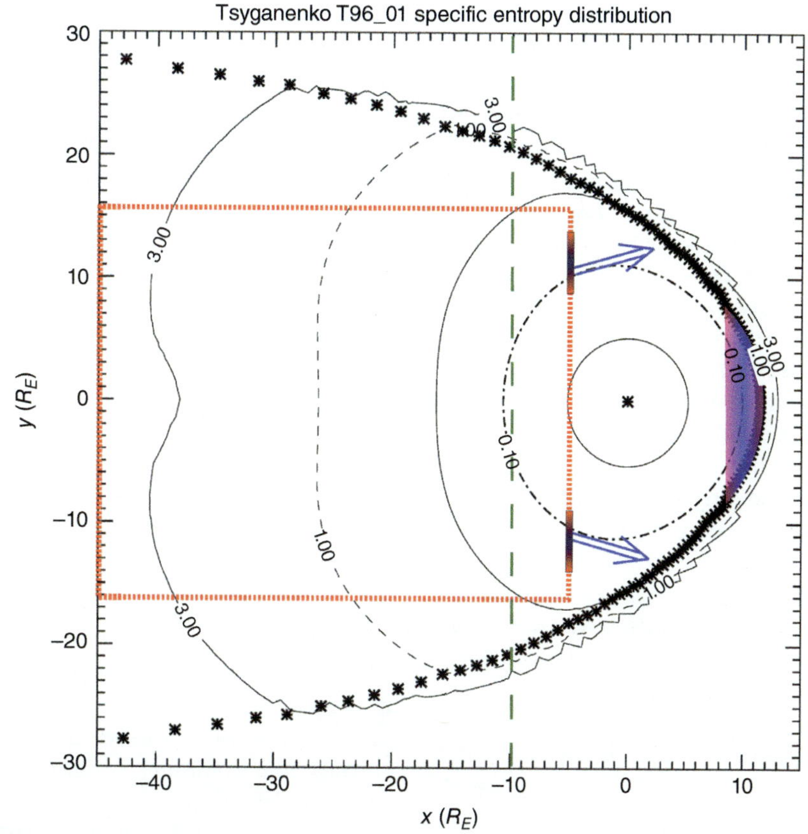

Figure 17.2 Contours of constant flux tube entropy [*Hall*, 2006] in the equatorial plane. The asterisks represent the magnetopause boundary, the red box outlines of the simulation domain, and the arrows sketch the outflow boundary condition used in section 3.2.

time scale of 30–60 min for typical isolated substorms? Considering a tail section of size 6 R_E by 12 R_E, an average B_z of about 50 nT in the midnight region, and a potential of about 50 keV for the sunward convection yields a time of about 50 min to deplete this tail section of its magnetic flux. This is well consistent with the typical duration of the growth phase. Assuming a constant reconnection potential (equals flux transport rate), CS thinning should accelerate during the growth phase because the typical time scale of the process is inversely proportional to the amount of magnetic flux in the reservoir.

17.3.2. Simulation Model

In order to demonstrate CS thinning by MFD in the midnight tail, we employ a three-dimensional mesoscale MHD simulation [*Otto*, 1990, 1995] of a section of the magnetotail. The finite difference simulation uses the full set of resistive MHD equations, and the simulation domain encompasses the magnetotail region defined by $-45\,R_E \le X \le -5\,R_E$, $-15\,R_E \le Y \le 15\,R_E$, and $0\,R_E \le Z \le 12\,R_E$, as indicated by the rectangular outline in Figure 17.2. For simplicity, the configuration uses a north-south symmetry which is applicable strictly only for a zero tilt angle of the magnetic pole and a purely northward or southward magnetic field.

Boundary conditions allow plasma flow tangential to all boundaries except for the equatorial plane and the sunward boundary. The boundary conditions at the equatorial plane assume symmetry of the magnetic field, plasma, and flow between the northern and southern hemispheres. The velocity boundary condition at the sunward boundary at $X = -5\,R_E$ is discussed below. The presented results use a grid with 123^3 nonuniformly spaced grid points. The best resolution of this nonuniform grid is chosen to be $\Delta x = 0.07$, $\Delta y = 0.06$, and $\Delta z = 0.03$ corresponding to a resolution between 200 and 450 km at the earthward boundary ($x = -5$, $y = 0$, $z = 0$). In the simulation all quantities are normalized to typical values, i.e., length scales to a typical $L_0 = 1\,R_E$, density ρ to $\rho_0 = m_0 n_0$ with the proton mass m_0 and a number density $n_0 = 1\,\text{cm}^{-3}$, magnetic field $\boldsymbol{B}$ to $B_0 = 20$ nT, velocity $\boldsymbol{u}$ to a typical Alfvén speed $V_A = B_0(\mu_0 \rho_0)^{-1/2} \sim 430$ km/s, time to the Alfvén transit time $t_0 = L_0/V_A \sim 14.4$ s, current density $\mathbf{j}$ to $j_0 = B_0/(\mu_0 L_0) \sim 2.5\,\text{nA/m}^2$, and pressure $\boldsymbol{p}$ to $p_0 = B_0^2/(2\mu_0) = 1.6 \times 10^{-10}$ Pa. Most presented results will either use physical units or include an interpretation in terms of physical units. Distances are measured in Earth radii R_E.

Figure 17.3 presents a sketch of the overall simulation domain and the magnetic field and flow conditions. The initial state is determined by a ballistic or frictional relaxation method [*Hesse and Birn*, 1993; *Adamson et al.*, 2011] using a T96 magnetic field with a dynamic pressure $p_{dyn}=1$ nPa, a magnetic index $D_{st}=-15$, $B_{y,IMF}=0$ nT, and $B_{z,IMF}=-1$ nT. In order to obtain an equilibrium configuration the T96 magnetic field is used to determine $\mathbf{j}\times\mathbf{B}$ forces. Starting at the sunward equatorial edge of the boundary, these forces are used to integrate the pressure distribution within this model. The resulting configuration is not in equilibrium; however, it serves to minimize the residual forces. Subsequently the ballistic relaxation method is used to obtain an equilibrium solution with very small residual forces.

The arrows in Figures 17.2 and 17.3 demonstrate that the simulations require azimuthal outflow at the sunward boundary to test the hypothesis of MFD. This outflow is nontrivial and can be applied not just at the equatorial edge of the sunward boundary. In order to construct this outflow boundary condition, we note that contours of constant flux tube entropy are approximately circular at the radial distance of about 10 R_E. We assume an approximation to the azimuthal convection of the form

$$\tilde{\mathbf{v}}_\varphi = v_0 f(r)\frac{\varphi-\pi}{\pi}\mathbf{e}_\varphi, \tag{17.2}$$

where v_0 provides the amplitude of the flow, $f(r)$ centers this flow at the radial distance of r_0 with a width of about $2R_E$, and φ is the azimuthal angle measured from the sunward direction. Equation 17.2 provides a divergent flow from the midnight sector. This convection is not applied within the simulation domain or at the equatorial boundary! Rather, we integrate field lines from the sunward boundary toward the equatorial plane, thereby mapping the expected electric field from the anticipated flow to the sunward boundary. At this boundary this electric field is used to determine the x and y components of the plasma velocity to provide the outflow condition at the sunward boundary. This formulation of the outflow condition allows to move magnetic field lines smoothly through the boundary without generating very large surface currents. Figure 17.4 presents this outflow condition in a plot of the velocity at the sunward boundary of the simulation 24 min after the outflow is switched on. The relaxation is applied up to time $t=300t_0$ after which the outflow boundary condition is applied. Furthermore the integral

$$\Phi_{out} = \int_{y,min}^{y,max} u_y B_z dy \tag{17.3}$$

along the equatorial edge of the sunward boundary ($X=-5\,R_E$, $Z=0$) determines the rate of magnetic flux transport (or potential) out of the simulation box. Here u_y is the y component of $\tilde{\mathbf{v}}_\varphi$ along this line, and v_0 in combination with $f(r)$ can be used to adjust this flux

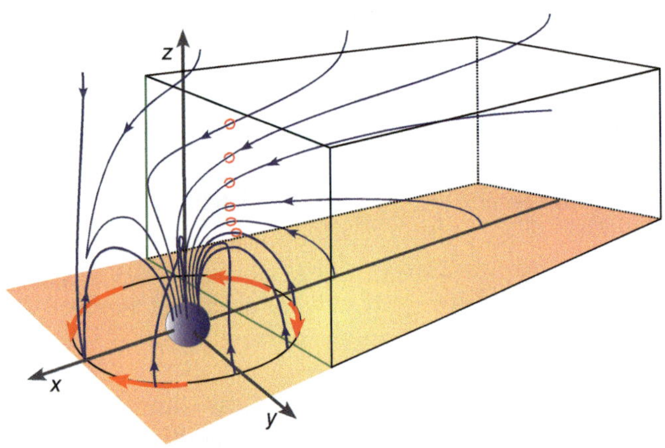

Figure 17.3 Magnetic flux transport from the night-side to the dayside after the IMF turns southward as indicated by the red arrows.

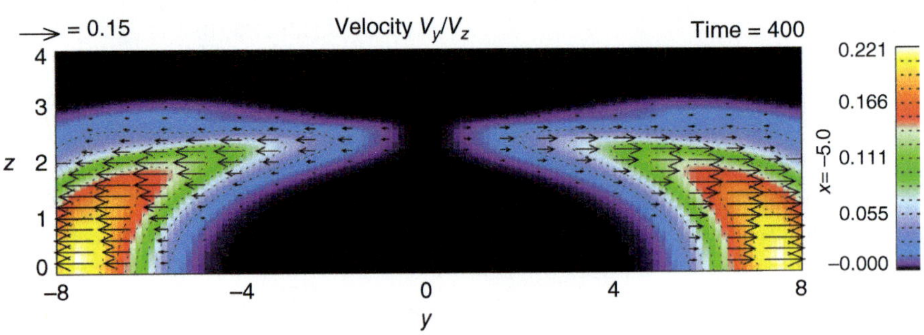

Figure 17.4 Velocity (in units of 430 km/s) at the sunward boundary at time 24 min for case 2. Color represents the x (sunward) component of the velocity, dotted lines are contour lines of this component, and arrows the y and z components. Distances are in R_E.

transport. For the presented results v_0 is regularly updated throughout the simulation and slowly increased by about a factor of 2 to maintain a reasonable level of the potential. The presented results are mostly based on three specific cases using maximum initial velocities of $50\,km/s$ (case 1), $75\,km/s$ (case 2), and $100\,km/s$ (case 3) at the sunward boundary. All cases assume an initial radial distance of the center of the flow channel of $r_0 = 10R_E$, which is reduced slowly to $9R_E$ at the end of the simulation to account for some erosion of the dayside magnetopause. The average potential over the course of each simulation run applied at the earthward boundary amounts to 35 kV (case 1), 50 kV (case 2), and 69 kV (case 3). The simulations use a small background resistivity of $\eta = 10^{-3}$ to balance nonphysical grid oscillations; however, this is switched off in the immediate vicinity of the equatorial plane to avoid the onset of magnetic reconnection.

17.3.3. Simulation Results

All simulation cases develop a divergent convection (from the midnight sector) in the near-Earth tail soon after the azimuthal outflow is applied. Figure 17.5 illustrates the convection just above the equatorial plane for case 2 and 24 min after the outflow is applied. The plot demonstrates the evolution of two azimuthal flow channels. There is almost no earthward convection, as expected from the constraint of flux tube entropy conservation. This flow is accompanied by a very slow convection converging toward the equatorial (tailward of about $X = -10\,R_E$).

Of central importance is the evolution of B_z in the equatorial plane. At $X \sim -10R_E$, B_z decreases from about 25 nT initially to a value of less than 1 nT 50 min after the sunward outflow is applied (Figure 17.6). The region of MFD extends from about $X \sim -8R_E$ to $X \sim -15R_E$ in the

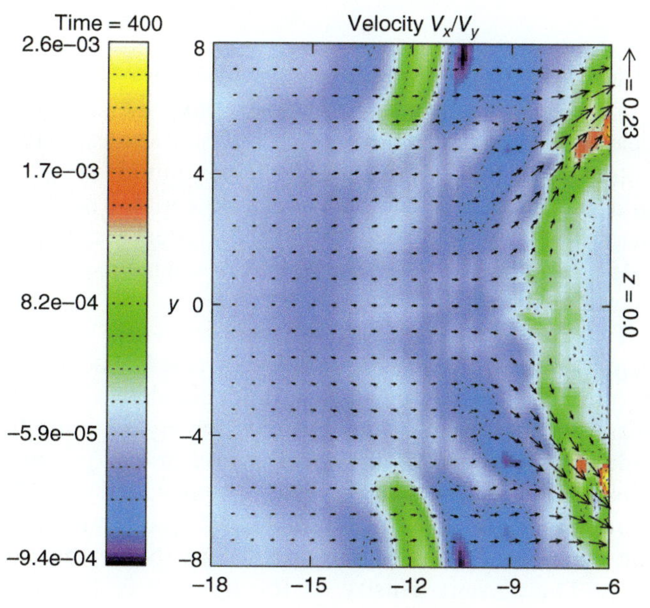

Figure 17.5 Velocity (in units of 430 km/s) for case 2 in the equatorial plane at $t = 24$ min. Color indicates the z component, which however is very small, as can be expected.

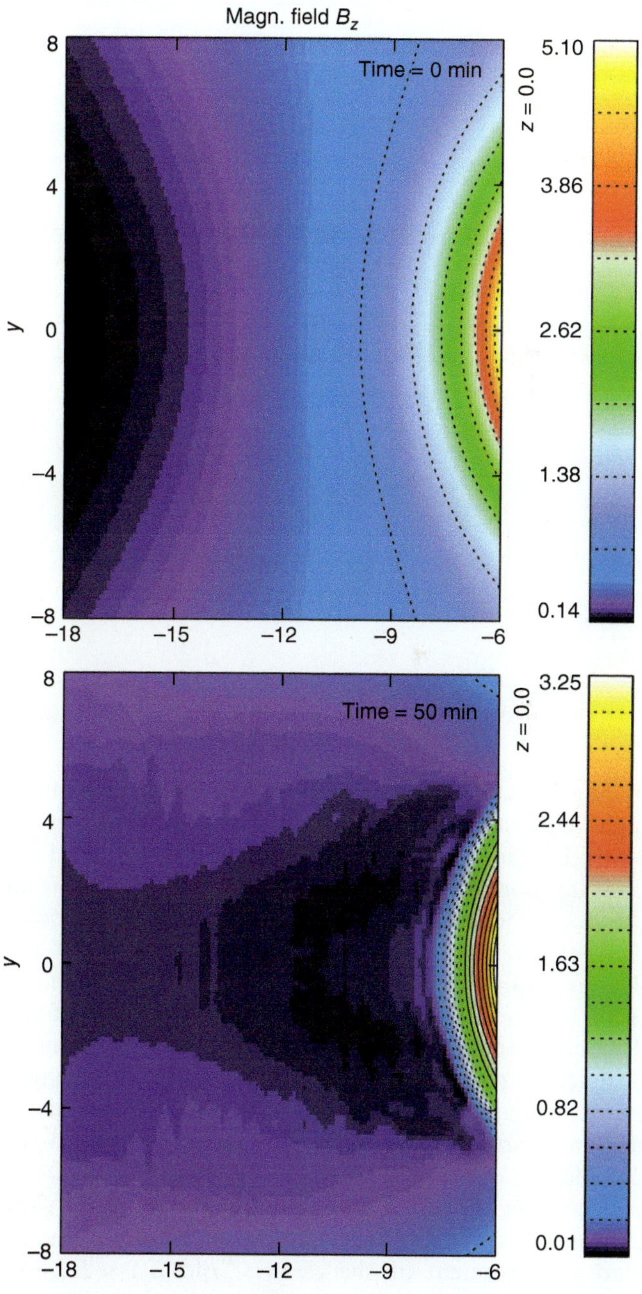

Figure 17.6 Color shows the magnetic field B_z component in the equatorial plane (in units of 20 nT) at $t = 0$ and $t = 50$ min after outflow boundary condition has been switched on for case 2.

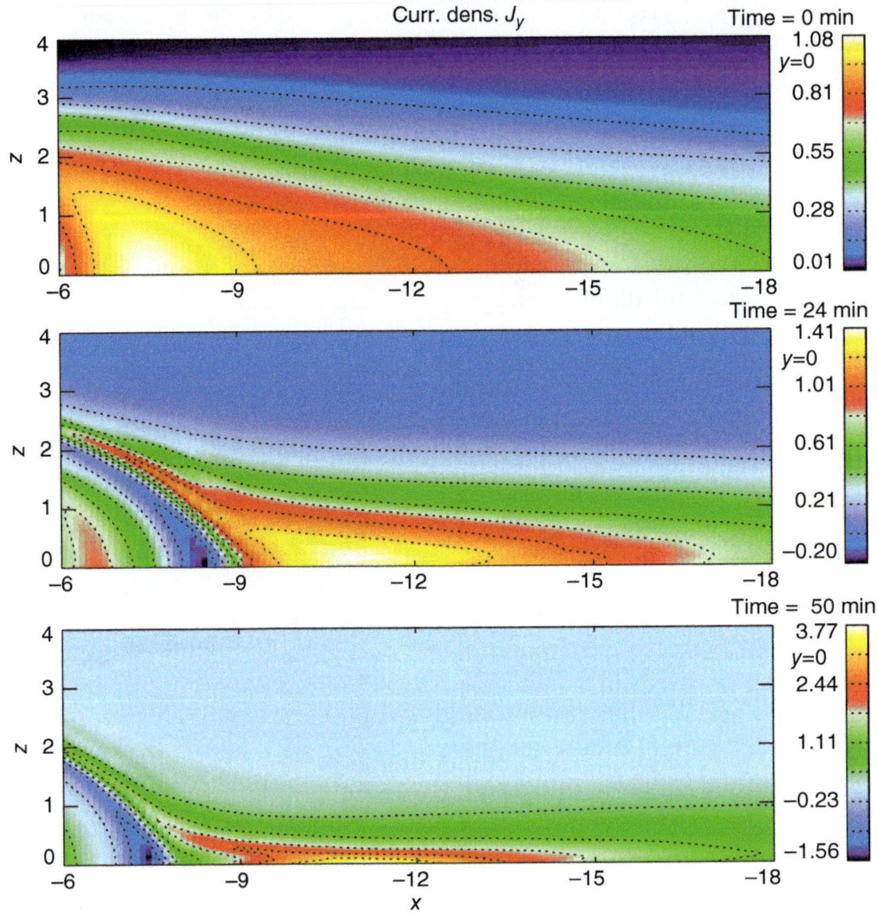

Figure 17.7 Cross-tail current density j_y component at the noon-midnight meridian at times $t=0$, 24, 50 min. Current density is normalized to 2.5 nA/m^2 for case 2.

midnight sector and covers about 6 h around local midnight, consistent with typical observations [*Newell et al.*, 1998]. Different from the initial B_z distribution, the final configuration at $t=50$ min shows a sharp transition from the small and strongly depleted B_z region to the dipolar field region. There is also a significant reduction of the dipolar field. At the plot boundary ($X=-6\,R_E$) B_z decreases from 102 to 65 nT, consistent with observations of the H component of the magnetic field at geosynchronous distances during the growth phase of substorms. This represents a major reduction of the closed magnetic flux in the near-Earth tail caused by the convection toward the dayside.

The strong reduction of B_z has a feedback on the flow channels. A reduced B_z implies a reduction of the associated y component of the electric field $E_y=u_xB_z$, and thus the associated potential decreases unless the flow channels widen or u_x increases. For the presented results we have increased the initial maximum velocity at the boundary to about twice the initial value, i.e., to 100 *km/s* (case 1), 150 *km/s* (case 2), and 200 *km/s* (case 3) during the simulation. However, despite this increase and the slow displacement of the flow channels from $r_0=10R_E$ to $r_0=9R_E$ (which taps into more closed flux), the magnetic field B_z decrease is so strong that the final values of the potential are reduced to about half their initial values or to about two thirds of the average potential. The total magnetic flux removed from the system is about 1.45×10^8 Wb close to the total magnetotail magnetic flux at midtail distances.

The resulting evolution of the cross-tail current density is illustrated in Figure 17.7 for case 2. The initial current density profile is almost identical to the T96 current distribution. The current density is initially reduced earthward of $8R_E$ or $9R_E$ and intensifies tailward of this region. Subsequently, an embedded thin CS develops with a half-width of about $0.2R_E$ at 50 min into the evolution. The CS is bifurcated at the earthward edge. The current density of the thin CS is 4 nA/m^2, 20–30 min into the simulation and increases to about 10 nA/m^2 or 4–6 times the initial value at $t=50$ min. These results agree well with a statistical study correlating the decrease of B_z and increase of the current density [*Petrukovich et al.*, 2007]. CS location and the

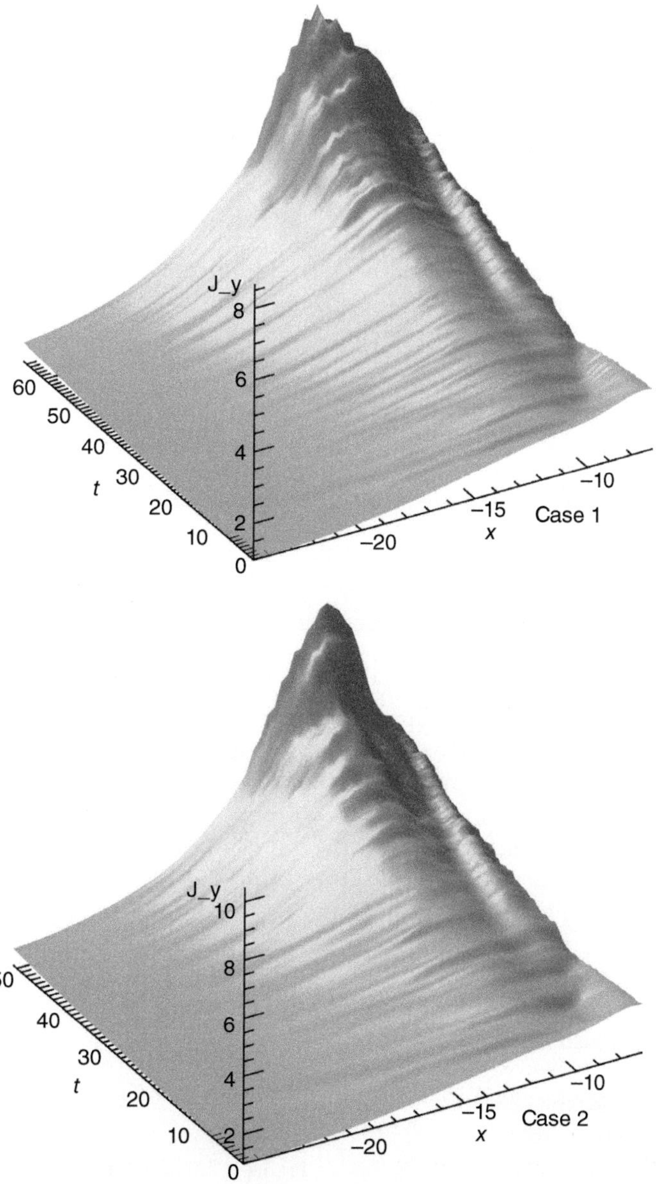

Figure 17.8 Evolution of the cross-tail current density (in nA/m²) in the midnight meridian as a function of time (in minutes) for cases 1, 2.

narrow transition to the dipolar region are consistent with recent THEMIS events [*Zhou et al.*, 2009; *Kubyshkina et al.*, 2011; *Sergeev et al.*, 2011], and the CS formation is also consistent with the characterization as an incompressible CS thinning [*Saito et al.*, 2011].

It is interesting to note that the current evolves in a very similar manner for the three cases, although on different time scales. The plots in Figure 17.8 show the evolution of the current density profile along the midnight meridian as a function of time for cases 1 and 2 (3 is very similar). The plots demonstrate that the current evolution for the different runs is almost identical except for different time scales. The maximum current density for all cases approaches about $10\,\mathrm{nA/m^2}$. The simulation of case 1 takes about 67 min to reach this current density, while case 2 takes about 50 min, and case 3 takes 34 min. The product of this time scale with the average potential is approximately the same for the three cases. This implies that the total amount of closed flux removed is approximately the same, which suggests that the time scale for the thin CS formation is inversely proportional to the applied flux transport rate, e.g., doubling the transport rate cuts the time to develop a thin CS approximately in half. Furthermore, the plots in Figure 17.8 indicate an acceleration of the thin CS formation toward the end of the runs even though the applied potential decreases.

This confirms a time scale for CS formation that is inversely proportional to the amount of closed magnetic flux available in the reservoir. These results are also consistent with a slow quasi-static evolution of the tail configuration, which is always approximately force balanced.

In order to shed additional light on the formation of the CS, it is assumed that three-dimensional effects in the midnight meridian are small. With this assumption we integrate the magnetic field to determine the flux function A_y. For any snapshot in time it is now possible to determine the thermal pressure and the cross-tail current density as a function of A_y. Figure 17.9 presents plots of $p(A_y)$ and $j_y(A_y)$ at four different times for case 3. Despite the large changes of $p(A_y)$ and $j_y(A_y)$, the figure illustrates that the maximum current density always occurs approximately at the steepest incline of $p(A_y)$. We believe the physical reason for the evolution of $p(A_y)$ is the reduction of B_z, which reduces the range of A_y for a fixed difference in p. Note that the thermal pressure p in the plasma sheet is mostly determined by the lobe magnetic field pressure and therefore does not change as much as the flux function.

The presented mechanism for CS thinning is specific to the circulation of magnetic flux in an Earth-like magnetosphere after the southward turning of the IMF. However, it should also operate whenever a persistent divergent flow depletes the magnetic field in the equatorial plane. This mechanism would apply to the giant magnetospheres if a substormlike circulation of magnetic flux applies which is currently unresolved. It may also apply in the transition region from the magnetodisc to the magnetotail where a divergent equatorial flow is expected. The next section discusses possible alternatives based on nonadiabatic processes for such magnetospheres that may also lead to MFD.

17.4. ENTROPY CHANGES AND CS FORMATION

In recent years magnetic flux transport, for instance, in the form of bursty bulk flows [*Angelopoulos et al.*, 1992], has been associated with localized flux tubes of depleted flux tube entropy [*Birn et al.*, 2004b; *Wolf et al.*, 2009; *Zhang et al.*, 2008, 2009]. These regions are often addressed as plasma bubbles. The basic physics of these bubbles is an interchange motion based on the low flux tube entropy, similar to a ballooning instability, which requires low-entropy flux tubes to move earthward. This interchange motion generates a very thin CS at the trailing edge of the low-entropy flux tube [*Birn et al.*, 2009; *Yang et al.*, 2011]. Both a localized entropy decrease and an entropy increase such as used by the work of *Lee et al.* [1998] can generate thin CSs in a stretched magnetic field configuration. Therefore, CS formation based on a redistribution of entropy is of potential interest for all magnetotail and magnetodisc configurations.

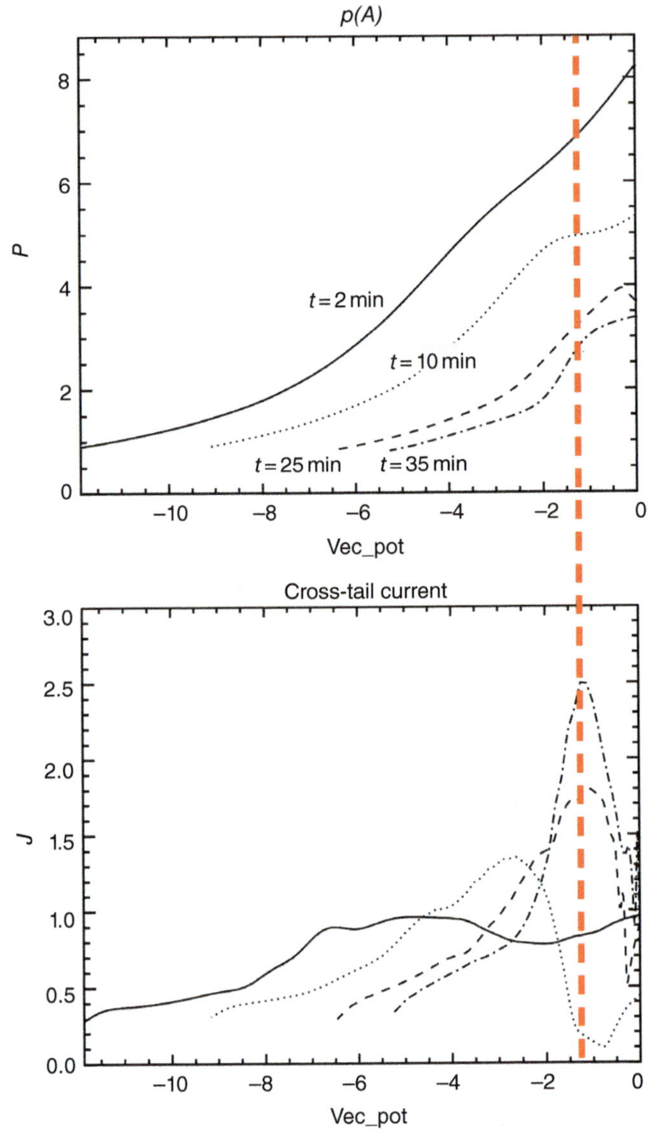

Figure 17.9 Thermal pressure $p(A_y)$ and cross-tail component of the current density $j_y(A_y)$ derived from case 3 at times 2 min, 10 min, 25 min, and 35 min after the outflow is switched on. The red line indicates that the maximum current density occurs coincident with the steepest pressure profile. Pressure and current density are in normalized units p_0 and j_0. The vector potential is in units of L_0B_0.

Processes that can alter flux tube entropy include magnetic reconnection, energy loss into the ionosphere, gradients in energetic particle drifts, or kinetic processes that change the pressure or its isotropy. It is expected that the plasma entropy increases with radial distance in the Jovian magnetosphere [*Bagenal and Delamere*, 2011], which requires nonadiabatic heating during the outward plasma transport. Here we present two examples of a slow localized nonadiabatic heating and cooling to demonstrate CS formation in a largely dipolar magnetic field.

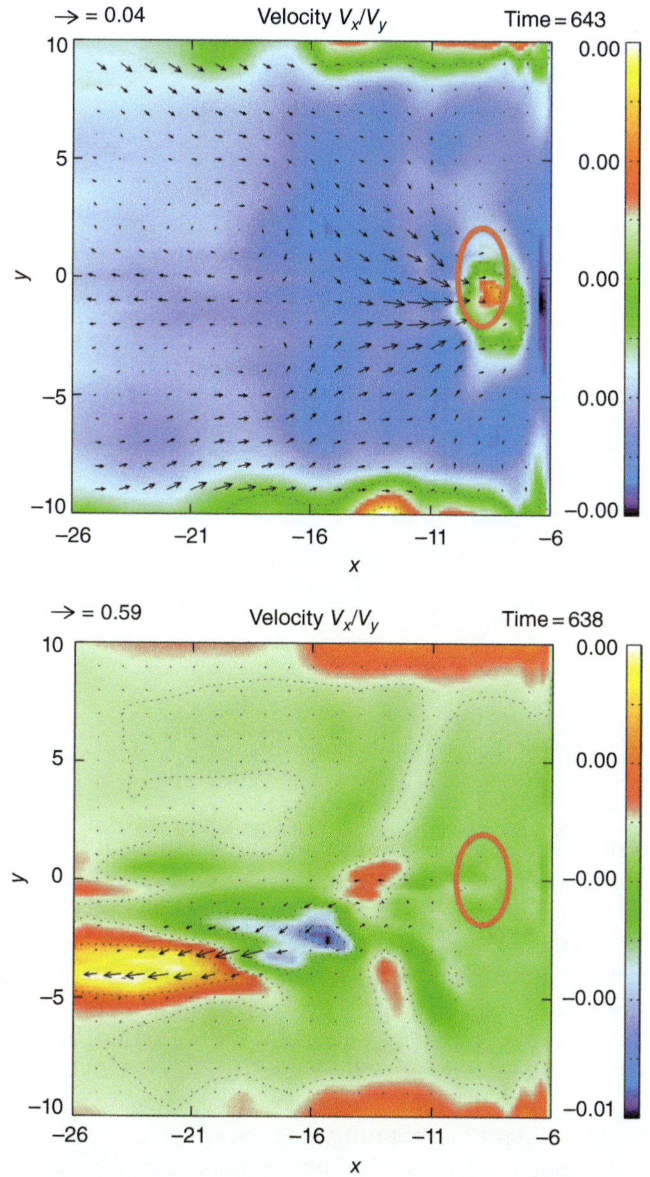

Figure 17.10 Equatorial convection for the cases of localized cooling (top) and heating (bottom) in the same format as Figure 17.5.

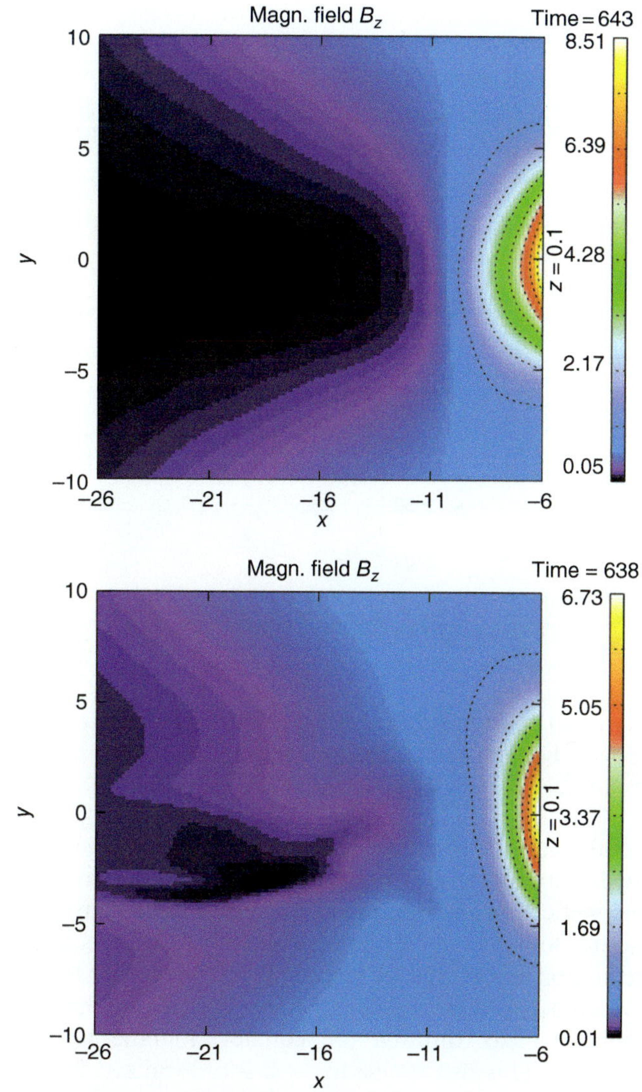

Figure 17.11 Magnetic field B_z in the equatorial plane for the cases of localized cooling (top) and heating (bottom) in the same format as Figure 17.6.

The following results make use of the same magnetotail configuration, but instead of applying a sunward outflow, localized heating or cooling is applied in the vicinity of $X=-10$ in the midnight tail. Compared to the ambient thermal energy density at this location the heating/cooling rate has a magnitude of $8\times10^{-3}t_0^{-1}$, i.e., it would take 120 Alfvén transit times, or 30 min, to heat the plasma to twice its initial temperature at this location. The results of Figure 17.10 show the equatorial velocity profile 80 min into the cooling (top) or heating (bottom) process. The red circle in the figures identifies the location of the localized cooling or heating. The case of localized cooling illustrates slow convection into the region where cooling occurs. This convection is also transporting magnetic flux into this region. The case of localized heating illustrates a narrow plasma jet away from the dipolar magnetosphere which is also associated with magnetic flux transport.

The local cause for this convection is a small perturbation of the thermal pressure which leads to a very small change of the local force balance. Alternatively, this convection can be understood by the change of flux tube entropy, e.g., cooling leads to a reduction of local entropy, allowing material to convect into the dipolar region. The evolution for these cases is very slow, and the faster flow in the narrow convection channel for the heating case develops only at the end of this run.

The effect of this transport on the equatorial magnetic flux is apparent in Figure 17.11. In comparison with the

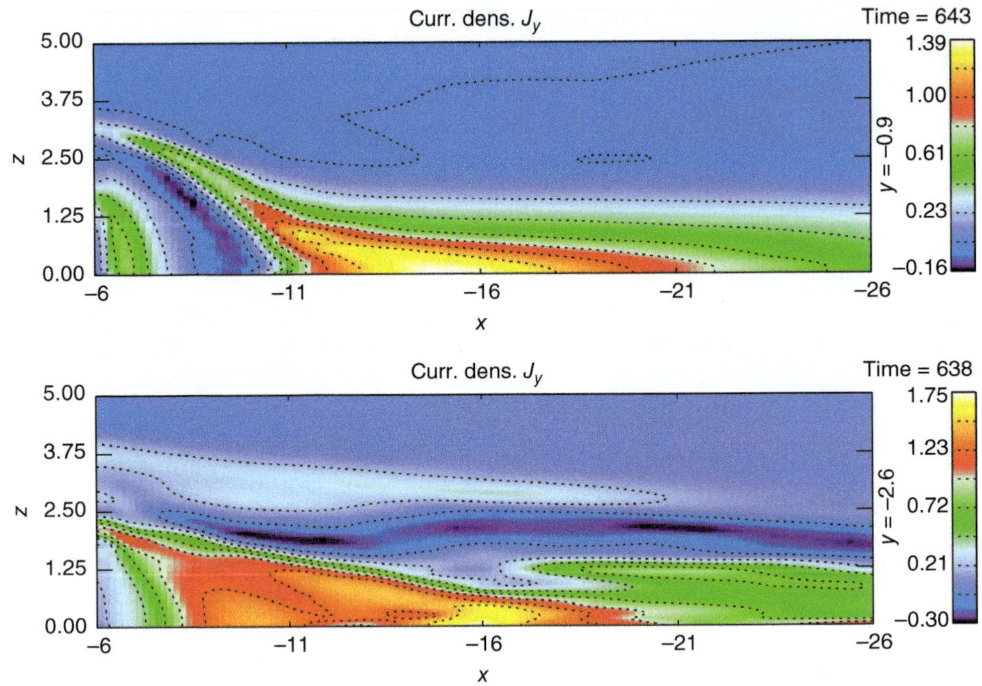

Figure 17.12 Cross-tail current density j_y in the equatorial plane for the cases of localized cooling (top) and heating (bottom) in the same format as Figure 17.7.

equilibrium distribution of B_z in Figure 17.6, the localized cooling and heating have strongly altered the magnetic flux distribution in the equatorial plane. The broad slow inward convection has generated an extended region of low B_z that extends close to the dipolar region. Different from the cases of sunward outflow, the total magnetic flux through the equatorial plane has not changed. This also applies to the case of localized heating. The narrow outward jetting plasma has generated a narrow channel of depleted B_z. Note that neither the localized heating nor the localized cooling in this almost dipolar field has caused the onset of reconnection in the simulation up to the times shown.

Figure 17.12 shows the cross-tail current in the midnight meridian plane for the localized cooling case (top) and in the center of the depleted B_z channel (bottom). Both cases show a significantly enhanced and thinned CS compared to the equilibrium current in Figure 17.7. These cases demonstrate that localized heating or cooling in a quasi-dipolar magnetic field can significantly alter the magnetic flux distribution and generate intense and thin CSs.

These examples use an equilibrium and parameters that strictly apply to the terrestrial magnetosphere. However, the basic physical mechanism of CS generation can be expected to apply to different magnetospheres as well. The examples ignore the inertial forces from the fast rotation of the giant magnetospheres as well as plasma anisotropy. Both can contribute to the physics of CS formation and to the onset of reconnection [*Vasyliunas,* 1983; *Kivelson and Southwood*, 2005]. However, none of these processes have been demonstrated in a quantitative model, and except for global simulations [*Jia et al.*, 2012; *Walker et al.*, 2011], there are currently no local simulation models that include self-consistent magnetospheric configurations with inertial forces [*Caudal*, 1986; *Zimbardo*, 1989; *Achilleos et al.*, 2010] for giant fast rotating magnetospheres.

Although CS formation by nonadiabatic processes is an important concept, studies have been unspecific on the details of the nonadiabatic processes. Models and simulations of plasma bubbles use either an ad hoc entropy reduction or magnetic reconnection to study plasma bubbles [*Birn et al.*, 2004b; *Hu et al.*, 2011; *Wolf et al.*, 2012]. However, the argument of CS formation by plasma bubbles is circular if such entropy-depleted flux tubes are created by reconnection in the first place.

17.5. SUMMARY AND DISCUSSION

The formation of CSs represents a fundamentally important and not-well-understood problem of planetary magnetotails and magnetodiscs. Various mechanisms can cause this formation; among those are persistent boundary perturbations, pressure diffusion at the plasma sheet boundary, and as discussed in detail, a local increase or decrease of the flux tube entropy and the MFD of the near-Earth field through flux circulation toward the dayside reconnection region. In general all of these mechanisms are capable of generating thin CSs as a result of a quasi-static evolution.

The growth phase of geomagnetic substorms is characterized by the requirement that magnetic flux has to circulate from the magnetotail to the dayside magnetopause. We have shown that the circulation and the duration of the CS formation are consistent with the mechanism of MFD in the midnight sector, which naturally causes a thin CS to evolve. This does not exclude other mechanisms which may operate simultaneously. The present study does not include transport of magnetic flux into the lobes, which can act as a driver for thin CS formation. While this assumption is not realistic for substorms, the approach allows to examine solely the physics and properties of MFD, which is shown to be highly efficient to generate thin CSs with properties that agree well with observations. A study including the additional influence of a driving electric field will be published elsewhere. Simulation results including such an electric field show an additional CS but leave the main results presented here unaltered. The main findings based on the presented results are as follows:

• Convection out of the midnight tail region generates a major depletion of the equatorial magnetic flux. This MFD has important influence also on the ionospheric mapping. The removal of this amount of closed magnetic flux suggests a significant equatorward shift of all plasma sheet and tail features when mapped into the ionosphere.

• The maximum velocities range from 50 to about 200 km/s; the process is slow and satisfies the conditions for force balance implying a quasi-static evolution.

• The time scale for the thin CS formation is determined by the available flux in the midnight reservoir and the applied potential. For realistic values of the potential the time scale for CS formation ranges between 0.5 and 1 h.

• CS formation develops naturally in the near-Earth tail between $8R_E$ and $15R_E$, consistent with observations.

• CS thinning accelerates toward the end of the process, confirming the argument that the time scale is inversely proportional to the amount of magnetic flux available in the reservoir.

• Toward the end of the growth phase, convection velocities in the convection channels need to increase significantly to balance the decrease of the B_z component in order to maintain a reasonable potential. This also implies the need of faster convection in the ionosphere with a narrower flow channel.

• The thin CS has a sharp boundary toward the still dipolar region; however, a moderate decrease of the geosynchronous magnetic field by up to 40% is also observed.

The presented model uses a number of assumptions that should be compared to observations and global simulations. The concept of flux tube entropy conservation can be violated, for instance, by energy loss into the ionosphere or by energetic particle drifts [*Heinemann*, 1999]. Estimating these losses indicates that a violation of entropy conservation may occur on time scales of about 10 h for Earth's magnetotail [*Hall*, 2006] such that flux tube entropy should be reasonably conserved during the substorm growth phase. Also, the amount of magnetic flux transport (1.5×10^8 Wb) is so large that there is insufficient flux in the magnetotail even if much of the closed flux would be convected into the near-Earth tail.

Properties of the MFD mechanism are very robust. For all parameter variations that have been tested, we found basically the same results with minor quantitative changes, for instance, a change of the radial distance of the assumed flow channel by $2R_E$ shifts the thin CS region by less than this change. Results of a mapping of the MFD process into the ionosphere have been submitted in a separate publication [*Hsieh and Otto*, 2014]. This mapping shows an equatorward motion of all plasma sheet features, including the R1 and R2 field-aligned currents, and a strong convergence between the maximum R1 and R2 current systems.

In addition to the substorm CS formation there is much evidence for sporadic or intermittent reconnection in Earth's magnetotail [*Angelopoulos et al.*, 1992; *Runov et al.*, 2011; *Fu et al.*, 2012; *Lee et al.*, 2012; *Sanchez et al.*, 2012] and in the magnetodisc of the giant magnetospheres [*Kronberg et al.*, 2008; *Ge et al.*, 2010; *Jackman and Arridge*, 2011; *Kasahara et al.*, 2013]. Bursty bulk flows and dipolarization fronts can occur at all times (although they appear more frequent during the substorm growth phase) in the magnetotail [*Baumjohann et al.*, 1990; *Angelopoulos et al.*, 1994]. These are likely the result of localized intermittent reconnection, which requires thin CSs to start operating. Similarly, the large variation of azimuthal plasma convection from an average rotation profile in the Jovian magnetodisc [*Woch et al.*, 2002] and events with super-corotational flows in the Kronian magnetodisc [*Masters et al.*, 2011] indicate significant deviations of the flow from an average equilibrium situation which are likely associated with reconnection. Furthermore, the large scatter from very low to rather large equatorial magnetic field B_z components in the premidnight and morning sectors and a persistent small value of B_z in the postmidnight sector [*Vogt et al.*, 2011] are indicative of intermittent and localized reconnection and thin CSs in the Jovian magnetosphere. Here boundary perturbations, divergent convection and MFD, or changes of the entropy content may be viable mechanisms for thin CS formation.

Possible scenarios for flux tube entropy changes are the pick-up of neutral material or the loss of particle energy through precipitation into the ionosphere. An azimuthal variation of the magnetodisc B_z also implies azimuthal gradients in the ∇B and curvatures drifts, which can deplete or enhance the local flux tube entropy. Our examples with a localized entropy source or sink on almost dipolar flux tubes demonstrate that a persistent flux tube entropy change results in thin CS formation adjacent to the location of localized heating or cooling. Observations indicate a radial increase of plasma entropy of an unresolved nature [*Bagenal and Delamere*, 2011]. However, studies for the giant magnetospheres should be extended to include inertial forces due to the fast rotation as well as allow for a gyrotropic pressure to account for variations of the parallel and perpendicular pressures, which are potentially important for CS formation in giant magnetospheres.

ACKNOWLEDGMENTS

Some of the presented results have been funded by the NASA G/Science grant NNX09AI09G. Computations were supported by the Arctic Region Supercomputer Center (ARSC) at the University of Alaska Fairbanks.

REFERENCES

Achilleos, N., P. Guio, and C. S. Arridge (2010), A model of force balance in Saturn's magnetodisc, *Monthly Notices R. Astron. Soc.*, *401*, 2349–2371. doi: 10.1111/j.1365-2966.2009.15865.x

Adamson, E., A. Otto, and K. Nykyri (2011), 3D mesoscale MHD simulations of a cusp-like configuration: Method and first results, *Ann. Geophys,* doi:10.5194/angeo-29-759-2011.

Akasofu, S.-I. (1964), The development of the auroral substorm, *Planet. Space Sci.*, *12*, 273.

Angelopoulos, V. (2008), The THEMIS Mission, *Space Sci. Rev.*, *141*(1–4), 5–34, doi:10.1007/s11214-008-9336-1.

Angelopoulos, V., W. Baumjohann, C. F. Kennel, F. V. Coroniti, M. G. Kivelson, R. Pellat, R. J. Walker, H. Luehr, and G. Paschmann (1992), Bursty bulk flows in the inner central plasma sheet, *J. Geophys. Res.*, *97*, 4027.

Angelopoulos, V., et al. (1994), Statistical characteristics of bursty bulk flow events, *J. Geophys. Res.*, *99*, 21,257, doi:10.1029/94JA01263.

Asano, Y., T. Mukai, M. Hoshino, Y. Saito, H. Hayakawa, and T. Nagai (2004), Statistical study of thin current sheet evolution around substorm onset, *J. Geophys. Res.*, *109*, A05213, doi:10.1029/2004JA010413.

Badman, S. V., and S. W. H. Cowley (2007), Significance of Dungey-cycle flows in Jupiter's and Saturn's magnetospheres, and their identification on closed equatorial field lines, *Ann. Geophys.*, *25*, 941–951, doi:10.5194/angeo-25-941-2007.

Bagenal, F., and P. A. Delamere (2011), Flow of mass and energy in the magnetospheres of Jupiter and Saturn, *J. Geophys. Res.*, 116, A05209, doi:10.1029/2010JA016294.

Baumjohann, W. (2002), Modes of convection in the magnetotail, *Phys. Plasmas*, *9*, 3665–3667, doi:10.1063/1.1499116.

Baumjohann, W., G. Paschmann, and H. Luehr (1990), Characteristics of high-speed ion flows in the plasma sheet, *J. Geophys. Res.*, *3801–3809*, doi:10.1029/JA095iA04p03801.

Becker, U., T. Neukirch, and K. Schindler (2001), On the quasi-static development of thin current sheets in magnetotail-like magnetic fields, *J. Geophys. Res.*, *106*, 3811–3826, doi:10.1029/2000JA900141.

Birn, J., M. Hesse, and K. Schindler (1998), Formation of thin current sheets in space plasmas, *J. Geophys. Res.*, *103*, 6843–6852, doi:10.1029/97JA03602.

Birn, J., J. C. Dorelli, M. Hesse, and K. Schindler (2004a), Thin current sheets and loss of equilibrium: Three-dimensional theory and simulations, *J. Geophys. Res.*, 109, A02215, doi:10.1029/2003JA010275.

Birn, J., J. Raeder, Y. Wang, R. Wolf, and M. Hesse (2004b), On the propagation of bubbles in the geomagnetic tail, *Ann. Geophys.*, *22*, 1773–1786, doi:10.5194/angeo-22-1773-2004.

Birn, J., M. Hesse, K. Schindler, and S. Zaharia (2009), Role of entropy in magnetotail dynamics, *J. Geophys. Res.*, *114*, A00D03, doi:10.1029/2008JA014015.

Boakes, P. D., S. E. Milan, G. A. Abel, M. P. Freeman, G. Chisham, and B. Hubert (2009), A statistical study of the open magnetic flux content of the magnetosphere at the time of substorm onset, *Geophys. Res. Lett.*, *36*, L04105, doi:10.1029/2008GL037059.

Borovsky, J. E., M. F. Thomsen, R. C. Elphic, T. E. Cayton, and D. J. McComas (1998), The transport of plasma sheet material from the distant tail to geosynchronous orbit, *J. Geophys. Res.*, *103*, 20,297–20,332, doi:10.1029/97JA03144.

Caudal, G. (1986), A self-consistent model of Jupiter's magnetodisc including the effects of centrifugal force and pressure, *J. Geophys. Res.*, *91*, 4201–4221, doi:10.1029/JA091iA04p04201.

Coroniti, F. V., and C. F. Kennel (1973), Can the ionosphere regulate magnetospheric convection?, *J. Geophys. Res.*, *78*, 2837, doi:10.1029/JA078i016p02837.

Cowley, S. W. H., S. V. Badman, E. J. Bunce, J. T. Clarke, J.-C. GéRard, D. Grodent, C. M. Jackman, S. E. Milan, and T. K. Yeoman (2005), Reconnection in a rotation-dominated magnetosphere and its relation to Saturn's auroral dynamics, *J. Geophys. Res.*, *110*, A02201, doi:10.1029/2004JA010796.

Delamere, P. A., and F. Bagenal (2010), Solar wind interaction with Jupiter's magnetosphere, *J. Geophys. Res.*, *115*, A10,201, doi:10.1029/2010JA015347.

Erickson, G. M., and R. A. Wolf (1980), Is steady convection possible in the earth's magnetotail, *Geophys. Res. Lett.*, *7*, 897–900.

Fu, H. S., Y. V. Khotyaintsev, A. Vaivads, M. André, and S. Y. Huang (2012), Occurrence rate of earthward-propagating dipolarization fronts, *Geophys. Res. Lett.*, *39*, L10101, doi:10.1029/2012GL051784.

Ge, Y. S., C. T. Russell, and K. K. Khurana (2010), Reconnection sites in Jupiters magnetotail and relation to Jovian auroras, *Planet. Space Sci.*, *58*, 1455–1469, doi:10.1016/j.pss.2010.06.013.

Hall, F. (2006), A mechanism for current sheet thinning in the growth phase of magnetospheric substorms, Ph.D. thesis, University of Alaska Fairbanks, Fairbanks, Alaska.

Heinemann, M. (1999), Role of collisionless heat flux in magnetospheric convection, *J. Geophys. Res.*, *104*, 28,397–28,410, doi:10.1029/1999JA900401.

Hesse, M., and J. Birn (1993), Three-dimensional magnetotail equilibria by numerical relaxation techniques, *J. Geophys. Res.*, *98*, 3973–3982, doi:10.1029/92JA02905.

Hill, T. W., et al. (2005), Evidence for rotationally driven plasma transport in Saturn's magnetosphere, *Geophys. Res. Lett.*, *32*, 14, doi:10.1029/2005GL022620.

Hsieh, M.-S., and A. Otto (2014), The influence of magnetic flux depletion on the magnetotail and auroral morphology during the substorm growth phase, *J. Geophys. Res Space Phys.*, *119*, 3430–3443, doi = 10.1002/2013JA019459.

Hu, B., R. A. Wolf, F. R. Toffoletto, J. Yang, and J. Raeder (2011), Consequences of violation of frozen-in-flux: Evidence from OpenGGCM simulations, *J. Geophys. Res.*, *116*, A06223, doi:10.1029/2011JA016667.

Jackman, C. M., and C. S. Arridge (2011), Statistical properties of the magnetic field in the Kronian magnetotail lobes and current sheet, *J. Geophys. Res.*, *116*, A05224, doi:10.1029/2010JA015973.

Jackman, C. M., and S. W. H. Cowley (2006), A model of the plasma flow and current in Saturn's polar ionosphere under conditions of strong Dungey cycle driving, *Ann. Geophys.*, *24*, 1029–1055, doi:10.5194/angeo-24-1029-2006.

Jackman, C. M., J. A. Slavin, and S. W. H. Cowley (2011), Cassini observations of plasmoid structure and dynamics: Implications for the role of magnetic reconnection in magnetospheric circulation at Saturn, *J. Geophys. Res.*, *116*, 10,212, doi:10.1029/2011JA016682.

Jia, X., K. C. Hansen, T. I. Gombosi, M. G. Kivelson, G. Tóth, D. L. DeZeeuw, and A. J. Ridley (2012), Magnetospheric configuration and dynamics of Saturn's magnetosphere: A global MHD simulation, *J. Geophys. Res.*, *117*, A05225, doi:10.1029/2012JA017575.

Kan, J. R. (1990), Tail-like reconfiguration of the plasma sheet during the substorm growth phase, *Geophys. Res. Lett.*, *17*, 2309–2312, doi:10.1029/GL017i013p02309.

Kan, J. R., T. A. Potemra, S. Kokubun, and T. Iijima (Eds.) (1991), Magnetospheric Substorms, of Geophys. Monogr. Ser., vol. 64 AGU, Washington, D. C..

Kasahara, S., E. A. Kronberg, T. Kimura, C. Tao, S. V. Badman, A. Masters, A. Retinò, N. Krupp, and M. Fujimoto (2013), Asymmetric distribution of reconnection jet fronts in the Jovian nightside magnetosphere, *J. Geophys. Res.*, *118*, 375–384, doi:10.1029/2012JA018130.

Kivelson, M. G., and D. J. Southwood (2005), Dynamical consequences of two modes of centrifugal instability in Jupiter's outer magnetosphere, *J. Geophys. Res.*, *110*, A12209, doi:10.1029/2005JA011176.

Kronberg, E. A., J. Woch, N. Krupp, and A. Lagg (2008), Mass release process in the Jovian magnetosphere: Statistics on particle burst parameters, *J. Geophys. Res.*, *113*, A10202, doi:10.1029/2008JA013332.

Kubyshkina, M., V. Sergeev, N. Tsyganenko, V. Angelopoulos, A. Runov, E. Donovan, H. Singer, U. Auster, and W. Baumjohann (2011), Time-dependent magnetospheric configuration and breakup mapping during a substorm, *J. Geophys. Res.*, *116*, A00127, doi:10.1029/2010JA015882.

Lee, D.-Y., H.-S. Kim, S. Ohtani, and M. Y. Park (2012), Statistical characteristics of plasma flows associated with magnetic dipolarizations in the near-tail region of $r < 12$ RE, *J. Geophys. Res.*, *117*, A01207, doi:10.1029/ 2011JA017246.

Lee, L. C., L. Zhang, G. S. Choe, and H. J. Cai (1995), Formation of a very thin current sheet in the near-earth magnetotail and the explosive growth phase of substorms, *Geophys. Res. Lett.*, *22*, 1137–1140, doi:10.1029/95GL01033.

Lee, L. C., L. Zhang, A. Otto, G. S. Choe, and H. J. Cai (1998), Entropy antidiffusion instability and formation of a thin current sheet during geomagnetic substorms, *J. Geophys. Res.*, *103*, 29,419–29,428, doi:10.1029/97JA02141.

Ma, X., and A. Otto (2013), Mechanisms of field-aligned current formation in magnetic reconnection, *J. Geophys. Res.*, *118*, 4906–4914, doi:10.1002/jgra.50457.

Masters, A., M. F. Thomsen, S. V. Badman, C. S. Arridge, D. T. Young, A. J. Coates, and M. K. Dougherty (2011), Supercorotating return flow from reconnection in Saturn's magnetotail, *Geophys. Res. Lett.*, *38*, L03103, doi:10.1029/2010GL046149.

McComas, D. J., and F. Bagenal (2007), Jupiter: A fundamentally different magnetospheric interaction with the solar wind, *Geophys. Res. Lett.*, *34*, L20106, doi:10.1029/2007GL031078.

McPherron, R. L. (1979), Magnetospheric substorms, *Rev. Geophys. Space Phys.*, *17*, 657.

McPherron, R. L., A. Nishida, and C. T. Russell (1987), Is near-Earth current sheet thinning the cause of auroral substorm onset? in *Quantitative Modeling of the Magnetosphere-Ionosphere Coupling Processes*, edited by Y. Kamide and R. A. Wolf, p. 252, Kyoto Sangyo Univ., Kyoto, Japan.

Mende, S. B., S. E. Harris, H. U. Frey, V. Angelopoulos, C. T. Russell, E. Donovan, B. Jackel, M. Greffen, and L. M. Peticolas (2008), The THEMIS array of ground-based observatories for the study of auroral substorms, *Space Sci. Rev.*, *141*, 357–387, doi:10.1007/s11214-008-9380-x.

Milan, S. E., A. Grocott, C. Forsyth, S. M. Imber, P. D. Boakes, and B. Hubert (2009), A superposed epoch analysis of auroral evolution during substorm growth, onset and recovery: Open magnetic flux control of substorm intensity, *Ann. Geophys.*, *27*, 659–668, doi:10.5194/angeo-27-659-2009.

Newell, P. T., V. A. Sergeev, G. R. Bikkuzina, and S. Wing (1998), Characterizing the state of the magnetosphere: Testing the ion precipitation maxima latitude (b2i) and the ion isotropy boundary, *J. Geophys. Res.*, *103*, 4739–4746, doi:10.1029/97JA03622.

Ohtani, S., A. T. Y. Lui, K. Takahashi, D. G. Mitchell, and T. Sarris (2000), Ion dynamics and tail current intensification prior to dipolarization: The June 1, 1985, event, *J. Geophys. Res.*, *105*, 25,233–25,246, doi:10.1029/2000JA000040.

Otto, A. (1990), 3D resistive MHD computations of magnetospheric physics, *Comput. Phys. Commun.*, *59*, 185.

Otto, A. (1995), Forced three-dimensional magnetic reconnection due to linkage of magnetic flux tubes, *J. Geophys. Res.*, *100*, 11,863.

Petrukovich, A. A., W. Baumjohann, R. Nakamura, A. Runov, A. Balogh, and H. Rème (2007), Thinning and stretching of the plasma sheet, *J. Geophys. Res.*, *112*, A10213, doi:10.1029/2007JA012349.

Runov, A., V. Angelopoulos, X.-Z. Zhou, X.-J. Zhang, S. Li, F. Plaschke, and J. Bonnell (2011), A THEMIS multicase study of dipolarization fronts in the magnetotail plasma sheet, *J. Geophys. Res.*, *116*, A05216, doi:10.1029/2010JA016316.

Saito, M. H., D. Fairfield, G. Le, L.-N. Hau, V. Angelopoulos, J. P. McFadden, U. Auster, J. W. Bonnell, and D. Larson (2011), Structure, force balance, and evolution of incompressible cross-tail current sheet thinning, *J. Geophys. Res.*, *116*, A10217, doi:10.1029/2011JA016654.

Sanchez, E. R., S. Wing, E. Spanswick, and E. Donovan (2012), Entropy conservation and rate of propagation of bubbles in the Earth's magnetotail: A case study, *J. Geophys. Res.*, *117*, A05226, doi:10.1029/2011JA017287.

Sanny, J., R. L. McPherron, C. T. Russell, D. N. Baker, T. I. Pulkkinen, and A. Nishida (1994), Growth-phase thinning of the near-Earth current sheet during the CDAW 6 substorm, *J. Geophys. Res.*, *99*, 5805–5816, doi:10.1029/93JA03235.

Schindler, K., and J. Birn (1982), Self-consistent theory of time-dependent convection in the earth's magnetotail, *J. Geophys. Res.*, *87*, 2263.

Schindler, K., and J. Birn (1993), On the cause of thin current sheets in the near-Earth magnetotail and their possible significance for magnetospheric substorms, *J. Geophys. Res.*, *98*, 15,477, doi:10.1029/93JA01047.

Sergeev, V., V. Angelopoulos, M. Kubyshkina, E. Donovan, X.-Z. Zhou, A. Runov, H. Singer, J. McFadden, and R. Nakamura (2011), Substorm growth and expansion onset as observed with ideal ground-spacecraft THEMIS coverage, *J. Geophys. Res.*, *116*, A00126, doi:10.1029/2010JA015689.

Sergeev, V. A., E. M. Sazhina, N. A. Tsyganenko, J. A. Lundblad, and F. Soraas (1983), Pitch-angle scattering of energetic protons in the magnetotail current sheet as the dominant source of their isotropic precipitation into the nightside ionosphere, *Planet. Space Sci.*, *31*, 1147–1155, doi:10.1016/0032-0633(83)90103-4.

Sergeev, V. A., P. Tanskanen, K. Mursula, A. Korth, and R. C. Elphic (1990), Current sheet thickness in the near-earth plasma sheet during substorm growth phase, *J. Geophys. Res.*, *95*, 3819–3828, doi:10.1029/JA095iA04p03819.

Sergeev, V. A., D. G. Mitchell, C. T. Russell, and D. J. Williams (1993), Structure of the tail plasma/current sheet at ~11 R_E and its changes in the course of a substorm, *J. Geophys. Res.*, *98*, 17,345–17,366, doi:10.1029/93JA01151.

Sergeev, V., Y. Nishimura, M. Kubyshkina, V. Angelopoulos, R. Nakamura, and H. Singer (2012), Magnetospheric location of the equatorward prebreakup arc, *J. Geophys. Res.*, *117*, 1212, doi:10.1029/2011JA017154.

Sharma, A. S., Y. Kamide, and G. Lakhina (Eds.) (2003), Disturbances in Geospace: The Storm-Substorm Relationship, of Geophys. Monogr. Ser., vol. 142 AGU, Washington, D. C.

Shevchenko, I. G., V. Sergeev, M. Kubyshkina, V. Angelopoulos, K. H. Glassmeier, and H. J. Singer (2010), Estimation of magnetosphere-ionosphere mapping accuracy using isotropy boundary and THEMIS observations, *J. Geophys. Res.*, *115*, 11,206, doi:10.1029/2010JA015354.

Slavin, J. A., et al. (2010), MESSENGER observations of extreme loading and unloading of Mercury's magnetic tail, *Science*, *329*, 665–668, doi:10.1126/science.1188067.

Slavin, J. A., et al. (2012), MESSENGER and Mariner 10 flyby observations of magnetotail structure and dynamics at Mercury, *J. Geophys. Res.*, *117*, 1215, doi: 10.1029/2011JA016900.

Sundberg, T., et al. (2012), MESSENGER observations of dipolarization events in Mercury's magnetotail, *J. Geophys. Res.*, *117*, 0, doi:10.1029/2012JA017756.

Thompson, S. M., M. G. Kivelson, K. K. Khurana, R. L. McPherron, J. M. Weygand, A. Balogh, H. Réme, and L. M. Kistler (2005), Dynamic Harris current sheet thickness from Cluster current density and plasma measurements, *J. Geophys. Res.*, *110*, A02212, doi:10.1029/2004JA010714.

Tsyganenko, N. A., and D. P. Stern (1996), Modeling the global magnetic field of the large-scale Birkeland current systems, *J. Geophys. Res.*, *101*, 27,187.

Vasyliunas, V. M. (1983), Plasma distribution and flow, in Physics of the Jovian Magnetosphere, edited by A. J. Dessler, pp. 395–453, Cambridge Univ. Press, New York.

Vasyliunas, V. M. (2011), The magnetospheres of Uranus and Neptune: What we know and what we don't know, proper presented at the EPSC-DPS Joint Meet. 2011, p. 1270.

Vogt, M. F., M. G. Kivelson, K. K. Khurana, R. J. Walker, B. Bonfond, D. Grodent, and A. Radioti (2011), Improved mapping of Jupiter's auroral features to magnetospheric sources, *J. Geophys. Res.*, *116*, A03220, doi:10.1029/2010JA016148.

Walker, R. J., K. Fukazawa, T. Ogino, and D. Morozoff (2011), A simulation study of Kelvin-Helmholtz waves at Saturn's magnetopause, *J. Geophys. Res.*, *116*, A03,203, doi:10.1029/2010JA015905.

Wiegelmann, T., and K. Schindler (1995), Formation of thin current sheets in a quasistatic magnetotail model, *Geophys. Res. Lett.*, *22*, 2057–2060, doi:10.1029/95GL01980.

Woch, J., N. Krupp, and A. Lagg (2002), Particle bursts in the Jovian magnetosphere: Evidence for a near-Jupiter neutral line, *Geophys. Res. Lett.*, *29*(7), 070,000–1, doi:10.1029/2001GL014080.

Wolf, R. A., Y. Wan, X. Xing, J.-C. Zhang, and S. Sazykin (2009), Entropy and plasma sheet transport, *J. Geophys. Res.*, *114*, A00D05, doi:10.1029/2009JA014044.

Wolf, R. A., C. X. Chen, and F. R. Toffoletto (2012), Thin filament simulations for Earth's plasma sheet: Interchange oscillations, *J. Geophys. Res.*, *117*, A02215, doi: 10.1029/2011JA016971.

Yang, J., R. A. Wolf, and F. R. Toffoletto (2011), Accelerated thinning of the near-Earth plasma sheet caused by a bubble-blob pair, *Geophys. Res. Lett.*, *38*, L01107, doi: 10.1029/2010GL045993.

Zaharia, S., and C. Z. Cheng (2003), Near-Earth thin current sheets and Birkeland currents during substorm growth phase, *Geophys. Res. Lett.*, *30*(17), 1883, doi: 10.1029/2003GL017456.

Zhang, J.-C., R. A. Wolf, S. Sazykin, and F. R. Toffoletto (2008), Injection of a bubble into the inner magnetosphere, *Geophys. Res. Lett.*, *35*, L02110, doi:10.1029/2007GL032048.

Zhang, J.-C., R. A. Wolf, R. W. Spiro, G. M. Erickson, S. Sazykin, F. R. Toffoletto, and J. Yang (2009), Rice convection model simulation of the substorm-associated injection of an observed plasma bubble into the inner magnetosphere: 2. Simulation results, *J. Geophys. Res.*, *114*, A08219, doi:10.1029/2009JA014131.

Zhou, X.-Z., et al. (2009), Thin current sheet in the substorm late growth phase: Modeling of THEMIS observations, *J. Geophys. Res.*, *114*, A03223, doi:10.1029/2008JA013777.

Zimbardo, G. (1989), A self-consistent picture of Jupiter's nightside magnetosphere, *J. Geophys. Res.*, *94*, 8707–8719, doi:10.1029/JA094iA07p08707.

18

Substorms: Plasma and Magnetic Flux Transport from Magnetic Tail into Magnetosphere

Gerhard Haerendel

18.1. INTRODUCTION

This review deals with the essence of the magnetospheric substorm, the return of magnetic flux into the magnetosphere after disconnection from the solar wind magnetic field. When the conditions for reconnection of the solar wind and Earth's magnetic fields are met at the frontside of the magnetosphere, open magnetic flux is added to the polar cap. The field lines are being stretched into a tail of hundreds of Earth radii length. At some point in time, reconnection somewhere in the near-Earth tail leads, on the one hand, to ejection of accumulated field into the antisolar direction and, on the other hand, to the return of a newly closed field loaded with energized plasma toward the inner magnetosphere. The phase from the beginning of new flux addition to the polar cap until the onset of a substorm, i.e., the massive shedding of energy into the atmosphere, is called the "growth phase". This is a practical phenomenological definition. However, neither does the flux addition by dayside reconnection have to end with the onset of a substorm nor does the onset coincide with the beginning of reconnection in the tail. Reconnection depends on sufficient thinning of the tail current sheet, while the substorm, though not necessarily its onset, requires arrival of magnetic flux and plasma at the outer boundary of the dipolar magnetosphere. This introduces in any case a delay of a few minutes. The flux accumulation in the tail, on the other hand, depends on the duration of the conditions allowing frontside reconnection. This can well extend beyond the substorm onset and lead to a great variability of the temporal profiles.

Max Planck Institute for Extraterrestrial Physics, Garching, Germany

The characterized sequence of events involves a number of very intriguing plasma physical processes all of which are still the subject of intense investigation. Here we will address the last step, the addition of the reconnected and earthward transported flux to the dipolar magnetic field and the conditioning of the magnetosphere for this to happen. Thus there are three fundamental transport processes involved: (1) thinning of the tail plasma sheet and accompanying recession of the outer boundary of the dipolar magnetosphere during the growth phase, (2) flux transport along the tail toward that boundary after onset of tail reconnection, and (3) penetration of plasma and magnetic flux into the dipolar magnetosphere. The last step has a clear onset signature through the appearance of a westward electric field. This review will be structured according to these three phases to be followed by (4) a short consideration of the further mass distribution after entry. Finally, it is rewarding to have a look at corresponding processes in the Jupiter and Saturn magnetospheres and tails, which are strongly dominated by the fast planetary rotations. The intention of this review is not to attempt completeness in referencing the rich pertinent literature but rather to elucidate some key aspects of the entry problem, albeit from a personal vantage point, and address the still open questions.

18.2. THINNING OF PLASMA SHEET AND RECESSION OF OUTER BOUNDARY OF DIPOLAR MAGNETOSPHERE

The first question arising in this context relates to the definition of the outer boundary of the dipolar magnetosphere on the nightside. This is not an easy undertaking, since the transition from a mildly distorted dipole field to a stretched tail field must be gradual. However, good

Magnetotails in the Solar System, Geophysical Monograph 207. First Edition.
Edited by Andreas Keiling, Caitríona M. Jackman, and Peter A. Delamere.

practical markers have been defined as "ion isotropy boundary" (IB) by *Sergeev et al.* [1983] or "ion precipitation energy flux maxima" (b2i) by *Newell et al.* [1998]. The argument behind these empirical definitions is that the hydrogen or proton arc is caused by pitch angle scattering owing to nonadiabatic motions in the stretched tail field and terminates where the field assumes dipolar geometry [*Samson et al.*, 1992]. On all-sky camera images that arc is often so faint as to be noticeable, but when observable, its homogeneity and continuity are indeed remarkable. While emissions at 5577 Å are normally attributed to electron precipitation, a good fraction of the diffuse aurora emission can be excited by proton precipitation [*Fukunishi*, 1975; *Eather*, 1967]. However, true electron arcs form during the growth phase. They are typically located 1° poleward of the equatorward boundary of the proton arc [*Deehr and Lummerzheim*, 2001; *Lessard et al.*, 2007]. The visibility of the arc tends to increase toward the end of the growth phase. Most important is its equatorward motion during the growth phase [*Deehr and Lummerzheim*, 2001; *Kadokura et al.*, 2002] and its brightening as the first visible signal of substorm onset [*Akasofu*, 1964]. Good images of its apparent growth in longitudinal extent before substorm onset are given by *Shiokawa et al.* [2009]. The arc consists of proton and electron precipitation, whereby the electron arc tends to be located approximately 100 km poleward of the equatorward border of the hydrogen arc, which is broadened through transport by charge-exchanged neutrals. The electron arc is by no means a unique structure. Splitting into or appearance of new parallel arcs is often observed, and it is not always the most equatorward structure that shows the first brightening [*Lyons et al.*, 2002]. Observations from FAST crossings [*Lessard et al.*, 2007] exhibit a peaked energy spectrum like the so-called inverted-V. In other cases diffuse or strongly field aligned soft electron fluxes dominate. Figure 18.1 shows proton and electron energy-time spectra of FAST which demonstrate the dual nature of the electron fluxes, dominantly arising from pitch angle scattering with narrow inverted-V arcs embedded. Ion and electron precipitations are strongly overlapping [see also *Lessard et al.*, 2007, Figure 15]. These features are relevant in the context of the underlying currents and flows (see below).

Little is known about the related field-aligned current system. In cases where the ionospheric conductivity is sufficiently large and measurable Hall currents are flowing overhead, ground magnetic perturbations systematically show negative H components on the dawnside and positive on the duskside [e.g., *McPherron*, 1970, 1973; *Kadokura et al.*, 2002]). The latter authors concluded accordingly that the growth-phase arc is tracing westward and eastward flow channels, respectively, in the sense of the DP2 return flows. Thus thinning of the plasma sheet proceeds under earthward approach of its inner edge, diversion of the flow along the poleward border of dipolar magnetosphere toward daytime, and mass loss through particle precipitation.

The continuity of the growth-phase arc, the absence of visible structures equatorward of it, and the balanced field-aligned current system together demonstrate a well-known result from magnetohydrodynamic (MHD) plasma physics when considering the plasma sheet above, namely that a plasma supported by a magnetic field is interchange stable if the field lines curve into the plasma [*Krall and Trivelpiece*, 1973]. The dipolar magnetosphere thus shields itself successfully against penetration of external plasma approaching from the tail, even when pushed earthward during the growth phase. The shrinking volume of the nightside magnetosphere is balanced by a sunward transport of magnetic flux, as shown by the DP2 equivalent current system (Figure 18.2). This flow is, at least in part, required by the need for replacing the magnetic flux eroded on the frontside by reconnection. This was originally proposed by *Coroniti and Kennel* [1973] [see also *Haerendel* [2010]].

Replacement of magnetic flux eroded on the frontside is only part of the story. There must be another driver. This is ultimately the solar wind. By draping the newly connected field lines into Earth's flaring tail, it drags plasma across the polar cap and exerts a transverse pressure force on the plasma sheet. The integrated force corresponds to a pull on Earth itself [*Siscoe*, 1966]. Owing to the lobe pressure the plasma sheet is progressively becoming thinner and its inner edge moves earthward, aided by the earthward directed tensions of strongly stretched but still closed magnetic field lines. This inward transport necessitates a DP2-type convection of magnetosphere and ionosphere (Figure 18.2). The generator of the related current system must be located at the interface of the tail lobe and plasma sheet, since the first essentially exerts the force, while the latter sheds energy into the ionosphere. The tail lobe and plasma sheet are in lateral pressure equilibrium. In addition, magnetic shear stresses transport momentum into the ionosphere and continue the polar cap convection into the trace of the plasma sheet. *Haerendel* [2009, 2011] identified the eastward directed magnetic gradient currents at the transition from the low-beta lobe field to the high-beta plasma sheet as the generator currents connecting to the field-aligned currents that transport these shear stresses. This is sketched in Figure 18.3, with generator, field-aligned, and ionospheric closure currents marked by green circles and arrows and flow directions in the plasma sheet and ionosphere by blue arrows. The equatorward flow from the polar cap is longitudinally diverted at the inner edge, because of the stability of the dipolar magnetosphere against interchange

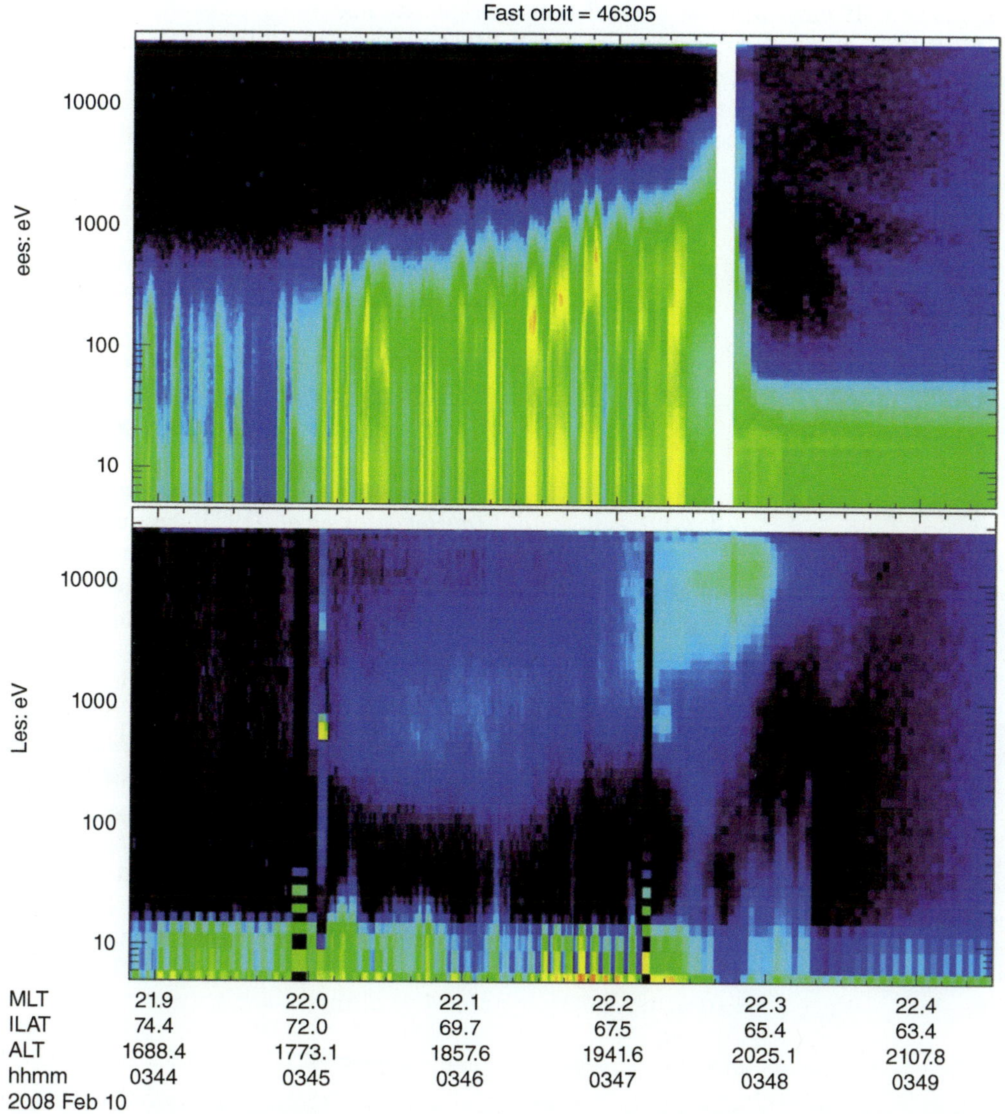

Figure 18.1 Electron and ion energy-time spectra measured by FAST at the crossing of a growth-phase arc at 22.2 MLT 26 min before substorm onset. (Courtesy of J. P. McFadden 2013.) Diffuse precipitation dominates for both species with narrow inverted-V-type spectra interdispersed. The graph covers the Invariant Latitude (ILAT) interval from 74.4° to 63.4° and a spacecraft altitude increasing from 1688 to 2108 km.

motions. However, the boundary slowly recedes inward. The field-aligned currents along this boundary, as shown in Figures 18.2 and 18.3, are shielding the convection from the dipolar magnetosphere. They are directed upward on the dawnside and downward on the duskside, just opposite of the situation at substorm onset (section 18.5.). In summary, we have identified two forces driving the plasma and field transport during the growth phase, suction from the eroded frontside magnetopause and solar wind/tail lobe pressure forces.

This seemingly straightforward scheme is more often than not disturbed by another transport phenomenon, the auroral streamers. *Nishimura et al.* [2010a] have found that faint auroral arcs, often structured by auroral rays, are launched from poleward boundary intensifications (PBIs) [*Zesta et al.*, 2002], propagate equatorward, and subsequently turn east- or westward while approaching the growth-phase arc. Most surprising was the discovery that substorm onsets (or auroral intensifications) often occurred when closest approach was reached between the two arcs [see also *Mende et al.*, 2011]. This led to the speculation that interaction between the two arcs may trigger the onsets, as opposed to the traditional thinking that they are caused by the arrival of reconnection-triggered flows. However, as shown by *Mende et al.* [2011], a significant number of substorm onsets do not exhibit the described trigger pattern. It must also be noticed that bursty bulk flows (BBFs) or dipolarization events (see next section) need only a few minutes from the reconnection site to the boundary of the dipolar magnetosphere, while the auroral streamers often appear up to 15 min prior to onset [*Zesta et al.*, 2002]. This time delay should not be surprising in the light of recent evidence for their origin outside $30R_E$, possibly at the distant neutral line [*Nishimura et al.*, 2013]. Therefore the question arises whether the arrival of auroral streamers near the growth-phase arc and substorm onset are causally related or whether both are expressions of the preconditioning by the same event, a final thinning of the plasma sheet. Clarification needs in the first place a better understanding of the origin and nature of the auroral streamers. This is a hot topic of present-day research which we will address in more detail in the next section.

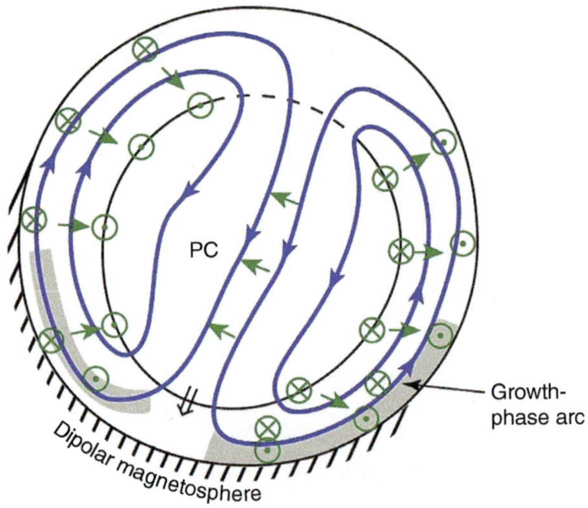

Figure 18.2 Idealized pattern of the plasma convection across the polar cap (inner circle) and of the return flow in a region of strongly stretched but closed magnetic field lines, poleward of the dipolar magnetosphere. Circles indicate the ionospheric entry and exit of field-aligned currents. Shaded regions represent growth-phase arcs.

18.3. MAGNETIC FLUX AND PLASMA TRANSPORT ALONG TAIL AFTER ONSET OF RECONNECTION

The discovery of BBFs by *Baumjohann et al.* [1990] and *Angelopoulos et al.* [1992] produced the missing link in *Dungey's* [1961] scenario of magnetic flux circulation from reconnection at the frontside of the magnetosphere, via reconnection in the tail, to magnetospheric convection back toward the dayside. Short flow bursts with high speeds were found to be accompanied by increases in B_z, so-called dipolarization fronts. They have been extensively studied in recent years due to the existence of the five THEMIS satellites and also by numerical simulations. Superimposed epoch analysis of six such events as shown in Figure 18.4 [*Runov et al.*, 2011] reveals a number of characteristic features. Most important besides the name-giving increasing B_z behind the front are the electron and ion temperature enhancements and density decreases of about 40% behind the front.

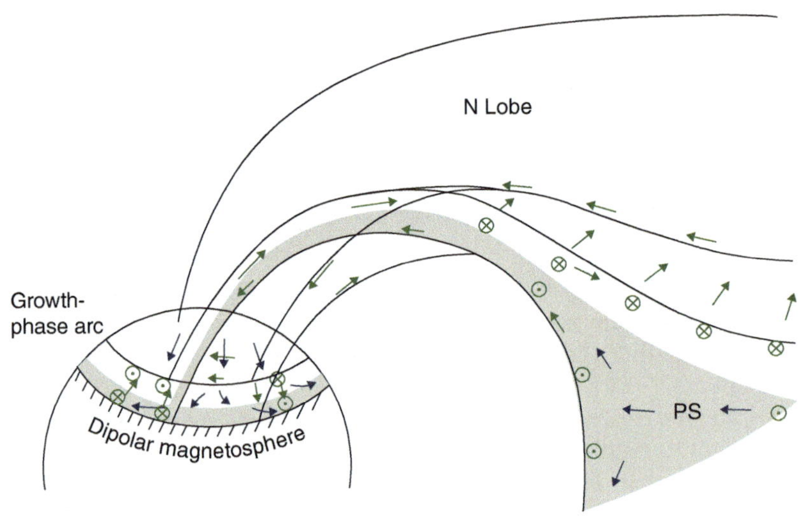

Figure 18.3 Concept of the sources of field-aligned currents driving the return flow during the growth phase (cf. Figure 18.2). The main generator is located in the boundary layer between the high-beta plasma sheet and the tail lobe. Generator, field-aligned, and ionospheric closure currents are marked by green circles and arrows and flow directions in plasma sheet and ionosphere by blue arrows. The field-aligned currents at the earthward edge are enforced by the stability of the outer surface of the dipolar magnetosphere.

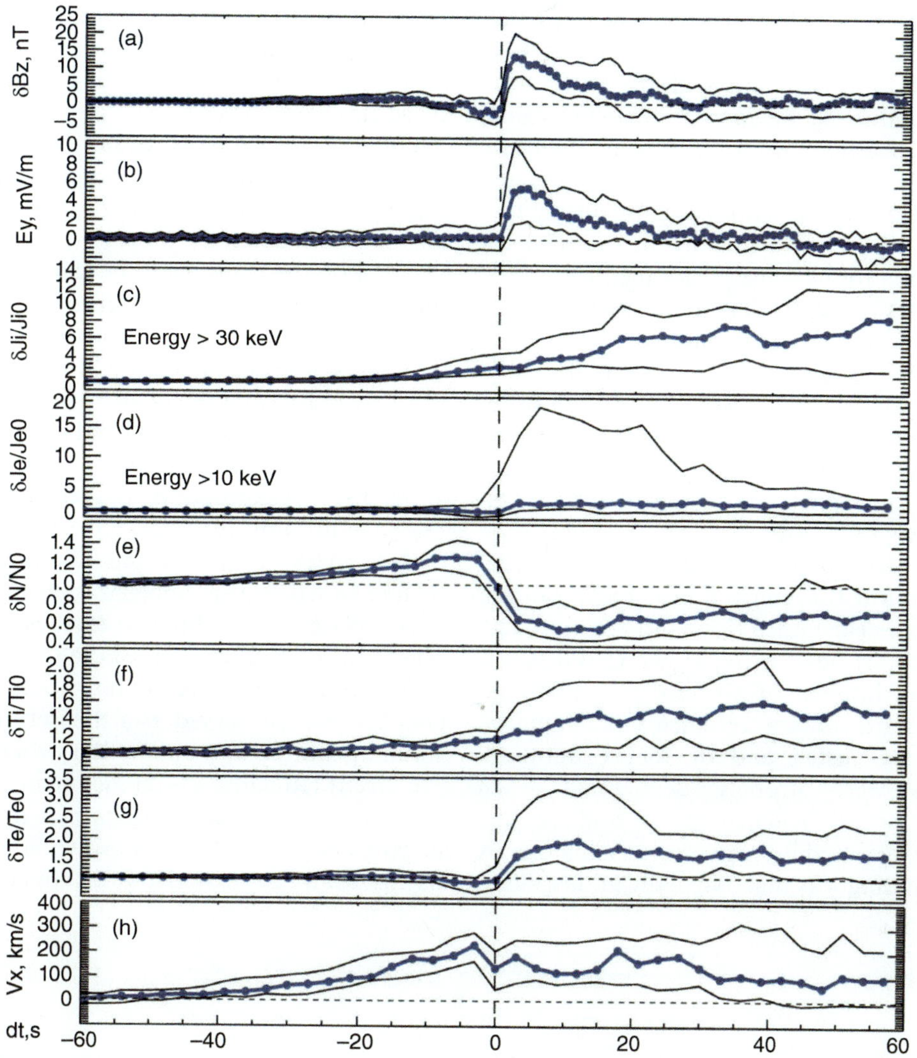

Figure 18.4 Superposed epoch analysis of magnetic and particle observations of six dipolarization fronts observed with the THEMIS spacecraft. Median (thick blue line) and lower quartiles (thin black lines) of (a) magnetic and (b) electric field components, (c) ion and (d) electron energy fluxes, normalized by the initial values, (e) density, (f) ion and (g) electron temperatures, and the *x* component of the ion velocity. From *Runov et al.* [2011].

Furthermore, preceding the front are density and ion velocity increases accompanied by slight decreases of B_z. The plasma pressure changes very little through the dipolarization event [*Dubyagin et al.*, 2010] whereas the magnetic pressure strongly increases. The plasma beta, typically above unity in front of the event, decreases significantly.

Dipolarization fronts are driven by magnetic tangential stresses generated by field line cutting and reconnection in the near-Earth tail. *Erickson and Wolf* [1980] had asked the question whether such a flux tube can proceed earthward under adiabatic contraction and concluded that the decreasing flux tube volume would cause an "absurdly" large pressure resisting further progression. Subsequent research, in particular following *Pontius and Hill* [1989] and *Pontius and Wolf* [1990], led to the concept of density depletions or bubbles which would be interchange unstable and could "fall" earthward, while enhancements would move in the opposite direction. *Sergeev et al.* [1996] found that flow bursts inside the plasma sheet had properties consistent with the bubble model of *Chen and Wolf* [1993]. *Sergeev et al.* [1996] and *Angelopoulos et al.* [1996] estimated a transverse scale of the BBFs of about $3R_E$. Since then, much work (observational and modeling) has been dedicated to the exploration of the behavior of bubbles and their relation to BBFs [*Birn et al.*, 2004, 2011; *Wolf et al.*, 2006; *Panov et al.*, 2010].

Two aspects are of particular interest, flow braking and reflections at the outer boundary of the dipolar magnetosphere. Flow braking had been introduced by *Haerendel* [1992] in an attempt to identify the forces driving the substorm wedge currents. This work had been carried further observationally by *Shiokawa et al.* [1997], who showed a fairly abrupt deceleration of the BBFs near $10R_E$. Numerical modeling by *Birn et al.* [1999, 2004] demonstrated the generation of vortexes when the flows were reflected at the inner boundary of the tail. A major issue is the behavior of the entropy parameter $pV^{5/3}$. *Dubyagin et al.* [2010] found statistically that it decreased behind dipolarization fronts while the local entropy $pn^{-5/3}$ strongly increased, suggesting that the flux tubes must have a ~40% reduction of particle content.

A critical question under the goal of this review is whether bubbles, after onset of reconnection, can penetrate up to the outer border of the dipolar magnetosphere. *Dubyagin et al.* [2011] found that the ability of BBFs to penetrate inward of $9R_E$ depended strongly on the flux tube entropy $pV^{5/3}$, with clear preference of those with lower values. Motivated by the finding of *Ohtani et al.* [2009] that tailward flows in the near-Earth tail are almost as common as earthward flows, *McPherron et al.* [2011] have studied fast flows with $|V_{\perp x}|>150$ km/s and $X>-11.9R_E$ for an interval of six months and found instead that earthward flows were much more common than tailward flows, faster, and of longer duration. Earthward flows were strongly decelerated when approaching the dipolar magnetosphere and became more azimuthally oriented toward the terminators. A most significant finding was that "virtually all fast flows reaching inside $12R_E$ in the local time sector 2100 to 0100 h can be associated with a sudden and persistent drop in the AL index indicating a substorm expansion phase has taken place."

While the above cited results are fully consistent with and thus seem to confirm the common working hypothesis that the BBFs or dipolarization fronts represent the means by which energy liberated by reconnection in the tail is carried earthward to be dumped in the substorm, there is still a substantial lack of understanding of the related auroral signatures. North-south aligned auroral structures have been observed both inside the oval during substorms [*Nakamura et al.*, 1993; *Elphinstone et al.*, 1996; *Henderson et al.*, 1998] and poleward of the growth-phase arc prior as well as during substorms [*Sergeev et al.*, 1999; Nakamura et al. 2001] and have been interpreted as auroral signatures of earthward flows inside the plasma sheet. Obviously, only the second group can be related to BBFs and dipolarization events. The first one must relate to the further earthward propagation of plasma already injected into the dipolar magnetosphere (see section 18.5.). The bubble model of *Chen and Wolf* [1993, 1999] has served as a theoretical basis for understanding motion and interaction with the ionosphere.

Current reduction inside the bubble owing to its lower plasma pressure in comparison with the ambient medium is postulated to lead to positive space charges on its dawnside and negative charge at its duskside (Figure 18.5),

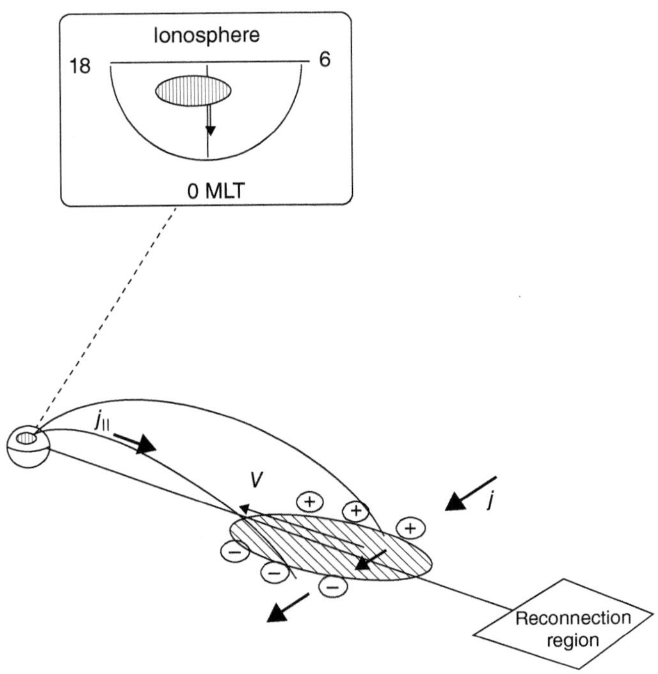

Figure 18.5 Illustration of the relationship between auroral streamers and plasma bubbles in the tail. The aurora is located to the west of the ionospheric trace of the bubble (inset) and propagating equatorward [*Nakamura et al.*, 2001].

connected with an enhanced westward electric field on the inside. Diversion of the outside current is seen as the origin of field-aligned currents. *Kauristie et al.* [2000] observed equivalent ionospheric current patterns suggesting the existence of a small Boström type I current wedge associated with transient fast flow events in the plasma sheet. *Nakamura et al.* [2001] interpreted the auroral streamer as trace of the upward current sheet. *Nishimura et al.* [2010b] showed that latitudinally narrow flow bursts on the polar cap, observed with the SuperDARN radar can precede PBIs followed by equatorward moving streamers, including those related to substorm onset, as discussed in the previous section.

Although there is plenty of evidence for the connection of fast earthward flows in the tail plasma sheet and auroral streamers and associated magnetic perturbations and flows, the causal connection remains uncertain. Field-aligned currents closing in the ionosphere are an expression of the transfer of force from above by magnetic shear stresses. Only when the source of the force is identified, can the current generation be considered as understood. The author has questioned the simple interpretation of the ionospheric phenomena as due to mapping of the flow bursts in the plasma sheet [*Haerendel*, 2011]. The somewhat simplistic argument of *Chen and Wolf* [1993], also taken up by *Sergeev et al.* [1996] and *Nakamura et al.* [2001], that a bubble, because of the reduced internal pressure, becomes electrically polarized and is accelerated by inertial currents of an increasing dawn-dusk electric field does not refer to the acting forces. Inertial currents are a consequence, not a driver of convection. The further argument that field-aligned currents are generated by the divergence of the cross-tail current at the lateral boundaries of a bubble (see Figure 18.5) suffers from the same deficiency. There are in fact two forces: the pressure gradient force of the tail lobe and the magnetic tensions set up in the reconnection process. The undisturbed cross-tail current represents the Lorentz force inside the plasma sheet balancing both forces. Inside the bubble it is reduced and in part replaced by the inertial current of the accelerated plasma. Therefore, lobe pressure and magnetic tensions jointly accelerate the bubble. Because of that there may indeed be an excess cross-tail current on the outside continued by field-aligned currents, but their closure remains unclear. In principle, this may happen further downstream in the tail plasma itself or in the ionosphere. In the first case, the closure current would accelerate the plasma; in the second case, it would transfer momentum to the ionospheric plasma and neutrals. The situation becomes clearer when the bubble becomes subject to flow braking. This is implicitly contained in the work of *Kauristie et al.* [2000] on transient plasma sheet flows. Two vortexlike flows associated with the propagating bubble are seen as sources of a pair of field-aligned currents, similar to the results of *Birn et al.* [2004]. Earthward flows are laterally deviated by the interaction with the ambient plasma and field. This implies loss of momentum or an earthward directed force and can thus serve as a field-aligned current generator. However, the vortex geometry is not easy to reconcile with the sometimes long extension and duration of some of the streamers, in particular when aligning along the growth-phase arc before substorm onset. These observations suggest quasi-stationary pumping of energy into an extended and persistent flow channel rather than the braking of a flow event. Clarification of the true relation between auroral streamers and plasma sheet flows is essential for a deeper understanding of the transport of field and plasma into the source region of the substorm phenomena.

18.4. PENETRATION OF PLASMA AND MAGNETIC FLUX INTO DIPOLAR MAGNETOSPHERE

The clearest indication of plasma entry is the negative bay, i.e., the striking negative excursion of the horizontal component of the magnetic field observed at ground, caused by a westward current overhead. This westward auroral electrojet (AEJ) can only be a load in a magnetosphere-ionosphere current system: $\underline{j} \cdot \underline{E} > 0$. Thus it requires the existence of a westward electric field E_w. This is equivalent to equatorward convection above the lower ionosphere. Since first auroral brightening at substorm onset occurs in the most equatorward arc, which before onset has been marking the outer border of the dipolar magnetosphere, the negative bay expresses the opening of a gate through this border. This means that the field-aligned currents shown adjacent to this border must suddenly reverse, flowing downward on the dawnside and upward on the duskside and closing by a westward current in the ionosphere. This westward electrojet has long since been understood as due to the combination of a westward electric field with a conductivity enhancement, caused by the auroral particle precipitation. The mechanism is expressed through the Cowling conductivity, which involves the setup of an equatorward electric polarization field and secondary Pedersen current nearly balancing the poleward Hall current.

An almost exponential increase of brightness of the equatorward arc [*Donovan et al.*, 2008] marks the conductivity enhancements providing the Cowling channel of the AEJ at the onset of a substorm. There is, however, a conflict. While the westward electric field signals equatorward plasma drift, the aurora tends to expand or propagate poleward. Figure 18.6 shows an example. The initially well defined arc extending over several hundred kilometers splits into a fold and quickly develops parallel and increasingly complex and smaller-scale mainly west-east oriented structures. While the westward edge, marked by the westward traveling surge (WTS), proceeds, new arcs seem to emanate from it in the eastward direction [*Oguti*, 1973; *Sergeev et al.*, 2000]. Soon after, north-south oriented structures appear bearing witness to the plasma entry. Before we address the implications of the conflict between arc and plasma motions, we look at the overall situation.

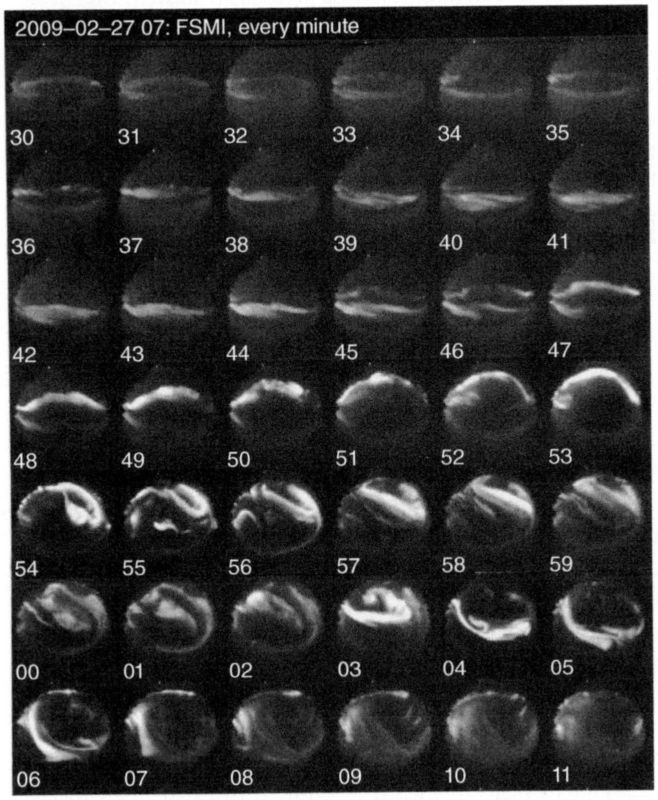

Figure 18.6 All-sky images of onset and breakup of a substorm above Ft. Smith (TEMIS GBO Network) on 27 February 2009. Courtesy of S. Mende and E. Donovan.

A summary picture of aurora, field-aligned currents, and plasma flows has been derived by Fujii et al. [1994] from many crossings of the substorm oval by the DE 2 spacecraft (Figure 18.7). It shows the locations of the downward and upward field-aligned currents at the dawnside and duskside, respectively, of the poleward border of the oval. They feed the auroral electrojet. While the bright poleward arc is a clear marker of the upward current leg of the AEJ, the downward field-aligned current region is only poorly marked because of the lack of an auroral signature. Measurements of the east-west magnetic perturbation fields across the whole auroral oval show that in its interior up- and downward directed currents are mainly balanced while about 30% can be attributed to net up- or downward currents, respectively, on the dusk and dawn flanks of the poleward border closing through the westward electrojet. It is only this component of the so-called region 1 field-aligned currents that belongs to a type I system after *Boström* [1964]. Its role is to impart an equatorward directed force to the ionospheric plasma and thus force an earthward convection inside the dipolar magnetosphere. Equatorward of the poleward arc and westward electrojet begins another plasma realm, the main auroral oval. It is dominated by pressure forces and the existence of a limited penetration depth into the dipolar magnetosphere, forcing the plasma to be diverted sunward along the morning and evening oval, respectively. We will address this region in the next section.

Part of the poleward expansion of the breakup arcs can be attributed to the sequential arrival of new energy

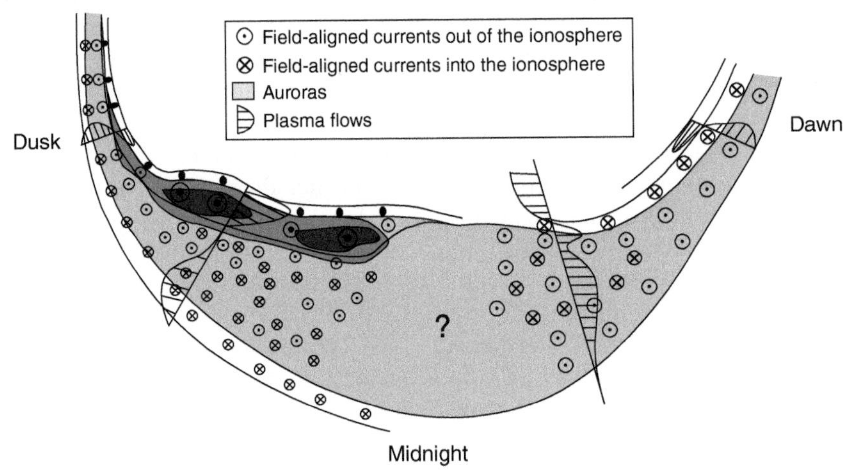

Figure 18.7 Distribution of field-aligned currents, plasma flows (electric fields), and auroras for a generic bulge-type substorm derived from Dynamics Explorer 2 passes during 35 substorms [*Fujii et al.*, 1994].

from high latitudes, e.g., in the form of dipolarization fronts [e.g., *Sergeev et al.*, 2000]. On the other hand, true poleward arc motions are quite common. That auroral arcs are not frozen in the background plasma has been experimentally verified by tracking artificially injected and visible plasma clouds in the neighborhood of auroral arcs. *Kelley et al.* [1975] documented an event with the plasma clouds moving southeastward while auroral arcs propagated poleward at high speed. Other demonstrations were achieved by simultaneous optical and radar tracking of arcs and plasma in the magnetic zenith [*Haerendel et al.*, 1993; *Frey et al.*, 1996]. *Haerendel* [1988, 2007] explained this "proper motion" between arc and plasma by a progressive erosion of magnetic shear stresses, "magnetic fractures," enabled by parallel potential drops above the arcs. This concept, although developed for arcs embedded in the evening auroral oval, may well apply also to the poleward expanding arcs. However, here the primary driving force is directed not along the arc but perpendicular to its dominant orientation, the very force driving the substorm current wedge. This is probably the reason for the absence of long homogeneous arcs as in the evening oval and creation of short and quickly changing forms. We will return to this difficult topic below.

Stimulated by the visual phenomena, the initial development of a wavy structure of the onset arc and the nonlinear growth into poleward protrusions (folds and spirals), several authors have attributed the poleward expansion to ballooning instabilities [*Miura et al.*, 1989; *Roux et al.*, 1991; *Pu et al.*, 1997; *Zhu et al.*, 2004; *Raeder et al.*, 2010]. *Haerendel* [1992, 2009] questioned the validity of this process with the argument that ballooning means increase of magnetic shear stresses at the expense of internal energy, whereas the substorm is just the opposite, a release of magnetic shear stresses stored in the tail. Another counterargument is that ballooning involves antisunward plasma motions, whereas the auroral observations relate to the propagation of arcs which are obviously decoupled from the sunward, i.e., equatorward moving plasma.

We are confronted with two major questions: (1) What is the origin of the substorm current system and the associated westward electric field? (2) Which process is behind the poleward motions of the aurora at breakup? There is one immediate conclusion: If poleward motions are proper motions enabled by decoupling of magnetospheric from ionospheric plasma, strong field-aligned currents must exist above the arcs. The observed high spatial and temporal structuring of the breakup arcs suggests that the substorm current wedge (Boström type I system) must be composed of filamentary and strongly varying field-aligned current elements. Zero-order theories of its origin naturally ignore this complication.

Basically three different mechanisms have been proposed for the generation of the substorm current wedge: current disruption [*McPherron et al.*, 1973; *Lui et al.*, 1990], flow braking [*Haerendel*, 1992; *Shiokawa et al.*, 1997; *Birn et al.*, 2004, 2011; *Birn and Hesse*, 2013], and tail lobe pressure acting on the plasma sheet [*Haerendel*, 2009]. (1) The first one has the deficiency that the current closing through the current wedge is generated at its exterior, while the force driving an earthward convection (equatorward in the ionosphere) can only be transmitted by magnetic shear stresses along field lines inside the wedge. Furthermore, the concept of current disruption [*Lui*, 1991, 1996] in an extended plasma is very problematic since the current is not forced to choose a certain path but is rather determined by the balance of forces. (2) Flow braking does work. Momentum is transferred to the ambient plasma, but quantitative evaluation showed that it is insufficient to drive the AEJ [*Shiokawa* et al., 1997; *Birn et al.*, 1999]. A viable mechanism seems to be vortex formation by the diversion of flow bursts, identified as entropy-depleted flux tubes or "bubbles," when approaching the outer boundary of the dipolar magnetosphere. It generates the right sense of field-aligned currents, as has been demonstrated in numerical models of *Birn et al.* [2004, 2011] and *Birn and Hesse* [2013]. However, it is based on flow braking and exchange of flow momentum with the environment. The quantitative agreement with the substorm currents has yet to be demonstrated. Furthermore, the observations of *McPherron et al.* [2011] (see section 18.3) exhibit a gradual slowing of the flow bursts inside about $10R_E$, whereas the poleward arcs are spatially concentrated events. In the likely initial source region below $10R^E$, the flow has become too slow to provide a substantial driving force. The fact that the brightness of the poleward arcs at substorm breakup exceeds by orders of magnitude the brightness of any auroral streamer which may be related to plasma sheet flows gives a strong hint that flow momentum is an insufficient source. Another process must be at work. (3) As the main driver of the substorm current wedge, *Haerendel* [2009, 2012] proposed the magnetic pressure force exerted by the tail lobe on the plasma piling up at the inner edge of the tail sheet. This concept respects the fact that the main source of energy dissipated in the substorm is residing not in the plasma sheet itself but in the stretched and compressed magnetic field connected to the polar cap and filling the tail lobe from well inside $10R_E$. The energy is thus flowing from this source into the plasma sheet plasma under compression and earthward pushing. The sense and magnitude of the current emerging from this generator were shown to be capable of accounting for observed auroral energy fluxes and electric fields, but the force by itself is rather static and cannot be easily reconciled with the auroral structures and dynamics at breakup. We must admit that so far we have not arrived at a concept shared by most involved researchers.

In spite of this pessimistic conclusion, we will pursue the scenario of *Haerendel* [2009] a little further, because it allows assessing the ionospheric feedback. The substorm current wedge was modeled as a relatively thin and high-beta plasma layer with an ionospheric trace of 33.4 km width in the north-south direction, as shown in Figure 18.8. The sheet current density (integral across the layer) ranged from 0.05 to 0.4 A/m depending on the plasma beta of the generator plasma. The auroral energy conversion was calculated by using the Knight conductance. The Pedersen and Hall conductivities were derived from the calculated auroral energy flux and parallel potential drop by using the empirical relations of *Robinson et al.* [1987]. Perhaps the most important result of this exercise was that, irrespective of the sheet current strength and the related increases of energy flux, conductivities, and electric polarization field E_x normal to the arc, the tangential electric field had always the same magnitude of about 15 mV/m. This means that the poleward arc(s) constitutes an obstacle that the plasma influx has to overcome in order to enter into the magnetosphere. The stronger the driving force, the stronger becomes the obstacle in terms of enhanced conductivity and dissipation.

The way the obstacle is being overcome is an important issue in this context. The layer of upward field-aligned current creates a strong magnetic shear. While the plasma transits the arc, it experiences a substantial reduction of this shear. The thereby released magnetic energy is converted into auroral particle energy. The traces of the magnetospheric part of a field line and of the decoupled ionospheric part are shown in Figure 18.9. After transit through the arc, the two parts have found a new connection (reconnection of type 2). The energy dumped into the arc covers the enhanced dissipation by the equatorward Pedersen current. The associated electric polarization field imposes an eastward flow in the ionosphere or westward Hall current. This is the basis of the prominence of the auroral electrojet on the dusk side. (The dawnside has been ignored in this treatment.)

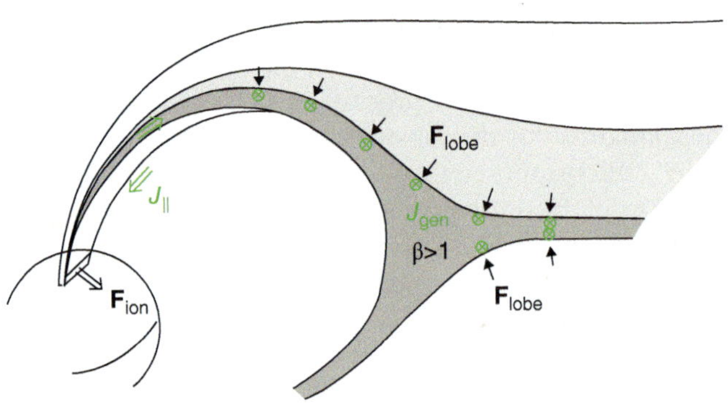

Figure 18.8 Sketch of the force transmission from the tail lobe via field-aligned currents to the ionospheric plasma and (by friction) to the neutral atmosphere. $\mathbf{F}_{lobe}$ designates the force exerted by the lobe field on the high-beta plasma layer, and $\mathbf{F}_{ion}$ the frictional force of the ionosphere on the equatorward drifting ions. Circles indicate the magnetic gradient currents, J_{gen}, flowing along the layer boundaries [*Haerendel*, 2009].

The here discussed model of *Haerendel* [2009] is, of course, strongly idealized. However, one can extract some more general information from it. Arcs represent obstacles to the plasma entry. Flow bursts arriving at the border of the dipolar magnetosphere, say at $7R_E$ or $8R_E$, inject new hot plasma which is subjected to the normal force exerted by the tail lobes. Field-aligned currents transfer the normal force to the conductivity enhancements created by the auroral particles, which become energized by the released tangential stresses. At the same time, the stress release motions in and above the auroral acceleration region are subject to the Kelvin-Helmholtz instability and generate waves, appearing as folds and rays [*Hallinan and Davis*, 1970; *Haerendel et al.*, 1996]. A new aspect in comparison with the elongated evening arcs embedded in a longitudinal convection is the presence of a force normal to the arc. Because of the high-conductivity region, it cannot be readily converted into normal plasma flows, as shown above. Reflected Alfvén waves carry normal momentum back to the generator plasma and, in all likelihood, will create smaller-scale structures in the source region appearing as complex auroral forms, as visible in Figure 18.6. Such a sequence of events, so far based on intuition without theoretical underpinning, would suggest that the observed auroral motions are a combination of the arc's proper motion and reactive motions of the generator plasma.

Furthermore, one finds a tendency for eastward growth and/or motions of the visible structures. They appear to

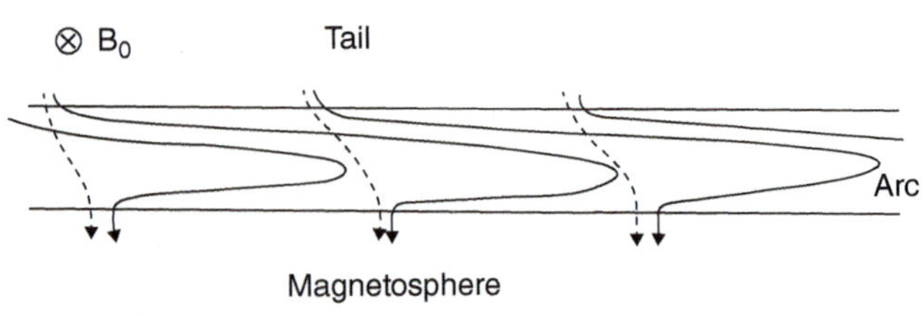

Figure 18.9 View of the plasma and field transit through the poleward arc and mapped into the ionosphere. The magnetospheric section of the field lines undergoes stress release motions indicated by full lines. The sections below the acceleration region ("fracture zone"), shown by dashed lines, move according to the electric field impressed on the ionosphere. Magnetic field lines become disconnected in this process and find new connections upon exit from the arc [*Haerendel*, 2009].

grow out of the westward traveling surge [*Oguti*, 1973; *Sergeev et al.*, 2000]. Furthermore, when brightening, they split and unfold in the eastward direction, thus creating left-handed loops, when viewed from above (see Figure 18.6 after 54 min). Interestingly, this is in conflict with the sense of rotation of the evening vortex which in the model of *Birn et al.* [2004] is the source of the upward current. The reason may lie in the existence of the earthward flow barriers in terms of the arc-related conductivity enhancements which convert normal stresses into tangential ones. There is an analogy with plate tectonics. For instance, where the Pacific and American plates meet, east-west driving forces generate north-south shear stresses and related fault lines.

Besides the poleward and eastward expansion of individual arcs, stepwise poleward auroral expansions (leaps) are quite common [*Sergeev et al.*, 2000]. If, as suggested here, the current driver is the normal force of the lobe magnetic field acting on the inner edge of the plasma sheet, the observed sequence of auroral motions would require an outward recession of the lobe-plasma sheet interface. This is indeed consistent with findings that the flow bursts or dipolarization fronts inject plasma into their higher-latitude wings even faster than the progression in the central current, as, for example, observed by *Runov et al.* [2009]. Consecutively arriving flow bursts would thus create new lobe-plasma interfaces at the transition to the dipolar magnetosphere with foot points at progressively higher latitudes. New current generators would replace their predecessors which had given up much of their free energy and had moved earthward.

So far we have avoided the difficult and vividly discussed question of what triggers the substorm onset. Is it instability of the configuration at the inner edge of the tail or is it caused by the arrival of BBFs? We will not get into the various observations from which arguments in either direction have been derived (e.g., *Angelopoulos et al.* [2008], *Liu et al.* [2011] *Mende et al.* [2009], and *Sergeev et al.* [2011] or *Lui* [1991], *Dubyagin et al.* [2003], *Rae et al.* [2009], and *Lui*, [2011]) but rather remind ourselves that the basic condition, already recognized by *Hones* [1979], is the thinning of the plasma sheet during the growth phase. This may, somewhat independently, create the necessary conditions for both near-Earth neutral line formation around $20R_E$ and current sheet collapse at the tail's inner edge [*Haerendel*, 2009, 2011]. The two events would occur not at exactly the same time but within a few minutes. For instance, the analysis of an isolated substorm by *Kubishkina et al.* [2011] and *Sergeev et al.* [2011] shows the sudden reduction of a very thin plasma sheet at distances as close as $6R_E$ coincident with substorm onset and start of earthward flows outside (!) $10R_E$.

18.5. FURTHER PROGRESSION INTO DIPOLAR MAGNETOSPHERE

Once the hurdle of the poleward arcs has been overcome by the intruding plasma sheet, it seems to break into narrow north-south directed flows. North-south aligned auroral forms inside the midnight sector of the substorm auroral oval have been known for a long time [*Rostoker et al.*, 1987; *Nakamura et al.*, 1993; *Henderson et al.*, 1994]. They seem to emanate from the poleward arc, although there is often a dark gap between this arc and the polar edge of the north-south structures (Figure 18.10). *Henderson et al.* [2002] interpreted them as upward field-aligned current signatures of the western edge of narrow flow channels. In addition, *Nakamura et al.* [1993] identified at the dawn flank "eastward propagating aurora" a few tens of kilometers wide eastward expanding forms developing into diffuse aurora.

It is not the objective of this chapter to describe all the complex auroral structures inside the oval during a substorm and their evolution. Only two striking observations shall be commented on, the already mentioned dark gap between the poleward arc and the north-south auroral forms and the absence of such forms on the dawnside of the oval, equatorward of the downward current source of the AEJ. The first indicates that the related equatorward motion is no longer owed to the westward electric field imposed at the poleward arcs but involves already new driving forces, in all likelihood pressure gradient forces. Owing to the existence of an inner border of the oval, marked by the region 2 currents, all flows are eventually turning sunward. The second observation seems to show that plasma entry is restricted to the dusk leg of the AEJ. This is surprising, since there are no poleward arcs and conductivity enhancements to create obstacles against inward transport. A hint comes from the data assembled by *Fujii et al.* [1994] showing the existence of a westward flow channel immediately poleward of the downward current region (see Figure 18.7). This indicates the existence of polar cap flows, tangential to the oval and proceeding until making their entry through the upward current leg of the AEJ. Some of the electric field patterns of the polar cap and oval presented by *Heppner and Maynard* [1987] can be interpreted in this sense. This leads to the remarkable consequence that the poleward arcs are not only obstacles but at the same time required for plasma entry, perhaps because shedding of internal energy lifts the pressure catastrophy arising from conservation of the flux tube entropy content.

The essence of the here-depicted scenario is that the poleward arc(s) of the auroral oval is the port of entry of plasma and magnetic flux from the tail into the dipolar magnetosphere [*Haerendel*, 2009, 2012]. Once it has entered, the plasma redistributes itself according to its

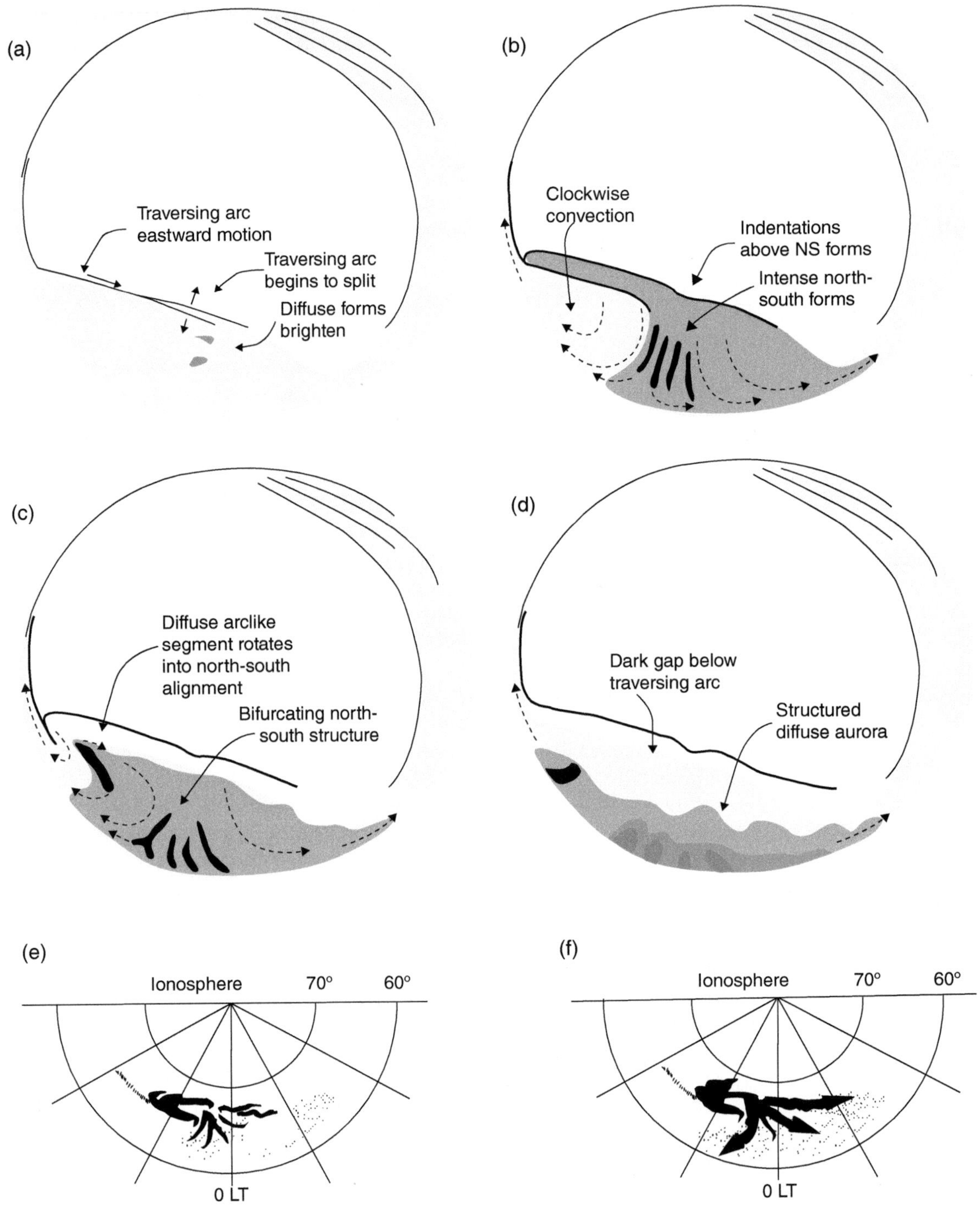

Figure 18.10 Schematic representations of north-south aligned auroral forms inside the auroral bulge during poleward expansion of a substorm. After *Henderson et al.* [1994] (a–d) and *Nakamura et al.* [1993] (e, f). Reprinted from *Henderson* [2012].

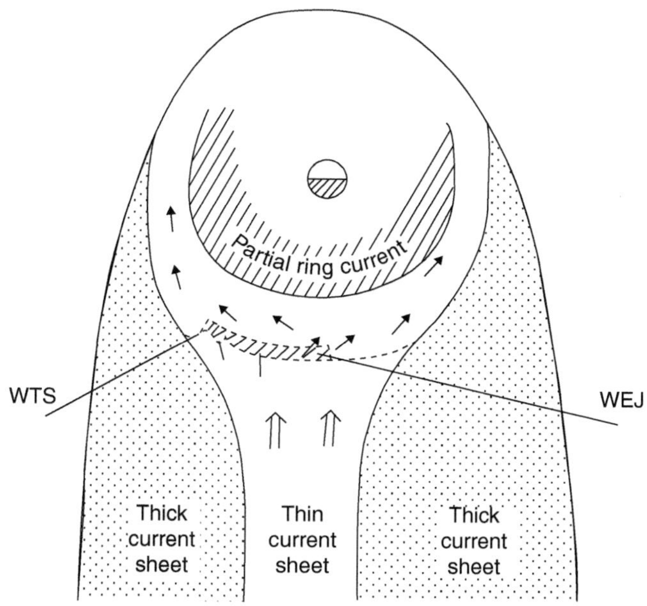

Figure 18.11 Injection of plasma from central current sheet into the outer magnetosphere and diversion into oval convection after current sheet collapse [*Haerendel*, 2009, 2012].

internal pressure forces, friction with the ionosphere, and shielding forces of the stronger internal low-beta dipole field (region 2 currents), eventually streaming sunward along the evening and morning sections of the auroral oval (Figure 18.11).

18.6. OBSERVATIONS RELATED TO PLASMA ENTRY INTO MAGNETOSPHERE OF JUPITER AND SATURN

None of the processes discussed in the preceding sections in the context of Earth's magnetosphere can be identified with the same accuracy and level of detail for the outer planets, neither with respect to the optical coverage of the aurora (by the Hubble Space Telescope (HST)) nor with respect to the magnetic field configurations and plasma flows (e.g., by the Galileo and Cassini missions). It is therefore quite natural that findings on the outer planets are more suspected to proceed similarly to those around Earth than proven to do so. However, a new quality is introduced by the strong role, even prominence, of the fast planetary rotation. It is therefore very rewarding, also for the understanding of the "home" situation, to search for similarities or correspondences. There are two possibilities for plasma entry from the tail side, one due to reconnection within mass-loaded and severely stretched internal flux tubes (Vasyliunas cycle [*Vasyliunas*, 1983]), the other one due to reconnection of the open field stretched into the tail (Dungey cycle [*Dungey*, 1961]) (Figure 18.12). Ignoring here many of the fascinating new features of the magnetospheres of Jupiter and Saturn, I will adhere to the four above-discussed aspects of plasma and field entry into the inner magnetosphere.

Searches for dependences of the Jovian and Kronian auroras on solar wind conditions have so far not revealed dependences on periods of northward IMF (both internal fields point southward at the magnetopause). The probable reason is that the IMF switching periods of tens of minutes to one hour are short compared with the times scales of appreciable flux addition to the huge tail [*Kurth et al.*, 2009]. However, solar wind interactions do exist and must also depend on IMF orientation, but more striking are the rather strong responses of aurora and radio emissions of both Jupiter and Saturn to solar wind pressure increases [*Gurnett et al.*, 2002; *Crary et al.*, 2005; *Nichols et al.*, 2007]. We will return to this below.

Sudden changes of the magnetic field, particle population, and aurora do occur on Jupiter and Saturn, suggesting the existence of substorms. A very striking one, appearing at Jupiter, has manifestations of all well-known substorm phases, growth phase, expansion phase with plasmoid ejections and fast earthward flows, and plasma penetrations into the magnetosphere. On the basis of energetic particle and magnetic field measurements, supported by auroral radio emissions [*Louarn et al.*, 1998], *Woch et al.* [1998], *Krupp et al.* [1998], and *Kronberg et al.* [2005, 2008] showed that the Jovian magnetosphere undergoes dramatic reconfigurations with a recurrence period of about two to four days. The *growth phase*, typically lasting for more than two days, is very different from that at Earth. It is not the tail lobe pressure or suction from the frontside what conditions plasma sheet for reconnection, but the centrifugal force acting on the continuously loaded plasma from Io and stretching the near-corotational magnetic field. Figure 18.13 summarizes a sequence of events as observed from about $100R_J$ [*Kronberg et al.*, 2005]. For a period of two to three days the plasma sheet is in a quiet state of subcorotation. Somewhere between the Galileo spacecraft and the planet, sufficient *thinning of the central current sheet* must have been reached to trigger reconnection, resulting in plasmoid ejection as well as inward plasma injection. At large radial distances one observes for many hours only fast outward particle streaming, while the magnetic field orientation agrees with that of an ejected plasmoid. The current sheet is gradually thinning and high-intensity beams of ions and electrons are streaming antisunward in the plasma sheet boundary layers. Eventually, the flow direction in the central plasma sheet turns toward the planet, and the magnetic field exhibits the southward tilt expected for the normal orientation. Here we have the corollary of the *bursty bulk flows*. Strong southward directed fields observed behind such inward flows have

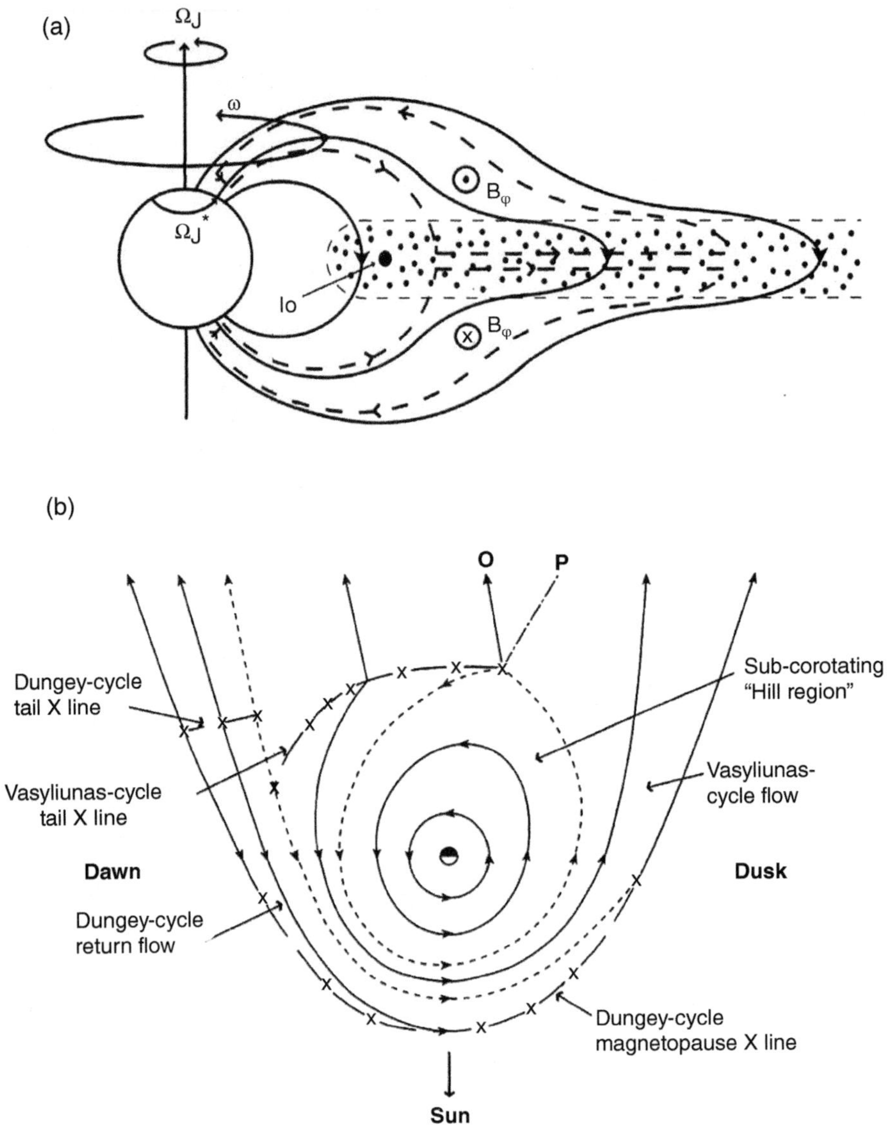

Figure 18.12 (a) Cross section of the configuration of the Jovian magnetosphere [*Cowley and Bunce*, 2001]. The plasma of the middle magnetosphere originates from Io. Currents are indicated by dashed lines. Out-of-plane magnetic field components due to rotation are also shown. (b) Polar view of the mapping of different regions of the northern auroral zone into the magnetosphere indicating flow lines from subcorotation in the inner part to those of the Vasyliunas and Dungey cycles, including the respective X lines. Reprinted from *Clarke* [2012].

been associated with *dipolarization fronts* [*Kasahara et al.*, 2011]. Figure 18.14 [*Vogt et al.*, 2010] shows a statistical distribution of the dominant B_θ signature taken as proxy for inward (red) and outward (blue) flows in the Jovian magnetotail. The white line indicates the most likely location of the X line. The planetary rotation superimposed on the inward flow leads to an increasingly deeper inward penetration toward the dawn sector.

Auroral and radio emissions correlated with reconnection events indicate the inflow of energy, but the conjugacy of the ionospheric phenomena and events in the distorted magnetosphere and stretched tail is not easily established. Isolated auroral spots are often observed poleward of the main oval in the dusk-to-midnight sectors, so-called nightside polar spots [*Grodent et al.*, 2004] as well as so-called polar dawn spots [*Radioti et al.*, 2008]. These emissions have been interpreted as signatures of reconnection events. Their locations just poleward of the main oval is very likely still on closed field lines. Hence they can be associated with *plasma entry*. The polar dawn spots cluster between 04 and 09 LT and exhibit a periodicity of two to three days [*Radioti et al.*, 2008]. The latter suggests that they are products of the above-described internal substorms. A somewhat different interpretation

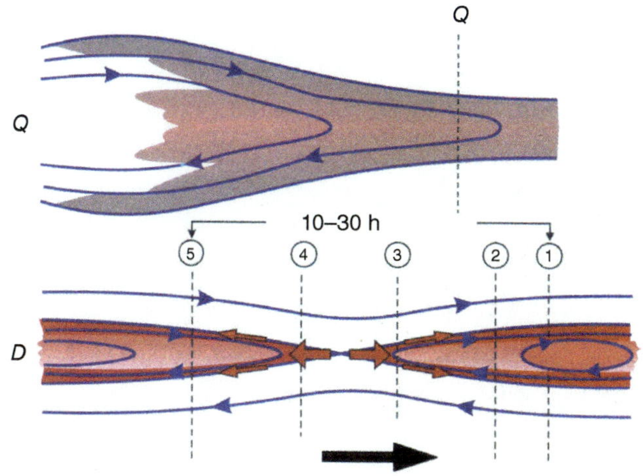

Figure 18.13 Sketch of two states of the Jovian magnetotail: top, "quiet" state; bottom, disturbed state. The numbered dashed lines refer to the magnetic and flow configurations at subsequent observations during a period of 10–30 h [*Kronberg et al.*, 2005].

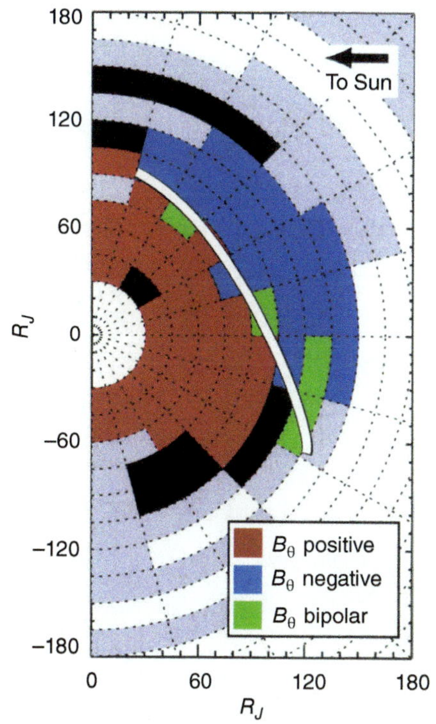

Figure 18.14 Distribution of the dominant B_θ component as proxy for the sense of inward (positive, red) or outward (negative, blue) flows derived from 249 events. The white line indicates the most likely location of the X line. From *Vogt et al.* [2010].

associates them with inward moving flow bursts or bubbles as sketched in Figure 18.5 without a spatial progression being actually observed [*Radioti et al.*, 2010].

The situation at Saturn is somewhat different. While on Jupiter the main auroral oval is thought to be caused by the breakdown of corotation of the middle magnetosphere [*Hill*, 1979; *Cowley and Bunce*, 2001], analysis of the same process for Saturn revealed that the associated field-aligned currents are too weak to account for the auroral emissions by field-parallel electron acceleration [*Cowley and Bunce*, 2003]. Instead, *Cowley et al.* [2004] suggested that the auroral oval corresponds to a ring of upward current at the boundary of open and closed field lines with the corresponding downward currents being located in the region of open field lines. This means that the Saturn aurora would be associated with the Dungey cycle, which, in combination with the planetary rotation, leads to a single cell flow on the polar cap entering the border of closed field lines in the dawn sector (Figure 18.15). During the high-latitude orbits of Cassini in 2008, *Talboys et al.* [2009] were able to measure the magnetic perturbations due to sheets of field-aligned currents near the open-close field boundary. Two types were observed. The most frequent ones consist of downward currents near or just equatorward of that boundary and upward ones extending further into the magnetosphere. The azimuthal field showed clearly the lagging nature (with respect to corotation) of the associated convection. A rarer type (25%) exists showing on the equatorward side of the above configuration an additional leading field signature, indicating superrotational flows or dragging of the ionosphere by high-altitude plasma. A natural interpretation is that they are driven by plasma of high angular momentum on flux tubes contracting inward after separation from ejected plasmoids. While the latter observations seem to support the *entry of plasma* from the tail into the magnetosphere, it cannot be decided whether they are related to the Dungey cycle by reconnection of open field lines or to the Vasyliunas cycle related to pinching off strongly stretched internal fields. However, the location of the regular aurora at the boundary of closed and open field lines and its dynamics are likely due to solar wind interactions in combination with the planetary rotation [*Bunce*, 2012].

Finally, we address the above-mentioned correlation between solar wind ram pressure and auroral activity and brightness on Jupiter and Saturn. During the Cassini flyby of Jupiter, *Nichols et al.* [2007] obtained a number of UV auroral images with HST. During the passage of a compression region the oval was significantly brighter over its whole length. "Active regions" were present in the postnoon and dusk sectors and broadened the oval substantially by ~5° in the poleward direction but also equatorward. During a HST-Cassini campaign at Saturn in 2004, spectacular observations of the impact of the solar wind on the aurora were obtained [*Clarke et al.*, 2005;

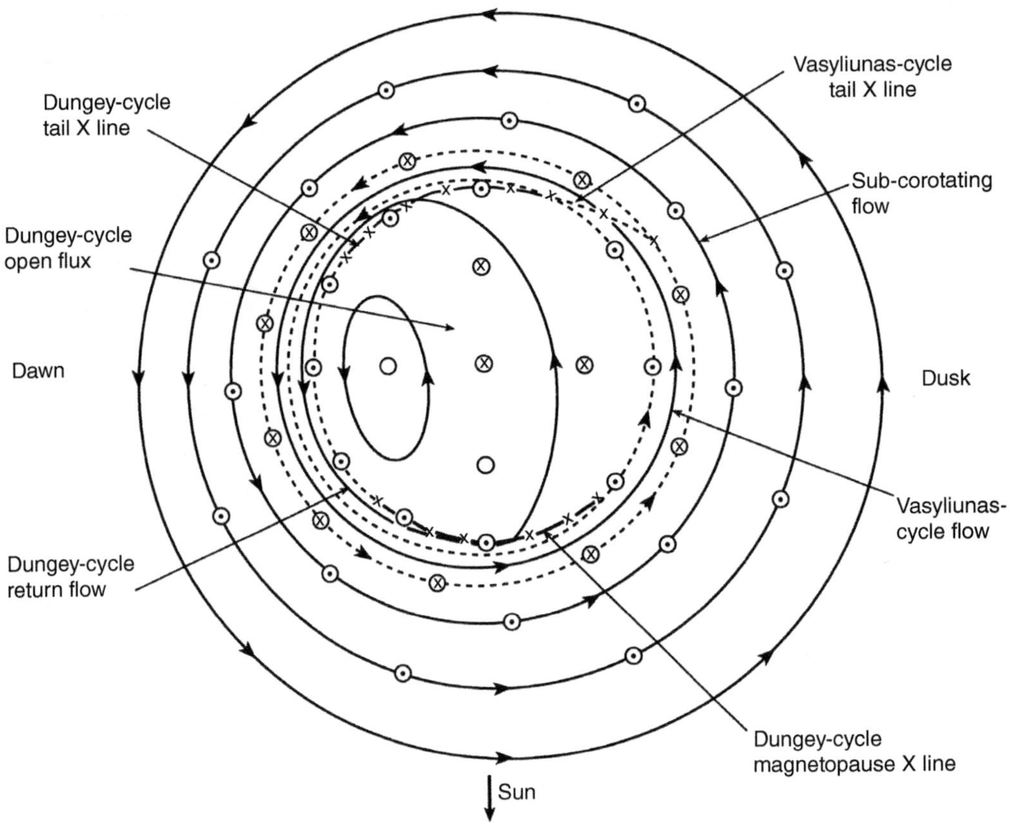

Figure 18.15 Sketch of the plasma flow in the Kronian ionosphere [*Cowley et al.*, 2004]. The outer circle corresponds to the colatitude of ~30° from the pole, which maps to about $3R_S$ in the equatorial plane. Circled dot and crosses indicate regions of upward and downward field-aligned currents, respectively. The Dungey cycle flow on the polar cap returns via dawn due to the planetary torque on the newly closed flux tubes.

Crary et al., 2005; *Kurth et al.*, 2005]. Under the compression of the magnetosphere by the arrival of a corotating interaction region (CIR), the size of the main auroral oval was reduced and a bright aurora starting from the dawn sector filled the polar cap. Such events clearly show that the dissipated energy had been stored by compression of the magnetospheric and tail fields and transferred to the injected plasma and further into the ionosphere by the generation of strong field-aligned current systems. Here we have a corollary to the compression of Earth's tail field during the growth phase (see section 18.4) in building up the energy to be converted during substorms. However, in Earth's case we have a clear dependence on the IMF B_z because of the shorter time scales of growth phase and substorms.

In summary we can state that substormlike events occur on both outer planets. At Jupiter they are mostly internal affairs involving the Vasyliunas cycle. However, the active aurora poleward of the main oval in response to solar wind pressure most likely also involves tail reconnection (Dungey cycle) and thus leads to solar wind plasma entry into the magnetosphere. Saturn's aurora shows stronger dependence on solar wind conditions, suggesting operation of the Dungey cycle. However, projections of auroral locations into the magnetosphere and tail are considerably more uncertain than at Earth. Allocations to open or closed field lines may change with new data. Under these conditions it is clear that sophisticated debates as to where and how a substorm is triggered are not yet on the agenda of magnetospheric research at the outer planets.

REFERENCES

Akasofu, S.-I. (1964), The development of the auroral substorm, *Planet. Space Sci.*, *12*, 273–282.

Angelopoulos, V., et al. (1992), Bursty bulk flows in the inner central plasma sheet, *J. Geophys. Res.*, *97*, 4027.

Angelopoulos, V., et al. (1996), Multipoint analysis of a bursty bulk flow event on April 11, 1985, *J. Geophys. Res.*, *101*, 4967–4989.

Angelopoulos, V., et al. (2008), Tail reconnection triggering substorm onset, *Science*, *321*, 931–935.

Baumjohann, W., G. Paschmann, and H. Lühr (1990), Characteristics of high-speed ion flows in the plasma sheet, *J. Geophys. Res.*, *95*, 3801–3809.

Birn, J., and M. Hesse (2013), The substorm current wedge in MHD simulations, *J. Geophys. Res.*, *118*, 3364–3376, doi:10.1002/jgra.50187.

Birn, J., M. Hesse, G. Haerendel, W. Baumjohann, and K. Shiokawa (1999), Flow braking and the substorm current wedge, *J. Geophys. Res.*, *104*, 19,895–19,904.

Birn, J., J. Raeder, Y. L. Wang, R. A. Wolf, and M. Hesse (2004), On the propagation of bubbles in the geomagnetic tail, *Ann. Geophys.*, *22*, 1773–1786.

Birn, J., R. Nakamura, E. V. Panov, and M. Hesse (2011), Bursty bulk flows and dipolarization in MHD simulations of magnetotail reconnection, *J. Geophys. Res.*, *116*, A01210, doi:10.1029/2010JA01116083.

Boström, R. (1964), A model of the auroral electrojets, *J. Geophys. Res.*, *69*, 4983–4999.

Bunce, E. (2012), Origins of Sarurn's auroral emissions and their relationship to large-scale magnetosphere dynamics, in *Auroral Phenomenology and Magnetospheric Processes: Earth and Other Planets*, Geophys. Monogr., edited by A. Keiling, E. Donovan, F. Bagenal, and T. Karlsson, vol. 197, pp. 397–410, AGU, Washington, D.C.

Chen, C. X., and R. A. Wolf (1993), Interpretation of high-speed flows in the plasma sheet, *J. Geophys. Res.*, *98*, 21,409.

Chen, C. X., and R. A. Wolf (1999), Theory of thin-filament motion in Earth's magnetotail and its application to bursty bulk flows, *J. Geophys. Res.*, *104*, 14,613–14,626.

Clarke, J. T. (2012), Auroral processes on Jupiter and Saturn, in *Auroral Phenomenology and Magnetospheric Processes: Earth and Other Planets*, Geophys. Monogr., edited by A. Keiling, E. Donovan, F. Bagenal, and T. Karlsson, Geophys. Monogr., vol. 197, pp. 113–121, AGU, Washington, D.C.

Clarke, J. T., et al. (2005), Morphological differences between Saturn's ultraviolet aurorae and those of Earth and Jupiter, *Nature*, *433*(7027), 717–719.

Coroniti, F. V., and C. F. Kennel (1973), Can the ionosphere regulate magnetospheric convection, *J. Geophys. Res.*, *78*, 2837–2851.

Cowley, S. W. H., and E. J. Bunce (2001), Origin of the main auroral oval in Jupiter's coupled magnetosphere-ionosphere system, *Planet. Space Sci.*, *49*, 1067–1088.

Cowley, S. W. H., and E. J. Bunce (2003), Corotation-driven magnetosphere-ionosphere coupling currents in Saturn's magnetosphere and their relation to the auroras, *Ann. Geophys.*, *21*, 1691–1707.

Cowley, S. W. H., E. J. Bunce, and R. Prangé (2004), Saturn's polar ionospheric flows and their relation to the main auroral oval, *Ann. Geophys.*, *22*, 1379–1394.

Crary, F. J., et al. (2005), Solar wind dynamic pressure and electric field as the main factors controlling Saturn's aurorae, *Nature*, *433*(7027), 720–722.

Deehr, C., and D. Lummerzheim (2001), Ground-based optical observations of hydrogen emission in the auroral substorm, *J. Geophys. Res.*, *106*(A1), 33–44.

Donovan, E., et al. (2008), Simultaneous THEMIS in situ and auroral observations of a small substorm, *Geophys. Res. Lett.*, *35*, L17S18, doi:10.1029/2008GL033794.

Dubyagin, S. V., V. A. Sergeev, C. W. Carlson, S. R. Marple, T. I. Pulkkinen, and A. G. Yahnin (2003), Evidence of near-Earth breakup location, *Geophys. Res. Lett.*, *30*, 1282, doi:10.1029/2002GL016569.

Dubyagin, S., V. Sergeev, S. Apatenko, V. Angelopoulos, R. Nakamura, J. McFadden, D. Larson, and J. Bonnell (2010), Pressure and entropy changes in the flow-braking region during magnetic field dipolarization, *J. Geophys. Res.*, *115*, A10225, doi:10.1029/2010JA015625.

Dubyagin, S. V., et al. (2011), Can flow bursts penetrate into the inner magnetosphere? *Geophys. Res. Lett.*, *38*, L08102, doi:10.1029/2011GL047016.

Dungey, J. W. (1961), Interplanetary magnetic fields and the auroral zones, *Phys. Rev. Lett.*, *6*, 47.

Eather, R. H. (1967), Secondary processes in proton auroras, *J. Geophys. Res.*, *72*, 11,481.

Elphinstone, R. D., J. S. Murphree, and L. L. Cogger (1996), What is a global substorm? *Rev. Geophys.*, *34*, 169–232.

Erickson, G. M., and R. A Wolf (1980), Is steady convection possible in the Earth's magnetotail? *Geophys. Res. Lett.*, *7*, 897–900.

Frey, H. U., G. Haerendel, D. Knudsen, S. Buchert, and O. H. Bauer (1996), Optical and radar observations of the motion of auroral arcs, *J. Atmos. Terr. Phys.*, *58*, 57–69.

Fujii, R., et al. (1994), Electrodynamic parameters in the nighttime sector during auroral substorms, *J. Geophys. Res.*, *99*(A4), 6093–6112.

Fukunishi, H. (1975), Dynamic relationship between proton and electron auroral substorms, *J. Geophys. Res.*, *80*, 553.

Grodent, D., J.-C. Gérard, J. T. Clarke, G. R. Gladstone, and J. H. Waite, Jr. (2004), A possible auroral signature of magnetotail reconnection process on Jupiter, *J. Geophys. Res.*, *109*, A05201, doi:10.1029/2003JA010341.

Gurnett, D. A., et al. (2002), Control of Jupiter's radio emission and aurorae by the solar wind, *Nature*, *415*, 985–987.

Haerendel, G. (1988), Cosmic linear accelerators, in *Proceedings of the Joint Varenna-Abastumani International School & Workshop on Plasma Astrophysics, Varenna, Italy*, ESA SP-285, vol. 1, pp., 37–44, European Space Agency, Paris.

Haerendel, G. (1992), Disruption, ballooning or auroral avalanche-on the cause of substorms, in *Substorms 1, Proceedings of the First International Conference on Substorms, Kiruna, Sweden, 23–27 March 1992* (ESA SP-335), pp. 417–420, European Space Agency, Paris.

Haerendel, G. (2007), Auroral arcs as sites of magnetic stress release, *J. Geophys. Res.*, *112*, A09214, doi:10.1029/2007JA012378.

Haerendel, G. (2009), Poleward arcs of the auroral oval during substorms and the inner edge of the plasma sheet, *J. Geophys. Res.*, *107*, A06214, doi:10.1029/ 2009JA014138.

Haerendel, G. (2010), Equatorward moving arcs and substorm onset, *J. Geophys. Res.*, *115*, A07212, doi:10.1029/2009JA015117.

Haerendel, G. (2011), Six auroral generators: A review, *J. Geophys. Res.*, *116*, A00K05, doi:10.1029/2010JA016425.

Haerendel, G. (2012), Auroral generators: A survey, in *Auroral Phenomenology and Magnetospheric Processes: Earth and Other Planets*, Geophys. Monogr. Ser. 197, edited by A. Keiling, E. Donovan, F. Bagenal, and T. Karlsson, pp. 347–354, AGU, Washington, D.C.

Haerendel, G., S. Buchert, C. L. LaHoz, B. Raaf, and E. Rieger (1993), On the proper motion of auroral arcs, *J. Geophys. Res., 98,* 6087–6099.

Haerendel, G., B. U. Olipitz, S. Buchert, O. H. Bauer, E. Rieger, and C. LaHoz (1996), Optical and radar observations of auroral arcs with emphasis on small-scale structures, *J. Atmos. Terr. Phys., 58,* 71–83.

Hallinan, T. J., and T. N. Davis (1970), Small-scale auroral arc distortions, *Planet. Space Sci., 18,* 1735–1745.

Henderson, M. G. (2012), Auroral substorms, poleward boundary activations, auroral streamers, omega bands, and onset precursor activity, in *Auroral Phenomenology and Magnetospheric Processes: Earth and Other Planets,* edited by A. Keiling, E. Donovan, F. Bagenal, and T. Karlsson, Geophys. Monogr., vol. 197, pp. 39–54, AGU, Washington, D.C.

Henderson, M. G., J. S. Murphree, and G. D. Reeves (1994), The activation of the dusk-side and the formation of north-south aligned structures during substoems, in *Proceedings of the Second International Conference on Substorms (ICS-2),* edited by J. R. Kan, J. D. Craven, and S.-I. Akasofu, pp. 37–42, Geophysical Institute, Univ. of Alaska Fairbanks, Fairbanks, AK.

Henderson, M. G., G. D. Reeves, and J. S. Murphree (1998), Are north-south aligned auroral structures an ionospheric manifestation of bursty bulk flows? *Geophys. Res. Lett., 25,* 3737–3740.

Henderson, M. G., L. Kepko, H. E. Spence, M. Connors, J. B. Sigwarth, L. A. Frank, H. J. Singer, and K. Yumoto (2002), The evolution of north-south aligned auroral forms into auroral torch structures: The generation of omega bands and Ps6 pulsations via flow bursts, in *Proceedings of the 6th International Conference on Substorms, March 25–29,* edited by R. M. Winglee, pp. 169–174, University of Washington, Seattle.

Heppner, J. P., and N. C. Maynard (1987), Empirical high-latitude electric field models, *J. Geophys. Res., 92,* 4467–4489.

Hill, T. W. (1979), Inertial limit on corotation, *J. Geophys. Res., 84,* 6554.

Hones, E. W., Jr. (1979), Transient phenomena in the magnetotail and their relation to substorms, *Space Sci. Rev., 23,* 393–410.

Kadokura, A., et al. (2002), Detailed analysis of a substorm event on 6 and 7 June 1989, 1, Growth phase evolution of nightside auroral activities and ionosheric convection toward expansion phase onset, *J. Geophys. Res., 107*(A12), 1479, doi:10.1029/2001JA009127.

Kasahara, S., et al. (2011), Magnetic reconnection in the Jovian tail: X-line evolution and consequent plasma sheet structures, *J. Geophys. Res., 116,* A11219, doi:10.1029/2011JA016892.

Kauristie, K., et al. (2000), A conjugate study of Wind and ground-based observations of transient plasma sheet flows, *J. Geophys. Res., 105,* 10,677.

Kelley, M. C., G. Haerendel, H. kappler, F. S. Mozer, and U. V. Fahleson (1975), Electric field measurements in a major magnetospheric substorm, *J. Geophys. Res., 80,* 3181–3195.

Krall, N. A., and A. W. Trivelpiece (1973), *Pinciples of Plasma Physics*, Mc Craw-Hill Book Co., New York, pp. 233–241.

Kronberg, E. A., J. Woch, N. Krupp, A. Lagg, K. K. Khurana, and K.-H. Glassmeier (2005), Mass release at Jupiter: Substorm-like processes in the Jovian magnetotail, *J. Geophys. Res., 110,* A03211, doi:10.1029/2004JA010777.

Kronberg, E. A., J. Woch, N. Krupp, A. Lagg P. W. Daly, and A. Korth (2008), Comparison of periodic substorms at Jupiter and Earth, *J. Geophys. Res., 113,* A04212, doi:10.1029/2007JA012880.

Krupp, N., J. Woch, A. Lagg, B. Wilken, S. Livi, and D. J. Williams (1998), Energetic particle bursts in the predawn Jovian magnetotail, *Geophys. Res. Lett., 25,* 1249–1252.

Kubyshkina, M., et al. (2011), Time-dependent magnetospheric configuration and breakup during a substorm, *J. Geophys. Res., 116,* A00127, doi:10.1029/2010JA015882.

Kurth, W. S., et al. (2005), An Earth-like correspondence between Saturn's auroral features and radio emissions, *Nature, 433*(7027), 722–725.

Kurth, W. S., et al. (2009), Auroral processes, in *Saturn from Cassini-Huygens,* edited by M. K. Dougherty et al., pp. 333–374, Springer Dordrecht, Heidelberg.

Lessard, M. R., et al. (2007), Ground and satellite observations of the evolution of growth phase auroral arcs, *J. Geophys. Res., 112,* A09304, doi:10.1029/2006JA011794.

Liu, J., V. Angelopoulos, M. Kubyshkina, J. McFadden, K.-H. Glassmeier, and C. T. Russell (2011), Revised timing and onset location of two isolated substorms observed by Time History of Events and Macroscale Interactions during Substorms (THEMIS), *J. Geophys. Res., 116,* A00I17, doi:10.1029/2010JA015877.

Louarn, P., A. Roux, S. Perraut, W. Kurth, and D. Gurnett (1998), A study of the large-scale dynamics of the Jovian magnetosphere using the Galileo plasma wave experiment, *Geophys. Res. Lett., 25*(15), 2905–2908.

Lui, A. T. Y. (1991), A synthesis of magnetospheric substorm models, *J. Geophys. Res., 96*(A2), 1849–1856.

Lui, A. T. Y. (1996), Current disruption in the Earth's magnetosphere: Observations and model, *J. Geophys. Res. 101*(A6), 13,067–13,088.

Lui, A. T. Y. (2011), Revisiting Time History of Events and Macroscale Interactions during Substorms (THEMIS) substorm events implying magnetic reconnection as the substorm trigger,*J. Geophys. Res. 116,* A03211, doi:10.1029/2010JA016078.

Lui, A. T. Y., A. Mankofsky, C.-L. Chang, K. Papadopoulos, and C. S. Wu (1990), A current disruption mechanism in the neutral sheet: A possible trigger for substorm expansions, *Geophys. Res. Lett., 17,* 745–748.

Lyons, L. R., I. O. Voronkov, E. F. Donovan, and E. Zesta (2002), Relation of substorm breakup arc to other growth-phase auroral arcs, *J. Geophys. Res., 107*(A11), 1390, doi:10.1029/2002JA009317.

McPherron, R. L. (1970), Growth phase of magnetospheric substorms, *J. Geophys. Res., 75*(28), 5592–5599.

McPherron, R. L., C. T. Russell, and M. P. Aubry (1973), Satellite studies of magnetospheric substorms on August 15, 1968, 9, Phenomenological model for substorms, *J. Geophys. Res., 78,* 3131–3149.

McPherron, R. L., T.-S. Hsu, J. Kissinger, X. Chu, and V. Angelopoulos (2011), Characteristics of plasma flows at the inner edge of the plasma sheet, *J. Geophys. Res., 116,* A00133, doi:10.1029/2010JA015923.

Mende, S., et al. (2009), Timing and location of substorm onsets from THEMIS satellite and ground based observations, *Ann. Geophys., 27,* 2813–2830.

Mende, S. B., H. U. Frey, V. Angelopoulos, and Y. Nishimura (2011), Substorm triggering by poleward boundary intensification and related equatorward propagation, *J. Geophys. Res.*, *16*, A00131, doi:10.1029/2010JA015733.

Miura, A., S. Ohtani, and T. Tamao (1989), Ballooning instability and diamagnetic hydromagnetic waves in a model magnetosphere, *J. Geophys. Res.*, *94*(A11), 15,231–15,242.

Nakamura, R., T. Oguti, T. Yamamoto, and S. Kokubun (1993), Equatorward and poleward expansion of the auroras during auroral substorms, *J. Geophys. Res.*, *98*, 5743–5759.

Nakamura, R., et al. (2001), Earthward flow bursts, auroral streamers, and small expansions, *J. Geophys. Res.*, *106*(A6), 10,791–10,802.

Newell, P. T., V. A. sergeev, G. R. Bikkuzina, and S. Wing (1998), Characterizing the state of the magnetosphere: Testing the ion precipitation maxima latitude(b2i) and ion isotropy boundary, *J. Geophys. Res.*, *103*(A3), 4739–4745.

Nichols, J. D., E. J. Bunce, J. T. Clarke, S. W. H. Cowley, J.-C. Gérard, D. Grodent, and W. R. Pryor (2007), Response of Jupiter's UV aurora to interplanetary conditions as observed by the Hubble Space Telescope during the Cassini flyby campaign, *J. Geophys. Res.*, *112*, A02203, doi:10.1029/2006JA012005.

Nishimura, Y., L. Lyons, S. Zou, V Angelopoulos, and S. Mende (2010a), Substorm triggering by new plasma intrusion: THEMIS all-sky imager observations, *J. Geophys. Res. 115*, A07222, doi:10.1029/2009JA015166.

Nishimura, Y., et al. (2010b), Preonset time sequence of auroral substorms: Coordinated observations by all-sky imagers, satellites, and radars, *J. Geophys. Res.*, *115*, A00108, doi:10.1029/2010JA015832.

Nishimura, Y., L. R. Lyons, X. Xing, V. Angelopoulos, E. F. Donovan, S. B. Mende, J. W. Bonnell, and U. Auster (2013), Identifying the magnetotail source region leading to preonset poleward boundary intensifications, *J. Geophys. Res.*, *118*, 4335–4340, doi:10.1002/jgra.50407.

Oguti, T. (1973), Hydrogen emission and electron aurora at the onset of the auroral breakup, *J. Geophys. Res.*, *78*, 7543–7547.

Ohtani, S., Y. Miyashita, H. Singer, and T. Mukai (2009), Tailward flows with positive B_z in the near-Earth plasma sheet, *J. Geophys. Res.*, *114*, A06218, doi:10.1029/2009JA014159.

Panov, E. V., et al. (2010), Multiple overshoot and rebound of a bursty bulk flow, *Geophys. Res. Lett.*, *37*, L08103, doi:10.1029/2009GL041971.

Pontius, D. H., and T. W. Hill (1989), Rotation driven plasma transport: The coupling of macroscopic motion and microdiffusion, *J. Geophys. Res.*, *94*, 15,041.

Pontius, D. H., Jr., and R. A. Wolf (1990), Transient flux tubes in the terrestrial magnetosphere, *Geophys. Res. Lett.*, *17*, 49–52.

Pu, Z. Y., et al. (1997), MHD drift ballooning instability near the inner edge of the near-Earth plasma sheet and its application to substorm onset, *J. Geophys. Res.*, *102*(A7), 14,397–14,406.

Radioti, A., D. Grodent, J.-C. Gérard, B. Bonfond, and J. T. Clarke (2008), Auroral dawn spots: Signatures of internally driven reconnection processes at Jupiter's magnetotail, *Geophys. Res. Lett.*, *35*, L03104, doi:10.1029/2007GL032460.

Radioti, A., D. Grodent, J.-C. Gérard, and B. Bonfond (2010), Auroral signatures of flow bursts released during magnetotail reconnection at Jupiter, *J. Geophys. Res.*, *115*, A07214, doi: 10.1029/2009JA014844.

Rae, I. J., et al. (2009), Near-Earth initiation of terrestrial substorm onset, *J. Geophys. Res.*, *114*, A00C09, doi:10.1029/2008JA013559.

Raeder, J., P. Zhu, Y. Ge, and G. Siscoe (2010), Open Geospace General Circulation Model simulation of a substorm: Axial tail instability and ballooning mode preceding substorm onset, *J. Geophys. Res.*, *115*, A00I16, doi:10.1029/2010JA015876.

Robinson, R., R. Vondrak, K. Miller, T. Dabbs, and D. Hardy (1987), On calculating ionospheric conductances from the flux and energy of precipitating electrons, *J. Geophys. Res.*, *92*, 2565–2569.

Rostoker, G., A. T. Y. Lui, C. D. Anger, and J. S. Murphree (1987), North-south structures in the midnight sector auroras as viewed by the Viking imager, *Geophys. Res. Lett.*, *14*, 407–410.

Roux, A., et al. (1991), Plasma sheet instability related to the westward traveling surge, *J. Geophys. Res.*, *96*(A10), 17,697–17,714.

Runov, A., et al. (2009), THEMIS observations on an earthward-propagating dipolarization front, *Geophys. Res. Lett.*, *36*, L14106, doi:10.1029/2009GL038980.

Runov, A., V. Angelopoulos, X.-Z. Zhou, X.-J. Zhang, S. Li, F. Plaschke, and J. Bonnell (2011), A THEMIS multicase study of dipolarization fronts in the magnetotail plasma sheet, *J. Geophys. Res.*, *116*, A05216, doi:10.10229/2010JA016316.

Samson, J. C., L. R. Lyons, P. T. Newell, F. Creutzberg, and B. Xu (1992), Proton aurora and substorm intensifications, *Geophys. Res. Lett.*, *19*, 2167–2170.

Sergeev, V. A., E. M. Sazhina, N. A. Tsyganenko, J. A. Lundblad, and F. Soraas (1983), Pitch-angle scattering of energetic protons in the magnetotail current sheet as the dominant source of their isotropic precipitation into the nightside ionosphere, *Planet. Space Sci.*, *31*, 1147.

Sergeev, V. A., V. Angelopoulos, J. T. Gosling, C.A. Catell, and C. T. Russell (1996), Detection of localized, plasma depleted flux tubes or bubbles in the midtail plasma sheet, *J. Geophys. Res.*, *101*(A5), 10,817–10,826.

Sergeev, V. A., et al. (1999), Development of auroral streamers in association with localized impulsive injections to the inner magnetotail, *Geophys. Res. Lett.*, *26*, 417–420.

Sergeev, V. A., et al. (2000), Plasma sheet ion injections into the auroral bulge: Correlative study of spacecraft and ground observations, *J. Geophys. Res.*, *105*(A8), 18,465–18,481.

Sergeev, V., et al. (2011), Substorm growth and expansion onset as observed with ideal ground-spacecraft THEMIS coverage, *J. Geophys. Res.*, *116*, A00126, doi:10.1029/2010JA015689.

Shiokawa, K., W. Baumjohann, and G. Haerendel (1997), Braking of high-speed flows in the near-Earth tail, *Geophys. Res. Lett.*, *24*, 1179–1182.

Shiokawa, K., et al. (2009), Longitudinal development of a substorm brightening arc, *Ann. Geophys.*, *27*, 1935–1940.

Siscoe, G. L. (1966), A unified treatment of magnetospheric dynamics with applications to magnetic storms, *Planet. Space Sci.*, *14*, 947–967.

Talboys, D. L., C. S. Arridge, E. J. Bunce, A. J. Coates, S. W. H. Cowley, M. K. Dougherty, and K. K. Khurana (2009), Signatures of field-aligned currents in Saturn's nightside magnetosphere, *Geophys. Res. Lett.*, *36*, L11119107, doi:10.1029/2009GL039867.

Vasyliunas, V. M. (1983), Plasma distribution and flow, in *Physics of the Jovian Magnetosphere*, edited by A. J. Dessler, pp. 395–453, Cambridge Univ. Press, Cambridge, U. K.

Vogt, M. F., M. G. Kivelson, K. K. Khurana, S. P. Joy, and R. J. Walker (2010), *J. Geophys. Res.*, *115*, A06219, doi:10.1029/2009JA015098.

Woch, J., N. Krupp, A. Lagg, B. Wilken, S. Livi, and D. J. Williams (1998), Quasi-periodic modulations of the Jovian magnetotail, *Geophys. Res. Lett.*, *25*, 1253–1256.

Wolf, R. A., V. Kumar, F. R. Toffoletto, G. M. Erickson, A. M. Savoie, C. X. Chen, and C. L. Lemon (2006), Estimating local plasma sheet $PV^{5/3}$ from single spacecraft measurements, *J. Geophys. Res.*, *111*, A12218, doi:10.1029/2006JA012010.

Zesta, E., E. Donovan, L. Lyons, G. Enno, J. S. Murphree, and L. Cogger (2002), Two-dimensional structure of auroral poleward boundary intensifications, *J. Geophys. Res.*, *107*(A11), 1350, doi:10.1029/2001JA000260.

Zhu, P., A. Bhattacharjee, and Z. W. Ma (2004), Finite-ky ballooning instability in the near-Earth magnetotail, *J. Geophys. Res.*, *109*, A11211, 249–258.

19

Injection, Interchange, and Reconnection: Energetic Particle Observations in Saturn's Magnetosphere

D. G. Mitchell[1], P. C. Brandt[1], J. F. Carbary[1], W. S. Kurth[2], S. M. Krimigis[1], C. Paranicas[1], Norbert Krupp[3], D. C. Hamilton[4], B. H. Mauk[1], G. B. Hospodarsky[2], M. K. Dougherty[5], and W. R. Pryor[6]

19.1. INTRODUCTION

There have been many papers written on the topic of particle injection at Saturn, at Jupiter, and of course at Earth. In the following, we will not discuss Earth at all and in fact will only go into detail regarding the phenomenology at Saturn, where the Cassini fields and particle data, with support from UV and IR auroral imaging data, have provided a number of well-correlated measurements of injections. While we are not discussing Jupiter, many of the observations of injections at Saturn likely inform the plasma and energetic particle observations and phenomena at Jupiter as well.

The term "injection" has probably been used too loosely, which has led to some confusion regarding what physics and which plasma dynamics are being described. In this chapter we will focus on two phenomena for which the term has been used, although even the two described here may have multiple interpretations. These two flavors of injections are (1) those interpreted to be the plasma heating manifestation of inward moving flux tubes associated with the interchange instability and (2) those interpreted as particle acceleration associated with the planetward transport and heating of plasma in the nightside postreconnection plasma sheet/current disk. It may be argued that these two categorizations are not distinct from one another because some regard the planetward transport and heating of plasma in flux tubes associated with the postreconnection plasma sheet as just another example of flux tube interchange. We would argue against that interpretation, since in the postreconnection plasma sheet the transition to an anomalous resistivity regime as the current sheet collapses is accompanied by development of a cross-sheet electric field that nonadiabatically energizes, heats, and transports the plasma particles via $E \times B$ drift in the planetward direction. This is quite different from flux tube interchange, in which the primary forces are buoyancy and field line tension, and the plasma is heated primarily adiabatically. Of course, an electric field is associated with the motion of the plasma planetward in interchange as well, but because the particles are for the most part adiabatic, they do not gain energy directly from that electric field.

So, what do these two types of injections look like, both in energetic particles and when imaged remotely, in energetic neutral atoms (ENAs)?

19.2. OBSERVATIONS

We will investigate the two types of injection through measurements of energetic ions and electrons by the magnetospheric imaging instrument (MIMI) ion and neutral camera (INCA), the MIMI low-energy magnetospheric measurement system (LEMMS), and the MIMI charge, energy, and mass sensor (CHEMS); ENAs from INCA; magnetic fields from the flux gate magnetometer (MAG); and plasma waves from the radio and plasma wave science instrument (RPWS). Descriptions of these instruments are given by *Krimigis et al.* (2004), *Dougherty et al.* (2004), and *Gurnett et al.* (2004). These injections are readily

[1] *Applied Physics Laboratory, Johns Hopkins University, Laurel, Maryland*

[2] *Department of Physics and Astronomy, University of Iowa, Iowa City, Iowa*

[3] *Max-Planck-Institut für Sonnensystemforschung, Göttingen, Germany*

[4] *University of Maryland, College Park, Maryland*

[5] *The Blackett Laboratory, Imperial College, London, United Kingdom*

[6] *Central Arizona College, Coolidge, Arizona*

Magnetotails in the Solar System, Geophysical Monograph 207. First Edition.
Edited by Andreas Keiling, Caitríona M. Jackman, and Peter A. Delamere.

identified in the thermal plasma observed by the Cassini Plasma Sensor (CAPS) as well and we do include CAPS electron data in our Figure 19.1, but this chapter is motivated by the characterization of injections of energetic particles, and including the CAPS data is generally beyond the scope of this work, and it will not be emphasized here. A complete treatment of particle injections at Saturn would include CAPS data. Many papers have been written on interchange injections at Saturn, and the signatures of such injections are very clear and well understood [*André et al.*, 2005; *Burch et al.*, 2005; *Hill et al.*, 2005; *Chen and Hill*, 2008; *Kennelly et al.*, 2013]. Fewer papers have appeared on plasma observations associated with current sheet collapse, with most of the focus being on the tailward side of the reconnection region. One exception in this latter category was described by *Bunce et al.* (2005). They discussed a plasma heating event observed just after Cassini Saturn Orbit Injection (SOI) in July 2004. Another event highlighted by *Thomsen et al.* (2013) may be an example of a high-speed flow moving planetward from such an event (discussed below).

19.3. CURRENT SHEET EVENTS; CHARGED PARTICLES

Figure 19.1 provides an overview, primarily of the energetic particle characteristics of the SOI event. This event is typical of a current sheet event. Cassini encountered the injection fairly far out at dipole L values between about 17 and 20. The bursty structure internal to the event showed no significant energy dispersion, and energetic particles were accelerated to energies of several hundred kilo-electron-volts. The lack of clear energy dispersion in this event is generally characteristic of events in the magnetosphere beyond about $12R_S$ to $15R_S$ and probably reflects the immersion of the spacecraft within the energization region (more on this later). The measured intensities at those energies were much higher than is typical for the outer magnetosphere suprathermal particle population. Furthermore, the event's particles have a harder energy spectrum than even the inner magnetospheric suprathermal particle population (excluding the durably trapped radiation belt ions and electrons).

The composition of the ions in this event is the same as that of the thermal plasma, that is, a mix of protons and oxygen. These particles were energized by a dynamic event that occurred on the nightside of the planet, possibly still taking place as it corotates over Cassini at about 0400 local time (LT). As suggested by *Bunce et al.* (2005), such events are probably the consequence of nightside reconnection and plasmoid release into the tail, with the subsequent dipolarization and collapse of the current disk central current sheet after the plasmoid was released and field line tension was no longer balanced by the centrifugal force that had been exerted by the rotating cold plasma. During this reconfiguration of the magnetic field, an electric field is generated and the particles are quickly energized, a process that has been investigated for decades in explaining events in Earth's magnetosphere. The energization is more efficient for the higher-energy and higher-mass ions because their first adiabatic invariant is not conserved. This mechanism only applies to the ions; the electrons are also energized, but because they ought to remain adiabatic, their energization must be through a different mechanism. Simple compression into a smaller volume as well as betatron acceleration may account for at least some of the electron heating. Plasmoid release and reconnection leave behind the reconnected flux tubes containing hot but very low density plasma. This entire region formed by the newly reconnected flux tubes is left occupied by depleted flux tubes, for which the corotational centrifugal forces are now much smaller than they were before the plasmoid release.

Figure 19.1 "Typical" current sheet collapse injection event. The lower two panels show MIMI LEMMS electron (upper) and ion (lower) spectrograms from just after Cassini SOI. More detail on this event regarding suprathermal particle populations appears above these lower panels. The top panel shows magnetic field magnitude, which exhibits irregular variations and is generally diamagnetically depressed relative to the surrounding medium. The second panel shows electron energy flux from 100 to 500 keV, derived by combining data from the CAPS ELS and the MIMI LEMMS sensors. Note that the energy density peaks in the energy range between 10 and 100 keV. There is evidence for remnants of the cold plasma population in filaments where the electron energy density extends down to 100 eV or less. The third panel from the top shows energetic ion energy flux from LEMMS. Ion pitch angle anisotropies (insets) are displayed for specific times at the top of the spectrogram. The open, egg-shaped patterns are characteristic of plasma convection in the usual, azimuthal (corotational) direction. The diagonal shapes are characteristic of fast planetward convective flow. The energetic ions are a mix of energetic protons and O^+, as determined by the MIMI CHEMS sensor. For brevity we show only O^+ between 50 and 230 keV as measured by telescope 3 from CHEMS, the telescope that for this spacecraft attitude records planetward flowing ions. The plasma flows as determined by the LEMMS and CHEMS ion anisotropy measurements are primarily in the azimuthal (corotation) direction, with only short intervals of strong planetward flow (especially between 0630 and 0640 UT).

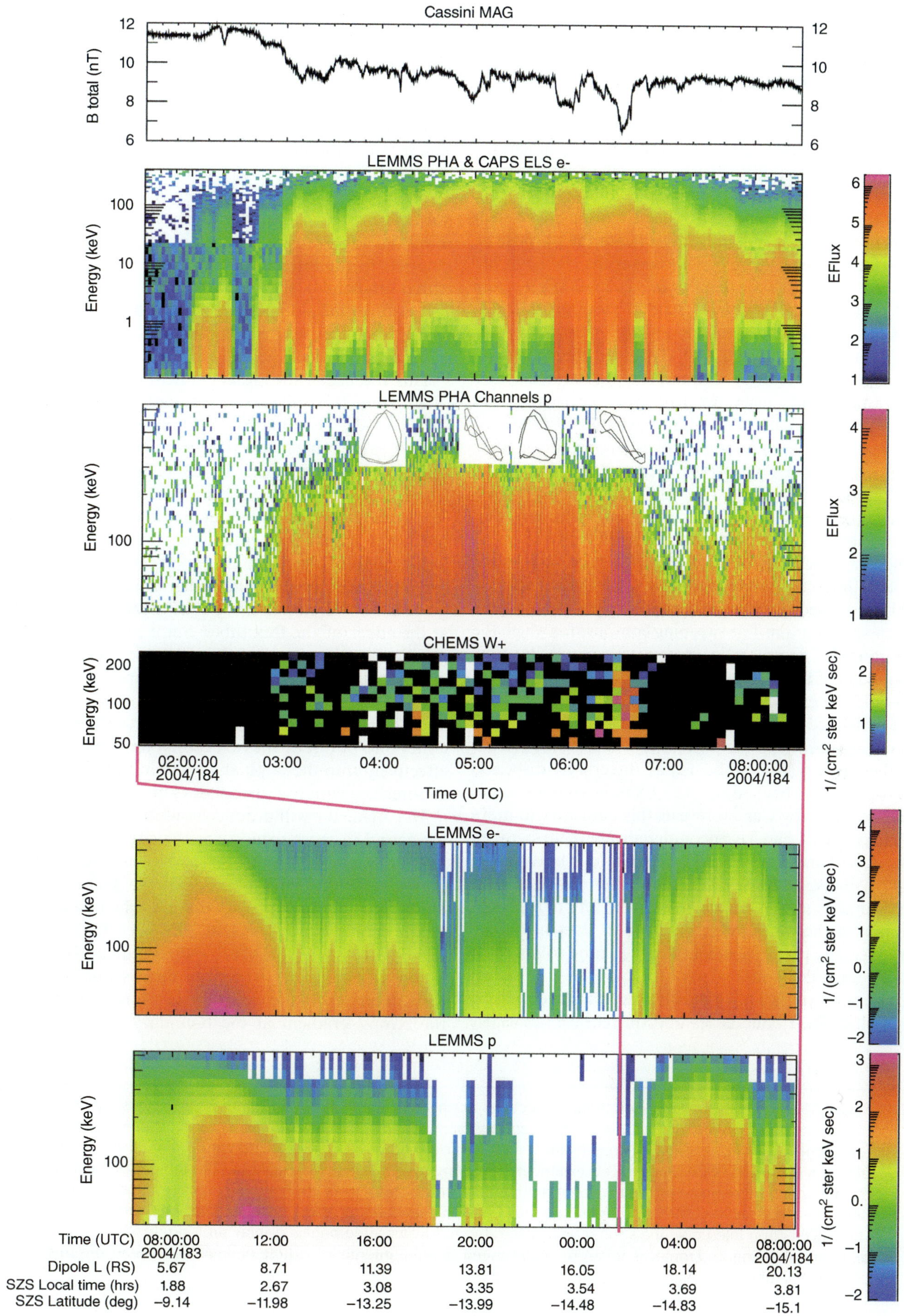

Cassini MAG
B total (nT)
12
10
8
6
LEMMS PHA & CAPS ELS e-
Energy (keV)
100
10
1
EFlux
6
5
4
3
2
1
LEMMS PHA Channels p
Energy (keV)
100
EFlux
4
3
2
1
CHEMS W+
Energy (keV)
200
100
50
02:00:00
2004/184
03:00
04:00
05:00
06:00
07:00
08:00:00
2004/184
Time (UTC)
1/ (cm2 ster keV sec)
2
1
LEMMS e-
Energy (keV)
100
1/ (cm2 ster keV sec)
4
3
2
1
0.
−1
−2
LEMMS p
Energy (keV)
100
1/ (cm2 ster keV sec)
3
2
1
0.
−1
−2
Time (UTC) 08:00:00 2004/183 12:00 16:00 20:00 00:00 04:00 08:00:00 2004/184
Dipole L (RS) 5.67 8.71 11.39 13.81 16.05 18.14 20.13
SZS Local time (hrs) 1.88 2.67 3.08 3.35 3.54 3.69 3.81
SZS Latitude (deg) −9.14 −11.98 −13.25 −13.99 −14.48 −14.83 −15.1

19.4. INTERCHANGE INJECTION EVENTS: CHARGED PARTICLES

Interchange events have been the subject of a great many studies at Earth, at Jupiter, and especially at Saturn, where they are thought to play the dominant role in radial plasma transport in Saturn's middle magnetosphere [e.g., *Mauk et al.*, 2009]. Many others have written about such events in considerable detail, and it is not the purpose of this work to repeat or review those results [*Burch et al.*, 2005; *Hill et al.*, 2005; *André et al.*, 2005, 2008; *Leisner et al.*, 2005; *Chen and Hill*, 2008; *Rymer et al.* 2009; *Kennelly et al.*, 2013]. *Mauk et al.* [2005], *Paranicas et al.* [2007], and *Müller et al.* [2010] discussed the dispersion of energetic particles as these events age and disperse via gradient and curvature drifts relative to the cold subcorotating plasma. We will not extend those detailed treatments here, but rather we will attempt to distinguish these events from the current sheet collapse injections introduced above. Figure 19.2 details several large-scale interchange events as they appear in energetic ions and electrons as well as in the magnetic field.

The events highlighted in Figure 19.2 are very recent flux tube interchange events. Older events are characterized by dispersed energies quite easy to spot in an energetic particle spectrogram: The sweeping contours for which the peak intensity decreases in energy with increasing time seen in each of the four lower panels of Figure 19.2 are just such "old" interchange events. The vertical features with sharply defined boundaries are good examples of "new" interchange events. In the bottom panel, it is also evident that there is very little in the way of discernable dispersed injection events beyond $L=15$. Examination of other periapsis passes by Cassini reveals this boundary to move around a bit; it can be encountered anywhere between about $L=10$ and $L=18$, with a typical value being about $L=12$. We take this as evidence that interchange, at least interchange as seen in the region inside ~$12R_S$, either is not common in the outer magnetosphere or does not accelerate energetic particles in confined flux tube volumes the way it does in the inner regions.

These new events, unlike the old, dispersed events, have very well defined boundaries both in particle intensities and in magnetic field strength. The magnetic field signature is as expected for a near-equatorial interchange flux tube in that the field interior to the event is larger in magnitude and visibly quieter as well [*Andre et al.*, 2007], with much less variability in strength than the field in the surrounding medium. This magnetic field behavior is in direct contrast to the diamagnetically depressed, highly variable field characteristic of the current sheet collapse event in Figure 19.1. Furthermore, whereas the field external to the interchange flux tubes clearly shows the radial gradient in the field magnitude as the spacecraft moves outward along its orbit, the field internal to each interchange flux tube is basically flat, with no indication of a radial gradient.

This absence of the radial gradient in $|B|$ internal to the interchange flux tubes contrasts with the radial gradient clearly evident external to the flux tubes. This is important, because it is the radial gradient of the magnetic field that dominates the drifts of the energetic particles and causes these events to disperse in energy as they age. Apparently, for these fresh events, the radial gradient is not yet established. There are very sharp gradients in $|B|$ at the edges of the flux tubes, and these gradients (which are much steeper than the radial gradient in the background field) should dominate the energetic particle drifts at the edges of the flux tubes. Since this gradient is perpendicular to the flux tube surface, the energetic particle gradient drifts for a cylindrical flux tube will be predominantly circular about the flux tube edge. So for 90° pitch angle particles, the drifts will not cause them to move either into or out of the flux tube volume but instead will effectively trap the population within the flux tube or in its interface with the surrounding medium.

This behavior will depend on whether these flux tubes are round in cross section or elongated radially, as may well be the case. Our estimate of their size is based on their azimuthal convection across the spacecraft in the background (sub)corotating plasma, which does not constrain their radial dimension. However, their shape will have implications for where these gradient drifting particles ultimately go; if the flux tube is round, they will drift in circles. If the flux tube is a long radial channel, the strong gradient at the boundary will be primarily

Figure 19.2 Typical flux tube interchange injections in energetic particles. The bottom panel covers from 0000 UT on day 80, 2006, to 1800 UT on day 81, featuring many ion injections, both new (vertical, nondispersed features) and older, dispersed injections (leaning to the left at higher energies). In the panels above, the interval between 0400 and 0800 on day 80 is expanded, revealing several interesting features. The top panel shows the B_z (dominant) component of the magnetic field; the second panel shows energetic electrons. In that panel, the high intensities at low energies decreasing between 0400 and 0500 result from sunlight background in the LEMMS electron telescope and should be ignored. The following three panels display proton intensities at three different pitch angles. From this it can be readily seen that the energetic particle energization inside the interchange flux tubes is strongly pitch angle dependent.

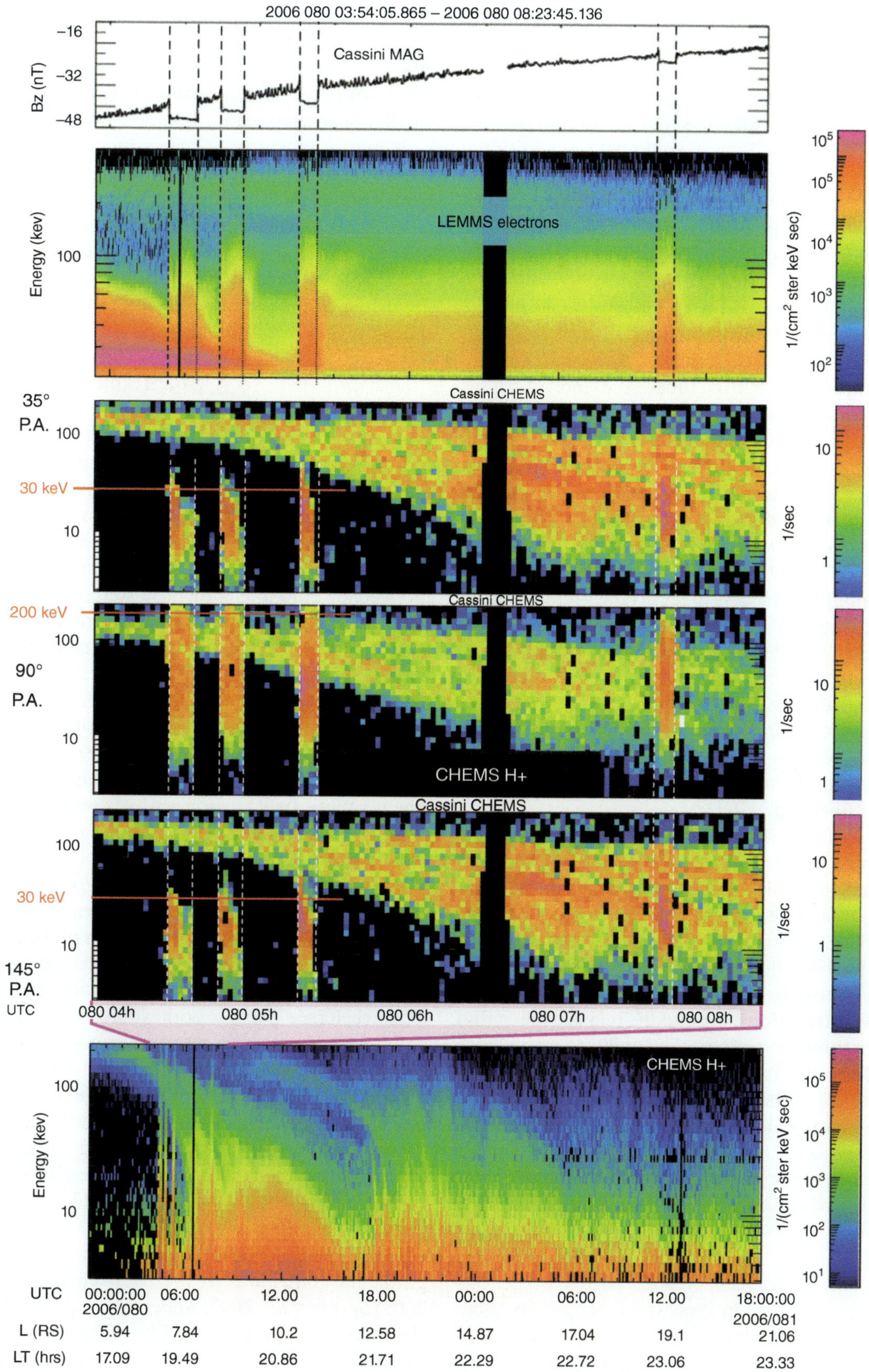

2006 080 03:54:05.865 – 2006 080 08:23:45.136
Cassini MAG
Bz (nT)
LEMMS electrons
Energy (kev)
1/(cm² ster keV sec)
Cassini CHEMS
35° P.A.
30 keV
1/sec
200 keV
90° P.A.
CHEMS H+
145° P.A.
UTC
080 04h
080 05h
080 06h
080 07h
080 08h
UTC
00:00:00 2006/080
06:00
12.00
18.00
00:00
06:00
12.00
18:00:00 2006/081
L (RS)
5.94
7.84
10.2
12.58
14.87
17.04
19.1
21.06
LT (hrs)
17.09
19.49
20.86
21.71
22.29
22.72
23.06
23.33

azimuthal, and the locally mirroring energetic particles will be transported radially along the edges of the interchange flux tube. This latter scenario would also result in a redistribution of the energetic particles in the magnetosphere, as it would provide a mechanism for radial transport (whose relative significance would require additional analysis).

Particles with smaller (closer to either parallel or antiparallel) pitch angles will experience both gradient and field line curvature drifts. The curvature drifts will presumably be very similar to the curvature drifts in the surrounding, preexisting medium. So small pitch angle particles may be much more freely able to drift onto and off of these "new" flux tubes.

Inside the interchange flux tubes, the energetic ions have been energized far more efficiently in the direction perpendicular to the magnetic field than in the parallel direction. The protons accelerated perpendicular to the magnetic field in these events reach energies of about 200 keV, whereas those in directions closer to parallel reach only about 30 keV. As noted by *Paranicas et al.* [2008], the dispersed energetic protons from older events apparently have direct, unimpeded access to these newly interchanged flux tubes. Examining either the 35° or the 145° pitch angle particles in the three events between 0430 and 0530, it would appear that the gradient and curvature drifting ions above about 50 keV are virtually unaffected by the flux tubes' presence. This would indicate that these new flux tubes have actually been at the L shell of the drifting ions for sufficient time that the ions have curvature drifted onto the flux tube field.

Another question raised by these data is, if interchange events accelerate ions to much higher energies perpendicular to B than parallel, why are the old, dispersed ion events in Figure 19.2 approximately isotropic (same intensity at all three pitch angles)? This question deserves further investigation, but we suggest that over the time it takes to disperse such events, pitch angle scattering via wave particle interactions eventually relaxes the initial pitch angle distribution from peaked at 90° to much more isotropic. Also note that below about 30 keV the pitch angle distribution in the new events becomes much more isotropic and even field aligned near the lower energy bound of the accelerated ion population. The older event(s) shown here may have also been larger, more energetic events for which even the 45° pitch angle particles reached 200 keV (but we favor pitch angle scattering as the explanation).

The ion intensities between 40 and 100 keV perpendicular to the magnetic field in these interchange flux tubes range between 10^3 and 10^4 (cm^2-s-sr-keV)$^{-1}$, as compared with 10^2–10^3 (cm^2-s-sr-keV)$^{-1}$ for the ion intensity in the current sheet collapse injection in Figure 19.1. However, outside the interchange flux tubes, the ambient energetic ion intensity in the same energy range falls to one to two orders of magnitude lower such that a large volume average intensity may be either higher or lower than the average intensity in the current sheet collapse injection, for which the accelerated ions fill the volume fairly uniformly. This means that when comparing the ENA intensity of these two types of events as measured by INCA, which from high altitude does not resolve structure at the scale of an interchange flux tube, the apparent intensity should directly reflect the heated ion population in the current sheet collapse events but will underestimate the peak intensities in the interchange flux tubes that contribute to the ENA in a region undergoing active interchange.

19.5. DISCUSSION

These two kinds of injection events (current sheet collapse and flux tube interchange) both accelerate charged particles to high energies, both transport magnetic flux tubes containing low plasma density radially inward, and both are consequences of the requirement for Saturn's rotating magnetosphere to rid itself of the stresses induced by cold plasma whose source is near Enceladus. In the inner to middle magnetosphere, the magnetic field is sufficiently strong that the centrifugal force of the rotating cold plasma does not distort it greatly, the azimuthal current associated with the cold plasma angular momentum is weak, and the field remains nearly dipolar. Still, the rotating plasma exerts a radial force, and unloaded flux tubes outside this region are "light" (the magnetic field tension is not offset by the cold plasma centrifugal force) and so conditions are ripe for flux tube interchange. The process results in a slightly more dipolar field on the light flux tubes that have participated in interchange and the outward displacement of the loaded flux tubes. Again, much has been written on this topic, and it is not our intention to expand upon it here.

Current sheet collapse and the accompanying processes that accelerate plasma both adiabatically and nonadiabatically are expected responses to plasmoid release in the tail. As the loaded flux tubes move farther out in the magnetosphere through interchange motion, the field magnitude decreases until a point is reached where the dipolar field gives way to a stretched field dominated by the azimuthal equatorial current sheet. This current sheet is continuous from about $L = 7$ out to the magnetopause [e.g., *Connerney et al.* 1983; *Bunce et al.*, 2007] and both the centrifugal (inertial) current and particle pressure gradients contribute to the total current [*Mauk et al.*, 2009; *Sergis et al.*, 2007; *Kellett et al.*, 2010]. According to the mechanism put forward by *Vasyliunas* (1983), as the rotating plasma moves farther from the planet, a point is reached when the field tension can no longer balance the inertial force of the plasma. The field will then reconnect across the current disk current sheet, and the cold plasma will be released tailward in a plasmoid (this does not occur on the dayside because the solar wind pressure helps confine the plasma within the magnetopause in the dayside outer magnetosphere).

When the plasmoid has been released, the tension in the reconnected field lines (still stretched) is no longer balanced by plasma centrifugal force, and they begin to snap back toward the planet. Their plasma content has been greatly reduced by the plasmoid release, but there will remain whatever plasma they contained in their off-equatorial extensions before reconnection took place. The current that separates the field reversal between the northern and southern hemispheres must collapse in this dipolarization process, and as it does so a self-consistent electric field is generated in the current sheet region that can accelerate the plasma that remains on the planetward retreating flux tubes. It is this heated plasma that we associate with the first type of event discussed above, namely, the current sheet collapse events.

These events, while covering a large region, are not uniform. There are fast planetward flows embedded in the event presented in Figure 19.1, for example, as discussed in the Figure caption. The event referenced earlier by *Thomsen et al.* [2013] is quite possibly just such a fast planetward flow; that event was observed at about $14R_S$ in the region we have found to be transitional between current sheet collapse and interchange. As such, it may be an important example of a direct transition of fast flows produced during current sheet collapse into planetward traveling interchange flux tubes. This will require further analysis and hopefully additional examples to place it on firmer ground.

As an aside, we would like to point out that the plasmoid released during this process may or may not retreat quickly tailward. Following reconnection of the last closed field line, conditions for continued reconnection involving open lobe magnetic flux may or may not prevail. If reconnection of open flux does proceed, then the field line tension associated with the newly reconnected open flux, whose other connection is to the solar wind, will exert tension in the tailward direction on the plasmoid, accelerating it tailward (the classical concept of plasmoid behavior). However, if lobe reconnection does not proceed, then the plasmoid (which was released because of the centrifugal force exerted by the rotation of the cold heavy plasma contained within it) as a whole will not experience a tailward force. The planetward portion of the plasmoid may move tailward in response to the stretched fields there, but the bulk of the plasma should experience no new forces on it. Therefore, it will not accelerate tailward but rather maintain constant velocity in the direction it was already moving, which is primarily tangent to the azimuthal trajectory it was recently describing. This being the case, the signature of such a recently released plasmoid will be difficult to distinguish from the general flow of the plasma throughout the tail. Eventually such a plasmoid will leave the magnetosphere; if it was released near the dusk flank, then by the time it travels across the tail it will have acquired a significant tailward component to its velocity. A plasmoid released postmidnight would presumably travel into the dawnside magnetopause and eventually be picked up in the low-latitude boundary and/or the sheath flow.

The two types of events (current sheet collapse and interchange) are thus generally located in different regions. Flux tube interchange takes place primarily in the inner and middle magnetosphere where the magnetic field is quasi-dipolar, and buoyancy drives the instability. Current sheet collapse takes place primarily in the outer magnetosphere, and the unopposed field line tension that results after a plasmoid is released through reconnection drives the instability.

However, the two types of events are very likely connected. When a plasmoid is released in the tail and the current sheet collapses, the flux tubes returning planetward contain only a tenuous, hot plasma. This planetward propagation will stop when the dipolarizing flux tubes reach the already existing dipolar region inside about $12R_S$ (in much the same way that bursty bulk flows stop at the dipolar region of Earth's field). This is where there is no longer really a stretched current sheet with stored energy that can collapse further, shutting off the current sheet collapse and thus also the electric field that both heats the particles and causes them to convect planetward. This situation is a perfect setup for flux tube interchange. The rather large region (one estimate would be ~4 h observation × ~100 km/s typical azimuthal velocity = $23R_S$ in azimuth) of reconnected, light flux tubes now sits just radially outward from the inner region, which is still loaded with cold, dense plasma. Buoyancy forces will therefore be strongest just at the longitude where the plasmoid release and current sheet collapse have occurred (of course, everything will continue to rotate azimuthally, driven by the ionosphere). Flux tube interchange is therefore most likely to happen in a (rotating) longitudinal sector where plasmoid release has just taken place, and the two types of events will therefore be, at least to some extent, correlated. This is not to say that one will not happen without the other, but there should be a greater probability for flux tube interchange directly following current sheet collapse.

Conversely, if flux tube interchange is already proceeding in the inner to middle magnetosphere at a particular (rotating) longitude, then cold plasma is being efficiently transported radially outward at that longitude. This could lead directly to plasmoid release at that longitude, since the enhanced outward transport of cold plasma will more quickly load the outer, stretched field with even more cold plasma, leading more imminently to reconnection and plasmoid release at that longitude.

19.6. ENERGETIC NEUTRAL ATOM INJECTIONS

A primary reason for this contribution to the conference is to use it as a platform for clearing up confusion regarding ENA observations of Saturn's magnetosphere. Again, in ENA observations abrupt brightenings are often referred to as injections and bright intensifications in general as "blobs." Here, we will relate these observations to the two injection types discussed above.

As the current sheet collapses, injections of charged particles are generally encountered in the outer magnetosphere, so it is in the outer magnetosphere that we would expect to see the ENA emissions from such events. Indeed, there is such a class of event seen in ENA with quite repeatable characteristics that fit very well the current sheet collapse scenario.

Figure 19.3 presents ENA images from 22 January 2009. Cassini was at about $14R_S$ just postnoon and at 62° latitude. This vantage point provides a good view of the nightside of Saturn's magnetosphere out to beyond the orbit of Titan (the outermost dashed contour). The large image shows a time integration of the ENA flux over an 8 min span from 0624 to 0632 UT for neutral hydrogen energies between 55 and 90 keV. The image shows a localized bright emission region very close to midnight local time. This emission is not fully resolved in angle but has a full width at half maximum (FWHM) of ~$3R_S$. This is a fairly typical location for the onset of current sheet collapse events as seen in ENA, but they can initially appear any place between this local time and about the dawn meridian and from about $15R_S$ to just beyond Titan's orbit. Regardless of where they first appear, they always rotate in the direction of the general magnetospheric corotational flow.

Above the large image in Figure 19.3 we have included six smaller images of the ENA emission at slightly lower energy, between 24 and 55 keV, spanning the time between about 0350 and 0810 UT. This sequence shows the progression from a weak but increasing emission just postmidnight and centered near $10R_S$ in the first frame to a more well-developed, spread-out emission at $10R_S$ that is rotating toward dawn. This is a fairly common behavior for events we identify as being associated with interchange processes and lends weight to the possibility that active interchange may predispose the outer magnetosphere to reconnection. By the third upper panel a new brightening at midnight and beyond $20R_S$ appears in juxtaposition with the $10R_S$ emission. By the sixth panel we can see that the "outer" emission has grown to cover ~4 h of local time and moved a bit planetward. This is likely very similar to the in situ current collapse event discussed in Figure 19.1. The fourth panel at the top is centered very close to the time of the main large image. From this we can see that the relative intensity of the new midnight enhancement beyond $20R_S$ and the rotating event centered near $10R_S$ is quite different at the two energies. At 55–90 keV, the $10R_S$ emission is much weaker than the outer midnight feature, whereas at 24–55 keV the two are comparable in intensity. To understand this difference, it is important to bear in mind how the ENA observation comes about. It is the line-of-sight integration of the charge exchange product between the local neutral gas density and the energetic ion intensity at the particular pitch angle that corresponds to the intersection of the line of sight with the local magnetic field. Given the high latitude of Cassini's location during this sequence, the ENA emission is generated by ions with pitch angles closer to field aligned than field perpendicular. From Figure 19.2 we can see that the ion acceleration for pitch angles within 30° or 40° of the magnetic field extends only to roughly 30 keV, whereas perpendicular to the field it extends to over 200 keV. So with our closer-to-parallel pitch angle vantage point for the images in Figure 19.3, flux tube interchange events would be expected to drop to relatively lower intensities at higher energies than would current sheet collapse events, for which the acceleration (or at least the observed distribution) is relatively independent of pitch angle, presumably as a consequence of pitch angle scattering in the weak, fluctuating magnetic field. And that is what is observed—the relative brightness of the two areas of emission at high energies favors the outer region (current sheet collapse) while they are more nearly equal at the lower energies.

This is not to say that the in situ ion intensities are equal for the two types of events for the two regions. The outer region would be expected to have higher volume-averaged ion intensity by virtue of the fact that the neutral gas density is lower at greater radial distances, so for the line-of-sight integral of the ENA flux to be comparable in value, the ion intensity must be higher. However, from Cassini's distance the ENA imager does not resolve structure at the angular size of an interchange flux tube, and so the ENA brightness again reflects the average ENA intensity from the region. So the ion intensities within the interchange flux tubes could easily be comparable to or higher than the ion intensities in the current sheet collapse events (at about 25–40 keV), and still the volume average over the ENA imager's resolution element could be lower. A lower average ion intensity combined with a higher neutral gas density in the inner dipolar region and vice versa in the outer current disk region can result in nearly equal ENA intensities from the two regions.

A clear example of a current sheet collapse event in relative isolation was presented by *Mitchell et al.* [2005,

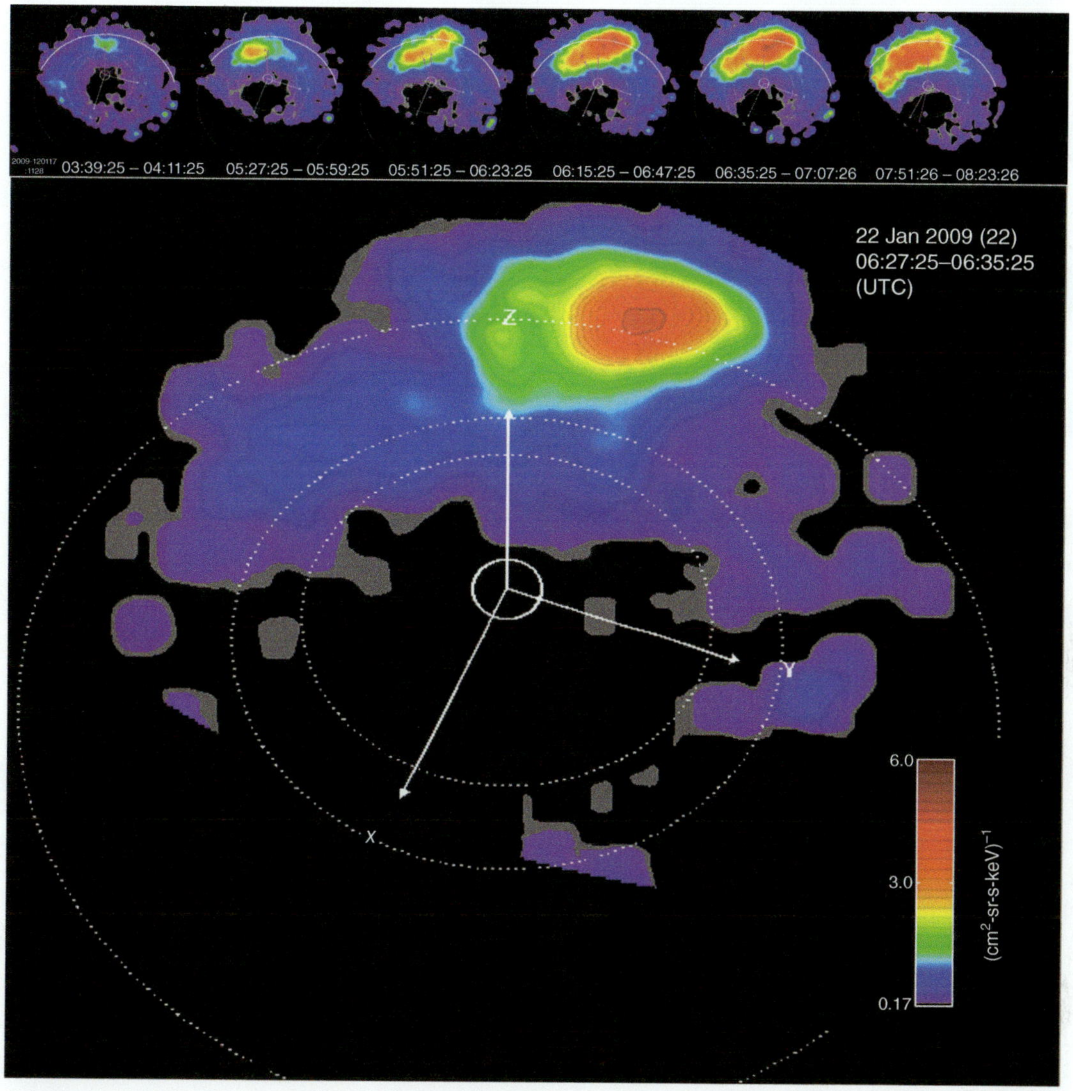

Figure 19.3 ENA image in 55–90 keV hydrogen at the onset of a current sheet collapse event. The orbits of Titan at $20R_S$, Rhea at $8.7R_S$, and Dione at $6.25R_S$ are included for reference. In the row of smaller images at the top, the progression of the event is shown in 24–55 keV H. There is ENA emission in the midnight to premorning LT region at about $10R_S$, with much stronger emission centered near midnight, a bit beyond Titan's orbit, beginning at about 0630 UT. In the small images the segment of Titan's orbit that passes through the bright emission has been enhanced for visibility.

Figures 2 and 3], who attributed it to current sheet disruption and showed a close correspondence with the onset and development of a particularly intense Saturn kilometric radiation (SKR) event. That event was characteristically dispersionless in energy and showed a fast rise in both hydrogen and oxygen over a range of energies. *Hill et al.* [2008] closely associated a similar event with the release of a fast tailward propagating plasmoid observed by Cassini at $44R_S$ and 0300 LT in the magnetotail. Another example by *Mitchell et al.* [2009, Figure 8] documents the association between an injection event and a solar wind pressure enhancement, SKR enhancement, and dawn auroral brightening. Another event by *Mitchell et al.* [2009, Figure 7] again associates an event that likely includes both current sheet collapse and interchange with a strong SKR enhancement and a well-associated, rotating dawnside auroral bulge.

Examples of especially well localized versions of the inner, interchange types of events have been associated with discrete, rotating auroral features (*Radioti et al.*, 2013). Figure 19.4 shows a keogram comparing the auroral observations of UVIS to the ENA observations of INCA, showing that the auroral features and the

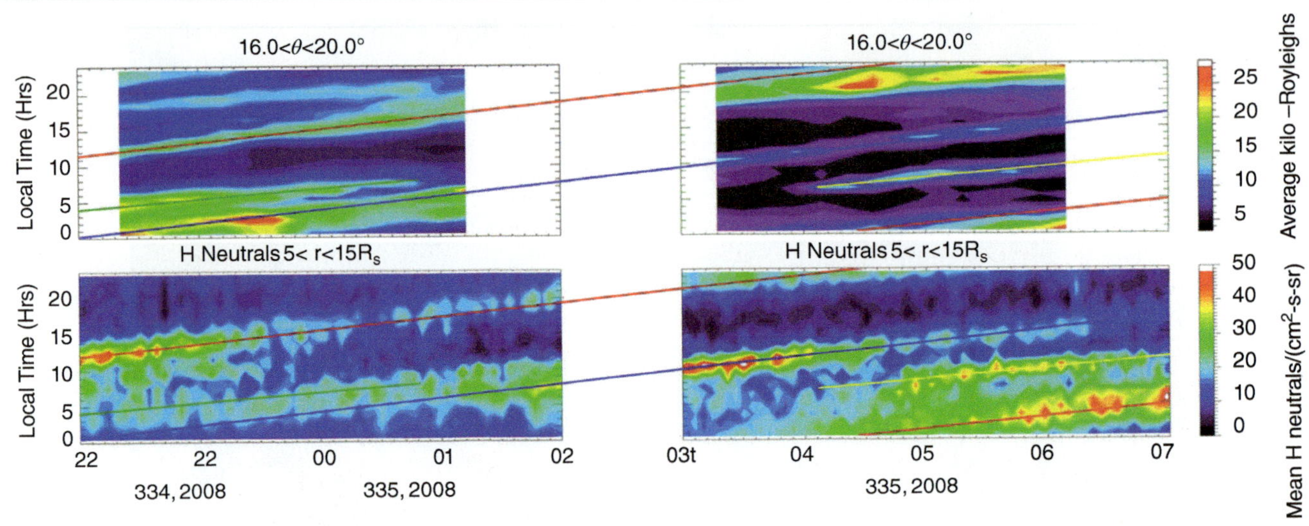

Figure 19.4 Keogram of UVIS auroral intensity between 16° and 20° colatitude (upper panels) and INCA ENA intensity between $5R_S$ and $15R_S$ (lower panels). The colored diagonal lines are repeated above and below to guide the eye. The slope of the diagonal reflects the rotation angular velocity for each feature. The slopes are not identical, indicating different rotation angular velocities for different pairs of features. Overall, the ENA and UV features are quite closely linked by common local times, as concluded by *Radioti et al.* [2013].

associated ENA emissions track each other very closely in local time over many hours.

These events have characteristics in ENA entirely consistent with the characteristics of the interchange events presented in Figure 19.2. In Figure 19.2, there were three successive interchange flux tubes seen in the span of 1h, observed at locations between $7.3 < L < 7.6$. For typical subcorotation plasma flows this corresponds to a region approximately $4R_S$ in azimuth containing three recent interchange events rotating past the spacecraft. *Radioti et al.* [2013] give the estimate for the extent of the observed ENA emitting region as $3R_S$ centered near $L = 8$, very comparable to the values for the event in Figure 19.2. As individual interchange events do age, cool, and spread out through drifts over time, we consider the observations of these long-lived rotating regions to indicate continued interchange happening repeatedly at preferred locations in the subcorotating system.

Of the two types of injections discussed above, the inner-to-middle magnetosphere interchange type of injection is by far the more common. It is rare that an ENA imaging sequence of a full rotation of Saturn goes by without at least one such event at some intensity occurring. During more active times, it is not uncommon for up to three such regions to be active simultaneously (e.g., in Figure 19.4 above), rotating in the corotation direction and typically intensifying as they rotate through the dusk meridian, through midnight, and dimming again at dawn. The current sheet collapse events in the outer magnetosphere are infrequent and are often (although based on scanty statistics) associated with solar wind compression events.

Although the association between ENA observations of rotating features and interchange events above is made through inference, we do have an example for which the ENA images of such an event are validated by direct observation of the same interchange events that produce the imaged ENA emission. In Figure 19.5a we show another example of a series of three interchange events, seen on day 301, 2004, during an inbound equatorial

Figure 19.5 (a) Example of three consecutive interchange events seen on day 301, 2004. The CHEMS sensor measured the energetic protons in these events, which show some energy dispersion consistent with the fact that they have aged several hours since they were freshly injected. The inset in the upper left shows Cassini's orbital geometry. The three interchange events are shown schematically at three different locations as they rotate about Saturn near $10R_S$, from the nightside through noon and back through dusk. The orbits of Rhea and Dione are included for reference. Cassini's orbit is close to Saturn's equatorial plane and during the data interval shown was inbound between about $12R_S$ and $10R_S$, moving from near noon to about 1300 LT. (b) ENA images from Cassini acquired before, during, and after the measurement of the three interchange events shown in (a) This annotated sequence demonstrates directly the correspondence between ENA images of rotating features near $10R_S$ and observed interchange events.

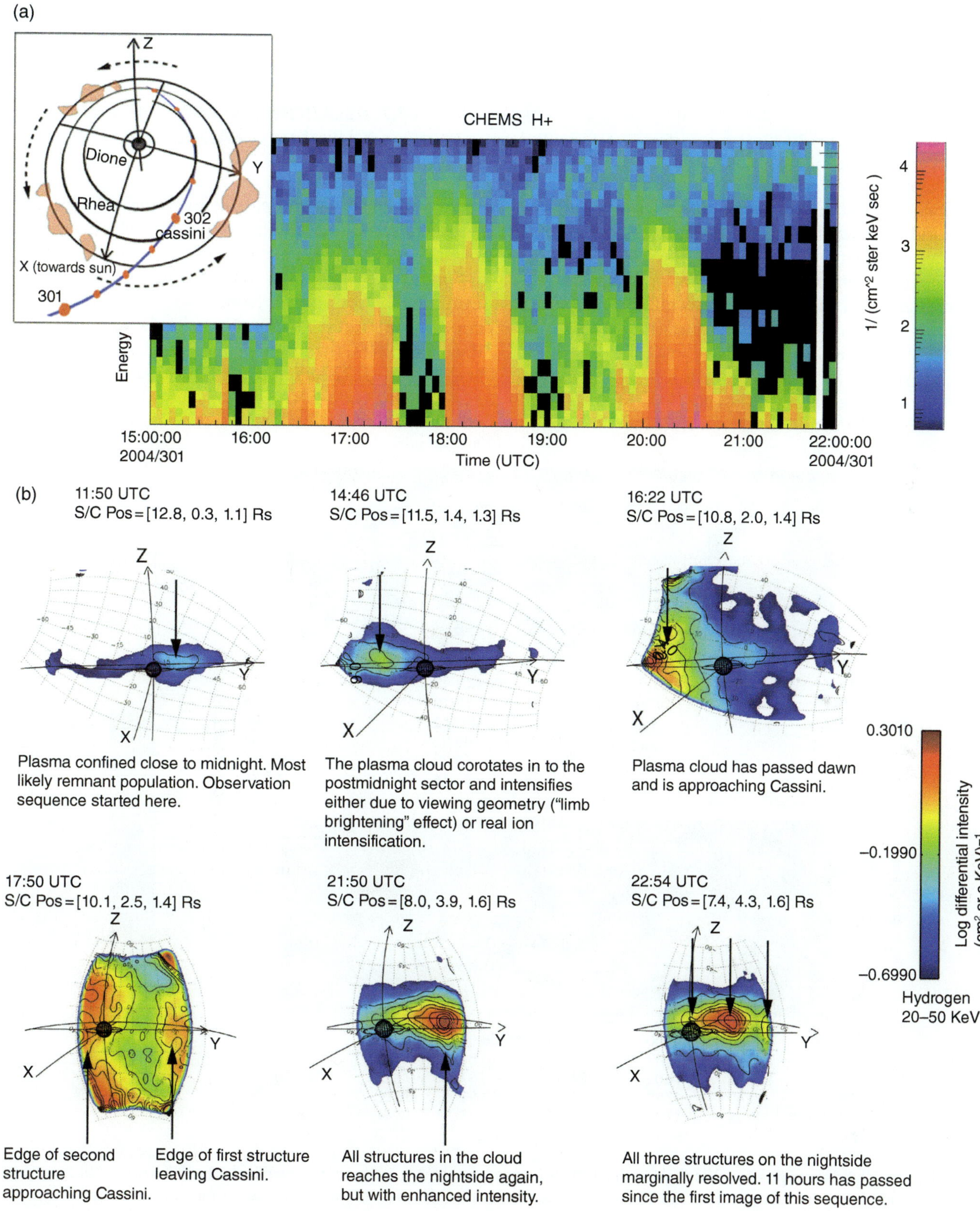
(a)
Z
Y
Dione
Rhea
302
cassini
X (towards sun)
301
CHEMS H+
Energy
1/ (cm^2 ster keV sec)
4
3
2
1
15:00:00
2004/301
16:00
17:00
18:00
19:00
20:00
21:00
22:00:00
2004/301
Time (UTC)
(b)
11:50 UTC
S/C Pos = [12.8, 0.3, 1.1] Rs
Plasma confined close to midnight. Most likely remnant population. Observation sequence started here.
14:46 UTC
S/C Pos = [11.5, 1.4, 1.3] Rs
The plasma cloud corotates in to the postmidnight sector and intensifies either due to viewing geometry ("limb brightening" effect) or real ion intensification.
16:22 UTC
S/C Pos = [10.8, 2.0, 1.4] Rs
Plasma cloud has passed dawn and is approaching Cassini.
17:50 UTC
S/C Pos = [10.1, 2.5, 1.4] Rs
Edge of second structure approaching Cassini.
Edge of first structure leaving Cassini.
21:50 UTC
S/C Pos = [8.0, 3.9, 1.6] Rs
All structures in the cloud reaches the nightside again, but with enhanced intensity.
22:54 UTC
S/C Pos = [7.4, 4.3, 1.6] Rs
All three structures on the nightside marginally resolved. 11 hours has passed since the first image of this sequence.
0.3010
–0.1990
–0.6990
Log differential intensity
(cm² sr s KeV)⁻¹
Hydrogen
20–50 KeV

Cassini orbit near noon local time. Although later in the mission INCA was typically run in ion mode for this orbital geometry, for this pass INCA was in ENA mode and was able to obtain a sequence of ENA images beginning about 6 h before the events passed over the spacecraft and extending until 4 h after the events had rotated past Cassini (see the inset in Figure 19.5a for the orbital geometry). In Figure 19.5b we present the INCA ENA images. As the text annotation of the individual images indicates, each of the ENA images corresponds quite naturally with the motion of the observed interchange events from an initial position on Saturn's nightside, through dawn and over the spacecraft just postnoon, and finally retreating through dusk back to the nightside. The time progression of the INCA images is just what would be expected for a high energetic ion intensity set of flux tubes rotating about Saturn at close to corotation speed at about $10R_S$ and gives us confidence in making the association between similar interchange events and the high inclination image sequences of rotating features seen near $10R_S$.

19.7. RELATIONSHIP WITH SATURN KILOMETRIC RADIATION

The current sheet collapse events of the outer magnetosphere are usually accompanied by some enhancement in Saturn kilometric radiation (SKR), which is radio emission from ~10 kHz to ~1500 kHz [*Lamy et al.*, 2008]. In Figure 19.6 we display RPWS electric field plasma wave power spectrograms from 3 to 3000 kHz for two periods of interest. The upper panel is for 11 through 13 January of 2009, the lower panel from a period 10 days later, 21 through 22 January 2009. Both were high-latitude intervals with good ENA imaging from Cassini, the second of which was the subject of Figure 19.3 and its

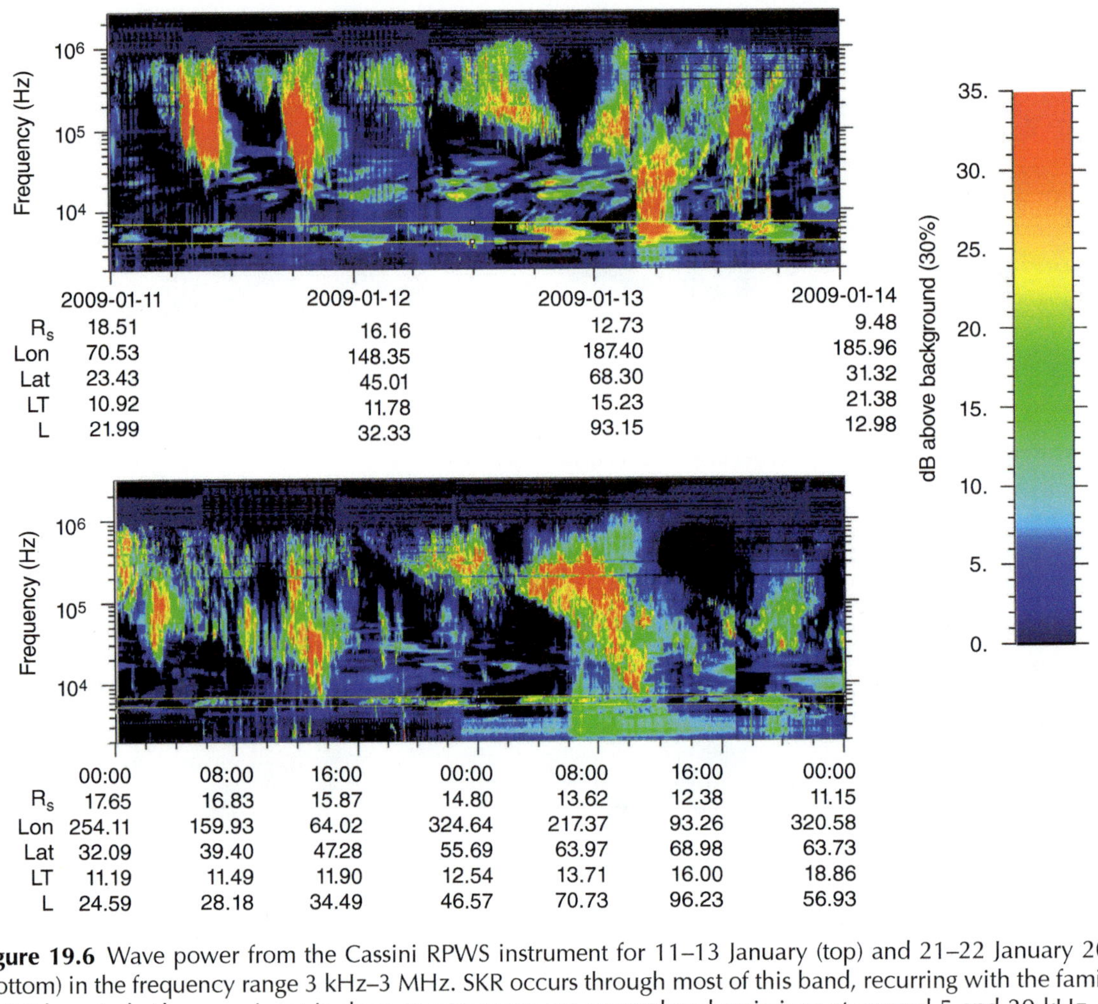

Figure 19.6 Wave power from the Cassini RPWS instrument for 11–13 January (top) and 21–22 January 2009 (bottom) in the frequency range 3 kHz–3 MHz. SKR occurs through most of this band, recurring with the familiar ~10.8 h period. Also prominent in these spectrograms are narrowband emissions at around 5 and 20 kHz. We have outlined the ~5 kHz narrowband emissions in yellow boxes, which we integrate to estimate the time profile of this emission in Figure 19.8. The other feature we will address n Figure 19.7 is the low-frequency extension of the SKR power between 10 and 30 kHz.

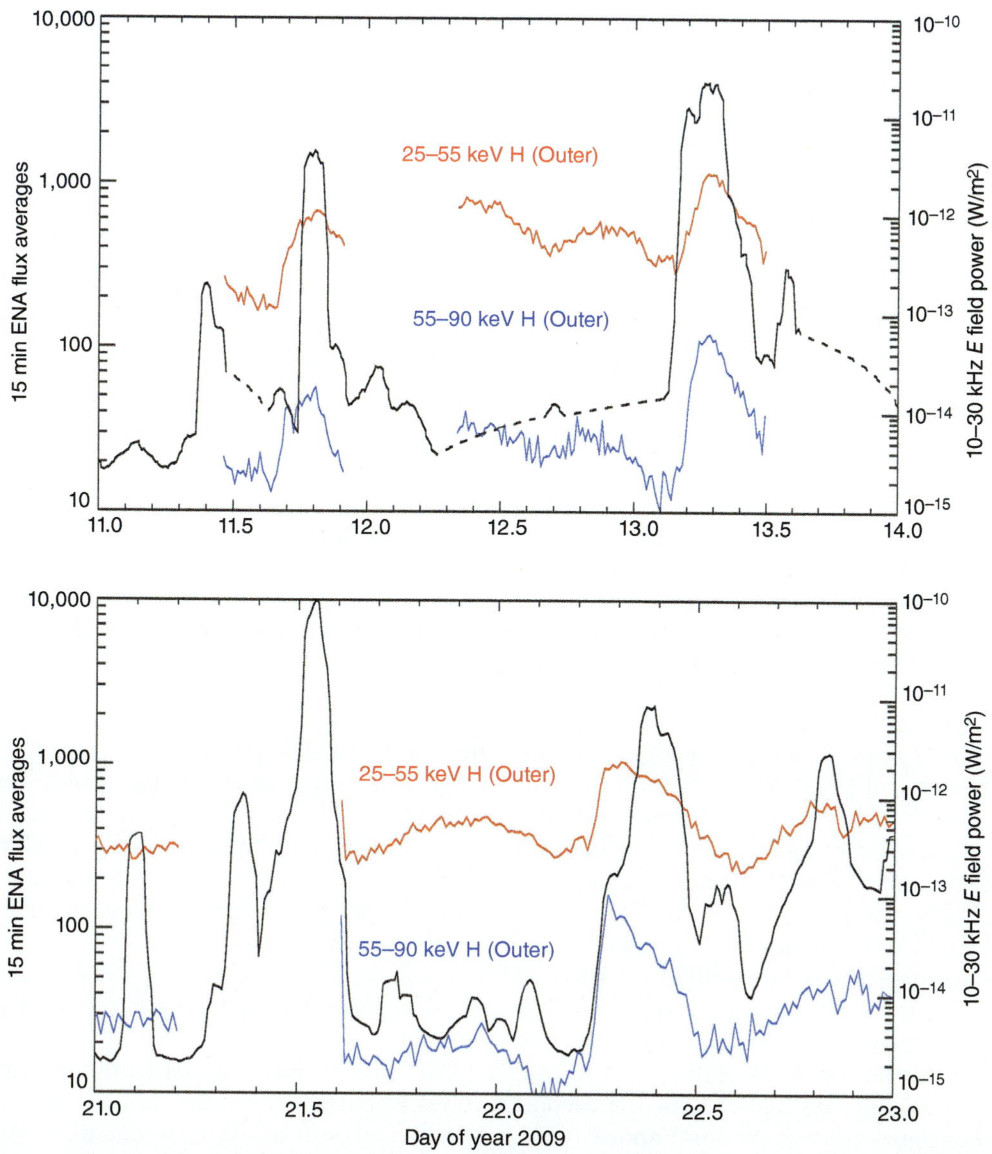

Figure 19.7 Comparisons of integrated INCA ENA fluxes for $14R_S$ to $25R_S$ radial distance with RPWS power integrated over the 10–30 kHz SKR wave band for the events discussed previously for 11–14 January 2009 (top) and 21–23 January 2009 (bottom). The red traces represent integrations of 25–55 keV ENA fluxes, while the blue traces represent 55–90 keV ENA fluxes. The SKR power has been normalized to $20R_S$ after correction for spacecraft range.

discussion. This frequency band fully contains the SKR emission (~10^4–10^6 Hz, occasionally extending below 10^4 Hz), and it also contains periodic bursts of narrowband emission, generally below but overlapping in frequency with the SKR at times. Horizontal yellow boxes identify the narrowband emission in the range between about 4 and 7 kHz in each interval.

For the ENA image emissions previously discussed in Figure 19.3 as well as for images from a very similar event viewed by INCA in ENA from 11 through 13 January 2009, Figure 19.7 compares integrated ENA intensity at 25–55 keV neutral hydrogen (red) and at 55–90 keV neutral hydrogen (blue) to the frequency-integrated SKR power from Figure 19.6 between 10 and 30 kHz. This is at the lower end of the SKR spectrum and was chosen because it shows a correlation with the ENA emission. It also corresponds to relatively high altitude SKR emission since the emission frequency is a function of field strength. The ENA fluxes have been corrected for slant viewing and integrated between $14R_S$ and $25R_S$, which is intended to capture the outer, current sheet collapse events. The RPWS power has been normalized to

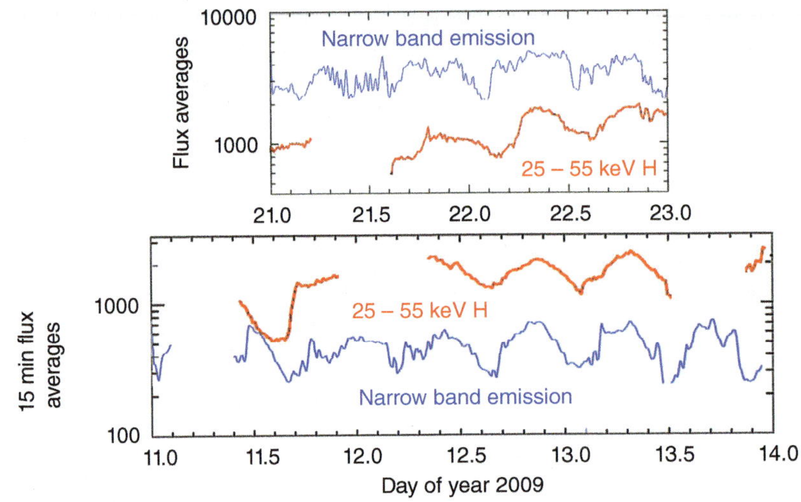

Figure 19.8 Comparison of integrated INCA ENA fluxes for $5R_S$ to $14R_S$ with RPWS power integrated over the 3–7 kHz narrowband emission wave band for the events discussed previously for 11–14 January 2009 (top) and 21–23 January 2009 (bottom). The red traces represent integrations of 25–55 keV ENA fluxes. The blue traces are estimates of the narrowband power (unitless in this Figure; we are simply trying to demonstrate the relationship with the ENA emission intensity in this radial range).

$20R_S$ to correct for spacecraft range changes during the time interval.

One notices a rough correspondence in comparing the ENA and RPWS curves. During the 11–13 period (top) the strongest ENA enhancements at both energies correspond quite closely with the abrupt increases in 10–30 kHz SKR power (the dashed lines correspond to interpolated intervals during which the 10–30 kHz SKR was not present, but the power in the chosen frequency range was elevated by enhancements in the narrowband wave power). Likewise the largest ENA enhancement between 21.0 days and 23.0 days agrees with the second largest SKR enhancement during the same approximate interval (The ENA measurement was missing during the largest SKR enhancement, and this interval did not have significant narrowband power within the 10–30 kHz band, which can be seen by inspection of Figure 19.6). The ENA and SKR profiles do not exactly match, of course, and this should not be expected owing to the rather crude way in which the integrations are performed and the effects of viewing geometry on both measurements.

In Figure 19.8 we compare estimates of the RPWS narrowband power within the yellow boxes drawn in the spectrograms in Figure 19.6 with the ENA intensity between 25 and 55 keV integrated over the distance range from $5R_S$ to $14R_S$. Here we can see that on both days the narrowband emission power estimate quite closely follows the ENA intensity in this radial distance range. This suggests that the two may be closely related, possibly through a common cause of the interchange instability. Although these two quantities appear in these two case studies to be closely related, we will follow up with a statistical look at this relationship to ascertain whether the phase relationship between the narrowband emission and the ENA emission at $5R_S$ to $14R_S$ is fixed or the apparent relationship is maintained at all in a general sense.

19.8. SUMMARY AND CONCLUSIONS

This chapter has discussed examples of two types of energetic particle injection events at Saturn from the perspective of both in situ energetic particle measurements and ENA imaging measurements. We provided examples based in part on the quality of the measurements and in part on the clarity of the events. With regard to the latter, we mean that the events were chosen primarily to show the features that we have found to characterize each kind of event. In one case, the event chosen has a clear-cut example of a current sheet collapse event that is closely associated with an event that we believe to be a flux tube interchange type of event (the event shown in Figure 19.3). Below, we summarize the features that we find distinguish the two types of energetic ion injections from one another.

Current Sheet Collapse

- **Location** $>12R_S$ to $15R_S$ out to $>20R_S$; initiated between midnight and dawn.
- **Size** $\sim 5R_S$ radial, 3–6 h LT azimuthal, rotating in direction of corotation.

• **Time Profile** Growing for several hours, peaking late morning LT, decaying by postnoon LT.

• **Energetic Ions** Enhanced flux, hard spectrum, some fast planetward flows, nearly isotropic pitch angles, little or no energy dispersion.

• **Energetic Electrons** Enhanced flux, hard spectrum, little or no energy dispersion.

• **Magnetic Field** Variable, diamagnetic depression consistent with current sheet/plasma sheet structure.

• **ENA** Enhanced postmidnight, well correlated with low-frequency SKR and dawn auroral brightening, planetward motion following initial onset, near corotation speed azimuthal propagation.

• **Aurora** Postmidnight onset with fast growth into dawnside bright emission. Structured emission may spread either or both poleward and equatorward from onset latitude.

• **SKR** Generally earlier onset than either aurora or ENA at higher frequency (above 100 kHz), growing to include lower frequencies below ~30 kHz at the time of visible onset in ENA. SKR emission depends on viewing geometry, so this may not always be evident.

• **Narrow Band** No association determined.

Flux Tube Interchange (Region of Enhanced Interchange Including Multiple Interchange Flux Tubes)

• **Location** $\sim 6 < L < 12R_S$ to $15R_S$, any LT, brighter and closer to planet on the nightside.

• **Size** $\sim 2R_S$ radial, 2–6 h LT azimuthal, rotating in direction of corotation.

• **Time Profile** Lasting for hours to multiple Saturn rotations (not individual interchange events, but regions of active interchange).

• **Energetic Ions** Enhanced flux, limited maximum energy, pitch angles peaked at 90° at higher energies, distinct energy dispersion as events age, absence of low-energy plasma in new events. Enhanced flux limited to distinct boundaries of flux tubes in new events, scale of flux tubes $\sim 0.5R_S$. Often occurring in clusters, such that a larger region is filled with multiple interchange flux tubes.

• **Energetic Electrons** Enhanced flux, distinct energy dispersion as events age. Enhanced flux limited to distinct boundaries of flux tubes in new events, scale of flux tubes $\sim 0.5R_S$.

• **Magnetic Field (Equator, Internal to Interchange Flux Tube)** Higher $|B|$ than surrounding medium, no radial gradient inside flux tube (though not yet shown to be characteristic of all such events), low fluctuations, more dipolar than surrounding medium.

• **ENA** Enhanced emission between about $6R_S$ and $10R_S$ peaked near $7R_S$ to $8R_S$, intensity reflects pancake pitch angle distribution at higher energies, correlated with narrowband radio emission and in some cases with auroral arcs equatorward of main oval, corotating at ~ plasma rotation speed, little radial motion (slightly planetward between noon and midnight, slightly anti-planetward between midnight and noon as established statistically by *Carbary et al.* 2008).

• **Aurora** For intense localized ENA emissions, well-correlated bright arc equatorward of main oval. For more diffuse ENA emission spread over many hours of LT, more diffuse auroral emission equatorward of main oval location.

• **SKR** Not generally correlated

• **Narrow Band** Close association between narrow band intensity and ENA intensity in some cases. More events must be studied to establish generality.

We have demonstrated through several case studies that there are two distinct types of energetic particle injection mechanisms at Saturn. The first type is a consequence of the centrifugal interchange instability in which flux tubes containing hot, tenuous plasma radially interchange with flux tubes containing cold, dense plasma as a result of centrifugal forces. It is associated with strong perpendicular heating of energetic ions, auroral arc formation close to or equatorward of 70° latitude, and possibly narrowband radio emission (this must be confirmed through a more extensive analysis). The second type develops farther from the planet during the collapse of the plasma sheet current sheet, which we propose follows the formation and release of a plasmoid caused by stretching of the field in the magnetotail by the same centrifugal force exerted by the corotating cold plasma. It is associated with isotropic particle heating, auroral brightening in the midnight-to-dawn sector, and SKR emission, especially in the low end of the SKR frequency range. The first type of injection occurs in the inner magnetosphere ($L \leq 12$) and can occur at any local time, although in ENA at least it is typically stronger on the night side, while the second type occurs in the outer magnetosphere, moving into $\sim 12R_S$ before stalling, primarily in the midnight-to-dawn sector. Each type of injection can be recognized by its signatures in energetic particles in situ (pitch angles, energy dispersion, scale) and in remote images in energetic neutrals atoms (location, persistence, spectral characteristics).

Based on the ENA events discussed in this chapter as well as similar events published previously, we can also comment on the lifetimes of the injections in terms of identifiable newly energized particle populations. The current sheet collapse types of events that begin in the midnight-to-dawn sector typically do not persist for a full rotation (~10.8 h). They typically reach a maximum intensity at some location between dawn and noon, and then the intensities beyond $\sim 15R_S$ dissipate and are no longer identifiable past about dusk. It can be somewhat confusing in that the peak intensity often moves fairly seamlessly into $\sim 10R_S$ (where we have argued that the mechanism is dominated by flux tube interchange), and that activity may indeed continue for a full rotation or more. The flux tube interchange type of event can persist for multiple Saturn rotations, over which time it does not propagate radially, but rather its motion is predominantly azimuthal. There is a whole range of behaviors and morphologies for the ENA emissions from this inner region, ranging from localized emissions that are relatively easy to follow over multiple rotations to much broader swaths of longitude that typically become dim on the dayside and brighter on the nightside but may also contain brighter patches whose motion can be followed more easily. So the lifetimes of these interchange type of events can be long, which we interpret as a continuously operating process embedded in the rotating plasma, rather than a one-time heating of particles and their subsequent motion.

ACKNOWLEDGMENTS

This research was supported in part by the NASA Office of Space Science under Task Order 003 of contract NAS5-97271 between NASA Goddard Space flight Center and the Johns Hopkins University and under NASA grant number NNX12AG81G. The research at The University of Iowa is supported by the National Aeronautics and Space Administration through contract 1279973 with the Jet Propulsion Laboratory.

REFERENCES

André, N., M. K. Dougherty, C. T. Russell, J. S. Leisner, and K. K. Khurana (2005), Dynamics of the Saturnian inner magnetosphere: First inferences from the Cassini magnetometers about small-scale plasma transport in the magnetosphere, *Geophys. Res. Lett., 32*, L14S06, doi:10.1029/2005GL022643.

André, N., et al. (2007), Magnetic signatures of plasma-depleted flux tubes in the Saturnian inner magnetosphere, *Geophys. Res. Lett., 34*, L14108, doi:10.1029/2007GL030374.

André, N., et al. (2008), Identification of Saturn's magnetospheric regions and associated plasma processes: Synopsis of Cassini observations during orbit insertion, *Rev. Geophys., 46*, doi:10.1029/2007RG000238.

Bunce, E. J., S. W. H. Cowley, D. M. Wright, A. J. Coates, M. K. Dougherty, N. Krupp, W. S. Kurth, and A. M. Rymer (2005), In situ observations of a solar wind compression-induced hot plasma injection in Saturn's tail, *Geophys. Res. Lett., 32*, L20S04, doi:10.1029/2005GL022888.

Bunce, E. J., S. W. H. Cowley, I. I. Alexeev, C. S. Arridge, M. K. Dougherty, J. D. Nichols, and C. T. Russell (2007), Cassini observations of the variation of Saturn's ring current parameters with system size, *J. Geophys. Res., 112*, A10202, doi:10.1029/2007JA012275.

Burch, J. L., J. Goldstein, T. W. Hill, D. T. Young, F. J. Crary, A. J. Coates, N. Andre, W. S. Kurth, and E. C. Sittler, Jr. (2005), Properties of local plasma injections in Saturn's magnetosphere, *Geophys. Res. Lett., 32*, L14S02, doi:10.1029/2005GL022611.

Carbary, J. F., D. G. Mitchell, P. Brandt, E. C. Roelof, and S. M. Krimigis (2008), Statistical morphology of ENA emissions at Saturn, *J. Geophys. Res., 113*, A05210, doi:10.1029/2007JA012873.

Chen, Y., and T. W. Hill (2008), Statistical analysis of injection/dispersion events in Saturn's inner magnetosphere, *J. Geophys. Res., 112*, A07215, doi:10.1029/2008JA013166.

Connerney, J. E. P., M. H. Acuña, and N. F. Ness (1983), Currents in Saturn's magnetosphere, *J. Geophys. Res., 88*, 8779–8789.

Dougherty, M. K., et al. (2004), The Cassini magnetic field investigation, *Space Sci. Rev., 114*, 331–383.

Gurnett, D. A., et al. (2004), The Cassini radio and plasma wave investigation, *Space Sci. Rev., 114*, 395–463.

Hill, T. W., A. M. Rymer, J. L. Burch, F. J. Crary, D. T. Young, M. F. Thomsen, D. Delapp, N. André, A. J. Coates, and G. R. Lewis (2005), Evidence for rotationally driven plasma transport in Saturn's magnetosphere, *Geophys. Res. Lett., 32*, L14S10, doi:10.1029/2005GL022620.

Hill, T. W., et al. (2008), Plasmoids in Saturn's magnetotail, *J. Geophys. Res., 113*, A01214, doi:10.1029/2007JA012626.

Kellett, S., C. S. Arridge, E. J. Bunce, A. J. Coates, S. W. H. Cowley, M. K. Dougherty, A. M. Persoon, N. Sergis, and R. J. Wilson (2010), Nature of the ring current in Saturn's dayside magnetosphere, *J. Geophys. Res., 115*, A08201, doi:10.1029/2009JA015146.

Kennelly, T. J., J. S. Leisner, G. B. Hospodarsky, and D. A. Gurnett (2013), Ordering of injection events within Saturnian SLS longitude and local time, *J. Geophys. Res., 118*, 1–7, doi:10.1002/jgra.50152.

Krimigis, S. M., et al. (2004), Magnetospheric imaging instrument (MIMI) on the Cassini mission to Saturn/Titan, *Space Sci. Rev., 114*, 233–329.

Lamy, L., P. Zarka, B. Cecconi, R. Prangé, W. S. Kurth, and D. A. Gurnett (2008), Saturn kilometric radiation: Average and statistical properties, *J. Geophys. Res., 113*. A07201, doi:10.1029/2007JA012900.

Leisner, J. S., C. T. Russell, K. K. Khurana, M. K. Dougherty, and M. Andre (2005), Warm flux tubes in the E-ring plasma torus: Initial Cassini magnetometer observations, *Geophys. Res. Lett., 32*, L14S08, doi:10.1029/2005GL022652.

Mauk, B. H., et al. (2005), Energetic particle injections in Saturn's magnetosphere, *Geophys. Res. Lett., 32*, L14S05, doi:10.1029/2005GL022485.

Mauk, B. H., et al. (2009), Fundamental plasma processes in Saturn's magnetosphere, in *Saturn from Cassini Huygens*, edited by M. Dougherty and S. M. Krimigis, pp. 281–331, Springer, Dordrect, Holland.

Mitchell, D. G., et al. (2005), Energetic ion acceleration in Saturn's magnetotail: Substorms at Saturn? *Geophys. Res. Lett., 32*, L20S01, doi:10.1029/2005GL022647.

Mitchell, D. G., et al. (2009), Recurrent energization of plasma in the midnight-to-dawn quadrant of Saturn's magnetosphere, and its relationship to auroral UV and radio emissions, *Planet. Space Sci., 57*, 1732–1742, doi:10.1016/j.pss.2009.04.002.

Müller, A. L., J. Saur, N. Krupp, E. Roussos, B. H. Mauk, A. M. Rymer, D. G. Mitchell, and S. M. Krimigis (2010), Azimuthal plasma flow in the Kronian magnetosphere, *J. Geophys. Res., 115*, A08203, doi:10.1029/2009JA015122.

Paranicas, C., et al. (2007), Energetic electrons injected into Saturn's neutral gas cloud, *Geophys. Res. Lett., 34*, L02109, doi:10.1029/2006GL028676.

Paranicas, C., D. G. Mitchell, D. C. Hamilton, S. M. Krimigis, B. H. Mauk, P. C. Brandt, J. F. Carbary, and A. M. Rymer (2008), Energetic Charged Particle Injections at Saturn, AGU, Fall Meeting, 2008AGUFMSM32A..03P.

Radioti, A., E. Roussos, D. Grodent, J.-C. Gérard, N. Krupp, D. G. Mitchell, J. Gustin, B. Bondfond, and W. Pryor (2013), Signatures of magnetospheric injections in Saturn's aurora, *J. Geophys. Res., 118*, 1922–1933, doi:10.1002/jgra.50161.

Rymer, A. M., et al. (2009), Cassini evidence for rapid interchange transport at Saturn, *Planet. Space Sci., 57*, 1779–1784, doi:10.1016/j.pss.2009.04.010.

Sergis, N., S. M. Krimigis, D. G. Mitchell, D. C. Hamilton, N. Krupp, B. H. Mauk, E. C. Roelof, and M. Dougherty (2007), Ring current at Saturn: Energetic particle pressure in Saturn's equatorial magnetosphere measured with Cassini/MIMI, *Geophys. Res. Lett., 34*, L09102, doi:10.1029/2006GL029233.

Thomsen, M. F., R. J. Wilson, R. L. Tokar, D. B. Reisenfeld, and C. M. Jackman (2013), Cassini/CAPS observations of duskside tail dynamics at Saturn, *J. Geophys. Res. Space Phys., 118*, 5767–5781, doi:10.1002/jgra.50552.

Vasyliunas, V. M. (1983), Plasma distribution and flow, in *Physics of the Jovian Magnetosphere*, edited by A. J. Dessler, pp. 395–402, Cambridge Univ. Press, New York.

20

Radiation Belt Electron Acceleration and Role of Magnetotail

Geoffrey D. Reeves

20.1. INTRODUCTION

While the process of energetic electron and ion trapping in Earth's magnetosphere is well understood, many outstanding questions about the origin and dynamics of the radiation belts remain to this day. Those questions and their implications for space weather were the primary motivation for the NASA Van Allen Probes mission launched in August 2012 [*Mauk et al.*, 2012]. In this chapter we consider the rapid evolution in our understanding of radiation belt dynamics that has occurred over the past decade, a period that might be called "the radiation belt revolution," when new observations challenged long-held assumptions and new theoretical frameworks emerged. This period also introduced a dramatic shift in radiation belt research when improvements in energetic particle measurements, the development of more realistic global magnetic field models, and advances in computational power enabled researchers to investigate radiation belt electrons in their canonical coordinate system: phase space density as a function of adiabatic magnetic invariants [*Schulz*, 1996]. We concentrate on the observational studies of relativistic electron phase space density gradients that have led to our current understanding of radiation belt dynamics and the role that the magnetotail plays in controlling those dynamics.

20.2. EARLY OBSERVATIONS

Earth's radiation belts were discovered by the Explorer 1 and 3 satellites in 1958 [*Van Allen et al.*, 1958]. The first source proposed for the origin of the radiation belts was decay of neutrons produced by the interaction of cosmic rays with Earth's atmosphere [*Singer*, 1958; *Kellog*, 1959; *Vernov et al.*, 1959]. Balloon and rocket measurements, however, showed that the loss rates of radiation belt electrons through precipitation into the atmosphere [e.g., *Winckler et al.*, 1962] would then demand improbably high rates of replenishment. The time dependence of outer belt electron fluxes was even more problematic [see, e.g., *Walt*, 1996], but, until spectral measurements were available, variations in detector count rates could potentially be explained by more subtle changes in spectral slope, an ambiguity that would resurface, for different reasons, in the radiation belt revolution. (In contrast, the proposal by *Kellog* [1959] that neutron albedo might be responsible for protons in the radiation belts is now widely accepted [*Selesnick et al.*, 2007].)

In the (remarkably few) years following their discovery much was learned about the radiation belts, including the inner and outer zone spatial structure, spectral information, pitch angle distributions, and characteristic response of the radiation belts to geomagnetic storms: a rapid decrease in electron fluxes during the main phase followed by a several-day reintensification. Two primary sources for radiation belt electrons were considered: either a solar wind source or an internal acceleration process. Since little was known about the structure of the magnetosphere at that time, "internal" was naturally taken to mean internal to the radiation belts themselves. (See the early review article by *Farley* [1963].) Unlike protons of solar origin, the rigidity of electrons with mega-electron-volt energies is too small to allow significant penetration of the geomagnetic field [*Størmer*, 1955] so attention turned to how electrons might be accelerated within the magnetosphere.

Northrop and Teller [1960] had worked out the equations of motion that govern adiabatic motion in the geomagnetic

Space Science and Applications Group, Los Alamos National Laboratory, Los Alamos, New Mexico

Magnetotails in the Solar System, Geophysical Monograph 207. First Edition.
Edited by Andreas Keiling, Caitríona M. Jackman, and Peter A. Delamere.

field and derived the invariants associated with electron gyromotion, μ, bounce between mirror points, J or K, and drift azimuthally around Earth, Φ or L^*. (We discuss adiabatic invariants at greater length in section 20.4.) Adiabatic motion, by definition, results in no net changes to the electron distribution but *Fälthammar* [1965] (and others) showed that acceleration of radiation belt electrons could occur through radial diffusion when field fluctuations occur on the time scale of electron drifts. He concluded that "when the first and second invariants are conserved, any radial motion of the particles is associated with a corresponding energy change. Some particles move outward and others inward; but, if there is a source in the outer magnetosphere and a sink farther in, there will be net inward transport and associated net energy gain." *Dungey* [1963] proposed that resonant interaction between electromagnetic whistler-mode waves and radiation belt electrons could form a sink for electrons by pitch angle scattering electrons into the atmospheric loss cone. He concluded that "the effect is found to be important between the inner and outer belts," the region that is called the slot region near $L \approx 2$. *Lyons and Thorne* [1973] showed, quantitatively, that the quiet time structure of the outer electron belt could be "explained on the basis of a balance between pitch angle scattering loss and inward radial diffusion from an average outer zone source". Such studies and others, including the comprehensive treatment by *Schulz and Lanzerotti* [1974], established the fundamental paradigm for radiation belt dynamics: acceleration by inward radial diffusion and loss by pitch angle scattering into the atmosphere. This model placed the magnetotail front and center as the source for radiation belt electrons through convection, substorm injections, or both.

20.3. RADIATION BELT REVOLUTION

In 1991 a dramatic event occurred in the radiation belts that took the entire community by surprise and refocused attention on the radiation belts. On 22 March a solar flare and coronal mass ejection produced an interplanetary shock that hit the magnetosphere at ~03:42 UT on 24 March. The Combined Release and Radiation Effects Satellite (CRRES) was fortuitously positioned at $L \approx 2.6$, in the radiation belt slot region, where, in less than a minute, it observed a four order of magnitude increase in the counting rates for >6 MeV electrons [*Blake et al.*, 1992]. Based on the timing of drift echoes in the event (Figure 20.1), *Blake et al.* concluded that the signatures in the >6, >9, and >13 MeV electron channels were actually produced by 15–50 MeV electrons. These observations were totally inconsistent with the paradigm of gradual radial diffusion from a magnetotail source. Subsequent modeling and analysis did, however, show that the source was the magnetotail (or at least the edge of the stable trapping boundary at high L shells) and the means of acceleration was inward radial transport. However, the very rapid inward transport of electrons was due to drift

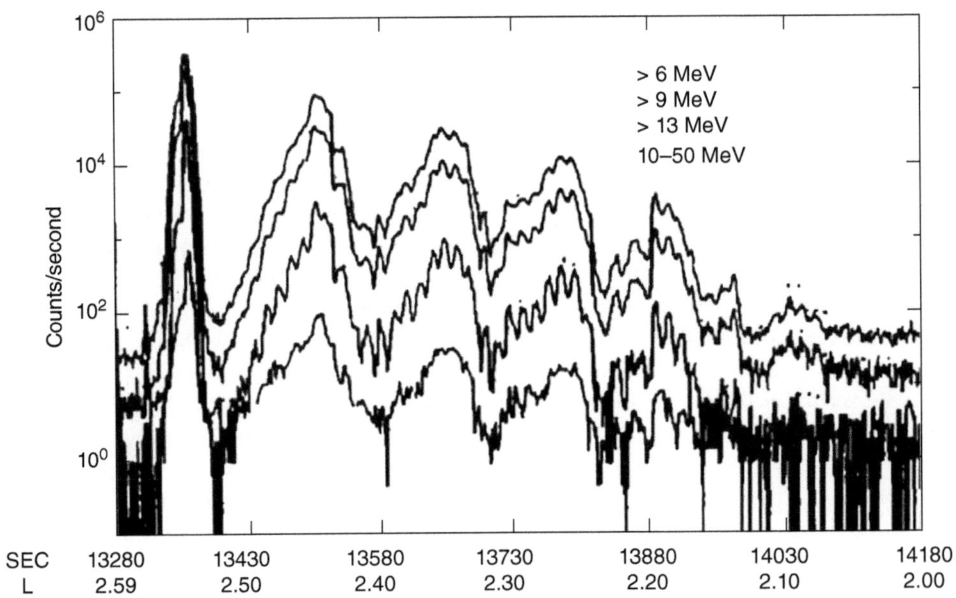

Figure 20.1 CRRES observations of relativistic electrons in the slot region following the impact of an interplanetary shock. Periodic enhancements are the signatures of the drift echoes of 15–50 MeV electrons producing background counts in the lower energy channels [from *Blake et al.*, 1992]. This event is widely recognized as starting a revolution in thinking about radiation belt dynamics.

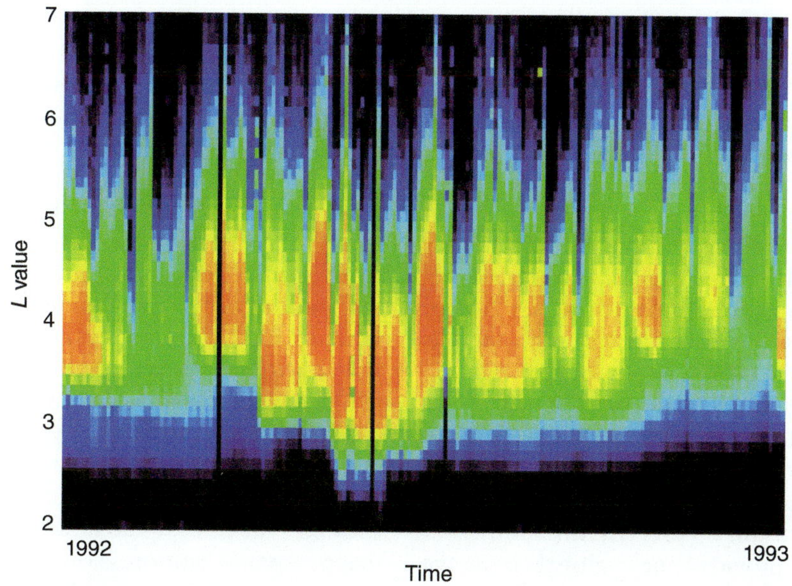

Figure 20.2 Plot of 2–6 MeV electron fluxes from the SAMPEX satellite. (Courtesy of S. Kanekal.) Frequent enhancements (and depletions) of radiation belt electrons are seen during geomagnetic storms but their characteristics are, at present, unpredictable.

resonance with the passage of a shock-induced magnetosonic wave [*Li et al.*, 1993]. Subsequent searches of previous observations showed that this process has occurred at other times in less dramatic fashion [*Gussenhoven et al.*, 1994], but it is not the cause for the vast majority of radiation belt enhancement events.

While the March 1991 event was anomalous (if not unique), it dramatically reinforced a growing picture of the radiation belts as a highly dynamic and unpredictable part of the magnetosphere. That picture was strongly confirmed by the Solar Anomalous Magnetospheric Particle Explorer (SAMPEX). Figure 20.2 shows the flux of >2 MeV electrons measured by SAMPEX in low Earth orbit (LEO) [*Baker et al.*, 1993]. Fluxes are color coded and plotted as a function of L shell and time. As Figure 20.2 suggests, the radiation belts are highly dynamic. The time scale to establish equilibrium is generally much longer than the time between dynamic events, so the state of equilibrium described by Lyons and Thorne is rarely, if ever, fully realized. Not only do the intensities change but the location of the peak of the radiation belts, the size and location of the slot region, and the whole spatial structure of the outer belts change, sometimes on time scales faster than a day. Also apparent in Figure 20.2 are dramatic dropouts of radiation belt electron fluxes that accompany some geomagnetic storms [e.g., *Onsager et al.*, 2002]. (It would later be shown that only about half of geomagnetic storms intensify the radiation belts while roughly 20% produce strong losses [*Reeves et al.*, 2003].) Plots such as this from CRRES also showed that the dynamics of the radiation belts as seen in

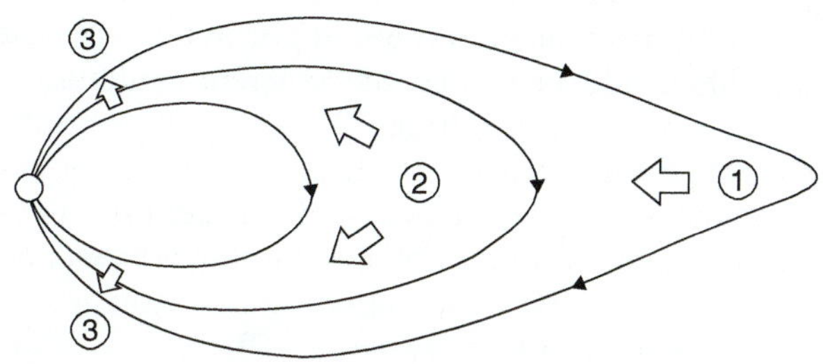

Figure 20.3 Schematic of the "recirculation" model for radiation belt acceleration [from *Nishida*, 1976].

different energy channels could be quite different. The rapid and profound changes in the radiation belts seen by SAMPEX and other satellites were generally consistent with the radial diffusion paradigm but they did provide challenging constraints.

Around this same time *Baker et al.* [1989] suggested that the "recirculation model" developed to explain electron observations at Jupiter [*Nishida*, 1976] might also apply in Earth's radiation belts and this modification of the standard radial diffusion paradigm was widely adopted by radiation belt scientists. In the recirculation model (Figure 20.3) electrons radially diffuse inward gaining energy while conserving the first two adiabatic invariants (step 1). In the slot region the pitch angle scattering rates exceed the radial diffusion rates and the particles are scattered toward the atmosphere (step 2). At low altitude the electrons are scattered across field lines without substantial loss of energy (step 3). They are then free to repeat the cycle. While the details were problematic (particularly step 3), the recirculation model had several advantages. It naturally accounted for the 2–3 day buildup of electron fluxes following the commencement of a storm and the very high energies seen at geosynchronous orbit. Since geosynchronous orbit lies close to the trapping boundary, magnetotail electrons cannot diffuse far to get there and therefore cannot gain much energy through radial diffusion alone. (Radial diffusion does not apply outside the trapping boundary where other, primarily convective, transport dominates.) Recirculation could explain acceleration up to arbitrarily high energies even at geosynchronous orbit.

Studies of the radiation belts also benefited from the growing network of satellites capable of measuring the belts. Energetic particle measurements from Geostationary Operational Environmental Satellites (GOES) and Los Alamos National Laboratory (LANL) instruments at geosynchronous orbit provided near-global coverage, albeit in a narrow range of L shells [*Reeves et al.*, 1998b]. Combining those measurements with data from other orbits (SAMPEX, global positioning system (GPS), high Earth orbit (HEO), and others) started to fill in a three-dimensional picture of the radiation belts [*Friedel et al.*, 1999]. Perhaps even more importantly, multisatellite studies could overcome the inherent limitation of satellite orbital periods (typically one half to several days) and provide measurements at multiple L shells on time scales less than an hour. Multisatellite analysis made new, detailed studies of space weather events possible. One of the first was a radiation belt enhancement event in November 1993 [*Li et al.*, 1997].

One such event, in January 1997, was exceptionally well studied [*Baker et al.*, 1998; *Goodrich et al.*, 1998; *Jordanova et al.*, 1998, 1999; *Kamide et al.*, 1998; *Reeves et al.*, 1998a, 1998b; *Hudson et al.*, 2000, 2001; *Lyons et al.*, 2000]. Following a procedure similar to that used by *Li et al.* [1997], *Reeves et al.* [1998a] binned data from 11 satellites in different orbits according to L shell. Near geosynchronous orbit, $L \approx 6.6$, the 2 MeV radiation belt electrons showed their "typical" behavior, a gradual buildup of fluxes over ~4 days (with a less typical shock compression early in the event). The behavior was startlingly different in the heart of the radiation belts at $L \approx 4.6$, where the GPS satellites cross the equatorial plane (Figure 20.4a). There, the 2 MeV electrons shot up by over two orders of magnitude in less than 12 h. Following the initial acceleration, fluxes remained remarkably constant and showed no evidence of the gradual buildup seen at geosynchronous orbit. In fact, the acceleration was so rapid that the radial profiles seen by GPS satellites moving away from the equator (i.e., outward in L) measured significantly different fluxes than on the inbound part of the orbit even though inbound and outbound measurements at the same L shell were separated by less than an hour (Figure 20.4b). Furthermore, the gradual buildup of fluxes seen near geosynchronous orbit appeared to be the result of outward radial diffusion from the newly accelerated population near $L \approx 4.6$ (Figure 20.4b). It is interesting to contrast these high-time-resolution measurements with observations from a single satellite. Polar with an ~18-h orbit captured the state of the belts before and after the acceleration but could not resolve the acceleration event itself.

Reeves et al., [1998a] drew two conclusions from the January 1997 event. First, the observations were not consistent with the recirculation model. One of the strengths of the recirculation model was that it naturally explained the gradual buildup to a delayed peak that is characteristic of radiation belt observations at geosynchronous orbit. For January 1997 the geosynchronous observations still showed that delay. (The adiabatic response to the strong magnetopause compression on 10–11 January was only temporary and limited to the outer L shells.) In the recirculation model that delay should be observed at all L shells as electrons complete the energization loop sketched in Figure 20.3. Instead at $L \approx 4.6$ the energization was not only much faster but also short-lived, with almost no change in flux after 12 January. Second, if the electrons were accelerated inside geosynchronous orbit and diffused outward, then the acceleration region would have to be confined to a remarkably narrow range of L shells, well inside geosynchronous orbit, in the heart of the radiation belts. If true, then the source of the MeV radiation belt electrons would not be the magnetotail, but rather the source would be lower-energy electrons in that same spatial region, somewhere between the plasmasphere and geosynchronous orbit.

At the same time that new missions and multisatellite studies were providing a new window on radiation belt dynamics, new theories of relativistic electron acceleration were being developed that included localized acceleration processes deep in the radiation belts. *Kennel and Petschek* [1966] (extending the work of *Dungey* [1963]) showed that the very low frequency (VLF) whistler-mode waves that pitch angle scattered electrons into the atmosphere could be produced by plasma distributions that had distribution functions with temperature anisotropies unstable to the growth of whistler-mode waves. The (suprathermal) electron temperature anisotropy excites whistler-mode waves which cause pitch angle scattering of electrons toward the loss cone. In the late 1990s *Summers et al.*, [1998] proposed that cyclotron resonant interactions at much higher (relativistic) energies can change electron momentum. At those energies gradients in the energy dimension implied that the flow of energy is from the waves to the particles providing a mechanism for accelerating electrons in situ from the local population of lower-energy electrons.

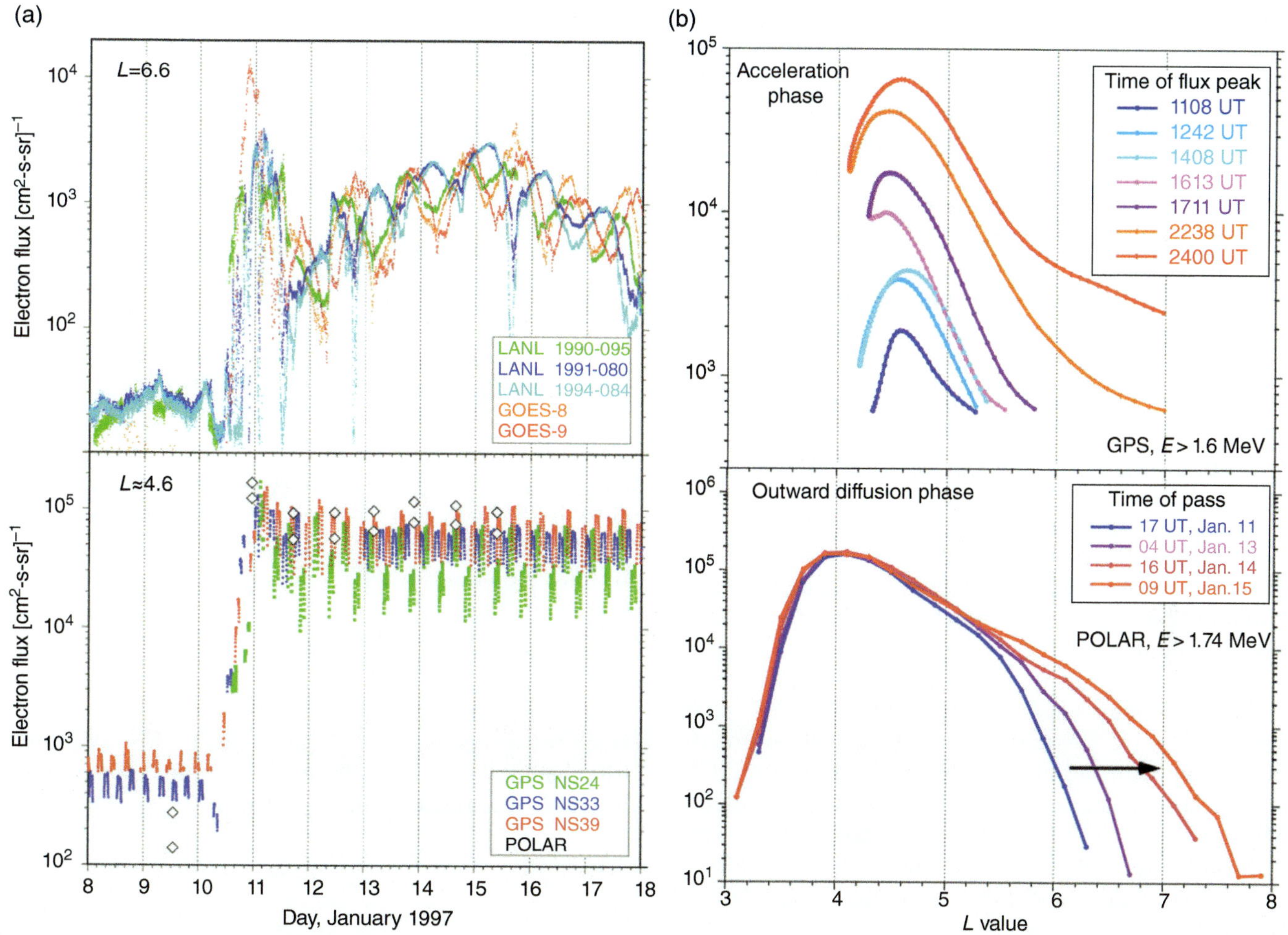

Figure 20.4 Multisatellite electron observations during the January 1997 radiation belt event [from *Reeves et al.*, 1998a]. Rapid acceleration in the heart of the radiation belts ($L \approx 4.6$) and geosynchronous orbit ($L \approx 6.6$) suggested local acceleration and subsequent outward radial diffusion but, without phase space density observations, were inconclusive.

This "local acceleration" scenario suggested a more complex and interdependent set of processes might be responsible for acceleration of radiation belt electrons to MeV energies. First electrons must be transported from the magnetotail to the inner magnetosphere by processes such as substorm injections and enhanced large-scale convection. In the process a fraction of the magnetotail electrons are energized to form a seed population of electrons with energies from a few to a few hundred kilo-electron-volts. Temperature anisotropies at lower energies (<30 keV) provide a source of free energy for the generation of whistler-mode waves while higher-energy electrons (≈100–500 keV) provide a "seed" population that can be accelerated to MeV energies by those same waves.

These theories of local acceleration were to undergo much future development [e.g., *Horne and Thorne*, 1998, 2003; *Horne et al.*, 2000, 2005a, 2005b; *Summers and Ma*, 2000; *Horne*, 2002; *Summers et al.*, 2002, 2004, 2007; *Albert and Young*, 2005; *Shprits et al.*, 2006a; *Omura et al.*, 2007, 2008; *Albert et al.*, 2009; *Thorne*, 2010]. One characteristic has remained constant though; the resonance between VLF waves and relativistic electrons takes place locally, on a given flux tube, without transporting the electrons radially. This local acceleration is therefore fundamentally different from radial-diffusive acceleration in both the energization process and the source population.

Unfortunately, the measurements needed to clearly distinguish between the two types of acceleration process were not yet available. A radially localized peak in electron fluxes (at a given energy) can be produced by either local or radial-diffusive acceleration. It was widely recognized in the radiation belt community that it would be necessary to convert measurements of electron flux j to phase space density f. The two quantities are related by the simple formula $f = j/p^2$. The challenge comes in converting flux from its natural coordinate system (position, energy, and pitch angle) to phase space density as a function of the adiabatic invariants (μ, K, L^*) that define the canonical coordinate system for relativistic electron motion [*Schulz*, 1996; see also GEM Inner Magnetosphere Storms Working Group report, 2000, http://www-ssc.igpp.ucla.edu/gemwiki/index.php/Documents_and_Reports]. In any realistic magnetic field the conversion from $j(\vec{r},E,\alpha)$ to $f(\mu,K,L^*)$ is both unique and time dependent for any point along a satellite's orbit. Fortunately, new global magnetic field models and advances in computational power were available to make that conversion.

20.4. PHASE SPACE DENSITY GRADIENTS

Radial diffusion occurs whenever various processes produce field fluctuations on the time scale of electron drifts. At some level, these fluctuations always exist so radial diffusion is always operating in Earth's radiation belts, either at rapid rates or slower rates. Since diffusion always moves a quantity from a region of higher density to a lower one, the radial gradients of phase space density (i.e., $f(L^*)$ at fixed μ and K) will be positive (increasing with radial distance) if electrons are diffusing inward from a magnetotail source population whereas they will be negative (decreasing with radial distance) in the outer part of the belt if the electrons are accelerated locally and the magnetotail/magnetopause acts as a sink.

Electron motion in a steady-state dipolelike magnetic field can be described by three periodic motions, gyration, bounce, and drift, each of which has an associated magnetic invariant which is conserved under adiabatic changes to the magnetic field. The first invariant, μ, is associated with the gyromotion around the magnetic field and is given by

$$\mu = \frac{p_\perp^2}{2m_0 B},$$

where m_0 is the rest mass and $p_\perp$ is the relativistic momentum in the direction perpendicular to the magnetic field B. It is the conservation of $p_\perp^2/B$ that leads to energization as a particle diffuses radially inward. Importantly, it is the perpendicular component of the momentum that changes so inward diffusion tends to create distributions that are strongly peaked near 90°.

The first invariant can be calculated based on local measurements of the electron fluxes and magnetic field. The second and third invariants are spatial integrals and therefore must be calculated using a magnetic field model. The second invariant is associated with the bounce motion between the northern and southern mirror points:

$$J = \oint p_\parallel \, ds = 2\int_{S_s}^{S_n} p_\parallel \, ds.$$

When μ is conserved, the second invariant can also be expressed in a convenient form that is independent of energy,

$$K = \frac{J}{2\sqrt{2m_0\mu}} = \int_{S_s}^{S_n} \sqrt{B_m - B(s)}\, ds,$$

where B_m is the magnetic field strength at the mirror point and s is the path length along the field line. The third invariant is an integral around the drift shell and between the mirror points,

$$\Phi = \oint \vec{B} \cdot d\vec{A},$$

which conserves the total magnetic flux enclosed within a drift shell. In radiation belt physics it is more common to use the L^* parameter [*Roederer*, 1970]:

$$L^* = \frac{2\pi M}{\Phi R_E},$$

where M is Earth's magnetic moment and R_E is the radius of Earth (6370 km).

In a dipole magnetic field $L^* = L$. However, when the field is not a pure dipole, L^* is not equivalent to L. In a realistic field, electrons with different pitch angles follow different drift paths, an effect known as drift shell splitting (Figure 20.5). Therefore, electrons with different pitch angles measured at a given point in space (such as a specific satellite location) have different third invariants Φ and therefore different L^*. That means that, while a given point in space can have a single value of L for all electrons on the field line, it cannot be assigned a single value of L^*. Furthermore, for a fixed point in space, all three invariants change when the magnetic field changes. In general, L^* for a fixed measurement point decreases with increasing geomagnetic activity, a feature known as the "Dst effect" [*Kim and Chan*, 1997].

Looked at from the other perspective, when phase space density is plotted as a function of L^* and L^* is calculated in a realistic field model, phase space density on that drift shell does not change when the magnetic field does. In the absence of acceleration or loss, L^* (for

Figure 20.5 Visualization of radiation belt drift shells based on numerical particle tracing in a realistic geomagnetic field [*Reeves et al.*, 2012]. The copper-colored sphere in the bottom left represents a satellite that measures electrons on a single field line at a given point. Mirror points for different equatorial pitch angles (in 10° steps) are shown spaced along the field. As electrons drift around Earth, conservation of μ and *K* requires that the electrons follow different drift paths as indicated by the color-coded field lines. Drift shell splitting is particularly apparent in the upper right. Electrons measured at the satellite with 30° pitch angles move along the copper-colored surface, which is the drift shell for those electrons. On that drift shell, μ, *K*, and *L** are constant but pitch angle, energy, and radial distance are not.

a given μ, *K*) defines a surface of constant phase space density, $f(\mu, K, L^*) = \text{const}$. Therefore, the only way to unambiguously identify radial diffusion, acceleration, or loss is to look for changes in phase space density, e.g., $f(\mu, K, L^*, t)$, that are the signatures of violation of one or more of the adiabatic invariants.

Figure 20.6 illustrates, schematically, the changes in phase space density profiles expected from radial diffusion (D_{LL}) and local acceleration (D_{EE}). Phase space density profiles are represented for three different times with the blue, magenta, and red lines, respectively. If the phase space density at the source and sink do not vary with time, then increasing radial diffusion will move electrons inward, flattening the gradient at high L^* where diffusion is strongest (Figure 20.6). In the inner regions ($L^* \approx 4$–5) the phase space density increases locally but the increase is due to the radial transport of electrons. Stated another way, the phase space density is not conserved, $f(\mu, K, L^*, t) \neq \text{const}$, because "new" electrons are added from the source region. Note, however, that regardless of how strong the radial diffusion is, the radial gradient will always be positive and monotonic from the sink (slot region) to the source (magnetotail).

VLF waves energize electrons locally through energy diffusion and produce a distinctively different change in phase space density profiles – a radially-localized peak (Figure 20.7) that grows from the background distribution to produce negative radial gradients at high L^*. Here again the phase space density increases in the inner regions

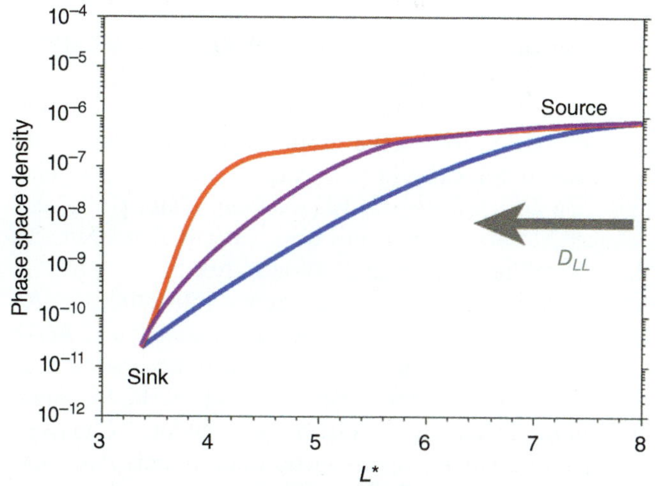

Figure 20.6 Schematic of radial diffusion from a source region (the magnetotail) to a sink region (the slot). An increase in radial diffusion will increase the transport rate and increase the phase space densities in the heart of the belts ($L^* \approx 4$–5).

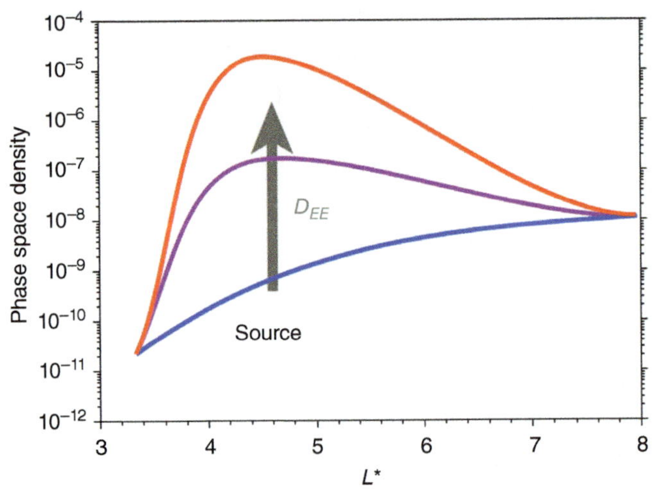

Figure 20.7 Schematic of local acceleration. Wave-particle interactions produce energy diffusion that accelerates electrons. In this case the source of the relativistic electrons is the local lower-energy populations.

($L^* \approx 4$–5) but the increase is due to the local acceleration of electrons. In this case, $f(\mu, K, L^*, t) \neq \text{const}$ because acceleration (energy diffusion) does not conserve μ.

20.5. PHASE SPACE DENSITY OBSERVATIONS

Numerical models such as T89 [*Tsyganenko*, 1989] made it possible to calculate the second and third invariants in a fairly realistic and activity-dependent global magnetic field. To test the proposal that electrons could be accelerated locally, in the heart of the radiation belts, and then diffuse outward, *Selesnick and Blake* [1998] calculated $f(\mu, K, L^*, t)$ using observations from NASA's Polar satellite. Figure 20.8 shows flux at $E = 0.68$–2.4 MeV and phase space density at $\mu = 150$–2400 MeV/G and $K = 0.5$ $G^{1/2}$ R_E. In a later work, *Selesnick and Blake* [2000] analyzed several other storms with associated relativistic electron enhancements. In that work, Selesnick and Blake calculated L^* (rather than the local L parameter) using a variety of magnetic field models that were available at the time. They noted that, for time-dependent models with large local time asymmetries, the values of L^* and L could be substantially different and they found that the presence or absence of a radial peak in phase space density was quite dependent on which magnetic field model was chosen.

Green and Kivelson [2004] also found evidence of a phase space density peak increasing locally in the region around $L^* \approx 5$ and extended the work of Selesnick and Blake. They provide a detailed analysis of the uncertainties in the calculation of the adiabatic invariants and therefore $f(\mu, K, L^*)$ as well as the effects of uncertainties in parameters that characterize the global magnetic field models. They also extended the studies to consider a range of μ and K values covering $\mu = 200$–2000 MeV/G and $K = 500$–5000 $G^{1/2}$ km. (The invariants can be expressed in a variety of units and a helpful reference to convert among them has been provided by *Taylor et al.*, [2004].) Their careful analysis showed that the growth of local peaks of phase space density was a robust feature of the observations and that internal acceleration by VLF waves was a *viable mechanism* but also concluded that "the pieces to the puzzle presented by Van Allen's first radiation belt measurements have narrowed but the complete picture has not yet fully emerged."

One limitation of the work by Green and Kivelson was the orbital characteristics of the Polar satellite. Polar was in an elliptical orbit with ~9R_E apogee, an orbital period of ~18 h, and, as the mission name suggests, a very high inclination orbit. As a result Polar made infrequent passes through the heart of the radiation belts (Figure 20.4a) and, more importantly, for large parts of its orbit Polar could only measure very small equatorial pitch angles, i.e., very field aligned electrons. In general, to get continuous coverage over a sufficiently large range of L^* required using very large values of K corresponding to equatorial pitch angles $\alpha < 35^\circ$. A scenario where phase space density peaks could arise from radial-diffusive acceleration is shown in Figure 20.9. Conservation of μ implies that radial diffusion increases the perpendicular component of momentum. Therefore, the strongest signatures of radial diffusion will be seen in electrons with equatorial pitch angles near 90°. If pitch angle diffusion ($D_{\alpha\alpha}$) occurs (or is strongest) in a narrow range of L shells, then, in combination with radial diffusion, the two processes can produce very different radial profiles of phase space density for different equatorial pitch angles [e.g., *Fennell and Roeder*, 2008]. Since pitch angle diffusion alone does not energize electrons, all the energy in this case comes from radial diffusion.

Hilmer et al., [2000] had examined equatorial electrons measured at geosynchronous orbit ($L \approx 6.6$) and GPS (equatorial crossing at $L \approx 4.2$). They found that, consistently, geosynchronous phase space densities were greater than or equal to GPS phase space densities and tended to increase sooner. They concluded that "fourteen of the 15 analyzable HSSWS (high speed solar wind stream) events … are consistent, we suggest, with a radial diffusion process transporting electrons of similar μ inward from a source region near or beyond $L = 6.6$." As with other studies, ambiguities remained. Hilmer et al. used omnidirectional fluxes and assumed that they were equal to the flux at 90°, $K = 0$, which would be true only

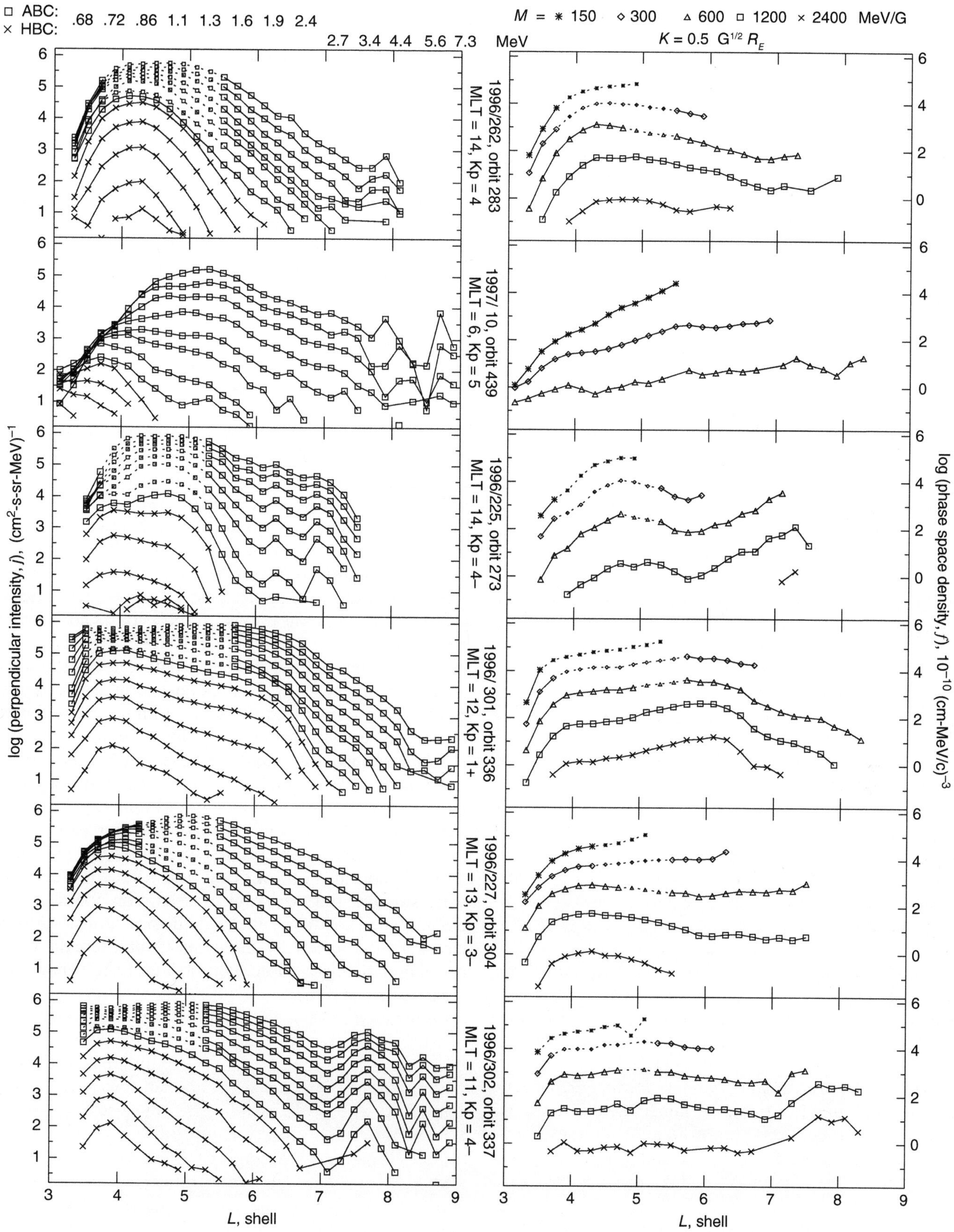

Figure 20.8 Observations from NASA'S Polar satellite for January 1997. Electron fluxes as a function of energy and L shell are shown on the left. Phase space densities as a function of μ and K are shown on the right [from *Selesnick and Blake*, 1998].

for a purely isotropic distribution. In addition, to get near-equatorial measurements, the equatorial crossing of the GPS orbit could be used giving the gradient between just two radially separated positions. The observations from Polar (and the schematic in Figure 20.6) show that a radial peak in phase space density can exist and still give positive gradients between 4.2 and 6.6, depending on the width and location of that peak.

Chen et al. [2007b] added observations from equatorial crossings of Polar to the LANL-GEO and GPS data. They chose an interval in 2002 when the equatorial crossing of Polar was well outside geosynchronous orbit and therefore able to determine radial gradients both inside and outside geosynchronous orbit. Chen et al. also used pitch-angle-resolved data from LANL-GEO. (The GPS detectors are inherently omnidirectional so an activity-dependent statistical pitch angle distribution model was used.) LANL-GEO provided continuous (4 min) observations from three satellites operating at different local times. Comparison of the phase space densities

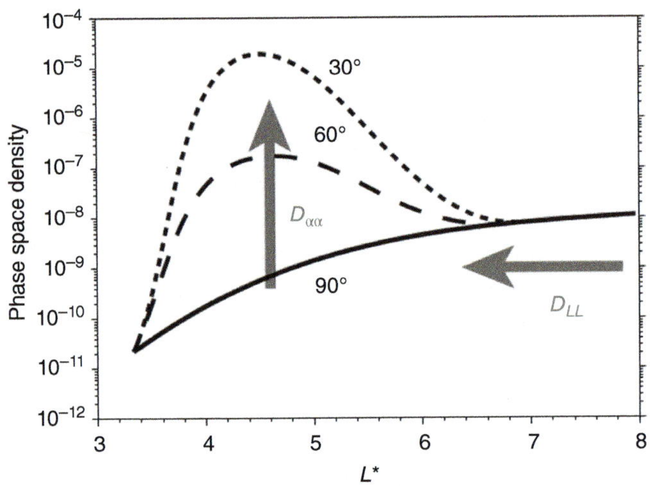

Figure 20.9 Schematic of the different radial profiles of phase space density that can be produced for different equatorial pitch angles (*K* values) through a combination of radial diffusion and localized pitch angle diffusion.

when the satellites were on the same L^* shell provided confidence in the global magnetic field model [*Chen et al.*, 2005, 2007a]. Data from Polar were only used when Polar was on closed radiation belt drift shells (i.e., inside the trapping boundary) but near the outer edge of the radiation belts. The results of *Chen et al.* [2007b] showed clear examples of peaked phase space density for near-equatorially mirroring electrons ($K=0.03\ G^{1/2}\ R_E$). For example, a storm in October 2002 showed growing phase space density measured by all three geosynchronous satellites ($L^* \approx 5$–6) while GPS ($L^* \approx 4$) and Polar ($L^* \approx 7$–9) remained at much lower values (Figure 20.10).

To some, the event and statistical studies by *Chen et al.* [2007b] and the numerous observational and theoretical studies that preceded pointed to an "emerging consensus that in addition to radial diffusion, some other mechanism accelerates electrons locally, at the most intense parts of the belts" [*Reeves*, 1998]. This so-called Roratonga Consensus was widely discussed but, it turns out, far from universally shared. *Selesnick and Blake* [2000] had shown, through modeling, that peaks in phase space density deep in the outer electron belt could also be produced by radial diffusion if the source at high L shells first increased and then decreased. This is illustrated in Figure 20.11. At time t_1, inward diffusion from a strong source at high L^* increases phase space density in the heart of the belts. If the source strength decreases at some later time, t_2, then radial diffusion continues to operate but the direction is reversed and a radial peak will exist until a new equilibrium can be established.

There is ample evidence that the strength of the source population does indeed vary, particularly during geomagnetic storms when many processes affect the high-L-shell electron populations: convection, substorm injections, electron microburst precipitation, magnetopause losses, and others. That and a growing body of work showing that very rapid acceleration could be produced by ULF-driven radial diffusion and transport (see, e.g., the review by *Elkington* [2006]) left the cause of relativistic electron acceleration and the role of the magnetotail in doubt.

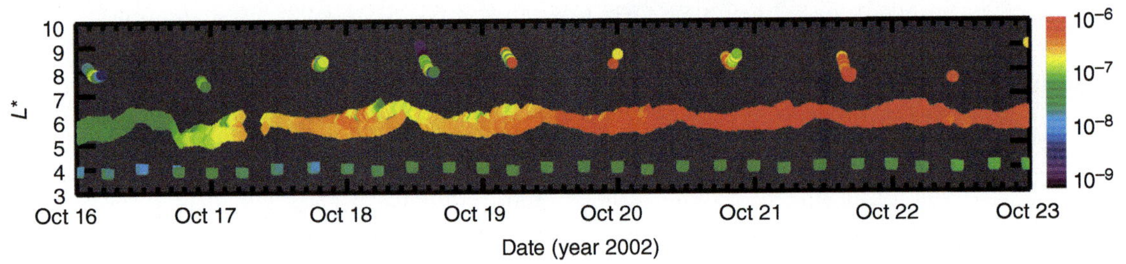

Figure 20.10 Equatorial phase space densities measured by Polar ($L^* > 7$), GPS ($L^* \approx 4$), and three LANL-GEO satellites ($L^* \approx 5$–7) [*Chen et al.*, 2007b].

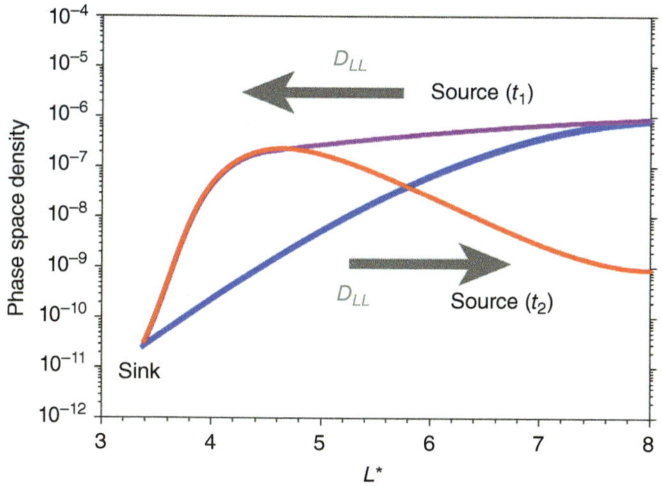

Figure 20.11 Schematic showing how peaks in phase space density can be formed by a combination of radial diffusion and a variable source at high L^*.

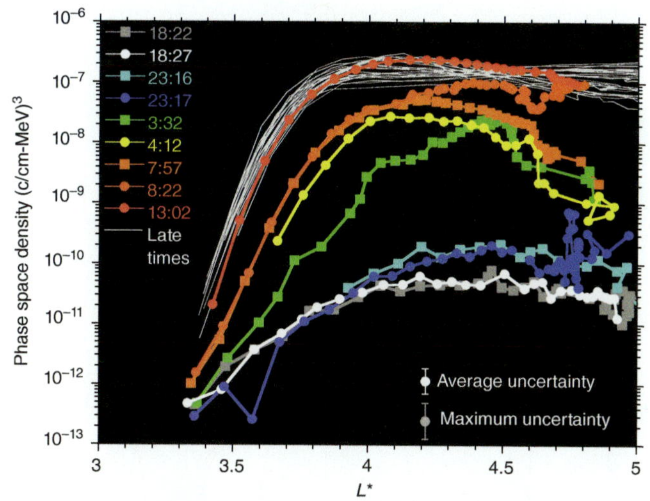

Figure 20.12 Radial profiles of phase space density measured by the two NASA Van Allen Probes satellites on 8–9 October 2013. A rapid increase in phase space density in a region centered near $L^* \approx 4.2$, a slower increase in the outer portion of the belt ($L^* > 4.7$), and the resulting radial peaks in phase space density are the signatures expected for local wave-particle acceleration [from *Reeves et al.*, 2013].

20.6. VAN ALLEN PROBES

NASA's Van Allen Probes mission (formerly Radiation Belt Storm Probes) was designed, in part, to overcome the observational limitations of previous missions and differentiate, clearly, between radial-diffusive acceleration and local wave-particle acceleration. The Van Allen Probes are a pair of satellites with a comprehensive set of particle and field instruments in an equatorial, elliptical orbit with apogee at $\approx 5.8 R_E$. Their orbital period is ≈ 9 h but is slightly different for the two satellites. This allows for the phasing in the orbits to vary with time. At times the satellites are perfectly out of phase with apogees separated by only 4.5 h, passing through the heart of the belts every 2–3 h. This unprecedented temporal resolution, near-equatorial orbit, and paired satellite configuration is providing views of radiation belt dynamics that were not previously possible.

Reeves et al. [2013] analyzed phase space density profiles for a storm that occurred on 8–9 October, early in the Van Allen Probes mission. The October 2013 storm produced a strong and rapid enhancement of the fluxes in the outer electron belt and the first opportunity to test whether or not local wave-particle interactions are a viable mechanism for relativistic electron acceleration. The 2 MeV electron fluxes at $L \approx 4.2$ increased by nearly three orders of magnitude in approximately 12 h. (The resemblance to the event in January 1997 in Figure 20.4a is remarkable.) The unique configuration of the Van Allen Probes was able to capture the rapid transformation of the belts but, unlike the January 1997 event, the observations could be converted into phase space densities and analyzed in detail.

Figure 20.12 shows profiles of phase space density as a function of L^* for fixed values of μ and K. Reeves et al. used the TS04 [*Tsyganenko and Sitnov*, 2005] magnetic field model to calculate the adiabatic invariants. In Figure 20.12 each radial pass is labeled by the time at which the satellite crosses $L^* = 4.2$. Here phase space densities are for $\mu = 3433$ MeV/G and $K = 0.11$ $G^{1/2}$ R_E. ($E \approx 2$–6 MeV and $50° < \alpha_e < 90°$ depending on the point on the orbit). Between 23:17 UT on 8 October and 03:12 UT on 9 October, phase space densities increased by approximately two orders of magnitude at $L^* = 4.2$ and continued increasing until about 1302 UT when the interplanetary magnetic field (IMF) turned northward.

On 8–9 October 2013 the Van Allen Probes were nearly perfectly out of phase in their orbits. This provided the fastest possible time resolution but also provided simultaneous measurements at high and low L^*. The inset in Figure 20.13 shows the orbits of the two satellites. RBSP-A (red) was inbound from apogee and RBSP-B (blue) was outbound from perigee. In Figure 20.13 the times (5 min resolution) of the satellite measurements are color coded for the times when the satellites were both in the outer electron belt (0722–0812 UT). During this time, RBSP-B measured a positive radial gradient at $L^* < 4.2$. At the same time, RBSP-A measured a negative radial gradient at $L^* > 4.2$, showing that the radial peak in phase space density was not an artifact of spatial-temporal aliasing. At 0812 UT, by coincidence, the orbital dynamics were such that both

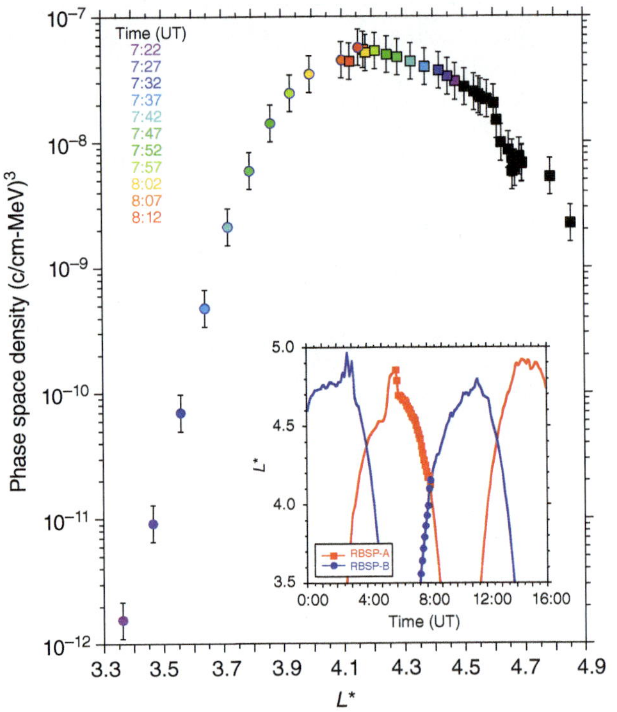

Figure 20.13 Two Van Allen Probes, RBSP-A and RBSP-B, measured simultaneously, a positive radial gradient at $L^* < 4.2$ and a negative radial gradient at $L^* > 4.2$. Points are color coded by the time of each measurement and error bars are given [from *Reeves et al.*, 2013].

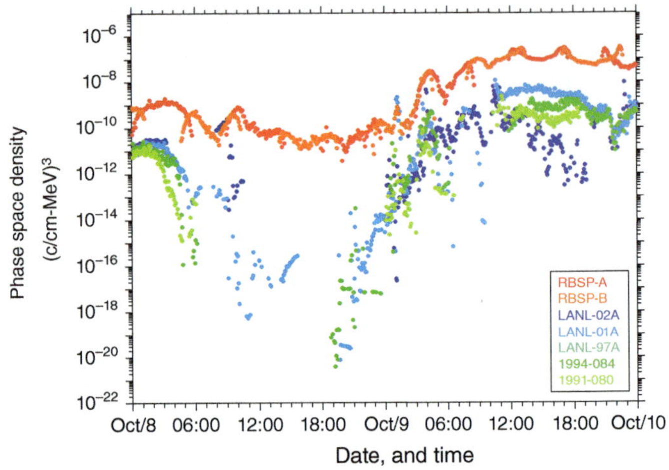

Figure 20.14 Phase space densities measured as a function of time by the two Van Allen Probes satellites ($R < 5.8R_E$) and five LANL geosynchronous satellites ($R \approx 6.6R_E$). Near-continuous geosynchronous measurements near the outer boundary of the radiation belts show that phase space density peaks are not produced, in this case, by a variable outer boundary [from *Reeves et al.*, 1998b].

spacecraft reached $L^* \approx 4.2$ at the same time (red points). Although the satellites were at quite different local times, the satellites should measure the same phase space density when they are on the same L^* (a "drift shell conjunction"). Any mismatches provide limits on uncertainties due to instrument calibration and also to uncertainties in the calculation of the invariant integrals. Reeves et al. used this (and other techniques) to quantify the average and worst case uncertainties as shown in Figures 20.12 and 20.13.

The Van Allen Probes measurements, by themselves, could not completely rule out the possibility of a variable source population at high L^* (Figure 20.11) even though the time between apogee passes was short (4.5 h). Figure 20.14 addresses this remaining ambiguity. Here the two Van Allen Probes (RBSP-A and RBSP-B) phase space densities are plotted in red and orange as a function of time whenever they are in the outer belt. Simultaneous phase space density measurements from five LANL geosynchronous satellites are plotted in greens and blues. (No data are plotted when the fluxes are too low to produce reliable pitch angle distributions.) At a few rare times, the inclination of the Van Allen Probes orbit and orientation of the field were such that they were at approximately the same L^* as a geosynchronous satellite ($R = 6.6R_E$). At those times the satellites measured approximately the same phase space densities. At all other times the Van Allen Probes ($R < 5.8R_E$) were at lower L^* but measured higher phase space density. Therefore, at least for this event, a variable source at high L^* cannot explain the peaks in phase space density leaving local wave-particle acceleration as the only viable process [*Reeves et al.*, 2013]. Further, although the precise wave-particle interaction mechanism deserves further study, chorus waves were observed during this period and model calculations show that chorus can indeed produce the observed spectral and pitch angle distributions [*Thorne et al.*, 2013].

20.7. DISCUSSION

In the traditional radial diffusion model of radiation belt electron acceleration the magnetotail provided a variable source of "seed" electrons that were further accelerated as they were transported inward to the heart of the radiation belts. Radiation belt electrons were gradually lost from the system through pitch angle diffusion into the atmospheric loss cone. This picture of a dynamic equilibrium between magnetotail source and atmospheric sink was challenged in the 1990s as new observations and theories pointed to a model in which seed electrons were accelerated locally, in the heart of the radiation belts. At the same time other observations and theoretical studies reinforced the evidence for radial-diffusive acceleration and the nature of the acceleration, and therefore the role of the magnetotail, remained ambiguous.

Recent results from NASA's Van Allen Probes mission have shown that, for at least one storm in October 2013, the growth of radial peaks in phase space density were fully consistent with local acceleration by wave-particle interactions and could not be explained by diffusive radial acceleration alone. Thus, although more complete and thorough investigations are ongoing, we can say that local acceleration of relativistic electrons in the radiation belts can be the dominant process for radiation belt acceleration.

What are the implications for magnetotail radiation belt dynamics? First, even though magnetotail electrons are not directly accelerated in the local acceleration scenario, they are still the source of thermal and energetic electrons in the inner magnetosphere and, hence, important for understanding the available seed population. Transport of electrons into the inner magnetosphere (by convection and substorms) and subsequent trapping produce distributions that are unstable to the growth of whistler-mode chorus (and other waves) that are thought to be responsible for local acceleration. Furthermore, the suprathermal distributions are responsible for Landau damping and therefore the spatial extent of the chorus waves [*Bortnik et al.*, 2007]. The spatial distribution of waves may, in turn, be responsible for determining which storms produce net enhancements of the radiation belts and which produce net decreases [*Reeves et al.*, 2011].

Observations from both the January 1997 and October 2013 storms suggest that electrons that are accelerated in the heart of the belts later diffuse outward [*Reeves et al.*, 1998a; *Shprits et al.*, 2006b; *Reeves et al.*, 2013]. Radial diffusion is always enhanced in storms and always acts to redistribute electrons throughout the inner magnetosphere. This, then, raises the intriguing possibility that electrons deep in the inner magnetosphere can, at times, be a source for (lower-energy) energetic particle populations in the near-Earth magnetotail, a two-way coupling through a remarkable sequence of energy transfer processes.

ACKNOWLEDGMENTS

This research was supported by RBSP-ECT funding under NASA's Prime contract no. NAS5- 01072 and by the Dynamic Radiation Environment Assimilation Model (DREAM) project at Los Alamos National Laboratory.

REFERENCES

Albert, J. M., and S. L. Young (2005), Multidimensional quasi-linear diffusion of radiation belt electrons, *Geophys. Res. Lett.*, *32,* L14110, doi:10.1029/2005GL023191.

Albert, J. M., et al. (2009), Three-dimensional diffusion simulation of outer radiation belt electrons during the 9 october 1990 magnetic storm, *J. Geophys. Res.*, *114*, A09214, doi:10.1029/2009JA014336.

Baker, D. N., et al. (1989), Relativistic electrons near geostationary orbit: Evidence for internal magnetospheric acceleration, *Geophys. Res. Lett.*, *16*, 6, doi:10.1029/GL016i006p00559.

Baker, D. N., et al. (1993), An overview of the solar, anomalous, and magnetospheric particle explorer (SAMPEX) mission, *IEEE Trans. Geosci. Remote Sens.*, *31*, 3, doi:10.1109/36.225519.

Baker, D. N., et al. (1998), Coronal mass ejections, magnetic clouds, and relativistic magnetospheric electron events: ISTP, *J. Geophys. Res. Space Phys.*, *103*, A8, doi:10.1029/97ja03329.

Blake, J. B., et al. (1992), Injection of electrons and protons with energies of tens of mev into l<3 on 24 march 1991, *Geophys. Res. Lett.*, *19*, 8, doi:10.1029/92gl00624.

Bortnik, J., et al. (2007), Modeling the propagation characteristics of chorus using CRRES suprathermal electron fluxes, *J. Geophys. Res.*, *112*, A08204, doi:10.1029/2006JA012237.

Chen, Y., et al. (2005), Multisatellite determination of the relativistic electron phase space density at geosynchronous orbit: Methodology and results during geomagnetically quiet times, *J. Geophys. Res. Space Phys.*, *110*, A10, doi:10.1029/2004ja010895.

Chen, Y., et al. (2007a), Multisatellite determination of the relativistic electron phase space density at geosynchronous orbit: An integrated investigation during geomagnetic storm times, *J. Geophys. Res. Space Phys.*, *112*, A11, doi:10.1029/2007ja012314.

Chen, Y., et al. (2007b), The energization of relativistic electrons in the outer Van Allen radiation belt, *Nature Phys.*, *3*, 9, doi:10.1038/Nphys655.

Dungey, J. W. (1963), Loss of van allen electrons due to whistlers, *Planet. Space Sci.*, *11*, 6, doi:10.1016/0032-0633(63)90166-1.

Elkington, S. R. (2006), A review of ulf interactions with radiation belt electrons, in *Magnetospheric ulf Waves: Synthesis and New Directions*, edited by K. Takahashi, P. J. Chi, R. E. Denton, and R. L. Lysak, pp. 177–193, AGU, Washington, D. C.

Fälthammar, C. G. (1965), Effects of time-dependent electric fields on geomagnetically trapped radiation, *J. Geophys. Res.*, *70*, 11, doi:10.1029/Jz070i011p02503.

Farley, T. A. (1963), The growth of our knowledge of the Earth's outer radiation belt, *Rev. Geophys.*, *1*, 1, doi:10.1029/RG001i001p00003.

Fennell, J. F., and J. L. Roeder (2008), Storm time phase space density radial profiles of energetic electrons for small and large k values: Scatha results, *J. Atmos. Solar-Terrest. Phys.*, *70*, 14, doi:10.1016/j.jastp.2008.03.014.

Friedel, R. H. W., et al. (1999), A multi-spacecraft synthesis of relativistic electrons in the inner magnetosphere using LANL, GOES, GPS, SAMPEX, HEO and POLAR, *Radiat. Measur.*, *30*, 5, doi:10.1016/S1350-4487(99)00230-9.

Goodrich, C. C., M. Wiltberger, R. E. Lopez, K. Papadopoulos, and J. G. Lyon (1998), An overview of the impacr of the January 10-11, 1997 magnetic cloud on the magnetosphere via global MHD simulation, *J. Geophys. Res.*, 25, 14, doi:10.1029/98GL01159.

Green, J. C., and M. G. Kivelson (2004), Relativistic electrons in the outer radiation belt: Differentiating between acceleration mechanisms, *J. Geophys. Res.*, *109*, A03, doi:10.1029/2003JA010153.

Gussenhoven, M. S., et al. (1994), Solar particle events as seen on CRRES, *Adv. Space Res.*, *14*, 10, doi:10.1016/0273-1177(94)90517-7.

Hilmer, R. V., et al. (2000), Enhancement of equatorial energetic electron fluxes near l=4.2 as a result of high speed solar wind streams, *J. Geophys. Res.*, *105*, doi:10.1029/1999JA000380.

Horne, R. B. (2002), The contribution of wave particle interactions to electron loss and acceleration in the Earth's radiation belts during geomagnetic storms, in *Review of Radio Science 1999–2002*, edited by W. R. Stone, pp. 801–828, Wiley, Hoboken, N. J.

Horne, R. B., and R. M. Thorne (1998), Potential waves for relativistic electron scattering and stochastic acceleration during magnetic storms, *Geophys. Res. Lett.*, *25*, 15, 289–296, doi:10.1029/98GL01002.

Horne, R. B., and R. M. Thorne (2003), Relativistic electron acceleration and precipitation during resonant interactions with whistler-mode chorus, *Geophys. Res. Lett.*, *30*, 10, doi:10.1029/2003GL016973.

Horne, R. B., et al. (2000), Proton and electron heating by radially propagating fast magnetosonic waves, *J. Geophys. Res.*, *105*, 27,597–27,610.

Horne, R. B., et al. (2005a), Timescale for radiation belt electron acceleration by whistler mode chorus waves, *J. Geophys. Res. Space Phys.*, *110*, A3, doi:10.1029/2004ja010811.

Horne, R. B., et al. (2005b), Wave acceleration of electrons in the Van Allen radiation belts, *Nature*, *437*, 8, doi:10.1038/nature03930.

Hudson, M. K., et al. (2000), Increase in relativistic electron flux in the inner magnetosphere: Ulf wave mode structure, *Adv. Space Res.*, *25*, 12, doi:10.1016/S0273-1177(99)00518-9.

Hudson, M. K., et al. (2001), Radiation belt electron acceleration by ulf wave drift resonance: Simulation of 1997 and 1998 storms, *Space Weather*, *125*, 289–296.

Jordanova, V. K., et al. (1998), Effect of wave-particle interactions on ring current evolution for january 10–11, 1997: Initial results, *Geophys. Res. Lett.*, *25*(15), 2971–2974, doi:10.1029/98GL00649.

Jordanova, V. K., et al. (1999), Ring current activity during the early bz<0 phase of the January 1997 magnetic cloud, *J. Geophys. Res.*, *104*(A11), 24895–24914, doi:10.1029/1999JA900339.

Kamide, Y., et al. (1998), Internally and externally triggered substorms: A case study of the January 10, 1997 events, in *Proceedings of the 4th International Conference on Substorms, ICS-4*, Hamanako, Japan, edited by S. Kokubun and Y. Kamide, Kluwer Academic Publishers, Boston.

Kellog, P. (1959), Possible explanation of the radiation observed by Van Allen at high altitude in satellites, *Nuovo Cimento*, *11*, 48.

Kennel, C. F., and H. E. Petschek (1966), Limit on stably trapped particle fluxes, *J. Geophys. Res.*, *71*, 1, doi:10.1029/JZ071i001p00001.

Kim, H. J., and A. A. Chan (1997), Fully adiabatic changes in storm time relativistic electron fluxes, *J. Geophys. Res. Space Phys.*, *102*, A10, doi:10.1029/97ja01814.

Li, X. L., et al. (1993), Simulation of the prompt energization and transport of radiation belt particles during the March 24, 1991 ssc, *Geophys. Res. Lett.*, *20*, 22, doi:10.1029/93gl02701.

Li, X. L., et al. (1997), Multisatellite observations of the outer zone electron variation during the November 3–4, 1993, magnetic storm, *J. Geophys. Res. Space Phys.*, *102*, A7, doi:10.1029/97ja01101.

Lyons, L. R., and R. M. Thorne (1973), Equilibrium structure of radiation belt electrons, *J. Geophys. Res.*, *78*, 2142.

Lyons, L. R., et al. (2000), Auroral disturbances during the January 10, 1997 magnetic storm, *Geophys. Res. Lett.*, *27*, 20, doi:10.1029/1999gl000014.

Mauk, B. H., et al. (2012), Science objectives and rationale for the radiation belt storm probes mission, *Space Sci. Rev.*, doi:10.1007/s11214-012-9908-y.

Nishida, A. (1976), Outward diffusion of energetic particles from the jovian radiation belt, *J. Geophys. Res.*, *81*, 10, doi:10.1029/JA081i010p01771.

Northrop, T., and E. Teller (1960), Stability of the adiabatic motion of charged particles in the Earth's field, *Phys. Rev.*, *117*, 1, doi:10.1103/PhysRev.117.215.

Omura, Y., et al. (2007), Relativistic turning acceleration of resonant electrons by coherent whistler mode waves in a dipole magnetic field, *J. Geophys. Res.*, *112*, A06236, doi:10.1029/2006JA012243.

Omura, Y., et al. (2008), Theory and simulation of the generation of whistler-mode chorus, *J. Geophys. Res.*, *113*, A04223, doi:10.1029/2007JA012622.

Onsager, T. G., et al. (2002), Radiation belt electron flux dropouts: Local time, radial, and particle-energy dependence, *J. Geophys. Res. Space Phys.*, *107*, A11, doi:10.1029/2001ja000187.

Reeves, G. D. (1998), Relativistic electrons and magnetic storms: 1992–1995, *Geophys. Res. Lett.*, *25*, 11, doi:10.1029/98gl01398.

Reeves, G. D., et al. (1998a), The global response of relativistic radiation belt electrons to the January 1997 magnetic cloud, *Geophys. Res. Lett.*, *25*, 17, doi:10.1029/98gl02509.

Reeves, G. D., et al. (1998b), The relativistic electron response at geosynchronous orbit during the January 1997 magnetic storm, *J. Geophys. Res.*, *103*, A8, doi:10.1029/97ja03236.

Reeves, G. D., et al. (2003), Acceleration and loss of relativistic electrons during geomagnetic storms, *Geophys. Res. Lett.*, *30*, 10, doi:10.1029/2002gl016513.

Reeves, G. D., et al. (2011), On the relationship between relativistic electron flux and solar wind velocity: Paulikas and blake revisited, *J. Geophys. Res. Space Phys.*, *116*, doi:10.1029/2010ja015735.

Reeves, G. D., et al. (2012), Dynamic radiation environment assimilation model: Dream, *Space Weather Int. J. Res. Appl.*, *10*, doi:10.1029/2011sw000729.

Reeves, G. D., et al. (2013), Electron acceleration in the heart of the Van Allen radiation belts, *Science*, doi:10.1126/science.1237743.

Roederer, J. G. (1970), *Dynamics of Geomagnetically Trapped Radiation*, Springer-Verlag, Berlin.

Schulz, M. (1996), Canonical coordinates for radiation belt modeling, in *Radiation Belt Models and Standards*, edited by

J.F. Lemaire, D. Heynderickx, and D.N. Baker, vol. 97, pp. 153–161, AGU, Geophysical Monograph, Washington, D.C.

Schulz, M., and L. J. Lanzerotti (1974), *Particle Diffusion in the Radiation Belts*, Springer-Verlag, New York.

Selesnick, R. S., and J. B. Blake (1998), Radiation belt electron observations following the January 1997 magnetic cloud event, *Geophys. Res. Lett.*, *25*, 14, doi:10.1029/98gl00665.

Selesnick, R. S., and J. B. Blake (2000), On the source location of radiation belt relativistic electrons, *J. Geophys, Res.*, *105*, 2607–2624.

Selesnick, R. S., et al. (2007), A theoretical model of the inner proton radiation belt, *Space Weather*, *5*, S04003, doi:10.1029/2006SW000275.

Shprits, Y. Y., et al. (2006a), Bounce-averaged diffusion coefficients for field-aligned chorus waves, *J. Geophys. Res.*, *111*, A10225, doi:10.1029/2006JA011725.

Shprits, Y. Y., et al. (2006b), Outward radial diffusion driven by losses at magnetopause, *J. Geophys. Res. Space Phys.*, *111*, A11, doi:10.1029/2006ja011657.

Singer, S. F. (1958), Trapped albedo theory of the radiation belt, *Physi. Revi. Lett.*, *1*, 5, doi:10.1103/Physrevlett.1.181.

Stømer, C. (1955), *The Polar Aurora*, Clarendon Press, Oxford.

Summers, D., and C. Ma (2000), A model for generating relativistic electrons in the earth's inner magnetosphere based on gyroresonant wave-particle interactions, *J. Geophys. Res.*, *105*, 2625–2639.

Summers, D., et al. (1998), Relativistic theory of wave-particle resonant diffusion with application to electron acceleration in the magnetosphere, *J. Geophys. Res.*, *103*, A9, doi:10.1029/98JA01740.

Summers, D., et al. (2002), Model of the energization of outer-zone electrons by whistler-mode chorus during the October 9, 1990 geomagnetic storm, *Geophys. Res. Lett.*, *29*, 24, doi:10.1029/2002gl016039.

Summers, D., et al. (2004), Competition between acceleration and loss mechanisms of relativistic electrons during geomagnetic storms, *J. Geophys. Res.*, *109*, A04221, doi:10.1029/2004JA010437.

Summers, D., et al. (2007), Timescales for radiation belt electron acceleration and loss due to resonant wave-particle interactions: 1. Theory, *J. Geophys. Res.*, *112*, A04206, doi:10.1029/ 2006JA011801.

Taylor, M. G. G. T., et al. (2004), Multisatellite measurements of electron phase space density gradients in the Earth's inner and outer magnetosphere, *J. Geophys. Res. Space Phys.*, *109*, A5, doi:10.1029/2003ja010294.

Thorne, R. M. (2010), Radiation belt dynamics: The importance of wave-particle interactions, *Geophys. Res. Lett.*, *37*, doi:10.1029/2010gl044990.

Thorne, R. M., et al. (2013), Van Allen probe evidence of relativistic radiation belt electron acceleration by magnetospheric chorus, *Nature*, *504*, 411–414.

Tsyganenko, N. A. (1989), A magnetospheric magnetic field model with a warped tail current sheet, *Planet. Space Sci.*, *37*, 183–192.

Tsyganenko, N. A., and M. I. Sitnov (2005), Modeling the dynamics of the inner magnetosphere during strong geomagnetic storms, *J. Geophys. Res.*, *110*, doi:10.1029/2004JA010798.

Van Allen, J. A., et al. (1958), Observation of high intensity radiation by satellites 1958 alpha and gamma, *Jet Propulsion*, *28*, 588–592.

Vernov, S. N., et al. (1959), Possible mechanism of production of terrestrial corpuscular radiation under the action of cosmic rays, *Soviet Phys. Doklady, 4*, 154.

Walt, M. (1996), Source and loss processes for radiation belt particles, in *Radiation Belt Models and Standards*, edited by J. F. Lemaire, D. Heynderickx and D. N. Baker, pp. 1–15, AGU, Washington, D.C.

Winckler, J. R., et al. (1962), A study of the precipitation of energetic electrons from the geomagnetic field during magnetic storms, *J. Geophys. Res.*, *67*, 10, doi:10.1029/JZ067i010p03717.

21

Substorm Current Wedge at Earth and Mercury

L. Kepko[1], K.-H. Glassmeier[2], J. A. Slavin[3] and T. Sundberg[4,5]

21.1. INTRODUCTION

The Hermean magnetosphere is an ideal laboratory for comparative studies of magnetospheric dynamics. Although it does not possess an ionosphere, it exhibits a similar set of solar-wind-driven phenomena that are observed within Earth's magnetosphere, including flux transfer events [*Russell and Walker*, 1985; *Slavin et al.*, 2009, 2010a, 2012a], Kelvin-Helmholtz oscillations [*Glassmeier and Espley*, 2006; *Sundberg et al.*, 2012a], and magnetospheric substorms [*Siscoe et al.*, 1975; *Baker et al.*, 1986; *Slavin et al.*, 2010b]. Within magnetospheric substorms, both Mercury and Earth exhibit tailward moving flux ropes and plasmoids [*Slavin et al.*, 2009, 2012b], tail loading and unloading [*Slavin et al.*, 2010b], and near-planet dipolarizations and injections [*Baker et al.*, 1986; *Christon et al.*, 1987; *Baker et al.*, 1986; *Sundberg et al.*, 2012b]. Other than differences in relative amplitude and time scale, substorms at Mercury and Earth are remarkably similar.

The similarity in substorm dynamics between the two magnetospheres allows a comparative examination of how the ionosphere influences substorm development. Earth's ionosphere is formed primarily by the absorption of solar extreme ultra violet (EUV) radiation by neutrals in the upper atmosphere [*Kelley*, 2009]. Additional ionization occurs due to the braking of energetic particles precipitating from the magnetosphere, enhancing ionospheric conductivity in the auroral zone (see *Strickland et al.* [1983] and review by *Pfaff* [2012]). In contrast, Mercury does not possess a neutral atmosphere. The direct impact of the solar wind and particles accelerated by magnetospheric processes onto the surface is known to sputter neutrals forming a sparse, dynamic exosphere [*Hunten et al.*, 1988; *Killen et al.*, 2007, 2001; *Zurbuchen et al.*, 2011]. These sputtered neutrals can subsequently become ionized, but they never form a consistent, large-scale ionosphere of the type observed at Earth. In addition, *Grard et al.* [1999] suggested that photoelectrons are emitted from the surface of Mercury, cause a low-conductance sheet at the surface, and thereby take part in charge exchanges with magnetosphere particles. Finally, Mercury's regolith is expected to be highly resistive [*Lammer and Bauer*, 1997; *Hill et al.*, 1976]. While Earth contains a conducting shell through which currents can flow and create resistance to magnetospheric motion, Mercury does not contain a shell for current closure, and field lines threading the surface flow freely. These different boundary conditions may impact how substorms store and release energy [*Glassmeier*, 1997, 2000] and provide a natural experiment to test the impacts of ionospheric line tying on the progression of magnetospheric activity.

At Earth, the ionosphere conducts electrical current in regions of high conductivity, dissipating energy through collisional Joule heating and linking different regions of geospace via coupling to field-aligned currents (see, e.g., *Thayer and Semeter*, 2004; *Strangeway*, 2012). In addition, regions of high conductivity can act to regulate and feed back upon magnetospheric motion by opposing the imposed force. In the magnetosphere, currents flow in

[1] *NASA Goddard Space Flight Center, Greenbelt, Maryland*
[2] *Institute of Geophysics and Extraterrestrial Physics, Technical University of Braunschweig, Braunschweig, Germany*
[3] *Department of Atmospheric, Oceanic and Space Sciences, University of Michigan, Ann Arbor, Michigan*
[4] *School of Physics and Astronomy, Queen Mary College, University of London, London, United Kingdom*
[5] *Center for Space Physics, Boston University, Boston, Massachusetts*

Magnetotails in the Solar System, Geophysical Monograph 207. First Edition.
Edited by Andreas Keiling, Caitríona M. Jackman, and Peter A. Delamere.

response to magnetic shear and in regions of enhanced pressure gradients that are generated by convective plasma flows. Dipolarizations, which represent transient, localized increases in magnetic field strength relative to the surrounding plasma, generate intense sheets of current to separate the disparate plasmas. These currents must close somewhere to satisfy current continuity; at Earth, these currents close primarily in the ionosphere, with some evidence for local closure.

Dipolarizations during substorms within Earth's magnetosphere are closely linked to the substorm current wedge (SCW) [*Birn et al.*, 1999; *Hesse and Birn*, 1991; *Sergeev et al.*, 2011]. The SCW remains a central component of substorm dynamics since its development over four decades ago [*McPherron et al.*, 1973; *Kepko et al.*, 2014]. The importance of the SCW lies in its ability to link ionospheric and magnetospheric processes during the dynamic collapse of the near-Earth current sheet under a single, coherent phenomenological picture. It is generated by the impact of high-speed plasma flows in the near-Earth region and maintained by the resultant pressure gradients and magnetic shear, and this high-pressure region of dipolarized magnetic field deflects subsequent flow bursts toward the dayside [*Birn et al.*, 2011]. Dipolarizations are observed at Mercury, with characteristics similar to dipolarizations at Earth, and currents must exist to separate the dipolarized region from the surrounding plasma. It is therefore natural to ask the question: Are dipolarizations at Mercury associated with a Hermean SCW? As we will show, the answer to this question relies primarily on the relative time scales of substorm processes and Alfvénic travel times.

In this chapter we review magnetospheric substorms and dipolarizations observed at both Earth and Mercury. We briefly discuss new insights into the physics of the SCW that have been revealed the past few years, primarily by advances in numerical simulation and coordinated, multipoint measurements during the THEMIS era [*Angelopoulos et al.*, 2008a; *Keiling et al.*, 2009]. The SCW is now known to be far more complicated than the simple wire circuit originally envisioned by *McPherron et al.* [1973].The insight into the drivers of the currents allows us to understand how different elements of the SCW may apply to Hermean substorms. Internal current closure via diamagnetic currents as suggested earlier by *Glassmeier* [2000] is probably not a major controlling factor. We conclude that the lack of an appreciable ionosphere at Mercury is not the major factor controlling substorm dynamics. Rather, it is the rapid time scales of the Hermean magnetosphere and comparatively slow Alfvénic travel times that lead to a self-regulating magnetosphere, rather than the coupled magnetosphere-ionosphere (M-I) system familiar at Earth.

21.2. BACKGROUND

21.2.1. Substorms at Mercury and Earth

The concept of magnetospheric substorms grew out of the "auroral substorm," a phenomenological picture developed to explain the repeatable pattern of auroral activity observed at high latitudes [*Akasofu*, 1964]. In situ measurements, primarily around geosynchronous orbit, combined with concurrent ground magnetic field measurements revealed that substorm phenomena extended out into the magnetosphere [*McPherron and Coleman*, 1971; *McPherron*, 1972]. These impulsive substorm phenomena rapidly reconfigure the magnetospheric magnetic field in localized regions, and this new configuration is communicated to the ionosphere via field-aligned currents. In addition to expanding the definition of substorms, these new observations formed the basis for the SCW model [*McPherron et al.*, 1973].

The term "substorm" now refers to a specific sequence of events that occurs in response to solar wind driving and energy input into the magnetosphere [*Rostoker et al.*, 1980]. A substorm occurs when there is unbalanced flux transfer between the dayside and nightside magnetosphere, as can happen during a sudden change in interplanetary magnetic field (IMF) orientation from north to south. The substorm that follows from this transient energy input consists of three distinct phases:

1. Growth phase (loading of energy in tail lobes)
2. Expansion phase (explosive release of energy)
3. Recovery phase (relaxation)

During the substorm growth phase, also known as the "loading" phase, the solar wind magnetic field connects with the magnetic field at the dayside magnetopause. This magnetic flux is transported from the dayside to the nightside, where it is stored in the tail lobes. Eventually, this flux must return to the dayside to complete the convection cycle. Magnetic reconnection occurs in the nightside plasma sheet, causing the tail lobes to "unload," leading to fast flows and flux transport away from the reconnection site. Figure 21.1 shows examples of loading and unloading observed at Mercury by MESSENGER and at Earth by ISEE-3 [*Slavin et al.*, 1992, 2010b]. Tail loading during the growth phase is observed as an increase in the magnetic field of the lobes, while the rapid unloading of the expansion phase appears as a rapid decrease in this lobe field. Although qualitatively similar, the loading/unloading cycle at Mercury and Earth differ substantially in terms of amplitude and duration.

Figure 21.1 demonstrates a number of key parameters of substorms, as observed in the tail lobes of both planets. During the loading phase (shaded in blue), the lobe field increases as magnetic flux is transferred from the dayside to the nightside. At Earth, the typical lobe field strength

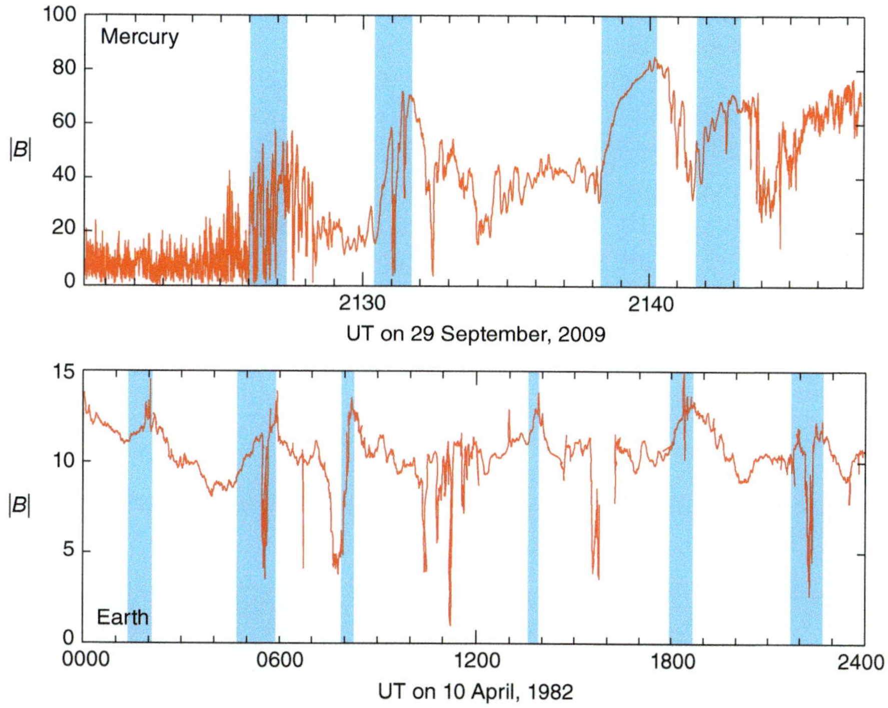

Figure 21.1 Examples of tail loading and unloading observed at Mercury (top) and Earth (bottom). Plotted are measurements of the total lobe field strength taken at (–1.7, 1.67, 0.0) R_M and (–80, 11, 11) R_E, respectively. Loading intervals are shaded in blue. Note the very different time scales and amplitudes. Adapted from *Slavin et al.* [1992, 2010b].

varies between ~10 and 20 nT near $20 R_E$, and a typical increase during the growth phase is 10–30% [*Hsu and McPherron*, 2004]. Tail loading is much more intense at Mercury, increasing the lobe field during the growth phase by a factor of 2 [*Slavin et al.*, 2010b]. Substorms at Mercury are also much more rapid, with loading/unloading events occurring every 60 s, as opposed to every 2–3 h at Earth [*Borovsky et al.*, 1993]. We discuss the time scales of substorm activity in more detail in section 21.2.5.

Reconnection that begins in the plasma sheet quickly reaches the lobe, and this stored magnetic flux is rapidly reconnected and convected both toward and away from the planet. The high-speed flow generated within the nightside plasma sheet by reconnection travels outward at speeds approaching the Alfvén velocity. At Earth, these speeds are typically between 500 and 1000 km/s. Flow speeds at Mercury have not been directly measured but are thought to be similar [*Slavin et al.*, 2012b]. Unimpeded by any obstacle, the tailward flow eventually rejoins the solar wind returning mass and magnetic flux in the form of plasmoids and flux ropes [*Slavin et al.*, 1993; *Moldwin and Hughes*, 1992]. The planet-directed flows run into an obstacle that slows the high-speed flow and deflects it toward the dayside. But the obstacle forcing the deceleration differs significantly between Mercury and Earth.

21.2.2. Flow Braking and Dipolarization

"Flow braking" refers to the rapid slowing of high-speed flows as they near Earth's inner magnetosphere [*Shiokawa et al.*, 1997; *Birn et al.*, 1999; *Panov et al.*, 2013]. The magnitude of Earth's dipole magnetic field at the equator is approximately 30,000 nT, 80 nT at nightside geosynchronous, and 10–20 nT in the midtail region where reconnection occurs. A flow originating in the midtail region would travel relatively unimpeded until it neared geosynchronous, where it would rapidly experience a factor of 16–64 increase in magnetic pressure. It is this strong and rapid increase in pressure in the inner magnetosphere that slows and deflects the incoming flow [*Shiokawa et al.*, 1997]. The magnetic field in the Hermean plasma sheet is slightly larger, at ~20–50 nT, but the equatorial surface field is less than 300 nT [*Anderson et al.*, 2008, 2010]. While at Earth flow braking occurs near and just beyond geosynchronous, the equivalent location for flow braking in the Hermean magnetosphere is near the planet's surface. The intriguing possibility that Hermean flow bursts directly impact Mercury's surface will not be discussed further in this chapter.

Dipolarization is a phenomenon closely related to flow braking. As the fast flow slows, the magnetic flux that is carried by the flow starts to "pile up." The process is

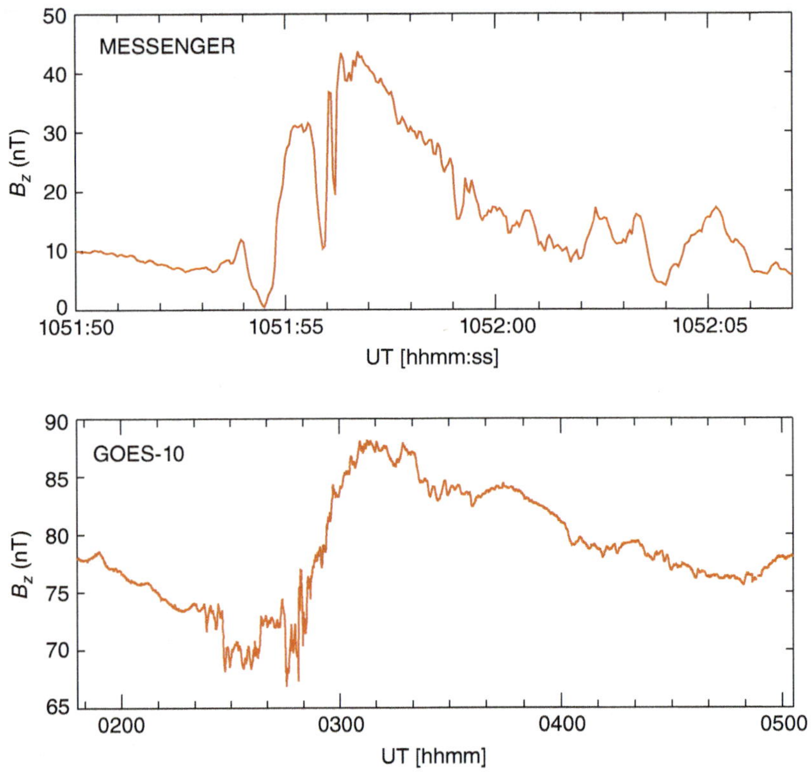

Figure 21.2 Examples of dipolarizations observed at Mercury (top) and Earth (bottom) with nightside, in situ magnetometers. Mercury observations were made by MESSENGER at (–1.41, 0.26, –0.06) R_M (equivalent to $X \sim -11 R_E$ at Earth), and Earth measurements were made by GOES-10 at geosynchronous orbit near local midnight. Both measurements were taken in the near-equatorial plasma sheet, where the magnetic effects of the SCW would appear primarily in the vertical (B_z) component. B_z is plotted, and growth-phase stretching is indicated by a decreasing B_z, while dipolarization is indicated by a sudden increase in B_z.

called dipolarization because the magnetic field line reconfigures from the stretched taillike field near the reconnection site to quasi-dipolar in the inner magnetosphere. While dipolarizations have been a well-known component of magnetospheric substorms at Earth, only recently have MESSENGER observations firmly established that dipolarization occurs at Mercury [*Sundberg et al.*, 2012b]. Examples of dipolarizations observed at Mercury and Earth are shown in Figure 21.2. Both events show the characteristics of dipolarization: an initial stretching of the field (decrease in B_z) during the growth phase, followed by a rapid increase in B_z at onset, often to values close to the dipole field value, followed by a slow relaxation as the accumulated stresses are dissipated. Although the dipolarizations in both magnetospheres are qualitatively similar, as with lobe loading and unloading they have strongly different amplitudes and occur on fundamentally different time scales. An important aspect of dipolarization is that the region of increased field strength is localized in azimuth, with stretched taillike field lines remaining on either side. This different configuration of magnetic field represents magnetic shear and requires a current sheet to separate the plasmas, as we will discuss in section 21.2.3.

The fast flows generated by reconnection during substorm expansion produce currents through two processes: the inertial current associated directly with the deceleration of the flow and pressure gradients built up by dipolarization. Both currents can be described by solving the momentum equation for the perpendicular component of the current, $j_\perp$ [e.g., *Glassmeier*, 2000]:

$$j_\perp = \frac{\vec{B}}{B^2} \times \rho \frac{d\vec{u}}{dt} + \frac{\vec{B}}{B^2} \times \nabla P. \tag{21.1}$$

The first term on the right-hand side (RHS) is the inertial current (j_{inrt}) and is created by time-dependent changes in the flow velocity. For planet-directed flows, the current flows dusk to dawn such that the $\vec{J} \times \vec{B}$ force is directed opposite the direction of the flow and is nonzero only while the flow is decelerating. The second term on the RHS of equation 21.1 is generated by the accumulated magnetic stresses and thermal pressure gradients created during flux pileup. At Earth, this current is at least a factor of 100 higher than the inertial current [*Birn et al.*, 1999] and represents the primary source for field-aligned currents during substorms. At Mercury, however, the inertial current is potentially an important source of current.

An estimate of the magnitude of the inertial current density at Mercury can be calculated directly using equation 21.1. Reasonable estimates of du/dt can be made at Earth, where in situ measurements have provided quantitative knowledge of how the flow burst speed changes as a function of distance within the braking region (e.g., *Baumjohann et al.*, 1990; *Schödel et al.*, 2001). Unfortunately, similar results do not yet exist at Mercury. As a first approximation, we relate the time scale of braking (dt) to the braking length L_x and the flow velocity u_x so that $dt = 2L_x/u_x$ [*Haerendel*, 1992]. The inertial current term now becomes

$$j_{\text{inrt}} = \frac{\rho_{\text{ps}} u_x^2}{2B_{\text{ps}} L_x} = 1.6 \times \frac{n_{\text{ps}}(\text{cm}^{-3}) u_x^2(\text{km/s})}{2B_{\text{ps}}(\text{nT}) L_x(\text{km})}, \qquad (21.2)$$

where j_{inrt} is in nA/m^2, and we have assumed the mass composition is entirely protons. Equation 21.2 can be used to estimate the magnitude of the inertial current during flow braking. Direct measurements of the flow speed have not been made at Mercury, but we can use a reasonable assumption of $u_x = 500$ km/s, comparable to values observed at Earth (see *Slavin et al.*, 2012b), and values of $n = 5$ cm^{-3} [*Raines et al.*, 2011] and $B = 80$ nT. The length scale over which braking occurs is not known at Mercury, but as a reasonable estimate we scale down the $\sim 1\text{R}_\text{E}$ length scale observed at Earth (see Table 21.1). Putting these numbers into equation 21.2, and using a braking length $L_x = 500$ km, we find $j_{\text{inrt}} = 16$ nA/m^2. *Slavin et al.* [1997] identified a field-aligned current sheet observed with the Mariner 10 magnetometer and estimated a current density of 700 nA/m^2, a factor of 40 higher than the inertial currents estimated here. However, we note that a flow speed of 1000 km/s yields j_{inrt} of 60 nA/m^2, while a further increase in density to 10 cm^{-3}, which may be more appropriate for the low-altitude braking region, yields j_{inrt} of 125 nA/m^2. In addition, our calculation assumed only protons, and it is known that Mercury's magnetosphere contains significant quantities of heavier species. An increase in the mass density would linearly increase the inertial current. These estimates should therefore be considered a lower limit on j_{inrt}. MESSENGER so far has not observed field-aligned current (FAC) signatures similar to the one reported by *Slavin et al.* [1997], so it remains to be seen whether smaller FACs, of the magnitude estimated here, are common in Mercury's magnetosphere.

Finally, we note that while we estimated the scale size and radial distance of flow braking at Mercury by scaling from values observed at Earth, the bulk plasma parameters (number density, flow speed, etc.) remain similar to the terrestrial environment. At Mercury, the ion inertial length, $d_i = c/\omega_{\text{pi}}$, in the braking region for a number density of 5 cm^{-3} is ~ 100 km, compared to the braking scale length of ~ 800 km. The flux pileup events observed by MESSENGER bears a striking resemblance to events observed at Earth yet are far more localized, approaching the scalesize of ion dynamics. Yet stable current sheets of thickness $< 10d_i$ have been routinely observed within Earth's magnetosphere (e.g., *Sergeev et al.*, 2011), and the leading edge of dipolarization fronts, stable over at least 10 R_E during propagation, are of the order of one ion inertial length [*Runov et al.*, 2011]. The ion inertial length decreases with the square root of the number density, and since the ion number density in the region of flow braking very near the Mercury's surface is likely to be greater than 5 cm^{-3}, a magnetohydrodynamic (MHD)–based description of braking dynamics is generally appropriate. Lastly, flux pileup and dipolarization, while fed by convective flows, is inherently a process dominated by the behavior of the magnetic field, not the particles, as seen by the behavior of B in the examples shown in Figure 21.2. Even a dipolarization event that began on the ion inertial scale would quickly expand into an MHD appropriate regime as flux piledup; otherwise, dipolarization would not be observed.

Table 21.1 Comparisons of scale lengths and locations of magnetospheric boundaries and substorm phenomena at Mercury and Earth.

	Mercury (R_M)	Scaled Mercury (R_M)	Earth (R_M)
Subsolar point	1.4[a]	1.5	12
X line	2[b]	16	20[c]
Braking region	<1.2[d]	0.75–1.25	6–10
Braking length	800 km[d]	800 km[d]	1
d_i (braking region)	100 km	100 km	100 km

Note: Scaling of Hermean distances to R_E assumes a factor of 8 scaling based on the distance of the subsolar point from the planet's center [*Ogilvie et al.*, 1977; *Siscoe et al.*, 1975]. Distances are shown as absolute values.

[a] *Winslow et al.* [2013].

[b] *Slavin et al.* [2012b].

[c] *Nagai et al.* [1998].

[d] Estimated.

21.2.3. Substorm Current Wedge

As defined within the context of substorms at Earth, the substorm current wedge refers to the diversion of cross-tail current at the edges of the dipolarized region into the ionosphere at the downward edge, across the

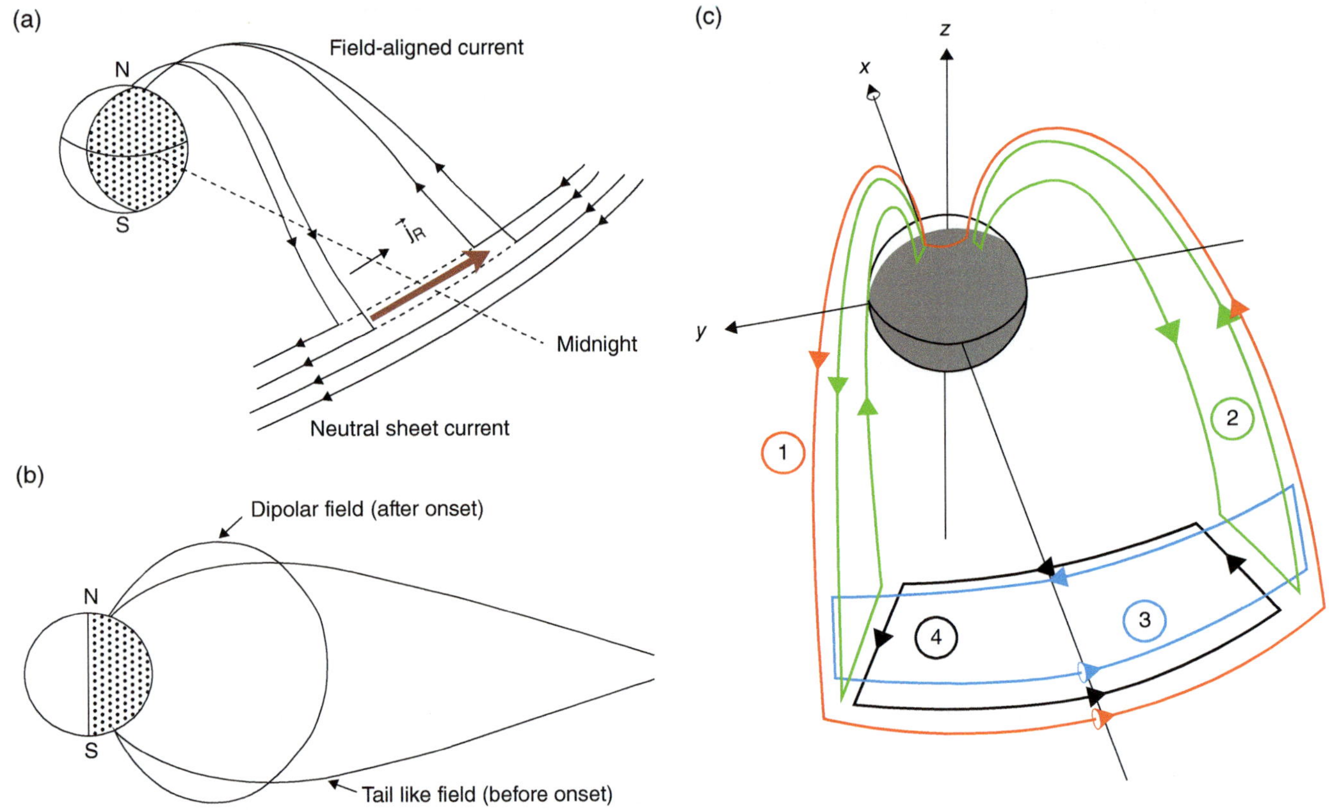

Figure 21.3 (a) Original substorm current wedge, showing a diversion of the cross-tail current into the ionosphere postmidnight and out of the ionosphere premidnight [*Glassmeier*, 1997]. The current can be considered a closed loop by closing the FACs across the magnetotail (red line arrow), with this loop superimposed upon the cross-tail current. (b) Time-dependent aspects of dipolarization. In a meridional plane, these loops represent the preonset (stretched) and postonset (dipolar) magnetic field configuration. It also shows the azimuthal structure, with the dipolar field inside the SCW, while outside the wedge the field lines retain the stretched configuration. (c) Updated current wedge derived from MHD simulations. [Only figure (c) is *Kepko et al.* [2014]. Courtesy of J. Birn.]

ionosphere from east to west, and out of the ionosphere at dusk (Figure 21.3a) [*McPherron et al.*, 1973]. The SCW is directly connected to dipolarization, as shown in Figure 21.3b. Outside the dipolarized region the plasma sheet still retains a stretched configuration, along with a cross-tail current required to support it. Inside the dipolarized region the cross-tail current is greatly reduced. Current continuity requires the cross-tail current to flow around the dipolarized region. At Earth, the current flows along field lines to divert through the ionosphere.

Although commonly described as "diversion" of the cross-tail current (Figure 21.3a), it is useful to think of the SCW as an independent current system superimposed upon and opposing the cross-tail current. One can then describe the SCW as driven by processes occurring within or at the boundaries of the dipolarized region. Figure 21.3c shows an updated picture of the substorm current wedge [*Birn et al.*, 1999; *Kepko et al.*, 2014]. Loop 1 shows the traditional SCW as an independent loop that in the equatorial plane opposes the cross-tail current. Loop 2 communicates the stress of flow diversion to the ionosphere while loop 4 contains the high-pressure dipolarized region. Loops 2 and 4 oppose each other in the equatorial plane, and the relative strengths of the currents control the azimuthal rate of expansion. Loop 3 is driven by off-equatorial pressure gradients and is not important for the discussion here.

The role of the SCW and equivalent current loops can be understood by examining $J \times B$ at different locations. The $J \times B$ force of loop 4 points radially outward in every direction and represents the expansive force of the high-pressure dipolar region. The ionospheric portion of loop 2 drives the ionospheric foot points to move outward in response to this expansion. This motion at the equator is opposed by $J \times B$ of loop 2. This represents ionospheric resistance to the flux tube motion. Higher conductivity increases the ionospheric resistance to motion, and the amount that loop 2 opposes the expansion of the dipolarized region is directly related to the ionospheric conductivity. Ignoring for the moment the effects of time-dependent coupling and ionospheric dissipation, for a fixed electric field, the strength of the current in loop 2 is directly related to the ionospheric conductivity. High conductivity leads to a large current, which is able to cancel the azimuthal expansion associated with loop 4. An ionosphere with low conductivity would not strongly oppose motion imposed by the magnetosphere, and loop 2 would carry comparatively little current. The exact closure and relative strengths of the current loops vary from event to event. But these basic principles can be used to understand the coupling.

21.2.4. Transient Coupling

The preceding discussion was based on a static description of a SCW that had already been established by the flows. Particularly for understanding the coupling at Mercury, it is important to understand the transient coupling that occurs as the current wedge is forming. The magnetosphere communicates transient movement to the ionosphere via Alfvén waves and the associated FACs carried by these waves [*Scholer*, 1970; *Southwood and Kivelson*, 1991]. We can understand the consequences of transient coupling by looking at how the Alfvén waves carrying the current are reflected.

Imagine a magnetospheric flow moving sunward from a reconnection site. This flow convects the equatorial portion of a field line toward the planet, eventually returning the flux to the dayside. The movement of this flux tube first occurs near the equator due to the enhanced electric field of the reconnection site. The portions of the field line close to the planet are not yet involved in this motion. This localized movement of the equatorward portion of the magnetic field introduces a kink, or magnetic perturbation, on the field line, which propagates down to the ionosphere as an Alfvén wave. The Alfvén wave carries the FAC that will flow across field lines in the inosphere to provide the $J \times B$ force to move the footprint in the direction of magnetospheric motion [e.g., *Strangeway et al.*, 2000]. What happens when this wave and associated FAC reaches the ionosphere depends on the relative conductivities of the flux tube and the ionosphere [*Mallinckrodt and Carlson*, 1978]. The transient interaction of Alfvén waves with a conducting boundary within the context of magnetospheric substorms has been well studied (see reviews by *Baumjohann and Glassmeier* [1984] and *Keiling and Takahashi* [2011]), and we provide only a brief review here.

One can define a reflection coefficient of the interaction of the wave with an ionosphere that dictates the amplitude of the imposed and reflected electric and magnetic fields. This reflection coefficient is defined as [*Scholer*, 1970; *Glassmeier*, 1983, 1997]

$$R = \frac{\Sigma_A - \Sigma_P}{\Sigma_A + \Sigma_P}. \tag{21.3}$$

The reflection coefficient depends on the Alfvén conductance (Σ_A) of the flux tube carrying the wave and the Pedersen (parallel to the applied electric field) conductance (Σ_P) of the ionosphere. The Alfvén waves carry FAC, and any partial reflection of the wave by the ionosphere will reduce the applied current. The buildup of the FAC occurs over several bounce times of the Alfvén wave, with higher ionospheric conductivity requiring more bounces. The net FAC after many bounces is given by *Glassmeier* [1997]:

$$j_{\parallel}^{\infty} = j_{\parallel,i} \cdot \frac{1-R}{1+R}. \tag{21.4}$$

Here $j_{\parallel,i}$ refers to the amplitude of the initial FAC, while $j_{\parallel}^{\infty}$ is the eventual FAC after many bounces.

The two limits of interest are $R \sim -1$ and $R \sim 1$, corresponding to a highly conducting ionosphere (Earth) and a poorly conducting ionosphere (Mercury), respectively (Table 21.2). A highly conducting ionosphere will resist the motion imposed by the magnetosphere that is being communicated by the Alfvén wave. The ionosphere reflects the incoming electric field, reversing the sign. This does more than cancel the magnetospheric motion; it reverses the motion. In essence, it is a "kickback" from the ionosphere that attempts to return the equatorial portion of the flux tube to its original position. The FAC builds up in response. At Earth, this is observed as the transient response Pi2 [*Baumjohann and Glassmeier*, 1984; *Keiling and Takahashi*, 2011] and is also loop 2 in Figure 21.3c. At the opposite limit, a poorly conducting ionosphere will readily move in response to magnetospheric motion. The reflection cancels the incoming wave and drives the FAC to zero. The current, and the associated $J \times B$ force, is not required to move the ionospheric footpoints.

Table 21.2 Reflection coefficients and the impact on an impinging Alfvén wave.

R Value	Definition	Impact
$R = 0$	Equal conductance	Ionosphere invisible
$R \sim -1$	Highly conducting ionosphere	Ionosphere opposes motion, builds current
$R \sim 1$	Nonconducting ionosphere	Ionosphere moves freely, cancels current

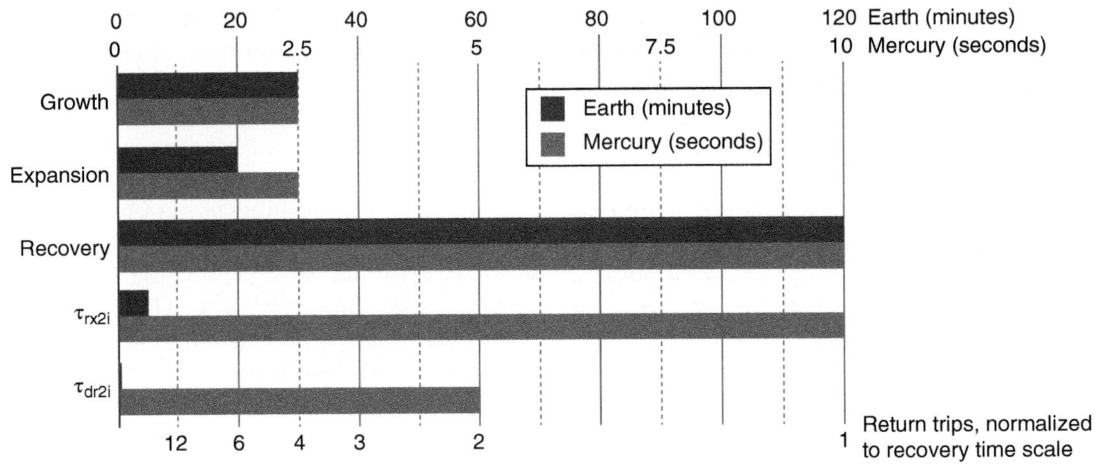

Figure 21.4 Timescales of substorm phenomena at Earth (minutes) and Mercury (seconds). The number of round trips from the specified location to the ionosphere and back is shown at bottom.

Equation 21.4 helps understand the time-dependent response of a system in which many Alfvén bounce times are possible. Additional complexity in the interaction arises when the magnetospheric motion occurs more rapidly than the communication time scale with the ionosphere or in situations where ionospheric resistance must be overcome. At Earth, field-aligned potential drops, for example, are one way in which magnetospheric motion can decouple from ionospheric motion. For Mercury, we will see that the substorms form and dissipate on time scales similar to the Alfvén transit time, in stark contrast to Earth.

21.2.5. Time Scales

Figure 21.4 shows the time scales for substorm-related phenomena at Earth and Mercury. Time scales (τ) are shown in minutes for Earth and seconds for Mercury. The cumulative time required to complete a substorm cycle is the combination of the length of the substorm growth, expansion, and recovery phases, $\tau_g + \tau_e + \tau_r$. For Earth, this is on average 2.5–3 h [*Borovsky et al.*, 1993], while at Mercury this varies from ~ 15 to 60 s for the MESSENGER flybys [*Slavin et al.*, 2012b]. The recovery time scale τ_r ~ 10 s at Mercury is based on the statistical characteristics of MESSENGER observations of dipolarization events [*Sundberg et al.*, 2012b].

At both planets, the substorm growth and expansion phases are approximately 25% the duration of the recovery phase. Although the relative time scales of substorm phases are comparable, the communication time scales between the magnetosphere and ionosphere differ significantly. Two time constants are shown in Figure 21.4: the roundtrip Alfvén transit time (T_A) between the reconnection site and the ionosphere (τ_{rx2i}) and the round-trip transit time between the dipolarization region to the ionosphere (τ_{dr2i}). These time constants represent the cumulative travel time for an Alfvén wave to travel along a field line from the source location to the ionosphere and return back to the original location.

The number of round trips for Alfvén waves with respect to the length of the substorm recovery phase is shown at the bottom of Figure 21.4. While Alfvén speeds between the two magnetospheres are comparable, Hermean substorms occur much closer to the planet than their terrestrial counterparts. The average location of the near-Earth neutral line is ~ 20 R_E to 25 R_E, while the average location of the X line at Mercury is near 2R_M. The Alfvén travel times are comparable to Hermean substorm time scales, while at Earth, communication times are much shorter than substorm times. This difference dictates that at Mercury the ionosphere has very little time to communicate with the magnetosphere, and vice-versa, during the course of a substorm.

21.3. DISCUSSION

The formation and evolution of the substorm current wedge are closely tied to the braking of flows convecting flux away from the reconnection site and the resultant near-planet flux pileup that creates the dipolarization. At Earth, the SCW plays a critical role in substorms, coupling magnetospheric to ionospheric motions, deflecting incoming plasma flows, and regulating the dissipation of pressure built up in the near-Earth magnetosphere during dipolarization. The braking, diversion, and pileup of these flows create a localized area of high pressure that is separated from the surrounding plasma by currents that continue to flow until the time that pressure equilibrium is reestablished or the magnetic shear is removed. At

Earth, the SCW exists long enough for the ionosphere and magnetosphere to reach equilibrium. Given that Mercury has no appreciable ionosphere, it is perhaps surprising, then, that the characteristic features of dipolarization are so similar between the two magnetospheres (Figure 21.2).

The key to understanding the implications of this dipolarization at both planets and how the associated currents flow in response is in comparing the relative time scales between substorm-related processes and Alfvénic travel times. At Earth, the Alfvén transit time from the near-geosynchronous dipolarization region to the ionosphere is $\sim 1-2$ min, compared to an average substorm expansion phase duration of 20 min. The formation of the substorm current wedge involves multiple bounces of current-carrying Alfvén waves. These are known as transient-response Pi2 pulsations and are observed with midlatitude nightside ground magnetometers as damped sinusoids on top of the growing background perturbation created by the SCW current [*Baumjohann and Glassmeier*, 1984; *Keiling and Takahashi*, 2011]. Over many bounces, these transient-response Pi2 establish the currents necessary for magnetospheric stresses transmitted to the ionosphere to reach equilibrium and set the ionosphere in motion.

In contrast, at Mercury the SCW forms and decays within the time it takes an Alfvén wave to travel to the surface and back [*Sundberg et al.*, 2012b]. The increased pressure inside the dipolarization region and magnetic shear at the edges have dissipated before the reflected signal returns. Even if Mercury had an ionosphere that was similar to Earth, the majority of the substorm cycle would be completed without ionospheric communication, in stark contrast to the terrestrial magnetosphere where several ionospheric bounces occur during each stage. What, then, is the role of the ionosphere during dipolarization?

There is strong evidence that at Earth the ionosphere controls the rate of decay of the SCW, i.e., how quickly the pressure gradient inside the dipolarized region smears out. Returning to Figure 21.3c, current loop 2 is associated with the gradient in pressure between dipolar and nondipolar field lines. It is also the loop that transmits ionospheric resistance to azimuthal expansion to the magnetosphere. The relative strengths of loops 2 and 4 control the azimuthal rate of expansion, but only once several bounces have occurred. Magnetic field signatures of the SCW at midlatitudes measure primarily the magnetic perturbation due to the field-aligned portions of the SCW. A typical signature is a 10–20 min rise in the perturbation field (the so-called midlatitude positive bay) with a damped sinusoidal perturbation with a period of 1–2 min riding on top, eventually damping at the point of maximum magnetic perturbation. The magnetic perturbation then slowly decays. These measurements indicate that the communication time scales between the magnetosphere and ionosphere are sufficiently rapid with respect to near-Earth dipolarization for the ionosphere to effectively reach an equilibrium condition where ionospheric flow is consistent with magnetospheric flow, or at the very least magnetospheric flow is deflected around the dipolarized region. In this picture, the ionosphere does not regulate large-scale magnetospheric convection but does affect the flow patterns near the planet, either by moving in the direction of motion or by forcing the magnetosphere to deflect flow further out. In either case, magnetospheric convection continues as needed to satisfy the return flux condition, which is easily accomplished because the distance between the dipolarized region and the source of the flows is large enough that flows are simply deflected further out. The generation and geometry of the SCW itself are established by the details of flow braking.

Further insight into the role of the SCW and ionospheric regulation of magnetospheric convection comes from studying intervals of steady magnetospheric convection (SMC). SMC events are intervals of quasi-steady convection in which the dayside and nightside flux transfer rates are roughly equivalent. This obviates the need for a growth phase, and the magnetosphere remains in a relatively stable, constant-recovery phase. During SMCs at Earth, the average total pressure within the inner magnetosphere is increased relative to other modes of magnetospheric convection [*Kissinger et al.*, 2012a]. This high-pressure region acts to deflect flows dawn and dusk further out than would normally occur during a substorm, and therefore the rapid braking and flux pileup associated with dipolarization during substorms do not occur. An interval of steady magnetospheric convection begins with a substorm and the establishment of a substorm current wedge [*Kissinger et al.*, 2012b]. During the SMC, the high-pressure region is bounded by FACs of the same sense as the SCW [*Sergeev et al.*, 1990].

SMCs are known to exist for up to several hours, much longer than a typical substorm recovery [*Sergeev et al.*, 1996]. If SMCs exist at Mercury, the long duration of SMCs would encompass many Alfvén bounce times and enable the generation of a quasi-steady SCW-like system. Recently, *Slavin et al.* [2012b] found evidence of steady magnetospheric convection in Mercury's magnetosphere. The lobe magnetic field strength was observed to remain relatively constant at $\sim 30-35$ nT while plasmoids and traveling compression regions were observed by MESSENGER, suggesting the tail flux was being replenished by transfer from the dayside. In a separate study, *Slavin et al.* [1997] found strong evidence of a FAC using Mariner 10 observations. The orientation of the current was in the sense of the SCW, being toward the planet postmidnight, and was observed approximately 3 min after a substorm dipolarization. Insufficient observations precluded determination of whether Mercury's magnetosphere was in an SMC state, however. But if a standing SCW system exists at Mercury, it will most likely occur during an interval of steady magnetospheric convection.

21.4. CONCLUSIONS

The lack of a conducting boundary at Mercury provides a natural experiment to examine the role of an ionosphere on regulating magnetospheric convection. The similarities in dipolarizations between the two planets is surprising since, at Earth, dipolarizations are closely tied to the SCW. Yet the purpose of the SCW is to communicate the stresses associated with dipolarization to the ionosphere and, with no ionosphere, it is not clear how a SCW at Mercury could exist.

For Mercury, the substorm cycle is extremely rapid and intense, and the Alfvén travel time is similar to this time scale. This leads to a magnetospheric system that responds primarily to local stresses and forces, with no input from any conducting boundary near the surface of, or even within, Mercury, since there is insufficient time for Alfvén waves to reflect back. Hermean dynamics are locally self-controlled and self-regulated by magnetospheric pressure gradients and associated diamagnetic currents [*Glassmeier*, 2000], which is quite different than the terrestrial system where the ionosphere has sufficient time to respond to changes and communicate resistance. At Mercury, currents will flow where they need to as required to dissipate the high pressure within the dipolarized region with respect to surroundings. Mercury may have radial portions of loop 2, and off equatorial loop 3, but these will be transient currents, not the long-duration currents observed at Earth. It is quite interesting that, without an ionosphere, we see very similar qualitative behavior of dipolarizations between Mercury and Earth. Future comparative studies between near-planet substorm effects at both planets may yield substantial new insight into the role of the ionosphere in controlling or regulating magnetospheric convection at Earth.

A larger question is why substorms exist at all at Mercury. With no ionospheric line tying, what inhibits convection? Again, the answer may be found in the relative time scales of loading vs. ionospheric communication. Loading occurs more rapidly than communication to the ionosphere via MHD waves. A related question is how the energy stored during the substorm growth phase is dissipated. At Earth, a substantial portion of the substorm energy is dissipated into the ionosphere via Joule heating. Without Joule heating, different mechanisms are required to dissipate energy from the tail. Energetic particles can be generated at MHD wave fronts [*Kan and Tamao*, 1988] and usually can travel much faster than MHD waves, thereby relaxing the communication problem [*Lysak and Lotko*, 1996; *Angelopoulos et al.*, 2008b]. Furthermore, energetic particle bombardment of the surface is associated with strong deposition of energy into the surface (e.g., *Vasyliunas* 1987). This energy deposition directly links magnetospheric substorms to physical and chemical modification processes at the planetary surface [e.g., *Orsini et al.*, 2007]. Energetic particles may play a much greater role within substorms at Mercury than at Earth, providing another opportunity for comparative studies.

ACKNOWLEDGMENTS

Portions of this work grew out of discussions L. Kepko had during "The Substorm Current Wedge" ISSI team meetings, and he is indebted to the team's insights into magnetosphere-ionosphere coupling during substorms. The work of KHG is financially supported by the German Ministerium für Wirtschaft und Technologie and the Deutsches Zentrum für Luft-und Raumfahrt under contract 50 QW 1101.

REFERENCES

Akasofu, S.-I. (1964), The development of the auroral substorm, *Planet. Space Sci.*, *12*, 273.

Anderson, B. J., M. H. Acuña, H. Korth, M. E. Purucker, C. L. Johnson, J. A. Slavin, S. C. Solomon, and R. L. McNutt (2008), The structure of Mercury's magnetic field from MESSENGER's first flyby, *Science*, *321* (5), 82.

Anderson, B. J., et al. (2010), The magnetic field of Mercury, *Space Sci. Rev.*, *152* (1), 307–339.

Angelopoulos, V., et al. (2008a), First results from the THEMIS mission, *Space Sci. Rev.*, *141* (1), 453–476.

Angelopoulos, V., et al. (2008b), Tail reconnection triggering substorm onset, *Science*, *321* (5891), 931–935.

Baker, D. N., J. A. Simpson, and J. H. Eraker (1986), A model of impulsive acceleration and transport of energetic particles in Mercury's magnetosphere, *J. Geophys. Res.*, *91*, 8742–8748.

Baumjohann, W., and K.-H. Glassmeier (1984), The transient response mechanism and Pi2 pulsations at substorm onset: Review and outlook, *Planet. Space Sci.*, *32*, 1361–1370, doi:10.1016/0032-0633(84)90079-5.

Baumjohann, W., G. Paschmann, and H. Luehr (1990), Characteristics of high-speed ion flows in the plasma sheet, *J. Geophys. Res.*, *95*, 3801.

Birn, J., M. Hesse, G. Haerendel, W. B. Aumjohann, and K. Shiokawa (1999), Flow braking and the substorm current wedge, *J. Geophys. Res.*, *104* (A), 19,895–19,904.

Birn, J., R. Nakamura, E. V. Panov, and M. Hesse (2011), Bursty bulk flows and dipolarization in MHD simulations of magnetotail reconnection, *J. Geophys. Res.*, *116* (A), 01210.

Borovsky, J. E., R. J. Nemzek, and R. D. Belian (1993), The occurrence rate of magnetospheric-substorm onsets: Random and periodic substorms, *J. Geophys. Res.*, *98*, 3807–3813.

Christon, S. P., J. Feynman, and J. A. Slavin (1987), Dynamic substorm injections: Similar magnetospheric phenomena at earth and Mercury, in *Magnetotail Physics*, edited by A. T. Y. Lui, pp. 393–402, The Johns Hopkins University Press, Baltimore.

Glassmeier, K.-H. (1983), Reflection of MHD-waves in the PC4-5 period range at ionospheres with non-uniform conductivity distributions, *Chapman Conference on Waves in Magnetospheric Plasmas*, *10*, 678–681.

Glassmeier, K.-H. (1997), The Hermean magnetosphere and its ionosphere-magnetosphere coupling, *Planet. Space Sci.*, *45*, 119–125.

Glassmeier, K.-H. (2000), Currents in Mercury's magnetosphere, in *Magnetospheric Current Systems*, Geophys. Monogr. 118, edited by S. Ohtani, p. 371, AGU, Washington, D. C.

Glassmeier, K.-H., and J. Espley (2006), ULF waves in planetary magnetospheres, in *Magnetospheric ULF Waves: Synthesis and New Directions*, Geophys. Monogr. 169, edited by K. Takahashi, p. 341, AGU, Washington, D. C.

Grard, R., H. Laakso, and T. I. Pulkkinen (1999), The role of photoemission in the coupling of the Mercury surface and magnetosphere, *Planet. Space Sci.*, *47* (1), 1459–1463.

Haerendel, G. (1992), Disruption, ballooning or auroral avalanche: On the cause of substorms, in *Proceedings of the International Conference on Substorms (ICS-1), Kiruna, Sweden, 23–27 March 1992, (ESA SP-337)*, pp. 417–420.

Hesse, M., and J. Birn (1991), On dipolarization and its relation to the substorm current wedge, *Journal of Geophysical Research (ISSN 0148-0227)*, *96*, 19,417.

Hill, T. W., A. J. Dessler, and R. A. Wolf (1976), Mercury and Mars: The role of ionospheric conductivity in the acceleration of magnetospheric particles, *Geophys. Res. Lett.*, *3*, 429–432.

Hsu, T.-S., and R. L. McPherron (2004), Average characteristics of triggered and nontriggered substorms, *J. Geophys. Res.*, *109*, 07208.

Hunten, D. M., T. H. Morgan, and D. E. Shemansky (1988), The Mercury atmosphere, in *Mercury*, edited by F. Vilas, C. R. Chapman, and M. S. Matthews, pp. 562–612, University of Arizona Press, Tucson, Arizona.

Kan, J. R., and T. Tamao (1988), On the propagation of auroral electron currents by MHD Alfvén waves, *Planet. Space Sci.*, *36*, 417–421.

Keiling, A., and K. Takahashi (2011), Review of Pi2 models, *Space Sci. Rev.*, *161* (1), 63–148.

Keiling, A., et al. (2009), Substorm current wedge driven by plasma flow vortices: THEMIS observations, *J. Geophys. Res.*, *114* (A), A00C22, doi:10.1029/2009JA014114.

Kelley, M. C. (2009), *The Earth's Ionosphere: Plasma Physics and Electrodynamics*, 2nd ed., Academic, Boston, Mass.

Kepko, L., R. L. McPherron, O. Amm, S. Apatenkov, W. Baumjohann, J. Birn, M. Lester, R. Nakamura, T. I. Pulkkinen, and V. A. Sergeev (2014), Substorm current wedge revisited, *Space Sci. Rev.*, in preparation.

Killen, R., et al. (2007), Processes that promote and deplete the exosphere of Mercury, *Space Sci. Rev.*, *132* (2), 433–509.

Killen, R. M., A. E. Potter, P. Reiff, M. Sarantos, B. V. Jackson, P. Hick, and B. Giles (2001), Evidence for space weather at Mercury, *J. Geophys. Res.*, *106* (E), 20,509–20,526.

Kissinger, J., R. L. McPherron, T. S. Hsu, and V. Angelopoulos (2012a), Diversion of plasma due to high pressure in the inner magnetosphere during steady magnetospheric convection, *J. Geophys. Res.*, *117* (A5).

Kissinger, J., R. L. McPherron, T. S. Hsu, V. Angelopoulos, and X. Chu (2012b), Necessity of substorm expansions in the initiation of steady magnetospheric convection, *Geophys. Res. Lett.*, *39* (15), L15105.

Lammer, H., and S. J. Bauer (1997), Mercury's exosphere: Origin of surface sputtering and implications, *Planet. Space Sci.*, *45* (1), 73–79.

Lysak, R. L., and W. Lotko (1996), On the kinetic dispersion relation for shear Alfvén waves, *J. Geophys. Res.*, *101* (A), 5085–5094.

Mallinckrodt, A. J., and C. W. Carlson (1978), Relations between transverse electric fields and field-aligned currents, *J. Geophys. Res.*, *83*, 1426.

McPherron, R. L. (1972), Substorm related changes in the geomagnetic tail: The growth phase, *Planet. Space Sci.*, *20*, 1521.

McPherron, R. L., and P. J. Coleman (1971), Magnetic field variations at ATS 1, in *Proceedings of ESRO Colloquium on Geostationary Magnetospheric Satellite*, pp. 119–144, ESRO pub #SP-60 ESRO Colloquium, Lyngby, Denmark.

McPherron, R. L., C. T. Russell, and M. P. Aubry (1973), Satellite studies of magnetospheric substorms on August 15, 1968. 9. Phenomenological model for substorms, *J. Geophys. Res.*, *78*, 3131.

Moldwin, M. B., and W. J. Hughes (1992), On the formation and evolution of plasmoids: A survey of ISEE 3 Geotail data, *J. Geophys. Res.*, *97*, 19,259.

Nagai, T., M. Fujimoto, Y. Saito, S. Machida, T. Terasawa, R. Nakamura, T. Yamamoto, T. Mukai, A. Nishida, and S. Kokubun (1998), Structure and dynamics of magnetic reconnection for substorm onsets with Geotail observations, *J. Geophys. Res.*, *103*, 4419.

Ogilvie, K. W., J. D. Scudder, V. M. Vasyliūnas, R. E. Hartle, and G. L. Siscoe (1977), Observations at the planet Mercury by the plasma electron experiment: Mariner 10, *J. Geophys. Res.*, *82*, 1807–1824.

Orsini, S., L. G. Blomberg, D. Delcourt, R. Grard, S. Massetti, K. Seki, and J. Slavin (2007), Magnetosphere-exosphere-surface coupling at Mercury, *Space Sci. Rev.*, *132*, 551–573, doi:10.1007/s11214-007-9222-2.

Panov, E. V., W. Baumjohann, R. Nakamura, O. Amm, M. G. Kubyshkina, K.-H. Glassmeier, J. M. Weygand, V. Angelopoulos, A. A. Petrukovich, and V. A. Sergeev (2013), Ionospheric response to oscillatory flow braking in the magnetotail, *J. Geophys. Res. Space Phys. 118*, 1529–1544, doi:10.1002/jgra.50190.

Pfaff, R. F. (2012), The near-Earth plasma environment, *Space Sci. Rev.*, *168* (1), 23–112.

Raines, J. M., J. A. Slavin, T. H. Zurbuchen, G. Gloeckler, B. J. Anderson, D. N. Baker, H. Korth, S. M. Krimigis, and R. L. McNutt (2011), MESSENGER observations of the plasma environment near Mercury, *Planet. Space Sci.*, *59* (1), 2004–2015.

Rostoker, G., S.-I. Akasofu, J. Foster, R. A. Greenwald, A. T. Y. Lui, Y. Kamide, K. Kawasaki, R. L. McPherron, and C. T. Russell (1980), Magnetospheric substorms: Definition and signatures, *J. Geophys. Res.*, *85*, 1663.

Runov, A., V. Angelopoulos, X. Z. Zhou, X. J. Zhang, S. Li, F. Plaschke, and J. Bonnell (2011), A THEMIS multicase study of dipolarization fronts in the magnetotail plasma sheet, *J. Geophys. Res.*, *116* (A5), A05216.

Russell, C. T., and R. J. Walker (1985), Flux transfer events at Mercury, *J. Geophys. Res.*, *90*, 11,067.

Schödel, R., W. B. Aumjohann, R. Nakamura, V. A. Sergeev, and T. Mukai (2001), Rapid flux transport in the central plasma sheet, *J. Geophys. Res.*, *106* (A), 301–314.

Scholer, M. (1970), On the motion of artificial ion clouds in the magnetosphere, *Planet. Space Sci.*, *18*, 977.

Sergeev, V., V. Angelopoulos, M. Kubyshkina, E. Donovan, X.-Z. Zhou, A. Runov, H. J. Singer, J. P. McFadden, and R. Nakamura (2011), Substorm growth and expansion onset as observed with ideal ground-spacecraft THEMIS coverage, *J. Geophys. Res.*, *116*. A00I26, doi:10.1029/2010JA015689.

Sergeev, V. A., W. Lennartsson, R. Pellinen, M. Vallinkoski, and N. I. Fedorova (1990), Average patterns of precipitation and plasma flow in the plasma sheet flux tubes during steady magnetospheric convection, *Planet. Space Sci.*, *38* (3), 355–363.

Sergeev, V. A., T. I. Pulkkinen, and R. J. Pellinen (1996), Coupled-mode scenario for the magnetospheric dynamics, *J. Geophys. Res.*, *101*, 13,047.

Shiokawa, K., W. B. Aumjohann, and G. Haerendel (1997), Braking of high-speed flows in the near-Earth tail, *Geophys. Res. Lett.*, *24* (1), 1179–1182.

Siscoe, G. L., N. F. Ness, and C. M. Yeates (1975), Substorms on Mercury, *J. Geophys. Res.*, *80*, 4359–4363.

Slavin, J. A., M. F. Smith, E. L. Mazur, D. N. Baker, T. Iyemori, H. J. Singer, and E. W. Greenstadt (1992), ISEE 3 plasmoid and TCR observations during an extended interval of substorm activity, *Geophys. Res. Lett.*, *19*, 825–828.

Slavin, J. A., M. F. Smith, E. L. Mazur, D. N. Baker, E. W. J. Hones, T. Iyemori, and E. W. Greenstadt (1993), ISEE 3 observations of traveling compression regions in the earth's magnetotail, *J. Geophys. Res.*, *98*, 15,425.

Slavin, J. A., J. C. J. Owen, J. E. P. Connerney, and S. P. Christon (1997), Mariner 10 observations of field-aligned currents at Mercury, *Planet. Space Sci.*, *45*, 133–141.

Slavin, J. A., et al. (2009), MESSENGER observations of magnetic reconnection in Mercury's magnetosphere, *Science*, *324* (5), 606.

Slavin, J. A., et al. (2010a), MESSENGER observations of large flux transfer events at Mercury, *Geophys. Res. Lett.*, *37* (2), 2105.

Slavin, J. A., et al. (2010b), MESSENGER observations of extreme loading and unloading of Mercury's magnetic tail, *Science*, *329* (5), 665.

Slavin, J. A., et al. (2012a), MESSENGER observations of a flux-transfer-event shower at Mercury, *J. Geophys. Res.*, *117*, A00M06, doi:10.1029/2012JA017926.

Slavin, J. A., et al. (2012b), MESSENGER and Mariner 10 flyby observations of magnetotail structure and dynamics at Mercury, *J. Geophys. Res.*, *117*, A01215, doi:10.1029/2011JA016900.

Southwood, D. J., and M. G. Kivelson (1991), An approximate description of field-aligned currents in a planetary magnetic field, *J. Geophys. Res.*, *96*, 67.

Strangeway, R. J. (2012), The equivalence of Joule dissipation and frictional heating in the collisional ionosphere, *J. Geophys. Res.*, *117*, A02310, doi:10.1029/2011JA017302.

Strangeway, R. J., R. Elphic, and W. Peria (2000), FAST observations of electromagnetic stresses applied to the polar ionosphere, in *Magnetospheric Current Systems*, *Geophys. Monogr. Ser.*, vol. 118, edited by S.-I. Ohtani et al., pp. 21–29, AGU, Washington, D.C.

Strickland, D. J., J. R. Jasperse, and J. A. Whallen (1983), Dependence of auroral FUV emissions on the incidence electron spectrum and neutral atmosphere, *J. Geophys. Res.*, *88*, 8051.

Sundberg, T., S. A. Boardsen, J. A. Slavin, B. J. Anderson, H. Korth, T. H. Zurbuchen, J. M. Raines, and S. C. Solomon (2012a), MESSENGER orbital observations of large-amplitude Kelvin-Helmholtz waves at Mercury's magnetopause, *J. Geophys. Res.*, *117* (A4), A04216.

Sundberg, T., et al. (2012b), MESSENGER observations of dipolarization events in Mercury's magnetotail, *J. Geophys. Res.*, *117*, A00M03, doi:10.1029/2012JA017756.

Thayer, J. P., and J. Semeter (2004), The convergence of magnetospheric energy flux in the polar atmosphere, *J. Atmos. Solar Terrest. Phys.*, *66*, 807–824, doi: 10.1016/j.jastp.2004.01.035.

Vasyliunas, V. M. (1987), Heat conduction limits on calorimetric effects at Mercury due to solar wind-magnetosphere interactions, *J. Geophys. Res.*, *92*, 13,658–13,660.

Winslow, R. M., B. J. Anderson, C. L. Johnson, J. A. Slavin, H. Korth, M. E. Purucker, D. N. Baker, and S. C. Solomon (2013), Mercury's magnetopause and bow shock from MESSENGER Magnetometer observations, *J. Geophys. Res. Space Phys.*, *118*, 2213–2227, doi:10.1002/jgra.50237.

Zurbuchen, T. H., et al. (2011), MESSENGER observations of the spatial distribution of planetary ions near Mercury, *Science*, *333* (6), 1862.

22

Review of Global Simulation Studies of Effect of Ionospheric Outflow on Magnetosphere-Ionosphere System Dynamics

M. Wiltberger

22.1. INTRODUCTION

Since the discovery of O^+ ions in the magnetosphere [*Shelley et al.*, 1972], it has been widely known that the ionosphere is an important source of magnetospheric plasma [*Sharp et al.*, 1985]. During active times observations have shown that O^+ can become a major component of the plasma sheet with O^+-to-H^+ number density ratios approaching unity [*Peterson et al.*, 1981] and energy density ratios also approaching unity [*Nosé et al.*, 2000]. During superstorms *Nosé et al.*, [2005] reported that the energy density ratios can become as large as 10–30 more O^+. *Kistler et al.* [2005] showed that these O^+ ions are a major contributor to the cross-tail and ring current systems in the magnetotail. The detailed study of Akebono observations by *Cully et al.* [2003] found correlations between O^+ densities and solar wind conditions, including dynamic pressure, electric field, the magnitude of the interplanetary magnetic field (IMF), and the direction of the IMF.

While not the focus of this review, significant research efforts have investigated the many different pathways that ions follow in acquiring energy during their exodus from the ionosphere [*Yau and Andre*, 1997; *Schunk*, 2007, and references therein]. These processes produce two main categories of outflow. One is the classical polar wind, composed principally of light ions (H^+, He^+) that are accelerated upward by an ambipolar electric field supported by thermal pressure gradients. The second category, sometimes referred to as generalized polar wind, includes outflow driven by process other than those that create the classical polar wind. One pathway involves acceleration of ions in the direction perpendicular to the magnetic field followed by mirror force lifting. This process acts first on heavy ions (O^+, N^+) due to their larger Lamour radii [*Lemaire and Scherer*, 1978; *Cladis*, 1986; *Demars et al.*, 1996; *Ganguli* , 1996]. Parallel acceleration processes associated with field-aligned currents (FACs) [*Ganguli*, 1996] and centrifugal acceleration [*Cladis and Francis*, 1992] can lead to ion conics and beams which are also classified as part of the "generalized polar wind." An additional pathway relies on the interaction of photoelectrons with the ambipolar electric field in the sunlit hemisphere. This interaction provides additional acceleration to the outflowing ions [*Khazanov et al.*, 1997]. As *Yau et al.* [2007] point out, it can be challenging observationally to distinguish between the classical and generalized polar wind so we shall use the term polar wind to refer to outflow of ions, mainly H^+, He^+, and O^+, from the ionosphere into the magnetosphere. The outflow flux and energy of the polar wind depend on the density of the ionospheric source population and the downward flow of electromagnetic power and precipitating particles.

In chapter 14, Wing and Johnson provide a comprehensive review of the pathways for entry and transport of solar wind ions into the magnetotail. The entry from the solar wind can involve double-cusp reconnection, Kelvin-Helmholtz instability, or kinetic Alfvén waves. Once in the magnetotail $E \times B$ and gradient curvature drifts play important roles in plasma transport and the specific entropy becomes an important parameter for tracking the transport of the solar wind ions. The pathway ions of ionospheric origin take into the

High Altitude Observatory, National Center for Atmospheric Research, Boulder, Colorado

Magnetotails in the Solar System, Geophysical Monograph 207. First Edition.
Edited by Andreas Keiling, Caitríona M. Jackman, and Peter A. Delamere.

magnetosphere depends primarily on their source region. Ions coming from the cusp subjected to transverse heating processes move up the field line as they are convected across the polar cap. As *Horwitz* [1986] and *Chappell et al.* [1987] describe, the resulting velocity filter effect results in higher energy and lighter ions traveling further down than their less energetic or heavier counterparts. On the nightside outflows from the aurora region can have direct access to the near-Earth plasma sheet. *Daglis et al.* [1994] found correlations between the strength of the auroral currents and the energy density of O^+ ions measured by AMPTE/CCE satellite mission. More recently *McFadden et al.* [2001] used FAST observations during substorms to connect ion outflows occurring on newly reconnected field lines in the nightside auroral region to enhancements in the plasma sheet density and changes in the ring current plasma composition during magnetic storms.

Several studies have indicated that O^+ could influence the substorm process by changing the growth phase [*Daglis et al.*, 1994] and effecting the initiation mechanism [*Baker et al.*, 1982; *Cladis and Francis*, 1992; *Lakhina*, 1995]. Microscale simulations of magnetic reconnection by *Shay and Swisdak* [2004] indicate that the reconnection rate in an H^+-O^+ plasma is lower than in a single fluid plasma with the rate controlled by the mass of the heaviest species present. They speculate that substorms will either need to occur more frequently or exhibit longer expansion phases to maintain global flux balance when O^+ is present in the magnetotail. In contrast, *Liu et al.* [2013] conducted a statistical study using Cluster data and concluded that adding O^+ to the magnetotail made it more difficult to trigger a substorm onset but the rate at which energy is dissipated in the magnetotail during the substorm is faster. Using ring current simulations driven by observations from geosynchronous satellite observations, *Liemohn and Jazowski* [2008] found a strong correlation between the O^+-to-H^+ ratio and the D_{ST} index, indicating that the polar wind may play an important role in the formation and evolution of the ring current.

Significant progress has also been made in the numerical modeling of the geospace system by coupling models of its various regions. Efforts by the Center for Integrated Space Weather Modeling (CISM) have produced the Coupled Magnetosphere-Ionosphere-Thermosphere (CMIT) Model [*Wang et al.*, 2004; *Wiltberger et al.*, 2004], which has undergone extensive validation [*Taktakishvili et al.*, 2007; *Burns et al.*, 2008; *Lei et al.*, 2008b; *Wang et al.*, 2010] and has allowed advances in understanding how the magnetosphere drives the ionosphere-thermosphere (IT) system [*Wang et al.*, 2007; *Lei et al.*, 2008a; *Wang et al.*, 2010]. Recent work within this group has focused on the development of a multifluid version of the magnetospheric component commonly referred to as the MultiFluid Lyon-Fedder-Mobarry (MFLFM) model [*Wiltberger et al.*, 2010]. The Space Weather Modeling Framework (SWMF) developed at the University of Michigan has assembled a similar set of coupled models, including a sophisticated polar wind module [*Tóth et al.*, 2005; *Glocer et al.*, 2009a]. Its developers have also validated the coupled models [*Yu and Ridley*, 2008; *Ridley et al.*, 2010; *Welling and Ridley*, 2010] and used them to investigate the fundamental physics of the geospace system [*Ridley et al.*, 2003, 2004; *Kivelson and Ridley*, 2008]. The OpenGGCM model [*Raeder*, 2003] also includes coupling between magnetosphere and IT models. It has been used to simulate substorms [*Raeder et al.*, 2001, 2008] and to study substorm onset [*Raeder et al.*, 2010], flux transfer events [*Raeder*, 2006], and driven reconnection at the magnetopause [*Dorelli*, 2004]. The Winglee group has pioneered the utilization of multifluid modeling techniques in studying Earth's magnetosphere [*Winglee*, 2000, 2003; *Winglee et al.*, 1998a, 2002]. This group has also used their code to simulate other planetary magnetospheres [*Winglee* et al., 1992, 2009b] and even studied a novel method for using mini-magnetospheres as a scheme for spacecraft propulsion [*Winglee et al.*, 2000]. In addition to the work conducted by the developers of these models, their availability at the Community Coordinated Modeling Center (CCMC) has allowed numerous investigators to conduct their own investigations on topics ranging from understanding the evolution of flux transfer events [*Daum et al.*, 2008] to ultralow-frequency (ULF) wave structure [*Samsonov et al.*, 2011] and even to investigations of fundamental questions about how the cross-polar cap potential (CPCP) saturates [*Borovsky et al.*, 2009]. Clearly the technique of coupling component models of geospace has been quite effective in advancing our understanding of the tightly coupled geospace system.

This chapter will focus on the efforts of three groups, Winglee, SWMF, and MFLFM, using multifluid simulations to study the impacts of polar wind outflow on the evolution of the magnetosphere-ionosphere system. The first portion of the chapter will follow the format of independently presenting the techniques used by each group used to conduct modeling of the impacts of ionospheric outflows on the geospace system. Section 22.2 describes the forms of the multifluid magnetohydrodynamic (MHD) equations each group uses as well as the numerical techniques employed by their simulations. Section 22.3 discusses the methods each group uses to include ionospheric outflow in their simulations. We then break from the approach of reporting each group independently in order discuss the impacts of the polar wind outflow on the different regions of geospace in section 22.4. The chapter concludes in section 22.5 with a brief review of the results and the future directions research is likely to proceed with these simulation techniques.

22.2. MAGNETOSPHERIC METHODS

In this review we focus on the results coming from three different groups using global-scale magnetospheric simulations to study the impacts of ionospheric outflows on the magnetosphere. While each group is using a multifluid MHD approach in solving for the plasma and magnetic field in the magnetosphere, they have different formulations of these equations and utilize vastly different numerical techniques to solve the set of partial differential equations (PDEs) on the computer. In this section we present a review of the equations and techniques drawn from key publications, but putting the equations into a common notation in order to facilitate comparison between these approaches.

Since all the groups are using multifluid MHD equations before presenting the formulations used within each model, we present a short derivation of the MHD ion momentum equation for a multifluid plasma. We start with the ion momentum equation

$$\frac{\partial \rho_\alpha \vec{u}_\alpha}{\partial t} = -\nabla\left(\rho_\alpha \vec{u}_\alpha \vec{u}_\alpha + \bar{I}P_\alpha\right) + n_\alpha q_\alpha\left(\vec{E} + \vec{u}_\alpha \times \vec{B}\right), \quad (22.1)$$

where the α subscript denotes the ion species; the typical convention of ρ, n, $\vec{u}$, and P is used for the mass density, number density, velocity, and pressure, respectively; and $\vec{B}$ and $\vec{E}$ are the magnetic and electric fields. The first step in replacing the electric field in the ion momentum equation is to utilize the generalized Ohm's law,

$$\vec{E} = -\vec{u}_e \times \vec{B} - \frac{1}{en_e}\nabla P_e + \eta \vec{J}, \quad (22.2)$$

where the subscript e donates the electron fluid velocity and $\vec{J}$ is the current density. The generalized Ohm's law can also be thought of as electron momentum equation when terms proportional to the electron mass are neglected. The second step in replacing the electric field is to use the definition of current density to determine the electron velocity,

$$\vec{u}_e = -\frac{1}{en_e}\vec{J} + \vec{u}_M, \quad (22.3)$$

where

$$\vec{u}_M = \frac{1}{en_e}\sum_\beta n_\beta q_\beta \vec{u}_\beta \quad (22.4)$$

is the charge-averaged ion velocity. Proceeding with the math results in

$$\frac{\partial \rho_\alpha \vec{u}_\alpha}{\partial t} = -\nabla\left(\rho_\alpha \vec{u}_\alpha \vec{u}_\alpha + IP_\alpha\right) + n_\alpha q_\alpha\left(\vec{u}_\alpha - \vec{u}_M\right) \times \vec{B} + \frac{n_\alpha q_\alpha}{n_e e}\left(\vec{J} \times \vec{B} - \nabla P_e\right) \quad (22.5)$$

as the final form of the ion momentum equation.

22.2.1. Winglee Magnetosphere Model and Methods

Winglee [1998] was the first to publish a study using multifluid MHD simulation to identify the location of the geopause. Basically the geopause is the boundary between where the plasma comes primarily from the ionosphere versus the solar wind [*Moore*, 1991; *Moore and Delcourt*, 1995]. *Winglee* [2000] extended the techniques in order to study plasma outflow in more detail, so those are the equations we present here. The units and naming convention are modified to match the conventions used in this text. For the ions,

$$\frac{\partial \rho_\alpha}{\partial t} = -\nabla \cdot \left(\rho_\alpha \vec{u}_\alpha\right), \quad (22.6)$$

$$\rho_\alpha \frac{d\vec{u}_{\alpha\parallel}}{dt} = -\left(\nabla P_\alpha\right)_\parallel - \frac{n_\alpha}{n_e}\left(\nabla P_e\right)_\parallel - \frac{GM_E \rho_\alpha}{R^2}\vec{r}_\parallel, \quad (22.7)$$

$$\rho_\alpha \frac{d\vec{u}_{\alpha\perp}}{dt} = en_\alpha\left(\vec{u}_\alpha \times \vec{B} - \sum_\beta \frac{n_\beta}{n_e}\vec{u}_\beta \times \vec{B}\right) + \frac{n_\alpha}{n_e}\left(\vec{J} \times \vec{B} - (\nabla)P_{e\perp}\right) - \left(\nabla P_\alpha\right)_\perp - \frac{GM_E \rho_\alpha}{R^2}\vec{r}_\perp, \quad (22.8)$$

$$\frac{\partial P_\alpha}{\partial t} = -\gamma \nabla \cdot \left(P_\alpha \vec{u}_\alpha\right) + (\gamma - 1)\vec{u}_\alpha \cdot \nabla P_\alpha. \quad (22.9)$$

The polytropic index (γ) is set to the typical fluid value of $\frac{5}{3}$. In addition to the traditional MHD forces, the Winglee group includes the effect of gravitational force and thus the last terms in the perpendicular 22.7 and parallel 22.7 velocity equations. This formulation also includes an explicit breakdown of the velocity equation into parallel and perpendicular components, which the other formulations do not typically present. Adding these two equations together yields the MHD ion momentum equation 22. 4 derived in the beginning of this section. It is also worth noting here that, consistent with the subsequent formulations, this approach results in the ion species having the same drifts across the field lines.

For the electrons, the first step is to assume quasi-neutrality so that the election density is simply

$$n_e = \sum_{\beta} n_\beta . \tag{22.10}$$

The electron velocity is found using the definition of current density presented in equation 22.3. The standard fluid equation

$$\frac{\partial P_e}{\partial t} = -\gamma\nabla\left(P_e\vec{u}_e\right)+(\gamma-1)\vec{u}_e\nabla\cdot P_e \tag{22.11}$$

is used to solve for the electron pressure.

For the calculation of the magnetic field, Faraday's law is used:

$$\begin{aligned}\frac{\partial\vec{B}}{\partial t} &= \nabla\times\vec{E} \\ &= \nabla\times\left(\vec{u}_M\times\vec{B}\right)-\frac{1}{en_e}\vec{J}\times\vec{B}+\frac{1}{en_e}\nabla P_e,\end{aligned} \tag{22.12}$$

with the Hall and electron pressure terms from the generalized Ohm's law maintained along with the tradition ideal MHD $\vec{u}\times\vec{B}$ term when replacing the electric field. The only term from the generalized Ohm's law that is neglected in this formulation is the term proportional to resistivity.

The computational solution of these equations is implemented using the classic Lax-Wendroff finite difference method with a flux-corrected transport scheme applied to the plasma parameters only. These shock-capturing methods allow for removal of nonphysical oscillations at sharp discontinuities such as the bow shock. The computational grid is Cartesian and extends $30R_E$ in the ±GSM (geocentric solar magnetospheric) X and Z directions. The front boundary of the grid is located at +35 R_E in the GSM X direction with the back boundary at –200 R_E. The inner radius of the computational grid is 3 R_E. The finest resolution of the simulation, approximately 0.4 R_E, is in the inner magnetosphere and dayside region. It gradually increases to a value of 3 R_E at the outer edges of the computational domain. Along the outer edges of the computational domain solar wind forcing, either from observations or idealized, along with outflow boundary conditions are used. While the details of the plasma boundary conditions for this model will be discussed in section 22.2.2, it is worth noting here that the simulation includes a resistive layer near the inner boundary which acts like the ionosphere [*Winglee et al.*, 1998b]. In this layer an artificial resistivity term is introduced that is a strong function of radial distance so that in effect it is only operating in the first few cells above the inner boundary. The resistivity can also be a function of latitude and longitude so that effects resulting from extreme ultraviolet (EUV) ionization and electron precipitation can be included in the simulation, although typically a weak uniform resistive layer is utilized.

22.2.2. SWMF and Methods

In their recent paper *Tóth et al.* [2012] provide a comprehensive review of the capabilities and numerical techniques available within the SWMF. This framework includes models for solar, magnetosphere, and ionosphere simulations. A key component of the framework is the Block-Adaptive Tree Solar-Wind Roe-Type Upwind Scheme (BATS-R-US) code that can solve various forms of the MHD equations, including Hall, anisotropic pressure, and semirelativistic forms. The multifluid formulation is most relevant to this review and *Glocer et al.* [2009b] provides a detailed derivation and explanation of this formulation which we will summarize here using our consistent set of units and naming conventions.

The full set of multifluid equations typically used for SWMF ionospheric outflow studies are

$$\frac{\partial\rho_\alpha}{\partial t} = -\nabla\cdot\rho_\alpha\vec{u}_\alpha + S_{\rho_\alpha}, \tag{22.13}$$

$$\begin{aligned}\frac{\partial\rho_\alpha\vec{u}_\alpha}{\partial t} &= -\nabla\left(\rho_\alpha\vec{u}_\alpha\vec{u}_\alpha+\bar{I}P_\alpha\right) \\ &\quad + n_\alpha q_\alpha\left(\vec{u}_\alpha-\vec{u}_M\right)\times\vec{B} \\ &\quad + \frac{n_\alpha q_\alpha}{n_e e}\left(\vec{J}\times\vec{B}-\nabla P_e\right)+S_{\rho_\alpha\vec{u}_\alpha},\end{aligned} \tag{22.14}$$

$$\frac{\partial P_\alpha}{\partial t} = -\nabla\cdot\left(P_\alpha\vec{u}_\alpha\right)-(\gamma-1)P_\alpha\nabla\cdot\vec{u}_\alpha + S_{P_\alpha}, \tag{22.15}$$

$$\frac{\partial\vec{B}}{\partial t} = \nabla\times\left(\vec{u}_M\times\vec{B}\right), \tag{22.16}$$

where the term q_α has been added to the naming convention to allow for consideration of higher charge state ions, e.g., He^{2+}, that were not considered in the equations of the Winglee method but are a straightforward extension of their equations.

Glocer et al. [2009b] correctly point out it is not possible to express the multifluid equations in a fully conservative form as is typically done in numerical solutions of hydrodynamics [*Hirsch*, 1988]. The utilization of the momentum density in equation 22.13 moves the formulation in this direction as opposed to using the parallel 22.7 and perpendicular 22.7 velocity equations as was done in the Winglee model. These equations expand the convective derivative term and the part not involving the partial derivative with time has been moved to the right-hand side of the equations in order to make the formulation consistent with the format used for comparison with the MFLFM set of equations. One major difference between this formulation and that used by the Winglee group is seen in Faraday's law 22.16 where the Hall and electron pressure terms have been neglected. *Glocer et al.* [2009a] use a multispecies version of the BATS-R-US code in which ions have separate continuity equations, but a single momentum equation is solved for the motions of all ion species.

The implementation above uses the pressure equation as was done by the Winglee model. The flexibility of the SWMF allows for the implementation of using hydrodynamic energy density formulation of the energy equation, namely,

$$\frac{\partial \mathcal{E}_\alpha}{\partial t} = -\nabla\cdot\left[\left(\mathcal{E}_\alpha + P_\alpha\right)\vec{u}_\alpha\right] + \left[n_\alpha q_\alpha\left(\vec{u}_\alpha - \vec{u}_M\right)\times\vec{B} + \frac{\rho_\alpha q_\alpha}{n_e e}\left(\vec{J}\times\vec{B} - \nabla P_e\right)\right]\cdot\vec{u}_\alpha + S_{\mathcal{E}_\alpha}, \tag{22.17}$$

with $\mathcal{E}_\alpha = \frac{1}{2}\rho_\alpha u_\alpha^2 + P_\alpha/(\gamma-1)$ as the plasma energy density definition. While not exactly conservative, this formulation allows the numerical solution to give roughly correct jump conditions across shocks and other discontinuities when the magnetic energy density is small compared to the thermal and kinetic energy densities.

These equations also contain source terms $S_{\rho,\rho\vec{u},P,\mathcal{E}}$ that could be used to include a variety of additional physical processes such as charge exchange and photoionization. Most are set to zero for Earth's magnetosphere but could be used for other systems. An exception to this is the inclusion of the frictional source term in the momentum equation in order to include the subgrid-scale two-stream instabilities that limit the relative velocities of ions parallel to the magnetic field.

For the electrons the SWMF uses the same definitions for density and velocity as the Winglee method. The SWMF has the ability to solve the full electron pressure equation as well as to set the electron pressure to some fraction of the total ion pressure. *Glocer et al.* [2009b] have used comparisons with inner magnetospheric simulations to adopt a value of 20% of the total ion pressure for the electrons.

The computational methods used to solve these equations are described in great detail by *Tóth et al.* [2012] so we only provide a brief summary here. The general technique used to solve the MHD equations is the utilization of advanced numerical methods with numerical flux functions that involve limiters which introduce dissipation in regions of strong gradients. In magnetospheric applications the computational domain extends from –224 R_E to 32 R_E in the GSM X direction and $\pm 128R_E$ in the GSM Y and Z directions. The code uses a Cartesian block adaptive grid which can be thought of as a regular course grid throughout the computational domain with finer resolution cells in regions of interest obtained by dividing the cells in half along each direction. The finest resolution in the computational domain is controlled by the level of nesting utilized for a particular study. For a magnetospheric calculation with ~1 million cells the smallest cell size is $0.25R_E$, and for ~3million cells the smallest cell size is $0.125R_E$. Similar to the Winglee model, along the outer edges of the computational domain solar wind conditions can be imposed. The code has the ability to include the dipole tilt of Earth's magnetic field. The framework can allow the core MHD component of the magnetosphere to be coupled to other physics-based simulations, including inner magnetosphere (IM) models, ionospheric electrodynamics (IE), and a polar wind outflow model (PWOM). The details of the PWOM and the inner plasma boundary conditions will be discussed in more detail in section 22.3.2. The IE module includes the capability to include changes due to EUV ionization and auroral zone processes in the conductance [*Ridley et al.*, 2004]. The auroral zone conductance is regulated by the location and strength of the FACs obtained from the magnetospheric calculation. The IM module allows coupling to simulations such as the Rice convection model (RCM) [*De Zeeuw et al.*, 2004].

22.2.3. MFLFM Model and Methods

The multifluid Lyon-Fedder-Mobbary model is an extension of the single-fluid LFM *Lyon et al.* [2004] numerical model that has been used to conduct numerous studies of the magnetosphere. *Wiltberger et al.* [2010] presented initial simulations with the MFLFM along with a description of the equations and numerical methods. In the consistent unit and naming convention of this chapter these equations are

$$\frac{\partial \rho_\alpha}{\partial t} = -\nabla\cdot\rho_\alpha\vec{u}_\alpha, \tag{22.18}$$

$$\frac{\partial \rho_\alpha\vec{u}_\alpha}{\partial t} = -\nabla\cdot\left(\vec{p}_\alpha\vec{u}_\alpha + \bar{I}P_\alpha\right) + \vec{F}_\alpha^d - \frac{n_\alpha q_\alpha}{n_e e}\nabla P_{e\parallel}, \tag{22.19}$$

$$\frac{\partial \mathcal{E}_\alpha}{\partial t} = -\nabla\cdot\left[\vec{u}_\alpha\left(\mathcal{E}_\alpha + P_\alpha\right)\right] + \vec{u}_\alpha\cdot\left(\vec{F}_\alpha^d + n_\alpha q_\alpha\vec{E}_\parallel\right), \tag{22.20}$$

$$\frac{\partial \vec{B}}{\partial t} = \nabla\times\left(\vec{u}_M\times\vec{B}\right), \tag{22.21}$$

with Lorentz force due to the first-order ion drift given by

$$\vec{F}_\alpha^d \equiv \hat{b}\times\left[\vec{p}_\alpha\cdot\nabla\vec{u}_\alpha+\nabla P_\alpha+\frac{\rho_\alpha\left(\vec{u}_M-\vec{u}_\alpha\right)\cdot\hat{b}}{B}\frac{\partial\vec{B}}{\partial t}\right.$$
$$\left.-\frac{\rho_\alpha}{\rho}\left(\sum_\beta\left(\vec{p}_\beta\cdot\nabla\vec{u}_\beta+\nabla P_\beta\right)+\nabla P_e-\vec{j}\times\vec{B}\right)\right]\times\hat{b}. \tag{22.22}$$

The utilization of the Lorentz force term in the momentum density equation 22.19 helps to illustrate that to lowest order all the ion species move in the perpendicular direction with the $\vec{E}\times\vec{B}$ velocity. This result is consistent with the formulations utilized by the other modeling groups. This formulation more clearly shows how the time rate of change magnetic field in the force equation 22.22 results in the transfer of momentum between species when the magnetic field direction varies. As equation 22.21 shows, this code also neglects the Hall and electron pressure terms in Faraday's law. This formulation also only includes a hydrodynamic energy density version of the energy equation.

The MFLFM uses the same definition of electron density and velocity as the SWMF and the Winglee model. For the electron pressure, the preferred mode of operation is to set it to be a fraction of the ion pressure as is done in the SWMF framework. In fact, the initial runs with the MFLFM have been done setting the electron pressure to zero. The code has the capability to solve the full electron pressure equation.

A detailed discussion of the high-order techniques used within the MFLFM are discussed by *Lyon et al.* [2004]. Basically the techniques involve the utilizations of a eighth-order differencing stencil along with the partial donor method limiter [*Hain*, 1987] to find the numerical fluxes needed to utilize the finite volume method of solving PDEs. The computational grid for the MFLFM utilizes a logically rectangular distorted spherical mesh with hexagonal cells that have nonorthogonal faces used to create a grid aligned with the known important features of the magnetosphere, e.g., bow shock, magnetotail. The computational domain for the MFLFM covers from $-300R_E$ to $+30R_E$ in the GSM X direction and $\pm 60R_E$ in the GSM Y and Z directions. The nonuniform mesh has a resolution that varies from ~0.25 R_E for low-resolution simulations to 0.125 R_E for high-resolution simulations in the inner magnetosphere. Along the outer edges of the computational domain the typical cell sizes are a few R_E wide. This code also has the capability for dealing with the tilt in Earth's dipole magnetic field. As part of the CISM project [*Goodrich et al.*, 2004] the code has been coupled with thermosphere-ionosphere simulations [*Wiltberger et al.*, 2004] and the Rice convection model of the inner magnetosphere [*Pembroke et al.*, 2012]. The magnetosphere ionosphere coupler solver (MIX) module [*Merkin and Lyon*, 2010] provides the electrodynamic coupling between the magnetosphere and ionosphere, including a conductance model capable of handling both EUV and electron precipitation-based enhancements.

22.3. OUTFLOW METHODS

In addition to using slightly different formulations of the multifluid MHD equations along with vastly different computational grids and numerical techniques to solve them, each of these groups also uses unique methods for including the outflow of plasma from the ionosphere into the magnetosphere. In fact, it is in including the plasma outflow from the ionosphere into the magnetosphere that the differences between the simulations become most pronounced. This section provides an overview of these methods by discussing information about the species, velocity, and fluxes seen in the simulation results as well as any similarities between the techniques.

22.3.1. Winglee Outflow

Outflow in the Winglee model is accomplished primarily through centrifugal acceleration of plasma whose density at the inner boundary of the magnetospheric volume is held fixed. As shown in section 22.2.1, the equations solved by this model include the effects of gravitational forces. In order to account for the fact that the inner boundary radius is at $3R_E$, the group elects to set the gravitational term to 30 m/s^2 instead of the typical 10 m/s^2 in order to mimic the gravitational forces the O^+ ions experience at altitudes below the inner boundary of the simulation domain. They do this because the gravitational force at $3R_E$ experienced by an O^+ ion would be quite small with normal gravitational acceleration. The authors report that this limits the O^+ outflow relative to the H^+ outflow for weak solar wind driving.

At the inner boundary of the simulation the H^+ density is held fixed at 400 cm^3 with an initial distribution that has the plasma density decreasing as R^{-6}. This density is chosen to fall within the typical plasmasphere/ionosphere observations such as those reported by *Parks* [1991]. The density of O^+ at this boundary is imposed as a fraction of the H^+ density, which can vary from 5 to 100%, but is typically held at values of 10%. The initial plasma distribution is assumed to be in equilibrium along the dipole field lines. Following the rough observational guidelines provided by *Parks* [1991], the authors set the temperature of the equatorial plasma to 60 eV and the polar cap plasma to 0.3 eV. The authors admit the equatorial value is a bit high, but they believe this is justified because of possible contributions from hotter ring current plasma

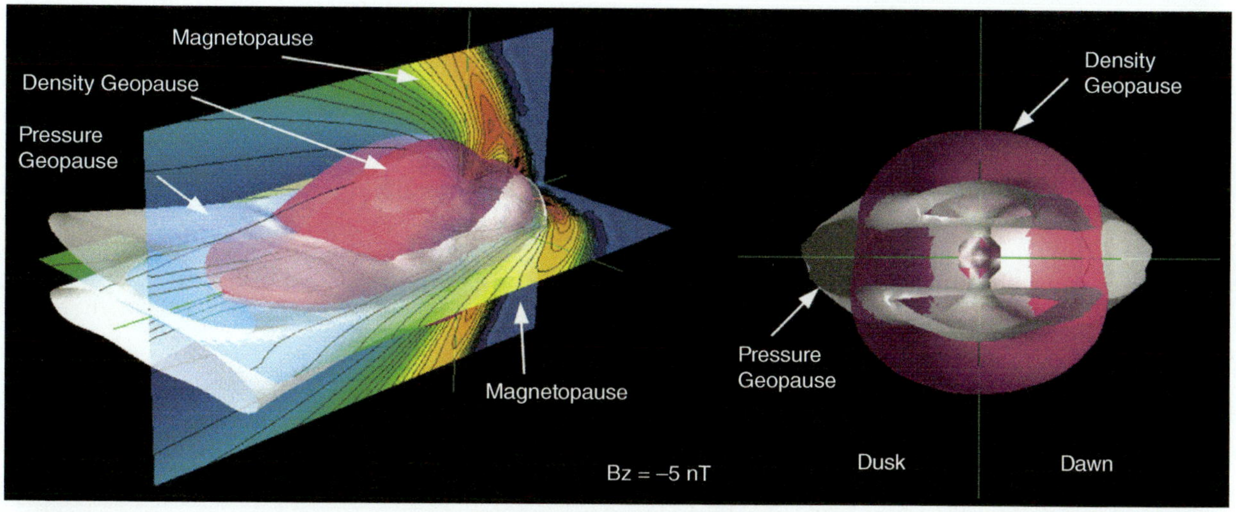

Figure 22.1 Location of the density and pressure geopauses during a period with southward IMF. The region where the pressure geopause is outside of the density geopause is indicative of the ionospheric plasma contributing significant amounts of hot plasma to the plasma sheet. Adapted from *Winglee* [2000].

populations. In simulations with idealized solar wind conditions, e.g., $n = 6\,cm^3$, $V_X = 400\,km/s$, during northward IMF the outflow fluence is typically 6×10^{26} ions/s and increases to 2×10^{27} ions/s for southward IMF. *Winglee et al.*, [2008] conducted a validation study comparing results of O^+ outflows in the simulation with observations from Polar and Akebono and finding relatively good agreement for realistic solar wind drivers. In more recent work, *Hornett et al.* [2008] developed a technique to dynamically alter the O^+ boundary density depending upon the magnitude of the aurora currents.

Winglee [2000] examined the pattern of outflow seen for idealized solar wind conditions. The model runs begin with the solar wind conditions mentioned above. The IMF starts with $B_X = -1.25$ nT and $B_Z = 1.25$ nT for a period of 2 h, after which time the B_Z is ramped southward to –1.25 nT over a period of 1 h. The IMF B_Z is then held constant for the next hour at 5, 0, or –5 nT. In order to compare the different distributions of plasma arising from the interaction of the solar wind driving with the inner boundary conditions, Winglee utilizes the concept of the geopause as was noted in the observational work of *Moore* [1991]. In addition to the traditional density geopause, Winglee utilized the same idea to create a pressure geopause. Figure 22.1 shows both of these surfaces for the southward IMF run. The cut plane shows contours of the total plasma pressure along which it is easy to pick out the magnetopause and bow shock. The magenta surface is the density geopause and the white surface is the pressure geopause. In regions where the pressure geopause extends outside the density geopause, the ionospheric plasma is less dense but is contributing more to the total pressure, and therefore the ionosphere is contributing significant amounts of hot plasma to the plasma sheet. While not shown here, the original figure by *Winglee* [2000] shows the configuration for northward IMF illustrating the opposite case, where the density in the lobes andplasmasphere is primarily of ionospheric origin.

22.3.2. SWMF Outflow Methods

The SWMF includes the most sophisticated method for including ionospheric plasma into the magnetosphere via the PWOM. Basically PWOM is a classical polar wind model whose outflows are primarily driven by pressure gradients and the ambipolar electric field [*Gombosi et al.*, 1985; *Glocer et al.*, 2007]. As part of the SWMF the PWOM is designed to couple with the multifluid version of the BATS-R-US global magnetosphere (GM) as well as the IE [*Ridley et al.*, 2004] and upper atmosphere (UA) [*Ridley et al.*, 2006] components. The initial work with this model has focused on coupling with the GM and IE modules, so in addition to providing a brief overview of the POWM, we discuss the methodology for coupling with the ionospheric electrodynamics and the magnetosphere.

The PWOM was initially developed to simulate the supersonic ion outflow along open magnetic field lines using the multifluid gyrotropic continuity, momentum, and energy equations. While the details of this model are beyond the scope of this chapter, we provide a brief description of the technique before focusing on the coupling with the magnetosphere. The origins of this work date back to work by *Gombosi et al.* [1985] and *Gombosi and Nagy* [1989]. While they focus on the application of

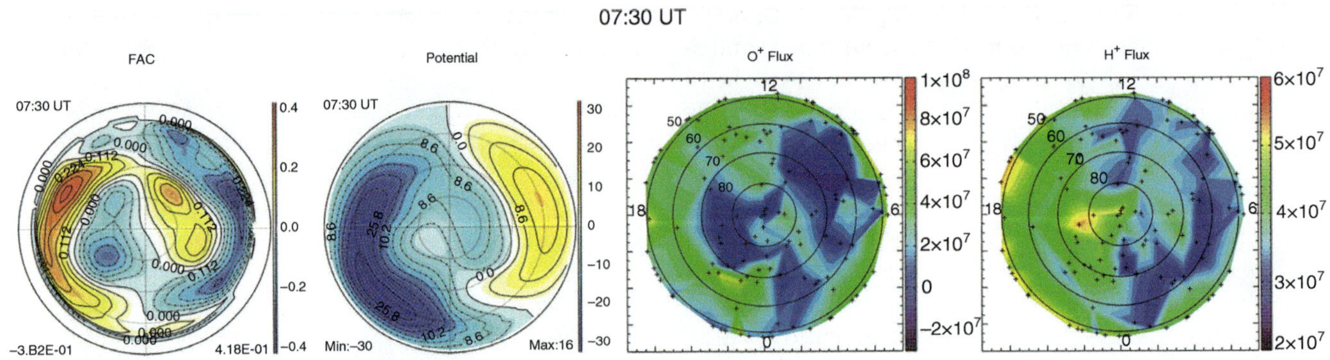

Figure 22.2 Ionospheric drivers and H^+ and O^+ fluxes during the main phase of the 4 May, 1998 geomagnetic storm. The left two panels show the FAC and CPCP potential at 150 km altitude. The right two panels show the fluxes at 1500 km for H^+ and O^+. Note that fluxes have different scales for each species. Also shown on the flux plots with + markers are the locations of the ionospheric foot points of the field lines used within PWOM mapped to 1500 km altitude. Adapted from *Glocer et al.* [2009a].

the PWOM to Saturn, *Glocer et al.* [2007] provide an excellent description of both the theoretical basis for the model and the numerical techniques used to solve the complete set of PDEs. *Glocer et al.* [2009a] cover the application of this technique to Earth's polar wind. The basic approach of the model is to solve for the motion of ions along a single fluid line starting at an altitude of about 250 km and extending a few Earth radii in altitude. The top boundary is well above the typical lower boundary of the GM module. The gyrotropic equations are solved along this field line, and in addition to the typical mass, velocity, and pressure terms seen in MHD, these equations include heat conductivity, gravity, ambipolar electric fields, and chemical reactions. These equations also depend on the cross-sectional area of the flux tube, which is a function of the magnetic field strength, and a simple dipole model is used to determine that area. The electrons are solved for using charge neutrality, steady electron velocity, and an energy equation that ensures conservation of the field-aligned currents and allows for contributions to the electron temperature from conduction, advection, heating, and energy transfer. In the case of Earth this allows for a complete description of the H^+, O^+, He^+, and e^- along a single field line.

One of the major challenges that the PWOM overcomes is the extension from a single field line model to one capable of simulating the entire polar cap. In order to accomplish this task, the PWOM uses a set of noninteracting field line models to cover the entire polar cap. These field lines are then allowed to convect under the influence of the corotation and convection electric fields. The convection electric field is driven by the field-aligned currents from the magnetosphere obtained from the GM module and computed in the IE module. Since each field line is independent, the model parallelizes trivially. The major challenge with this approach comes with passing information from the PWOM to the GM module. Since the footpoints of field lines are moving around under the influence of the ionospheric electric fields, even an initially uniform distribution will rapidly become a highly unstructured mesh after a short interval of strong driving. The code uses the method of Delaunay triangulation at each exchange between PWOM and the GM modules to interpolate between the unstructured PWOM grid and the grid points of the structured GM inner boundary region. When coupling with the full multifluid version of the GM model as in *Glocer et al.* [2009b], the values from the H^+ and O^+ fluids in the PWOM model are used to set the density, velocity, and temperature of these fluids in the GM simulation. It is also important to note that the mass coupling between PWOM and the GM is done with no feedback from the GM on PWOM and only out owing velocities are utilized in the coupling domain. If, as was done by *Glocer et al.* [2009a], the multispecies version of the GM is used, then the velocity information from each species is combined to produce a single velocity in the combined momentum equation. Since the multispecies version has separate continuity equations for each ion, the density information from the PWOM is used for each species included in the GM module. Regardless of the GM module used, the coupling with the PWOM includes feedback from the magnetosphere through alteration of the FACs obtained from the GM module.

Glocer et al. [2009a] presents case studies of the outflow seen during a strong storm (minimum $D_{ST} \sim -200$) on 4 May 1998 and a very strong storm (minimum $D_{ST} \sim -400$) on 31 March 2001. While we discuss the impacts of these outflows on the state of the magnetosphere in more detail later, Figure 22.2 shows the structure of the outflow during the main phase of the 4 May 1998 geomagnetic storm. In this figure the ionospheric conditions, namely the FAC structure and the cross-polar cap potential, near the base of the field lines at an altitude of 150 km are displayed on the left. The resulting H^+ and O^+ outflow patterns at an altitude of 1500 km, well below the lower boundary of the global magnetospheric solution, are shown on the right. This simulation was done with 125 distinct field lines, and the locations of those points are mapped into the outflow plots, illustrating the need for the triangulation techniques we previously discussed. Another key point to note when examining this figure is that the ranges for the H^+ and O^+ flux color bars are different. The O^+ range includes downflows and a maximum value that is 40% larger than used in the H^+ color bar. The O^+ outflow shows the expected increase in regions with strong upward FACs, the in uence of convection being more difficult to isolate. The O^+ flux at this altitude shows significant variability, including flows that are owing back toward Earth. The H^+ fluxes at this altitude do not show any downward flows and are generally less dynamic than those seen in the O^+ flows. The flux values are generally consistent with the statistical properties for flows seen during these geomagnetic conditions as reported by the *Abe et al.* [1993] data from the Akebono spacecraft. One notable missing feature in these results is the lack of any strong O^+ outflows consistently seen in the noon sector and typically associated with transverse ion heating in the cusp ion fountain [*Abe et al.*, 1996]. According toAbe et al., this limitation can be addressed by including a better model for electron heat flux at the top of field lines for the polar wind simulations, although including a transverse acceleration process would be needed to accurately capture the correct cusp outflow physics.

In addition to the PWOM, two types of outflow can be included in the SWMF. *Yu and Ridley* [2013] define the location of the cusp and auroral zone and then specify the amount of O^+ flux coming from these regions based upon observations. This technique is quite similar to the approach used by the MFLFM group and is discussed in more detail in section 22.3.3. Another source of outflow in SWMF simulations originates from the fact that the SWMF specifies and density value for plasma at the inner boundary of magnetospheric domain. As *Welling and Liemohn* [2014] discusse, this boundary condition results in outflow in the cusp, polar cap, and auroral zone. In the cusp and polar cap the major force responsible for outflow is related to the pressure gradient while Lorentz forces dominate the outflow in the auroral zone. This outflow source will be present in all simulations with the fixed inner boundary density.

22.3.3. MFLFM Outflow Methods

The MFLFM group has utilized two main methods for including plasma outflow into the magnetospheric domain. The first has been to use the statistical observations of where ionospheric outflow has been known so to occur to specify the region within the grid where the outflow occurs and then choose values for the flux coming from that region. The second approach has been to develop an empirical model for outflow using the observed correlations between Alfvénic Poynting flux and electron precipitation with the location and intensity of outflow reported by *Strangeway et al.* [2005]. The Poynting flux and electron precipitation parameters are easily obtained from the MFLFM. This approach has the advantage of reproducing the intensity and location of outflow, but it sacrifices knowledge about the acceleration process that would be gained from a full polar wind model. Another advantage of specifying the outflow is that it allows for easy construction of numerical experiments capable of isolating the impacts of outflows coming from regions such as the cusp and auroral zone.

In the work of *Wiltberger et al.* [2010] and *García et al.* [2010] the first approach of specifying the region for O^+ outflow was used along with observational evidence for ranges of O^+ flux. A broad range of fluxes have been observed, with *Lennartsson et al.* [2004] reporting fluxes of $10^7 cm^{-2} s^{-1}$ from Polar data and *Strangeway et al.* [2005] reporting fluxes in the range $10^6 - 10^9 cm^{-2} s^{-1}$ in FAST observations. It is important to note that while outflow is typically reported as a flux or fluence, to be used as a boundary condition for MHD simulation, this must be broken down into density, velocity, and temperature. Using observations from Akebono, Interball 2, and Cluster *Bouhram et al.* [2004] reported on the altitude profile of velocity and temperature of O^+ ions in the vicinity of the cusp. At $2.8R_E$ geocentric altitude they reported a parallel velocity ranging from 20 to 100 km/s. The temperature data was less noisy and they reported a typical value of 10 eV at $1.5R_E$ geocentric altitude. *Wiltberger et al.* [2010] identified the typical location of the cusp and specified the flux at values of 10^7 and $10^9 cm^{-2} s^{-1}$ by fixing the temperature to 10 eV and varying the density and outflow velocities to achieve the desired flux. This allowed them to examine the differences between fast and slow parallel velocities and the impacts to the magnetotail. *Garcìa et al.*, [2010] took a similar approach, except they constrained their outflow to be coming from the nightside auroral zone and held the outflow velocity fixed at 40 km/s and changed the density values to regulate the amount of flux coming from the auroral zone. An issue of concern here is the utilization of observations of outflow in the cusp to specify the parameters in the auroral zone.

Obviously, specifying a fixed region for outflow to occur is very limiting and will not work well for simulating real-world events. Using the observational correlations

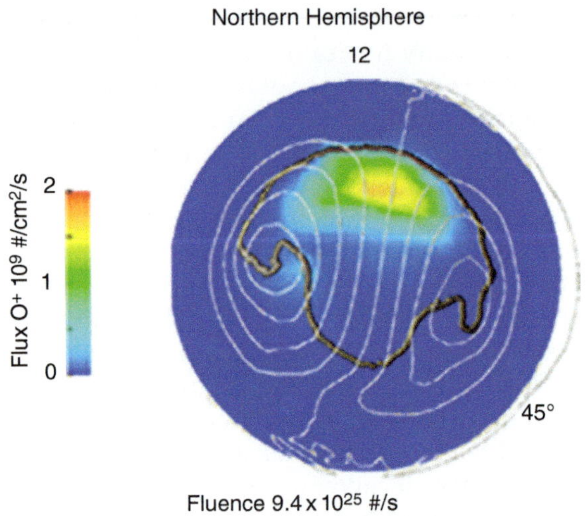

Figure 22.3 Flux of O^+ ions during southward IMF phase of CME passage resulting from the utilization of the MFLFM empirical outflow model. The color indicates the flux O^+, the white contour lines represent the cross-polar cap potential, and the thick gray line is the open/closed magnetic field boundary. Adapted from *Brambles et al.* [2010].

between outflow ion flux and both the DC Poynting flux and electron number flux reported by *Strangeway et al.* [2005] and *Zheng et al.* [2005], a causally regulated empirical model for outflow was developed by *Brambles et al.* [2010]. The model assumes a linear relationship between the DC Poynting flux and O^+ flux to specify the amount of outflow. Since the DC flux is relatively indiscriminate in determining the location, additional limiting masks based upon electron flux and velocity are used to limit the outflow to statistically relevant regions. It is because of this limitation that *Brambles et al.* [2011] chose to use the AC Poynting flux in the model instead of the DC Poynting flux in the advanced version of the empirical outflow model. In order to turn this flux into MHD boundary conditions, the model assumes a velocity and temperature with the density being chosen to make the flux dictated by the empirical outflow model. A range of velocities and temperatures have been used with the velocities ranging from 3 to 50 km/s and the temperatures ranging from 1 to 100 eV. Figure 22.3 shows the O^+ flux from a simulation of the CME passage that drove a modest geomagnetic storm on 31 August 2005. The region of outflow is clearly associated with the cusp with virtually no outflow coming from the nightside auroral zone. The outflow is polarward of the open/closed field line boundary and has a peak flux of a few $10^9 cm^{-2}s^{-1}$ over the course of the storm.

It is also worth noting here that the MFLFM does not specify a value for the plasma density at the inner boundary of the magnetospheric domain, unlike what was done in the Winglee and SWMF simulations. This means that only the source of plasma flowing into the computation grid will be that specified by either the empirical outflow model or cusp/auroral zone specifications.

22.4. IMPACTS

In this section we break from the previous methodology of describing the work of the groups separately and discuss the impact outflow has on the dynamics of geospace, including relevant result from the groups as needed. The first topic covers the impact of outflow on the strength of the cross-polar cap potential and the origin of the changes caused by outflow. The second topic addresses how outflow affects the dynamics of the magnetotail, including changes to the substorm process and the ring current.

22.4.1. Convection Dynamics

A key measure of the coupling between the solar wind and the magnetosphere-ionosphere system is the cross-polar cap potential. Simulations that have included ionospheric outflow have shown that this process affects the CPCP and the interaction between the solar wind and the magnetosphere. Each group has presented results in which the inclusion of outflow reduced the CPCP when compared with simulations that did not have outflow. *Winglee et al.* [2002] first reported this effect by showing a significant reduction in CPCP, shown in Figure 22.4, as the ratio of O^+ to H^+ at the inner boundary is increased. Since this ratio regulates the amount of O^+ that makes it into the magnetosphere, the higher the ratio, the more significant the reduction. Approximately 70 kV of reduction is seen between the simulation with limited O^+ outflow and one with strong O^+ outflow. The low-outflow simulation had a ratio of 0.05 and the high-outflow simulation had a ratio of 1.0. In the larger ratio run the O^+ flux is about 4 times larger than it was in the run with the smaller ratio. The reduction in CPCP brings the simulation results into better agreement with the values predicted by the assimilative mapping of ionospheric electrodynamics (AMIE) [*Richmond*, 1992] technique. The authors speculate that the cause for the reduction in the CPCP is the mass loading of the magnetosphere by the O^+ ions. The basic argument is that the increase in inertia in the inner magnetosphere caused by the ions is not compensated for by significant momentum transfer from these ions, and therefore the mass loading reduces the convection speed and results in a lower CPCP.

The MFLFM group reports seeing similar reductions in the CPCP when outflow is included in the simulations.

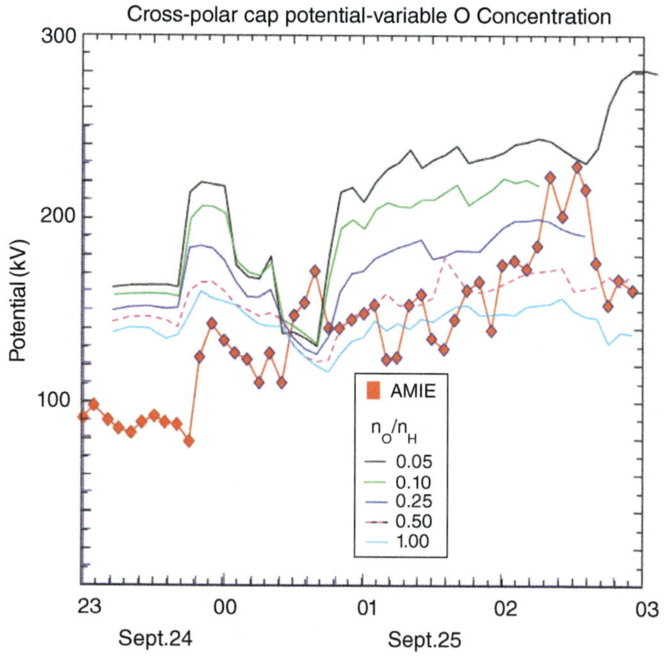

Figure 22.4 Inuence of increased O^+ outflow on the cross-polar cap potential. As the ratio of O^+ to H^+ density at the inner boundary is increased, the amount of O^+ outflowing into the magnetosphere increases and the cross-polar cap potential decreases. Adapted from *Winglee et al.* [2002].

Wiltberger et al. [2010], using the specified cusp outflow method, saw a reduction in the CPCP of roughly 50 kV in simulations with the strongest outflow when compared to single-fluid simulations without any outflow. The magnitude of the reduction of the CPCP was reduced to roughly 10 kV in simulations with the outflow flux two orders of magnitude smaller. In contrast to the Winglee results, the MFLFM includes an empirical model for determining the conductance, and the authors report seeing an increase in conductance driven by the increase in low-altitude plasma density near the inner boundary of the magnetospheric domain, which leads to higher electron fluxes and therefore larger conductances. This larger conductance contributes to a reduction in the region 1 FACs and therefore a lower CPCP. The authors comment that the Winglee mass loading effect may also be occurring and that a possible technique to determine the relative importance of these two factors would be to run MFLFM simulations with the empirical conductance model disabled. *García et al.* [2010] saw much smaller reductions in the CPCP for simulations with fixed outflow coming soley from the auroral zone.

Perhaps the most intriguing results coming from the MFLFM group with respect to the CPCP are those reported by *Brambles et al.* [2010] for a coronal mass ejection (CME) driven magnetic storm on 31 August 2005. In simulations for this event with modest outflow, they report seeing a slight increase in CPCP. In simulations conducted with stronger outflow, similar to those reported by *Winglee et al.* [2002] and *Wiltberger et al.* [2010], they report seeing a reduction in the CPCP. In both cases they report seeing a reduction in the region 1 FAC strengths. The explanation for the difference between the two outflow cases and the reason for the reduction in CPCP in the stronger outflow simulation rely on how the outflow alters the interaction of the magnetosphere with the solar wind. As discussed by *Lopez et al.* [2010], the interaction between the currents and magnetic fields in the magnetosheath can regulate the CPCP. A weaker region 1 FAC implies a weaker $\vec{J} \times \vec{B}$ force in the magnetosheath and a less effective diversion of the flow in the magnetosheath. It is expected that, with other factors being the same, the less effective flow diversion will result in an increase in flux delivered to the dayside and a higher CPCP as the result of either an increase in the reconnection line length or the reconnection rate at the magnetopause. This is what is observed in the simulation with the weaker outflow. In the second simulation, with the stronger outflow, the interaction is more complicated but still relies on understanding how the outflow alters the solar wind–magnetosphere interaction. In this case the stronger outflow leads to a dramatic increase in the strength of the ring current resulting in a blunter magnetosphere. The wider profile results in a bow shock location as much as 20% further from Earth than the baseline case. As discussed by *Merkine et al.* [2003], a blunter magnetosphere results in an enhanced flow diversion around the magnetopause and therefore a reduction of the CPCP. *Lopez et al.* [2010] also reports that this configuration will result in a shorter reconnection length and therefore a smaller CPCP. In the simulation with the stronger outflow the inflation effect is strong enough to overwhelm the reduction caused by the observed weaker region 1 FACs. The authors speculate that, while this simulation lacks any H^+ polar wind, the reduction in the CPCP due to storm time enhancements in the O^+ ring current is likely to be a nearly universal process.

Simulations with the SWMF including the PWOM also see a reduction in the cross-polar cap potential. *Glocer et al.* [2009a] report a reduction in the CPCP at all times when compared with storm simulations that did not include the outflow. Interestingly, while the simulations with outflow show dramatic improvements in the agreement with the D_{ST} index and other magnetospheric measurements, the inclusion of outflow actually reduces the agreement with the AMIE reconstruction of the CPCP. The authors note that, while the mass loading effect on the CPCP is likely to be occurring, they also believe that the alteration of the electrodynamic coupling through modifications of the auroral conductance

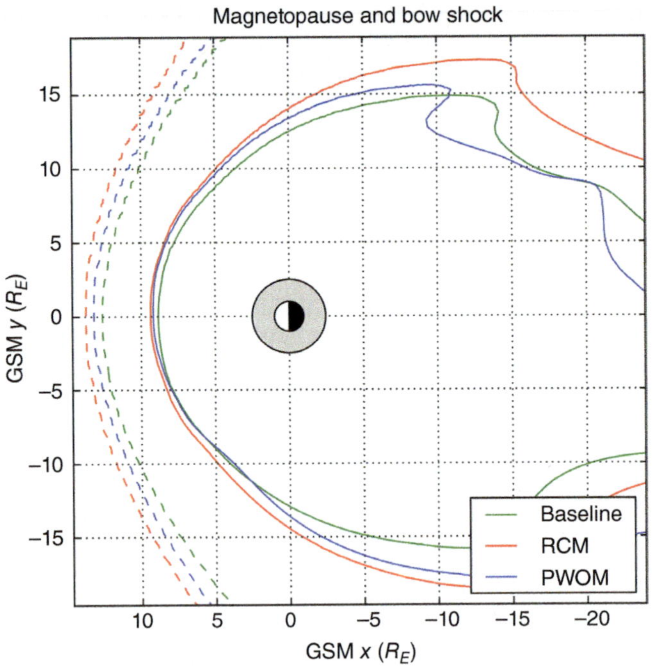

Figure 22.5 Location of the magnetopause (solid lines) and the bow shock (dashed lines) for three numerical experiments. The baseline simulation (green) has no outflow. The RCM simulation (red) includes the ring current model but no outflow. The PWOM simulation (blue) includes both the ring current model and outflow. Adapted from *Welling and Zaharia* [2012].

and improved FACs is contributing to the observed reduction of the CPCP in the simulation results. *Welling and Zaharia* [2012] used the SWMF construct in a series of numerical experiments to test the hypothesis proposed by the MFLFM group that the reduction in the CPCP is due to the increased ring current altering the shape of the magnetopause and therefore changing the solar wind–magnetosphere interaction as shown in Figure 22.5. In their experiment Welling and Zaharia conduct three runs, a baseline simulation with no outflow, a simulation without outflow but coupled to the RCM ring current model [*Toffoletto et al.*, 2003], and a final simulation with outflow from the PWOM module and the RCM model. All simulations were run for the same idealized solar wind conditions. The RCM simulation has the widest magnetopause and the furthest bow shock distance, with the PWOM having a slightly thinner magnetopause and correspondingly closer bow shock. Interestingly, the RCM simulation shows a reduction in the CPCP of nearly 7% while the PWOM simulation shows a reduction of 28%. The authors conclude that, while the inflation effect is occurring, it is not sufficient to fully account for the full reduction of the CPCP seen in their simulation results with outflow. They speculate that in addition to the mass loading effect the interaction of the cold plasma with the dayside reconnection site might be reducing the reconnection rate, resulting in a lower CPCP.

22.4.2. Magnetotail Dynamics

As previously mentioned, the presence of O^+ in the magnetotail is expected to affect the dynamic evolution of the magnetosphere. Global simulations with outflow present a unique opportunity to assess the importance of these effects since the presence, magnitude, and location of the outflow can be controlled. In this section we begin by examining results from each group's simulations of substorms and sawtooth events under conditions with both idealized solar wind conditions and those driven by actual observations. Next we report on the impact outflow has on the ring current in the coupled simulations.

We begin with the work of *Wiltberger et al.* [2010] which used the MFLFM with idealized solar wind conditions and outflow coming from the cusp. In these simulations the outflow is turned on as the IMF switches from northward (+5 nT) to southward (−5 nT) in a solar wind with constant velocity (400 km/s) and density (5 cm^3). Figure 22.6 compares the results of a simulation with and without outflow. In this case the outflow is coming from the cusp region at a velocity of 20 km/s with a total flux of 10^9cm^{-2}s^{-1} and a temperature of 10 eV. The baseline simulation produces a substorm with a plasmoid beginning to form at 0250 ST, as can be seen in the upper right panel of Figure 22.6. After the substorm completes its recovery phase, the simulation shows no further substorm activity with the nightside and dayside reconnection locations remaining relatively constant, indicative of a quasi-steady equilibrium. This behavior is fairly typical for LFM simulations with steady solar wind conditions and is thought to be similar to a steady magnetospheric convection (SMC) event [*McPherron et al.*, 2005] in the magnetosphere. In the simulation with outflow the evolution of the first substorm is not affected significantly due to the fact that the O^+ coming from the cusp has not had time to reach the magnetotail. The subsequent evolution of the magnetotail is dramatically different, as can be seen in the lower right-hand panel of Figure 22.6. A second plasmoid is clearly forming at 0445 ST indicating that this simulation is producing a second substorm event. The authors suspect that this is caused by the presence of O^+ ions altering the reconnection rate in the midtail region. This alteration prevents the simulation from reaching equilibrium, resulting in the onset of a second substorm cycle. After this cycle completes, the simulation does not produce any further substorms and falls into a SMC state with the reconnection line shifted tailward, reflecting a new balance point for the reconnection rate that has been altered by the O^+ ions.

The onset of the second substorm in the MFLFM simulations turns out to be fairly sensitive to the parameters chosen for the outflow flux. Increasing the outflow speed to 50 km/s while keeping the total flux the same results, in

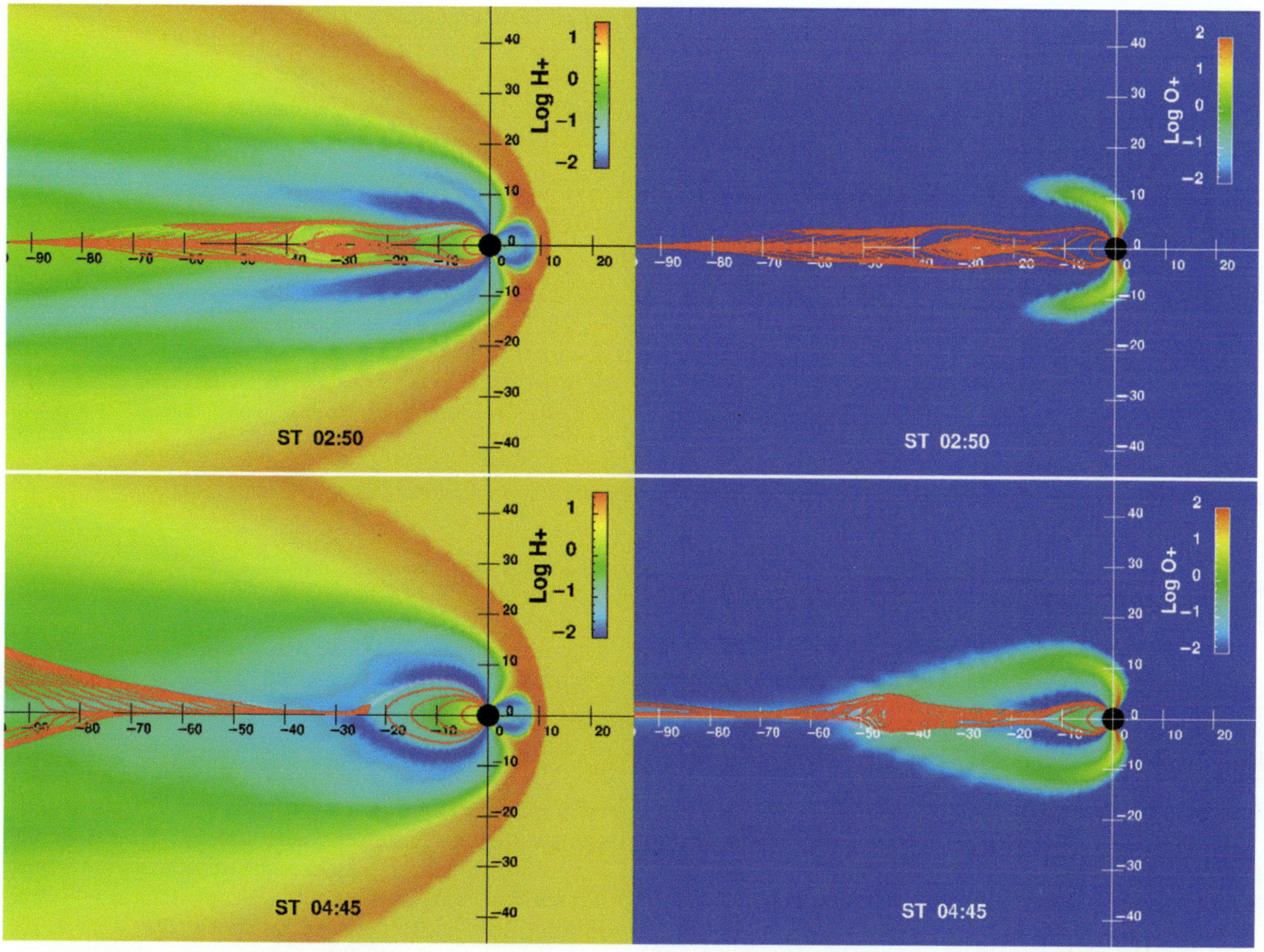

Figure 22.6 Comparison of the magnetospheric configuration for a run without (left) and with (right) O^+ outflow. The panels on the left show the H^+ density while those on the right show the O^+ density. In all panels red magnetic field lines are drawn from fixed locations along the *x* axis. The top panels show the configuration during the first substorm interval. In the lower panels a second substorm is only seen in the simulation with outflow. Adapted from *Wiltberger et al.* [2010].

outflow that flows well past the X-line location established after the first substorm onset. As a result, the O^+ ions never interact with the X line and the simulation goes into an SMC state. Reducing the flux resulted in O^+ ions landing much further earthward in the magnetotail. The O^+ density level never reaches magnitudes needed to dramatically alter the shape of the magnetotail so this simulation also entered a SMC state.

Brambles et al. [2011] used the MFLFM with the empirically regulated outflow model to more fully examine the evolution of the magnetotail with O^+ outflow from the ionosphere. In addition to the previously mentioned substorm and SMC states, the magnetosphere is also known to enter a state with quasi-periodic substorms commonly referred to as sawtooth intervals [*Henderson et al.*, 2006; *Cai and Clauer*, 2009]. It is not known what causes the magnetosphere to enter into a sawtooth or SMC state so *Brambles et al.* [2011] conducted a study to determine what role O^+ ions may play in the onset of sawtooth intervals. One of the characteristic observational features of sawtooth events is the periodic dipolarizations of the magnetic field seen in geostationary satellite observations. Figure 22.7 shows the magnetic field inclination angle extracted from the simulation near geosynchronous orbit. The baseline simulation without any outflow does not show any significant oscillations, while the simulation with stronger solar wind velocity and causally regulated O^+ outflow clearly shows multiple dipolarizations associated with the onsets of multiple substorm events. The addition of the causally regulated location and intensity of outflow has altered the dynamics of MI coupling resulting in the sawtooth

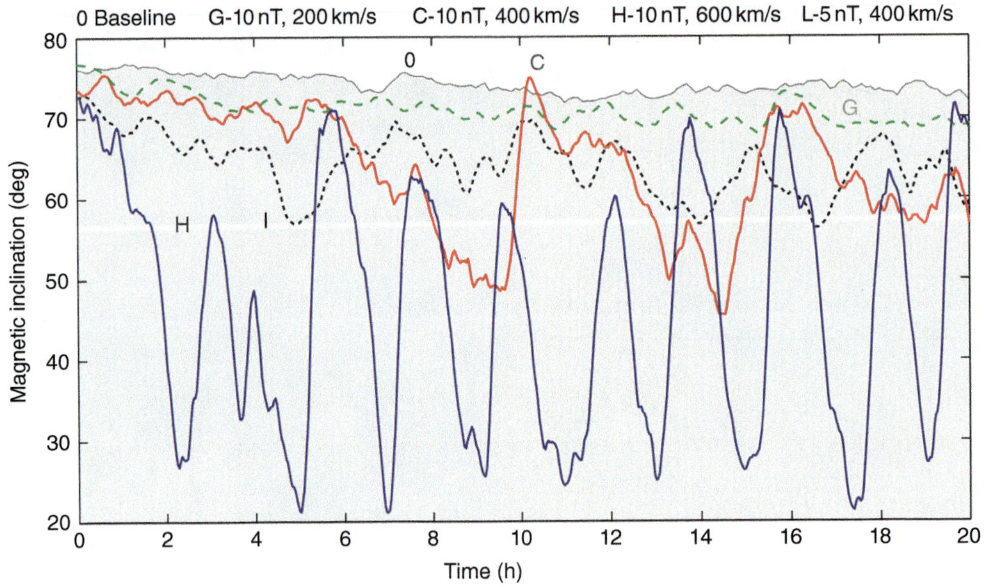

Figure 22.7 Magnetic inclination angle near geosynchronous orbit. Line colors reflect simulation runs with differing O^+ outflow intensities and solar wind conditions. The baseline simulation (0, grey) contains no outflow and has no periodic dipolarizations. The simulation which contains outflow and strong driving (H, blue) has the periodic dipolarizations associated with onsets in the sawtooth interval. From *Brambles et al.* [2011].

configuration that was not seen in the simple cusp outflow model used by *Wiltberger et al.* [2010].

Brambles et al. [2011] conducted a wide parameter study to investigate the role of outflow intensity and solar wind conditions on the sawtooth intervals. In case C, seen as the red line in Figure 22.7, they reduced both the intensity of the outflow and the solar wind velocity, resulting in a sawtooth state with a period between events much longer than seen in case H. A further reduction in the solar wind velocity and outflow intensity results in the simulated magnetosphere once again entering a SMC state, as seen in case G, illustrated by the green line in Figure 22.7. The authors conducted a series of runs with the same solar wind driving conditions but altered the intensity of the outflow. They found that the sawtooth period was a strong, nearly exponential, function of total fluence of O^+ ions. The authors conclude that, while they have shown that ionospheric outflow can produce sawtooth events in MFLFM simulations, they do not know if it is the only or most important mechanism driving the magnetosphere into this state. *Ouellette et al.* [2013] examined the dynamics of one of the sawtooth intervals from the *Brambles et al.* [2011] work in detail and found that there is a feedback loop in the system whereby the O^+ ions released during one expansion phase end up triggering the next substorm.

Yu and Ridley [2013] conducted simulations with the multifluid version of the SWMF, but instead of using the PWOM, they defined O^+ outflow to be coming from either the cusp or nightside auroral zones with a velocity of 50 km/s and total flux of $10^8 cm^{-2}s^{-1}$. This flux is an order-of-magnitude lower than the *Wiltberger et al.* [2010] study. The IMF is held constant at -5 nT with the solar wind velocity and density held constant at 400 km/s and $5 cm^3$, respectively. These values are the same as the *Wiltberger et al.* [2010] study with the temporal evolution being different. The authors conclude that cusp ions landing near the reconnection site in the magnetotail dramatically disrupt the magnetotail while those from the auroral zone appear to stabilize the magnetotail. In the simulations with cusp outflow a substorm not present in the simulation with only auroral outflow is observed. The authors further note that what is important is not only the location of ions, which determines their role in altering magnetotail dynamics, but also the ratio of parallel to perpendicular velocity at the inner boundary, which determines where the ions land in the magnetotail, and it is that location that has the strongest influence on the magnetotail dynamics.

García et al. [2010] used the MFLFM to examine the role of O^+ ions originating from the nightside aural oval. They used the idealized solar wind conditions with a velocity of 400 km/s and density of $5 cm^3$ with a IMF of 5 nT southward. They hold the O^+ outflow velocity fixed at 40 km/s and vary the density from 10 to $100 cm^3$ in the auroral zone to obtain fluxes that range from 4×10^7 to $4 \times 10^8 cm^{-2}s^{-1}$. In order to reduce the complications caused by the interaction of the empirical precipitation

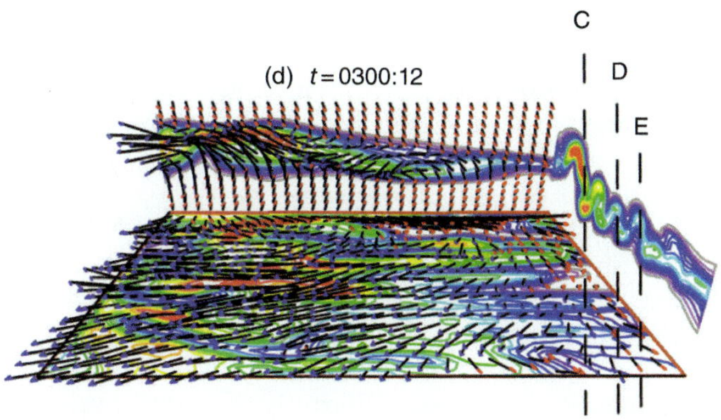

Figure 22.8 Flows and O^+ pressure contours for a substorm occurring in simulation with idealized solar wind conditions. The frame shown is shortly after the onset of the substorm. The kinks seen in the cross-tail direction (marked with C, D, E) are on the scale of the O^+ gyroradius. Adapted from *Winglee et al.* [2009a].

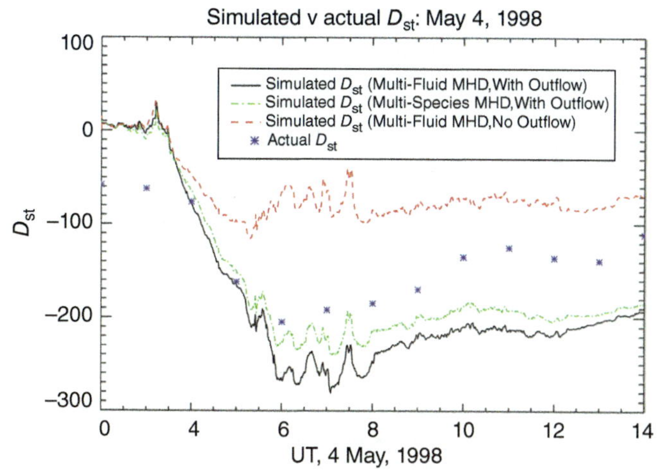

Figure 22.9 The D_{ST} index for the 4 May 1998 storm event from observations (blue stars) and a baseline simulation without outflow (red dashed line) and a multifluid simulation with outflow from the PWOM model (black solid line). Adapted from [*Glocer et al.*, 2009b].

model, the authors chose to conduct these studies with a constant ionospheric conductance. Not surprisingly, the total mass of the plasma sheet increases as the flux of O^+ ions from the auroral zone increases. Interestingly, both the width of the magnetotail and the distance to the nightside X line increase with larger auroral flux values. In agreement with the work of *Yu and Ridley* [2013], the authors only see significant alteration of the magnetotail dynamics when the magnitude of the flux is significantly above the statistical observations.

Winglee et al. [2009a] and *Harnett et al.* [2010] used the Winglee multifluid simulation to examine the role of O^+ ion in the substorm process. *Winglee et al.* [2009a] used idealized solar wind conditions with IMF due southward at –5 nT. In a slight modification to the previous techniques the ratio of O^+ to H^+ density is allowed to vary over the polar cap with a value of 5% being used in the auroral regions and ramping it down to zero at the poles and equator. Onset of the substorm in this simulation occurs at 0258 ST, and Figure 22.8 shows the O^+ pressure as an earthward moving flux rope enters the near-Earth region. We see that the O^+ plasma pressure in the plasma sheet is greater than 0.25 nPa in regions around the flux rope. The authors suggest that, while the O^+ does not initiate the substorm, the flux rope passage energizes the O^+ ions more than the H^+ due to their larger gyroradius, allowing them to play a major role in the substorm injection process and in the generation of the partial ring current. *Harnett et al.* [2010] simulated a substorm that occurred on 26 February, 2008 that was well observed by the THEMIS satellite constellation. In the simulation the heavy ions arrive in the magnetotail approximately 10 min prior to the substorm onset, and these ions contributed to enhanced reconnection and fast earthward flows. In agreement with the idealized simulations they further conclude that the substorm is internally triggered. This conclusion disagrees with the one reached by *Angelopoulos et al.* [2008] that onset was initiated by tail reconnection.

One final but important effect of outflow on the magnetotail is its impact on the ring current. All of the groups report that the O^+ outflow results in significantly enhanced ring current strengths. Figure 22.9 [*Glocer et al.*, 2009b] shows a comparison between the observed D_{ST} index and the simulations with and without outflow from the PWOM. In the case without outflow the simulation underestimates the D_{ST} index by approximately a factor of 2 during the course of the magnetic storm. Including outflow in the multifluid simulation results in slight overprediction of the magnitude of the D_{ST} index. The increase in the simulated D_{ST} index by *Brambles et al.* [2010] was not as strong, but it did improve agreement with the observed values. *Winglee et al.* [2009a] did not present any comparisons with the D_{ST} index in their results but did state that O^+ ions will contribute to an enhanced ring current.

22.5. SUMMARY AND FUTURE DIRECTIONS

Several groups have developed multifluid global-scale simulations of Earth's magnetosphere. These simulations all start from the same physics of idealized MHD for multiple ion species, although they have different formulations of these equations. The Winglee group includes the effects of gravity in their formulation and the SWMF and MFLFM groups do not typically include these terms. In contrast, while all the groups have the capability to solve for the electron pressure, the SWMF and MFLFM groups typically set the electron pressure to a fraction of the ion pressure. The formulation by the SWMF group explicitly includes source terms which could be used to include processes such as charge exchange, but these source terms are typically set to zero. The computational grids and numerical techniques used to solve the multifluid equations differ dramatically between the groups.

A major source of difference between the groups is the approaches used to include ionospheric outflow into the magnetospheric domain. The Winglee group defines the density of H^+ and O^+ at the inner boundary of the magnetospheric domain and alters the gravitational constant so that the O^+ outflow is limited relative to the H^+ outflow. Another key factor controlling outflow in their simulations is the ratio of O^+ to H^+ at the inner boundary. The main method for including ionospheric outflow in the SWMF group is through the utilization of the PWOM. This model uses a set of noninteracting field line simulations of the polar wind distributed throughout the polar cap to simulate the outflow of ions, including H^+ and O^+ , into the magnetosphere. Each field line is a full gyrotropic simulation of the ions, including the effects of heat conductivity, gravity, and the ambipolar electric field. In addition to the PWOM it is possible to specify the inflow conditions of ions coming from regions such as the cusp and auroral zone. The main method for including ionospheric outflow in the MFLFM simulations is through the use of an empirical model. This model is based on using the observed correlations between Poynting flux and electron precipitation number flux with O^+ outflow flux to specify both the intensity and location of the outflow. This group also makes extensive use of the outflow parameter in regions such as the cusp and auroral zone as part of their investigations.

One of the common impacts seen across the simulations was a reduction in the cross-polar cap potential. The Winglee group attributes this reduction to mass loading of the magnetosphere resulting in weaker convection. All groups agree that this occurrs, but the MFLFM and SWMF groups have additional contributions to the reduction of the CPCP. The MFLFM group proposed that the increase in the ring current alters the shape of the magnetopause, altering the interaction of the magnetosphere with the solar wind. The SWMF group tested this hypothesis with their simulations and found that it is occurring, but not to a degree sufficient to explain the full reduction in the CPCP. Alteration of the conductance patterns due to the interaction of the outflow with enhancements of electron precipitation through their empirical model was suggested by the MFLFM group as another possible origin of the reduction in the CPCP.

All the groups reported increased ring currents in their simulations with outflow included. The SWMF and MFLFM groups show increases in simulated D_{ST} indexes and bring the results into greater agreement with the observations for simulated events.

The inclusion of outflow can significantly alter the dynamics of the magnetotail. The SWMF and MFLFM groups showed that the inclusion of prescribed outflow from the cusp region can result in new substorm activity that was not present in simulations without outflow. The impact of the cusp outflow depends heavily on the location, intensity, and very significantly the parallel velocity used to specify the outflow coming from the cusp. The MFLFM group also found that, when using their empirical model for outflow, the simulation produces sawtooth events for steady solar wind conditions. Under the same conditions without the outflow the simulation went into a SMC state. They also found that the period of the sawtooth was heavily dependent upon the fluence of the outflow coming from the ionosphere.

In the opinion of this author there are three major directions for future development in coupled simulations of magnetosphere-ionosphere-thermosphere system. Both the SWMF and MFLFM groups can couple their codes with 3D simulations of the ionosphere-thermosphere system. These simulations will provide an important and dynamic lower boundary condition to the outflow studies. Likely questions to address include how the outflow is altered when high-density patches are formed in the ionosphere and what are the effects of features such as sub auroral polarization stream (SAPS) on the outflow. The inclusion of outflow from the ionosphere into the magnetosphere is the first step in being able to include a fully dynamic plasmasphere model into the coupled models of geospace. The plasmasphere will involve flow in both directions across the typical lower boundary of the magnetospheric domain. Inclusion of the plasmasphere is essential for getting the density in the inner magnetosphere correct, which is in turn needed to get the correct ULF wave speeds essential to radiation belt modeling. The final area for improvement will be in continued advances in the polar wind outflow methods, be they methods for including transverse ion heating in the PWOM or more realistic empirical models in the MFLFM.

In conclusion, it is clear that fully coupled global-scale simulations that include both electrodynamic coupling and outflow are proving to be excellent tools for addressing fundamental questions about the interaction of the solar wind with the magnetosphere-ionosphere-thermosphere system. Another obvious future challenge will be to extend these modeling capabilities to the challenges presented by the multiple sources present in the giant magnetospheres throughout the system. Key among these challenges will be dealing with the large H^+ sources of planetary bodies and the numerous ions that can come from moons in the systems.

ACKNOWLEDGMENTS

This material is based upon work supported by CISM funded by the STC Program of the National Science Foundation under agreement number ATM-0120950 and by NASA grants NNH12AU10I and NNH11AR82I. The National Center for AtmosphericResearch is sponsored by the National Science Foundation. The author wishes to thank Roger Varney for the many helpful conversations that took place during the preparation of this chapter.

REFERENCES

Abe, T., B. A. Whalen, A. W. Yau, S. Watanabe, E. Sagawa, and K. I. Oyama (1993), Altitude profile of the polar wind velocity and its relationship to ionospheric conditions, *Geophys. Res. Lett.*, *20*, 2825, doi:10.1029/93GL02837.

Abe, T., S. Watanabe, B. A. Whalen, A. W. Yau, and E. Sagawa (1996), Observations of polar wind and thermal ion outflow by Akebono/SMS, *J. Geomag. Geoelec.*, *48*(3), 319–325.

Angelopoulos, V., et al. (2008), Tail reconnection triggering substorm onset, *Science*, *321*, 931–931, doi:10.1126/science.1160495.

Baker, D. N., E. W. Hones, Jr., D. T. Young, and J. Birn (1982), The possible role of ionospheric oxygen in the initiation and development of plasma sheet instabilities, *Geophys. Res. Lett.*, *9*(12), 1337, doi:10.1029/GL009i012p01337.

Borovsky, J. E., B. Lavraud, and M. M. Kuznetsova (2009), Polar cap potential saturation, dayside reconnection, and changes to the magnetosphere, *J. Geophys. Res.*, *114*, 03224, doi:10.1029/2009JA014058.

Bouhram, M., B. Klecker, W. Miyake, H. Rème, J. Sauvaud, M. Malingre, L. Kistler, and A. Blagau (2004), On the altitude dependence of transversely heated O distributions in the cusp/cleft, *Ann. Geophys.*, *22*, 1787.

Brambles, O. J., W. Lotko, P. A. Damiano, B. Zhang, M. Wiltberger, and J. Lyon (2010), Effects of causally driven cusp O^+ outflow on the storm time magnetosphere-ionosphere system using a multifluid global simulation, *J. Geophys. Res.*, *115*(A), doi:10.1029/2010JA015469.

Brambles, O. J., W. Lotko, B. Zhang, M. Wiltberger, J. Lyon, and R. J. Strangeway (2011), Magnetosphere sawtooth oscillations induced by ionospheric outflow, *Science*, *332*(6034), 1183–1186, doi:10.1126/science.1202869.

Burns, A. G., W. Wang, M. Wiltberger, S. C. Solomon, H. Spence, T. L. Killeen, R. E. Lopez, and J. E. Landivar (2008), An event study to provide validation of TING and CMIT geomagnetic middle-latitude electron densities at the F2 peak, *J. Geophys. Res.*, *113*, 05310, doi:10.1029/2007JA012931.

Cai, X., and C. R. Clauer (2009), Investigation of the period of sawtooth events, *J. Geophys. Res.*, *114*, A06201, doi:10.1029/2008JA013764.

Chappell, C. R., T. E. Moore, and J. H. J. Waite (1987), The ionosphere as a fully adequate source of plasma for the earth's magnetosphere, *J. Geophys. Res.*, *92*, 5896–5910.

Cladis, J. B. (1986), Parallel acceleration and transport of ions from polar ionosphere to plasma sheet, *Geophys. Res. Lett.*, *13*, 893–896, doi:10.1029/GL013i009p00893.

Cladis, J. B., and W. E. Francis (1992), Distribution in magnetotail of O(+) ions from cusp/cleft ionosphere: A possible substorm trigger, *J. Geophys. Res.*, *97*, 123–130, doi:10.1029/91JA02376.

Cully, C. M., E. F. Donovan, A. W. Yau, and G. G. Arkos (2003), Akebono/suprathermal mass spectrometer observations of low-energy ion outflow: Dependence on magnetic activity and solar wind conditions, *J. Geophys. Res.*, *108*(A), 1093, doi:10.1029/2001JA009200.

Daglis, L. A., S. Livi, E. T. Sarris, and B. Wilken (1994), Energy density of ionospheric and solar wind origin ions in the near-Earth magnetotail during substorms, *J. Geophys. Res.*, *99*, 5691, doi:10.1029/93JA02772.

Daum, P., J. A. Wild, T. Penz, E. E. Woodfield, H. Rème, A. N. Fazakerley, P. W. Daly, and M. Lester (2008), Global MHD simulation of flux transfer events at the high-latitude magnetopause observed by the Cluster spacecraft and the SuperDARN radar system, *J. Geophys. Res.*, *113*, doi:10.1029/2007JA012749.

De Zeeuw, D. L., S. Sazykin, R. A. Wolf, T. I. Gombosi, A. J. Ridley, and G. Tóth (2004), Coupling of a global MHD code and an inner magnetospheric model: Initial results, *J. Geophys. Res.*, *109*(A), 12,219, doi:10.1029/2003JA010366.

Demars, H. G., A. R. Barakat, and R. W. Schunk (1996), Effect of centrifugal acceleration on the polar wind, *J. Geophys. Res.*, *101*(A), 24,565–24,572, doi:10.1029/96JA02234.

Dorelli, J. C. (2004), A new look at driven magnetic reconnection at the terrestrial subsolar magnetopause, *J. Geophys. Res.*, *109*(A12), doi:10.1029/2004JA010458.

Ganguli, S. B. (1996), The polar wind, *Rev. Geophys.*, *34*(3), 311, doi:10.1029/96RG00497.

García, K. S., V. G. Merkin, and W. J. Hughes (2010), Effects of nightside O^+ outflow on magnetospheric dynamics: Results of multifluid MHD modeling, *J. Geophys. Res.*, *115*, doi:10.1029/2010JA015730.

Glocer, A., T. I. Gombosi, G. Tóth, K. C. Hansen, A. J. Ridley, and A. Nagy (2007), Polar wind outflow model: Saturn results, *J. Geophys. Res.*, *112*(A1), doi:10.1029/2006JA011755.

Glocer, A., G. Tóth, T. Gombosi, and D. Welling (2009a), Modeling ionospheric outflows and their impact on the magnetosphere, initial results, *J. Geophys. Res.*, *114*(A), 05216, doi:10.1029/2009JA014053.

Glocer, A., G. Tóth, Y. Ma, T. Gombosi, J.-C. Zhang, and L. M. Kistler (2009b), Multifluid block-adaptive-tree solar wind Roe-type upwind scheme: Magnetospheric composition and dynamics during geomagnetic storms—Initial results, *J. Geophys. Res.*, *114*(A), 12,203, doi:10.1029/2009JA014418.

Gombosi, T. I., and A. F. Nagy (1989), Time-dependent modeling of field-aligned current-generated ion transients in the polar wind, *J. Geophys. Res.*, *94*(A1), 359–369, doi:10.1029/JA094iA01p00359.

Gombosi, T. I., T. E. Cravens, and A. F. Nagy (1985), A time-dependent theoretical model of the polar wind: Preliminary results, *Geophys. Res. Lett.*, *12*(4), 167–170, doi:10.1029/GL012i004p00167.

Goodrich, C. C., A. L. Sussman, J. G. Lyon, M. A. Shay, and P. A. Cassak (2004), The CISM code coupling strategy, *J. Atmos. Solar Terr. Phys.*, *66*(1), 1469–1479, doi:10.1016/j.jastp.2004.04.010.

Hain, K. (1987), The partial donor method, *J. Comp. Phys.*, *73*, 131.

Harnett, E. M., R. M. Winglee, A. Stickle, and G. Lu (2008), Prompt ionospheric/magnetospheric responses 29 October 2003 Halloween storm: Outflow and energization, *J. Geophys. Res.*, *113*, A06209, doi:10.1029/2007JA012810.

Harnett, E. M., R. M. Winglee, and T. Lerud (2010), Multiscale-multifluid simulations of the 26 February 2008 substorm: Evidence for internal triggering of a substorm, *J. Geophys. Res.*, *115*, A12238, doi:10.1029/2010JA015672.

Henderson, M. G., et al. (2006), Substorms during the 10-11 August 2000 sawtooth event, *J. Geophys. Res.*, *111*, 06206, doi:10.1029/2005JA011366.

Hirsch, C. (1988), *Numerical Computation of Internal and External Flows*, vol. II, Wiley, New York.

Horwitz, J. L. (1986), The tail lobe ion spectrometer, *J. Geophys. Res.*, *91*, 5689, doi:10.1029/JA091iA05p05689.

Khazanov, G. V., M. W. Liemohn, and T. E. Moore (1997), Photoelectron effects on the self-consistent potential in the collisionless polar wind, *J. Geophys. Res.*, *102*(A), 7509–7522, doi:10.1029/96JA03343.

Kistler, L. M., et al. (2005), Contribution of nonadiabatic ions to the cross-tail current in an O+ dominated thin current sheet, *J. Geophys. Res.*, *110*(A), 06213, doi:10.1029/2004JA010653.

Kivelson, M. G., and A. J. Ridley (2008), Saturation of the polar cap potential: Inference from Alfvén wing arguments, *J. Geophys. Res.*, *113*(A), 05214, doi:10.1029/2007JA012302.

Lakhina, G. S. (1995), Excitation of plasma sheet instabilities by ionospheric O+ ions, *Geophys. Res. Lett.*, *22*(2), 3453–3456, doi:10.1029/95GL03427.

Lei, J., A. G. Burns, T. Tsugawa, W. Wang, S. C. Solomon, and M. Wiltberger (2008a), Observations and simulations of quasiperiodic ionospheric oscillations and large-scale traveling ionospheric disturbances during the December 2006 geomagnetic storm, *J. Geophys. Res.*, *113*(A), 06310, doi:10.1029/2008JA013090.

Lei, J., J. P. Thayer, J. M. Forbes, E. K. Sutton, R. S. Nerem, M. Temmer, and A. M. Veronig (2008b), Global thermospheric density variations caused by high-speed solar wind streams during the declining phase of solar cycle 23, *J. Geophys. Res.*, *113*(A11), A11303, doi:10.1029/2008JA013433.

Lemaire, J., and M. Scherer (1978), Field aligned distribution of plasma mantle and ionospheric plasmas, *Eur. Geophy. Soc.*, *40*, 337–342.

Lennartsson, O. W., H. L. Collin, and W. K. Peterson (2004), Solar wind control of Earth's H+ and O+ outflow rates in the 15-eV to 33-keV energy range, *J. Geophys. Res.*, *109*(A), 12,212, doi:10.1029/2004JA010690.

Liemohn, M. W., and M. Jazowski (2008), Ring current simulations of the 90 intense storms during solar cycle 23, *J. Geophys. Res.*, *113*(A), doi:10.1029/2008JA013466.

Liu, Y., L. M. Kistler, C. G. Mouikis, B. Klecker, and I. Dandouras (2013), Heavy ion effects on substorm loading and unloading in the Earth's magnetotail, *J. Geophys. Res.*, *118*(5), 2101–2112, doi:10.1002/jgra.50240.

Lopez, R. E., R. Bruntz, E. J. Mitchell, M. Wiltberger, J. G. Lyon, and V. G. Merkin (2010), Role of magnetosheath force balance in regulating the dayside reconnection potential, *J. Geophys. Res.*, *115*(A), 12,216, doi:10.1029/2009JA014597.

Lyon, J. G., J. A. Fedder, and C. M. Mobarry (2004), The Lyon-Fedder-Mobarry (LFM) global MHD magnetospheric simulation code, *J. Atmos. Solar Terr. Phys.*, *66*, 1333, doi:10.1016/j.jastp.2004.03.020.

McFadden, J. P., Y.-K. Tung, C. W. Carlson, R. J. Strangeway, E. Moebius, and L. M. Kistler (2001), FAST observations of ion outflow associated with magnetic storms, in *Space Weather*, edited by P. Song et al., pp. 413–421, AGU, Washington, D. C.

McPherron, R. L., T. P. O'Brien, and S. M. Thompson (2005), Solar wind drivers for steady magnetospheric convection, in *Multiscale Coupling of Sun-Earth Processes*, vol. 1, edited by A. T. Y. Lui, Y. Kamide, and G. Consolini, p. 113, Elsevier B. V., Amsterdam, The Netherlands.

Merkin, V. G., and J. G. Lyon (2010), Effects of the low-latitude ionospheric boundary condition on the global magnetosphere, *J. Geophys. Res.*, *115*(A), 10,202, doi:10.1029/2010JA015461.

Merkine, V. G., K. Papadopoulos, G. Milikh, A. S. Sharma, X. Shao, J. Lyon, and C. Goodrich (2003), Effects of the solar wind electric field and ionospheric conductance on the cross polar cap potential: Results of global MHD modeling, *Geophys. Res. Lett.*, *30*(2), 2180, doi:10.1029/2003GL017903.

Moore, T. E. (1991), Origins of magnetospheric plasma, *Rev. Geophys.*, *29*, 1039–1048.

Moore, T. E., and D. C. Delcourt (1995), The geopause, *Rev. Geophys.*, *33*(2), 175–209.

Nosé, M., S. Ohtani, A. T. Y. lui, S. P. Christon, R. W. McEntire, D. J. Williams, T. Mukai, Y. Saito, and K. Yumoto (2000), Change of energetic ion composition in the plasma sheet during substorms, *J. Geophys. Res.*, *105*, 23,277, doi:10.1029/2000JA000129.

Nosé, M., S. Taguchi, K. Hosokawa, S. P. Christon, R. W. McEntire, T. E. Moore, and M. R. Collier (2005), Overwhelming O+ contribution to the plasma sheet energy density during the October 2003 superstorm: Geotail/EPIC and IMAGE/LENA observations, *J. Geophys. Res.*, *110*(A), doi:10.1029/2004JA010930.

Ouellette, J. E., O. J. Brambles, and J. G. Lyon (2013), Properties of outflow-driven sawtooth substorms, *J. Geophys. Res.*, *118*, 3223–3232, doi:10.1002/jgra.20309.

Parks, G. K. (1991), Physics of space plasmas - an introduction, Addison-Wesley Publishing Co., Redwood City, CA.

Pembroke, A., F. Toffoletto, S. Sazykin, M. Wiltberger, J. Lyon, V. Merkin, and P. Schmitt (2012), Initial results from a dynamic coupled magnetosphere-ionosphere-ring current model, *J. Geophys. Res.*, *117*(A2), doi:10.1029/2011JA016979.

Peterson, W. K., R. D. Sharp, E. G. Shelley, R. G. Johnson, and H. Balsiger (1981), Energetic ion composition of the plasma sheet, *J. Geophys. Res.*, *86*, 761–767, doi:10.1029/JA086iA02p00761.

Raeder, J. (2003), Global magnetohydrodynamics: A tutorial, in *Space Plasma Simulation*, edited by J. Büchner et al., Springer Verlag, Berlin, Heidelberg, and New York.

Raeder, J. (2006), Flux transfer events: 1. Generation mechanism for strong southward IMF, *Ann. Geophys.*, *24*(1), 381–392, doi:10.5194/angeo-24-381-2006.

Raeder, J., Y. Wang, and T. J. Fuller-Rowell (2001), Geomagnetic storm simulation with a coupled magnetosphere-ionosphere-thermosphere Model, in *Space Weather*, edited by H. Singer, pp. 377–384, AGU, Washington, D. C.

Raeder, J., D. Larson, W. Li, E. L. Kepko, and T. Fuller-Rowell (2008), OpenGGCM simulations for the THEMIS mission, *Space Sci Rev.*, *141*, 535, doi:10.1007/s11214-008-9421-5.

Raeder, J., P. Zhu, Y. Ge, and G. Siscoe (2010), Open Geospace General Circulation Model simulation of a substorm: Axial tail instability and ballooning mode preceding substorm onset, *J. Geophys. Res.*, *115*, A00I16, doi:10.1029/2010JA015876.

Richmond, A. D. (1992), Assimilative mapping of ionospheric electrodynamics, *Adv. Space Res.*, *12*(6), 59–68, doi:10.1016/0273-1177(92)90040-5.

Ridley, A., T. Gombosi, and D. DeZeeuw (2004), Ionospheric control of the magnetosphere: Conductance, *Ann. Geophys.*, *22*(2), 567–584, doi:10.5194/angeo-22-567-2004.

Ridley, A. J., A. D. Richmond, T. I. Gombosi, D. L. De Zeeuw, and C. R. Clauer (2003), Ionospheric control of the magnetospheric configuration: Thermospheric neutral winds, *J. Geophys. Res.*, *108*(A), 1328, doi:10.1029/2002JA009464.

Ridley, A. J., Y. Deng, and G. Tóth (2006), The global ionosphere-thermosphere model, *J. Atmos. Solar Terr. Phys.*, *68*(8), 839–864, doi:10.1016/j.jastp.2006.01.008.

Ridley, A. J., T. I. Gombosi, I. V. Sokolov, G. Tóth, and D. T. Welling (2010), Numerical considerations in simulating the global magnetosphere, *Ann. Geophys.*, *28*(8), 1589–1614, doi:10.5194/angeo-28-1589-2010.

Samsonov, A. A., D. G. Sibeck, N. V. Zolotova, H. K. Biernat, S.-H. Chen, L. Rastaetter, H. J. Singer, and W. Baumjohann (2011), Propagation of a sudden impulse through the magnetosphere initiating magnetospheric Pc5 pulsations, *J. Geophys. Res.*, *116*(A), 10,216, doi:10.1029/2011JA016706.

Schunk, R. W. (2007), Time-dependent simulations of the global polar wind, *J. Atmos. Solar Terr. Phys.*, *69*(1), 2028–2047, doi:10.1016/j.jastp.2007.08.009.

Sharp, R. D., W. Lennartsson, and R. J. Strangeway (1985), The ionospheric contribution to the plasma environment in near-earth space, *Radio Sci.*, *20*, 456–462, doi:10.1029/RS020i003p00456.

Shay, M. A., and M. Swisdak (2004), Three-species collisionless reconnection: Effect of O+ on magnetotail reconnection, *Phys. Rev. Lett.*, *93*(1), 175,001, doi:10.1103/PhysRevLett.93.175001.

Shelley, E. G., R. G. Johnson, and R. D. Sharp (1972), Satellite observations of energetic heavy ions during a geomagnetic storm, *J. Geophys. Res.*, *77*, 6104–6110, doi:10.1029/JA077i031p06104.

Strangeway, R. J., R. E. Ergun, Y.-J. Su, C. W. Carlson, and R. C. Elphic (2005), Factors controlling ionospheric outflows as observed at intermediate altitudes, *J. Geophys. Res.*, *110*(A), 03221, doi:10.1029/2004JA010829.

Taktakishvili, A., M. M. Kuznetsova, M. Hesse, M.-C. Fok, L. Rastätter, M. Maddox, A. Chulaki, T. I. Gombosi, and D. L. De Zeeuw (2007), Buildup of the ring current during periodic loading-unloading cycles in the magnetotail driven by steady southward interplanetary magnetic field, *J. Geophys. Res.*, *112*, 09203, doi:10.1029/2007JA012317.

Toffoletto, F., S. Sazykin, R. Spiro, and R. Wolf (2003), Inner magnetospheric modeling with the Rice Convection Model, *Space Sci. Rev.*, *107*(1), 175–196, doi:10.1023/A:1025532008047.

Tóth, G., et al. (2005), Space Weather Modeling Framework: A new tool for the space science community, *J. Geophys. Res.*, *110*(A), 12,226, doi:10.1029/2005JA011126.

Tóth, G., et al. (2012), Adaptive numerical algorithms in space weather modeling, *J. Comp. Phys.*, *231*(3), 870–903, doi:10.1016/j.jcp.2011.02.006.

Wang, W., M. Wiltberger, A. G. Burns, S. C. Solomon, T. L. Killeen, N. Maruyama, and J. G. Lyon (2004), Initial results from the coupled magnetosphere-ionosphere-thermosphere model: Thermosphere-ionosphere responses, *J. Atmos. Solar Terr. Phys.*, *66*(1), 1425–1441, doi:10.1016/j.jastp.2004.04.008.

Wang, W., A. G. Burns, M. Wiltberger, S. C. Solomon, and T. L. Killeen (2007), An analysis of neutral wind generated currents during geomagnetic storms, *J. Atmos. Solar Terr. Phys.*, *69*, 159, doi:10.1016/j.jastp.2006.06.014.

Wang, W., J. Lei, A. G. Burns, S. C. Solomon, M. Wiltberger, J. Xu, Y. Zhang, L. Paxton, and A. Coster (2010), Ionospheric response to the initial phase of geomagnetic storms: Common features, *J. Geophys. Res.*, *115*(A), 07321, doi:10.1029/2009JA014461.

Welling, D. T., and M. W. Liemohn (2014), Outflow in global magnetohydrodynamics as a function of a passive inner boundary source, *J. Geophys. Res.*, *119*(4), 2691–2705, doi:10.1002/2013JA019374.

Welling, D. T., and A. J. Ridley (2010), Validation of SWMF magnetic field and plasma, *Space Weather*, *8*(3), 03002, doi:10.1029/2009SW000494.

Welling, D. T., and S. G. Zaharia (2012), Ionospheric outflow and cross polar cap potential: What is the role of magnetospheric inflation? *Geophys. Res. Lett.*, *39*(2), 23,101, doi:10.1029/2012GL054228.

Wiltberger, M., W. Wang, A. G. Burns, S. C. Solomon, J. G. Lyon, and C. C. Goodrich (2004), Initial results from the coupled magnetosphere ionosphere thermosphere model: Magnetospheric and ionospheric responses, *J. Atmos. Solar Terr. Phys.*, *66*(1), 1411–1423, doi:10.1016/j.jastp.2004.03.026.

Wiltberger, M., W. Lotko, J. G. Lyon, P. Damiano, and V. Merkin (2010), Influence of cusp O+ outflow on magnetotail dynamics in a multifluid MHD model of the magnetosphere, *J. Geophys. Res.*, *115*(1), doi:10.1029/2010JA015579.

Winglee, R. M. (1998), Multi-fluid simulations of the magnetosphere: The identification of the geopause and its variation with IMF, *Geophys. Res. Lett.*, *25*(2), 4441–4444, doi:10.1029/1998GL900217.

Winglee, R. M. (2000), Mapping of ionospheric outflows into the magnetosphere for varying IMF conditions, *J. Atmos. Solar Terr. Phys.*, *62*(6), 527–540, doi:10.1016/S1364-6826(00)00015-8.

Winglee, R. M. (2003), Circulation of ionospheric and solar wind particle populations during extended southward interplanetary magnetic field, *J. Geophys. Res.*, *108*(A), 1385, doi:10.1029/2002JA009819.

Winglee, R. M., J. D. Menietti, and H. K. Wong (1992), Numerical simulations of bursty radio emissions from planetary magnetospheres, *J. Geophys. Res.*, *97*, 17,131, doi:10.1029/92JA01521.

Winglee, R. M., S. Kokubun, R. P. Lin, and R. P. Lepping (1998a), Flux rope structures in the magnetotail: Comparison between Wind/Geotail observations and global simulations, *J. Geophys. Res.*, *103*, 135–150, doi:10.1029/97JA02415.

Winglee, R. M., A. T. Y. Lui, R. P. Lin, R. P. Lepping, S. Kokubun, G. Rostoker, and J. C. Samson (1998b), Magnetospheric-ionospheric activity during an isolated substorm: A comparison between WIND/Geotail/IMP 8/ CANOPUS observations and modeling, *Geospace Mass and Energy Flow: Results from the International Solar-Terrestrial Physics Program*, *104*, 181.

Winglee, R. M., J. Slough, T. Ziemba, and A. Goodson (2000), Mini-magnetospheric plasma propulsion: Tapping the energy of the solar wind for spacecraft propulsion, *J. Geophys. Res.*, *105*(A), 21,067–21,078, doi:10.1029/1999JA000334.

Winglee, R. M., D. Chua, M. Brittnacher, G. K. Parks, and G. Lu (2002), Global impact of ionospheric outflows on the dynamics of the magnetosphere and cross-polar cap potential, *J. Geophys. Res.*, *107*(A), 1237, doi:10.1029/2001JA000214.

Winglee, R. M., W. K. Peterson, A. W. Yau, E. Harnett, and A. Stickle (2008), Model/data comparisons of ionospheric outflow as a function of invariant latitude and magnetic local time, *J. Geophys. Res.*, *113*(A), 6220, doi:10.1029/2007JA012817.

Winglee, R. M., E. Harnett, and A. Kidder (2009a), Relative timing of substorm processes as derived from multifluid/multiscale simulations: Internally driven substorms, *J. Geophys. Res.*, *114*, A09213, doi:10.1029/2008JA013750.

Winglee, R. M., D. Snowden, and A. Kidder (2009b), Modification of Titan's ion tail and the Kronian magnetosphere: Coupled magnetospheric simulations, *J. Geophys. Res.*, *114*(A), 5215, doi:10.1029/2008JA013343.

Yau, A. W., and M. Andre (1997), Sources of ion outflow in the high latitude ionosphere, *Space Sci. Rev.*, *80*(1), 1–25, doi:10.1023/A:1004947203046.

Yau, A. W., T. Abe, and W. K. Peterson (2007), The polar wind: Recent observations, *J. Atmos. Solar Terr. Phys.*, *69*(16), 1936–1983, doi:10.1016/j.jastp.2007.08.010.

Yu, Y., and A. J. Ridley (2008), Validation of the space weather modeling framework using ground-based magnetometers, *Space Weather*, *6*(5), 05002, doi:10.1029/2007SW000345.

Yu, Y., and A. J. Ridley (2013), Exploring the influence of ionospheric O+ outflow on magnetospheric dynamics: Dependence on the source location, *J. Geophys. Res.*, *118*(4), 1711–1722, doi:10.1029/2012JA018411.

Zheng, Y., T. E. Moore, F. S. Mozer, C. T. Russell, and R. J. Strangeway (2005), Polar study of ionospheric ion outflow versus energy input, *J. Geophys. Res.*, *110*(A), 7210, doi:10.1029/2004JA010995.

INDEX

Magnetotails in the Solar System, Geophysical Monograph 207. First Edition.
Edited by Andreas Keiling, Caitríona M. Jackman, and Peter A. Delamere.